47th American Association of Swine Veterinarians Annual Meeting 2016

Standing on the Shoulders of Giants: Collaboration & Teamwork

New Orleans, Louisiana, USA
27 February - 1 March 2016

ISBN: 978-1-5108-2174-3

Printed from e-media with permission by:

Curran Associates, Inc.
57 Morehouse Lane
Red Hook, NY 12571

CURRAN ASSOCIATES INC.
proceedings.com

Some format issues inherent in the e-media version may also appear in this print version.

Printed by Curran Associates, Inc. (2016)

American Association of Swine Veterinarians
830 26th Street
Perry, IA 50220-2328
USA

Phone: 515-465-5255
Fax: 515-465-3832

aasv@aasv.org

Additional copies of this publication are available from:

Curran Associates, Inc.
57 Morehouse Lane
Red Hook, NY 12571 USA
Phone: 845-758-0400
Fax: 845-758-2633
Email: curran@proceedings.com
Web: www.proceedings.com

Proceedings of the 47th AASV Annual Meeting
Standing on the Shoulders of Giants:
Collaboration & Teamwork
American Association Swine Veterinarians
February 27 – March 1, 2016 • New Orleans, Louisiana

Table of Contents

Howard W. Dunne Memorial Lecture

Alex Hogg Memorial Lecture

Sunday: Research Topics

Sunday Concurrent Session #1: Student Seminar

Sunday Concurrent Session #2: Industrial Partners

Sunday Concurrent Session #3: Industrial Partners

Sunday Concurrent Session #4: Industrial Partners

Poster Session: Veterinary Students

Poster Session: Research Topics

Poster Session: Industrial Partners

Monday General Session: Standing on the Shoulders of Giants: Collaboration and Teamwork

Monday Concurrent Session #1: Swine Enteric Coronaviruses

Monday Concurrent Session #2: PRRS, *Mycoplasma* and Influenza

Monday Concurrent Session #3: Antibiotic Use: 2017 and Beyond

Tuesday General Session: Emerging Disease Response

Howard Dunne Memorial Lecture

Emergence of "*Brachyspira hampsonii*" in western Canada: A collaborative success

John C. S. Harding, DVM, MSc, Dipl ABVP

In remembrance and honor of Howard W. Dunne

Dr. Howard W. Dunne received his veterinary degree from Iowa State University in 1941. He was in engaged in private practice for one year and industry for four years before entering graduate school at Michigan State University. After receiving his PhD in 1951, he spent two years as deputy chief of veterinary microbiology at the Chemical Corps Biological Laboratories in Camp Detrick, MD. In 1953, he joined the faculty of Pennsylvania State University, where he headed the veterinary science research program and the large animal diagnostic laboratory. Dr. Dunne was internationally known for his work on swine diseases, particularly hog cholera and the swine reproductive viruses then referred to as the SMEDI syndrome. Dr. Dunne authored more than 90 scientific publications, and is perhaps best known for initiating and editing the textbook *Diseases of Swine*, which is still widely considered the "bible" of swine disease.

In 1971, the American Veterinary Medical Association presented Dr. Dunne with the XII International Veterinary Congress Prize to recognize his contributions to the international understanding of veterinary medicine. At the 1974 Congress of the International Pig Veterinary Society (IPVS), Dr. Dunne was elected president of the 1976 IPVS Congress, to be held in Ames, Iowa. His sudden death at the age of 61 in September of 1974 saddened the swine veterinary community, and the 1976 IPVS Congress was dedicated to his memory.

At the suggestion of Dr. Allen Leman, a financial gift made by the 1976 IPVS Congress organizers to the fledgling American Association of Swine Practitioners was combined with additional association funds to endow this prestigious annual lecture in remembrance and honor of Dr. Howard W. Dunne.

HOWARD DUNNE MEMORIAL LECTURERS

1978	Michael Muirhead
1979	W. J. (Bill) Smith
1980	N.C. Nielsen
1981	Allen Leman
1982	Michael Wilson
1983	Edward Bohl
1984	Richard Ross
1985	Kent Kislingbury
1986	Ralph Vinson
1987	Gary Dial
1988	Thomas Alexander
1989	Steven Henry
1990	Stanley Curtis
1991	Roy Schultz
1992	Joseph Connor
1993	Camille Moore
1994	Robert Morrison
1995	Rodney Johnson
1996	Paul Yeske
1997	Barbara Straw
1998	Mike Daniel
1999	Larry Rueff
2000	David Reeves
2001	Max Rodibaugh
2002	Tim Blackwell
2003	Lonnie King
2004	Robert Desrosiers
2005	Scott Dee
2006	Gregg BeVier
2007	Tom Burkgren
2008	Timothy J. Loula
2009	John U. Thomson
2010	James F. Lowe
2011	Locke A. Karriker
2012	R. Dean Boyd
2013	Craig Rowles
2014	Daryl Olsen
2015	Gregory W. Stevenson

Emergence of "*Brachyspira hampsonii*" in western Canada: A collaborative success

John C. S. Harding, DVM, MSc, Dipl ABVP (Swine Health Mgmt)
Western College of Veterinary Medicine, University of Saskatchewan, Saskatoon, Canada

It is an honor to present the 2016 Howard Dunne Memorial Lecture and I thank Dr. George Charbonneau and the programming committee for the invitation. The theme of this meeting is "collaboration," and my presentation aims to highlight various aspects of our research program that has evolved at the University of Saskatchewan in response to the discovery of an emerging pathogen "*Brachyspira hampsonii*." My paper begins with a review of the literature pertaining to "*B. hampsonii*," then highlights various aspects of our multidisciplinary collaborative research program aimed at finding solutions to combat dysentery. Thirdly, I have provided a commentary on the essential elements of collaboration based on my experience working in diverse research groups during the last two decades. These concepts are applicable to anyone working with organizations or groups of individuals with varying backgrounds. Finally, I will close with a tribute to Dr. Dunne and other pioneers in our profession.

Re-emergence of swine dysentery in North America

Swine dysentery is a severe, economically significant disease of grow-finish pigs associated with moderate to severe mucoid or mucohemorrhagic diarrhea and typhlocolitis. Although economic assessments are sparse, the most recent estimates suggest the cost of dysentery is between $7 and $16 (USD) per pig[1,2] due to elevated morbidity, mortality and medications, and loss of grow-finish performance. The first clinical description of swine dysentery was published in 1921,[3] although it was not until 1971 that an intestinal spirochete, *Treponema hyodysenteriae* (now renamed *Brachyspira hyodysenteriae;* Bhyo), was recognized as the cause.[4-6] The disease has global distribution and was commonly diagnosed in commercial piggeries in North America until the 1990's when industry consolidation, health technologies, and the gaining popularity of higher health farms possibly played roles in its diminution. That being said, some strains of Bhyo are more virulent than others, and some strains may inhabit the intestinal tract but cause no disease at all.[7] Thus, the absence of clinical signs cannot be interpreted as absence of the organism in animal populations.

In late 2009, after a decade or more of relative quiescence, pigs with mucohemorrhagic colitis unrelated to Bhyo were identified in two Saskatchewan farms. These farms had no epidemiological linkages, were geographically separated by more than 200 km, and had no neighboring pig farms within 5 km. Initial diagnostic testing conducted at the Prairie Diagnostic Services Inc., Saskatoon, suggested the presence of an intestinal spirochete in lesional colonic tissues. Uncertain with the meaning of these results, the herd veterinarians independently contacted us to assist with obtaining a definitive diagnosis. Genus specific (*nox*) PCR[8] followed by sequencing of the amplicon revealed the presence of a potentially novel species of *Brachyspira* in affected tissues. The organism temporarily named *Brachyspira* sp. Sask30446.[9,10]

In 2011, Primus *et al.* from the University of Minnesota also reported the discovery of "novel, strongly-hemolytic *Brachyspira* spp." associated with bloody diarrhea.[11] The Iowa State University Veterinary Diagnostic Laboratory had also reported a resurgence of swine dysentery associated with Bhyo beginning in the mid-2000s, as well as the isolation of an atypical, strongly hemolytic *Brachyspira* spp.[12] Subsequent phylogenetic analyses of representative atypical/novel *Brachyspira* strains obtained from all three institutions confirmed that they were genetically similar.

"*Brachyspira hampsonii*"

Two years after the initial published reports of *B.* spp. Sask30446-associated dysentery in western Canada, Gebhart *et al.* in her lead lecture at the 2012 IPVS, announced the new species was provisionally named "*B. hampsonii*."[13] A more detailed characterization of the organism, along with a description of two distinct clades, was subsequently published in peer-review.[14] Interestingly, Chander *et al.*[14] proposed the reclassification of a strongly beta-hemolytic strain of *Serpulina* sp., P280/1, originally isolated from a UK pig with "porcine colitis syndrome" and later shown to cause mucoid diarrhea in gnotobiotic pigs,[15] as "*B. hampsonii*." If appropriate, this is the earliest isolate of "*B. hampsonii*" to the author's knowledge.

Experimental reproduction

Burrough *et al.* was the first to experimentally reproduce the characteristic disease associated with "*B. hampsonii*" (clade II reported as *B.* sp. Sask30446; clade I as strongly hemolytic *B. intermedia)* in mice[16] and subsequently pigs.[17] Similar reproduction of mucohemorrhagic diarrhea and colitis was subsequently confirmed using Canadian "*B. hampsonii*" reference strains: clade I (30599) and clade II (30446).[18,19]

Distribution

While primarily a North American disease, two cases of diarrhea in pigs associated with "*B. hampsonii*" clade I have been reported in Europe.[20,21] At the time of writing, there are no reports outside North America of the isolation of, or disease caused by, "*B. hampsonii*" clade II in pigs. However, the organism has been isolated from a number of waterfowl including Mallard ducks and Graylag geese in Spain[22] and Lesser Snow geese in the Canadian arctic.[23] Although these findings raise the possibility of transmission between waterfowl and pigs, "*B. hampsonii*" clade II strain KL180, a Canadian isolate cultured from a snow goose nesting at the Karrak Lake Queen Maud Gulf Bird Sanctuary in Nunavut, Canada did not cause disease when used to inoculate susceptible pigs.[23] Similar results were obtained when pigs were inoculated with pig-passaged KL180 in two successive trials, suggesting this non-pathogenic strain does not readily revert to virulence (Harding, unpublished).

To better understand the global epidemiology and evolution of "*B. hampsonii*," a new multilocus sequence typing (MLST) approach has been developed in which 4 genetic groups (GG) have been proposed.[24] Of the 81 pig and waterfowl isolates included in this analysis GG I and III contain pig and waterfowl isolates from N.A. and EU respectively, whereas GG II only contains clade II pig isolates from N.A., and GG4 contains only waterfowl isolates from Spain.[24]

Diagnosis

Multiple assays are available to enable accurate and timely diagnosis of swine dysentery, including the identification and/or isolation of "*B. hampsonii*," *B. hyodysenteriae* or both. Routine methods include bacterial culture and PCR conducted on feces, rectal swabs or colonic tissues collected from affected pigs. Clade-specific real-time PCR assays are published,[18,19] and have been used quantitatively to help distinguish clinical from subclinically affected animals,[10] an important distinction because "*B. hampsonii*" can be detected in the feces of healthy pigs following experimental challenge[18,19] or natural exposure.[10] A genus-specific PCR,[8] originally intended to aide in the identification of cultured isolates can also be used for the direct detection of *Brachyspira* spp. in tissue or feces, including unspeciated and less virulent *Brachyspira* such as *B. murdochii* and *B. innocens*. Matrix-assisted laser desorption ionization time-of-flight mass spectrometry (MALDI-TOF MS) has been used successfully to distinguish *Brachyspira* spp. although its ability to distinguish "*B. hampsonii*" clades is not yet perfect.[25] However, the methodology will likely become more popular, particularly in larger laboratories as caseload and throughput increases. For formalin fixed paraffin embedded colonic tissues, fluorescent *in-situ* hybridization (FISH) has been successfully used to identify both "*B. hampsonii*" and Bhyo.[26] FISH on formalin fixed feces has also been developed and has a similar level of detection as PCR used on fresh feces but lower than bacterial culture which remains the most sensitive assay.[27] In Saskatoon, we have developed protocols for PCR detection of "*B. hampsonii*" clades I and II in paraffin embedded formalin fixed tissues (Hill, unpublished).

Treatment and control

Control and elimination of swine dysentery is largely dependent on mass antimicrobial therapy. Commonly used products include tiamulin, tyvalosin, lincomycin, tylosin, salinomycin and in the USA, carbadox. Although the relative efficacy of these products is difficult to accurately judge and relies mainly on field testimonial, tiamulin water soluble at 0.006% and 0.0018% is reported to result in a cessation of clinical disease within 24h and "*B. hampsonii*" shedding within 72h of treatment initiation.[28] Antimicrobial sensitivity testing is extremely laborious and not performed routinely. There is only one recently published report of MICs for North American *Brachyspira* species,[29] which included ten *B. intermedia* isolates and 11 unspeciated *Brachyspira* isolates, some of which may or may not be "*B. hampsonii*". Because sensitivity testing techniques are not standardized between labs, clinical breakpoints are not established for any *Brachyspira*, spp., and sensitivity profiles from different laboratories and regions are difficult to directly compare, veterinarians are advised to carefully evaluate the efficacy of any chosen antibiotic based on cessation of clinical signs in field applications.

Diets

It is reported that diets low in both soluble non-starch polysaccharides (NSP) and resistant starch (RS) offer some protection against swine dysentery caused by Bhyo.[30] While the mechanisms are unclear, grain-processing methods also appear to be important, whereas overall microbial activity (measured by ATP levels) or the production fermentation byproducts such as volatile fatty acids (VFAs) that lower the pH of the hind gut are not. Regardless of the mechanism(s), Australian research indicates that diets using rice, sorghum, corn as the sole cereals appear to be more protective than diets using wheat, barley and dehulled oats.[30] The heavy reliance on wheat and barley in western Canada may certainly be a risk factor for clinical SD expression, but does not completely explain SD's re-emergence in the mid 2000s as these ingredients have been used for decades. The use of distiller's grains (DDGS) in swine diets does, however, coincide with the re-emergence of SD across North America. In a challenge experiment comparing the clinical expression of SD associated with Bhyo and "*B. hampsonii*" clade II in diets containing either 0% or 30% DDGS, similar proportions of pigs in all groups developed mucoid or mucohemorrhagic diarrhea[31] indicating that the relationship with DDGS is far from certain.

Collaborative "*B. hampsonii*" research program at the University of Saskatchewan

Swine researchers at the Western College of Veterinary Medicine have studied novel infectious diseases for many years, and welcomed the opportunity to investigate "*B. hampsonii*" when presented. From a single case of non-Bhyo mucohemorrhagic diarrhea presented in October 2009, our research has expanded into a multidisciplinary, collaborative program involving five laboratory groups. We have raised ~1.7M CAD in extramural funding to support a wide variety of short and long-term objectives with the overarching aim of understanding mechanisms and enhancing the control of "*B. hampsonii*" associated colitis and diarrhea on commercial farms (Figure 1). Our program also

Figure 1: Overview of the "*Brachyspira hampsonii*" collaborative program at the University of Saskatchewan.

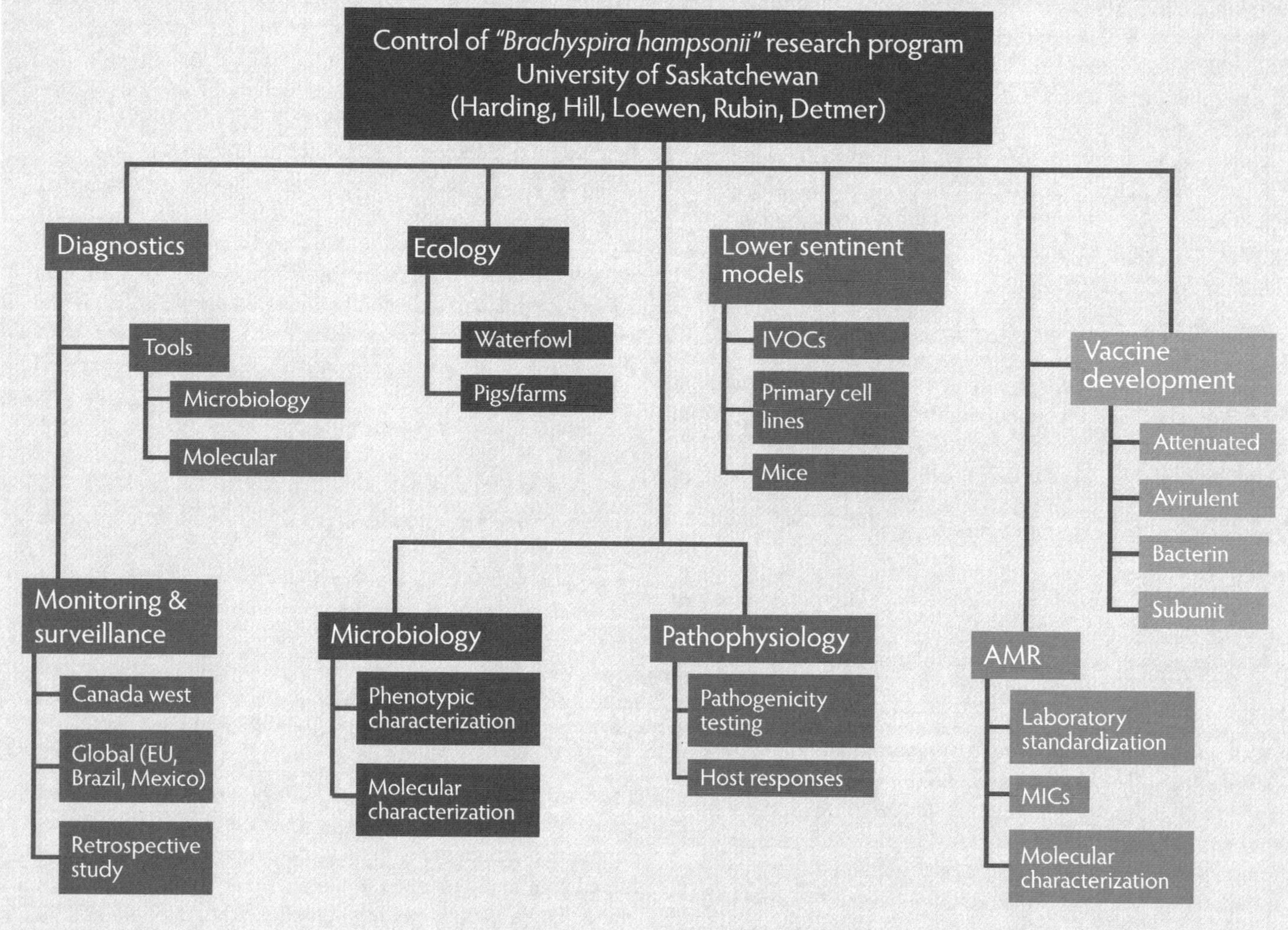

includes the only full service *Brachyspira* diagnostic laboratory in western Canada, which provides farm-level surveillance data for the Canada West Swine Health Intelligence Network (CWSHIN). A brief description of past and ongoing research activities in our various laboratory groups follows.

Diagnostic tools

Establishing species-specific, real-time quantitative PCR assays was priority and are now used routinely for diagnostic and research materials. Our standard protocol for working up diagnostic cases includes species-specific PCR assays for Bhyo, *B. pilosicoli*, and "*B hampsonii*", in addition to genus-specific (*nox*) PCR (followed by sequencing if positive) and *Brachyspira* culture (if samples are not frozen). The latter is done using CVS and/or BJ agar. Positive cultures are confirmed using phase contrast microscopy, and identified by *nox* PCR and sequencing. Positive cultures are also grown in JBS broth and frozen in glycerol at -80°C for future use. We have biochemically-characterized species in the past but this is not performed on routine diagnostic cases. Because various *Brachyspira* spp. produce a spectra of beta-hemolysis from strong to weak, we have recently developed a hemolytic assay that is more objective than traditional visualization. We have not adopted fluorescent *in site* hybridization (FISH) for diagnostic cases because our diagnostic service aims to identify any *Brachyspira* in the sample, and approximately 25% of samples contain a species other than Bhyo or "*B. hampsonii*". We have completed whole genome sequencing on a number of pathogenic and non-pathogenic "*B. hampsonii*" isolates and are analyzing the data to identify genomic modifications potentially associated with virulence.

Monitoring and surveillance

Diagnostic results are entered into a secure database that enables the tracking of cases by *Brachyspira* spp., farm, system, province and year. Anonymous summary data is reported to Canada West Swine Health Intelligence Network (CWSHIN) quarterly. Since 2009, we have processed over 850 samples (a sample being feces, tissues or extracted DNA) from over 120 farms in Western Canada. The samples were collected from clinical cases, as part of a trace back in related farms, or as part of a monitoring program following herd elimination. One of more species of *Brachyspira* has been isolated or detected in about 70% of farms, the most frequent being "*B. hampsonii*" (Table 1).

We have other ongoing surveillance projects. One project involves collaborators in Mexico, Brazil and Europe with the objective of better understanding the global distribution of "*B. hampsonii*" outside of North America. The other project is a retrospective study using PCR on paraffin embedded formalin fixed tissues that aims to determine when "*B. hampsonii*" was first identified in western Canada.

Microbiology

Detailed phenotypic characterization of clade II "*B. hampsonii*" (strain 30446) was undertaken within a year of its first isolation and a Canadian reference strain deposited in the international Depositary Authority of Canada (IDAC) in Winnipeg. Detailed phenotypic characterization of clade I (strain 30599) is ongoing. More interesting is a comparison of whole genome sequences of virulent and non-virulent "*B. hampsonii*" clade II isolates. This work has identified a number of bacterial genomic regions that may be associated with virulence.

Epidemiology and ecology

We conducted a cross-sectional study in one of the two Saskatchewan index farms with a clinical history of "*B. hampsonii*" clade II colitis.[32] Our goal was to identify critical control points associated with the on-farm transmission of the organism to help develop more effective control strategies. Some results were unexpected. After removing all feed and water medications that had been used to control mucohemorrhagic diarrhea during the previous year, the farm remained free of clinical signs. Culture positive rectal swabs were more prevalent in grower pigs (21%) but "*B. hampsonii*" clade II was not isolated from this stage of production. The organism was only isolated from 3/96 (3%) finisher pigs. *B. murdochii* was the most frequently isolated *Brachyspira* spp. on the farm and was recovered from all stages of production except for suckling pigs, the only age group in which no *Brachyspira* was isolated. Approximately 30% (24/81) of the isolates on the farm were unspeciable, and many of these fell into three phylogenetic clusters based on their partial *nox* sequence. Strongly beta-hemolytic isolates were found in each of these three novel clusters, as well as isolates clustering as *B. murdochii*. This study indicated that an ecology of strong and weak beta-hemolytic *Brachyspira* spp. may exist on farms, some of which may be beneficial to gut health.

There are many non-pig *Brachyspira* reservoirs, including some waterfowl.[33, 34] Because there exist multiple migratory flyways across North America, and several of the farms initially confirmed positive had access to natural waterways where waterfowl congregate (dugouts, reservoirs, river systems), we were interested in determining if "*B. hampsonii*" could be isolated from snow geese. Cloacal swabs collected from lesser snow geese located in four regions of the Canadian Arctic were cultured on selective media. A *Brachyspira* spp. was isolated from 15/170 samples (8.8%) including two clade I and two clade II "*B. hampsonii*" isolates.[23] Of 12 pigs inoculated with "*B. hampsonii*" clade II strain KL180 (a snow goose strain), 5 developed a transient non-mucohemorrhagic watery diarrhea for 1-2 days and 11 shed KL180 for an average of 4.5 days (range 1-12d) during the 10-14 day post inoculation period.[23] Similar results were found in two subsequent pig passages of strain KL180 (Harding *et al.*, unpublished), confirming that this snow goose strain is unlikely to cause mucohemorrhagic diarrhea in pigs.

Pathophysiology and virulence testing

After confirming that Canadian clade I and II reference strains, 30446 and 30599 respectively, were pathogenic to pigs,[18,19] we turned our attention to KL180[23] and several strains of moderate to strongly beta-hemolytic *B. murdochii* that were common to several western Canadian pork systems (Ek *et al.*, in preparation). For these experiments, 6-week old pigs were orally inoculated

Table 1: Identity of *Brachyspira* species in diagnostic cases from western Canada (2009 to September 30, 2015).

Species	Positive/tested (%)	# species/sample	Sample types submitted
Positive (any Brachy spp.)	432/804 (53.7%)	1 - 320/804 (39.8%)	Feces – 613/808 (75.9%)
B. hyodysenteriae	84/804 (10.5%)	2 - 95/804 (11.8%)	Tissue – 166/808 (20.5%)
"*B. hampsonii*"	210/804 (26.1%)	3 - 13/804 (1.6%)	Swabs – 12/808 (1.5%)
clade I	89/804 (11.1%)		
clade II	132/804 (16.4%)		
B. pilosicoli	62/804 (7.7%)	4 - 2/804 (0.25%)	DNA – 17/808 (2.1%)
B. murdochii	105/804 (13.1%)		
B. intermedia	27/804 (3.4%)		
B. innocens	27/804 (3.4%)		
Unspeciated/novel	31/804 (3.9%)		

Source: WCVM Molecular Research Laboratory

thrice with the species of interest, then monitored daily for 14-21 days following inoculation for fecal consistency and fecal shedding. This animal model has been reasonably robust in this age group except in summer, when the incidence of mucohemorrhagic diarrhea in positive control pigs is decreased. This suggests that cold room temperature stress may be a risk factor for "*B. hampsonii*" colitis.

A substantial portion of the research conducted by our team in the last 2-3 years has focused on the pathophysiology "*B. hampsonii*" colitis, and most has yet to be published. It was shown in the early 1980's that Bhyo causes a malabsorptive diarrhea associated with the failure of sodium and chloride absorption from the intestinal lumen to the blood.[35] However, the exact transport mechanism used, and other virulence mechanisms are not fully understood. Using the modern laboratory tools now available, we have assessed the electrogenic and electroneutral secretory and absorptive responses in colonic tissue of pigs experimentally challenged with Bhyo or "*B. hampsonii*" at their peak clinical signs. Our objective is to identify the key cellular mechanisms associated with mucohemorrhagic diarrhea, which will subsequently assist in the identification of specific virulence factors or proteins that induce these changes. At the time of writing, the animal work and most of the laboratory work is completed, and analysis well underway.

Lower sentinent models

To better characterize the pathophysiology of Bhyo and "*B. hampsonii*", we have challenged well over 200 susceptible pigs with various *Brachyspira* species and strains. This work is very expensive and laborious, especially since we typically collect fecal samples daily from each pig. In the last 2 years, we have developed three lower sentinent models used to study pathogenicity and host responses. Although various mouse models have been reported by other researchers, all differ in terms of mouse strain, diet, inocula used, and few provide a detailed clinical description of fecal consistency post-inoculation.[36-38] To this end, we have just completed work comparing the clinical, microbiological, and pathological responses in two strains of laboratory mice and two types of mouse chow following oral challenge with "*B. hampsonii*" and Bhyo. This work will be used to screen the vaccine prototypes we are working on, prior to intended pig studies.

Two *in vitro* models are in development by our team. The advantages of *in vitro* models are reduced cost and the ability to investigate host responses at multiple time points in the same specimen. Porcine *in vitro* organ culture (IVOC) has been used to study host responses to *Salmonella typhimurium,* but only over short periods of time.[39] Our goal was to modify the technique in order to maintain IVOC viability for several days because we were unsure how long it would take "*B. hampsonii*" to induce the specific cellular responses critical to the pathogenesis of disease. Using procedures of Dame *et al.* (2010),[40] we have successfully cultured polarized colonic mucosa from healthy pigs for 5 days, and have recently completed an experiment investigating host responses following "*B. hampsonii*" infection (Costa, in preparation). We anticipate a future experiment to identify potential differences in the host transcriptome following infection of colonic IVOC with pathogenic and non-pathogenic "*B. hampsonii*" strains.

In addition to IVOCs, we have developed the protocols to successfully harvest, culture, and vitrify primary colonic and small intestinal epithelial cell lines from healthy donor pigs (Loewen, unpublished). These will be used in Ussing chamber studies to study colonic host responses following exposure to any number of potential virulence factors or proteins that are discovered in our pathophysiology studies described above.

Vaccine development

Efficacious vaccines for "*B. hampsonii*" are not presently available but are a research priority for our team, and will help reduce production losses and antimicrobial usage for farms unable to perform herd elimination. We are taking four different approaches to vaccine development, but our work is too preliminary to report herein. We have prepared and tested a parenteral "*B. hampsonii*" strain 30446 bacterin which induces a systemic IgG response measured by "*B. hampsonii*" strain 30446 ELISA, but our animal experiment was inconclusive because an insufficient number of animals, including non-vaccinated, developed mucohemorrhagic diarrhea. We have also tested an orally administered avirulent "*B. hampsonii*" strain KL180 and are developing an attenuated *in vitro* passaged "*B. hampsonii*" strain 30446 prototype. Finally, we are purifying biologically active proteins that may be used as prototype subunits.

Antimicrobial sensitivity testing

Establishing MICs for *Brachyspira* spp. is extremely laborious and not routinely performed in commercial diagnostic labs. It requires a series of five to seven agar plates or broth media containing serially diluted concentration of an antibiotic. One series is needed for each antibiotic of interest. The media is inoculated with a *Brachyspira* sp., incubated anaerobically and the MIC is defined as the lowest concentration in which no bacterial growth is detected. In addition to understanding the MICs of *Brachyspira* strains isolated in western Canada, we are interested in developing standardized broth and agar dilution methods, and characterizing the genetic determinants of resistance to antimicrobials used to treat *Brachyspira*-associate colitis in western Canada.

Key elements of fostering collaboration and creating team science

If I have seen further, it is by standing on the shoulders of giants.
Issac Newton (1676) and Bernard of Chartres (12th century)

I have been very fortunate to be part of several interdisciplinary research groups, as these have provided some of the most exciting opportunities of my career. Interdisciplinary collaboration forms the basis of "team science", as detailed at the far right hand side of Figure 2, because no single research group has sufficient expertise to make sustained, timely and substantial progress in solving complex research questions. Successful collaboration requires substantial effort, and the bigger and more diverse the team, the more effort it requires. Whether in research or business, collaboration exists as a continuum as aptly illustrated by Bennett *et al.* (Figure 2).[41] That being said, individuals choose the degree to which they work in collaboration based on their interests and desire to work with others.

While our *Brachyspira* project is better described as a multidisciplinary collaboration, team members have expertise in a wide variety of veterinary disciplines including swine production and infective diseases, animal models, microbiology, molecular biology, pathology, physiology, biochemistry, and epidemiology and biostatistics. Our industry partner, Boehringer Ingelheim, has expertise in vaccinology. At any one time we have over a dozen senior researchers, technicians and students involved in various experiments. In spite of each senior researcher having a number of other important projects on the go, we have developed a successful collaboration and are making forward progress. Although we are all incredibly motivated and build on our collective achievements, it has not always been easy and we are not a perfect team. Management of a collaborative team is dynamic process; leaders must be responsive to the members' sentiment, learn from mistakes and strive to continuously improve. With this as background, in the section below I offer what I believe are the five most important elements of building successful collaborations, based on my experience working in a number of inter-disciplinary, team science programs. In my view, these elements are equally applicable to team building in science, business or in the community.

1. Leadership

All collaborative teams require a *champion* to lead the team, but this person does not need to be the brightest or most technically skilled member. There are often advantages of a distributed leadership style, particularly where the research objectives are technically complex and/or multifaceted. The leader must be transparent and fair, must understand personalities and inter-generational differences, and efficiently deal with the administrative tasks. Egocentric leadership must be avoided as it alienates and stifles communication and trust. A successful leader values diversity, credits team members for their achievements, and allows others to step up to fill gaps when needed.

2. Recruitment of team members

Although team members must have the capability and desire to contribute to the end goal, hiring solely based on technical skills and expertise is clearly not sufficient for team membership. I hire slowly and live by Jim Collin's motto:[42] *"get the right people on the bus, wrong people off the bus, and the right people in the right seats".* I expect all members of my team to work hard and build efficiencies, collaborate, and assist others when required. I find the right people are self-motivated, willing to step up before and when asked, and think about others. I can trust their decisions while holding them accountable. By contrast, I do not tolerate personal agendas, grandstanding or victimism. Although termination is not always easy or desirable, it is inevitable as soon as there is a loss of trust or persistent disruptive behavior. When required, fire as fast as possible.

Figure 2: The collaborative research continuum*

What is a Scientific Research Team?
...think of it as a continuum...

Low — *Level of interaction and integration* — ***High***

Independent research	Collaboration	Integrated research team
• Investigator works largely independently on a research problem with his or her lab.	• Each group member brings expertise to address the research problem. • Group members work on separate parts of the research problem, which are later integrated. • Data sharing or brainstorming among lead investigators varies from limited to frequent.	• Each team member brings specific expertise to address the research problem. • Team meets regularly to discuss team goals, individuals' objectives, and next steps. • Team shares leadership responsibility, decision-making authority, data, and credit.

* Reproduced with permission. Bennett et al. National Institutes of Health, 2010. Available at http://teamscience.nih.gov

3. Organizational culture

The organizational culture refers to the values and behaviors that contribute to the working environment. Developing a unified *organizational culture* is one of the keys to developing a cohesive and sustainable collaboration. This is very difficult, but exceptionally important for collaborative teams that are comprised of multiple subgroups working on different diverse objectives. Successful collaborations should seek and welcome diversity, as it brings new ideas, different approaches, and varying expertise to the project. Although it is not uncommon to see a strong leader hold these diverse subgroups together for a period of time, the collaboration is at risk of failure when the leader moves on, or the cultural differences amongst the subgroups become unsurpassable. In situations where different organizational cultures are present and cannot be easily unified, leadership must work to recognize the differences, emphasize the value of working together, and ensure competing interests do not develop or become disruptive.

4. Communication

Poor communication is the weakness of many organizations and is the default for collaborative teams that are focused and pushing deadlines. It is extremely destructive, yet creeps naturally into busy professional and personal lives. It is the leadership's responsibility to establish a process to ensure information flows vertically as well as horizontally throughout the organization. This will help motivate and empower team members at all levels. Communication is particularly important for subgroups that are geographically separated or have complementary, non-competing areas of expertise. To avoid working in silos, communication must be regular; pre-arranged works best in my experience. The frequency and mode must be commensurate with the objectives, scope, and timelines of the project. In theory, communication is easier today than in our parents' generation because of the many electronic tools that are presently available. While face-to-face meetings at central locations may be periodically desired, the communication strategy for diverse teams requires more than this. Communication must start early in the project, and be efficient and productive in order to create value for all team members.

5. Motivate with purpose

All members must clearly understand and be motivated towards a single overarching objective. Individuals within the team may be motivated towards smaller specific objectives in the short term, but these must be viewed as milestones. Moreover, while the road towards the end goal may be long and winding, it should be clearly defined at the outset, and be reemphasized so that no one loses sight of it. Measure and celebrate achievements along the way, encourage excellence, and provide downtime for building comradery.

Tribute to Howard Dunne

I, like many younger AASV members, did not know Dr. Dunne yet we celebrate his achievements with this memorial lecture. In a much broader context, Dr. Dunne should epitomize the numerous pioneers of veterinary science, some of which have passed and others that are still contributing to their field of expertise. In some ways, the purpose of a memorial lecture is not only to recognize the achievements of one person long passed, but to remind ourselves that we make scientific progress only because we build on the achievements of others who have come before. Thus, as a profession striving to work within the realm of evidence-based medicine, the theme selected for this meeting *"standing on the shoulders of giants"* is most commendable.

In swine medicine, the list of giants is very long. In terms of *Brachyspira*, the list includes notable individuals such as Harris, Taylor and Hampson, as well as less prominent individuals such as Whiting. These individuals were pioneers in a time when there were no smart phones, personal computers, social media, internet, email, or modern laboratory tools such as RT-qPCR and next generation sequencing. Science was much different then, yet their discoveries made a lasting impact on our profession. Even though science seems more complex now and requires more specialization, more collaboration, and more funding to the tackle the problems most relevant to industry and society, it provides the foundation for evidence-based medicine. So as my lecture fades from your memory, I would simply ask that whether you are reading a paper, reviewing a grant proposal, or researching a topic online for a specific client, strive to be a pioneer; strive to make a lasting difference in your professional and personal lives. After all...

"Ah, but a man's reach should exceed his grasp,
or what's a heaven for?"
Robert Browning, 1855

Acknowledgements and funding

I wish to acknowledge the numerous principal investigators, students and technicians at the University of Saskatchewan who have contributed to the work reported herein: Janet Hill, Matt Loewen, Susan Detmer, Joe Rubin, Yanyun Huang, Brendan O'Connor, Maneul Chirino, Matheus Costa, Cole Enns, Lisa Johnson, Jason Perez, Brandon Keith, Ruwini Gamage, Mo Patterson, Champika Fernando, Roman Nosach, Courtney Ek, Michelle Sniatynski, Holly Lemieux, Steve van Ravenstein and Heather Kittrell.

Our research is funded by competitive grants from the Canadian Swine Health Board, Alberta Livestock and Meat Agency, Agriculture and Agrifood Canada, Saskatchewan Ministry of Agriculture and Food, Genome Canada, Boehringer Ingelheim, Novartis Animal Health, and the University of Saskatchewan. The University of Saskatchewan has filed patent applications pertaining to the "*B. hampsonii*" clade II Canadian type strain (*Brachyspira* sp.sask30446) titled: "Isolated *Brachyspira* and methods and compositions for expanding and isolating *Brachyspira*" in Canada (serial number 2842068) and USA (serial number 14/233414) on 7/18/2012. The application are published and available online.

References

1. Waddilove J. Swine dysentery eradication programmes - practical experiences in the UK. *Proc. Carlos Pijoan Symposium on Swine Dysentery; A. D. Leman Swine Conference.* St. Paul, MN. Sept 18, 2011:31-44.

2. Henry S. The clinical past - thinking about swine dysentery, a historic perspective. *Proc. Carlos Pijoan Symposium on Swine Dysentery; A. D. Leman Swine Conference.* St. Paul, MN. September 18, 2011:1-5.

3. Whiting RA, Doyle LP, Spray RS. Swine Dysentery. *Bulletin.* 1921:3-15.

4. Taylor DJ and Alexander TJL. The production of dysentery in swine by feeding cultures containing a spirochaete. *Br. Vet. J.* 1971;127:58-61.

5. Harris DL, Glock RD, Christensen CR, Kinyon JM. Swine dysentery-1. Inoculation of pigs with *Treponema hyodysenteriae* (new species) and reproduction of the disease. *Vet. Med. Small Anim. Clin.* 1972;67:61-64.

6. Glock RD and Harris DL. Swine dysentery-II. Characterization of lesions in pigs inoculated with *Treponema hyodysenteriae* in pure and mixed culture. *Vet. Med. Small Anim. Clin.* 1972;67:65-68.

7. Thomson JR, Smith WJ, Murray BP, Murray D, Dick JE, Sumption KJ. Porcine enteric spirochete infections in the UK: surveillance data and preliminary investigation of atypical isolates. *Anim. Health Res. Rev.* 2001;2:31-36.

8. Rohde J, Rothkamp A, Gerlach GF. Differentiation of porcine *Brachyspira* species by a novel nox PCR-based restriction fragment length polymorphism analysis. *J. Clin. Microbiol.* 2002;40:2598-2600.

9. Harding JCS, Chirino-Trejo M, Fernando C, Jacobson M, Forster Z, Hill JE. Detection of a novel *Brachyspira* species associated with haemorrhagic and necrotizing colitis. *Proc. Int. Pig Vet. Soc. Congr.* Vancouver, Canada. July 18-21, 2010:740.

10. Harding J, Chirino-Trejo M, Vermette C, Fernando C, Jacobson M, Forster Z, Hill JE. Detection of a novel *Brachyspira* species associated with haemorrhagic and necrotizing colitis. *Proc. West. Can. Assoc. Swine Vet. Conf.* Saskatoon, SK. October 15-16, 2010:65-70.

11. Primus A, Oliveira S, Gebhart CJ. Identification of a new potentially virulent *Brachyspira* species affecting swine. *Proc. Amer. Assoc. Swine Vet. Conf.* Phoenix, AZ. March 5-8, 2011:109-110.

12. Schwartz K. *Brachyspira*: what is happening in Iowa and why. *Proc. West. Can. Assoc. Swine Vet. Conf.* Saskatoon, SK. October 20-21, 2011:25-33.

13. Gebhart CJ, Chander Y, Vannucci F, Kelly M. Bacterial enteric diseases of growing pigs: dysentery and proliferative enteropathy. *Proc. Int. Pig Vet. Soc. Congr.* Jeju, Korea. June 10-13, 2012:48-50.

14. Chander Y, Primus A, Oliveira S, Gebhart CJ. Phenotypic and molecular characterization of a novel strongly hemolytic *Brachyspira* species, provisionally designated "*Brachyspira hampsonii*". *J. Vet. Diagn. Invest.* 2012;24:903-910.

15. Neef NA, Lysons RJ, Trott DJ, Hampson DJ, Jones PW, Morgan JH. Pathogenicity of porcine intestinal spirochetes in gnotobiotic pigs. *Infect. Immun.* 1994;62:2395-2403.

16. Burrough E, Strait E, Kinyon J, Bower L, Madson D, Schwartz K, Frana T, Songer JG. Comparison of atypical *Brachyspira* spp. clinical isolates and classic strains in a mouse model of swine dysentery. *Vet. Microbiol.* 2012;160:387-394.

17. Burrough ER, Strait EL, Kinyon JM, Bower LP, Madson DM, Wilberts BL, Schwartz KJ, Frana TS, Songer JG. Comparative virulence of clinical *Brachyspira* spp. isolates in inoculated pigs. *J. Vet. Diagn. Invest.* 2012;24:1025-1034.

18. Rubin JE, Costa MO, Hill JE, Kittrell HE, Fernando C, Huang Y, O'Connor B, Harding JCS. Reproduction of mucohaemorrhagic diarrhea, colitis and shedding indistinguishable from swine dysentery following experimental inoculation with *"Brachyspira hampsonii"* strain 30446. *PLoS One.* 2013;8:e57146.

19. Costa MO, Hill JE, Fernando C, Lemieux HD, Detmer SE, Rubin JE, Harding JC. Confirmation that *"Brachyspira hampsonii"* clade I (Canadian strain 30599) causes mucohemorrhagic diarrhea and colitis in experimentally infected pigs. *BMC Vet. Res.* 2014;10:129.

20. Mahu M, de Jong E, De Pauw N, Vande Maele L, Vandenbroucke V, Vandersmissen T, Miry C, Pasmans F, Haesebrouck F, Martel A, Boyen F. First isolation of *"Brachyspira hampsonii"* from pigs in Europe. *Vet. Rec.* 2014;174:47 doi:10.1136/vr.101868.

21. Rohde J, Habighorst-Blome K, Seehusen F. *"Brachyspira hampsonii"* clade I isolated from Belgian pigs imported to Germany. *Vet. Microbiol.* 2014;168:432-435.

22. Martínez-Lobo FJ, Hidalgo Á, García M, Argüello H, Naharro G, Carvajal A, Rubio P. First Identification of *"Brachyspira hampsonii"* in Wild European Waterfowl. *PLoS One.* 2013;8:e82626.

23. Rubin JE, Harms NJ, Fernando C, Soos C, Detmer SE, Harding JCS, Hill JE. Isolation and characterization of *Brachyspira* spp. including "*Brachyspira hampsonii*" from Lesser Snow Geese (Chen caerulescens caerulescens) in the Canadian arctic. *Microb. Ecol.* 2013;66:813-822.

24. Mirajkar NS, Bekele AZ, Chander YY, Gebhart CJ. Molecular epidemiology of novel pathogen *"Brachyspira hampsonii"* reveals relationships between diverse genetic groups, regions, host species, and other pathogenic and commensal *Brachyspira* species. *J. Clin. Microbiol.* 2015;53:2908-2918.

25. Warneke HL, Kinyon JM, Bower LP, Burrough ER, Frana TS. Matrix-assisted laser desorption ionization time-of-flight mass spectrometry for rapid identification of *Brachyspira* species isolated from swine, including the newly described *"Brachyspira hampsonii". J. Vet. Diagn. Invest.* 2014;26:635-639.

26. Burrough ER, Wilberts BL, Bower LP, Jergens AE, Schwartz KJ. Fluorescent in situ hybridization for detection of *"Brachyspira hampsonii"* in porcine colonic tissues. *J. Vet. Diagn. Invest.* 2013;25:407-412.

27. Wilberts BL, Warneke HL, Bower LP, Kinyon JM, Burrough ER. Comparison of culture, polymerase chain reaction, and fluorescent in situ hybridization for detection of *Brachyspira hyodysenteriae* and *"Brachyspira hampsonii"* in pig feces. *J. Vet. Diagn. Invest.* 2015;27:41-46.

28. Wilberts BL, Arruda PH, Warneke HL, Erlandson KR, Hammer JM, Burrough ER. Cessation of clinical disease and spirochete shedding after tiamulin treatment in pigs experimentally infected with *"Brachyspira hampsonii". Res. Vet. Sci.* 2014;97:341-347.

29. Clothier KA, Kinyon JM, Frana TS, Naberhaus N, Bower L, Strait EL, Schwartz K. Species characterization and minimum inhibitory concentration patterns of *Brachyspira* species isolates from swine with clinical disease. *J. Vet. Diagn. Invest.*;23:1140-1145.

30. Pluske JR, Pethick DW, Hopwood DE, Hampson DJ. Nutritional influences on some major enteric bacterial diseases of pigs. *Nutrition Research Reviews.* 2002;15:333-371.

31. Wilberts BL, Arruda PH, Kinyon JM, Frana TS, Wang C, Magstadt DR, Madson DM, Patience JF, Burrough ER. Investigation of the impact of increased dietary insoluble fiber through the feeding of distillers dried grains with solubles (DDGS) on the incidence and severity of *Brachyspira*-associated colitis in pigs. *PLoS One.* 2014;9:e114741.

32. Patterson AH, Rubin JE, Fernando C, Costa MO, Harding JCS, Hill JE. Fecal shedding of *Brachyspira* spp. on a farrow-to-finish swine farm with a clinical history of " *Brachyspira hampsonii*"-associated colitis. *BMC Vet. Res.* 2013;9:DOI: 10.1186/1746-6148-1189-1137.

33. Råsbäck T, Jansson DS, Johansson KE, Fellström C. A novel enteropathogenic, strongly haemolytic spirochaete isolated from pig and mallard, provisionally designated *'Brachyspira suanatina'* sp. nov. *Environmental Microbiology.* 2007;9:983-991.

34. Jansson DS, Johansson KE, Olofsson T, Rasback T, Vagsholm I, Pettersson B, Gunnarsson A, Fellstrom C. *Brachyspira hyodysenteriae* and other strongly beta-haemolytic and indole-positive spirochaetes isolated from mallards (Anas platyrhynchos). *J. Med. Microbiol.* 2004;53:293-300.

35. Argenzio RA, Whipp SC, Glock RD. Pathophysiology of swine dysentery: colonic transport and permeability studies. *J. Infect. Dis.* 1980;142:676-684.

36. Hutto DL and Wannemuehler MJ. A comparison of the morphologic effects of *Serpulina hyodysenteriae* or its beta-hemolysin on the murine cecal mucosa. *Vet. Pathol.* 1999;36:412-422.

37. Hutto DL, Galvin JE, Wannemuehler MJ. Morphologic and temporal characterisation of lesions in an enhanced murine model of *Serpulina hyodysenteriae* infection. *J. Med. Microbiol.* 1998;47:275-280.

38. Nibbelink SK and Wannemuehler MJ. An enhanced murine model for studies of *Serpulina (Treponema) hyodysenteriae* pathogenesis. *Infect. Immun.* 1992;60:3433-3436.

39. Collins JW, Coldham NG, Salguero FJ, Cooley WA, Newell WR, Rastall RA, Gibson GR, Woodward MJ, Ragione RMl. Response of porcine intestinal in vitro organ culture tissues following exposure to *Lactobacillus plantarum* JC1 and *Salmonella enterica* serovar Typhimurium SL1344. *Appl. Environ. Microbiol.* 2010;76:6645-6657.

40. Dame MK, Bhagavathula N, Mankey C, DaSilva M, Paruchuri T, Aslam MN, Varani J. Human colon tissue in organ culture: preservation of normal and neoplastic characteristics. *In Vitro Cell Dev. Biol. Anim.* 2010;46:114-122.

41. Bennett LM, Gadlin H, Levine-Finley S. Collaboration and Team Science: A Field Guide. National Institutes of Health.

42. Collins J. Good to Great: Why Some Companies Make the Leap... and Others Don't. William Collins, USA. 2001.

Alex Hogg Memorial Lecture

Whose shoulders are we standing on?

Peggy Anne Hawkins, DVM, MS

In remembrance and honor of Alexander Hogg

Alexander Hogg
1920-2006

A native of Scotland, Alex (pronounced "Alec") Hogg immigrated to the United States with his family when he was six years old. After reaching adulthood, he served in the US Navy during World War II before entering veterinary school at Kansas State University. He received his DVM degree in 1950, and was a mixed animal practitioner in Iowa for twenty years. Dr. Hogg returned to college – this time, Iowa State University – to complete a Masters Degree in 1972. He spent the following years serving and educating swine veterinarians and producers in his roles as Nebraska Swine Extension veterinarian and professor at the University of Lincoln-Nebraska. Upon "retirement" from the university in 1990, Alex worked as a technical services consultant for MVP Laboratories. Ever an enthusiastic learner, he graduated from the University of Illinois Executive Veterinary Program in 1995.

An active and longstanding member of the AASV, Dr. Hogg joined what was then the American Association of Swine Practitioners shortly after it was formed. He served as secretary/treasurer of the association from 1974 to 1976, and was elected president in 1979.

The many awards and honors bestowed upon him over the years included several presented by the AASV: The Howard Dunne Memorial Award in 1985, the Meritorious Service Award in 1996, and in 2005, the first-ever AASV Foundation Heritage Award for a lifetime of achievement in veterinary medicine.

ALEX HOGG MEMORIAL LECTURERS

2011 Joseph F. Connor

2012 Lisa Tokach

2013 John T. Waddell

2014 Mark J. Engle

2015 C. Scanlon Daniels

Whose shoulders are we standing on?

Peggy Anne Hawkins, DVM, MS
Northfield, Minnesota

Introduction

This lecture generally begins with an explanation of the speaker's connection to Dr Alex Hogg. I met Dr Hogg in 1986 or 87. I was the laboratory technician for Dr Lauren Christian, Swine Breeding professor at Iowa State University. Dr Christian introduced me to Dr Hogg. I thought they were joking with me. Dr Hogg the hog vet from Nebraska! I was embarrassed. More recently I have gotten to know Mary Lou Hogg. I appreciate the chance to know more about Alex through my friendship and collaborations with her. Also, I have read many of the *Hogg Sense* newsletters. These tidbits of pork production information were the result of collaboration with swine veterinarians, animal science professors and others. Alex's shoulders were a foundation worth standing on. This particular Hogg Lecture is a celebration of the early AASP/V women, a discussion about personality types, a look at generational values and collaborative opportunities for today's swine veterinarian.

The shoulders of women

So, there is a difference between boys and girls, men and women. I mean, besides the obvious physical differences. For example, golf. When I learned to play golf, I was pleased to see that women tee off in front of the men. Sometimes this distance is so great the guys have difficulty driving the ball that far. For those who don't play golf, all I can say, is that there are special considerations for the guy that doesn't make it past the ladies' tee.

In my own life experience, my dad told me in 1975 that girls don't go to college so we compromised and I attended the Medical Institute of Minnesota in Minneapolis in the Veterinary Medical Technician program (now part of Argosy University, Eagan, MN). I worked for my hometown veterinary clinic (Glenwood, Iowa) for two years then disobeyed my dad by attending Iowa State University and got a BS degree in Animal Science. [Before I go any further, I have to say, my dad was very proud of me and my accomplishments. I appreciate the fear that a father must have when his daughter wants to do something that he could not understand.] There were five women in the Animal Science (AnSci) Class of 1982. I had declared AnSci-PreVet. I was told more than once by my professors that as a woman, veterinary medicine might be too difficult. I almost believed it when I didn't get accepted on my first application. The school provided a meeting with Dr Roger Hogle to see how close to being accepted you were so you could decide to try again or choose another career path. Dr Hogle spent a lot of time trying to find my name on the green and white striped computer print-out. It wasn't alphabetical; names were listed in the order of your application score. Finally, Dr Hogle refolded the print-out and started from the top. He seemed relieved and a bit excited to say; "Here you are, on the first page!" I was close. I had been accepted to the US Peace Corps, so, I decided to take that challenge knowing I would continue my quest to be a veterinarian when I returned home.

I spent three years in Lesotho, Africa as a Peace Corps Volunteer teaching junior high and high school students. What you need to know about that experience is the effect it had on me to see junior high students, seventh graders, change before my eyes. I thought the students were getting used to my Iowa accent and they understood me better. The home economics teacher explained that corn meal (*papa*) and cabbage (*meroho*) were the staples of the Basotho diet. As the fat and protein deficiencies were fulfilled at lunch, the students began to stay awake in class, follow instructions better and their attention span increased. For two more years I paid close attention along with the other teachers as we saw our students blossom. This experience solidified my desire to become a food animal veterinarian. A similar experience is documented in a paper titled: *The role of nutrition in children's neurocognitive development, from pregnancy through childhood.* 2013.[1]

Knowing I was close to getting into the veterinary college, my second application was an affirmation of reasons that the selection committee would accept me into the Veterinary College: technician experience, Peace Corps experience, and a recommendation from Dr Christian – how could they resist? I was accepted into the ISU Class of 1991 which was 45% women. A year later, as a veterinary school sophomore, an opportunity to pursue the Master of Science degree in Livestock Production (emphasis in pork and poultry) came along. With Drs Christian, Rothschild, Draper, Sebraneck and Cox as my major professor and committee, the challenge was on. In December 1992, I was awarded the MS degree. Perhaps the influence of swine classes during that degree pushed me from the broader food animal career to become a swine veterinarian. But the truth is, I always felt more welcomed by the swine veterinarians, researchers, producers/farmers and pigs than I did in other food animal situations. This was my experience, so I was wondering what caused other women to become swine veterinarians. In the 1970's for example, it was difficult for women to enter the college. It was felt that women would likely be part-time veterinarians. They would take time off to raise a family. Some instructors felt a woman would take a place in veterinary school that should go to a man who would make it a full-time profession. A study published in July 2010 shows there were three reasons that have led to the increase in women veterinary students: 1.) In 1972 a federal amendment outlawed discrimination against female students. 2.) Male students decided not to apply to graduate schools because of the growing number of women applicants, and 3.) The number of women attaining the bachelor's (BS) degree exceeded the number of men with BS degrees.[2] Demographics aside, who were the first women that joined the AASP? The first names that came to my mind were Drs Rexanne Struve, Barb Straw, Cate Dewey, Beth Lautner and Fay Vittetoe. With much help from Dr Sue Schulteis, I was able to gather some facts. The AASP got its start in July 1969 (changed to AASV in 2001). Drs Fay Vittetoe joined in 1975 and Barb Straw 1976 neither are current AASV members. They may have been some of the first female members of the association. The records are not completely digitalized from before 2000. Of current continuous members, Dr Beth Lautner has been a member since 1978 followed closely by Dr Teddi Wolff in 1980.[3] A review of the entire AASV membership data, for the years 2000, 2005, 2010 and 2015, shows an increasing number of women at each five-year interval (Table 1). Comparing the percentage of women in the active roster of the AVMA to that of the AASV USA membership shows a stark difference. (Figure 1, Figure 2)[3,4] The overall, 2015, AASV membership is about 2% of the current AVMA active membership which includes members with overseas addresses. If, you look only at AASV membership living in the USA, then the AASV (882) is 1% of the AVMA (77,870) membership. The following are interesting leadership parallels between the two organizations: The AVMA membership flipped to having more women than men sometime between 2005 and 2010. I don't expect that to happen in the AASV any time soon. The AVMA started in 1863 and 133 years later, the first AVMA women president Dr Mary Beth Leininger was elected in 1996. It took the AASV 33 years to elect a women president: Dr Lisa Tokach (2002). More recently, Drs Tara Donovan (2012) and Michelle Sprague (2014) have served our association. The AVMA elected a second women president, Dr. Bonnie Beaver in 2004 and a third Dr René Carlson in 2011. Based on the fact that the AASV has a much lower percentage of women members (23%) as compared to the AVMA (60%), I believe the AASV is ahead of the AVMA in its acceptance of women in leadership roles.

The USA represents about 66% of the AASV membership and Canada is second with 11% averaged across the four years of the database. Mexico is third averaging about 3% of the total AASV membership. Canadian membership over the last 15 years shows that both total membership and number of women has been fairly steady. (Figure 3)[3] There is more AASV membership data to present but space is limited in these proceedings. Additional information will be part of the presentation.

The contributions of the first female swine veterinarians to our profession are numerous. Dr Rexanne Struve made her mark in the specific pathogen free (SPF) programs by delivering baby pigs through caesarian section. And, developing the surgical "bubble" to ensure the offspring were not exposed to the sow during the procedure. Dr Barbara Straw collaborated with many authors/contributors to get multiple editions of *Diseases of Swine* published. On numerous occasions, Dr Cate Dewey has collaborated on research projects and currently has projects in Eastern Africa to add to her contributions to swine production medicine. As for the two who have been on the roster the longest, Dr Beth Lautner is the Director of APHIS' National Veterinary Services Laboratories in Ames, Iowa. Beth collaborates daily with influential policy makers. Dr Teddi Wolff recently retired after many years as a Technical Service Veterinarian. These women role models and others contributed to our profession and that means a lot to me. I would like to share a personal example. I asked Rexanne once how she managed being a wife, a mom, and having a veterinary practice. She said that the first time she had back to back emergency calls and didn't have time to write a "babysitters" list for her husband, she was worried. She came home from the chain of emergencies and noticed her husband had cared for their child just fine. From Rexanne I learned about work-life balance and, it is not babysitting when the children are your own.

I became a student member of AASV in 1989 with full membership in 1991. The data was not entered electronically at that time. However, using the 2015 subset (1,325 members) and the Membership year by Gender, I was able to determine that today there are 22 women and 336 men who have been members of AASV since 1991 and before. I was certainly not the first woman swine veterinarian, however, I probably was the first woman to be the veterinarian for a large, top 20, integrator. I was the staff veterinarian for White Oak Mills in Elizabethtown, PA in 1992-1994. During those years we went from 18 thousand to 20 thousand sows and implemented 100 percent artificial insemination of the sow herd. That was my first job as a swine veterinarian. I had wonderful mentors: Drs Bob Graybill and Tim Trayer in PA; Drs Howard Hill and Kent Schwartz from Iowa State University; Drs John Waddell, Roy Schultz and Dreyfus Froe also come to mind from those days. One of my most memorable moments was being asked to wear a dress instead of coveralls in a shower-in sow farm owned by a Mennonite family. Good collaboration requires a willingness to accept different cultural customs. As a Peace Corps Volunteer, I learned as long as the custom does not interfere with my own ethics, it is best to go along. That I wear a dress was not an unacceptable request. I am sure none of the men

Table 1: AASV membership by year and gender

Year	Men	Women	Total membership
2000	1,342	153	1,495
2005	1,099	170	1,269
2010	1,047	226	1,273
2015	1,031	294	1,325

Figure 1: AVMA active membership by gender and year (due to database update, data for 2000 was not available so 2001 was used instead).

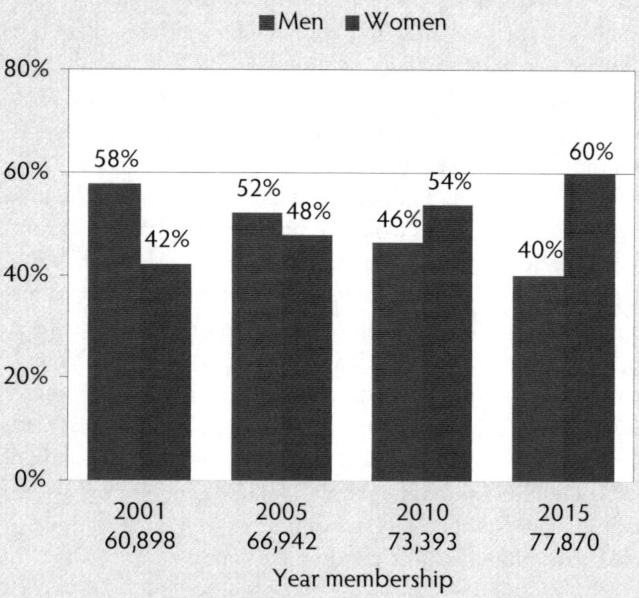

Figure 2: AASV USA membership by gender and year.

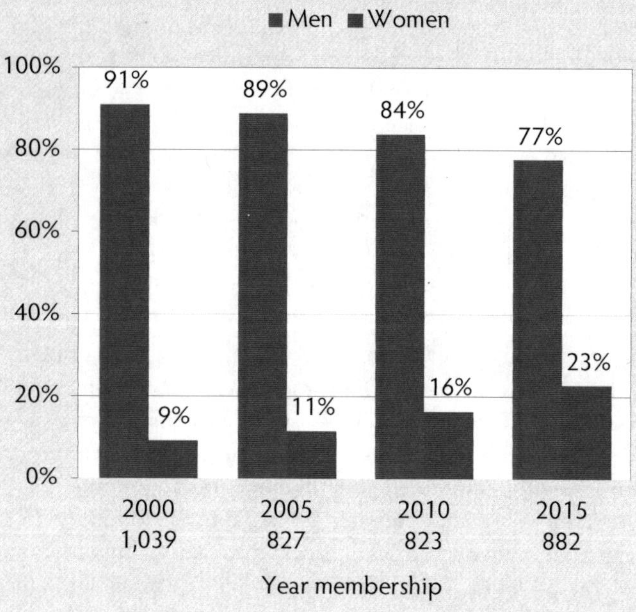

Figure 3: AASV Canadian membership by gender and year.

in the audience were ever confronted with the choices I had to make that day. The dress was a very modest one piece style that covered all the right places. It was a little short, just above my knees. I was confused by a white piece of cloth made of jersey (t-shirt) material. Since it was put on the shelf to the left of the dress next to other undergarments, I decided it should be worn as a slip (petticoat) under the skirt of the dress. I walked out of the shower room into the barn – tube socks to the knee, skirt an inch or two above the knee. Adding barn boots to the ensemble really made the outfit. Once inside I met a young lady tending to the piggys (that is what they call piglets in Pennsylvania). To my surprise, she was wearing the white jersey material as an apron on the top of the skirt. Well, not much I could do about it now, besides, no one could see under my skirt. No harm – until! There came an opportunity to post a piggy for diagnostic sampling. There was one place in the barn that was big enough for everyone to watch. The problem is, I had to post the pig on the floor. Oh, I can post a pig on the floor without a problem. This time, I had to squat down keeping the dress tucked behind my knees and allow the skirt to fall over the front of my legs. Now, that apron I was wearing under the dress, it was visible. No one said anything about the apron. We were all very professional. The piggy had an E. coli infection that improved with vaccination implementation. I wonder sometimes what they thought of the educated woman veterinarian that didn't know how to wear a dress.

About the mid 1990's, some of the AASV women thought it might be nice to have a gathering of just women during the annual meeting. With the help of the AASV staff (Tom) the women arranged to have breakfast together to talk about issues that women swine veterinarians could relate to. Mostly, we talked about children, babies and how to manage family and work. We soon realized, though we have some specific issues as women, our work-life balance issues were no different than our male colleagues. I didn't like the idea of separating from the other members of our Association. At the time most of my mentors were men. They treated me as an equal. Separating from the herd so to speak didn't make sense. In recent years, some of the pharmaceutical companies have sponsored meetings specifically inviting women swine veterinarians. The differences between those meetings and the hunting/fishing/golfing trips include accommodation for the entire family (not just the spouse) and activities like going shopping, the spa or family entertainment. There are differences between men and women. Within the AASV, those differences are recognized, internalized and celebrated. Everyone is given a chance to make it as a swine veterinarian based on his/her own merits.

Personality types and Myers-Briggs

Part of collaboration and teamwork revolves around the personalities of the people working on a project together. During my employment at a couple of large companies (Pfizer Animal Health and Monsanto) we often took time to do team building exercises. One of the recurring activities was taking some version of the Myers-Briggs Type Indicator (MBTI®) personality type test.[5] We learned something about the way people think. It helps to explain why co-workers might handle their part of the project different than the way you might have handled it. We are fortunate that in this day and age, we can go on-line and take a free Type test. It will take about 15 to 20 minutes to complete. There are two internet sites that provide free evaluations. Choose one site and get your 4-letter personality type:

`http://www.humanmetrics.com/personality/type`

Very easy – truly free - get the answer straight away

`http://www.16personalities.com/free-personality-test`

Quick – truly free – you do not have to give your email just say "no thanks". Multiple Languages!!

If you have participated in a MBTI exercise and already have your 4-letter preferences, use those results. By the time we get to this part of the AASV meeting, I hope that most of you will have taken the Type test and will know which of the sixteen possible combinations you are.

So, what is personality type testing all about? It is a series of questions/situations where you choose one answer over another answer. The answers are not right or wrong answers. The Type test measures your preferences. It tracks what you prefer to do or think. Using your preference each question gets a score. The final scores determine your results. There are four personality areas with two sides to each area. These are not opposites, they are just different ways we look at things. The breakdown is as follows:

1. Introverted (turning inward) or Extraverted (turning out) – **I or E** - Attitude
2. Intuitive (intuition) or Sensing (sensation) – **N or S** – Information gathering
3. Feeling (story) or Thinking (data) – **T or F** – Decision making
4. Perceiving (empathy) or Judging (situational)- **P or J** – Relating to the world/Lifestyle

Myers-Briggs personality types are used for career counseling, team building, leadership training, and personal development. People in Sales may use what they have learned to approach customers differently based on preferences exhibited by the customer. Knowing your type and the types of the people you work with can be a fun way to get to know each other. These are personality types not traits. There are tendencies toward certain types among gender, occupations and geographical regions. Individual results can change over time depending on how you feel the day you take it. Generally the dominant function of Introvert versus Extravert remains the same. The rest may depend on the actual questions and your current situation or state of mind. My results are ESFJ. Extravert-Sensing-Feeling-Judging. That makes me friendly, outgoing, reliable, conscientious, organized, practical, seek to be helpful and please others, enjoy being active and productive. I think that describes the way I see myself fairly well.

If your team is having difficulties working together, take a step back and do a team building exercise like the MBTI test. This will give the members of your team something to talk about besides the project. Collaborations are easier when we see each other as individuals.

Generational influence

Just as our personalities affect the way we interact with the people and the world around us, the year of your birth has an effect on you. You are defined by your generation. Now we will look at the way Lancaster and Stillman have sorted out the generational behaviors.[6] Your MB type is the result of the way you respond to questions. Lumping behavior based on the generation of your birth relies on stereotypes. Generational values are based on the year of your birth, geographic area, cultural, political and economic influences.[6] The actual birth year from generation to generation is flexible and overlaps.

What generation are you (Table 2)?[7] The Greatest Generation; Traditionalist (The Silent Generation); Baby Boomers; Generation X; Millennials (Generation Y or Gen Next); and Generation Z. I will concentrate the discussion to Traditionalist, Boomers, Gen X and Millennials. Things that influenced Traditionalists include the Great Depression, Roaring 20's, WWI and II, Korean War and the GI Bill. This led to a patriotic, loyal, waste-not/want-not attitude. It fostered faith in institutions leading to one company careers. The companies had a military-like top down structure. Boomers were influenced by Suburbia, TV, Vietnam/Watergate, Protests & Human rights movements and Drugs & Rock 'n Roll. This fostered a generation that was idealistic, competitive and who questioned authority. Generation X grew-up with Sesame Street, Game Boy,

Personal computers, the divorce rate tripled and many were latch key kids. GenX came through their childhood eclectic, resourceful, self-reliant, distrustful of institutions and highly adaptive to change and technology. Millennials have been influenced by the rapid expansion of technology, natural disasters, violence, gangs and diversity. Millennials are realistic, cyber literate, personal safety is their primary concern, and thanks to 24/7 real time news, they are more globally concerned.

I took some liberties with the data set. We do not have birth dates for members, but we do have veterinary college graduation dates. The earliest graduation date is 1942. The names were removed from the data set, so I don't know who that is for sure. I do know that they are male. Table 3 gives us a rough estimate of the generational group breakdown of AASV members from North America. There are many examples that show how people from different generations respond to career and work (Table 4).

Generational collaboration is the subject of many articles. The best advice on how multiple generations should work together: learn the differences, embrace the differences and give each person what they need. **Traditionalists** make good teachers, project leaders and coaches. **Baby Boomers** need flexibility in their projects. Give them authority to make decisions and show respect. Challenge them to keep growing. Be mindful that they will burn out if workload is not managed. **Gen Xers** do not want to burn out like boomers and they do not want to be over supervised. Manage them by coaching. They need feedback and credit for results. Push them to keep learning new things and to take on new projects. **Millennials** like flexibility in work hours and dress code. Treat them like professionals and they will respond like professionals. Mentor millennials and keep them focused with speed, customization and interactivity.

In short, show respect to Traditionalist who teach by storytelling; offer praise and recognition to Boomers for a job well done; define the parameters of projects for Gen Xers, coach them, then stand back and let them get it done; marvel at the resourceful, technology savvy Millennials, let them know why the project is important and let them complete the task in their own way.

Collaboration opportunities

Since 2012 I have volunteered to represent Swine Practice on the Committee on Environmental Issues (CEI) at the AVMA. This year I also volunteered to be on the Regulatory Advisory Panel (RAP). The CEI is a long standing, policy generating committee while the RAP is a pilot project to see if there is a better way to address policy issues. The idea is to be more nimble and to get more members engaged in the work of the AVMA. More involvement sounds like a good idea. Consider that swine veterinarians make up about 1% of the entire AVMA membership. These entities are great opportunities to collaborate at a high level. Many of the policies directly affect food animals and wildlife. Our companion animal colleagues may not be as capable in making policies about livestock as we are. So far, the AVMA has not developed the process by which they will get more members vetted and involved. In any case, we need to be ready and willing to serve on these committees, panels, working groups and taskforce teams. I would suggest that owners of clinics, managers of staff veterinarians and anyone in-charge add an item to the work goals of your employees or direct reports: be available to help the AASV fill these positions. This is a good way to reach out to the younger generations to make their job more meaningful. The AASV staff will let us know, we just need to be ready to volunteer.

There is one more way to give meaning to being a swine veterinarian and to show how collaborative we can be. Most of you are aware of Operation Main Street (OMS). It is a program supported by the National Pork Board to get pig farmers and pig veterinarians speaking to audiences about pork production. From high school students, to veterinary colleges, to Rotary Clubs, Eagles, Elks, and Lions, these are all places where we can talk about what we do, how we do it, why it is important and show the audience what progress we are making. Not everyone wants to speak in front of strangers. Be open to the idea of adding OMS training to work goals. Then, add a goal to do 1 talk every 2 or 3 months, even twice a year would be a good start. All you have to do is take the OMS presentation and add your own story to it. The NPB and state pork boards provide handouts and often pay mileage.

Table 2: The Generations*: What generation are you?

Generation name	Births starting	Births ending	Youngest age	Oldest age
Greatest generation	1910	1925	90	105
Traditionalists	1923	1944	71	92
Baby boomers	1945	1964	51	70
Generation X	1961	1981	34	54
Millennials	1975	1995	20	40
Generation Z	1995	2015	0	20

* Modified from MT Robinson; The Generations[6]
These are western cultural generations. Other geographical locations will have their own definitions based on major cultural, political, and economic influences. Some years overlap because the data are still being collected to determine a natural break.

Table 3: Histogram of 2015 AASV members from Canada, USA and Mexico using Graduation Year to estimate which Generation they represent. (Total: 985).

Generation	Graduation year	Frequency
Greatest (1910-1925)	1942-1950	4
Traditionalist (1923-1944)	1951-1960	14
Boomer (1945-1964)	1961-1988	458
Gen X (1961-1981)	1989-2005	296
Millennials (1975-1995)	2006-2014	213

Table 4:

	Traditionalist	Baby boomers	GenerationX	Millennials
Key word	Loyal	Optimist	Skepticism	Realistic
Career goals	Build a legacy	Build a stellar career	Build a portable career	Build parallel careers
Career path	Job changing has stigma	Job changing puts you behind	Job changing is necessary	Doesn't need to be a straight line
Institutions	Deserve loyalty	Deserve change	Are suspect	Should be judged on their own merit
Rewards	Job well done	Money, Title, Recognition	Freedom!	Work that has meaning
Feedback	No news is good news	Once a year whether needed or not	So, how am I doing?	From virtual coach to touch of a button

Summary

The AASV is an international organization. This presentation was mostly about the USA with a little more about Canada and Mexico. Overall, our organization should be proud of the way women have been and continue to be included. We are a strong organization built of strong membership and strong leadership in our office. Drs Burkgren, Schulteis and Snelson set the tone of cooperation and collaboration. Our board members and elected officers also make us a great organization. Our President-elect is a French-Canadian, male, Boomer, and our immediate past president is an American, woman Gen Xer. Collaboration between the genders and the generations are strengths of the AASV.

Whose shoulders are we standing on? Your shoulders, and mine! The AASV is an inclusive organization. Past performance is the greatest indicator of future behavior. Collaboration is a skill the AASV membership has exhibited in the past. It is the way of our future.

Acknowledgement

Thank you to Dr Sue Schulteis who went the extra mile to get a data set to me. Thanks to my two bosses, Drs Mike Strobel and Mark Werner for letting me continue to be a swine veterinarian **and** a GMP microbiologist. Thanks to all the great mentors and role models in the words of our colleague, Dr Roy Schultz; "I am proud to be a swine veterinarian." Finally, to my two sisters, T. Leigh and Patti Kaye, you keep me grounded. Because of you and the rest of the family, I always have a place to call home.

References

1. Nyaradi A et al. The role of nutrition in children's neurocognitive development, from pregnancy through childhood. *Front Hum Neurosci.* 2013; Mar 26:7:97

2. Lincoln AE, The shifting supply of men and women to occupations: feminization in veterinary education. *Social Forces.* 2010. Vol 88. No5. P1969-1998

3. Personal communication/collaboration, Dr Sue Schulteis, Associate Director, AASV, Perry, Iowa.

4. Personal communication, Laura Pikel, Marketing & Research Associate, AVMA

5. Kroeger, O. Thuesen, JM. Rutledge, H. Type Talk at Work: How 16 Personality Types Determine Your Success on the Job. New York: Dell Pub., 2002. 192. Print

6. Lancaster, L. Stillman, D. When Generations Collide: Traditionalists, Baby Boomers, Generation Xers, Millennials : Who They Are, Why They Clash, How To Solve The Generational Puzzle At Work. New York : Harper Business, 2002. Print.

7. Robinson MT. The Generations. 2015 Available at: www.careerplanner.com. Accessed 17 November 2015.

Vaccine efficacy of live-attenuated virus, whole-virus inactivated and alphavirus vectored subunit vaccines against antigenically distinct H3N2 swine influenza A viruses

Eugenio J. Abente[1], BA, PhD; Nicola S. Lewis[2], BSc, BVetMed, PhD, MRCVS; Mark A. Mogler[3], BSc, PhD; Daniela Rajao[1], DVM, PhD; Jefferson Santos[4], BSc, MSc; Daniel Perez[4], BSc, PhD; Amy L. Vincent[1], DVM, PhD
[1]Virus and Prion Diseases Research Unit, National Animal Disease Center, ARS-USDA, Ames, Iowa; [2]Department of Zoology, University of Cambridge, Cambridge, United Kingdom; [3]Harrisvaccines, Ames, Iowa; [4]Poultry Diagnostic and Research Center, University of Georgia, Athens, Georgia

Introduction

Subtype H3 swine influenza A virus (IAV-S) is an important pathogen in swine. The main intervention strategy for preventing infection is vaccination with the goal of inducing neutralizing antibodies against the surface glycoprotein hemagglutinin (HA).[1,7] However, due to antigenic drift and shift, vaccine components must be periodically updated. Previous work by our group and others employed antigenic cartography, a computational tool used in this case with hemagglutination-inhibition assay data, and identified two major antigenic clusters, cyan and red, and several outlier groups, including a green outlier that we later characterized as a third antigenic group among H3N2 viruses circulating in pigs in the United States from 1998 to 2013.[2,5] Studies examining human IAV found that the antigenic changes from over forty years of human seasonal H3 IAV that resulted in vaccine strain updates were largely driven by single amino acid substitutions.[4] The substitutions predicted to be associated with changes among antigenic clusters and outliers in H3 from US swine populations were shown to require 1-4 amino acid changes in the HA, and coincided with the same antigenic positions identified for human IAV.[2,5] The clinical importance of these differences in antigenicity is not fully understood, nor is it known how efficacious different vaccines are at protecting against antigenically distinct viruses. Additionally, differences in efficacy with cross-protection has been observed when comparing whole-inactivated virus with oil-and-water adjuvant (WIV) and live-attenuated influenza virus (LAIV),[3,11] demonstrating that the mechanism of vaccine type or delivery route can play a role in the breadth of vaccine-induced protection. In this study we compared the effectiveness of protection of different vaccine platforms including WIV, LAIV[8] and an alphavirus vectored vaccine[10] against challenge strains that were antigenically distinct based on the predicted amino acid substitutions defined as antigenic determinants in H3 swine IAV.

Materials and methods

Three animal experiments were carried out to examine protection provided by WIV, LAIV and an alphavirus vectored vaccine that expressed the HA of the virus of interest (Table 1). For all three studies pigs were challenged with 2 ml intratracheally and 1 ml intranasally of 1×10^5 50% tissue culture infectious dose per ml ($TCID^{50}$/ml) of the challenge virus, nasal swabs were collected at 0, 1, 3, and 5 days post-infection (DPI) from primary pigs. At 5 DPI, primary pigs were humanely euthanized and necropsied for macroscopic lesion evaluation, when bronchoalveolar lavage fluid (BALF) and trachea and lung tissue were collected. Standard virological techniques including virus isolation, virus titer, hemagglutination inhibition (HI) assays, lung lesion scoring and mucosal antibody quantitation were all performed as described previously.[11] Results were analyzed using analysis of variance (ANOVA), with $P \leq 0.05$ considered significant.

Animal experiment 1

Animal experiment 1 was conducted to investigate the efficacy of a cyan antigenic virus WIV and LAIV against a red antigenic virus. Thirty-two crossbred healthy pigs (n = 8 per WIV group, n = 8 per LAIV group, n = 8 per no vaccine and challenge, n = 8 per negative control group) were obtained from a high-health herd free of IAV and PRRSV and vaccinated with either WIV or LAIV and challenged with a red antigenic virus. WIV consisted of two doses of 128-256 hemagglutinin units (HAU) of ultraviolet (UV)-inactivated IAV combined with 20% commercial adjuvant (Emulsigen D; MVP Technologies, Omaha, NE, USA) delivered by the intramuscular route. The LAIV consisted of two doses of $1 \times 10^5 TCID^{50}$/ml (2 ml) delivered intranasally.

Animal experiment 2

Animal experiment 2 was carried out to investigate the efficacy of alphavirus vectored vaccines that expressed the HA of green or red antigenic viruses as monovalent and bivalent vaccines against homologous and heterologous challenge. The alphavirus vectored platform to deliver HA antigen as vaccine has been described previously.[10] Eighty-five crossbred healthy pigs (n = 10 per red antigenic virus HA vaccine and homologous challenge, n = 10 per red antigenic virus HA vaccine and heterologous challenge, n = 10 per green antigenic virus HA vaccine and homologous

Table 1: Animal experiment designs

Group (# of pigs)	Vaccine (antigenic phenotype)	Challenge virus (antigenic phenotype)
Animal experiment 1		
WIV/C (8)	OH/04 (cyan)	IN/13 (red)
LAIV/C (8)	OH/04 (cyan)	IN/13 (red)
C (8)	-	IN/13 (red)
NEG (8)	-	-
Animal experiment 2		
VEEV/C (10)	NY/11 (red)	NY/11 (red)
VEEV/C (10)	NY/11 (red)	IA/14 (light green)
VEEV/C (10)	IA/14 (light green)	IA/14 (light green)
VEEV/C (10)	IA/14 (light green)	NY/11 (red)
VEEV/C (10)	NY/11 (red) + IA/14 (light green)	NY/11 (red)
VEEV/C (10)	NY/11 (red) + IA/14 (light green)	IA/14 (light green)
VEEV/C (10)	GFP	NY/11 (red)
VEEV/C (10)	GFP	IA/14 (light green)
VEEV (5)	GFP	-
Animal experiment 3		
WIV/C (10)	IA/14 (light green)	IA/14 (light green)
WIV/C (10)	IA/14 (light green)	NY/11 (red)
LAIV/C (10)	IA/14 (light green)	IA/14 (light green)
LAIV/C (10)	IA/14 (light green)	NY/11 (red)
C (10)	-	IA/14 (light green)
C (10)	-	NY/11 (red)
NEG (5)	-	-

C, challenged; WIV, whole-virus inactivated vaccine; LAIV, live attenuated influenza vaccine; VEEV, alphavirus vectored vaccine expressing the HA of the virus of interest; GFP, green fluorescent protein employed as a sham vaccine; NEG, non-vaccinated, non-challenged (negative control); OH/04, A(H3N2)/turkey/OH/313053/2004; IN/13, A(H3N2)/swine/Indiana/A01260254/2013; NY/11, A(H3N2)/Swine/New York/A01104005/2011; IA/14, A(H3N2)/swine/Iowa/A01480656/2014

challenge, n = 10 per green antigenic virus HA vaccine and heterologous challenge, n = 10 per GFP sham vaccine and green antigenic virus challenge, n = 10 per a green antigenic virus and red antigenic virus bivalent HA vaccine and red antigenic virus challenge, n = 10 per a green antigenic virus and red antigenic virus bivalent HA vaccine and green antigenic virus challenge, n = 10 per GFP sham vaccine and red antigenic virus challenge, n = 5 per GFP negative control group) were obtained from a high-health herd free of IAV and PRRSV and vaccinated with either monovalent or bivalent vaccine and challenged with homologous and heterologous virus. The dose of vaccine was 1×10^8 alphavirus particles delivered as 2 ml by IM.

Animal experiment 3

Animal experiment 3 was designed to investigate the efficacy of a green antigenic virus WIV and LAIV against homologous and heterologous (red antigenic virus) challenge. Sixty-four crossbred healthy pigs (n = 10 per WIV and homologous challenge, n = 10 per LAIV and homologous challenge, n = 10 per WIV and heterologous challenge, n = 10 per LAIV and heterologous challenge, n = 10 per control group and green antigenic virus challenge, n = 9 per control group and red antigenic virus challenge, n = 5 per control group) free of IAV and PRRSV were administered WIV or LAIV and challenged with homologous and heterologous virus. The WIV and LAIV were administered to the pigs as described above. The LAIV consisted of the OH/04 attenuated backbone expressing the IA/14 HA.

Antigenic cartography

The quantitative analyses of the antigenic properties of swine and human influenza A H3N2 viruses were performed using antigenic cartography as previously described for human H3, equine H3,

Figure 1: 3D antigenic map of vaccine and challenge viruses. Relative position of vaccine, challenge and reference viruses; vaccine and challenge viruses are shown as larger spheres. Reference viruses corresponding to antigenic clusters are color-coded following the same color scheme used previously.[5] Antigenic distances are noted in antigenic units (AU; 1 antigenic unit corresponds to a 2-fold dilution of antiserum in the HI assay).

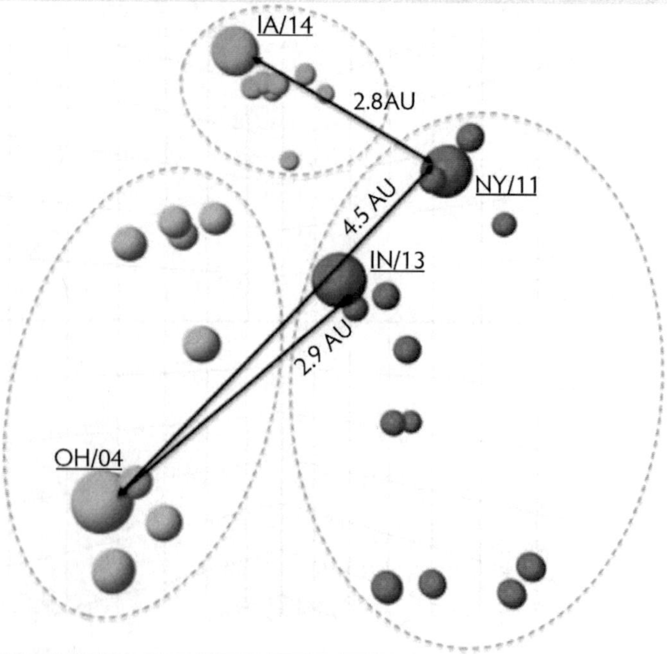

and swine influenza A H3 and H1 viruses.[5,6,9] Relative antigenic distances of vaccine and challenge viruses were quantified in the map and reported in antigenic unites (AU). One AU is equal to a two-fold loss in HI activity.

Results and discussion

The main strategy used to prevent or reduce morbidity of IAV for swine producers is to employ commercially available and farm-specific autogenous vaccines to prevent IAV clinical disease in swine.[1,7] Key components of a successful vaccine strain selection program include a comprehensive understanding of the antigenic distance of circulating strains from vaccine strains and early detection of antigenically-drifted viruses against which the current vaccine would be less efficacious. These findings would warrant an update of the vaccine formulation if epidemiologic evidence suggests circulation and spread of the variant have occurred. Studies conducted by our group and others demonstrated that amino acids at positions 145, 155, 156, 158, 159, and 189 were associated with antigenic cross-reactivity of H3 IAV-S in the United States[2,5] and this could facilitate rapid detection of the emergence of an antigenically novel virus. Animal experiments are on going and the results from sample analyses are pending, but these experiments will provide data that can inform how to best implement a successful vaccine program for both commercial vaccine producers and pork producers that use autogenous vaccines. The significance of these studies is that it may now be possible to define the number of genetic variants required to be included in multivalent vaccines to cover the breadth of antigenic variants of H3 swine IAV co-circulating in the US.

References

1. Agriculture] UUSDo: Swine 2006, Part II: reference of swine health and health management practices in the United States, 2006. 2007.

2. Feng Z, Gomez J, Bowman AS, Ye J, Long LP, Nelson SW, et al.: Antigenic characterization of H3N2 influenza A viruses from Ohio agricultural fairs. *J Virol* 2013:87(13):7655-7667.

3. Kitikoon P, Gauger PC, Anderson TK, Culhane MR, Swenson S, Loving CL, et al.: Swine influenza virus vaccine serologic cross-reactivity to contemporary US swine H3N2 and efficacy in pigs infected with an H3N2 similar to 2011-2012 H3N2v. *Influenza Other Respir Viruses* 2013:7 Suppl 4:32-41.

4. Koel BF, Burke DF, Bestebroer TM, van der Vliet S, Zondag GC, Vervaet G, et al.: Substitutions near the receptor binding site determine major antigenic change during influenza virus evolution. *Science* 2013:342(6161):976-979.

5. Lewis NS, Anderson TK, Kitikoon P, Skepner E, Burke DF, Vincent AL: Substitutions near the hemagglutinin receptor-binding site determine the antigenic evolution of influenza A H3N2 viruses in U.S. swine. *J Virol* 2014:88(9):4752-4763.

6. Lewis NS, Daly JM, Russell CA, Horton DL, Skepner E, Bryant NA, et al.: Antigenic and genetic evolution of equine influenza A (H3N8) virus from 1968 to 2007. *J Virol* 2011:85(23):12742-12749.

7. Ma W, Richt JA: Swine influenza vaccines: current status and future perspectives. *Anim Health Res Rev* 2010:11(1):81-96.

8. Pena L, Vincent AL, Ye J, Ciacci-Zanella JR, Angel M, Lorusso A, et al.: Modifications in the polymerase genes of a swine-like triple-reassortant influenza virus to generate live attenuated vaccines against 2009 pandemic H1N1 viruses. *J Virol* 2011:85(1):456-469.

9. Smith DJ, Lapedes AS, de Jong JC, Bestebroer TM, Rimmelzwaan GF, Osterhaus AD, et al.: Mapping the antigenic and genetic evolution of influenza virus. *Science* 2004:305(5682):371-376.

10. Vander Veen RL, Loynachan AT, Mogler MA, Russell BJ, Harris DL, Kamrud KI: Safety, immunogenicity, and efficacy of an alphavirus replicon-based swine influenza virus hemagglutinin vaccine. *Vaccine* 2012:30(11):1944-1950.

11. Vincent AL, Ma W, Lager KM, Richt JA, Janke BH, Sandbulte MR, et al.: Live attenuated influenza vaccine provides superior protection from heterologous infection in pigs with maternal antibodies without inducing vaccine-associated enhanced respiratory disease. *J Virol* 2012:86(19):10597-10605.

Novel reassortant human-like H3N2 and H3N1 influenza A viruses detected in pigs are virulent and antigenically distinct from endemic viruses

Daniela Rajao[1], DVM, PhD; Phillip Gauger[2], DVM, PhD; Tavis Anderson[1], PhD; Nicola Lewis[3], BSc, BVetMed, PhD, MRCVS; Eugenio Abente[1], PhD; Mary Lea Killian[4], MSc; Daniel Perez[5], BSc, PhD; Troy Sutton[6], BSc, PhD; Jianqiang Zhang[2], PhD; Amy Vincent[1], DVM, PhD

[1]National Animal Disease Center/USDA/ARS, Ames, Iowa; [2]Iowa State University, Ames, Iowa; [3]University of Cambridge, Cambridge, United Kingdom; [4]National Veterinary Services Laboratories/USDA/APHIS, Ames, Iowa; [5]University of Georgia, Athens, Georgia; [6]University of Maryland, College Park, Maryland

Introduction

Swine are important hosts in the evolution of influenza A viruses (IAV) because they can become infected with viruses from multiple species.[1] Although it is relatively uncommon for a swine lineage virus to become established in humans, spillover of human seasonal viruses to pigs has largely contributed to the great diversity of IAV circulating in swine populations through the maintenance of human-origin viral segments.[2,3] Since 2012, novel H3 viruses with the hemagglutinin (HA) from contemporary human seasonal viruses were identified through the United States Department of Agriculture (USDA) swine influenza A virus surveillance system. Here, we characterized two of these novel swine human-like H3N2 and H3N1 viruses and compared their pathogenicity and transmission in pigs to a human H3N2 with common HA ancestry. In a follow up study, we evaluated the role of gene segments on their pathogenesis using reverse genetics-generated reassortants between the swine human-like H3N1 and the seasonal human H3N2.

Materials and methods

The HA, NA, and M genes of the human-like H3N1 and H3N2 swine isolates submitted to the USDA surveillance system by the submitting NAHLN veterinary diagnostic lab were analyzed in a maximum likelihood (ML) tree using the RAxML (v7.4.2) program[4] in conjunction with representative sequences of North American swine and human viruses downloaded from GenBank and GISAID. Two swine isolates, A/Swine/Missouri/A01476459/2012 (H3N2; Sw/MO/12) and A/Swine/Missouri/A01410819/2014 (H3N1; Sw/MO/14), were obtained from the repository held at the USDA National Veterinary Service Laboratories to be used in an *in vivo* pathogenesis study. The H3N2 virus was isolated from a breeding herd during the winter of 2012, and the H3N1 virus was isolated from an epidemiologically linked location during the winter of 2013. A human H3N2 isolate A/Victoria/361/2011 (A/VIC/11) with genetically similar HA was included as a control. Two-way hemagglutination inhibition (HI) assays were performed using a panel of reference antiserum against reference swine and human H3N2 viruses, including Sw/MO/12, Sw/MO/14, and A/VIC/11, to quantify the antigenic interrelationships between them.

Thirty-five 3-week-old crossbred healthy pigs (n = 10 per challenge group; n = 5 for negative control group) were obtained from a high-health herd free of IAV and PRRSV and challenged with 2 ml intratracheally and 1 ml intranasally of 1×10^5 TCID50/ml of each virus. At 2 days post-infection (dpi), five contact pigs were placed in separate raised decks in the same room as each inoculated group to evaluate air borne transmission. Nasal swabs were collected at 0, 1, 3, and 5 dpi from primary pigs and at 0 to 5, 7, and 9 days post-contact (dpc) from indirect contact pigs for virus isolation as previously described.[5] At 5 dpi, primary pigs were humanely euthanized and necropsied for macroscopic lesion evaluation, when bronchoalveolar lavage fluid (BALF) and trachea and lung tissue were collected. BALF was tested by virus isolation.[5] Lungs and trachea tissue sections were scored for microscopic lesions.[5] Indirect contact pigs were euthanized at 15 dpc. In a follow up study, eight reassortant viruses between Sw/MO/14 and VIC/11 (Table 1) generated by reverse genetics (rg)[6] were used (n = 10 per challenge group; n = 5 for negative control group) following the same procedures as described above. Results were analyzed using analysis of variance (ANOVA), with $P \leq 0.05$ considered significant.

Results and discussion

Phylogenetic analysis of the HA genes of the novel human-like H3N2 and H3N1 isolates demonstrated that they were most closely related to human seasonal H3N2 strains from 2010-2011 (Figure 1A), suggesting they were of similar ancestry, and that the first detected isolate (Sw/MO/12) most likely evolved from a relatively recent spillover event of a human-seasonal virus into pigs. These human-like viruses have been detected in three generations of reassorted genome constellations: the neuraminidase (NA) gene initially identified was closely related to human N2 genes that circulated in 2010-2011, then the human N2 was replaced by an N1 from classical H1N1 viruses, and then replaced by the N2 of the swine 2002 N2 lineage (Figure 1B,C). Initially, all internal genes of five human-like viruses (the first H3N2 and four H3N1) were closely related to H1N1pdm09 viruses, but more recent swine human-like H3N2 have a combination of internal genes of the TRIG lineage with the M gene of H1N1pdm09 lineage (data not shown). After the introduction of the H1N1pdm09 into pigs,

Table 1: Viruses generated by reverse genetics (rg) using A/VIC/11 and Sw/MO/14.

Virus	HA gene origin	NA gene origin	Internal genes origin
Sw/MO/14rg	Sw/MO/14	Sw/MO/14	Sw/MO/14
VIC11-HA/NA	A/VIC/11	A/VIC/11	Sw/MO/14
VIC11-HA	A/VIC/11	Sw/MO/14	Sw/MO/14
VIC11-NA	Sw/MO/14	A/VIC/11	Sw/MO/14
A/VIC/11rg	A/VIC/11	A/VIC/11	A/VIC/11
MO14-HA/NA	Sw/MO/14	Sw/MO/14	A/VIC/11
MO14-HA	Sw/MO/14	A/VIC/11	A/VIC/11
MO14-NA	A/VIC/11	Sw/MO/14	A/VIC/11

HA= hemagglutinin; NA= neuraminidase

Figure 1: Maximum likelihood phylogeny of the HA (A) and NA (B, C) genes of the swine human-like H3 viruses. Numbers above or below branches indicate bootstrap support (%): bootstrap values ≤ 50% are not shown. The tree is midpoint rooted for clarity and all branch lengths are drawn to scale: scale bar indicates nucleotide substitutions per site.

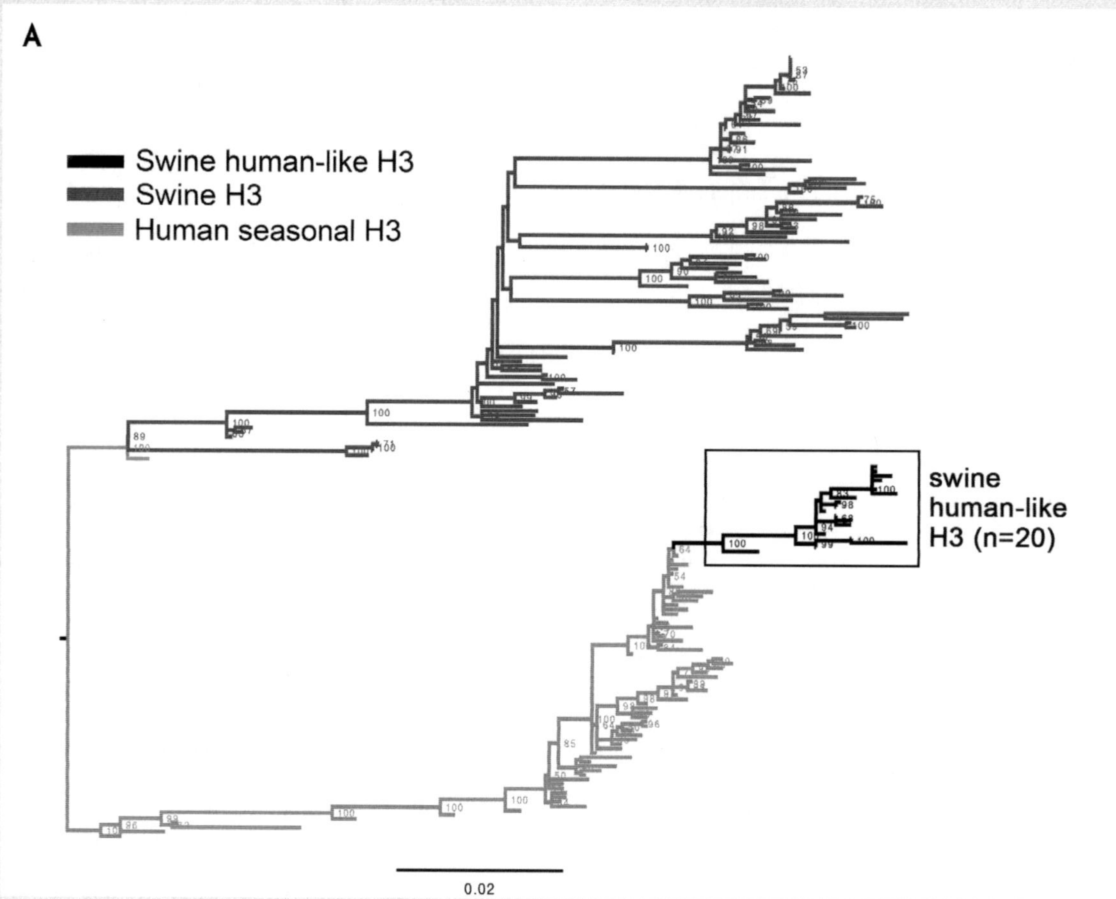

Figure 1: Maximum likelihood phylogeny of the HA (A) and NA (B, C) genes of the swine human-like H3 viruses. Numbers above or below branches indicate bootstrap support (%): bootstrap values ≤ 50% are not shown. The tree is midpoint rooted for clarity and all branch lengths are drawn to scale: scale bar indicates nucleotide substitutions per site.

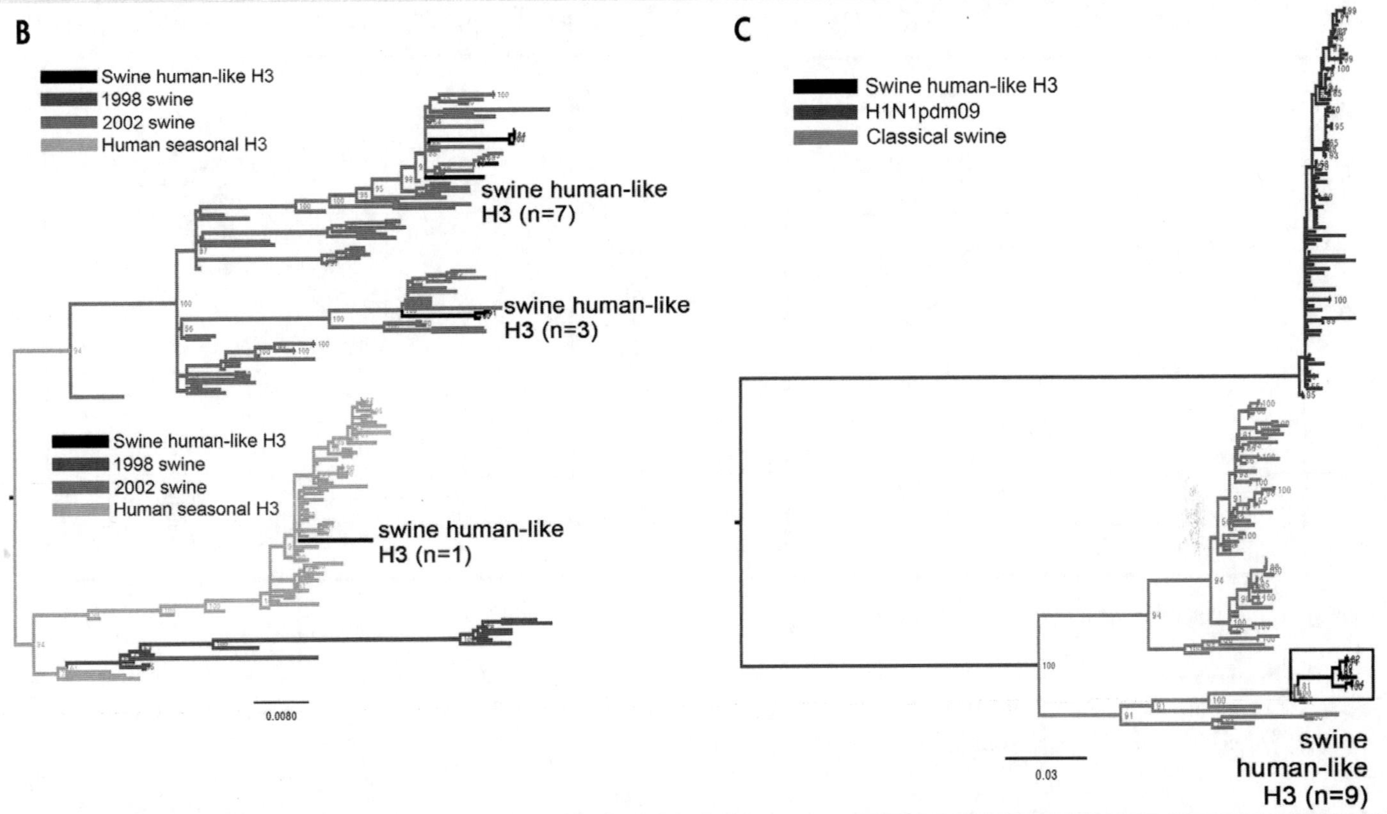

reassortants containing the H1N1pdm09 M gene have become predominant in North American swine IAV, with other H1N1pdm09 internal genes increasingly being detected.[7]

The two human-like viruses chosen for the *in vivo* study were antigenically distant from all swine H3 viruses that currently circulate in the US and from vaccine strains used in pigs, and they also showed antigenic drift from most recent human seasonal H3N2 strains (Figure 2). Therefore, the immune response elicited by the available commercial swine vaccines is highly unlikely to result in cross-protection against these novel H3 viruses. Additionally, the potential risk of these emerging swine IAV to humans should be considered.

The pathogenicity and transmission in pigs of the first detected human-like H3N2 and a second generation human-like H3N1 were compared to a human H3N2 with common HA ancestry. Our results demonstrated that the human-like viruses efficiently infected pigs, caused moderate to severe pneumonia and resulted in airborne transmission to indirect contacts, contrary to the results observed for the prototypic human A/VIC/2011 H3N2 virus (Table 2; Figure 3). The swine human-like H3N1 showed the highest average lung viral titers and degree of pathology, the initial swine human-like H3N2 virus showed an intermediate level of pathology and virus titers in the lungs, while the human H3N2 virus resulted in no significant viral titers and lung pathology (Table 2). Wholly human IAV have been shown to cause milder respiratory disease in comparison to swine-adapted viruses, such as the H1N1pdm09.[8,9] The novel field isolates studied here contained all internal genes from H1N1pdm09, leading to a speculation that these genes may be associated with the fitness of these viruses in the swine host.

To further investigate the role of gene segments on these novel viruses pathogenesis and adaptation to the swine host, reverse genetics-generated reassortants between the swine human-like H3N1 and the seasonal human H3N2 were tested in pigs. Our results suggest that the Sw/MO/14 HA alone conferred the ability to replicate in the lungs regardless of the NA or internal genes paired with it (Table 3). However, the HA combined with the other genes (NA and/or internal genes) were critical for ability to replicate in nasal epithelium and transmit to indirect contacts (Figure 4). While the HA from Sw/MO/14 contributed to replication in the lower respiratory tract, pairing the A/VIC/11 HA or HA and NA with the H1N1pdm09-lineage internal genes from the Sw/MO/14 virus resulted in significantly higher nasal

Figure 2: Graph illustrating the antigenic distances between the human-like swine H3 viruses Sw/MO/12 **(A)** and Sw/MO/14 **(B)** and a panel of reference human and swine H3N2 viruses. The viruses used in this study, Sw/MO/12 H3N2, Sw/MO/14 H3N1, and A/VIC/11, are represented by larger circles.

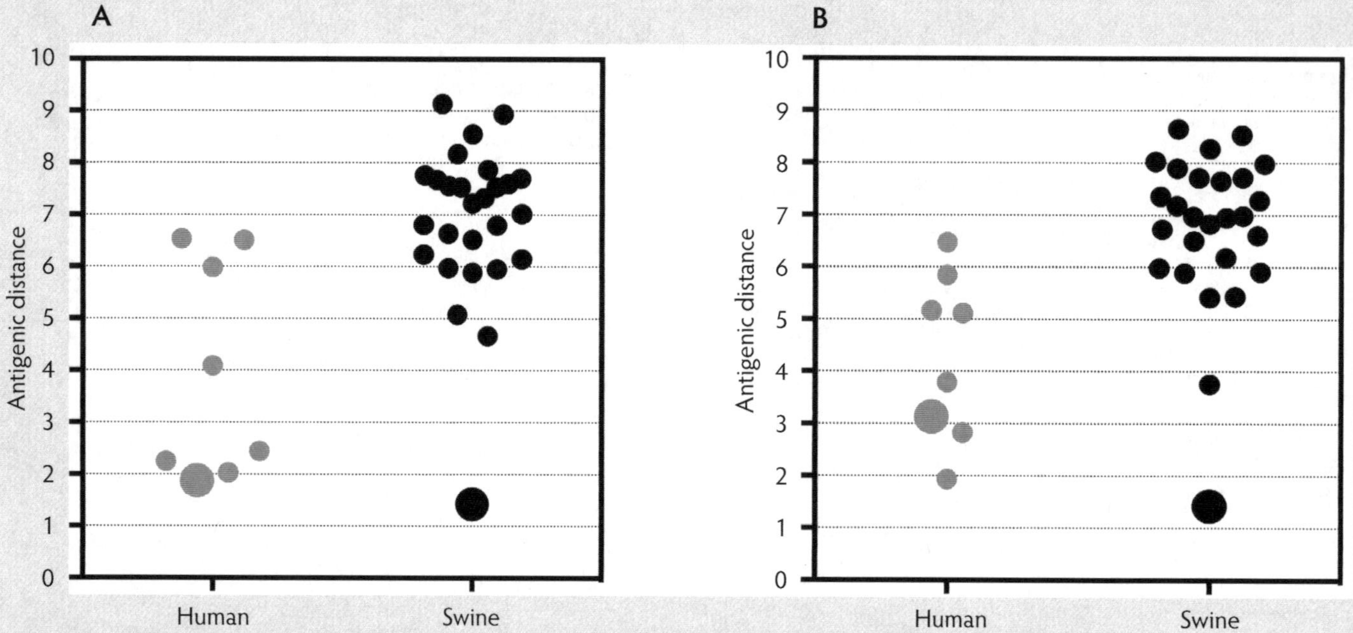

Table 2: Lung pathology and virus titers of pigs challenged with wild-type A/VIC/11, Sw/MO/12, or Sw/MO/14, and non-challenged controls (NC). Results shown as means ± standard error.

Group	Macroscopic pneumonia (%)	Microscopic pneumonia score (0-22)	Log_{10} virus titer ($TCID_{50}$) in BALF
NC	$0.0 \pm 0.0^{a,x}$	0.2 ± 0.1^{a}	0.0 ± 0.0^{a} (0/5)^{y}
A/VIC/11	0.0 ± 0.0^{a}	0.8 ± 0.2^{a}	0.6 ± 0.4^{a} (2/8)
Sw/MO/12	4.2 ± 1.0^{b}	4.9 ± 0.5^{b}	3.6 ± 0.2^{b} (10/10)
Sw/MO/14	12.0 ± 0.8^{c}	9.4 ± 0.6^{c}	5.1 ± 0.2^{c} (10/10)

x Different lower case letters within the same column indicate significant differences ($P \leq 0.05$).
y The number of virus-positive pigs/total number of pigs tested is indicated in parentheses.

shedding compared to the whole human virus (Table 3; Figure 4). These findings indicate that the HA played a critical role in the adaptation of these novel viruses to swine, with different phenotypes in the upper compared to lower respiratory tract, but the combination and balance between viral genes was also essential. Because these human-like viruses seem to be fully adapted to pigs and the US swine population likely has limited immune protection against them, effective surveillance and close monitoring of the evolution of these human-origin viruses in pigs are critical for vaccine preparedness and to improve preventive measures in the swine industry.

Acknowledgements

Dr. Rajao was a CNPq-Brazil fellowship recipient; Drs. Anderson and Abente were supported in part by an appointment to the USDA-ARS administered by ORISE. This study was supported by USDA-ARS and USDA-APHIS. We gratefully acknowledge the pork producers, swine veterinarians, and laboratories for participating in the USDA Influenza A Virus Swine Surveillance System. We thank Michelle Harland and Gwen Nordholm for technical assistance; and Jason Huegel, Tyler Standley, and Jason Crabtree for assistance with animal studies.

Figure 3: Virus titers in nasal swabs of **(A)** primary pigs at 1, 3, and 5 days post infection (dpi) with Sw/MO/12, Sw/MO/14 or A/VIC/11, and of **(B)** their respective indirect contact pigs at 4, 5, 7, and 9 days post contact (dpc). Results shown as means and standard error of the means. Numbers of infected pigs/total number of pigs are indicated in parentheses. Different lowercase letters between groups within the same sampling day indicate significant differences ($P \leq 0.05$).

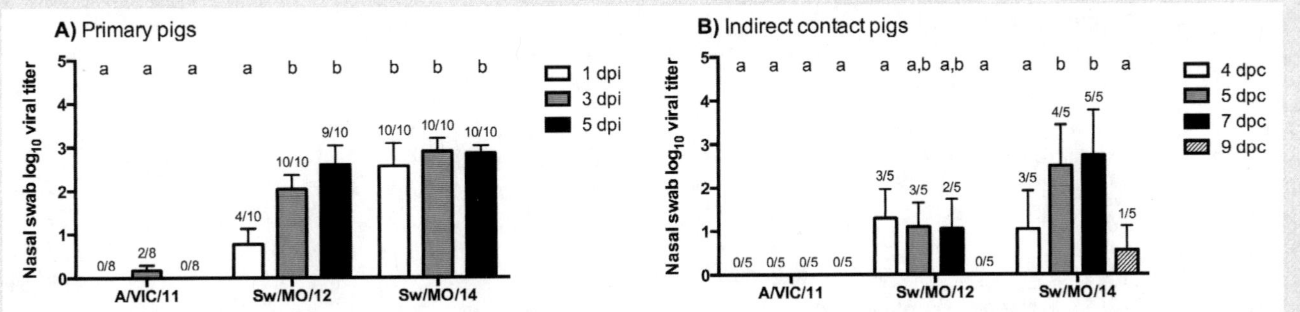

Table 3. Lung pathology and virus titers of pigs challenged with reverse genetics-generated viruses between A/VIC/11 and Sw/MO/14, and non-challenged controls (NC). Results shown as means ± standard error.

Group	Macroscopic pneumonia (%)	Microscopic pneumonia score (0-22)	Log_{10} virus titer ($TCID_{50}$) in BALF
NC	0.3 ± 0.2[a,x]	0.1 ± 0.1[a]	0.0 ± 0.0[a] (0/5)[y]
Sw/MO/14rg	6.3 ± 1.9[b]	6.4 ± 1.0[b]	4.1 ± 0.3[b] (10/10)
VIC11-HA/NA	0.5 ± 0.4[a]	0.1 ± 0.1[a]	0.0 ± 0.0[a] (0/10)
VIC11-HA	0.0 ± 0.0[a]	0.2 ± 0.1[a]	0.0 ± 0.0[a] (0/10)
VIC11-NA	2.7 ± 0.7[a]	2.3 ± 0.7[b]	3.6 ± 0.4[b,c] (9/10)
A/VIC/11rg	0.4 ± 0.2[a]	0.6 ± 0.2[a]	0.4 ± 0.3[a] (2/10)
MO14-HA/NA	1.0 ± 0.3[a]	0.5 ± 0.2[a]	1.7 ± 0.4[c] (7/10)
MO14-HA	1.1 ± 0.5[a]	0.9 ± 0.2[a]	2.6 ± 0.4[c] (9/10)
MO14-NA	1.4 ± 0.6[a]	0.4 ± 0.1[a]	0.4 ± 0.2[a] (2/10)

x Different lower case letters within the same column indicate significant differences ($P \leq 0.05$).
y The number of virus-positive pigs/total number of pigs tested is indicated in parentheses.

References

1. Webster R.G., W.J. Bean, O.T. Gorman, T.M. Chambers and Y. Kawaoka. 1992. Evolution and ecology of influenza A viruses. *Microbiol Rev*. 56:152-179.

2. Anderson T.K., B.A. Campbell, M.I. Nelson, N.S. Lewis, A. Janas-Martindale, M.L. Killian and A.L. Vincent. 2015. Characterization of co-circulating swine influenza A viruses in North America and the identification of a novel H1 genetic clade with antigenic significance. *Virus Res*. 201:24-31.

3. Nelson M.I., D.E. Wentworth, M.R. Culhane, A.L. Vincent, C. Viboud, M.P. LaPointe, X. Lin, E.C. Holmes and S.E. Detmer. 2014. Introductions and evolution of human-origin seasonal influenza a viruses in multinational Swine populations. *J Virol*. 88:10110-10119.

4. Stamatakis A. 2006. RAxML-VI-HPC: maximum likelihood-based phylogenetic analyses with thousands of taxa and mixed models. *Bioinformatics*. 22:2688-2690.

5. Vincent A.L., W. Ma, K.M. Lager, J.A. Richt, B.H. Janke, M.R. Sandbulte, P.C. Gauger, C.L. Loving, R.J. Webby and A. Garcia-Sastre. 2012. Live attenuated influenza vaccine provides superior protection from heterologous infection in pigs with maternal antibodies without inducing vaccine-associated enhanced respiratory disease. *J Virol*. 86:10597-10605.

6. Hoffmann E., G. Neumann, Y. Kawaoka, G. Hobom and R.G. Webster. 2000. A DNA transfection system for generation of influenza A virus from eight plasmids. *Proc Natl Acad Sci U S A*. 97:6108-6113.

7. Anderson T.K., M.I. Nelson, P. Kitikoon, S.L. Swenson, J.A. Korslund and A.L. Vincent. 2013. Population dynamics of cocirculating swine influenza A viruses in the United States from 2009 to 2012. *Influenza Other Respir Viruses*. 7:42-51.

8. Vincent A.L., K.M. Lager, K.S. Faaberg, M. Harland, E.L. Zanella, J.R. Ciacci-Zanella, M.E. Kehrli, Jr., B.H. Janke and A. Klimov. 2010. Experimental inoculation of pigs with pandemic H1N1 2009 virus and HI cross-reactivity with contemporary swine influenza virus antisera. *Influenza Other Respi Viruses*. 4:53-60.

9. Landolt G.A., A.I. Karasin, L. Phillips and C.W. Olsen. 2003. Comparison of the pathogenesis of two genetically different H3N2 influenza A viruses in pigs. *J Clin Microbiol*. 41:1936-1941.

Figure 4: Virus titers in nasal swabs of primary pigs at 1, 3, and 5 days post infection (dpi) with reverse genetics generated parental viruses **(A)** Sw/MO/14rg or **(B)** A/VIC/11rg, and reassortant viruses with surface genes exchanged on the parental backbones (**A:** VIC11-HA/NA, VIC11-HA and VIC11-NA in the Sw/MO/14rg backbone; **B:** MO14-HA/NA, MO14-HA and MO14-NA in the A/VIC/11rg backbone). Results shown as means and standard error of the means. Numbers of infected pigs/total number of pigs are indicated in parentheses. Different lowercase letters within the same sampling day indicate significant differences ($P \leq 0.05$).

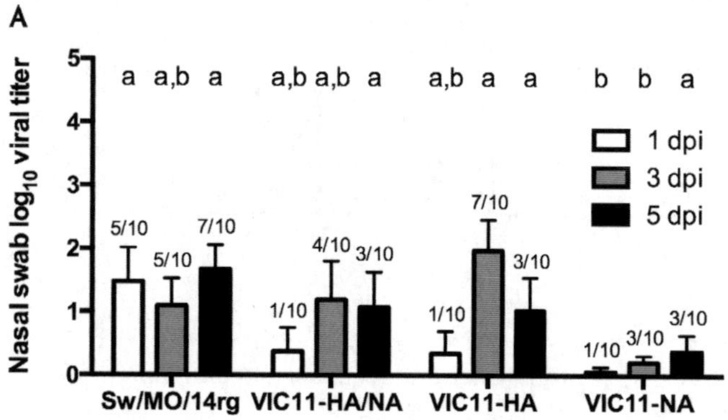

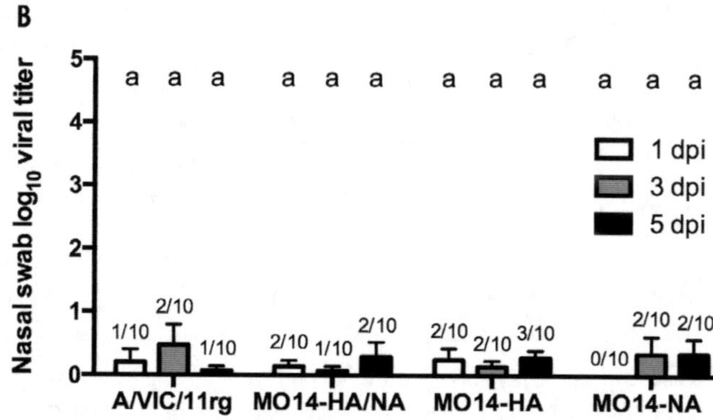

Effect of sow vaccination and maternally-derived antibodies on influenza infection

F. Chamba[1], DVM, MS; A. Diaz[1], DVM, MS, PhD; M. Allerson[2], DVM, MPH, PhD; M. Culhane[1], DVM, PhD; R. Morrison[1], DVM, PhD; P. Davies[1], DVM, PhD; A. Perez[1], DVM, PhD; M. Torremorell[1], DVM, PhD

[1]Department of Veterinary Population Medicine, College of Veterinary Medicine, University of Minnesota, Saint Paul, Minnesota; [2]Holden Farms, Inc. Northfield, Minnesota

Introduction

Influenza is an economically significant disease in US swine herds. Most control measures include vaccination in breeding herds using prefarrow, mass or a combination of prefarrow and mass vaccination protocols. Commercial and autogenous, and mono and/or multivalent vaccines are licensed and available in the US.[1] Furthermore, weaning age pigs are important in the maintenance of influenza A virus (IAV) infections in breeding herds as well as in the transmission of the virus to wean-to-finish facilities.[2] Moreover, maternally-derived antibodies (MDA) has been associated with improved clinical presentation in pigs after weaning and in certain instances, decrease on transmission.[3]

However, there is still a lack of understanding about the effect on sow vaccination on the level of influenza in prior to wean pigs, and also about the role that MDA have on infection dynamics in pigs after weaning, in particular under field conditions. In order to further the understanding of these questions, we conducted two different studies to 1) evaluate the effect of sow vaccination protocol and type of vaccine on the detection of influenza in groups of pigs at weaning, and 2) assess the effect of MDA at weaning on influenza infection dynamics in pigs after weaning.

Materials and methods

In study 1, participating breeding herds were selected based on their influenza vaccination protocol and herd status within 12 months prior to the start the study. There were 52 farms enrolled in a cohort study. Thirty nasal swabs were collected on a monthly basis during 6 months in each farm from pigs prior to wean. Sampling timeframe was between January and November 2013. One piglet was conveniently selected from each of 30 litters and nasal swabs collected. Swabs were tested in pools of 3 to detect IAV matrix gene by RT-PCR at the University of Minnesota Veterinary Diagnostic Laboratory. Generalized linear mixed logistic regression models (GLMM) were used with influenza RT-PCR results (Positive/Negative) at group level as the outcome, and vaccination protocol and type of vaccine as fixed predictors. Farm and season (time) were incorporated as random predictors. A group of pigs was considered positive if at least one pool tested positive. Best models were selected based on AIC criterion using SAS 9.4® (SAS Foundation, Cary, NC).

In study 2, a cohort of 132 out of 2200 pigs were identified at weaning and sampled weekly by collecting nasal swabs during 15 weeks in a wean-to-finish farm. Pigs were selected randomly and individually identified. Blood samples were also collected at weaning and every 4 weeks. Sera were assayed for antibodies by hemagglutination inhibition (HI) test including 2 field (H1N1 Gamma 1 and H3N2 Cluster IV A) and 4 vaccine (H1N2 Delta 1, H1N1 Delta 2, H1N1 Gamma 1 and H3N2 Cluster IV A) strains. Nasal swabs were tested by RT-PCR targeting the IAV matrix gene. A sample was considered positive if the cycle threshold (ct) value was less than 38. Time to first infection after weaning was considered the outcome and HI titers against each strain were considered as predictors. Analysis was done with a Kruskal-Wallis test in SAS using *proc npar1way*.

Results

In study 1, sixty three percent of herds vaccinated against IAV. Within vaccinated herds, 45%, 36% and 19% had prefarrow, mass and prefarrow plus mass vaccination protocols, respectively. Also, 53%, 31% and 16% used commercial, autogenous or a combination of commercial and autogenous vaccines, respectively. One farm did not report the type of vaccine used. Overall, 48% of the herds tested positive at least once. Moreover, 44% of vaccinated and 58% of non-vaccinated herds tested positive. Also, 25% of groups of pigs at weaning tested positive with vaccinated herds having 16% of positive groups compared to the 40% of the non-vaccinated herds. The odds of detecting IAV positive groups at weaning were decreased by either prefarrow or mass vaccination protocol. There were no differences between protocols. Commercial vaccines decreased the odds of detecting positive groups at weaning, and a similar trend was observed for autogenous vaccines. There were no significant differences between types of vaccines.

In study 2, two waves of influenza infection were identified. The first wave was dominated by an H1N1 Gamma 1 virus and peaked at week 2 post weaning. Ninety percent of pigs tested positive at least once during the first infection wave. Second wave of infection was dominated by an H3N2 Cluster IV A virus and peaked at week 7. Seventy two percent of pigs retested positive at least once by week 8. A total of 75% of the pigs had tested positive previously in non-consecutive weeks and were considered reinfected. Mean time to first infection was 1.5 weeks after weaning and mean duration of this infection was 1.6 weeks. Moreover, 17% and 0% of pigs at weaning had HI titers ≥ 40 against the field isolates H1N1 Gamma 1 and H3N2 Cluster IV A respectively. Also, 68%, 61%,

33% and 27% of pigs at weaning had HI titers ≥ 40 against vaccine strains H1N1 Gamma 1, H1N1 Delta 2, H1N2 Delta 1, and H3N2 Cluster IV A respectively. Finally, there were no significant differences in the time to infection after weaning regarding to the HI titer against to each strain at weaning.

Conclusions and discussion

Sow vaccination decreased the odds of detecting positive groups at weaning with either of the protocols as well as type of vaccines evaluated in comparison to no vaccination. There were no significant differences between protocols or type of vaccines. Overall, our first study indicates that IAV sow vaccination may be an effective tool at decreasing the prevalence of IAV in pigs at weaning.

Our results in study 2 indicated co-circulation of distinct IAV isolates in a wean-to-finish cohort. Also, it seems that the rapid spread of IAV observed shortly after weaning was related to the low levels of MDA found at weaning. However, lack of antibodies against the H3N2 isolate at weaning did not explain why the second wave did not appear earlier. Furthermore, IAV appeared to infect all pigs in this followed cohort. Finally, this second study provides a deeper insight into the duration and persistence of influenza infections in post weaning farms and highlights the complexity and importance of understanding the effect of MDA levels on influenza infection dynamics after weaning.

Acknowledgements

Study 1 was funded with support from the Swine Disease Eradication Center industrial partners: Zoetis, Newport, Merck and Novartis.

Study 2 was funded by National Pork Board (NPB) and Minnesota Pork Board (MPB).

References

1. Sandbulte, M.R., Spickler, A.R., Zaabel, P.K. & Roth, J.A. Optimal Use of Vaccines for Control of Influenza A Virus in Swine. *Vaccines 3*, 22-73 (2015).

2. Allerson, M.W., Davies, P.R., Gramer, M.R. & Torremorell, M. Infection dynamics of pandemic 2009 H1N1 influenza virus in a two-site swine herd. *Transbound Emerg Dis 61*, 490-9 (2014).

3. Allerson, M. et al. The impact of maternally derived immunity on influenza A virus transmission in neonatal pig populations. *Vaccine 31*, 500-505 (2013).

Airborne transmission of highly pathogenic avian influenza virus. Is the swine industry ready?

C. Alonso[1], DVM, MS; B. Olson[2], PhD; P. Raynor[2], PhD; D. Patanayak[1], PhD; P. Davies[1], BVSc, PhD; M. Torremorell[1], DVM, PhD
[1]Department of Veterinary Population Medicine, College of Veterinary Medicine, University of Minnesota, Saint Paul, Minnesota; [2]Division of Environmental Health Sciences, School of Public Health, University of Minnesota, Saint Paul, Minnesota

Introduction

Since first detected in the United States in December 2014, the poultry industry experienced severe outbreaks of the H5N2 highly pathogenic avian influenza (HPAI) virus during the spring of 2015. Only in the state of Minnesota, more than 9 million birds were killed and many more were affected throughout the Midwest.[1] Field observations of the rapid spread of cases suggested the potential for airborne spread. Airborne transmission of influenza virus in pigs and people is better understood since the virus is shed through the respiratory tract. However, airborne transmission of avian influenza virus (AIV), which is predominantly shed in feces, is more controversial. HPAI transmits primarily by the fecal-oral route when infected feces or contaminated materials via fomites are ingested by susceptible hosts.[2] There is evidence of other pathogens shed in feces that are also able to transmit through the air;[3,4] however more information is needed to determine the contribution of the airborne route in the spread of enteric pathogens such as AIV. Based on the hypothesis that HPAI virus particles can be suspended in the air and remain infectious while airborne, and given the on-going threat that HPAI represents for the virus to jump species and affect swine, our objectives were three-fold: 1) to detect and assess the viability of airborne HPAI virus in air samples collected inside and outside affected flocks; 2) to assess airborne particles deposition on surfaces outside affected flocks; 3) and to characterize particle size distribution of HPAI virus in positive farms.

Materials and methods

Three turkey flocks located in MN and 3 layer flocks in IA and NE were enrolled in the study during the spring of 2015. Air samples were collected inside and outside of affected flocks at 5m, 70-100m and 500-1000m approximately, using 2 types of samplers: a) cyclones (cyclonic air collector (CC, 200 lpm), and b) size selective air samplers [8 staged Andersen cascade impactor (ACI, 28.3 lpm) and 5 staged high volume Tish cascade impactor (TISH, 1,100 lpm)]. These size selective air samplers, previously validated under experimental conditions, are capable to collect particles ranging from 0.01 to 10 microns in size. Environmental samples from surfaces in locations at high risk of direct exposure to the air exhausted from the layer farms were taken to evaluate the risk of environmental virus deposition. Relative humidity (%RH), temperature (temp) and air speed as well as total particles were measured during the sampling periods. All samples were tested by RT-PCR and positive samples were inoculated on embryonated eggs to assess viability. Total quantity of viral RNA copies per volume of air collected, total airborne particles, %RH and temp were analyzed at the different sampling points.

Results

Six flocks with confirmed HPAI H5N2 infections were included in the study. A total of 138 sampling events, including 321 individual samples, were analyzed by quantitative RT-PCR. Overall HPAI samples tested positive in 5 out of the 6 flocks tested. Sixty seven percent of the sampling events inside, and 45% and 4% at 5m and 70-150m, respectively, tested positive (CT values 26-35) (Table 1).

Thirty seven percent of air samples, collected at 500-1000m, were suspect indicating low quantity of genetic material. Thirty eight percent of samples collected inside(18/48), 52% (14/27) at 5 m, and 25% (1/4) at 70-100m outside the facilities were positive for virus isolation demonstrating the presence of viable virus on those air samples.

Deposition of airborne HPAI RNA was also shown in several locations surrounding the facilities including farm fixtures as well as experimental testing material (35% positive samples). However virus isolation did not yield viable virus. HPAI virus was also detected associated to particles of all size ranges measured inside and outside (5m) the sampling locations in quantities ranging from 1.2×10^3 (particles collected at 5 m outside between 0.4-0.7µm) to 1.4×10^6 (particles collected inside > 9µm) (Figure 1). Overall, viable virus was detected in particles > 2.1 µm.

Conclusions and discussion

Results from this study indicate that infectious HPAI virus can be aerosolized from infected flocks and remain airborne. The transport of infectious virus airborne particles and their deposition on surfaces around infected premises appear to be a risk for the spread of HPAI to other locations. Our results also indicate that there is an immediate need to develop strategies and technologies to contain bioaerosols in order to limit the spread of epidemics in food animal production systems. Because of the similarities in the production systems between poultry and pigs,

Table 1: RT-PCR results of sampling events in turkey and layer farms inside and at outdoor distances ranging from 5m to 1000m.

	RT-PCR[a]	Inside	5m	70-150m	500-1000m	Sub_total
Total	Positive	26 (67%)	18 (45%)	1 (3.5%)	0 (0%)	45 (33%)
	Suspect	7 (18%)	11 (27.5%)	23 (79%)	11 (37%)	52 (38%)
	Negative	6 (15%)	11 (27.5%)	5 (17.5%)	19 (63%)	41 (30%)
	Total	**39 (100%)**	**40 (100%)**	**29 (100%)**	**30 (100%)**	**138 (100%)**

[a] Ct values < 35 (positive); 35 - < 40 (suspect); > 40 (negative)

Figure 1: Particle size distribution of HPAI airborne particles collected inside layer and turkey facilities.

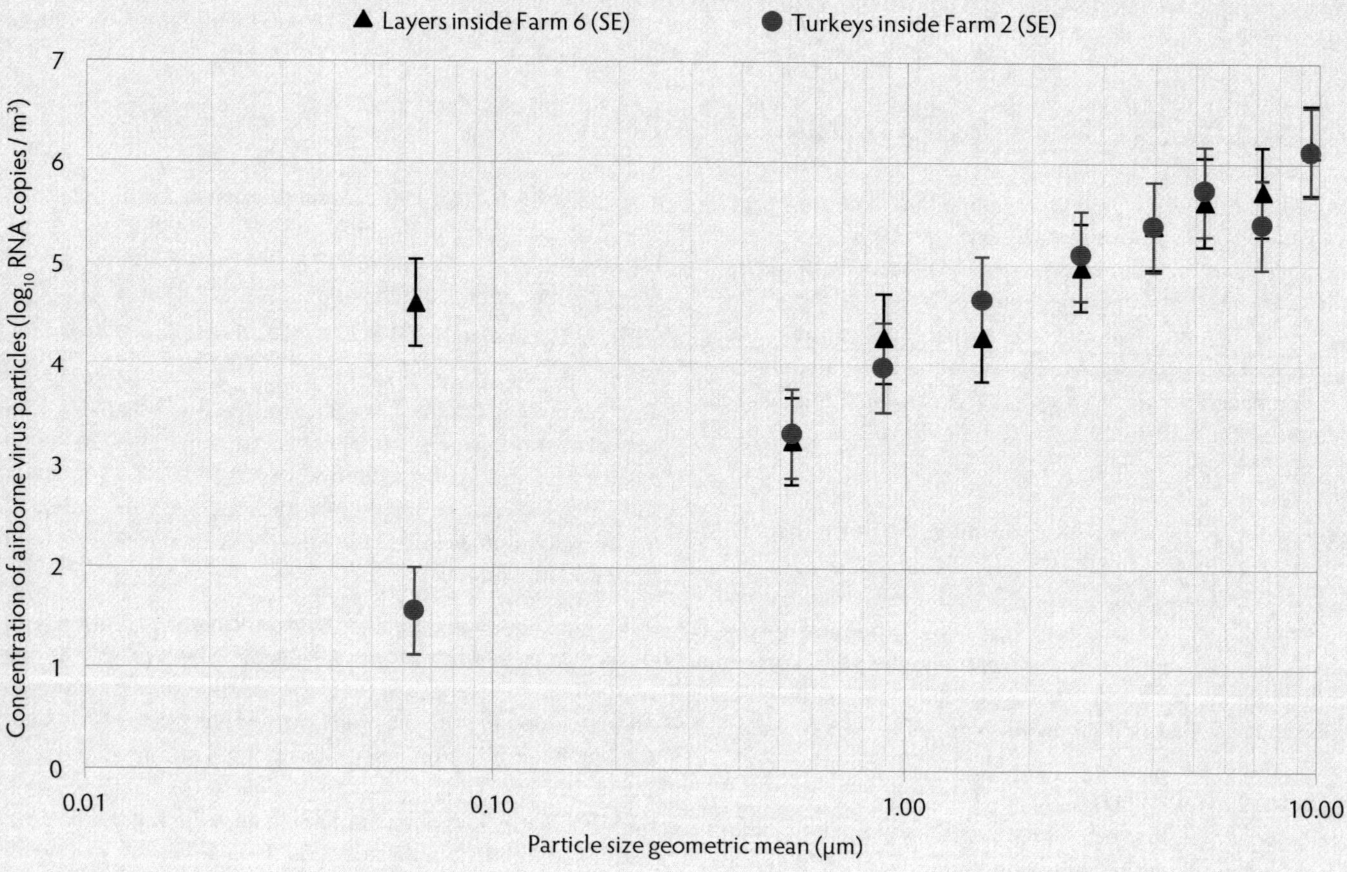

this information is crucial for the preparedness and response of the swine industry against this and other highly devastating pathogens.

Acknowledgements

Project founded by a Coop Agreement with APHIS/USDA.

References

1. https://www.bah.state.mn.us/avian-influenza.

2. Pantin-Jackwood, MJ., Swayne, DE., (2009) Pathogenesis and pathobiology of avian influenza virus infection in birds. Rev - Off Int Epizoot.;28: 113–136.

3. Alonso, C., Goede, D., Morrison, R., Davies, P., Rovira, A., Marthaler, D., et al. (2014) Evidence of infectivity of airborne porcine epidemic diarrhea virus and detection of airborne viral RNA at long distances from infected herds. Veterinary research, doi: 10.1186/s13567-014-0073-z.

4. Bonifait, L., Charlebois, R., Vimont, A., Turgeon, N., Veillette, M., Longtin, Y., et al. (2015) Detection and quantification of airborne Norovirus during outbreaks in healthcare facilities. Clinical infectious diseases.

Particle size association of PRRS and PED viruses in aerosols from acutely infected pigs under field conditions

C. Alonso[1], DVM, MS; B. Olson[2], PhD; P. Raynor[2], PhD; S. Goyal[1], PhD; P. Davies[1], BVSc, PhD; M. Torremorell[1], DVM, PhD

[1]Department of Veterinary Population Medicine, College of Veterinary Medicine, University of Minnesota, Saint Paul, Minnesota; [2]Division of Environmental Health Sciences, School of Public Health, University of Minnesota, Saint Paul, Minnesota

Introduction

Diseases that spread via aerosols are the most costly and difficult to control.[1] Porcine reproductive and respiratory syndrome virus (PRRSV) and porcine epidemic diarrhea virus (PEDV) are airborne pathogens able to spread rapidly and cause devastating financial losses. Despite this evidence, there is limited field information on the size of particles with which airborne viruses are associated while airborne. This association will determine the virus contribution to the airborne route to explain the in between-farm transmission, as well as the survivability and infectivity of the viruses in aerosols.[2] Thus, the objective of this study was to characterize the size distribution of the particles that transport PRRSV and PEDV generated by acutely infected pigs under field conditions, using 2 size selective air collectors, inside and outside swine facilities.

Materials and methods

Farms in Minnesota experiencing active outbreaks of PRRSV or PEDV were enrolled in the study during the winter of 2014-2015. Two and 3 replicates of air samples/ room were collected for 1 hour using an 8 staged Andersen cascade impactor (ACI, 28.3 lpm) and for 15 min using a 5 staged high volume Tish cascade impactor (TSI, 1,100 lpm), respectively. Both air samplers, previously validated under experimental conditions, are capable to collect particles as function of particle size ranging from 0.01 to 10 microns. Samples were collected inside and at 5 meters outside the swine facilities. Temperature, relative humidity (%) as well as oral fluids, were also collected for diagnostic confirmation at the time of air sampling. Additionally, an optical particle counter was used to analyze total airborne particles during the sampling periods. Samples were analyzed using quantitative PRRSV and PEDV RT-PCR. Concentrations of RNA copies/m^3 were analyzed in the log scale. Log of total quantity of virus RNA per volume of air, and total particle counts by size and type of virus were analyzed for the study.

Results

A total of 4 nursery farms participated in the study. Positive status was confirmed by oral fluids and a total of 210 air samples were analyzed by quantitative RT-PCR. PEDV farms had no clinical signs at the time of sampling; however PEDV was detected in the air and it was in higher quantities than PRRSV (Figure 1). PEDV was detected in all particle sizes measured inside (ACI air sampler) and outside the sampling locations (TIS air sampler). Inside PEDV was detected in quantities ranging from 2.8 logs (particles between 0.7-1.1μm) to 6.25 logs (particles ranging between 4.7-5.8μm) (Figure 2). PRRSV, was detected in all size ranges, except for particles < 2.1μm (limit of qPCR detection was added to the calculation) in quantities ranging from 2.4 logs (particles between 2.1-3.3μm) to 5.4 logs (particles > 9.0μm) (Figure 2). All samples collected outside with the TIS air sampler were negative for PRRSV.

Conclusions and discussion

Results from this study indicate that PEDV and PRRSV can be found in a wide range of particles sizes while airborne under field conditions, however, PRRSV seems to be essentially associated to larger particles. PRRSV airborne particles were not detected outside the swine facilities. This information is important to further our understanding of airborne transmission and to evaluate the efficacy and impact of biosecurity measures such as air filtration on the risk of pathogen introduction in filtered farms.

Acknowledgements

This project was founded by the University of Minnesota Signature Research Program.

References

1. Hyslop NS (1971) J Am Vet Med Assoc 159(11):1500-7.
2. Zuo et al. (2013) *Aerosol Science and Technology*, 47:373–382.

Figure 1: Virus concentration in air samples collected inside and outside swine facilities per virus type (viral copies/m³ of air).

Total virus concentration in air samples by virus type

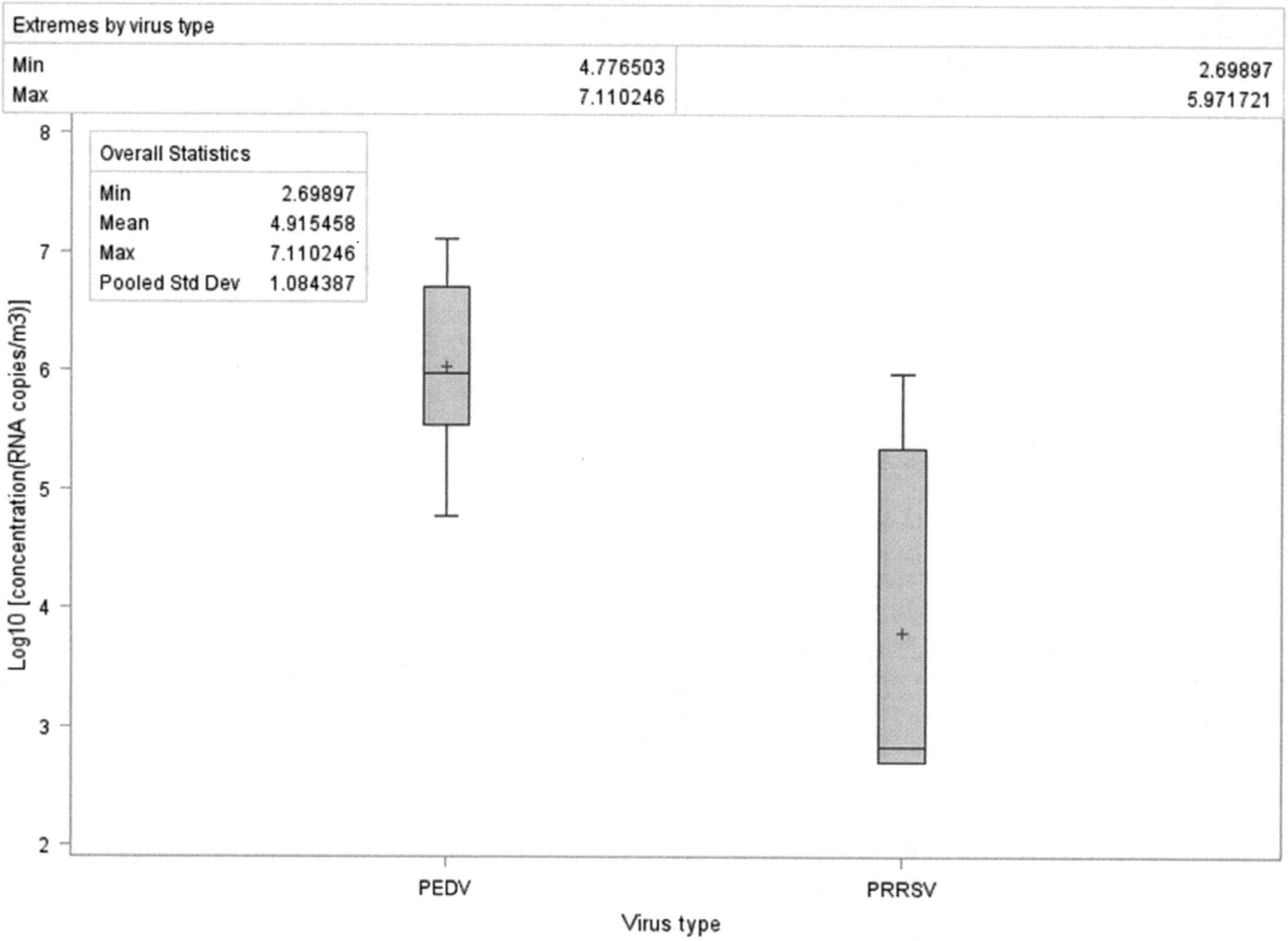

Figure 2: Particle size distributions (viral copies/m³ of air) for PRRSV and PEDV detected by the Andersen cascade impactor (ACI) sampler inside the study farms.

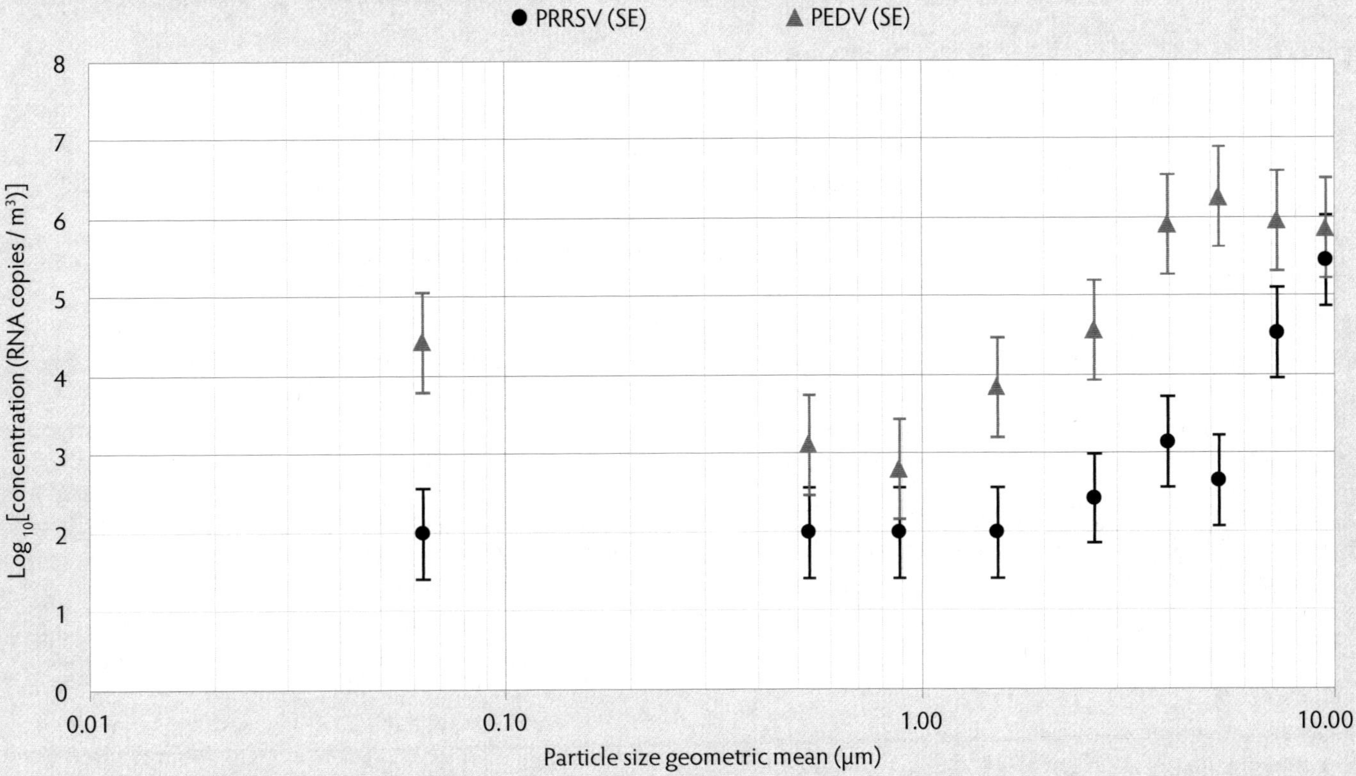

Porcine circovirus (PCV) as delivery virus vector to express porcine reproductive and respiratory syndrome virus (PRRSV) epitopes

P. E. Piñeyro[1], DVM, MVSc, DVSc, PhD; S. P. Kenney[1], PhD; L. G. Giménez-Lirola[2], PhD; T. Opriessnig[2,3], DVM, PhD; S. R. Matzinger[1], PhD; D. Tian[1]; C. L. Heffron[1], BS; X. J. Meng[1], MD, PhD
[1]Virginia Tech University, Blacksburg, Virginia; [2]Iowa State University, Ames, Iowa;
[3]The Royal (Dick) School of Veterinary Studies, United Kingdom

Introduction

Two types of porcine circovirus (PCV), with similar genomic organization, have been identified thus far. The non-pathogenic PCV1 has a low prevalence in swine herds, while the PCV2 is highly prevalent in most swine-producing countries and is associated with clinical porcine circovirus associated disease (PCVAD). Porcine reproductive and respiratory syndrome (PRRS) is caused by PRRS virus (PRRSV). PCV2 and PRRSV is the most common co-infection observed in the field, causing immense economic losses to the pork industry worldwide.

Objectives

The specific aims of this study were 1) to generate chimeric viruses expressing neutralizing epitopes of PRRSV *in vitro* using the backbone of the non-pathogenic PCV1 or PCV1-2a vaccine strain ($_{vs}$) and 2) to evaluate infectivity and immunogenicity of PCV1-PRRSV$_{EPI}$ PCV1-2a-PRRSV$_{EPI}$ chimeric viruses *in vivo.*

Methods

Four different B-cell linear epitopes derived from PRRSV strain VR2385, including GP2 II (aa 40-51), GP3 I (aa 61-72), GP5 I (aa 35-46), and GP5 IV (aa 187-200), were cloned individually in frame into the C-terminus of the PCV1 and PCV1-2a capsid gene. The infectivity and co-expression of the PCV1 and PCV1-2a-capsid protein and PRRSV epitopes were evaluated *in vitro* by IFA. The infectivity and immunogenicity was further evaluated *in vivo* in two independent experiments by inoculation of a total of 42 specific-pathogen-free (SPF) pigs. **Experiment 1**: 21pigs, randomly assigned into seven groups of three pigs each, including two positive control groups (PCV1 and PRRSV), a negative control (MEM-treated group), and four groups, one for each of the PCV1-PRRSV$_{EPI}$ chimeric viruses. **Experiment 2**: performed in the same fashion as PCV1-PRRSV$_{EPI}$ chimeric but using PCV1-2a-PRRSV$_{EPI}$ chimeric viruses. Pig serum samples were collected prior to inoculation and weekly thereafter for a period of 7 weeks. The serum and tissue viral DNA loads were tested and quantified by Taqman® qPCR. Anti-PCV1 IgG titers were evaluated by IFA, and specific anti-PCV2 and anti-PRRSV$_{EPI}$ IgG antibodies were evaluated by ELISA. Serum virus neutralization assay was performed to evaluate the neutralizing activity against PRRSV-VR2385

Results

Four PCV1-PRRSV$_{EPI}$ and four PCV1-2a-PRRSV$_{EPI}$ showed to be infectious *in vitro* and co-expressed PCV1cap and PCV2cap respectively as well as the respective PRRSV epitopes. Animal studies in the PCV1-PRRSV$_{EPI}$ chimeric experiment showed that three chimeric viruses produced viremia and replicated in lung and tracheobronchial lymph nodes; while in the PCV1-2a-PRRSV$_{EPI}$ chimeric experiment, two chimeric viruses showed viremia and were detected in tissues. Anti-PCV1 IgG antibodies were detected in PCV1 wild type and PCV1-PRRSV$_{EPI}$ chimeric viruses by dpi 14 and remained seropositives throughout the study (Figure 1). Anti-PCV2 IgG antibodies were detected in PCV1-2a control as well as PCV1-2a-PRRSV$_{EPI}$ chimeric viruses by dpi 28 and remained seropositives throughout the study (Figure 2). Anti-PRRSV-VR2385 neutralizing antibodies were detected in the PCV1-PRRSV$_{EPI}$GP3I, PCV1-PRRSV$_{EPI}$GP5I, and PCV1-PRRSV$_{EPI}$GP5IV chimeric viruses-infected groups at 28 dpi (Figure 3). Two of the four of the PCV1-2 chimeric viruses, PCV1-2a-PRRSV$_{EPI}$GP3IG and PCV1-2a-PRRSVEPIG-P5IV, elicited neutralizing antibodies against PRRSV VR2385 as well as PCV2 (straPCV2a, PCV2b, and mPCV2b) (Figure 4).

Discussion

These results provided a proof of concept for further exploring the use of the non-pathogenic PCV1as delivery system to generate immunity against PRRSV or PCV1-2a as a platform for the development of a bivalent modified live vaccine.

References

1. Piñeyro PE, Kenney SP, Giménez-Lirola LG, Opriessnig T, Tian D, Heffron CL, Meng XJ. Evaluation of the use of non-pathogenic porcine circovirus type 1 as a vaccine delivery virus vector to express antigenic epitopes of porcine reproductive and respiratory syndrome virus. Virus Res. 2015.

2. Piñeyro PE, Kenney SP, Giménez-Lirola LG, Heffron CL, Matzinger SR, Opriessnig T, Meng XJ. Expression of antigenic epitopes of porcine reproductive and respiratory syndrome virus (PRRSV) in a modified live-attenuated porcine circovirus type 2 (PCV2) vaccine virus (PCV1-2a) as a potential bivalent vaccine against both PCV2 and PRRSV. Virus Res. 2015.

Figure 1: Anti-PCV1 IgG antibodies and anti-PRRSV N antibodies in specific-pathogen-free (SPF) pigs experimentally infected with PCV1-PRRSV$_{EPI}$ chimeric viruses.

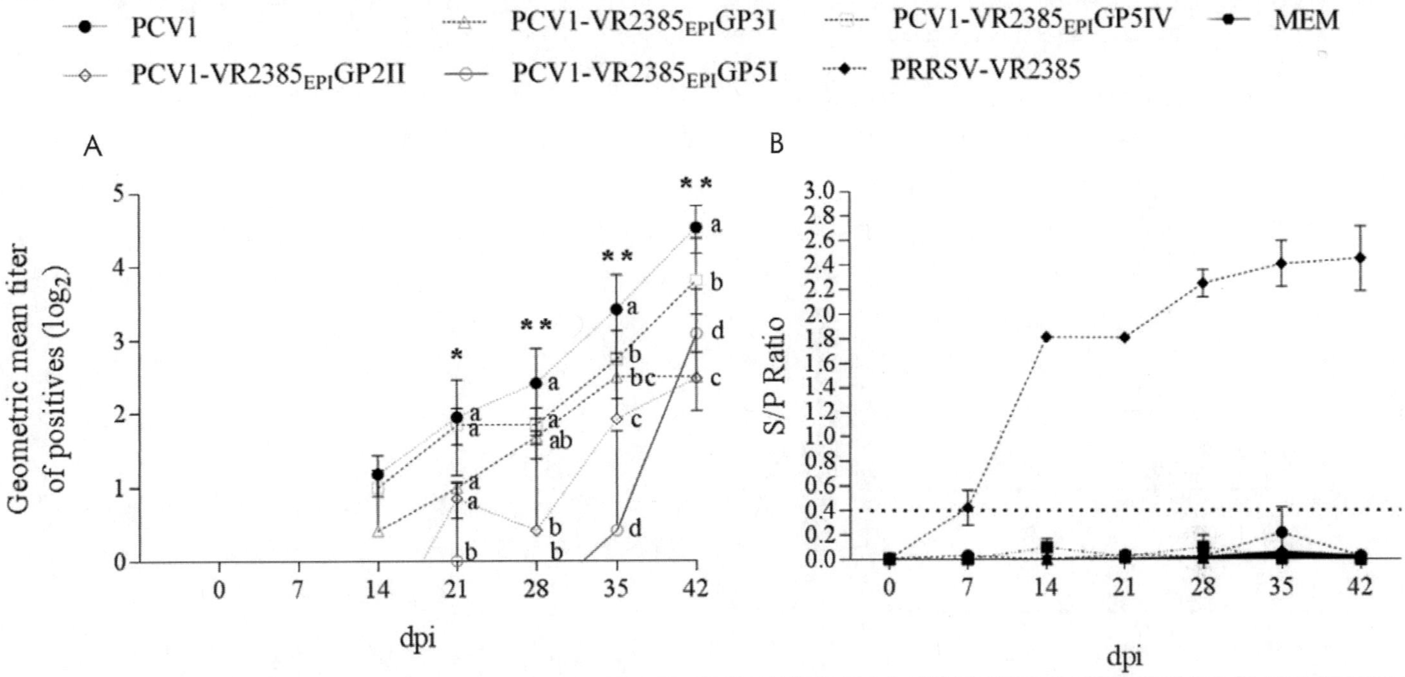

Figure 2: Anti-PCV2 IgG antibodies and anti-PRRSV N antibodies in specific-pathogen-free (SPF) pigs experimentally infected with PCV1-2a-PRRSV$_{EPI}$ chimeric viruses.

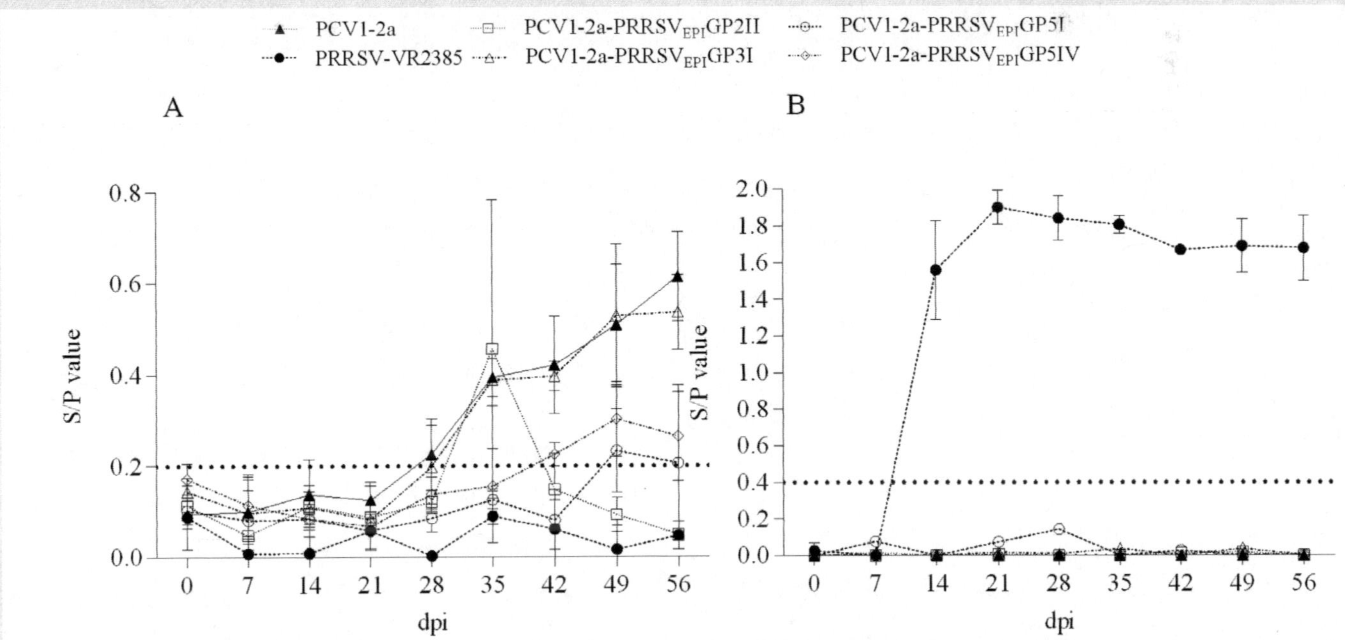

Figure 3: Kinetics of anti-PRRSV neutralizing antibody response in pigs experimentally infected with each of the four PCV1-VR2385$_{EPI}$ chimeric viruses as well as with the PRRSV VR2385 virus.

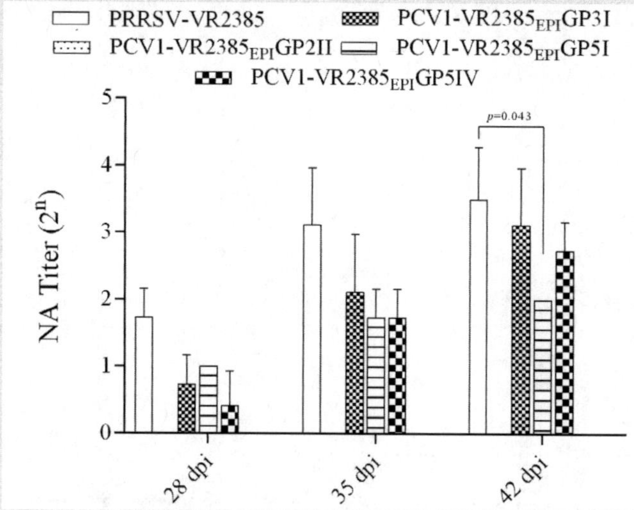

Figure 4: Kinetics of anti-PCV2 and anti-PRRSV neutralizing antibody responses in pigs experimentally infected with each of the four PCV1-2a-PRRSV$_{EPI}$ chimeric viruses as well as with the PRRSV VR2385 virus.

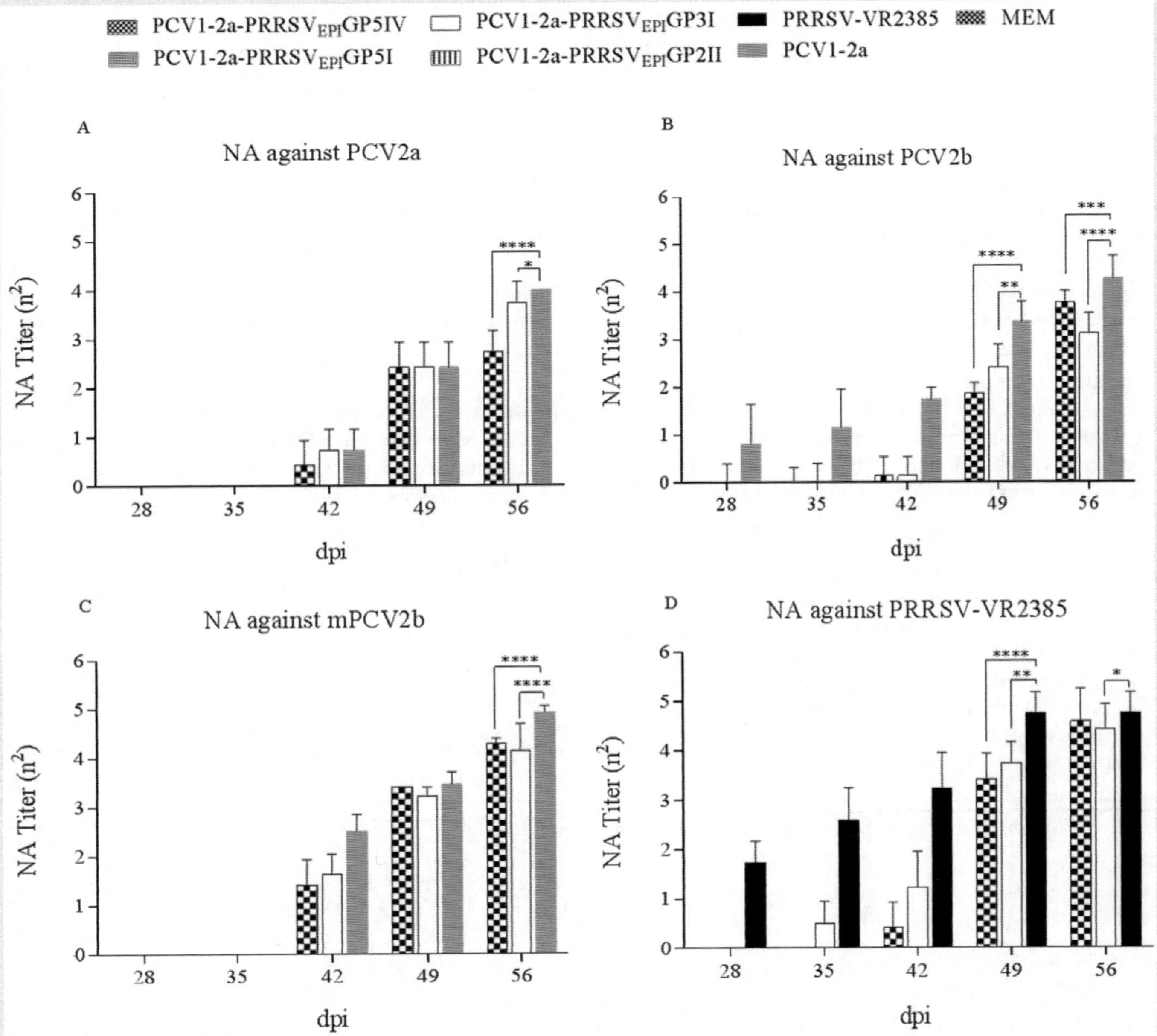

Developing sampling guidelines for oral fluid-based PRRSV surveillance

M. Rotolo; L. Gimenez-Lirola; Y. Sun; S. Bade; C. Wang; D. Baum;
P. Gauger; M. Hoogland; R. Main; J. Zimmerman
Iowa State University, Ames, Iowa

Introduction

Oral fluids (OF) are a convenient surveillance sample because they are easily collected and can be tested for nucleic acids and/or antibodies for PRRSV and a variety of pathogens. We are currently developing statistically-based guidelines for sample size, frequency, and location.

Materials and methods

Two studies were conducted to map the spatial and temporal aspects of PRRSV infection and further the development of oral fluid sampling guidelines. Study 1 - In one WTF barn on each of 10 production sites, OF samples were collected from 6 equidistant pens (~25 pigs per pen) every 2 weeks for 18 weeks. This provided a total of 600 OF samples. Study 2 - In 3 wean-to-finish barns on one finishing site, OF samples were collected weekly from every occupied pen (108 pens; ~25 pigs per pen) for 8 weeks. This provided a total of 972 OF samples. OF samples were completely randomized and then tested for PRRSV RNA, IgG, and IgA. To date, statistical analyses have been done to examine spatial autocorrelation, compare detection based on systematic spatial vs random sampling, and compare sampling from the same pens vs alternate pens at each time point. Additional analyses are currently in progress.

Results

Analyses showed that the disease status of a pen in a barn was highly influenced by the disease status of other pens in the same barn, i.e., the presence of ≥ 1 positive pens increased the probability of detecting another positive pen (Table 1). For most swine veterinarians, this is a statement of the obvious, but spatial autocorrelation has previously not been quantified at the barn level and has important implications for surveillance.

Analysis has also shown that systematic spatial sampling was as good as or better than random sampling. Sampling the same pens at each time point was more effective than changing the pens sampled at each time point.

Discussion

The results of the analyses performed to date show that systematic spatial sampling is a viable approach for routine surveillance. Sample size calculations are in progress, but frequency of sampling is more important that sample size. That is, fewer samples collected at regular 2-week intervals are more useful than more samples collected at long intervals. Overall, the results suggest that a simple, but reliable, oral fluid-based sampling strategy can be developed.

Table 1: Probability for future pen PRRSV status based on current pen status (ELISA)

	No. of positive pens in the barn (among 6 pens sampled)					
	0	1	2	3	4	5
Probability that a positive pen will remain positive at the next sampling	0.47	0.58	0.68	0.77	0.84	0.89
Probability that a negative pen will test positive at the next sampling	0.25	0.34	0.45	0.55	0.66	0.75

Table 2: Probability for future pen PRRSV status based on current pen status (PCR)

	No. of positive pens in the barn (among 6 pens sampled)					
	0	1	2	3	4	5
Probability that a positive pen will remain positive at the next sampling	0.12	0.21	0.33	0.49	0.65	0.78
Probability that a negative pen will test positive at the next sampling	0.04	0.07	0.12	0.21	0.34	0.5

Evaluation of the pathogenesis differences of the US PEDV prototype and S-INDEL-variant strains and examination of the cross-protective immunity of two strains in weaned pigs

Qi Chen, DVM, MS; Molly Stafne, BS; Melisa Spadaro, DVM; Holly Salzbrenner, BS; Phillip Gauger, DVM, PhD; Michael Welch; Joseph Thomas, BS; Paulo Arruda, DVM, PhD; Luis Gimenez-Lirola, PhD; Drew Magstadt, DVM, MS; Darin Madson, DVM, PhD, DACVP; Jianqiang Zhang MD, PhD

Department of Veterinary Diagnostic and Production Animal Medicine, Iowa State University, Ames, Iowa

Introduction

Porcine epidemic diarrhea virus (PEDV) was detected in the United States (US) for the first time in April 2013.[1] Currently at least two genetically different PEDV strains have been identified and co-circulate in US swine (US PEDV prototype strain and S-INDEL-variant strain).[2-4] In a previous study, we have experimentally confirmed that the US S-INDEL-variant strain is less pathogenic than the US prototype strain in 5-day-old piglets.[5] However, pathogenicity of PEDV can be age-dependent.[6-7] The objectives of the current study were to 1) evaluate the pathogenesis differences of two US PEDV strains in weaned pigs; and 2) examine the cross-protection efficacy between two strains in weaned pigs.

Materials and methods

Eighty-five 3-week-old pigs were purchased from a conventional breeding farm and delivered to the Iowa State University Laboratory Animal Resources facilities. All pigs were intramuscularly injected with a dose of Excede® upon arrival and confirmed negative for PEDV, PDCoV, and TGEV by virus-specific PCRs on rectal swabs and negative for PEDV antibody by a virus-specific indirect fluorescent antibody (IFA) assay on serum samples at the ISU VDL. Pigs were blocked by weight and then randomly divided into 7 groups with 15 or 10 pigs per group (Table 1). Pigs were orogastrically inoculated with virus-negative culture medium (N), PEDV prototype isolate USA/IN19338/2013 (P-prototype), or PEDV S-INDEL-variant isolate USA/IL20697/2014 (V-variant) at Day 0 (D0) followed by challenge at D28. Inoculation at each point used 10ml of inoculum per pig, containing 10^4 $TCID_{50}$/ml for virus inocula. Seven groups were designated according to 1st inoculation/2nd challenge: P/V (15 pigs), V/V (15 pigs), N/V (15 pigs), P/P (10 pigs), V/P (10 pigs), N/P (10 pigs), N/N (10 pigs). At D4, 5 pigs from the P/V, V/V, and N/V groups were necropsied for comparison of gross and microscopic lesions between two PEDV strains. Five pigs from all 7 groups were necropsied 6 days after the 2nd challenge (D34) to evaluate cross-protection efficacy in term of gross and microscopic lesions. The remaining 5 pigs per group were kept until 4 weeks after 2nd challenge (D56) to evaluate post-challenge antibody response.

Clinical observations were recorded. Rectal swabs were collected at D0, 2, 4, 7, 10, 14, 21, 28, 30, 32, 34, 38, 42, 49 and 56 and

Table 1: Experimental design

Group	Day 0 (Inoculation)	D4 (Necropsy)	D28 (Challenge)	D34 (Necropsy)	D56 (Necropsy)
G1 P/V (n = 15)	Prototype	n = 5	Variant	n = 5	n = 5
G2 V/V (n = 15)	Variant	n = 5	Variant	n = 5	n = 5
G3 N/V (n = 15)	Neg	n = 5	Variant	n = 5	n = 5
G4 P/P (n = 10)	Prototype	n = 0	Prototype	n = 5	n = 5
G5 V/P (n = 10)	Variant	n = 0	Prototype	n = 5	n = 5
G6 N/P (n = 10)	Neg	n = 0	Prototype	n = 5	n = 5
G7 N/N (n = 10)	Neg	n = 0	Neg	n = 5	n = 5

Notes:
P=Prototype PEDV isolate USA/IN19338/2013; V=Variant PEDV isolate USA/IL20697/2014; N=Neg Control.
Inocula: 10^4 $TCID_{50}$/ml, 10 ml per pig for either the Prototype or the Variant isolates; or 10 ml culture medium per pig for Neg Control.

tested by a quantitative PEDV N gene-based real-time RT-PCR. Serum samples were collected at D0, 7, 14, 21, 28, 35, 42, 49 and 56 and tested by indirect fluorescent antibody (IFA) assay and virus neutralization (VN) assay using two PEDV strains as indicator viruses, respectively. Small intestines were collected for histopathology evaluation and immunohistochemistry (IHC) staining.

Results and conclusion

Following the 1st inoculation of 3-week-old pigs, the PEDV prototype strain-inoculated groups shed higher amount of virus in feces than the S-INDEL-variant strain-inoculated groups during D0-7 (Figure 1). Fecal shedding in the prototype strain-inoculated groups reached the peak level at D4 and then gradually declined. In contrast, Fecal shedding in the variant strain-inoculated groups gradually increased in the first a few days and reached the peak level at D7 (Figure 1). For IHC staining on D4, prototype strain-inoculated pigs had respective average IHC scores of 4 and 4 in distal jejunum and ileum, which were significantly higher than the variant strain-inoculated pigs that had average IHC scores of 1.4 and 1.4 in distal jejunum and ileum, respectively. Interestingly, after the 2nd challenge at D28 (pigs were 7-week-old at that time), the N/V group shed more viruses in feces than the N/P group during D32-38 (4-10 days after the 2nd challenge, Figure 1). At D34 (6 days post the 2nd challenge), N/V group had higher average IHC scores in distal jejunum and ileum than N/P group.

After the 1st inoculation, all pigs inoculated with the prototype strain (P/P and P/V groups) or S-INDEL-variant strain (V/V and V/P group) developed IFA and VN antibodies starting from D7-14 regardless of which virus strain was used as the indicator virus. PEDV antibody titers in these four groups increased slightly after the 2nd challenge and then were maintained throughout the end of the study (D56). The N/P and N/V groups were PEDV antibody negative until D42 (14 days post the 2nd challenge) when IFA and VN antibodies became detectable and were maintained through D56. The N/N group remained PEDV antibody negative from D0-56. After the 2nd challenge at D28, fecal viral shedding in the P/P, V/P, V/V, and P/V groups were all less than detected in the N/V and N/P groups (Figure 1).

In summary, it appears that US PEDV prototype strain achieved higher levels of fecal virus shedding than S-INDEL-variant strain in 3-week-old pigs but the opposite was observed in 7-week-old pigs. However, this observation needs to be further confirmed. US PEDV prototype strain provided protection against challenge with both homologous and heterologous strains in weaned pigs; US PEDV S-INDEL-variant strain provided protection against homologous strain challenge and at least partial protection against heterologous (prototype) strain challenge in weaned pigs.

Figure 1: Virus shedding in rectal swabs of various inoculation groups during D0-D56. (A) Mean Ct values. (B) Mean Log_{10}(Genomic Copies/ml).

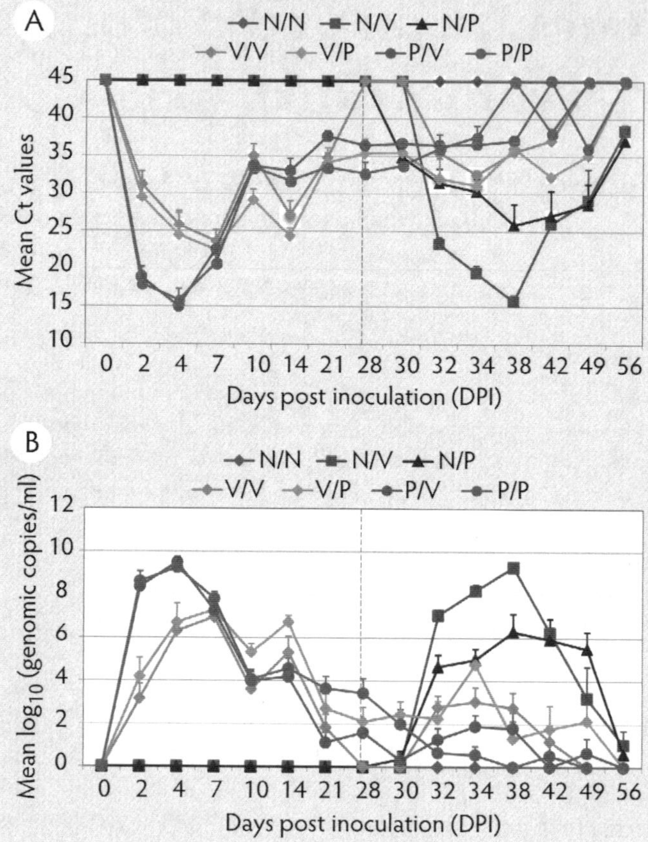

Acknowledgement

This study was supported by the Iowa Pork Producers Association. We are grateful to the Iowa State University Veterinary Diagnostic Laboratory faculty and staff for assistance with some testing. We also thank the Iowa State University LAR staff for animal care.

References

1. Stevenson GW, Hoang H, Schwartz KJ, Burrough EB, Sun D, Madson D, Cooper VL, Pillatzki A, Gauger P, Schmitt BJ, Koster LG, Killian ML, and Yoon KJ. (2013). Emergence of porcine epidemic diarrhea virus in the United States: clinical signs, lesions, and viral genomic sequences. *Journal of Veterinary Diagnostic Investigation* 25(5): 649-654.

2. Chen Q, Li G, Stasko J, Thomas JT, Stensland WR, Pillatzki AE, Gauger PC, Schwartz KJ, Madson D, Yoon KJ, Stevenson GW, Burrough ER, Harmon KM, Main RG, and Zhang J. (2014). Isolation and characterization of porcine epidemic diarrhea viruses associated with the 2013 disease outbreak among swine in the United States. *Journal of Clinical Microbiology* 52(1), 234-243.

3. Wang L, Byrum B, and Zhang Y. (2014). New variant of porcine epidemic diarrhea virus, United States, 2014. *Emerging Infectious Diseases* 20(5), 917-919.

4. Vlasova A, Marthaler D, Wang Q, Culhane M R, Rossow KD, Rovira A, Collins J & Saif LJ. (2014). Distinct characteristics and complex evolution of PEDV strains, North America, May 2013-February 2014. *Emerging Infectious Diseases* 20(10), 1620-1628.

5. Chen Q, Gauger PC, Stafne MR, Thomas JT, Madson DM, Huang H, Zheng Y, Li G, and Zhang J. Pathogenesis comparison between the United States porcine epidemic diarrhea virus prototype and S-INDEL-variant strains in conventional neonatal piglets. *Journal of General Virology* (submitted)

6. Thomas JT, Chen Q, Gauger PC, Gimenez-Lirola LG, Sinha A, Harmon KM, Madson DM, Burrough ER, Magstadt DR, Salzbrenner H, Welch MW, Yoon KJ, Zimmerman JJ, and Zhang J. (2015). Effect of porcine epidemic diarrhea virus infectious doses on infection outcomes in naïve neonatal and weaned pigs. *PLoS ONE* 10(10):e0139266.

7. Jung K, Annamalai T, Lu Z & Saif LJ. (2015). Comparative pathogenesis of US porcine epidemic diarrhea virus (PEDV) strain PC21A in conventional 9-day-old nursing piglets vs. 26-day-old weaned pigs. *Veterinary Microbiology* 178, 31-40.

An evaluation of porcine epidemic diarrhea virus survival in individual feed ingredients in the presence or absence of a liquid antimicrobial

Scott Dee[1]; Casey Neill[1]; Travis Clement[2]; Aaron Singrey[2]; Jane Christopher-Hennings[2]; Eric Nelson[2]
[1]Pipestone Applied Research, Pipestone Veterinary Services, Pipestone, Minnesota;
[2]Animal Disease Research and Diagnostic Laboratory, South Dakota State University, Brookings, South Dakota

Introduction

Porcine epidemic diarrhea virus (PEDV) is an enveloped single-stranded positive sense RNA virus belonging to the Order *Nidovirales*, the family *Coronaviridae* and the genus *Alphacoronavirus*.[1] Following detection in the US swine population during May, 2013 the virus spread rapidly across the country.[2] Initial risk factors proposed for the spread of PEDV between herds included infected pigs, contaminated transport and aerosols.[3-5] PEDV infection of nursery swine results in reduced performance and fecal shedding for out to 24 day post-infection (DPI).[3] In regards to the role of contaminated transport vehicles, data collected from the US suggests that collection points, such as harvest facilities and livestock auction markets are contaminated and can be a source of contamination of transport vehicles that return to pig farms.[4] Finally, PEDV RNA has been detected by PCR in aerosol samples out to 16 km from infected swine facilities and under experimental conditions may contain viable PEDV.[5] However, while pig-to-pig transmission of PEDV has been proven,[3] controlled transmission studies providing proof of concept regarding the role of aerosols and transport have not been published at this time. In 2014, new risk factors for PEDV were identified: contaminated feed and feed ingredients. Initial reports indicated the ability of PEDV to survive in dry feed for 7 days and in wet feed for 28 days when stored at room temperature.[6] Proof of concept that contaminated complete feed could serve as a route of PEDV transmission to naïve pigs was published,[7] with a follow-up report identifying spray-dried porcine plasma as a possible ingredient-specific risk.[8] Both of these studies collected contaminated feed material from farms or feed suppliers and performed controlled challenge studies involving naïve pigs. In both cases, use of swine bioassay demonstrated infection in naïve piglets following ingestion of contaminated feed bin material[7] or commercial plasma,[8] therefore raising awareness of these risks. Recently, transmission of PEDV via ingestion of contaminated complete feed has been repeated and the minimum infectious dose calculated at $5.6 \times 10^1 TCID_{50}/mL$ (Ct = 37).[9]

In regards to management strategies to reduce the risk of PEDV spread between herds, protocols of transport sanitation and air filtration have been validated using standard approaches [Dee S, unpublished data, 2013-2015]; however, the means to biosecure feed is a new paradigm for the swine industry. Recently, a liquid antimicrobial (LA) product containing formaldehyde and propionic acid has been proposed as a means to mitigate risk (SalCURB®, Kemin Industries, Des Moines, IA, USA).[10] This study demonstrated the ability to prevent PEDV infection in pigs consuming contaminated complete feed treated with LA.[10] However, information is not currently available regarding PEDV survival in other ingredients in swine diets or whether LA treatment can inactivate PEDV at the ingredient level. Therefore, a study was conducted to measure PEDV viability across a panel of ingredients commonly encountered in swine diets in the presence or absence of a LA presently used to maintain feed and feed ingredients as *Salmonella*-negative for up to 21 days. It was hypothesized that while PEDV survival is ingredient-specific, LA would be efficacious, independent of ingredient type.

Methods

A detailed summary of the Materials and Methods are described in Dee SA, Neill C, Clement T, Singrey A, Christopher-Hennings J and Nelson E. An evaluation of PEDV survival in individual feed ingredients in the presence or absence of a liquid antimicrobial. *Porcine Health Management* 2015,1:9.

Results

All feed ingredient samples were PCR negative on day 0 of the study. Successful PEDV inoculation was confirmed, as all day 1 samples were PCR-positive. The mean Ct of treated samples was 23.74 (SD = 5.9) on 1 DPI and 37.23 (SD = 2.5) on 30 DPI. Only SBM (Ct = 28.22) and meat and bone meal (Ct = 33.21) remained PCR positive at 30 days PI, all other treated samples had Ct values of ≥ 38 indicating PEDV RNA was not detected, including the treated positive control complete feed samples. When analyzed by t-test, the difference in mean Ct of treated ingredients was significant at $P < 0.0001$. In contrast, mean Ct values across non-treated ingredients on day 1 averaged 23.46 (SD = 5.5) and 20.78 (SD = 3.8) on day 30. When analyzed by t-test, this difference was not significant at $P = 0.1143$. Finally, when analyzed by t-test, the mean Ct of 30 DPI treated samples (Ct = 37.23) was significantly different ($P < 0.0001$) than the mean Ct of 30 DPI non-treated samples (20.78).

Viable PEDV was detected on 1 DPI and all sampling days (7, 14 and 30 DPI) from non-treated SBM, DDGS, RBCs, meat and bone meal, lysine HCL and D/L methionine. Select samples of non-treated choice white grease (1 and 7 DPI), threonine (1, 7 and 14 DPI) and limestone (1 and 7 DPI) were also VI positive. In contrast, all non-treated samples of corn, the 3 VTMs, spray-dried plasma, purified plasma, intestinal mucosa and dry choline chloride were VI negative on all 4 sampling days. Both the complete feed positive control samples and the virus stock controls were VI positive on all sampling days. All complete feed negative control samples and all LA treated ingredients were VI negative across all sampling points.

Samples selected for swine bioassay testing consisted of non-treated ingredients which were PCR positive/VI negative on 7, 14, and 30 DPI. This included corn, all 3 VTMs, intestinal mucosa, soy oil, choline chloride, spray-dried plasma, purified plasma and SalCURB® -treated ingredients. In addition, ingredients which VI-positive on 7 DPI but VI negative on 14 and 30 DPI (choice white grease and limestone) and threonine (VI negative 30 DPI) were tested as well. Following completion of the bioassay, viable PEDV was only detected in piglets in the choline unit and the choice white grease unit. Affected animals displayed evidence of mild diarrhea, shed PEDV in feces and samples of small intestine were PCR-positive at necropsy. All other piglets inoculated with the aforementioned feed ingredients, LA treated ingredients and the negative control piglets remained healthy and all rectal swabs and intestinal tract samples were negative by PCR.

Based on the magnitude of the SBM titers at 30 DPI (mean = 15,360 FFU/mL), it was decided to continue to store these non-treated samples as described and test at 30 days intervals for an additional 180 days. At 180 DPI, mean Ct and mean FFU/mL across the 2 samples of SBM were 20.11 and 90 FFU/mL, respectively. Throughout this period of time, the mean temperature was -0.13°C (range = -25°C to 22° C, 95% CI = -1.83 to 1.57°C, median = 0.00°C and SD = 11.7). Samples tested at 210 DPI were PCR-positive (21.84) but VI-negative. Differences in the mean temperature of sampling periods 1- 90 DPI were significantly different ($P < 0.0001$) than those recorded during days 91-120 DPI versus 121-210 DPI . As a control, the complete feed positive control was conducted and samples remained VI positive out to day 45 DPI (Ct 24.50, 160 FFU/mL).

Discussion

The objective of this study was to measure PEDV viability across feed ingredients commonly encountered in swine diets in the presence or absence of a liquid antimicrobial treatment. Under the conditions of this study, our results suggest that PEDV survival is ingredient-specific, as 50% of the non-treated ingredients harbored viable PEDV over some period of time while the remaining non-treated ingredients were VI and bioassay negative at 7, 14 and 30 DPI. The extended survivability of PEDV in SBM was a novel observation and may have been influenced by environmental temperature, based on the observed decrease in titer as temperatures increased over time. In addition, other novel information from this study included the recovery of viable PEDV from DDGS, 3 synthetic amino acids and dry choline chloride (all non-treated). A by-product of the ethanol industry, wet DDGS are stored outside/uncovered while dry DDGS are stored in open warehouses, raising the risk of post-processing contamination. As they are widely used in swine diets and their turnover at mills is short (1-2 days), if contaminated, as under the conditions of our study, virus may remain viable and the ingredient could pose a risk to complete feed during mixing. Similarly, the observed variability in PEDV survival across the 3 amino acids raises the question of whether ingredient chemistry, for example, D/L methionine, a sulfur-containing amino acid, may have influenced titer magnitude and duration. In addition, the detection of viable virus in both synthetic amino acids and choline chloride raises the awareness of both domestic risk and transboundary risk, due to their widespread global trade.

Just as intriguing was the inability of viable PEDV to be recovered from certain ingredients, such as spray-dried porcine plasma, particularly since infection of naïve piglets through the ingestion of PCR-positive porcine plasma has been published.[8] Surprisingly, we were not able to reproduce these results, despite employing two viability assays and purposely inoculating plasma samples with PEDV at a Ct of 16.34 (400,000 FFU/total dose), a viral load far greater than previously published.[8] Even at day 1 PI, we were not able to recover viable PEDV by VI and samples collected at 7, 14 and 30 DPI were negative by bioassay. Follow-up questioning confirmed that the plasma had not been chemically altered at the mill. While negative results were obtained with intestinal mucosa samples and purified plasma, the risk of other by-products, such as meat and bone meal and RBCs continue to raise awareness that certain by-products, if contaminated, may harbor viable PEDV for extended periods.

Similar to plasma, no evidence of viable PEDV was recovered from the 3 different VTMs tested. This is interesting as international shipments of VTMs were originally considered a possible source of PEDV introduction to the US. Finally, in regards to the management of "high risk" ingredients, under the conditions of this study, the formaldehyde-based liquid antimicrobial successfully reduced viral load and inactivated PEDV, independent of ingredient. These data confirm previously published reports of the efficacy of liquid antimicrobial biosecurity of PEDV-contaminated complete feed.[10]

As with all research studies, this project had its share of acknowledged strengths and limitations. Regarding the strengths of the study, we selected 18 ingredients commonly found in commercial swine diets and designed an experiment where the only variables in the study were the individual ingredients, and the presence or absence of the liquid antimicrobial. Specifically, we inoculated equal amounts of each ingredient with an equivalent quantity of virus in an attempt to mimic published viral loads associated with field cases of PEDV in feed. All ingredients were stored in identical containers, exposed to the same environment, sampled over periods representative of mill turnover times and tested in a single laboratory, involving consistent, trained personnel and validated assays for the detection of PEDV. We organized samples in a manner to prevent cross-contamination by insuring that the individual sample storage containers were never opened from the time of PEDV inoculation until testing occurred at the laboratory. While this may have resulted in some variability of the PCR assay since a new sample set was submitted each time, the fact that our negative control samples remained free of contamination throughout the entire project validated these protocols. Finally, we used multiple metrics (PCR, VI and bioassay) to document viral load across a total of 320 samples. At the time of this writing, this is the only such study we are aware of that has used this approach to assess PEDV viability in a large variety of ingredients.

In regards to limitations, while we did involve the use of a large number of samples, we were limited to 18 ingredients and we only conducted 2 replicates per ingredient. Furthermore, these results were derived from gram quantities of feed and may not directly equate to the vast quantity of tonnage used in actual swine production or feed manufacturing facilities. The bioassay protocol, while helpful at identifying ingredients containing low levels of virus presented limitations, was limited by cost thereby requiring pooling across sample days, i.e., 7, 14 and 30 DPI, instead of the testing of daily samples. Despite the need to pool samples, the Ct values across tested inoculums ranged from 22.95-31.06, clearly indicative of a high quantity of PEDV. Finally, the fact that the samples were stored under a specific set of environmental conditions for a defined period does not allow for extrapolation of results across other climates or over longer periods of time, with the exception of the supplementary testing of SBM and complete feed.

In closing, based on the information from this study, a theory to explain how PEDV infection of a swine farm via contaminated feed may actually occur in the field can now be proposed, based on the hypothesis that infection via contaminated feed is a multi-factorial event:

1. Ingredients capable of harboring viable PEDV for an extended period become contaminated.
2. Contaminated ingredients then contact a supportive matrix (SBM) during the preparation of complete feed.
3. Delivery of PEDV-positive feed introduces the virus to the farm.

While further work is needed, including replication under varying conditions of season, differing viral loads, a larger number of ingredient types, and alternative intervention methods, the risk of contaminated feed and feed ingredients as a vehicle for PEDV has now been well established. With this information, veterinarians and feed industry experts can now begin to target interventions to reduce risk. In addition, increased efforts at the mill level, focusing on improved biosecurity of facilities and feed transport vehicles may also reduce risk with the end result being a lowered incidence of feed-related PEDV cases across the national herd.

Finally, this paper provides the first proof of concept data that PEDV can survive for extended periods in select feed ingredients. We hope the new information generated from this project improves the understanding of pathogen risk through feed at both the domestic and international levels, thereby stimulating further research efforts in this critical area.

References

1. Saif LJ, Pensaert MB, Sestak K, Yeo S, Jung K: Coronaviruses. In *Diseases of Swine*. 10th Edition. Edited by Zimmerman JJ, Karriker LA, Ramierez A, Schwartz KJ, Stevenson GW. Ames: Wiley and Sons; 2012:501-524.

2. Chen Q, Ganwu L, Stasko J, Thomas JT, Stensland WR, Pillatzki AE, Gauger PC, Schwartz KJ, Madson D, Yoon KJ, Stevenson GW, Burrough ER, Harmon KM, Main RG, Zhang J: Isolation and characterization of Porcine Epidemic Diarrhea Viruses associated with the 2013 disease outbreak among swine in the United States. *J Clin Microbiol* 2014, 52:234-243.

3. Madson DM, Magstadt DR, Arruda PH, Hoang H, Sun D, Bower LP, Bhandri M, Burrough ER, Gauger PC, Pillatzki AE, Stevenson GW, Wilberts BL, Brodie J, Harmon KM, Wang C, Main RG, Zhang J, Yoon KJ. Pathogenesis of porcine epidemic diarrhea virus isolate (US/Iowa/18984/2013) in 3-week-old weaned pigs. *Vet Microbiol* 2014, 174:60-68.

4. Lowe J, Gauger P, Harmon K, Zhang J, Connor J, Yeske P, Loula T, Levis I, Dufrense L, Main R. Role of transportation in spread of porcine epidemic diarrhea virus infection, United States. *Emerg Infect Dis* 2014, 972-874.

5. Alonso C, Goede DP, Morrison RB, Davies PR, Rovira A, Marthaler DG, Torremorell. Evidence of infectivity of airborne porcine epidemic diarrhea virus and detection of airborne viral RNA at long distances from infected herds. *Vet Res* 2014, 45:73.

6. Goyal S. Interventions to control PEDV in feed and feed ingredients. http://www.pork.org/pedv-2014-research/pedv-feed. (2014). Accessed 23 March 2015.

7. Dee S, Clement T, Schelkopf A, Nerem J, Knudsen D, Hennings J, Nelson E. An evaluation of contaminated complete feed as a vehicle for porcine epidemic diarrhea virus infection of naïve pigs following consumption via natural feeding behavior: Proof of concept. *BMC Vet Res* 2014, 10:176.

8. Pascik J, Berhane Y, Ojkic D, Maxie G, Embury-Hyatt C, Swekla K, Handel K, Fairles J, Alexandersen S. Investigation into the role of potentially contaminated feed as a source of the first-detected outbreaks of porcine epidemic diarrhea in Canada. *Transbound Emerg Dis* 2014, 61:397-410.

9. Schumacher L, Woodworth JC, Zhang J, Gauger PC, Chen Q, Welch M, Salzebrenner H, Thomas J, Main R, Dritz S, Cochrane R, Jones C. Determining the minimum infectious dose of PEDV in a feed matrix. *Proceedings of the 2015 Midwest Animal Science Conference* Des Moines, IA, USA, Abstract 160.

10. Dee S, Neill C, Clement T, Hennings J, Nelson E. An evaluation of a liquid antimicrobial (SalCURB®) for reducing the risk of porcine epidemic diarrhea virus infection of naïve pigs during consumption of contaminated feed. *BMC Vet Res* 2014, 10:220.

Effect of thermal mitigation on porcine epidemic diarrhea virus (PEDV) contaminated feed

Roger A. Cochrane[1], MS; Loni L. Schumacher[2], DVM; Steve S. Dritz[2], DVM, PhD; Jason C. Woodworth[3], PhD; Anne R. Huss[1], PhD; Charles R. Stark[1], PhD; Joel M. DeRouchey[3], PhD; Mike D. Tokach[3], PhD; Robert D. Goodband[3], PhD; Jianfa Bia[4], PhD; Qi Chen[5], MS; Jianqiang Zhang[5], PhD; Phil C. Gauger[5], DVM; Roger G. Main[5], DVM, PhD; Cassandra K. Jones[1], PhD
[1]Grain Sciences and Industry, College of Agriculture, Kansas State University, Manhattan, Kansas; [2]Department of Diagnostic Medicine/Pathobiology, College of Veterinary Medicine, Kansas State University, Manhattan, Kansas; [3]Animal Sciences and Industry, College of Agriculture, Kansas State University, Manhattan, Kansas; [4]Veterinary Diagnostic Lab, College of Veterinary Medicine, Kansas State University, Manhattan, Kansas; [5]Veterinary Diagnostic & Production Animal Medicine, College of Veterinary Medicine, Iowa State University, Ames, Iowa

Introduction

In a review by Nitikanchana (2014), a theoretical temperature × time relationship was proposed to reduce the infectivity of PEDV in complete feeds based on data extrapolated primarily from PEDV environmental survival studies.[1] Typical swine feed pelleting conditioner retention times and temperatures encompass the theoretical temperature × time relationship proposed. We are unaware of any available direct research confirming this time × temperature relationship using conditioner and pellet mills that are present in modern feed manufacturing facilities.

Although it would be highly uncommon to set target pellet mill conditioning temperatures below 68°C, it is possible that feed may be produced below these operating limits. This commonly would occur during start-up of the pellet mill or during a pellet mill die plug, which is often resolved without manual intervention by turning off steam to the conditioner in an effort to unclog the die. This can lead to significant quantities of pelleted feed being produced that has not reached the target conditioning temperature. If PEDV particles were in the feed, the resulting feed conditioned at a lower temperature may still have infectious PEDV particles that may potentially contaminate the pellet cooler, post-pellet feed handling equipment and bins at the feed mill, trucks, and bins, feed lines, and feeders at the farm. Therefore, the objectives of this study were to 1) determine the impact of conditioning time and temperature of pelleted complete feed at two PEDV dosages on PEDV inactivation as measured by RT-PCR and bioassay, and 2) determine the efficacy of increasing pellet mill conditioning temperature on PEDV-survivability as measured by qRT-PCR (quantitative reverse transcriptase PCR) and bioassay.

Materials and methods

The feed used was a corn soybean meal-based swine diet (Table 1) and was manufactured at the Kansas State University O.H. Kruse Feed Technology Innovation Center in Manhattan, Kansas. A subsample of this feed was obtained prior to inoculation and confirmed negative by real-time PCR (RT-PCR) for the presence of PEDV RNA at the Kansas State University Research Park Molecular Diagnostics Development Laboratory in Manhattan, Kansas.

Experiment 1

The low dose (1×10^2 $TCID_{50}$/g) and high dose (1×10^4 $TCID_{50}$/g) concentrations in the feed were used based on the results obtained from previous research. Within each dose, the batches of inoculated feed were processed at 1 of 3 condition times (45, 90, 180 sec) with 1 of 3 processing temperatures (68.3, 79.4, and 90.6°C). Thus, treatments were arranged in a 2 (PEDV dose) × 3 (conditioning time) × 3 (processing temperature) arrangement. Additionally, 3 batches of unprocessed (meal based) feed were used as control treatments. These batches consisted of feed inoculated with virus-free tissue culture medium, feed inoculated with the low dose PEDV and feed inoculated with the high dose PEDV concentration.

Conditioning times were chosen to represent a typical conditioning time, an extended time using a typical conditioner, and an extended time that would require a modified conditioner. The temperature range reflects a typical nursery diet processing temperature on the low end and the maximum achievable temperature under typical commercial production of swine feed on the high end.

For each concentration, aliquots (500 ml) of the PEDV dilutions were uniformly blended into 4.5 kg batches of feed. Feed from each of the conditioning time × temperature combinations (9 batches within each PEDV dose) was processed using a pilot-scale single pass conditioner and pellet mill (Model CL5, CPM, Waterloo, IA). Prior to pelleting the first batch, non-inoculated feed was processed until the exit temperature of the feed was stable at the target temperature. In between each inoculated batch, a minimum of 5 kg of virus free feed was processed. The objective of this was twofold. The first objective was to provide a flush of the equipment and prevent virus carry over. The second was to stabilize the pelleting temperature so the contaminated feed

Table 1: Diet composition used in Exp. 1 and 2

Item	Negative control
Ingredient, %	
Corn	79.30
Soybean meal, 46.5 CP	15.70
Choice white grease	1.00
Monocalcium phosphate	1.40
Limestone, ground	1.15
Salt	0.50
L-Threonine	0.03
Trace mineral premix[1]	0.15
Sow add pack[2]	0.50
Vitamin premix[3]	0.25
Phytase[4]	0.02
Total	100.00
Formulated analysis, %	
Dry matter	91.4
Crude protein	17.1
Crude fiber	3.7
Ether extract	3.5
Ca	0.78
P	0.52

[1] Each kilogram contains 26.4 g Mn, 110 g Fe, 110 g Zn, 11g Cu, 198 mg I, and 198 mg Se.
[2] Each kilogram contains 220,000 mg choline, 88 mg biotin, 660 mg folic acid, 1,980 mg pyridoxine.
[3] Each kilogram contains 4,400,000 IU vitamin A, 660,000 IU vitamin D_3, 17,600 IU vitamin E, 1,760 mg menadione, 3,300 mg riboflavin, 11,000 mg pantothenic acid, 19,800 mg niacin, 15.4 mg vitamin B_{12}.
[4] High Phos 2700 GT, DSM Nutritional Products, Parsippany, NJ.

was processed under uniform temperature conditions. The low dose batches were pelleted prior to the high dose batches. Within dose, the highest temperature and longest retention time was pelleted first and then temperature was reduced to achieve the medium and low temperature batch before moving to the middle retention time. The same progression was used between the medium and low retention time such that the final batch within each dose was the low temperature and shortest retention time treatment.

Three 100 g samples of each batch of feed were added to 400 ml of cold phosphate buffered saline (PBS, pH 7.4) in 500 ml bottles, thoroughly mixed and stored at 4°C for approximately 12 hours. The feed suspension supernatants were then collected and evaluated using a PEDV N-gene based real-time RT-PCR and aliquots were harvested and frozen at -80°C until use in the pig bioassay.

Experiment 2

A subsample of the feed was obtained prior to inoculation for each replication and confirmed negative by qRT-PCR for the presence of PEDV RNA at the Kansas State University Research Park Molecular Diagnostics Development Laboratory in Manhattan, Kansas. The negative control treatment was formed by mixing a 50 kg batch of the swine diet in a 0.11 m^3 electric paddle mixer (H. C. Davis Sons Manufacturing, model# SS-L1; Bonner Springs, KS). A mixing time of 5 min was determined to achieve a uniform mix with a coefficient of variation of less than 10%.

A 500 ml aliquot of stock PEDV was mixed into a 4.5 kg batch of swine feed using procedures established in a prior experiment.[2] Briefly, feed and virus were mixed using a manual, bench-top laboratory scale stainless steel paddle mixer previously validated for mixer efficiency. The stock virus was slowly added during mixing, and a wet mix time of 2.5 min was used to confirm a consistent PEDV-contaminated mix, which served as the PEDV feed inoculum.

The PEDV feed inoculum (4.5 kg of feed + 500 ml of stock virus) was added to 45kg of PEDV-free swine diet to form the positive control in the same 0.11 m^3 electric paddle mixer and the entire batch of positive control feed was then mixed for 5 min, discharged for 10 min into biohazard containers, and finally held at -2°C until the start of the thermal processing portion of the study.

PEDV-free feed was pelleted using the same pilot-scale single pass conditioner and pellet mill (Model CL5, CPM, Waterloo, IA). The pellet mill was heated by pelleting PEDV-free feed with a conditioning temperature of 71°C for 60 min. The steam valve was then turned off until the conditioning temperature dropped below 37°C to mimic procedures commonly used to resolve a plug in the conditioner or pellet die. Next, PEDV-inoculated feed was removed from the freezer and added to the pellet mill hopper. Once PEDV-inoculated feed started passing through the pellet mill, steam was slowly added, and five pelleted samples were collected at targeted hot mash conditioner temperatures of 37,46,54,62, and 71°C (± 1.2°C) with feed pelleted using a 30 sec conditioning time. These conditioning temperatures were selected based on a previously-determined prediction equation for the specific pellet mill to result in hot pellet temperatures of 40, 52, 63, 73, and 80°C, respectively.

Bioassay

The Iowa State University Institutional Animal Care and Use Committee reviewed and approved the pig bioassay protocol. Eighty one crossbred, 10 d-old pigs (33 pigs for Exp. 1 and for 48 pigs for Exp. 2) of mixed sex were sourced from a single commercial, crossbred farrow-to-wean herd with no prior exposure to PEDV. Upon arrival, piglets were ear tagged, weighed and administered a dose of cefitiofur (Excede, Zoetis, Florham Park, NJ). Also, upon arrival fecal swabs were obtained and confirmed negative for PEDV, porcine delta coronavirus (PDCoV) and transmissible gastroenteritis virus (TGEV) using a RT-PCR assay. To further confirm PEDV negative status, serum was collected and confirmed negative for PEDV antibody by an indirect fluorescent antibody (IFA) assay and TGEV antibody by ELISA conducted at the Iowa State University Veterinary Diagnostic Laboratory (ISU-VDL). Pigs were allowed 2 d of adjustment to the new pens before the bioassay began.

In Exp. 1, a total of 22 rooms (66 pigs) were randomly assigned to treatments corresponding to the 2 × 3 × 3 factorial design with each room representing each treatment.

In Exp. 2, a total of 16 rooms (48 pigs) were assigned to treatment groups with 1 negative control room and 15 challenge rooms. Each pig from the negative control room was given a 10 mL aliquot of inoculum created from the PEDV-free feed collected from the electric mixer during each of replicate 1, 2, and 3. Different from the negative control room, each pig in each challenge room was given an aliquot of inoculum from the same replicate, and one room was representative of a treatment in a single replicate.

In both experiments, rectal swabs were collected on d -2, 0, 2, 4, 6, and 7 d post inoculation (dpi) from all pigs and tested for PEDV RNA by quantitative qRT-PCR. Pigs were humanely euthanized at 7 dpi, and fresh small intestine, cecum, and colon samples were collected at necropsy along with an aliquot of cecal content. One section of formalin-fixed proximal, middle, distal jejunum and ileum was collected per pig for histopathology (Chen et al., 2014).[3] Infectivity of cecal content was evaluated for PEDV by qRT-PCR.

Statistical analysis

Statistical analysis was carried out using SAS (SAS Institute, Inc., Cary, NC). In Exp 1, ANOVA was performed to evaluate PEDV RNA feed Ct values, Villus height, crypt depth, villous height to crypt depth ratio, and immunohistochemistry. In Exp. 2, data of the effects of conditioner temperature on villus height, crypt depth, and villus height to crypt depth ratio were analyzed as a completely randomized design using PROC MIXED in SAS (SAS Institute, Inc., Cary, NC) with pig as the experimental unit by a pairwise comparison. Results for treatment criteria were considered significant at $P \leq 0.05$ and marginally significant from $P > 0.05$ to $P \leq 0.10$.

Results and discussion

Experiment 1

There was no PEDV RNA detected in the unprocessed PEDV-free feed. When the low-dose PEDV media (Ct 20) was mixed with the feed, the resulting feed Ct value was 31 and when the high PEDV dose media (Ct of 13) was mixed with feed the resulting Ct value was 24 (Table 2).

As expected, fecal shedding of PEDV was not detected in rectal swabs from negative control pigs for the duration of the study (Table 2). Fecal swabs from pigs fed the low-and high-PEDV dose positive control treatment (inoculated, but non-processed feed) were PEDV-positive from 2 dpi through the end of the study at 7 dpi. Cecum contents at 7 dpi and IHC determined 7 dpi were also positive for the positive control pigs (Table 2). However, if either the low-or high-dose PEDV feed was processed at any of the 9 possible conditioning time x temperature combinations, no PEDV RNA was detected in fecal swabs or cecum contents at 7 dpi.

The villous height for pigs challenged with the non-inoculated feed was higher ($P < 0.01$) compared to the height in pigs challenged with the low or high dose PEDV unprocessed feed (Table 3). Porcine epidemic diarrhea virus IHC immnoreactivity was not visible in the cytoplasm of villous enterocytes of low or high dose challenged pigs from any of the time and temperature pellet treatment combination for the duration of the study.

Experiment 2

When PEDV-inoculated feed was processed at five different conditioning temperatures (37, 46, 54, 62, and 71°C), the respective mean cycle threshold (Ct) values as detected by qPCR were 32.5, 34.6, 37.0, 36.5, and 36.7, respectively (Table 4). All 9 of the feed samples conditioned at 37, 46, and 54°C had detectible PEDV RNA; whereas 8 of the 9 processed at 143.6 or 159.8°F had detectible PEDV RNA. It was observed that Ct value increased as conditioning temperature increased to 130°F, with little change thereafter.

As in Exp. 1, fecal shedding of PEDV was not detected in rectal swabs or cecum contents from pigs fed the PEDV-negative control for the duration of the study (Table 5). Of the 9 total pigs gavaged with aliquots from the PEDV-positive diet conditioned at 100°F, a fecal swab from 1 pig (Room 7, Replicate 2) yielded detectible PEDV RNA at 2 dpi, and all 3 pigs in Room 7 had fecal swabs and cecum contents with detectible viral particles by 4 through 7 dpi. In addition, 3 pigs gavaged with aliquots from the treatment conditioned at 46°C had detectible fecal PEDV RNA at 2 to 7 dpi, and all pigs were in the same room (Room 8, Replicate 2). No pig challenged with feed conditioned at or above 130°F had detectible PEDV RNA in fecal swabs or cecum content for the duration of the study.

The pigs challenged with the feed conditioned at 37°C had shorter ($P < 0.05$) villous heights than pigs challenged with any other treatment (Table 6). Furthermore, pigs challenged with feed conditioned at 37, 54, or 62°C had deeper ($P < 0.05$) crypt depths than pigs challenged with the negative control or feed conditioned at 71°C. This led to pigs challenged with feed conditioned at 37°C to have a lower ($P < 0.05$) villus height: crypt depth ratio than pigs challenged with the negative control or feed conditioned at 54, 62, and 71°C. Porcine epidemic diarrhea virus IHC immunoreactivity was visible in the cytoplasm of villous enterocytes of pigs challenged with the two lowest processed temperatures: 37 and 46°C when harvested at 7 dpi.

In summary, infected feed processed at common manufacturing temperatures and conditioner retention times effectively inactivated the PED virus at a low and high concentration and did

Table 2: Effects of porcine epidemic diarrhea virus (PEDV) dose, pelleting temperature, and conditioning retention time on PEDV detection from feed, pig fecal swabs and cecum contents, Exp. 1[1]

		PEDV N-gene Real Time-PCR, cycle threshold (Ct)							
				Fecal swabs					Cecum contents
PEDV dose, temp.,[3] and time[4]		Tissue culture	Feed	0 dpi[2]	2 dpi	4 dpi	6 dpi	7 dpi	7 dpi
Unprocessed virus-free feed[5]		–	–	–	–	–	–	–	–
Low dose inoculated feed[6]		20.0	31.0	–	22.4	18.2	18.8	24.1	26.7
	45s		42.6	–	–	–	–	–	–
68.3°C	90s		39.5	–	–	–	–	–	–
	180s		45.0	–	–	–	–	–	–
	45s		36.7	–	–	–	–	–	–
79.4°C	90s		39.7	–	–	–	–	–	–
	180s		42.3	–	–	–	–	–	–
	45s		39.7	–	–	–	–	–	–
90.6°C	90s		37.4	–	–	–	–	–	–
	180s		35.9	–	–	–	–	–	–
High dose inoculated feed[7]		13.0	24.0	–	23.0	15.3	20.4	24.3	24.0
	45s		30.2	–	–	–	–	–	–
68.3°C	90s		29.7	–	–	–	–	–	–
	180s		30.2	–	–	–	–	–	–
	45s		30.1						
79.4°C	90s		29.5	–	–	–	–	–	–
	180s		30.2	–	–	–	–	–	–
	45s		30.1						
90.6°C	90s		30.6	–	–	–	–	–	–
	180s		30.0	–	–	–	–	–	–

1 An initial tissue culture containing a low dose and high dose of PEDV was used to inoculate batches of feed. Three feed samples per batch were collected and diluted in PBS. The supernatant from each sample was then collected for pig bioassay. The supernatant was administered one time via oral gavage on d 0 to each of three pigs per treatment (10 ml per pig). Thus, each value represents the mean of 3 pigs per treatment. Pigs were initially 10 d old and 3.6 kg BW.

2 Day post inoculation.

3 Temperature of feed exiting the conditioner.

4 Retention time. The amount of time required for feed to pass through the conditioner.

5 A cycle threshold (Ct) of > 45 was considered negative for presence of PEDV RNA.

6 For low dose feed, PEDV (1×10^3 $TCID_{50}$/ml) was diluted into feed to provide a dose of 1×10^2 $TCID_{50}$/g of feed.

7 For high dose feed, PEDV (1×10^5 $TCID_{50}$/ml) was diluted into feed to provide a dose of 1×10^4 $TCID_{50}$/g of feed.

Table 3: Effects of porcine epidemic diarrhea virus (PEDV) dose, pelleting temperature, and conditioning retention time on morphologic and immunohistochemistry evaluation of small intestine from pigs, Exp. 1[1]

Item		Morphology[2] Villus height, µm	Crypt death, µm	Villus height to crypt depth ratio	Immunohisto-chemistry (IHC)[3]
Unprocessed virus-free feed		495.7	101.7	4.9	0
Low dose inoculated feed[4]		414.3	91.0	4.6	0.3
68.3°C	45s	481.6	101.3	4.8	0
	90s	489.3	108.1	4.6	0
	180s	504.4	115.6	4.4	0
79.4°C	45s	508.6	108.9	4.7	0
	90s	476.4	103.6	4.6	0
	180s	441.8	103.6	4.3	0
90.6°C	45s	443.2	97.8	4.6	0
	90s	495.7	101.7	4.9	0
	180s	441.8	103.6	4.3	0
High dose inoculated feed[7]		309.3	112.6	3.1	1.7
68.3°C	45s	423.1	105.3	4.0	0
	90s	429.4	118.3	3.7	0
	180s	389.2	100.0	4.0	0
79.4°C	45s	432.8	117.3	3.7	0
	90s	390.6	102.5	3.7	0
	180s	448.7	104.5	4.3	0
90.6°C	45s	383.1	119.5	3.2	0
	90s	446.4	102.8	4.4	0
	180s	408.9	105.7	3.9	0
	SEM	29	10.1	0.4	0.2

1 An initial tissue culture containing a low dose or high dose of PEDV was used to inoculate batches of feed. Three feed samples per batch were collected and diluted in PBS. The supernatant from each sample was then collected for pig bioassay. The supernatant was administered one time via oral gavage on d 0 to each of three pigs per treatment (10 ml per pig). Thus, each value represents the mean of three pigs per treatment necropsied at 7 d post infection. Pigs were initially 10 d old and 3.6 kg BW.

2 Intestinal cross-sections were fixed in formalin and stained with hematoxylin and eosin for evaluation.

3 Three sections of ileum were evaluated and averaged into one categorical value per pig. Categorical values were assigned for each pig (0 = no signal, 1 = mild, 2 = moderate, 3 = abundant, 4 = diffuse) and reported as the mean from 3 pigs per treatment, thus the mean of 9 values.

4 For low dose feed, PEDV (1 × 10^3 $TCID_{50}$/ml) was diluted into feed to provide a dose of 1 × 10^2 $TCID_{50}$/g of feed.

5 Temperature of feed exiting the conditioner.

6 Retention time. The time feed was inside the conditioner.

7 For high dose feed, PEDV (1 ×10^5 $TCID_{50}$/ml) was diluted into feed to provide a dose of 1 × 10^4 $TCID_{50}$/g of feed.

Table 4: Effect of increasing pelleting temperature on porcine epidemic diarrhea virus (PEDV) infected feed as analyzed by qRT-PCR (quantitative reverse transcriptase PCR) Exp. 2[1]

	Mash conditioner temperature				
Positive treatment[2]	**37.8°C**	**46.1°C**	**54.4°C**	**62.8°C**	**71°C**
% positive[3]	100.0 (9/9)	100.0 (9/9)	100.0 (9/9)	88.9 (8/9)	88.9 (8/9)
Cycle threshold (Ct)[4]	32.5	34.6	37.0	36.5	36.7

[1] 500 ml of tissue culture containing 4.5 × 10^6 $TCID_{50}$/ml of PEDV was inoculated into a 4.5 kg batch of feed, then added to 45 kg of PEDV negative feed to form the positive treatment. One feed sample per temperature per replicate × 3 replicates was collected, divided into 3 aliquots and diluted in PBS to form supernatants. Thus each feed supernatant value per treatment represents the mean of three replicates × 3 repetitions.
[2] Pellet mill die was warmed-up for approximately 1 hour before the steam was turned off. Steam was slowly added and the five processed samples were collected.
[3] Means represent the percent of samples that had detectible RNA by PEDV qPCR analysis.
[4] Mean cycle threshold (Ct) value of samples with detectible PEDV RNA below 45.

Table 5: Influence of processed porcine epidemic diarrhea virus (PEDV) inoculated feed on qRT-PCR (quantitative reverse transcriptase PCR) cycle threshold (Ct) of feed, fecal swabs and cecum contents of pigs Exp. 2[1]

		Fecal swabs, Ct					
Item	**KSU feed inoculum, Ct**	**0 dpi**[2]	**2 dpi**	**4 dpi**	**6 dpi**	**7 dpi**	**7 dpi Cecum content, Ct**
Processed feed, %[3]							
Negative	-[4]	-		-	-	-	-
37.8°C	100.0 (9/9)	-	11.1 (1/9)	33.3 (3/9)	33.3 (3/9)	33.3 (3/9)	33.3 (3/9)
46.1°C	100.0 (9/9)	-	33.3 (3/9)	33.3 (3/9)	33.3 (3/9)	33.3 (3/9)	33.3 (3/9)
54.4°C	100.0 (9/9)	-	-	-	-	-	-
62.8°C	88.9 (8/9)	-	-	-	-	-	-
71.1°C	88.9 (8/9)	-	-	-	-	-	-
Processed feed, Ct[5]							
Negative	-	-	-	-	-	-	-
37.8°C	32.5	-	15.8	27.5	16.6	17.8	16.9
46.1°C	34.6	-	24.5	15.2	15.4	17.9	18.8
54.4°C	37.0	-	-	-	-	-	-
62.8°C	36.5	-	-	-	-	-	-
71.1°C	36.7	-	-	-	-	-	-

[1] 500 ml of tissue culture containing 4.5 × 10^6 $TCID_{50}$/ml of PEDV was inoculated into a 4.5 kg batch of feed, then added to 45 kg of PEDV negative feed to form the positive treatment. One feed sample per temperature per replicate × 3 replicates was collected, divided into three aliquots and diluted in PBS to form supernatants. Thus each feed supernatant value per treatment represents the mean of three replicates × 3 repetitions.
[2] Day post inoculation.
[3] Means represent the percent of samples that had detectible RNA by PEDV qPCR analysis (< 45 Ct).
[4] No detectible PEDV RNA (Ct > 45).
[5] Mean cycle threshold (Ct) value of samples with detectible PEDV RNA below 45.

Table 6: Morphologic and immunohistochemistry evaluation of small intestine from pigs that were challenged with porcine epidemic diarrhea virus (PEDV) inoculated feed processed at increasing temperatures Exp. 2[1]

Item	Morphology[2]			Immunohisto chemistry (IHC)[3]
	Villus height, µm	Crypt depth, µm	Villus height to crypt depth ratio	
Processed feed				
Negative	395.3 ± 27.1[a]	245.1 ± 18.3[a]	2.1 ± 0.2[b]	0
37.8°C	299.8 ± 15.6[b]	188.2 ± 10.6[b]	1.6 ± 0.1[a]	0.7
46.1°C	374.4 ± 15.6[a]	217.0 ± 10.6[ab]	1.7 ± 0.1[ab]	0.6
54.4°C	380.3 ± 15.6[a]	190.9 ± 10.6[b]	2.0 ± 0.1[b]	0
62.8°C	387.3 ± 15.6[a]	199.5 ± 10.6[b]	1.9 ± 0.1[b]	0
71.1°C	385.9 ± 15.6[a]	218.8 ±10.6[a]	1.8 ± 0.1[b]	0

1 17 oz. of tissue culture containing 4.5×10^6 $TCID_{50}$/ml of PEDV was inoculated into a 4.5 kg batch of feed, then added to 45 kg of PEDV negative feed to form the positive treatment. One feed sample per temperature per replicate × 3 replicates was collected, divided into three aliquots and diluted in PBS to form supernatants. Thus each feed supernatant value per treatment represents the mean of three replicates × 3 repetitions.

2 Intestinal cross-sections were fixed in formalin and stained with hematoxylin and eosin (H&E) for evaluation.

3 Three sections of ileum were evaluated and averaged into one categorical value per pig. Categorical values were assigned for each pig (0 = no signal, 1 = mild, 2 = moderate, 3 = abundant, 4 = diffuse) and reported as the mean from 3 pigs per negative treatment and mean from 9 pigs per remaining treatments.

a,b Means within column lacking a common superscript are different ($P < 0.05$).

not cause detectible clinical or microscopic disease in 10 d old pigs. This suggests that thermal processing is one step that can be used to reduce the risk of PEDV transmission. If target processing temperatures are not reached, such as during resolution of a plugged pellet mill, there is a risk of transmitting PEDV through contaminated pelleted feed. Our data suggest that feed should be conditioned at temperatures above 54°C in order to minimize the risk of PEDV transfer. Thus, care should be taken if suspected PEDV-contaminated feed is conditioned below the target temperature to prevent downstream cross-contamination.

Acknowledgment

Appreciation is expressed to the National Pork Board for financial support (awards #14-159 and #14-273).

References

1. Nitikanchana, S. Potential Alternatives to Reduce Porcine Epidemic Diarrhea Virus (PEDV) Contamination in Feed Ingredients. 2014: http://www.asi.k-state.edu/species/swine/research-and-extension/PEDV%20contamination%20in%20feed%20ingredients_Feb%2026.pdf (Accessed 31 December 2014.

2. Woodworth J, Stark C, Hesse R, Main R, Zhang J, Tokach M, Gauger P, and Dritz S. Determining the impact of conditioning time and temperature in pelleted diets on porcine epidemic diarrhea virus (PEDV) survivability in complete swine diets-. *National Pork Board*.2014:NPB#14159. http://research.pork.org/FileLibrary/ResearchDocuments/14-159-WOODWORTH-KSt.pdf (Accessed 14 March 2015).

3. Chen Q, Ganwu L, Stasko J, Thomas J, Stensland W, Pillatzki A, et al. Isolation and characterization of porcine epidemic diarrhea viruses associated with the 2013 disease outbreak among swine in the United States. *Journal of Clinical Microbiology*. 2014;52(1):234-243.

PEDV antibody responses in fecal and oral fluid specimens

L. G. Giménez-Lirola; J. B. Kraft; K. Woodard; C. Wang; D. Baum; J. Zimmerman; R. Main
Iowa State University, Ames, Iowa

Introduction

The management of PEDV will depend upon the development of PEDV antibody assays able to detect infected animals and assess their level of protective (lactogenic) immunity. In an intestinal infection such as PEDV, the crucial confrontation of the pathogen with antibodies occurs within the environment of the gut lumen; thus, serum antibody levels are only an indirect reflection of the immune response against PEDV within the gut. On the other hand, fecal antibodies, as a mirror of mucosal immunity, could be an effective tool both in surveillance and for the prognostic assessment of immunity against PEDV.

Materials and methods

The immediate aims of this study were to 1) create a PEDV indirect ELISA capable of detecting antibody in feces, and using the new test, 2) describe the antibody (IgG and IgA) kinetics in pen-based feces and compare to antibody profiles in serum and pen-based oral fluids in commercial populations. The ultimate goal of this research is to characterize the maternal fecal antibody profiles associated with protection of piglets against PEDV infection.

Aim 1: Two PEDV recombinant spike (S) protein-based indirect ELISAs (IgG, IgA) were created to detect antibody in different specimens. In brief, a truncated PEDV S protein was selected as the ELISA antigen target by evaluating the serum antibody response of PEDV-infected pigs against different recombinant polypeptides derived from the major structural proteins (nucleocapsid, membrane, spike, and envelope) using Luminex® technology.

Aim 2: A wean-to-finish facility was monitored for PEDV at 2 week intervals beginning at placement. At ~13 weeks of age, this verified PEDV-negative population was exposed to PEDV using standard field exposure techniques. Pen-based feces (n = 72), pen-based oral fluids (n = 72), and pig serum samples (n = 360) from day post exposure (DPE -18, -1, 6, 13, 27, 41, 55, 69, 83, 97, and 111 were randomized and tested by RT-PCR and the two PEDV S ELISAs (IgG, IgA).

Results

PEDV was detected in oral fluids and feces at the first sampling post exposure (DPE 6). PEDV IgG and IgA antibody was detected in serum, oral fluid and feces collected on DPE 13. Substantially higher levels of serum IgG (peak at DPE 13) than IgA (peak at DPE 27) were found. However, IgA rather than IgG, was the most abundant antibody isotype in both oral fluid (peak at DPE 97) and fecal (peak at DPE 41) specimens.

Discussion

Preliminary results suggest that IgA antibody in fecal and oral fluids may facilitate PEDV monitoring in commercial pig populations. Future work should focus on establishing the correlation between mucosal IgA and lactogenic immunity against PEDV.

Stochastic modeling to determine number of laryngeal swab sample pools and sample collections needed for low *Mycoplasma hyopneumoniae* group prevalences

C. Fitzgerald[1]; A. Sponheim[2], DVM; E. Fano[2], DVM, PhD; D. Polson[2], DVM, PhD; M. Pieters[3], DVM, PhD
[1]Iowa State University, Ames, Iowa; [2]Boehringer Ingelheim Vetmedica, Inc., Saint Joseph, Missouri;
[3]University of Minnesota, CVM, Saint Paul, Minnesota

Introduction

Ante-mortem detection of *Mycoplasma hyopneumoniae* (Mhp) can be difficult during the early stages of an infection. There is a need for accurate diagnostic sampling protocols to confidently determine if populations are negative (≤ 1% prevalence), remain negative (≤ 1% prevalence) over time, and to detect early infection to prevent spread. In a recent study, the authors demonstrated with the use of stochastic modeling and diagnostic sensitivities (DxSe) that increased sample size resulting from ante-mortem laryngeal swab sampling combined with pooling allowed for a higher herd detection rate by PCR while pursuing the most economical approach, in a high prevalence population.[1] To further test this model in a Mhp low prevalence population scenario, a second study was designed. The objective of this study was to determine ante-mortem laryngeal swab sampling protocols for detection of Mhp by PCR at low prevalence levels.

Materials and methods

A stochastic sampling model developed in a spreadsheet program (Excel, Microsoft) was used to determine the number of animals along with the number of times to sample herds in order to detect one positive animal in a pool when the rest of the pool was negative in a Mhp low prevalence scenario (low prevalence and high Ct). The DxSe used in this project have been described previously.[2] The high (H = 36) Ct tables had 3:1 pools and 5:1 pools. For the 3:1 pools, three DxSe were run: 79.1% (for 99% lower confidence limit (LCL)), 81.9% (for 95% LCL), and 90% (for the mean). For the 5:1 pools three DxSe were run: 58.5% (for the 99% LCL), 61.8% (for the 95% LCL), and 72% (for the mean). For each adjusted DxSe value, the model was run for a specific number of individuals sampled from a 2,500 population size (N individuals = 120, 90, 60, and 30) and a percent prevalence (Percent prevalence = 1%, 2%, 3%, 4%, and 5%). For each pool, adjusted DxSe, N individuals and percent prevalence, two values were recorded from the stochastic model: a detection probability of ≥ 99% and a detection probability of ≥ 95%. For each detection probability, a minimum of 100 iterations (in groups of 10 iterations) were run and the highest value, indicative of the number of samples needed, was recorded.

Results and discussion

Results for 99% LCL and 95% detection probability are reported in Table 1. Additional tables and interpretation training will be available on a future website. To the authors' knowledge this is the first Mhp sampling table of its kind. This table takes into account DxSe for the laryngeal swab sample procedure and provides guidance for determining number of animals and number of samplings required to economically detect Mhp in low prevalence scenarios. A further aim of the study is to use the model to help evaluate the financial implications associated with missing Mhp positive animals in low prevalence scenarios.

References

1. Sievers et al. Comparison of ante-mortem sampling procedures to post-mortem bronchial sampling for the detection of *Mycoplasma hyopnuemoniae* by PCR. AASV. 2015; 115-116.

2. Sponheim et al. Influence of pooling on *Mycoplasma hyopneumoniae* polymerase-chain reaction diagnostic assay results in expected low prevalence scenarios. Allen D. Leman Swine Conference. 2015; 29.

Table 1: Number of sample collections needed at varying Mhp group prevalences by number of laryngeal swab samples collected (99% LCL, 95% detection probability, ≥ 2,500 population size).

N Individuals	**30**	**60**	**90**	**120**
N 3:1 pools	**10**	**20**	**30**	**40**
Percent prevalence				
1%	15	8	6	4
2%	8	4	3	3
3%	6	3	2	2
4%	4	2	2	2
5%	3	2	2	2
N 5:1 pools	**6**	**12**	**18**	**24**
Percent prevalence				
1%	21	11	7	6
2%	11	5	4	3
3%	7	4	3	2
4%	5	3	2	2
5%	4	3	2	2

Detection of *Actinobacillus pleuropneumoniae* ApxIV toxin antibody in serum and oral fluid specimens from pigs inoculated under experimental conditions

W. Gonzalez[1], DVM; L. G. Giménez-Lirola[1], PhD; A. Holmes[1], BS; S. Lizano[3], PhD; C. Goodell[3], PhD; K. Poonsuk[1], DVM, MS; P. Sitthicharoenchai[1], DVM, MS; Y. Sun[2], MS; J. Zimmerman[1], DVM, PhD
[1]College of Veterinary Medicine, Iowa State University, Ames, Iowa; [2]College of Liberal Arts and Sciences, Iowa State University, Ames, Iowa; [3]IDEXX Laboratories, Westbrook, Maine

Introduction

Actinobacillus pleuropneumoniae (APP) is an important bacterial respiratory pathogen of swine causing acute fibrinohemorrhagic and necrotizing pleuropneumonia.[12] The prevention and control of APP in commercial production settings is based on serological monitoring. ELISAs have been developed to detect specific serum antibodies against a variety of APP antigens, including long-chain lipopolysaccharides (LPS), capsular polysaccharide (CP), and Apx exotoxins (ApxI, II, III, IV).[4,6,9] Although serum is the historic choice for antibody detection and APP surveillance, a variety of antibody-based assays for oral fluid specimens have been reported, e.g., assays for African swine fever virus,[8] *Erysipelothrix rhusiopathiae,*[5] influenza A virus,[10] porcine circovirus type 2,[11] and porcine reproductive and respiratory syndrome virus.[7] The objective of the present study was to describe and compare antibody responses specific for ApxIV in serum and oral fluid from pigs inoculated with APP under experimental conditions using a commercial ApxIV antibody ELISA.

Methods

Four groups of pigs (~14-weeks-old; 6 per group) were used. Animals tested negative for APP by two commercial ELISA assays (APP ApxIV Ab ELISA, IDEXX Laboratories, Inc., Westbrook ME, USA; Swinecheck® Mix-APP, Biovet, Inc., St. Hyacinthe, Quebec, Canada) at the Iowa State University Veterinary Diagnostic Laboratory (Ames, Iowa). Pigs were individually housed in order to collect individual pig oral fluids. Two weeks post-placement, animals were inoculated with APP serovars 1 (ATCC 27088), 5 (ATCC 33377), 7 (ATCC WF83) or 12 (ATCC 9799/84). Pigs were mildly sedated to reduce stress and facilitate handling and exposed intranasally (2 ml) and by direct application (3 ml) to the tonsils. Blood samples were collected weekly and oral fluids were collected daily from DPI -14 through DPI 56. To confirm infection, serum samples were completely randomized and tested by long-chain lipopolysaccharide (LPS) ELISAs for serovars [(1(9, 11), 5a-5b, 7(4), and 12] (Service de Diagnostic, University of Montreal, Québec, Canada) and for ApxIV antibodies using a serum antibody ELISA (ApxIV Ab, IDEXX Laboratories, Inc., Westbrook, ME). Oral fluid samples were tested using the serum ApxIV ELISA using a protocol that optimized the detection of ApxIV antibody in oral fluids. Statistical analyses were performed to compare the ApxIV responses among and within serovars over time.

Results and discussion

The production of clinical signs was marked and prolonged in pigs exposed to APP serovars 1 or 7 and mild or absent in pigs exposed to APP serovars 5 or 12. One pig inoculated with serovar 1 was found dead the morning of DPI 3. All animals inoculated with serovars 1 or 7 were treated with ampicillin from DPI 4 through 11.

The LPS ELISA serum response showed that all pigs exposed to serovars 1 and 7 were positive from DPI 14 through DPI 56, with some animals demonstrating a LPS-specific IgG response as early as 7 DPI. In contrast, a weak and transient LPS ELISA response was observed in animals inoculated with serovar 5. In serovar 12, a single animal was LPS ELISA positive on DPI 35, 42, and 49. These results are compatible with previous studies reporting variable success at infecting pigs with APP under experimental conditions.[1]

Comparisons of the serum ApxIV ELISA S/P results from DPI 0 showed a significant increase ($P < 0.05$) in ApxIV antibody 14 DPI (IgM) and 21 DPI (IgG) in animals inoculated with serovars 1 and 7. This pattern was consistent with previous reports[2,3] and also followed the pattern observed for the LPS ELISA response in these animals. The results of this study suggested that the S/P cutoff of the commercial serum ApxIV IgG ELISA is too high. However, a limitation of the current study was the small number of animals. As a consequence, it was not possible to conduct a valid statistical analysis of test performance, e.g., receiver operating characteristic analysis, to re-calculate the ELISA cutoff and associated test sensitivity and specificity. Comparisons of the oral fluid ApxIV ELISA S/P results from DPI 0 showed a significant increase ($P < 0.05$) in ApxIV IgG S/P ratios at ≥ 21 DPI in pigs inoculated with serovar 1, with no detectable increase in IgM or IgA. In serovar 7, significantly higher IgG S/P ratios were observed at ≥ 14 DPI. In this group, a significant increase ($P < 0.05$) in the ApxIV IgM response was observed at 14 DPI and in the IgA response at DPI 21 and then at ≥ 35 DPI. This work suggests that ELISAs based on ApxIV antibody detection in oral fluid samples could be effective in population monitoring for APP.[13]

References

1. Costa G et al. Evaluation of *Actinobacillus pleuropneumoniae* diagnostic tests using samples derived from experimentally infected pigs. Vet. Microbiol. 2011; 148, 246-251.

2. Dreyfus A et al. Use of recombinant ApxIV in serodiagnosis of *Actinobacillus pleuropneumoniae* infections, development and prevalidation of the ApxIV ELISA. Vet. Microbiol.2004; 99: 227-238.

3. Eamens G et al. Evaluation of serovar-independent ELISA antigens of *Actinobacillus pleuropneumoniae* in pigs, following experimental challenge with *A. pleuropneumoniae*, *Mycoplasma hyopneumoniae* and *Pasteurella multocida*. Aus. Vet. J. 2012; 90: 490-498.

4. Gottschalk M et al. Evaluation of long chain lipopolysaccharides (LC-LPS) of *Actinobacillus pleuropneumoniae* serotype 5 for the serodiagnosis of swine pleuropneumonia. Vet. Microbiol. 1994; 38, 315-327.

5. Giménez-Lirola et al. Improving ante -mortem diagnosis of *Erysipelothrix rhusiopathiae* infection by use of oral fluids for bacterial, nucleic acid, and antibody detection. J. Microbiol. Methods. 2013; 92: 113-121.

6. Inzana and Fenwick. Serologic detection of *Actinobacillus pleuropneumoniae* in swine by capsular polysaccharide-biotin-streptavidin enzyme-linked immunosorbent assay. J. Clin. Microbiol.2001; 39: 1279-1282.

7. Kittawornrat et al. Detection of porcine reproductive and respiratory syndrome virus (PRRSV) antibodies in oral fluid specimens using a commercial PRRSV serum antibody ELISA. J. Vet. Diagn. Invest. 2012; 24: 262-269.

8. Mur et al. Potential use of oral fluid samples for serological diagnosis of African Swine Fever. Vet. Microbiol.2013; 165: 135-139.

9. Nielsen R. Evaluation of an indirect enzyme-linked immunosorbent assay (ELISA) for detection of antibodies to the Apx toxins of *Actinobacillus pleuropneumoniae*. Vet. Microbiol. 2000; 71: 81-87.

10. Panyasing et al. Kinetics of influenza A virus nucleoprotein antibody (IgM, IgA, IgG) in serum and oral fluid specimens from pigs infected under experimental conditions. Vaccine. 2013;31: 6210-6215.

11. Prickett et al. Prolonged detection of PCV2 and anti-PCV2 antibody in oral fluids following experimental inoculation. Transbound. Emerg. Dis. 2011; 58:121-127.

12. Shope R. Porcine contagious pleuropneumonia. I. Experimental transmission, etiology and pathology. J. Exp. Med.1964; 119: 357-368.

13. Gonzalez W et al. Detection of *Actinobacillus pleuropneumoniae* ApxIV toxin antibody in serum and oral fluid specimens from pigs inoculated under experimental conditions. Vet. Microbiol. (*submitted*).

Discovery of a divergent lineage pestivirus in piglets with congenital tremors and reproduction of disease following experimental challenge

B. Arruda[1], DVM, PhD; P. Arruda[1], DVM, MS, PhD; D. Magstadt[1], DVM, MS; K. Schwartz[1], DVM, MS; T. Dohlman[1], DVM, MS, DACT; J. Schleining[1], DVM, MS, DACVS-LA; A. Patterson[2], DVM, PhD; C. Visek[2]; J. Victoria[2], PhD

[1]Department of Veterinary Diagnostic and Production Animal Medicine, College of Veterinary Medicine, Iowa State University, Ames, Iowa; [2]Boehringer Ingelheim Vetmedica, Inc., Ames, Iowa

Introduction

Congenital tremors (CT; *myoclonia congenital*) is a sporadic but widely distributed disease of neonatal piglets characterized by action-related repetitive myoclonus. Tremors vary in severity within and between affected litters with some piglets having only fine muscle fasciculations while others have pronounced repetitive contractions that make it difficult for the piglet to stand, walk, or nurse. The first reports of clinical disease were made by Kinsley (1922) in the United States,[1] Payen and Fournier (1934) in France,[2] and Hindmarsh (1937) in Australia.[3]

The syndrome of CT defines the clinical signs that may be observed in fetal infections with classical swine fever virus (Type AI),[4] inherited disorders (Type AIII and Type AIV),[5,6] and trichlorfon toxicity (Type AV).[7] However, most contemporary outbreaks have been attributed to an unidentified virus (Type AII) based on the elimination of classical swine fever virus, genetic variability in swine herds, and current production practices.[8]

To investigate the cause of CT Type AII, the objectives of this study were to: 1) detect potential pathogen(s) in samples from piglets with CT and 2) develop an infection model to reproduce CT.

Materials and methods

Pestivirus identification

Next-generation sequencing.

Porcine tissues from three unrelated diagnostic investigations of CT were obtained. Samples were processed for MiSeq based sequencing through library generation using the NextEra XT library preparation kit. *De novo* assembled sequences were analyzed by BLASTn and BLASTx. ClustalW alignment was used for phylogenetic analysis of the NS3 and Npro gene. Neighbor-joining phylogenetic trees were generated using MEGA 6.0 software.

Quantitative real-time polymerase chain reaction (qRT-PCR). A qRT-PCR targeting the NS3 region was developed. A cut-off for positive samples was established at Ct values lower than 36. Two sample sets were collected from farms with congenital tremors. The first set consisted of 6 affected and 2 unaffected pre-suckle piglets, serum from five sows, and 5 affected and 2 unaffected post-suckle piglets. The second set consisted of 5 affected piglets suckle status unknown and serum from five sows with affected piglets.

Sow inoculation model

Eight crossbred sows were obtained from a commercial source with no known previous history of CT. Serum from all sows was negative for PCV2a, PCV2b, PRRSV, PPV1, PPV5 and the novel pestivirus by qRT-PCR prior to shipment and inoculation. Individual sows were assigned to one of three separately-housed groups [sham-inoculated at 45 days gestation (n = 1) and 62 days gestation (n = 1), pestivirus-inoculated at 45 days gestations (n = 3), and pestivirus-inoculated at 62 days gestation (n = 3)].

Terminal serum from a viremic pig was thawed at 37°C. Total nucleic acid was extracted and screened by PCR for the presence of PCV2a, PCV2b, PRRSV, PPV1, PPV5 and the novel pestivirus. Following anesthetic induction, an aseptic laparotomy was performed. The uterus was exteriorized and a sterile handheld ultrasound transducer was used to image each fetal unit. Each vesicle was inoculated with 0.25mL of inoculum (PBS or pestivirus-serum). The inoculum was also administered directly to the sow via an intranasal (2mL) and intravenous (2mL) route.

Clinical observations, sample collection, and necropsy

Fecal material, blood and nasal swabs were collected from sows at days post-farrowing (DPI) 2, 7, 10 and 14 and then weekly until farrowing. Serum, nasal swabs and fecal swabs were collected from piglets at birth. In a subset of piglets (n = 7), blood from the umbilical cord was collected. Videos of piglets were taken daily. Four blinded investigators reviewed the videos and assigned a tremor severity score to each piglet: 0 – absent, 1 –fine muscle fasciculation, 2 – mild tremor, 3 – moderate tremor, 4 – severe tremor with pronounced hopping. Piglets receiving a score of ≥ 0.75 were considered to be affected.

At necropsy on days post-farrowing (DPF) 2, sow serum, piglet serum, cerebrum, cerebellum, brainstem, spinal cord, kidney, mesenteric lymph node, tracheobronchial lymph node, thymus, heart, and spleen were collected. In a subset of piglets, whole blood (EDTA tubes; n = 20) and cerebrospinal fluid (n = 26) were collected.

Results

Pestivirus identification

Next-generation sequencing.

A novel virus closely related to a Chinese bat pestivirus, and now known to be more closely related to a recently reported provisionally named atypical porcine pestivirus,[9] was identified by three independent congenital tremor disease investigations.

Quantitative real-time polymerase chain reaction (qRT-PCR). Animals with congenital tremors from two unrelated farms were positive for the pestivirus by qRT-PCR while the virus was not detected in unaffected piglets.

Sow inoculation model

Sow observations and samples.

One sham-inoculated sow at 45 days gestation developed a high fever and aborted all fetuses on DPI 3 and 4. A sow from the group to be virus-inoculated at 45 days of gestation was found not to be pregnant at time of inoculation. No sows regardless of group displayed clinical signs, developed a detectable viremia, or shed the virus at levels detectable by qRT-PCR.

Piglet observations and samples

Sham-inoculated piglets did not have clinical signs consistent with CT. A majority of piglets that were pestivirus-inoculated as fetuses had clinical signs consistent with CT. The prevalence of congenital tremors in pestivirus-inoculated litters ranged from 57% to 100% on DPF 2.

Viral RNA was not detected in samples from sham-inoculated piglets. Viral RNA was detected in all piglets with CT and all but one of the pestivirus-inoculated unaffected piglets. Pestivirus RNA was detected in serum, nasal swabs, feces, terminal serum, cerebrum, cerebellum, brainstem, spinal cord, kidney, mesenteric lymph node, tracheobronchial lymph node, thymus, heart, and spleen by qRT-PCR in pestivirus-inoculated. In addition, pestivirus RNA was detected in umbilical cord blood, whole blood, and CSF from a subset of piglets. The average Cq of serum, nasal swabs, CSF, mesenteric lymph node, tracheobronchial lymph node, spleen and umbilical cord blood was less than 26.

Discussion

The clinical syndrome of CT was first recognized nearly 100 years ago; yet, most contemporary outbreaks are attributed to an unidentified virus. Through the use of next-generation sequencing, a novel agent originally identified to be closely related to a bat pestivirus and now more closely related to a recently published divergent pestivirus provisionally named atypical porcine pestivirus[9] was detected in samples of piglets with CT from three separate disease investigations.

Serum and central nervous system tissue from piglets with CT from two unrelated farms contained viral RNA detected by qRT-PCR suggesting that the virus has a systemic distribution while clinically impacting central nervous system function. This is supported by the tissue distribution of viral RNA in pestivirus-inoculated piglets most of which had clinical signs consistent with CT. However, the mechanism of central nervous systemic dysfunction is currently unknown. Despite IN and IV inoculation, sows did not develop a detectable viremia or shed the virus at levels detectable by qRT-PCR; therefore, sows were not infected following challenge or the available diagnostic tests were insufficient to detect infection.

The development of virus isolation techniques, validated antibody assays, *in situ* detection techniques and refined molecular tools will undoubtedly lead to better understanding of pathogenesis and epidemiology of this virus.

Acknowledgements

The authors wish to thank Drs. William Hollis and Mark Brinkman for their assistance in the congenital tremor disease investigations.

References

1. Kinsley AT. Dancing Pigs? *Vet Med.* 1922;17: 123.
2. Payen B, Fournier P. Pocelets "trembleurs." *Rec Med Vet.* 1934;110: 84-86.
3. Hindmarch WL. Trembling in young pigs. *Aust Vet J.* 1973;13: 249-251.
4. Harding JDJ, Done JT, Darbyshire JH. Congenital tremors in piglets and their relation to swine fever. *Vet Rec.* 1966;79: 388-390.
5. Harding JDJ, Done JT, Harbourne JF, Randall CJ, Gilbert FR. Congenital tremors type A III in pigs: A hereditary sex-linked cerebrospinal hypomyelinogenesis. *Vet Rec.* 1973;92: 527-529.
6. Patterson DSP, Sweasey D, Brush J, Harding JDJ. Neurochemistry of the spinal cord in British saddleback piglets affected with congenital tremor, type A-IV, second form of hereditary cerebrospinal hyopmyelinogenesis. *J Neurochem.* 1973;21: 397-406.
7. Knox B, Askaa J, Basse A, Bitsch V, Eskildsen M, Mandenep M, et al. Congenital ataxia and tremor with cerebellar hypoplasia in piglets borne by sows treated with NeguvonTM vet. (Metrifonate, trichlorfon) during pregnancy. *Nord Vet Med.* 1978;30: 538-545.
8. Bolin SR (1992). Congenital Tremors Virus. In: Leman AD, Straw BE, Mengeling WL, D'Allaire S, Taylor DJ editors. Diseases of swine. London: Wolfe Publishing Ltd. pp. 247-249.
9. Hause B, Collin EA, Peddireddi L, Yuan F, Chen Z, Hesse RA, et al. Discovery of a novel putative atypical porcine pestivirus in pigs in the United States. *J Gen Virol.* 2015. doi: 10.1099/jgv.0.000251.

The effects of tulathromycin injectable solution on reducing the transmission of swine respiratory pathogens from sows to wean pigs

Timothy Pearson[1], BS; Seth Krantz[2], DVM; Lucina Galina Pantoja[3], DVM, PhD
[1]College of Veterinary Medicine, University of Tennessee, Knoxville, Tennessee; [2]Tosh Pork, LLC, Henry, Tennessee; [3]Zoetis, Florham Park, New Jersey

Introduction

Diseases resulting from swine respiratory pathogens cause significant economic impact to the swine industry. *Mycoplasma hyopneumoniae, Haemophilus parasuis, Actinobacillus suis, Mycoplasma hyorhinis, Streptococcus suis, Pasteurella multocida,* and *Actinobacillus pleuropneumoniae* are some of these pathogens. The prevalence of infection by *M. hyopneumoniae* at weaning can be a predisposing factor for the development of respiratory disease in later stages of production.[1] It is probable that the presence of these other respiratory pathogens at weaning also contribute to disease in the finishing stage of production. In 2013, it was demonstrated that a single tulathromycin injection in sows, pre-farrowing, reduces the shedding of *M. hyopneumoniae* from sows to weaned pigs. The research showed a resulting reduction of *M. hyopneumoniae* positive pigs at weaning.[2] With this study, we aimed to expand upon these results by looking at a two-dose tulathromycin treatment, other respiratory pathogens, a uniform parity distribution, and a more sensitive sampling technique. Tulathromycin, a macrolide antibiotic of the subclass triamilide, is an effective antibiotic to treat several respiratory pathogens. Tulathromycin acts by disrupting protein synthesis which, at low concentrations is bacteriostatic, but at higher concentrations it is bactericidal. Due to its high volume of distribution, resulting in high systemic availability, tulathromycin concentrations can be higher in the tissues (i.e. the respiratory tract) than in the plasma. The purpose of this study was to evaluate the effect of a two-dose treatment of tulathromycin injectable solution in sows on the colonization of respiratory pathogens in piglets at weaning.

Materials and methods

This research was conducted at a 3,000 sow commercial breed-to-wean farm. The sows in this study all shared a common feed source and were porcine epidemic diarrhea virus (PEDV), swine influenza virus (SIV), and porcine reproductive and respiratory syndrome virus (PRRSV) negative. All sows in this study were PIC C29 females.

Treatment groups

Seventy-two sows were randomly selected for this study from a pool of sows with an expected farrow date within one week. Those sows were evenly stratified into three different groups based on parity: Parity 0 (P0), Parity 1-2 (P1-2), and Parity 3+ (P3+). Twenty-four sows were allocated to each parity group. The sows within each parity grouping were then randomly assigned by an Excel® randomization program to either a treatment or control group and evenly allocated between four farrowing rooms. All sows were randomly assigned to a treatment and farrowing stall location. Six sows per parity group were allocated per room, resulting in 18 trial sows per room. Of the six sows per parity group in a room, three were assigned to receive tulathromycin injections and three were assigned to the control group (Table 1) (Figure 1). The 36 treatment sows received two intramuscular injections (IM) of tulathromycin injectable solution (Draxxin®, Zoetis), based on the label dose (1 mL/88 lbs). Sows were not weighed individually. Estimated weights were utilized based on the parity group in order to have uniformly treated groups. The P0 group was estimated to weigh 400 lbs (181.8 kg) and received 4.5 mL of tulathromycin. The P1-2 group was estimated to weigh 440 lbs (200 kg) and received 5 mL of tulathromycin. The P3+ group was estimated to weigh 570 lbs (259.1 kg) and received 6.5 mL of tulathromycin. The 36 control sows were each given 5 mL of phosphate buffered saline as an IM injection in the same location as the treated sows. All injections were administered in either the left or right cervical region just caudal to the ear. The first injection was administered on the left side of the sow at approximately 3 days pre-farrow. The second injection was administered on the right side of the sow at 7 days post-farrow. At the time of each treatment, a sample was collected by a novel laryngeal scraping technique on each sow.[3] All sows were sampled at 3 days pre-farrow, 7 days post-farrow, and the day prior to litters being weaned. Five piglets from each sow were randomly selected. These piglets were sampled by laryngeal swab at 7 days of age and the day prior to weaning (day 20).

All animals in this study were raised and handled on farms that are Pork Quality Assurance Plus certified and adhere to standards set forth by the National Pork Board.

Sampling technique

The sow samples in this study were collected by a laryngeal scrape technique. A sample was obtained by first restraining the sow, then a speculum was placed to keep the mandible open. A smooth, long handled, spoon was used to gather the mucous at the back of the larynx and then transferred onto a swab.[3] The sample was stored in Stuart transport media tubes and frozen for shipping. The piglets were restrained and sampled by a laryngeal

Table 1: This table represents the distribution of sows in the farrowing rooms and their treatment in the study.

Parity	Room 1	Room 2	Room 3	Room 4	
P0	3 control	3 control	3 control	3 control	24 sows
	3 treated	3 treated	3 treated	3 treated	
P1-2	3 control	3 control	3 control	3 control	24 sows
	3 treated	3 treated	3 treated	3 treated	
P3+	3 control	3 control	3 control	3 control	24 sows
	3 treated	3 treated	3 treated	3 treated	
Total	**9 control**	**9 control**	**9 control**	**9 control**	
	9 treated	**9 treated**	**9 treated**	**9 treated**	
	18 Sows per Room	18 Sows per Room	18 Sows per Room	18 Sows per Room	**72 sows total**

swab technique. The swab was used to collect the mucus directly from the larynx and then stored in a similar manner to the sow samples.

Diagnostic analysis

Samples were submitted to the University of Minnesota Veterinary Diagnostic Laboratory for *M. hyopneumoniae, H. parasuis, A. suis,* and *M. hyorhinis* testing by real-time polymerase chain reaction (PCR). All sows laryngeal scrape samples were individually tested for each pathogen. Laryngeal swabs from individual piglets were pooled by litter. The primary objective was to determine the percentage of litters that were PCR positive for *M. hyopneumoniae, H. parasuis, A. suis,* and *M. hyorhinis* in treated pigs compared to control pigs. Next, the number of piglets found positive for respiratory pathogens was analyzed based on parity of the sow.

Statistical analysis

Chi square tests and analysis of variance (ANOVA) were utilized to detect variations in the percentage of positive litters between groups and parities with $P \leq 0.05$ considered significant.

Results and discussion

Five sows had to be eliminated from the study. Three of the sows did not survive to the end of the study due to death or euthanasia. Two of the sows were removed from the study to be nurse sows for other fallback litters. Their piglets were fostered onto other sows outside of the study removing them from the sampling group. All samples were collected by laryngeal scrape rather than tonsillar or nasal swabs. This method was chosen due to the increased sensitivity to detect these pathogens which cause the most issues in the lower airway. The laryngeal scraping technique was found to have the same sensitivity for detecting *M. hyopneumoniae* as the tracheo-bronchial catheter technique.[3] However, the laryngeal scrape technique was used because it is an easier sampling method and can be easily taught to a farm employee for surveillance of these pathogens. In order to determine a pathogen prevalence of 20% with a laryngeal scrape sensitivity of 50%, a sample size of 29 animals would be enough to find positive cases in a population of 1,000 sows.[4] This minimum sample size was exceeded in the study by having 34 sows in the control group and 33 sows in the treated group. All samples tested PCR negative for *M. hyopneumoniae* prior to treatment, thus this pathogen was not further evaluated in the study. The respiratory pathogen colonization results for the sows at 3 days pre-farrow are represented in Figure 2, which is the baseline status for the sows. Both groups had the same percentage of sows positive for *H. parasuis* and *A. suis*. However, the treatment group had a higher percentage of sows positive for *M. hyorhinis*, immediately before treatment was applied. There was a statistically significant difference in colonization of specific respiratory pathogens in piglets from the tulathromycin treated group as compared to piglets from the control group. Seven day old piglets from tulathromycin treated sows showed a significant reduction in colonization of *A. suis* ($P = 0.004$) and *H. parasuis* ($P = 0.005$). The percentage of positive piglets for all pathogens was lower, compared to the control piglets at 7 days of age (Figure 3). The samples for the sows at day 7 post-farrow were misplaced at the lab so there are no results to report from that sample. Weaning age piglets from tulathromycin treated sows, also showed a statistically significant reduction in colonization of *A. suis* ($P = 0.0002$) and *M. hyorhinis* ($P = 0.0007$) (Figure 4). This is important because, while the treated sows had a higher percentage of positive animals for *M. hyorhinis* at the beginning of the study, the percentage of positive piglets from the treated litters by weaning was significantly lower compared to the control piglets. The results for the treated sows at weaning showed a reduction for *H. parasuis* and *M. hyorhinis* compared to the pre-farrow sample. Both groups showed a reduction in the colonization of *A. suis* compared to the pre-farrow sample. In this study, the sow

Figure 1: This is an example of where the sows were allocated in two of the farrowing rooms. Each grid represents a farrowing stall in the room.

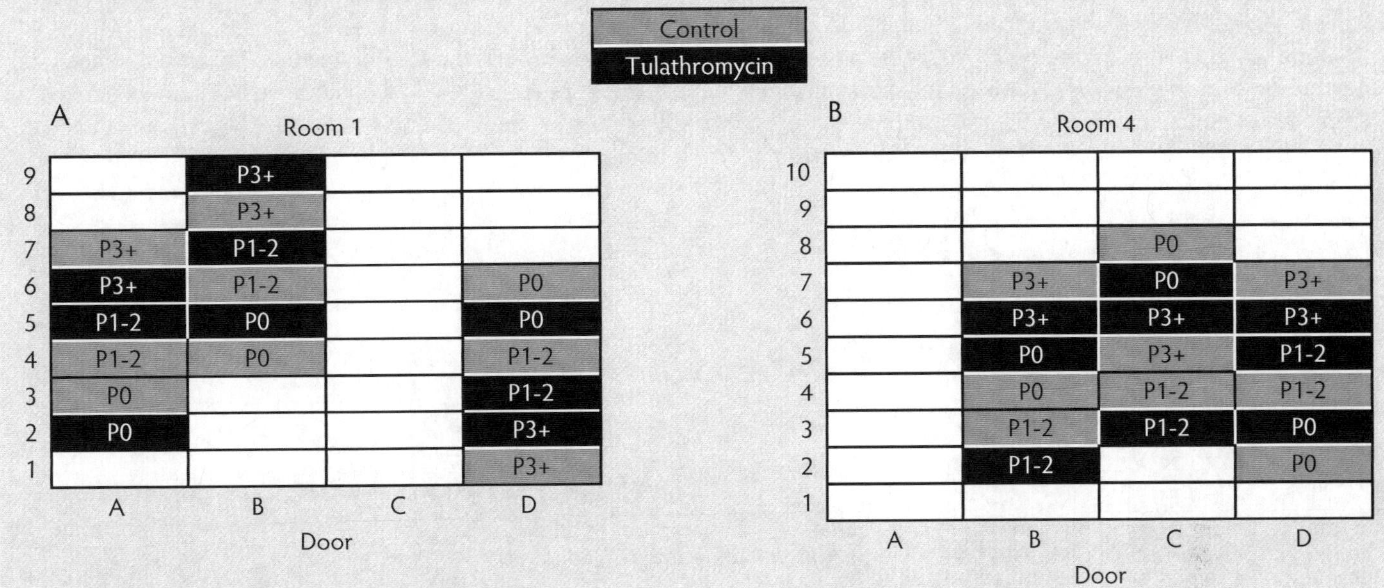

parity did not influence pathogen prevalence. The use of injectable tulathromycin in the sows to reduce transmission of *M. hyorhinis* and *A. suis* in wean piglets proves to be a viable option to improve the health status of wean piglets. In any production system, pathogen reduction of weaned pigs can lead to improved health through marketing with resulting improvements in growth performance and throughput. Significant reductions in medication costs for control and/or treatment of respiratory disease during the growing phase represents major financial advantages for swine producers. Antibiotic free production systems are becoming increasingly more common and uncontrolled respiratory disease in these settings can be detrimental to animal well being. Implementing mass antibiotic treatment eliminates the profit potential for this type of production. This study offers a solution to control respiratory disease while maintaining the antibiotic free status of the growing pigs.

Acknowledgements

The authors would like to sincerely thank the following individuals for their editorial contributions and sample collection assistance: Dr. Laura Greiner, Dr. Wesley Lyons, Ms. Katie Sherman and the staff of the Tosh Huntland sow unit.

References

1. Fano E, Pijoan C, Dee S. Effect of *Mycoplasma hyopneumoniae* colonization at weaning on disease severity in growing pigs. *Can J Vet Res*. 2007;71:195-200.

2. Holst S, Yeske P, Leuwerke B, Swalla R, Davies P, Pieters M. Effect of pre-farrow administration of tulathromycin injectable solution on *Mycoplasma hyopneumoniae* prevalence in suckling pigs at birth and weaning. *Proc AASV*. San Diego, California. 2013;89-90.

3. Sievers C, Dalquist L, Sponheim A, Fano E, Wetzell T. Comparison of ante-mortem sampling procedures to post-mortem bronchial sampling for the detection of *Mycoplasma hyopneumoniae* by PCR. *Proc AASV*. Orlando, Florida. 2015;115-116.

4. EpiTools epidemiological calculators. Available at: **http:epitools.ausvet.com.au/content.php?page=PrevalenceSS**. Accessed 15 November 2015.

Figure 2: Percent of individual PCR positive sows 3 days pre-farrow (n = 34 control and 33 tulathromycin treatment). This graph represents the pathogen prevalence of the sow population at the beginning of the study and immediately prior to treatment. There was an even distribution of positive sows in each treatment group for *Haemophilus parasuis* and *Actinobacillus suis*. However, there was a higher percentage of positive sows in the treatment group for *Mycoplasma hyorhinis*.

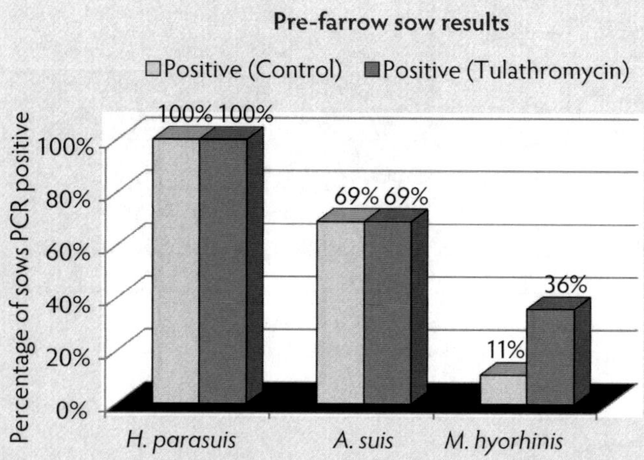

Figure 3: Percent of PCR positive piglet pools on day 7 of age (n = 34 control and 33 tulathromycin treatment). For all pathogens, the treated group had a lower percentage of positive piglets as compared to the control; however, these differences were statistically significant only for *Haemophilus parasuis* and *Actinobacillus suis*.

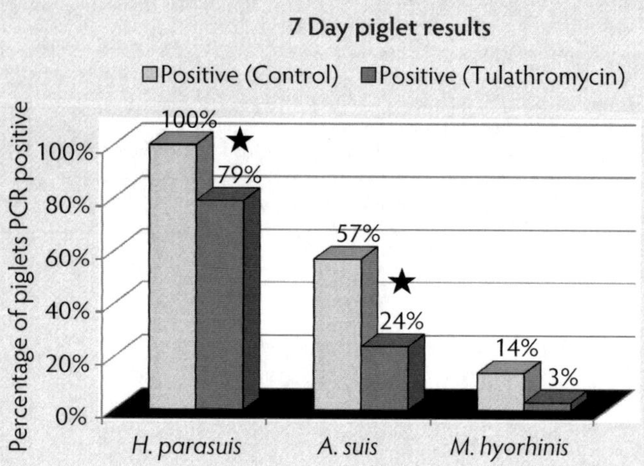

★ Significant differences found at $P \leq 0.05$

Figure 4: Percent of PCR positive piglet pools on day 20 of age (n = 34 control and 33 tulathromycin treatment). By time of weaning, all piglet samples tested positive for *Haemophilus parasuis*. For *Actinobacillus suis* and *Mycoplasma hyorhinis*, the treated group still had a significantly lower percentage of positive piglets as compared to the control. This is most notable for *M. hyorhinis*, in which the treated sows had a higher percentage of positives than the control sows, at the beginning of the study.

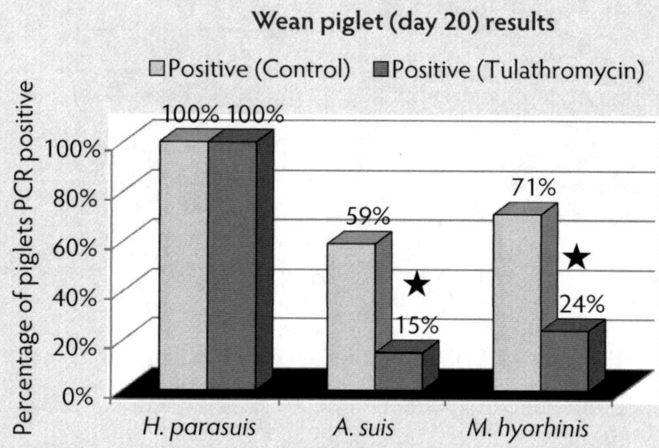

★ Significant differences found at $P \leq 0.05$

Correlation between aggression and cognitive bias in gestating sows

S. A. Manoharan; K. M. Horback, PhD; T. D. Parsons, VMD, PhD, DACAW
University of Pennsylvania School of Veterinary Medicine, Kennett Square, Pennsylvania

Introduction

Poor welfare is often measured in terms of physical attributes such as skin lesions, lameness, and body condition score. Objective assessments of an animal's mental state are more difficult and thus somewhat rare in the study of farm animal welfare. The introduction of new sows into a group results in aggressive interactions while the dominance hierarchy is established. The welfare of sows decreases if they experience fear, injury or pain during these fighting episodes. Recent debates about the psychological needs of sows encouraged us to explore the use of a measure of welfare purported to reflect the animal's psychological health called cognitive bias.[1] To observe the behavioral responses of individual sows during the cognitive bias training and testing, and see how the cognitive outcomes correlate with natural occurring behaviors such as aggression at the time of mixing.

Materials and methods

At the Swine Teaching and Research Center of the University of Pennsylvania Veterinary School, gestating sows (PIC 1050: Landrace-Yorkshire) are housed in pre-implantation, dynamic groups and fed via electronic sow feeding stations (Compident 7, Schauer Agrotronic). Twenty-six sows (parity 2) were trained a "go/no-go" operant task in a 7 × 2.4m arena for four days. The positive "go" cue (bowl with 200g of corn-soy feed) was placed in the far left or right corner of the field, while the negative "no-go" cue was placed of the same feed bowl in the opposite corner. Approaching the positive cue was rewarded with access to the feed, while approaching the negative field elicited a punishment of a red flag wave and no access to feed. The location of positive and negative cues was randomized for each sow. On the 5th and 6th day, sows were given ambiguous trials with a bowl placed equidistant from the positive and negative bowl placements. A cognitive bias index (CBI) was calculated for these sows based on their latency to approach the first ambiguous cue. It was scaled such that 0 would reflect a failure to approach the ambiguous cue, whereas 1 represents an approach comparable to a positive cue. Individual responses to cognitive bias test were compared to the behavior of each sow when mixed into the large gestation pen three days post-breeding. Behaviors recorded include rate of chasing, initiating and receiving bites, and frequency/duration of fighting. To determine significant relationships between variables, Spearman's rank correlations (r_s) were conducted, with alpha = 0.05 (SPSS® V. 23).

Results and discussion

Sows exhibited different rates of learning of the cognitive bias task, with "no go" proving more difficult to master than "go". Those sows that struggled during the learning phase were more likely to be aggressive at mixing (Table 1). Sows which chased others more often when mixed into pen after first farrow also approached the negative cue more often ($r_s = 0.45$, $P = 0.03$) and more quickly ($r_s = -0.44$, $P = 0.03$). Sows which chased others for longer durations also approached the negative cue more often ($r_s = 0.46$, $P = 0.02$) and more quickly ($r_s = -0.46$, $P = 0.02$). Sows which bit other sows more often during mixing also approached the negative cue more often ($r_s = 0.43$, $P = 0.03$). Similarly, the sows which interpreted the ambiguous cue as positive (CBI scores > 0.6) also engaged in more fights during pen mixing ($r_s = 0.5$, $P < 0.05$) and initiated more bites to other sows during mixing ($r_s = 0.5$, $P < 0.05$). Our data suggests that cognitive outcomes correlate with aggressive behavior during the establishment of social hierarchy. Aggressive animals were more challenged by the learning paradigm, but were more likely to treat an ambiguous stimulus as positive. Sows with a positive cognitive bias may simply be "risk-takers" (e.g. less risk-adverse). While the cognitive bias test has been used in some animal welfare studies to determine differences in affective state, we argue that here, this psychological test may be a more appropriate method of assessing behavioral phenotypes (i.e., bold-shy continuum). As such, we are able to use this information to modify group housing by taking into account these personality differences and housing sows of similar behavior phenotypes together.

Table 1: Spearman's correlation coefficient (*P*-value) between aggressive behaviors at pen mixing and incorrectly approaching a negative cue.

	Approach negative cue (frequency)	Approach negative cue (speed)
Chase other sow (frequency)	0.45 (0.03)	-0.44 (0.03)
Chase other sow (duration)	0.46 (0.02)	-0.46 (0.02)
Bite other sow (frequency)	0.43 (0.03)	-0.4 (0.05)

Acknowledgements

Supported by Pennsylvania Pork Producers Council, New York Farmers, and Swiss Village Foundation.

References

1. Mendl M et al. Cognitive bias as an indicator of animal emotion and welfare: Emerging evidence and underlying mechanisms *Appl Anim Behav Sci*. 2009; 118:161-181.

Evaluation of diagnostic sampling techniques used in the farrowing house to increase sensitivity of detection and sequencing of influenza A virus in swine (IAV-S)

Ryan Strobel[1]; Brett O'Brien[2], VMD, MS; Joel Nerem[3], DVM; Spencer Wayne[3], DVM, PhD; Bob Morrison[1], DVM, MBA, PhD; Ana Alba-Casals[1], PhD

[1]College of Vet Med, University of Minnesota, Saint Paul, Minnesota; [2]Zoetis, Florham Park, New Jersey; [3]Pipestone Veterinary Services, Pipestone, Minnesota

Introduction

Influenza A virus in swine (IAV-S) is widespread throughout the US and is one of the major infectious agents that contributes to respiratory disease in swine.[1] Current industry standards utilize two sampling methods to detect and sequence IAV-S in piglets in the farrowing house: oral fluids (OF) and nasal swabs (NS). NS sensitivity can be low because acutely infected pigs must be sampled. OF sequencing can be challenging due to interfering factors that lower the viral concentration.[2] From a herd health standpoint, detecting and sequencing IAV-S is crucial to downstream health and having the potential to wean influenza negative piglets.

Objectives

1) To compare four different testing techniques in farrowing for both isolating IAV-S and sequencing capabilities on a site basis as well as NS and Oropharyngeal (OP) on individual samples.
2) Establish a sequence detection cutoff from cycle time (Ct) on polymerase chain reaction (PCR) for OP technique.

Materials and methods

Seven sow units were selected based on history of endemic IAV-S status as defined by at least one positive PCR result on monthly oral fluid (OF) sampling within the previous six months of the trial start date. Four different sampling methods were compared on 10 different sampling days in the farrowing house: oropharyngeal (OP), nasal swab (NS), oral fluids (OF) and tissues. Tissue sampling involved harvesting lungs (pooled by four) from a maximum of eight non-select pigs (e.g. lame) that were 10-21 days of age. The OF collection procedure consisted of hanging 1-4 cotton ropes in the farrowing crates immediately prior to weaning. NS and OP were collected on 15 pigs (10-21 days of age) that showed clinical signs of acute IAV-S infection (bilateral serous nasal discharge and/or fever). For NS technique viral transport swabs (BD, Sparks, MD) were inserted into both nares. For OP, viral transport swabs were inserted to the back of the oral cavity on the same pigs as NS. NS and OP were pooled by three and results were compared directly to each other using a Kappa analysis. Established Ct cutoff values for sequencing were used at Ct < 25 for NS and tissues and Ct < 20 for OF. A sequence was attempted (virus isolation) on all OP samples that were PCR positive. Receiver operating characteristic (ROC) curve analysis was used to establish a Ct cutoff for OP swabs positive to PCR.

Results

Detection of IAV-S at a site was 100% for OF (9/9), 90% for both OP and NS (9/10) and 70% for tissues (7/10). When comparing pooled samples, tissues had the highest sequence rate at 73% (8/11, with a 95% Confidence Interval (CI) 42-92%) followed by NS at 58% (15/26, CI 38-75%), OP at 43% (13/30, CI 26-61%) and 0% for OF. On a site basis, however, OP was the most successful at yielding at least one sequence at a sampling time at 70% (7/10) followed by NS and tissue at 60% (6/10) and zero for OF. Table 1 shows the average Ct from each sampling day as well as the percent of positive virus isolation results. Between NS and OP on individual pigs there was a fair agreement on IAV-S detection with a kappa value at .36 ($P = 0.15$). Estimated test sensitivity for IAV-S detection was 0.63 for NS and 0.77 for OP. In this study, the established Ct cutoff for OP sampling to yield an IAV-S sequence positive was Ct < 25.

Discussion

Novel OP technique was the most successful at isolating IAV-S and obtaining a sequence on a site basis in this study and on the basis of these results should be considered as an alternative method to more effectively detect IAV-S in the farrowing house. Having the capability to detect and sequence IAV-S allows producers to more accurately assess their intervention strategies and improve monitoring.

References

1. Gauger P, Vincent A, Harmon K, Zhang J. 2015. Influenza A viruses in swine: diversity, diagnostics and interpretation. *Proc AASV, Pg. 493-501.*

2. Janke B. 2012. Influenza Virus Infections in Swine: Optimizing Diagnostic Efforts. *Proc ISU disease conference*, Pg. 22-26.

3. Ohrt K, Harris C, Kolb J, Edler R. 2015. Comparison of sampling methods for detection of influenza A virus. *Proc AASV, Pg. 281.*

Table 1: Average cycle times (CT) from each sampling day on positive polymerase chain reaction (PCR) and percent positive on virus isolation (VI) results.

Sampling day	NS (Ave.CT)	NS VI %positives	OP (Ave.CT)	OP VI %positives	OF (Ave. CT)	OF VI %positives	Tissue (Ave. CT)	Tissue VI %positives
1	26.89	83%	25.1	40%	26.7	0%	18.35	100%
2	28.47	0%	24.68	25%	24.48	0%	19.54	100%
3	31.15	0%	34.5	0%	28.16	0%	Neg	0%
4	25.08	66%	25.15	33%	27.85	0%	35.04	0%
5	28.3	33%	28.57	33%	N/A	N/A	20.35	100%
6	37.55	0%	Neg	0%	25.84	0%	38.43	100%
7	27.22	50%	22.33	100%	30.53	0%	Neg	0%
8	34.28	40%	32.07	0%	31.09	0%	Neg	0%
9	23.96	60%	26.49	40%	25.87	0%	23.71	50%
10	Neg	0%	33.23	20%	37.79	0%	Neg	0%
Average:	29.2	55.3%	28.01	41.57%	28.7	0%	25.9	90%

NS = Nasal Swab, OP = Oropharyngeal, OF = Oral Fluids, VI = Virus Isolation

Assessing and ranking sow farm biosecurity within a production system

Brent Sexton[1], BS; Elissa Toohill[2], DVM; Clayton Johnson[2], DVM; Darrell Neuberger[3], DVM
[1]Iowa State University, College of Veterinary Medicine, Ames, Iowa; [2]The Maschhoffs, LLC, Carlyle, Illinois; [3]Zoetis, Florham Park, New Jersey

Introduction

Production companies operate on strict budgets which mandate that investments in biosecurity measures offer adequate returns on investment. A ranking system of the relative sow farm biosecurity is beneficial in a number of ways: helps prioritize capital allocation among farms; provides a method to objectively identify and measure factors that contribute to a farm's biosecurity; ensures each individual farm has acceptable biosecurity measures in-place, taking into consideration the various factors that interact with that farm; and ensures judicious use of the production system's resources. The objective of this project was to customize a biosecurity assessment tool and use it to rank farms within the production system in order to determine which farms have the greatest need of capital improvements to enhance the biosecurity of the individual farm, of the other farms within the area the farm is located, and for the entire swine company as a whole.

Materials and methods

A two-part biosecurity assessment form focusing on location and farm practices was the starting point for this project. That form was adapted for use within a large swine company in order to assess over 70 sow farms across several states. The location assessment focused on geographic factors that would affect the spread of disease such as presence of windbreaks, terrain, and distance from roads, as well as local and regional swine density, distance to closest farm, and relation to nearby farms. The farm practices assessment was broken into several sections: non-moveable risks, moveable risks, health management procedures, internal risks, and pig and semen transport. Non-moveable risks focused on farm facilities and related practices concerning: entry onto the farm, gilt isolation, animal loading and unloading areas, and perimeter fences. The moveable risks pertained to non-fixed factors, such as: waste management, water supply, and feed delivery. Health management procedures focused on the farm's adherence to vaccination protocols, medication storage and health equipment hygiene. Internal risks involved both the removal and disposal of dead animals, as well as pest control. The final category, pig and semen transport, focused on the vehicles and trailers used to move pigs and semen onto and off of the farm. These categories work to provide a more comprehensive look at biosecurity risks associated with both the location and the individual farm.

The assessment was converted to an excel file which allowed for several sites to be completed simultaneously. The assessment was also modified to eliminate redundant questions and standardized for use across several regions. These assessments were filled out jointly by a trained employee and service managers who may oversee as many as eight different farms. The meetings were held in person or via telephone with a shared computer screen in order to improve accuracy. The assessments were typically completed in under an hour and a half. The responses were transcribed to the original assessment then submitted to an independent company for scoring. The scoring was weighted so that factors which have greater significance to the biosecurity of the farm earned more points than less significant factors.

In addition to the location and site assessments, a supplemental assessment was created to focus on assets or management practices that were not identified by the original assessments. This third assessment helped better differentiate and compare biosecurity factors between sites. The creation of the supplemental assessment required the development of the questions, potential responses, and appropriate scores. Compiling the results from the three assessments on a spreadsheet allows the farms to be sorted by 34 different variables such as by region and by specific production practices. Each assessment's results were run through a correlation analysis by an additional independent third party to identify the strength of the relationship between the scores and the farm's porcine reproductive and respiratory syndrome (PRRS) break-rates.

Results

The site and location assessments were each scored out of a total of 1,000 possible points and the supplemental assessment was scored out of a total of 320 possible points, with higher scores being more favorable for all three assessments. The highest scoring farms on the location portion were those in geographic areas of low swine density. The location assessment had a spread of over 700 points out of 1,000 from the highest scoring farm to the lowest scoring farm. The highest scoring farms on the site assessments were those doing well in the non-moveable risk category which identifies a farms' physical set up, such as perimeter fences, loading facilities, and isolation facilities. The non-moveable risk section alone saw a variation of over 200 points, while the entire site assessment had a variation of just 300 points out of 1,000. The relatively small variation in points is reflective of the similar operating procedures among the farms in the system. Many questions regarding health protocols, transportation, and nutrition had the same or very similar responses across the entire system. In contrast to the relatively low variation between scores on the site assessment, the supplemental assessment had a variation of over 200 points, out of just 320 points. According to the correlation analysis, both the location assessment and the supplemental assessment had a correlation of .79 with the PRRS break-rate; and the site assessment, which had more uniformity among farms, had a correlation of .63 with the PRRS break-rate.

Discussion

To optimize biosecurity in a cost effective manner, capital improvement funds must be distributed judiciously, and this project provides a tool to help achieve that goal. By using this tool, the user can arrange and rank the farms by several parameters and factors. This allows the lower scoring farms, which theoretically have the highest probability of a disease outbreak, to be identified. Improving the biosecurity of these farms will not only reduce the chances of an outbreak of disease on the farm, but also reduce the risk of any disease outbreak spreading to nearby farms. The results of this project have been used to justify altering management practices and making capital improvements. The results have also been used to identify areas of focus when training employees on the importance of biosecurity compliance. While the assessments have not been statistically validated to definitively rank farm biosecurity, they do allow a production system to identify strengths and weaknesses among its farms and ultimately prioritize resource allocation based upon an objective analysis.

Effect of administering tulathromycin to IAV-S positive weaned pigs to control swine respiratory disease

Kathleen Wood[1], BS; Brett O'Brien[2], VMD, MS; Sam Holst[3], DVM; Abigail Redalen[2], DVM; Andrea Hanson[3], PhD
[1]North Carolina State University College of Veterinary Medicine, Raleigh, North Carolina;
[2]Zoetis, Florham Park, New Jersey; [3]Swine Vet Center, Saint Peter, Minnesota

Introduction

Influenza A virus in swine (IAV-S) is an acute, highly transmissible respiratory disease that can affect all phases of swine production. Mortality associated with IAV-S is low; however, morbidity is high and can result in reduced body weight/condition, uneven growth, and prolonged finishing time.[1] IAV-S is characterized by excessive pulmonary inflammation and hypersecretion, as well as impaired immunity to secondary pathogens. Macrolide antibiotics, specifically tulathromycin, have demonstrated anti-inflammatory properties beneficial for treating swine respiratory disease (SRD).[2] These immunomodulating properties may benefit animals with respiratory inflammation caused by SRD, including IAV-S. The objective of this study was to evaluate the health and performance impact of administering tulathromycin (Draxxin 25®) to IAV-S positive weaned pigs.

Materials and methods

Pigs were sourced from 2 IAV-S positive sow farms and double-stocked into a 2000-head wean-to-finish barn. Weaned pig IAV-S status was confirmed via on-site FluDETECT testing and IAV-S PCR on nasal swabs and oral fluids. Upon arrival, 909 pigs were weighed, sexed, ear-tagged and placed into pens. Pens were blocked by gender, weight, and sow farm source. After pen placement, pigs were randomly assigned to either the tulathromycin (T01) or control (T02) group based on weight, gender, and source. T01 pigs received a 2.5 mg/kg intramuscular injection of tulathromycin; T02 pigs were handled similarly but did not receive an injection. Tissue samples were collected from mortalities during the first 2 weeks post-placement and submitted for diagnostic testing to characterize bacterial pathogens present. During the first 7 days post-placement, pigs were individually evaluated daily and assigned a clinical activity score (CAS) (0-normal, 1-mild depression, 2-moderate depression, 3-severe depression). Caregivers and CAS scores were blinded to treatment groups to avoid bias. Rectal temperatures were recorded during this time from a subset of pigs displaying clinical signs of IAV-S (lethargy, fever, anorexia, nasal discharge). Oral fluids were collected on 1, 3, 5, and 7 weeks post-placement from 5 designated trial pens and submitted for IAV-S PCR. Mortality, individual treatments, and pigs removed to hospital pens were logged throughout the trial. Final weights were recorded on roughly Day 49 post-placement. All samples were analyzed at the University of Minnesota Veterinary Diagnostic Laboratory. Least square means for measured performance data were compared using Fisher's Exact test, Chi Square test, or ANOVA using JMP 7.0 software (SAS, Cary, NC).

Results and discussion

Oral fluids were IAV-S PCR positive at placement and at 1 and 5 weeks post-placement; confirmed secondary bacterial pathogens included *P. multocida, M. hyorhinis*, and *H. parasuis*. As shown in Table 1, treatment with tulathromycin was associated with significant increases in ADG ($P = 0.002$) and final weight ($P = 0.001$). Mortality, mean rectal temperature, and mean CAS were not significantly different. Additionally, there was a trend for fewer pigs removed to hospital pens in T01 compared to T02 ($P = 0.085$). Results from this study indicate that treatment with tulathromycin was associated with improved nursery performance. Administering tulathromycin to IAV-S positive weaned pigs is a potential tool to help mitigate the post-weaning impact of IAV-S.

References

1. Neumann E, Ramirez A, Schwartz K. Influenza. *Swine Disease Manual* (4th ed.). Perry, Iowa. 2009;55-56.

2. Duquette SC, Fischer CD, Williams AC, Saman S, Feener TD, Bhargava A, Reti KL, Muench GP, Morck DW, Allison J, Lucas MJ, Buret AG. Immunomodulatory effects of tulathromycin on apoptosis, efferocytosis, and proinflammatory leukotriene B4 production in leukocytes from Actinobacillus pleuropneumoniae-or zymosan-challenged pigs. *Am J Vet Res.* 2015;76(6):507-519.

Table 1: Performance data

	No. allocated	No. for analysis	Start wt, lb	Final wt, lb	ADG, lb/d	Removed to hospital pen, %	Mortality, %	Mean rectal temp, °F	Mean CAS
T01	454	414	15.3[a]	62.5[a]	0.97[a]	7.5[a]	0.96[a]	103.3[a]	0.4[a]
T02	455	400	15.2[a]	60.7[b]	0.94[b]	11.0[a]	0.74[a]	103.5[a]	0.4[a]

a,b Values within the same column with different superscripts are significantly ($P \leq 0.05$) different

Characterization of shedding and viremia in naïve or vaccinated replacement gilts given 1-7-4 RFLP PRRS live virus inoculation

T. Engle[1]; S. Clark[1]; W. Burton[2]; J. S. Pittman[2]
[1]Virginia-Maryland College of Veterinary Medicine, Blacksburg, Virginia;
[2]Smithfield, Hog Production Division, North Region, Waverly, Virginia

Introduction

The estimated economic impact of porcine reproductive and respiratory syndrome virus (PRRSV) in the United States is approximately $66.75 million for breeding herds and $493.57 million for growers.[1,2] In the spring of 2014, an emergent and highly virulent strain of, characterized with 1-7-4 Restriction Fragment Length Polymorphism (RFLP),[3] devastated swine operations in North Carolina and other regions of the country. Breeding farms affected by the 1-7-4 strain experienced reproductive failure as well as sow (10-25% annualized) and piglet (40-100%) mortality, regardless of previous immune status and vaccination strategies.[4] Growing pigs experienced respiratory disease and increased mortality (10-50%).[4] Although current industry strategies for reducing morbidity and mortality associated with PRRSV involve vaccinating naïve animals, these methods may not produce sufficient immune stimulation for all circulating and emerging isolates.[5] Diversity in PRRSV isolates and the variable heterologous level of cross-protection produced by current PRRSV vaccines reduces vaccine effectiveness. Some research has suggested alternative vaccination methods such as planned exposure through presenting infected animals or tissues and aborted fetuses to naive animals to stimulate long-term immunity; however, these approaches have not been well described and may be unreliable in producing uniform infections.[1,6] Other approaches for exposure-based immune stimulation involve inoculating naïve animals with serum from PRRSV infected pigs (Live Virus Inoculation (LVI)) to ensure exposure to resident strains; though these methods have not been well documented in relation to safety or efficacy.[5] With recent PRRSV outbreaks movement of naïve gilts into positively populated farms is an increasing concern for sow farm stability. Positively populated sow farms can wait until the herd tests negative to bring in new naïve gilts, or immunize incoming naïve gilts, which may not provide sufficient cross-protection and cause the farm to destabilize. Investigative strategies to stabilize naïve gilts introduced to positive populated farms utilize Live Virus Inoculation (LVI) containing endemic circulating PRRSV isolates recovered from destination sow farms to provide homologous immune protection. The objective of this project was to explore strategies involving LVI to characterize shedding duration, mortality, and time to reach negative herd status using single dose LVI or PRRSV Modified Live Virus vaccine (MLV) prior to LVI.

Materials and methods

Gilts used for experimental inoculation were selected from traditional PRRSV-negative multiplication gilt flow systems. Selection criteria and housing assignment for gilts was based on available and current commercial sow farm flows; either sow farm A or sow farm B. Enrolled gilts were not specifically tested for presence of PRRSV at the beginning of the study, however all gilts originated from farms where PRRSV status was tested and confirmed negative every 2 weeks prior to and proceeding shipment. PRRSV naïve gilts (11 to 20 weeks old) were housed at separate off-site Gilt Development Units (GDU: A and B corresponding to associated sow farm) each containing 450 gilts. Natural circulating 1-7-4 PRRSV was isolated from weaned pig blood from corresponding destination sow farms and used to prepare GDU specific LVI. Pooled serum samples were submitted to South Dakota State University and analyzed for PRRSV via quantitative PCR (qPCR), sequenced for PRRSV to confirm 1-7-4 strain identification, and a Tissue Culture Infectious Dose^{50} (TCID^{50}) was performed. Samples were also tested for porcine circovirus type 2 (PCV) via PCR. The LVI was diluted to 10^3 TCID$_{50}$ ml^{-1} (1 ml dose) using 1X phosphate buffered saline. Gilts at GDU A were given 1 ml LVI intramuscularly (IM) upon arrival (Day 0), while gilts at GDU B gilts were given 2 mls PRRSV MLV (Ingelvac PRRS MLV, Boehringer Ingelheim Vetmedica, Inc., St Joseph, MO) IM upon arrival (Day 0), and 1ml LVI IM 28 days later. Blood (n = 45) and oral fluid (OF) (n = 10: GDU A or n = 12: GDU B) samples were collected every 2 weeks from the same ear-tagged gilts (blood) and pens (OF), respectively (Table 1). Samples were matched such that an OF sample was collected from pens where five blood samples were also collected. Blood samples were sent to Iowa State University Veterinary Diagnostic Laboratory (ISU-VDL) for pooled (by 5 based on pen grouping) analysis for PPRSV via qtPCR and individual ELISA. Oral fluid samples were sent to ISU-VDL for individual analysis for PPRSV via qtPCR and ELISA. Selected samples at particular time points were sequenced to confirm that LVI and MLV were the prominent virus during post-exposure phases of viremia. Mortality was recorded daily throughout the study.

Results

All gilts seroconverted to PRRSV by PCR 7 days after day 0 MLV and LVI injections. All serum and OF were PCR positive

Table 1: Timeline for experimental challenges for both LVI and MLV prior to LVI.

Live virus inoculation GDU A	Timeline	Modified live vaccine† and live virus inoculation GDU B
	Day -28	Received 450 gilts, vaccinated with MLV
	Week -3	Collected blood samples from 45 tagged gilts, collected 12 OF samples, sequenced selected samples
	Week -2	
	Week -1	
Received 450 gilts, inoculated with LVI	**Day 0**	Collected blood samples from 45 tagged gilts collected 12 OF samples, inoculated with LVI
Collected blood samples from 45 tagged gilts, collected OF samples, sequenced selected samples	**Week 1**	Collected blood samples from 45 tagged gilts, collected 12 OF samples, sequenced selected samples
	Week 2	Collected 12 OF samples
Collected blood samples from 45 tagged gilts	**Week 3**	
	Week 4	Collected 12 OF samples
	Week 5	
Collected 10 OF samples	**Week 6**	Collected 12 OF samples
Collected 10 OF samples	**Week 8**	Collected blood samples from 45 tagged gilts, collected 12 OF samples, sequenced selected samples
Collected blood samples from 45 tagged gilts	**Week 10**	Collected blood samples from 45 tagged gilts, collected 12 OF samples
Collected 10 OF samples		
Collected blood samples from 45 tagged gilts, collected 10 OF samples	**Week 12**	Collected 12 OF samples
STUDY ENDED		

† Ingelvac PRRS®MLV, Boehringer Ingelheim Vetmedica, Inc., St Joseph, MO
LVI = Live virus inoculation; MLV = Modified live virus; OF = Oral fluids

in all gilts through 8 weeks post-LVI; however, only serum samples from GDU B were negative at 10 weeks post-LVI. By 12 weeks post-LVI all serum and OF samples were PCR negative in both GDU A and B (Figures 1 and 2). GDU A mean PCR ct values 7 days post-LVI were 23.6 and 26.7 for serum and OF, respectively. GDU B average PCR ct values 7 days post-LVI were 30.4 and 31.2 for serum and OF, respectively. The cumulative mortality rate was 6.7 and 1.6%, for GDU A and B respectively (Figures 3 and 4).

Conclusion

Several confounding factors including facility differences, non-contemporary treatments, timeline limitations, and origin of LVI may potentially affect the results of this study. These experimental challenges were conducted in a commercial field setting, which restricted experimental design allowing potential confounding factors to occur. Each treatment used a GDU specific LVI, which are genetically different viral strains. It was difficult to prevent the spread LVI across treatment groups; though this mimics transmission in large commercial populations. The use of MLV in naïve gilts given prior to LVI may reduce viral load, as noted by the lower ct values in GDU B at 7 days post LVI as compared to GDU A. In addition, decreased mortality was also observed in GDU B, which may be associated to using MLV prior to LVI. This research indicates that giving MLV prior to LVI may not reduce the shedding duration in a population of naïve gilts, though may reduce ct values and decrease mortality in naïve gilts given MLV prior to LVI. Further research is needed to determine the effectiveness of using these methods as a strategy for immunizing and stabilizing naïve gilts.

Figure 1: Serum and oral fluid PCR results and average ct values for LVI (time of LVI shown in grey box).

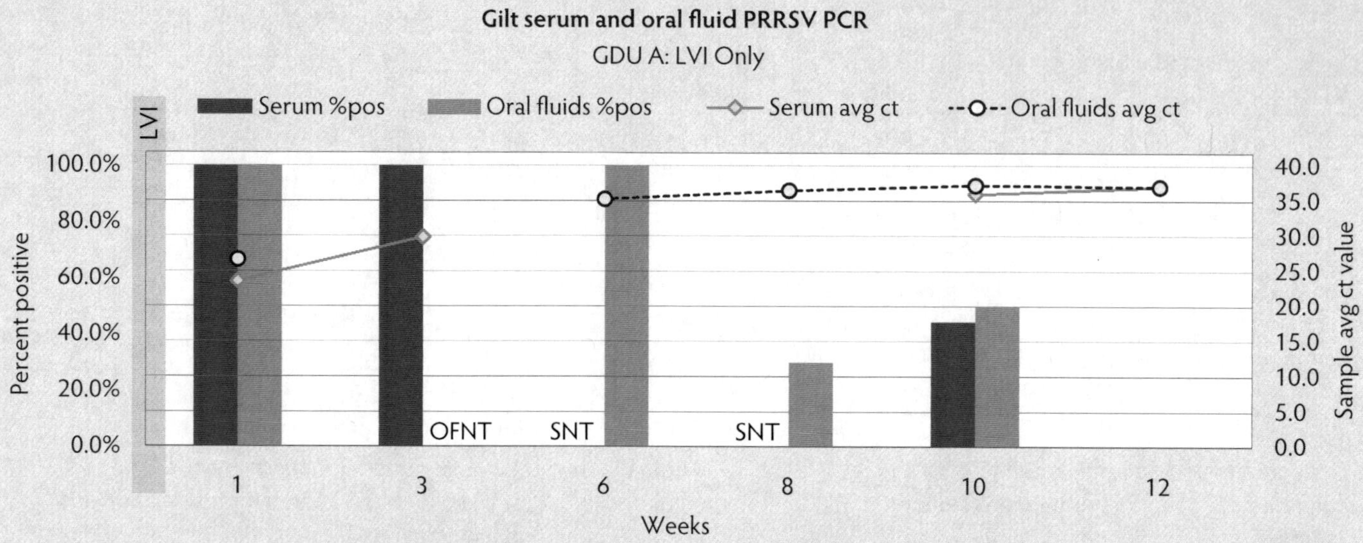

Week	Serum				Oral fluids			
	Avg. ct	Range ct.	%Pos	(Pos/total)	Avg. ct	Range ct.	%Pos	(Pos/total)
0								
1	23.6	(22.4, 24.9)	100.0%	9/9	26.7	(25.0, 28.1)	100.0%	10/10
3	29.9	(27.8, 32.2)	100.0%	9/9				
6					35.2	(33.5, 37.8)	100.0%	10/10
8					36.5	(35.0, 37.0)	30.0%	3/10
10	36.0	(31.3, 37.0)	44.4%	4/9	37.2	(35.8, 38.8)	50.0%	5/10
12	37.0	(37.0)	0.0%	0/9	37.0	(37.0)	0.0%	0/10

OFNT = Oral fluid not tested
SNT = Serum not tested
LVI = Live virus inoculation
MLV = Modified live virus
OF = Oral fluids
ct = cycle threshold value

Acknowledgements

The authors would like to thank the owners and staff of the farms where this research was conducted.

References

1. Opriessnig T., Baker R. B., and Halbur P. G. 2007. Use of an experimental model to test the efficacy of planned exposure to live porcine reproductive and respiratory syndrome. Clin.Vaccine Immunol. 14: 1572-1577.

2. Corzo C. A., Mondaca E., Wayne S., Torremorell M., Dee S., Davies P., and Morrison R. B. 2010. Review: Control and elimination of porcine reproductive and respiratory syndrome virus. Virus Research 154:185-192.

3. Johnson A., Smith C., Almond G. 2015. Update of porcine reproductive and respiratory syndrome virus 1-7-4 strain in North Carolina. Proc AASV Orlando, Florida; 513-515.

4. Pittman, JS, personal communication, 2015.

5. Batista L., Pijoan C., Torremorell M. 2002. Experimental injection of gilts with porcine reproductive and respiratory syndrome virus (PRRSV) during acclimatization. J Swine Health Prod. 10:147-150.

6. Christianson W. T., Choi C., Collins J. E., Molitor T. W., Morrison R. B., Joo H. 1993 Pathogenesis of porcine reproductive and respiratory syndrome virus infection in mid-gestation sows and fetuses. Can J Vet Res 57:262-268.

Figure 2: Serum and oral fluid PCR results and average ct values for MLV prior to LVI (time of MLV and LVI shown in grey boxes).

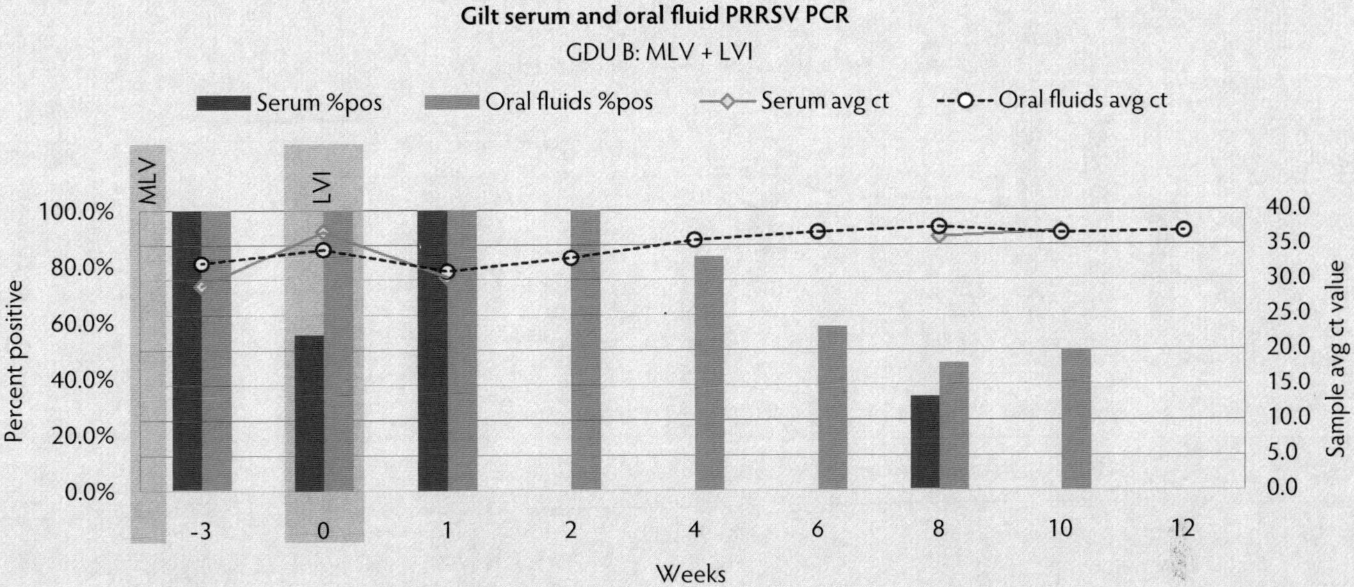

Week	Serum Avg. ct	Serum Range ct.	Serum %Pos	Serum (Pos/total)	Oral fluids Avg. ct	Oral fluids Range ct.	Oral fluids %Pos	Oral fluids (Pos/total)
-3	29.1	(27.6, 30.5)	100.0%	9/9	32.4	(29.5, 35.4)	100.0%	12/12
0	36.9	(32.3, 40.0)	55.6%	5/9	34.4	(30.9, 38.1)	100.0%	12/12
1	30.4	(27.0, 34.3)	100.0%	9/9	31.2	(29.2, 32.4)	100.0%	12/12
2					33.1	(31.7, 35.0)	100.0%	12/12
4					35.7	(33.0, 37.3)	83.3%	10/12
6					36.8	(35.0, 37.7)	58.3%	7/12
8	36.2	(33.3, 37.0)	33.3%	3/9	37.5	(37.0, 38.7)	45.5%	5/12
10	37.0	(37.0)	0.0%	0/9	36.7	(34.0, 39.1)	50.0%	6/12
12					37.0	(37.0)	0.0%	0/12

SNT = Serum not tested

Figure 3: Weekly and cumulative mortality for LVI.

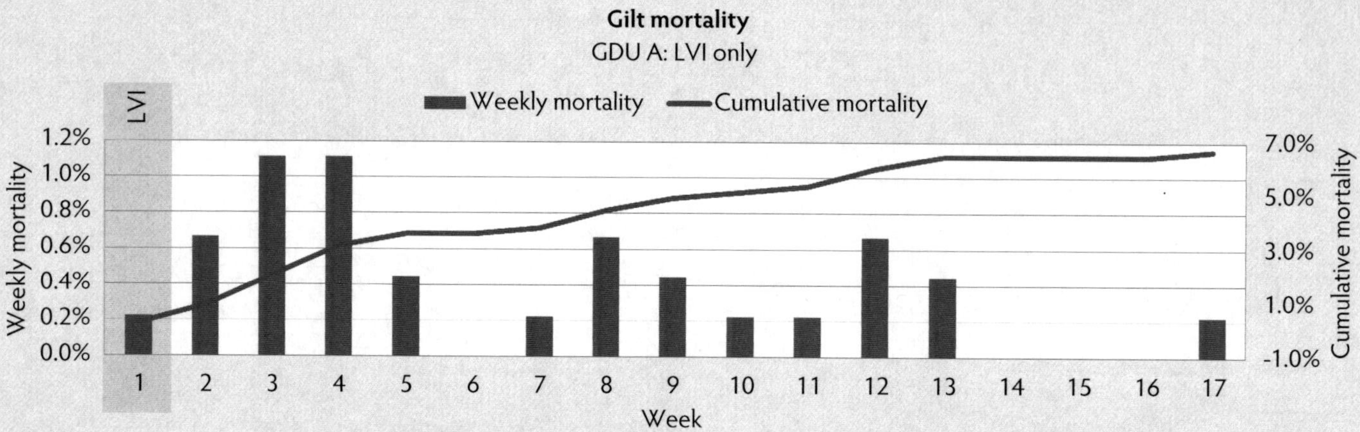

Figure 4: Weekly and cumulative mortality for MLV prior to LVI.

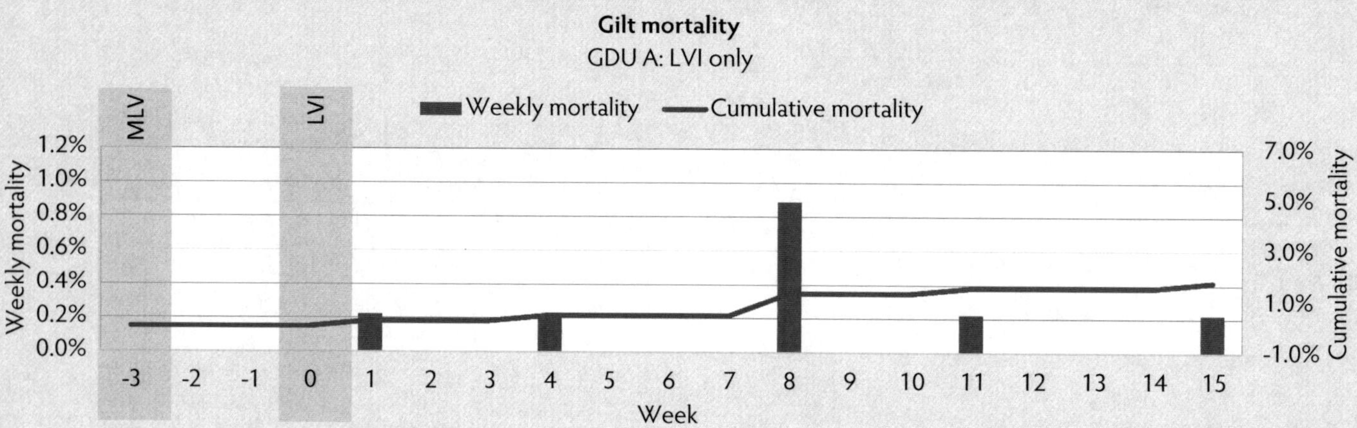

Evaluation of a second iron supplementation and the use of zinc oxide in the prevention of iron-deficient anemia in young pigs

Eric Perrin, BSc; Amanda Perri, BSc, MSc; Robert Friendship, DVM, MSc; Terri O'Sullivan, DVM, PhD
Department of Population Medicine, Ontario Veterinary College, University of Guelph, Canada

Introduction

A recent study has shown that standard iron supplementation protocols used on commercial swine farms may not be adequate.[1] The authors reported that many piglets at weaning had low hemoglobin (Hb) values, with an even higher prevalence of anemia 3 weeks into the nursery.[1] Due to competitive absorption between zinc, iron and other metals, pharmacological doses of zinc oxide (ZnO) in the nursery have been hypothesized to be a contributing factor to the higher prevalence of anemia in nursery pigs.[1,2]

Objectives

1) To assess if an additional injection of iron (at 2-weeks of age) improves Hb levels in piglets at weaning and subsequently in the nursery.

2) To assess the relationship between ZnO levels used in nursery starter rations and Hb concentration in nursery pigs.

Materials and methods

Trial 1: Nursing piglets (n = 186), over 2 cohorts on a commercial farm in southwestern Ontario were enrolled at 2-weeks of age. Prior to enrollment each piglet had already received an initial injection of gleptoferron (200 mg IM) by the producer at 3-days of age. The piglets were randomly assigned to one of two groups: 1. Treatment with gleptoferron 200 mg IM or 2. Control (no injection). Individual body weights (kg) and whole blood samples for measuring Hb were taken at 2 and 6 weeks of age.

Trial 2: At a different commercial farm 294 weaned piglets, over 3 cohorts, were divided into 2 groups. Group 1: fed 500 ppm ZnO coated pellets and Group 2: fed 3000 ppm mixed ZnO pellets. The diets were nutritionally identical commercial nursery rations, except for their respective ZnO concentration. The rations were offered as the sole diet to the respective groups from weaning (21-days of age) to 3-weeks post weaning at which time an individual blood sample was taken from each pig to measure Hb. Multi-level linear regression was used for both trials to assess potential associations.

Results

Trial 1: Piglets in the treatment group had higher Hb levels compared to piglets in the control group ($P < .01$). However, there was no difference in ADG between the two groups ($P > .05$).
Trial 2: Pigs in Group 1 had lower Hb levels at 3-weeks post weaning vs. pigs in Group 2 ($P < .05$).

Discussion

The administration of an additional injection of 200 mg of gleptoferron at 2-weeks of age increased piglet Hb. While no effect on ADG was observed in this study there may be other unmeasured benefits such as the promotion of optimal immunity that a higher Hb status may have. Hence, a second injection of iron may still have health benefits beyond improving ADG and future research is needed to evaluate this. Additionally, future research should focus on following iron supplemented pigs into the finishing barn to evaluate any unrealized effects. Pigs fed 500 ppm ZnO had lower Hb compared to pigs fed 3000 ppm by approximately 2 units which was antagonistic to our hypothesis as it was expected that the high ZnO levels would interfere with iron uptake. Further research, using a larger sample size, is needed to further investigate the effect of ZnO concentration on Hb values in pigs.

References

1. Kubik, A., et al. An investigation of iron deficiency and anemia in piglets. Centralia Swine Research Update. 34th edition. 2015. Available at: www.centraliaswineresearch.ca/proceedings/2015/centralia2015proceedings.pdf PI-9.

2. Jensen-Waern, et al. (1998). Dietary zinc oxide in weaned pigs--effects on performance, tissue concentrations, morphology, neutrophil functions and faecal microflora. Research in Veterinary Science, 64(3), 225–231.

Viral ecology determined by metagenomic sequencing within multi-sourced swine markets located in the southeastern United States

Olivia Myers[1], BS; Josh Duff[2], DVM; Chad Smith[2], DVM; Gene Nemechek[3], DVM; Ben Hause[4], PhD
[1]North Carolina State University College of Veterinary Medicine, Raleigh, North Carolina;
[2]Maxwell Foods, LLC, Goldsboro, North Carolina; [3]Zoetis Animal Health, Florham Park, New Jersey;
[4]Kansas State University Veterinary Diagnostic Lab, Manhattan, Kansas

Introduction

Commingled markets are often located near live production sites, and represent a collection point for diseases that could spread to the region. Surveillance of markets, where pigs from multiple sources are commingled, utilizing metagenomics sequencing could be useful in identifying emerging/reemerging pathogens. Metagenomic sequencing can be used to identify viruses without *a priori* sequence information. This new technology is different from traditional sequencing methods because it allows thousands of base pairs to be sequenced without the use of a known primer. This permits for a more comprehensive view of all viruses present in a sample. When compared to the traditional Sanger method of sequencing, metagenomic sequencing creates almost one million reads of shorter sequences and is cheaper per base pair. The Sanger method cost approximately $0.30 per base pair while metagenomic sequencing costs around $0.02 per base pair. Metagenomic sequencing has been used to identify several emerging pathogens in swine such as enterovirus G and a novel influenza virus in swine.[1,2] Metagenomic sequencing can also be used to identify bacterial genomes although that was not the goal of this study. The objective of this study was to identify viruses found in pigs sent to markets in the southeast region of the United States.

Materials and methods

Nasal and rectal samples were taken using BD™ Universal Viral Transport System swabs from pigs at five different markets in the southeast region of the United States. Our goal was to collect samples from five pigs from five different producers at each site. Rectal samples and nasal samples were then pooled by five for a total of 50 pooled samples. Samples were stored at -80°C until all five markets had been sampled, and then sent to Kansas State Veterinary Diagnostic lab for metagenomics sequencing. Before sequencing, the samples were filtered through 0.2 µm filter and treated with nucleases to remove bacteria. The sequenced samples were then compared to known sequences using a BLAST analysis. A second round of sampling, following the same protocol, was conducted approximately one month later at the same five sites.

Results

16 different viruses were identified in the first round of sampling and 36 different viruses were found in the second round of sampling. 48% of the nasal pools tested positive for viral material as did 88% of the rectal pools. Seneca Valley virus was identified in less than 10% of both the nasal and rectal pooled samples in the first sampling. In the second sampling, Seneca Valley virus was found in 28% of both the nasal and rectal pooled samples. Upon virus isolation, SVV found in this study is most closely related to Brazilian isolates associated with clinical disease in neonatal pigs in 2014 and 2015, and less related to Canadian isolates found in 2011 and historical US isolates (Figure 1). Figures 2 and 3 report noteworthy viruses identified in the first round of sampling. The noteworthy viruses identified in the first sampling included porcine astrovirus, enterovirus G, hemagglutinating encephalomyelitis virus, porcine parvovirus, swine norovirus GII-18, and torque teno suis virus. All of these viruses, with the exception of swine norovirus, were identified in the second

Figure 1: Dendrogram representing the relatedness of SVV P1 isolates. SVV isolates found in markets during this study are most closely related to Brazilian SVV P1 isolates found in 2014-2015 (97% identity). Isolates found in this study are also related to Canadian isolates (95% identity) and previous US isolates (87-92% identity).

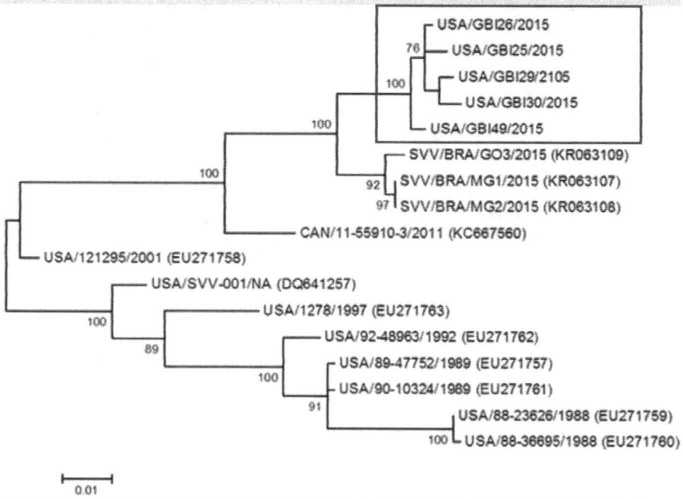

Figure 2: Percentage of nasal pools positive for established swine pathogens in first sampling.

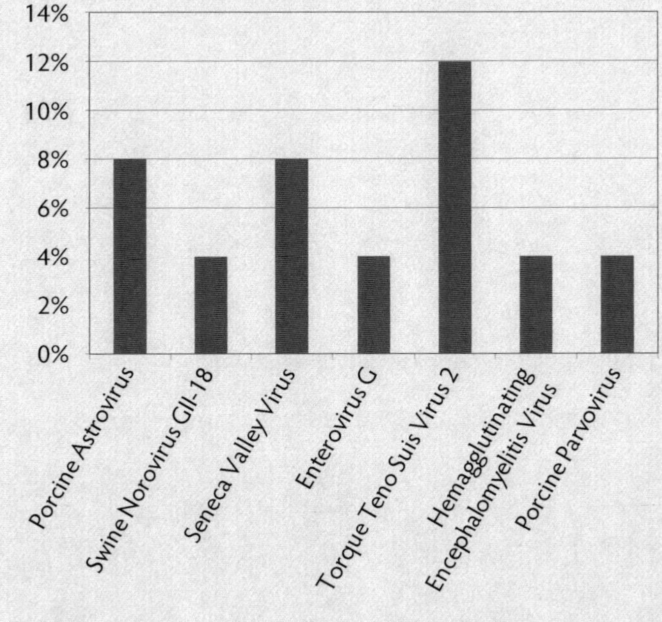

Figure 3: Percentage of rectal pools positive for established swine pathogens in first sampling.

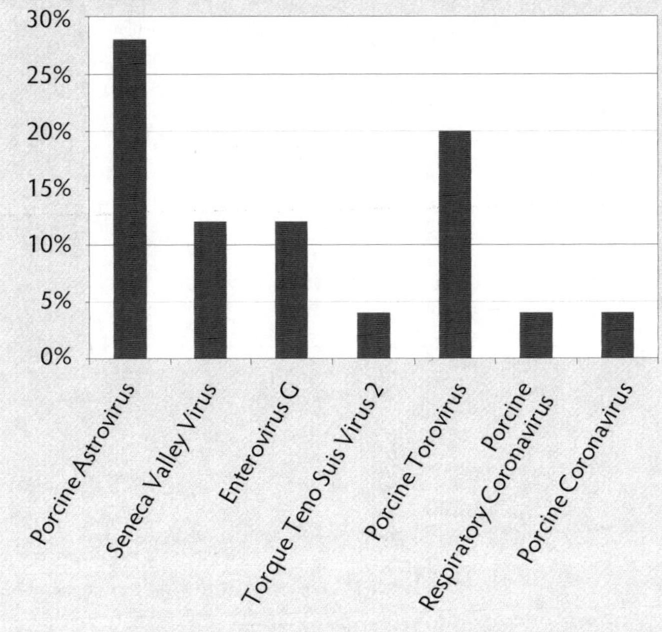

sampling with the addition of, porcine circovirus type 1 and type 2, influenza A virus, pestivirus, porcine deltacoronavirus, swine calicivirus, and porcine epidemic diarrhea virus. Figures 4 and 5 show noteworthy viruses found in the second round of sampling.

Discussion and implications

One of the important assets of metagenomic sequencing is its use in surveillance of emerging/reemerging pathogens. However, one of the drawbacks of this technology is that many clinically irrelevant diseases and environmental pathogens were found in the samples. Many of the viruses identified in the sequenced samples have not been well studied, and their role in swine disease is not well understood. Additionally, some of the viruses identified in our samples have zoonotic potential. The presence of Seneca Valley Virus, a reportable virus, increased between the first and second samplings. Seneca Valley Virus is significant because it is clinically indistinguishable from Food-and-Mouth disease. Visible lesions were not observed while collecting samples from animals that tested positive for SVV. Other notable viruses that were identified in the second round of sampling that were not found in the first round include PEDV and IAV-S. Metagenomic sequencing could be useful in confirming difficult cases that traditional diagnostic methods fail. This would be applicable if clinical signs and disease complexes are not matching test results from traditional diagnostics. Whole viral genome can be recovered by use of this technology and compared to determine strain differences to known pathogens. It should be noted that this study only provides a small snapshot of the viruses present in markets at a given time. Periodic surveillance of markets, as well as live production sites, could help identify potential emerging/reemerging pathogens.

References

1. Hause BM, Ducatez M, Collin EA, Ran Z, Liu R, Sheng Z, Armien A, Kaplan B, Chakravarty S, Hoppe AD, Webby RJ, Simonson RR, Li F. Isolation of a novel swine influenza virus from Oklahoma in 2011 which is distantly related to human influenza C viruses. *PLoS Pathog.* 2013: 9(2): e1003176.

2. Anbalagan S, Hesse RA, Hause BM. First identification and characterization of porcine enterovirus G in the United States. *PLoS ONE.* 2014: 9(5): e97517.

Figure 4: Percentage of nasal pools positive for established swine pathogens in second sampling.

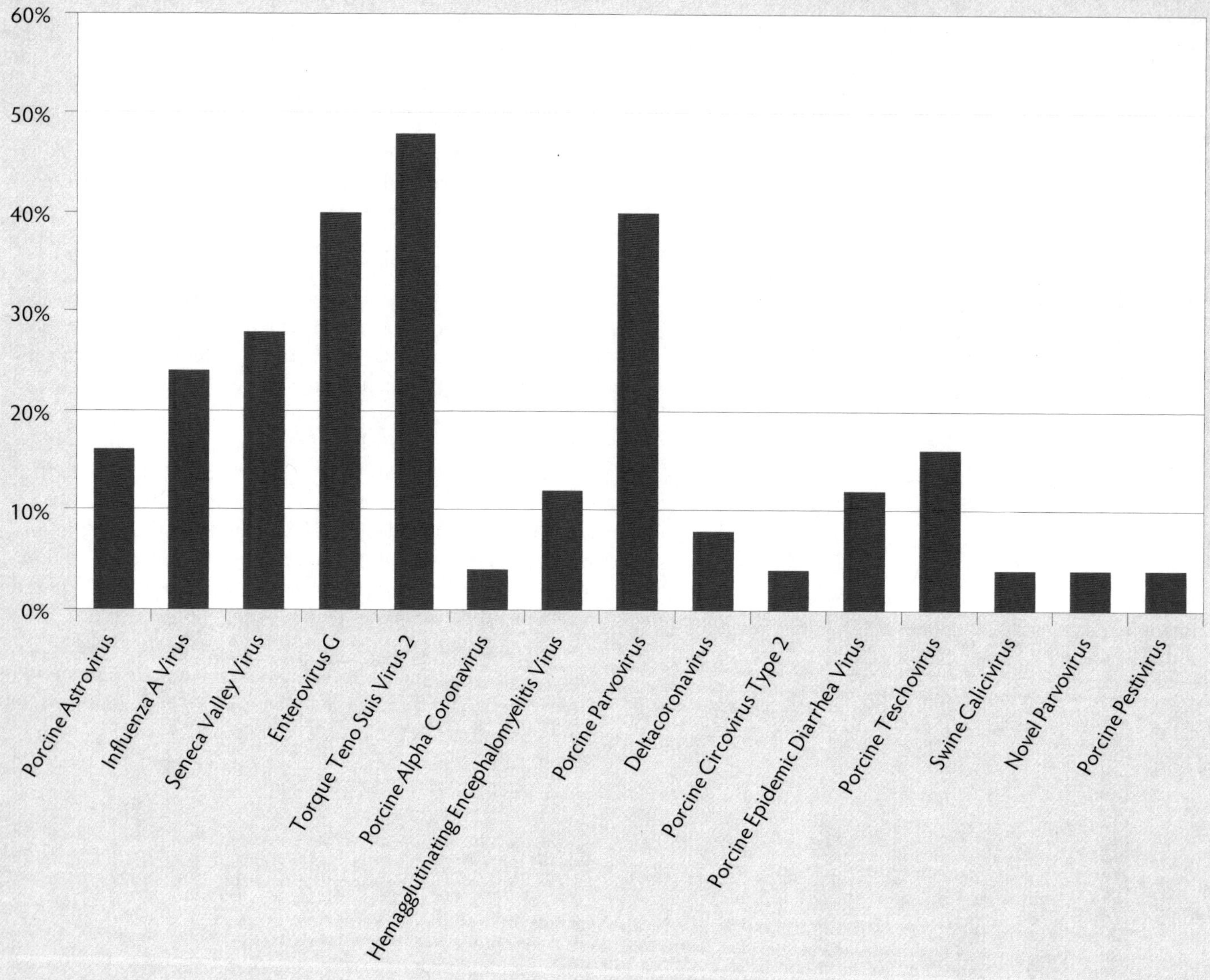

Figure 5: Percentage of rectal pools positive for established swine pathogens in second sampling.

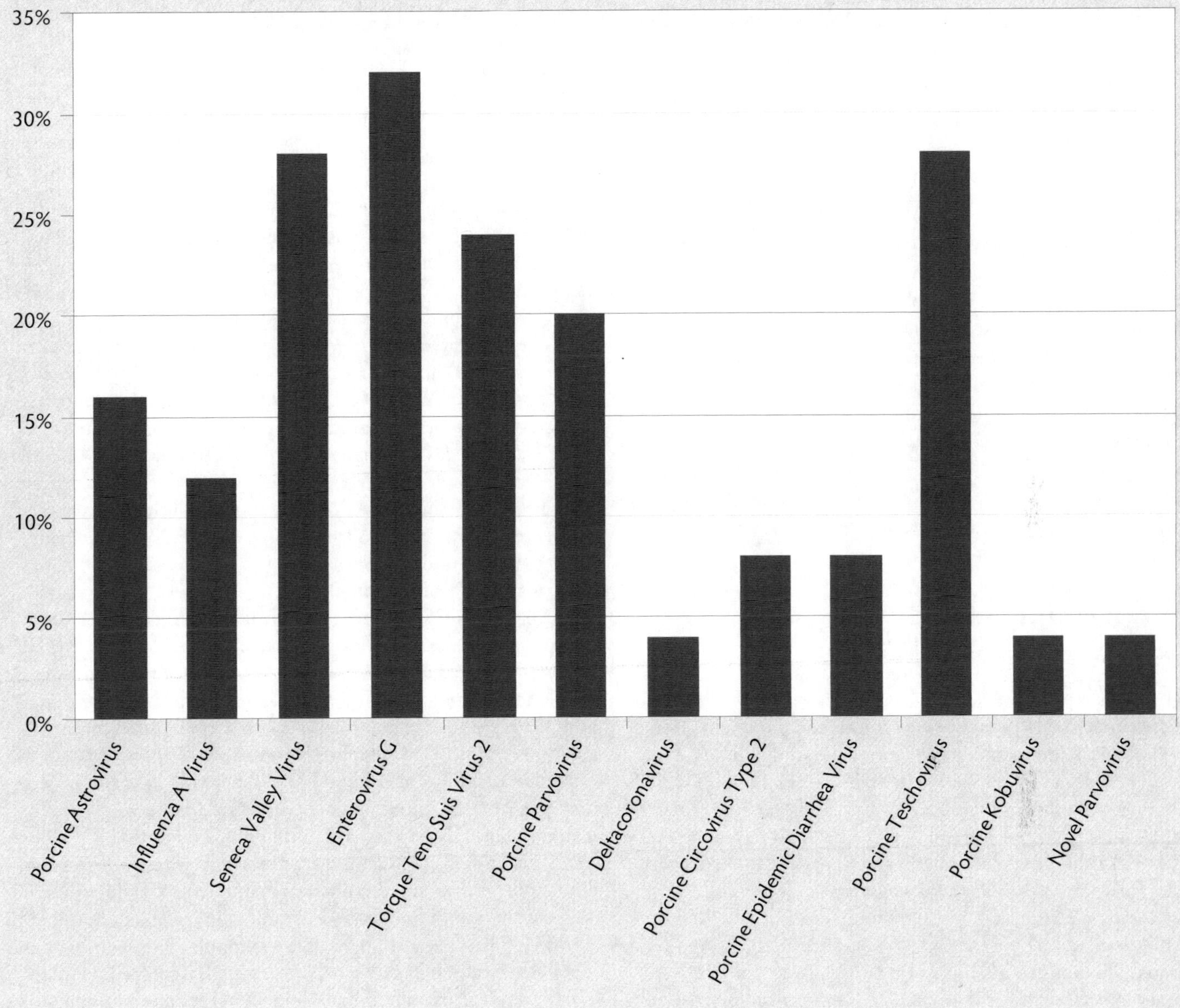

Effect of herd closure on the pattern of *Mycoplasma hyosynoviae* and *Mycoplasma hyorhinis* infection

A. Anderson[1], BS; W. Lyons[2], DVM; J. O. Geiger[2], DVM, MS; M. J. Clavijo[2], DVM, PhD
[1]University of Minnesota College of Veterinary Medicine, Saint Paul, Minnesota;
[2]Pig Improvement Company (PIC), Hendersonville, Tennessee

Introduction

Mycoplasma hyosynoviae (MHS) causes arthritis in pigs 3-5 months of age affecting up to 50% of the group.1 *Mycoplasma hyorhinis* (MHR) causes polyserositis and arthritis in pigs 3-10 weeks of age.1 Infection with these pathogens compromises animal welfare, creates continuous treatment costs and impacts performance.2 There is a lack of information regarding the duration of shedding and changes in the dynamics of infection of MHR and MHS when herd closure is applied. Such knowledge will provide further insight into the possibility of eliminating these agents from swine herds. The objective of this study was to determine whether the implementation of herd closure can have an effect on the prevalence of MHS and MHR infection.

Materials and methods

The study was conducted at a 1,150 farrow-to-finish sow farm with previous history of MHS and MHR associated lameness in grow-finish pigs. The herd was undergoing a *Mycoplasma hyopneumoniae* (MHP) elimination project, based on herd closure and strategic medication. At the time of the study, clinical evidence of MHR and MHS associated disease was not observed. A total of 462 oropharyngeal swabs (OS) and oral fluid (OF) samples were collected at five different times and tested by MHS and MHR PCR. To determine the initial pattern of MHS and MHR infection within the herd, OS were randomly collected from 60 sows, 60 gilts, 60 due to be weaned piglets, 30 70 day old pigs and 30 130 day old pigs 32 days after herd closure was initiated. To monitor MHS and MHR shedding from the herd, a total of 90 OF samples were collected from sows, gilts, pre- and post-weaning pigs at 93, 145, and 185 days after the start of herd closure (Table 1). Before the herd re-opened, a final sampling was carried out at 221 days post herd closure. The sampling included the collection of 60 OS and 12 OF samples from randomly selected sows (p1), as well as 60 OS from randomly selected due to weaned piglets.

Results and discussion

At baseline, MHS was detected from OS in 68% of the sows, 37% of the gilts and 7% of the due to be weaned piglets sampled (Table 1). Similarly, 22% of the sows, 32% of the gilts and 8% of the piglets were MHR PCR-positive. The detection of both MHS and MHR in OS from due to be weaned piglets suggests that pigs may become infected with these agents from their dams. In post-weaning, 13% of the 70 day old pigs and 16% of the 130 day old pigs were PCR-positive for MHS. In contrast, 97% of the 70 day old pigs and 43% of the 130 day old pigs were PCR-positive for MHR. The prevalence of MHR and MHS within the herd has been similarly observed in previous studies supporting the occurrence of high colonization (> 85%) of MHR in the nursery and the presence of MHS colonization in piglets at 3 weeks of age, which may play a role in clinical presentation downstream of the sow farm.[2,3] The subsequent monitoring at 93 days post herd closure revealed that OF samples from sows, gilts and post-weaning pigs tested PCR-positive for both MHS and MHR. MHR was no longer detected in OF at days 145 and 185 post herd closure, while MHS was detected in a high percentage (> 73%) of the samples. At 221 days post herd closure, MHR was detected in 5% of OS from p1 sows and 8.3% of OF from sows in pens F and MHS was detected in 3.3% of pre-weaning piglets and 10% of sows in OS and in 33.3% of OF. Low sample size, antibiotic treatment and intrinsic differences between sample groups are some potential explanations as to why MHR was not detected in OF at 145 and 185 days post herd closure. A numerical decrease was observed between the baseline and final sampling, suggesting a possible effect of herd closure, recent antibiotic treatment or a combination of both over bacterial shedding. Detection of MHR and MHS before the herd re-opened indicates that the duration of shedding for these two pathogens could potentially be longer than that reported for MHP.[4] However, it is important to note that success of herd closure depends on whole herd bacterial exposure prior to the start of herd closure. In this study, possibly not all animals were exposed to MHR and MHS before herd closure, as seen in the baseline results. This could have caused negative pigs to gradually become infected, thus extending the shedding period. Further research is needed to determine the shedding period in pigs infected with MHR and MHS. Also, methods for efficacious exposure to both pathogens should be evaluated.

References

1. Minion, F.C., Thacker, L. E. 2012. Mycoplasmosis. In Zimmerman, J.J., et al. (Eds.), *Diseases of Swine (10th Edition)* (pp. 788-791). NJ: John Wiley & Sons, Inc.

2. Schwartz, J. et al. 2014. Dynamics *of Mycoplasma hyosynoviae* detection and clinical presentation. Paper presented at 2014 American Association of Swine Veterinarians (AASV) Conference: Our Oath in Practice, Dallas, TX. Retrieved from AASV Library Catalog. Pp 115-116.

3. Clavijo, M.J. et al. 2012. Dynamics of infection of *Mycoplasma hyorhinis* at two commercial swine herds. Paper presented at 2012 Allen D. Leman Swine Conference, Saint Paul, MN. Retrieved from AASV Library Catalog. Pp 91-92.

4. Pieters M, Pijoan C, Fano E, et al. An assessment of the duration of *Mycoplasma hyopneumoniae* infection in an experimentally infected population of pigs. *Vet Microbiol.* 2009;134:261-266.

Table 1: Prevalence of *M. hyorhinis* and *M. hyosynoviae* by sample type and age.

Sample date	Age of pigs sampled	Number of samples	Type of sample	Percent positive samples for MHR* (%)	Percent positive samples for MHS* (%)
32 days post herd closure (Initial baseline)	Piglets (20 days)	60	OS	8.3	6.7
	Nursery (70 days)	30	OS	96.7	13.3
	Finisher (130 days)	30	OS	43.3	16.7
	Gilts	60	OS	31.6	36.7
	Sows	60	OS	23.3	68.3
93 days post herd closure	Finisher (154 days)	15	OF	46.7	33
	Gilts	8	OF	25	100
	Sows	7	OF	0	85.7
145 days post herd closure	Gilts	11	OF	0	72.7
	Sows	9	OF	0	88.8
185 days post herd closure	Nursery (70 days)	12	OF	0	0
	Finisher (130 days)	6	OF	0	100
	Gilts	9	OF	0	100
	Sows	13	OF	0	92
221 days post herd closure	Piglets (16 days)	60	OS	0	3.3
	Sows	60	OS	5	10
		12	OF	8.3	33.3

* OS: oropharyngeal swabs; OF: oral fluids
* MHR: *Mycoplasma hyorhinis;* MHS: *Mycoplasma hyosynoviae*

Vesicles, hoof lesions, and severe lameness in commercial finishing pigs in the United States

Scott Radke[1], BS; Abbey Canon[1], DVM, MPH; AJ Smit[2], DVM; Pablo Piñeyro[1], DVM, MVSc, DVSc, PhD; Daniel Linhares[1], DVM, MBA, PhD, FSVM; Derald Holtkamp[1], DVM, MS; Kim Gerardy[1], BS; Christopher Rademacher[1], DVM

[1]Iowa State University College of Veterinary Medicine, Ames, Iowa; [2]Orange City Veterinary Clinic, Orange City, Iowa

Problem, herd, and time period

The farm manager observed severe, acute lameness in 50% of commercial market hogs in a 1,200 head wean-to-finish facility in the Midwest on July 15, 2015, five days following the first loadout. Pigs marketed on the first loadout were not lame.

Case history, data collection, and diagnostics

On July 28, 2015, the herd veterinarian observed nasal and oral vesicles in about 25% of the herd. In addition, approximately 80-90% of the finisher pigs showed interdigital or coronary band vesicular lesions, ulcerations and lameness associated with animals that were reluctant to rise while squealing, shaking, and refusing to move. These lesions were indistinguishable from other vesicular diseases such as foot-and-mouth disease (FMD), vesicular stomatitis virus (VS), swine vesicular disease (SVD), and vesicular exanthema of swine (VES). A foreign animal disease (FAD) investigation was initiated immediately. Swabs of vesicular lesions and serum were collected by the Foreign Animal Disease Diagnostician on July 28, 2015, and tested at the Foreign Animal Disease Diagnostic Laboratory (FADDL) at Plum Island Animal Disease Center by RT-PCR along with either Antibody Enzyme-linked Immunosorbent Assay (AB-ELISA) or Immunofluorescence Antibody Assay (IFA). All samples were negative for FMD, VS, SVD, and VES. Senecavirus A (Seneca Valley Virus or SVA) was detected by RT-PCR on July 29, 2015, at the FADDL at Plum Island. Following negative results for a FAD at the FADDL at Plum Island, additional samples of vesicular fluid, vesicle tissue scrapings, whole blood, serum, nasal swabs, and oral fluids were submitted to the Iowa State University Veterinary Diagnostic Laboratory (ISU VDL) for further diagnostic work-up. Senecavirus A was detected in all samples. Oral fluid sampling indicated that the pigs were negative for porcine reproductive and respiratory syndrome virus (PRRSV) and swine influenza virus (SIV) but positive for porcine circovirus type 2 (PCV2) (32.0-35.2) and *Mycoplasma hyopneumoniae* (30.6-31.7). Because of the severe lameness and discomfort caused by the lesions, Oral-Pro Sodium Salicylate 48% was administered at an anti-inflammatory dose and was reported by the producer to help. The herd was evaluated following the outbreak on August 10, 2015. Clinical lameness was observed in 50-60% of the remaining 450 animals, with the majority possessing multifocal hemorrhagic lesions throughout the nail bed of the hoof (Figure 1). Two individuals had acute vesicular lesions on the snout and coronary bands; ulcerated and granulated lesions were also observed in clinically lame animals (Figure 2). Carpal hygromas were commonly observed in lame animals from limb dragging because of the pain inflicted by the hoof lesions along with occasional sloughing of the hoof (Figures 3, 4). On August 10, oral fluid and serum samples

Figure 1: Nail bed hemorrhage

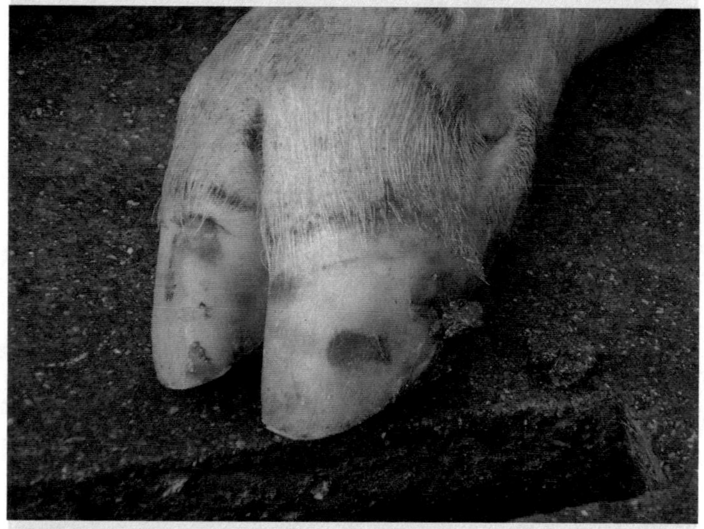

Figure 2: Ulcerated snout lesion

Figure 3: Carpal hygroma

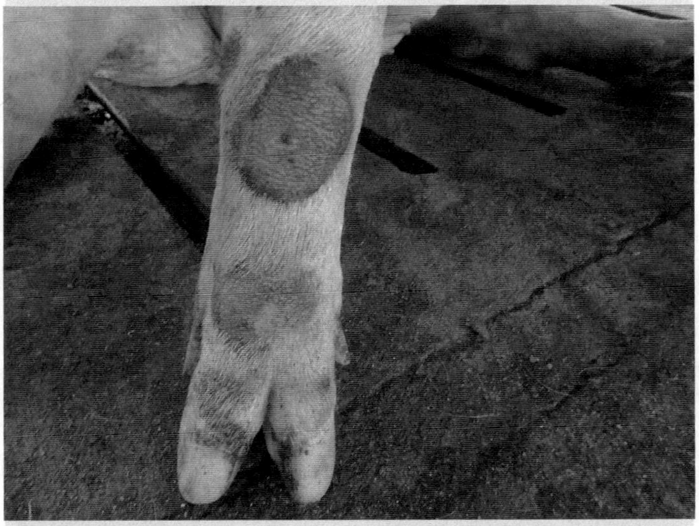

Figure 4: Sloughed hoof

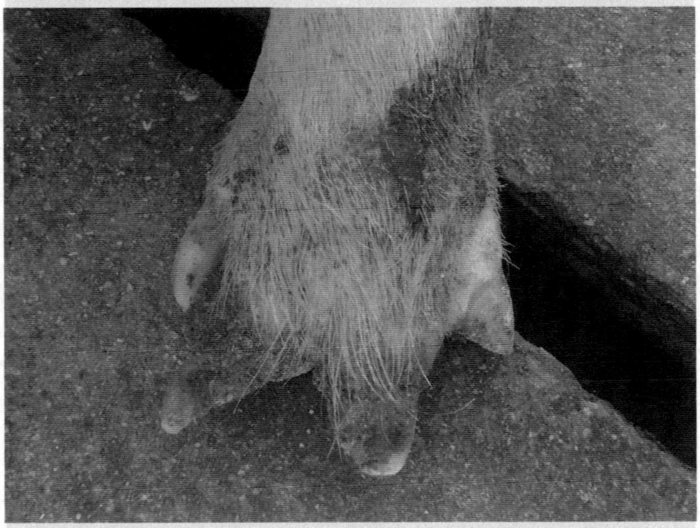

were collected from both clinically lame and asymptomatic animals. Vesicular swabs, scrapings and nasal and rectal swabs were taken from two acutely affected animals. Five animals were transported to the ISU VDL for necropsy. An outbreak investigation was performed to identify the most likely routes of introduction and provide recommendations to improve biosecurity. The loadout and use of a loadout crew five days preceding the break might have introduced the virus to the farm or triggered the clinical manifestation of the virus as the truck was dirty while the driver walked from the dirty trailer into the barn. The load out crew loaded trailers for other producers as well. *M. hyopneumoniae* detected at ISU VDL may have also exacerbated the clinical signs of senecavirus A if it worked as a cofactor of disease. *M. hyopneumoniae* was not only detected, but pigs were clinically affected by *M. hyopneumoniae* by the significant amount of coughing that was observed.[1] Serial samples of serum, nasal and rectal swabs, and tonsil scrapings were taken 36, 43 and 50 days post clinical onset from ten individuals that were moved to an isolated location following the outbreak investigation in order to attempt to determine how long the virus could be shed.

Significant findings and recommendations

Senecavirus A, a picornavirus, was detected by RT-PCR at both the FADDL and ISU VDL from pigs that presented with vesicular lesions clinically indistinguishable from FMD, VS, SVD, and VES. The virus was also isolated at the ISU VDL. Senecavirus A is a non-enveloped single strand RNA virus belonging to the same family as FMD and SVD.[2] No other gross or histopathological lesions than cutaneous vesicles and ulceration along the the coronary bands as well as the oral mucosa and hemorrhage in the nail beds were observed upon necropsy. Senecavirus A was detected in the tonsils, vesicular fluid, nasal, rectal and vesicular swabs, skin scrapings, and tissue, and in the serum during the acute phase of disease. Vesicular fluid as well as tissue scrapings from the vesicular lesions taken initially were shown to have the highest concentration, as measured indirectly by the CT of the PCR, of virus while oral fluids had the next highest concentration with average cycle time (CT) of 17.3,13.8 and 21 respectively (Figure 5).

Discussion

The clinical presentation of this case closely mirrors previous cases described of Idiopathic Vesicular Disease of Swine (IVDS) in Indiana in 2004, and Manitoba in 2007. All these cases share the same gross features; vesicular lesions on both the oral mucosa as well as coronary bands associated with high morbidity (approximately 80%) but no increase in mortality. Although these sporadic cases of IVDS showed a similar clinical pattern, no common predisposing or trigger factor has been identified certain. However, pigs in Indiana were moved one week prior to breaking, so stress due to transportation may play a role.[3] Although SVA was detected in the Manitoba case, at that time, there was no known association of the virus with vesicular lesions.[4] Recently several outbreaks of transient mortality in neonatal pigs associated with lameness in sows have been describe in Brazil and United States. Lesions in breeding herd animals in these cases resemble to those described in finishing animals.[5,6] In this current case the presence of vesicular disease has been associate with the presence of SVA. In addition other causes of vesicular disease such as FMD, VS, SVD, and VES have been ruled out. No other predisposing factor such as high stress events, transport or concurrent disease showed correlation with the presence of SVA and clinical disease. In this case, amongst multiple samples evaluated, SVA was recovered during the acute phase from vesicular fluid and scrapings which showed to be the samples with the lowest CT values for SVA diagnosis and holding the highest viral concentration. Oral fluids appeared to have the next lowest CT values and would make an excellent ante mortem sample for Senecavirus A in animals without classical vesicular lesions.

Figure 5: PCR CT for Senecavirus A sample types

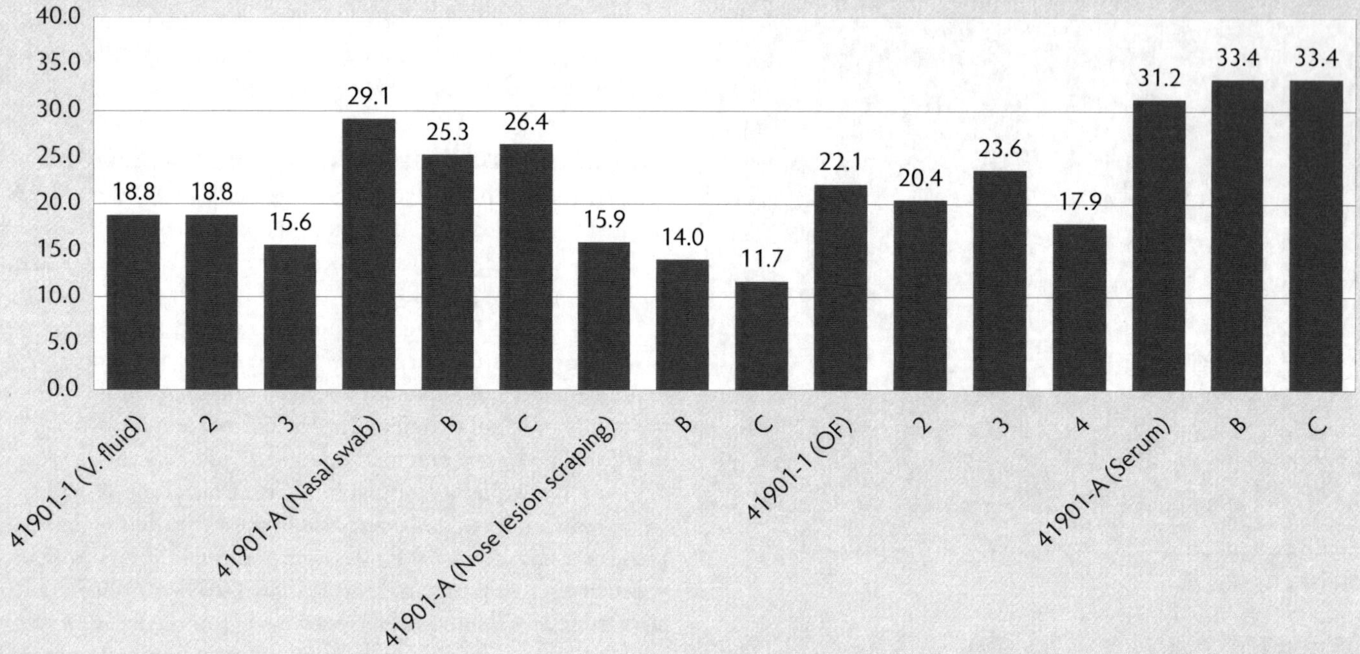

Take home message to the practicing veterinarian

The first observed clinical sign in finishing pigs was severe, acute lameness, with high morbidity but no impact on mortality. Only upon closer inspection were vesicular and hoof lesions identified. It is imperative that we, as veterinarians, are spending time training our farm employees on what the clinical presentation of vesicular disease looks like, so that we are notified immediately when they see something similar to this case. It is imperative that herd veterinarians observing these lesions follow suspect FAD investigation protocols by contacting their state or federal veterinarian who will initiate a FAD outbreak investigation to properly identify the lesions with the appropriate diagnostic tests. Veterinarians need to be able to communicate with not only the state veterinarian, but also producers and packers. Above all, with multiple cases of Senecavirus A occurring throughout the United States, veterinarians cannot allow themselves to become complacent and allow a FAD with similar lesions, such as FMD, into swine production systems.

Acknowledgements

All photos courtesy of the Swine Medicine Education Center. Table results courtesy of ISU VDL.

References

1. Canon A, Gerardy K, Rademacher C, Linhares D, Smit AJ, Bomgaars DC, Holkamp D. 2015. SVV Disease outbreak investigation. PRRS Outbreak Investigation Program, Iowa State University.

2. Rademacher C. 2015. Case reports from Senecavirus A. Allen D. Leman Swine Conference. 22 September.

3. Amass SF, Schneider JL, Miller CA, et al. 2004. Idiopathic vesicular disease in a swine herd in Indiana. *J Swine Health Prod* 12(4):192-196.

4. Pasma T, Davidson S, Shaw SL. Idiopathic vesicular disease in swine in Manitoba. *The Canadian Veterinary Journal*. 2008;49(1):84-85.

5. Linhares D. 2015. Epidemic transient neonatal losses in Brazil and Senecavirus A. James D. Mckean Swine Disease Conference. 6 November.

6. Piñeyro P.E, Rademacher, C.J, Derscheid R.J, Guo B, Yoon K.J, Zhang J, Harmon K.M, Magstadt, D.R, Gauger P.C, Chen Q, Arruda P.H.E., Stevenson G.W, Burrough E.R, Schwartz K.J, Cooper V.L, Arruda B.L, Baum D, Zimmerman J, Gimenez-Lirola L, Correia-Lima-Linhares D, Halbur P.G, Main R.G, Madson D.M. Senecavirus A vesicular disease outbreak in Midwest show and commercial pigs American Association of Veterinary Laboratory Diagnosticians (AAVLD) annual conference meeting pp158. Providence, RI October 2015.

Evaluation of *Salmonella* Typhimurium fecal shedding in pigs vaccinated with *Salmonella* Choleraesuis-Typhimurium vaccine via drinking water and challenged four weeks later

Q. Steichen[1], MS; R. A. Smiley[2], MS; B. Fergen[2], PhD; D. Jordan[2], DVM, PhD; K. F. Lechtenberg[3], DVM, PhD; T. J. Kaiser[2], DVM, MS

[1]Kansas State University College of Veterinary Medicine, Manhattan, Kansas; [2]Boehringer Ingelheim Vetmedica, Research & Development, St. Joseph, Missouri; [3]Midwest Veterinary Service, Oakland, Nebraska

Introduction

Salmonella enterica serovar Typhimurium (*ST*) and *Salmonella* Choleraesuis are prevalent pathogens in swine. *ST* is a primary cause of enteritis and subclinical production losses in growing or finishing swine and contributes to environmental and carcass contamination.[1] Due to the zoonotic potential, intervention programs for *ST* have been established across the world attempting to reduce carcass contamination with the ultimate goal of reducing human salmonellosis cases.[2,3] The objective of this study was to evaluate *ST* fecal shedding of pigs vaccinated with a commercial, attenuated live culture (ALC) *Salmonella* Choleraesuis-Typhimurium vaccine when challenged with virulent *ST* four weeks post-vaccination.

Materials and methods

Eight litters of six pigs (fecal enrichment culture-negative for *Salmonella* prior to vaccination) at two weeks of age were blocked by litter and assigned to one of two treatment groups, 3 pigs/litter to each group. Both treatments were administered through the drinking water: Placebo group (n = 24) and Vaccine group (n = 24). The vaccine was the bivalent commercial vaccine (Enterisol® Salmonella T/C) administered at the minimum immunizing dose established in studies approved by the USDA, and the placebo was a lyophilized presentation containing the media used in vaccine formulation without ALC bacteria. During the vaccination phase, pigs were housed by treatment to avoid unintentional exposure of ALC vaccine to the Placebo group. For the challenge phase, pigs were re-penned individually with treatments comingled in the same room and littermates within close spatial proximity. Four weeks after treatment administration, all pigs were challenged intranasally with 2 mL of virulent *ST* (4×10^8 CFU per dose). The trial was conducted in an Animal Biosafety Level 2 facility under the Midwest Veterinary Services Animal Use Protocol.

Fecal samples were collected daily from challenge through 14 days post-challenge (DPC) and then three times weekly until 84DPC. Fecal samples were tested for shedding via modified enrichment cultures with a lower limit of detection of ~4000 CFU/gram for frozen samples. Pigs were qualitatively considered positive or negative for *ST*. The number of positive samples by pig in the Vaccine group was compared to the number of positive samples by pig in the Placebo group using mitigated fraction analysis. Mitigated fraction is an estimate of disease intervention regularly used by the USDA to access vaccine efficacy.[4]

Results and discussion

During the 12-week challenge phase (44 samples per pig), the mean number of positive samples per pig for the Placebo group was 26.22 samples/pig and for the Vaccine group was 15.87 samples/pig, which was a significant improvement with a mitigated fraction of 78.8% and lower confidence interval of 51.52% (Table 1; Figure 1). All Placebo pigs were positive from 3DPC to 6DPC, and ≥ 78.3% of Placebo pigs continued to shed *ST* until 31DPC; then at least 47.8% of Placebo pigs shed *ST* from 33DPC to 49DPC (Figure 2). For Vaccine pigs, 82.6% were positive the day after challenge which steadily declined to ≤ 13.0% from 47DPC through the end of the study.

Fecal shedding is a primary mechanism for contamination of the environment and transmission to animals and humans, and

Table 1: Number of positive samples per pig.

Group	N	Mean	Median	Standard Deviation	Min.	Max.	Mitigated Fraction	95% Confidence Limits
Placebo	23	26.22	26	5.85	14	39		
Vaccine	23	15.87	15	5.38	5	31	0.788	0.5152;1.0000

Figure 1: Positive fecal samples by pig post-challenge. Each row is a pig, B is *Salmonella* Typhimurium cultured and "-" is negative.

Group	Days Post-Challenge																																												Total Positive Samples
	1	2	3	4	5	6	7	8	9	10	11	12	13	14	17	19	11	24	26	28	31	33	35	38	40	42	45	47	49	52	54	56	59	61	63	66	68	70	73	75	77	80	82	84	
	B	B	B	B	B	B	-	-	B	-	-	B	B	B	B	-	-	-	B	-	-	B	-	-	-	-	-	-	-	-	B	-	-	-	-	-	-	-	-	-	-	-	-	-	14
	B	B	B	B	B	B	-	B	B	B	-	-	-	B	-	B	-	B	B	B	-	-	-	-	-	B	B	-	-	-	-	-	-	-	-	B	B	-	-	-	-	-	B	-	19
	-	B	B	B	B	B	B	B	-	-	-	-	-	B	B	-	-	B	B	B	B	B	-	B	-	-	-	-	-	-	-	-	-	-	-	-	-	-	-	-	B	B	B	B	19
	B	B	B	B	B	B	B	B	B	B	-	B	B	-	-	B	-	B	B	B	B	B	-	B	-	-	-	-	-	-	B	-	-	-	-	-	-	-	-	-	-	-	-	-	20
	B	B	B	B	B	B	B	B	B	B	B	B	B	B	B	B	B	B	B	-	B	-	-	-	-	B	-	B	-	-	-	-	-	-	-	-	-	-	-	-	-	-	-	-	22
	B	B	B	B	B	B	B	B	B	B	B	B	B	B	B	B	B	-	-	-	-	B	B	B	-	-	-	B	B	-	-	-	-	-	-	-	-	-	-	-	-	-	-	-	22
	B	B	B	B	B	B	B	B	B	B	B	B	B	B	B	B	B	B	B	B	-	-	-	-	-	B	B	-	-	-	-	-	B	-	-	-	-	-	-	-	-	-	-	-	23
	B	B	B	B	B	B	B	B	B	B	B	B	B	B	B	B	B	B	B	B	B	-	-	-	B	-	-	-	-	-	-	-	-	-	-	-	-	-	-	-	B	-	-	-	23
	B	B	B	B	B	B	B	B	B	B	B	-	B	-	B	B	B	B	B	B	B	-	B	B	B	B	-	B	-	-	-	-	-	-	-	-	-	-	-	-	-	-	-	-	24
	B	B	B	B	B	B	B	B	-	B	-	B	B	-	B	B	B	B	-	B	B	-	B	-	B	-	B	B	B	-	-	-	-	B	B	-	-	-	-	-	-	-	-	-	24
	B	B	B	B	B	B	B	B	B	B	B	B	B	B	B	B	B	B	B	B	B	B	B	-	B	B	-	-	-	-	-	-	-	-	-	-	-	-	-	-	-	B	-	-	26
Placebo	B	B	B	B	B	B	B	B	B	B	B	B	B	-	B	B	B	B	B	B	B	B	B	B	-	B	-	B	B	-	-	-	-	-	-	-	-	-	-	-	-	-	-	-	26
	B	B	B	B	B	B	B	B	B	B	B	B	B	B	B	B	-	B	B	B	B	B	-	-	B	-	B	-	B	-	B	-	-	B	-	-	-	-	-	-	-	-	-	-	26
	B	B	B	B	B	B	B	B	-	-	-	B	B	B	B	B	B	B	B	B	B	B	B	B	B	B	B	B	B	-	-	-	-	-	B	-	-	-	-	-	-	-	-	-	27
	-	B	B	B	B	B	B	B	B	B	B	B	B	B	B	-	B	B	B	B	B	B	B	B	B	B	B	B	-	B	-	-	-	-	B	-	-	-	-	-	-	-	-	-	28
	B	B	B	B	B	B	B	B	B	B	-	B	B	B	B	B	B	B	-	-	B	B	B	B	-	B	-	B	-	-	-	-	B	B	B	-	-	B	-	-	-	-	-	B	28
	B	B	B	B	B	B	B	B	B	B	B	B	B	B	B	B	B	B	B	B	B	B	B	B	B	B	B	B	B	B	-	-	-	-	-	-	-	-	-	-	-	-	-	B	31
	B	B	B	B	B	B	B	B	B	B	B	B	B	B	B	B	B	B	B	B	B	B	B	B	B	B	B	-	B	-	-	B	B	-	-	-	-	-	B	-	-	-	-	-	31
	B	B	B	B	B	B	B	B	B	B	B	B	B	B	B	-	B	-	B	B	B	B	B	-	B	-	-	B	-	-	-	-	-	B	-	B	-	B	-	B	B	B	B	B	31
	B	-	B	B	B	B	B	B	B	B	B	B	-	-	B	-	B	B	B	B	B	-	-	B	B	B	B	B	B	-	B	-	B	B	B	B	B	-	-	B	-	B	B	-	32
	B	B	B	B	B	B	B	B	B	-	B	B	B	B	B	B	B	-	B	B	B	B	B	B	B	B	-	B	B	-	-	-	-	-	-	B	B	B	B	B	B	B	-	-	33
	B	B	B	B	B	B	B	B	B	B	B	B	B	B	B	B	B	B	B	B	B	B	B	B	B	B	B	B	B	-	B	B	-	-	B	-	-	-	-	B	-	B	B	-	35
	B	B	B	B	B	B	B	B	B	B	B	B	-	B	B	B	B	B	B	B	B	B	B	B	B	B	B	B	B	B	-	-	B	B	B	B	B	B	B	B	B	-	-	B	39
	-	-	-	-	B	-	-	-	-	-	-	-	B	-	B	B	-	-	-	B	-	-	-	-	-	-	-	-	-	-	-	-	-	-	-	-	-	-	-	-	-	-	-	-	5
	B	-	B	B	B	-	-	-	-	-	B	-	-	-	-	-	B	-	B	-	-	B	-	-	-	-	-	-	-	-	-	-	-	-	-	-	-	-	-	-	-	-	B	-	9
	-	B	B	B	B	-	-	B	-	-	B	B	-	-	-	B	-	-	-	-	B	-	-	-	-	-	-	-	-	-	-	-	-	-	-	-	-	-	-	-	-	-	-	-	9
	-	B	B	B	B	B	B	B	B	B	-	-	B	-	-	B	-	-	-	-	-	-	-	-	-	-	-	B	-	-	-	-	-	-	-	-	-	-	-	-	-	-	-	-	12
	B	-	B	B	-	-	-	B	B	B	-	B	B	-	B	B	-	B	-	-	-	B	-	-	-	-	-	-	-	-	-	-	-	-	-	-	-	-	-	-	-	B	-	-	13
	B	B	-	B	B	B	B	B	-	-	B	B	-	B	B	-	-	-	-	-	-	-	-	-	B	-	B	-	-	-	B	-	-	-	-	-	-	-	-	-	-	-	-	-	14
	B	B	B	B	B	B	B	B	B	B	B	-	B	-	B	-	-	-	-	-	-	-	-	-	-	-	-	-	-	-	-	-	-	-	-	B	-	-	-	-	-	-	-	-	14
	B	B	-	-	-	B	B	B	B	B	B	B	-	B	-	B	-	-	-	B	B	-	-	B	-	-	-	-	-	-	-	-	-	-	-	-	-	-	-	-	-	-	-	-	14
	-	B	-	-	B	B	B	B	-	B	-	B	B	-	B	B	-	-	B	-	B	B	-	-	-	-	-	-	-	-	B	-	-	-	-	-	-	-	-	-	-	-	-	-	14
	B	-	B	B	B	B	-	-	B	-	B	B	-	B	-	-	-	-	-	-	-	-	B	B	B	-	-	-	-	-	-	-	-	B	-	-	B	-	-	-	-	-	-	-	14
	B	B	B	B	B	B	-	B	B	B	B	B	B	-	-	-	B	-	B	-	-	-	-	-	B	-	-	-	-	-	-	-	-	-	-	-	-	-	-	-	-	-	-	-	15
Vaccine	B	-	B	B	B	B	B	B	B	B	-	B	-	B	-	-	-	-	B	B	-	-	B	B	-	-	-	-	-	-	-	-	-	-	-	-	-	-	-	-	-	-	-	-	15
	B	-	-	-	B	B	B	B	-	B	B	B	B	B	B	-	-	-	-	B	B	-	-	-	B	B	-	-	-	-	-	-	-	-	-	-	-	-	-	-	-	-	-	-	15
	B	B	B	B	B	B	-	B	B	B	B	B	-	-	-	-	-	-	B	-	-	-	-	-	-	-	-	-	-	B	-	B	-	-	-	-	-	-	-	-	B	-	-	B	16
	B	-	-	-	-	-	B	B	B	-	B	B	B	-	B	B	B	B	-	B	B	-	-	B	B	-	-	-	-	B	-	-	-	-	-	-	-	-	-	-	-	-	-	-	16
	B	-	-	B	-	B	-	B	B	B	B	B	B	B	B	-	-	-	B	-	B	B	B	-	B	-	-	-	-	-	-	-	-	B	-	-	-	-	-	-	-	-	-	-	17
	B	B	-	B	B	B	B	-	B	-	-	B	B	B	B	B	B	B	B	B	-	-	B	B	-	-	-	-	-	-	-	-	-	-	-	-	-	-	-	-	-	-	-	-	18
	B	B	B	B	-	B	-	B	B	B	B	B	B	-	B	B	-	B	B	B	B	B	-	-	-	-	-	-	-	-	-	-	-	-	-	B	-	-	-	-	-	-	-	-	19
	B	-	B	B	B	B	B	B	B	B	B	-	B	B	-	B	-	-	B	-	B	-	-	B	-	-	B	-	-	-	-	-	-	-	B	-	-	-	B	-	-	-	-	-	19
	B	B	-	-	-	B	B	B	-	B	B	-	-	B	B	-	-	B	B	B	B	B	B	-	-	B	-	-	-	-	-	B	-	-	-	B	-	-	B	-	B	-	-	-	20
	B	B	B	B	B	-	-	B	B	B	B	B	B	B	-	-	-	B	-	B	B	B	-	B	B	-	B	-	-	-	-	-	B	-	-	-	-	-	-	-	-	-	B	-	21
	B	B	B	B	B	B	-	B	B	B	B	B	B	B	B	B	-	B	B	B	B	B	B	-	B	B	B	-	-	-	-	-	-	-	-	-	-	-	-	-	-	B	-	-	25
	B	B	B	B	B	B	B	B	B	B	B	B	B	B	B	-	B	B	B	B	B	-	B	B	B	B	B	B	B	-	B	B	B	-	B	-	-	-	-	-	-	-	-	-	31

shedding dynamics and patterns are affected by challenge exposure dose.[5] In general, enteric pathogens tend to have sporadic fecal shedding patterns as was observed for *ST* in this study. While expectations are not that vaccination will eliminate shedding, the Vaccine group had, for the most part, ceased excretion of *ST* within two to six weeks post-challenge exposure while several pigs in the Placebo group continually shed through the 12-week challenge phase. The cease in shedding may become one important implementation strategy to help reduce environmental contamination during the lifespan of the animals. The lower detection limit of enrichment cultures was approximately 4000 CFU/gram, thus the sporadic positive fecal samples later in the challenge phase may indicate a low level of colonization in both groups. Additionally, since pigs were housed on slats during the challenge phase, the cleaning of the pens was minimized to lessen aerosolization of challenge organism throughout the room. This study-related practice may have led to recirculation of the challenge organism to the pigs individually housed by pen. Regardless, the vaccine reduced the consistent shedding earlier in the challenge phase, thus reducing the pathogen load in the environment which ultimately will reduce recirculating *Salmonella* in the pigs in a production setting.

Figure 2: Percent of pigs shedding *Salmonella* Typhimurium by day post-challenge.

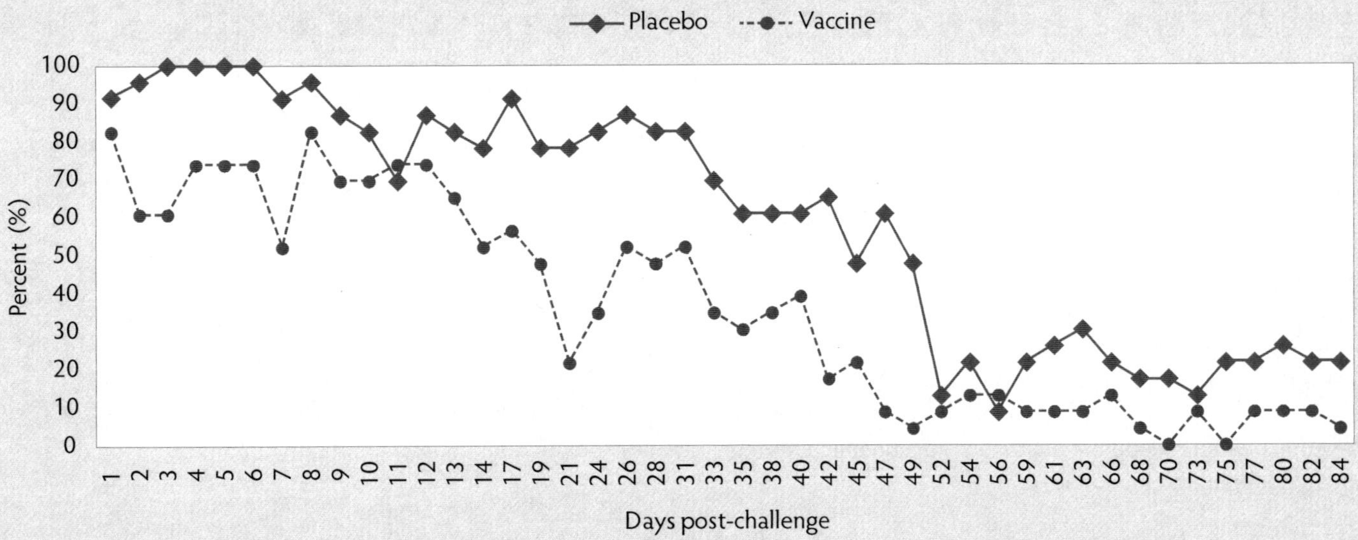

Two pigs were removed from the study after vaccination. Prior to challenge, a Vaccine group pig was removed from the study 20 days after vaccination for humane reasons (severe joint abscess). On 3DPC, a Placebo group pig was removed and was considered a challenge-related mortality by the attending veterinarian who observed, during gross necropsy, moderate intestinal autolysis, watery diarrhea, enlarged spleen and fibrinous peritonitis (normal lungs).

The apparent reduction of shedding observed in the vaccinated group is highly relevant. This preliminary data suggests that the use of a commercially available ALC *Salmonella* Choleraesuis-Typhimurium vaccine clinically reduces *ST* fecal shedding in pigs post-challenge. At the time of submission for publication, the USDA had not reviewed this information and has not approved a shedding claim.

References

1. Botteldoorn N et al. *Salmonella* on pig carcasses: positive pigs and cross contamination in the slaughterhouse. *Journal of Applied Microbiology*. 2003:95.

2. Wong DMA Lo Fo et al. Epidemiology and control measures for *Salmonella* in pigs and pork. *Livestock Production Science*. 2002;76.

3. Wegener HC et al. *Salmonella* control programs in Denmark. *Emerging Infectious Diseases*. 2003;9(7).

4. Siev D. An estimator of intervention effect on disease severity. *Journal of Modern Applied Statistical Methods*. 2005;4(2).

5. Ivanek R et al. *Salmonella* fecal shedding and immune responses are dose- and serotype- dependent in pigs. *PLoS ONE*. 2012;7(4): e34660. doi:10.1371/journal.pone.0034660.

Evaluation of the effect of shipping and handling of oral fluid samples on influenza A in swine by PCR

Victoria Foerster[1], BS; Greg Cline[2], DVM; Wayne Chittick[3], MS; Fred Sylvia[3], BASc; Marlin Hoogland[4], DVM
[1]Iowa State University CVM; [2]Boehringer Ingelheim Vetmedica Inc.; [3]Health Management Center (HMC);
[4]Smithfield Hog Production Division, Midwest Region

Introduction

Influenza is both a costly and difficult disease to control in the swine industry. Oral fluid (OF) samples are a useful method to detect influenza virus in swine (IAV-S) using polymerase chain reaction (PCR).[1] Best management practices for collecting and submitting oral fluid samples for IAV-S testing do not exist to the best of the investigators' knowledge.

Objective

This study evaluated the effects of time, temperature, and the addition of stabilizing or transport media on detection of IAV-S in oral fluids by real-time PCR.

Materials and methods

Five hundred milliliters of oral fluids were collected and confirmed to be negative for IAV-S by PCR. A culture of swine origin gamma clade H1N2 influenza virus was sequenced and propagated to arrive at a final concentration of 7.2 logs/ml. The oral fluids and H1N2 culture were mixed to make an 11:1 negative oral fluid to IAV-S master mix with PCR cycle threshold (Ct) of 23.23. The master mix was then further diluted to target four mixtures with Ct values of 25, 28, 34 and 37 by adding the appropriate amount of oral fluids. Each dilution was tested in quadruplicate at three temperatures (22°C, 4°C and -80°C) and with two types of transport media (phosphate buffered saline (PBS) (Gibco 1X Thermo Fisher Scientific, Waltham, MA) and viral transport media (VTM) (Eagle minimum essential media Sigma-Aldrich, St. Louis, MO)) in a 1:1 ratio. Samples were extracted and prepared for PCR at time 0 and plus 24, 48, and 96 hours. All extracted material had real-time PCR performed the same day with two complete replicates per PCR plate. All treatments were represented on a single PCR plate to minimize plate to plate variation. A multiple regression analysis was used to analyze the data collected.

Results

Of the 576 expected positive PCR samples, 98.2% tested positive. This left 10 negative samples which were all samples that had the weakest IAV-S concentration at the onset. Table 1 describes the negative samples. Mean Ct values were also evaluated. Figure 1 shows the mean Ct value of dilution 3 over time without transport media. There was little difference in virus degradation between the temperature treatments of 4°C and -80°C, but 22°C showed much more virus degradation resulting in greater Ct values. Least squares mean analysis showed that using no media or VTM were not significantly different from each other, but PBS use resulted in lower Ct values.

Table 1: IAV-S negative PCR samples testing parameters

Sample(s)	Media	Temperature	Time
1-4	None	22°C	96 hrs
5	VTM	-80°C	96 hrs
6	VTM	-80°C	48 hrs
7	VTM	4°C	48 hrs
8	VTM	22°C	48 hrs
9	VTM	22°C	0 hrs
10	PBS	22°C	48 hrs

Figure 1: Mean Ct value of dilution 3 over time at 22°C.

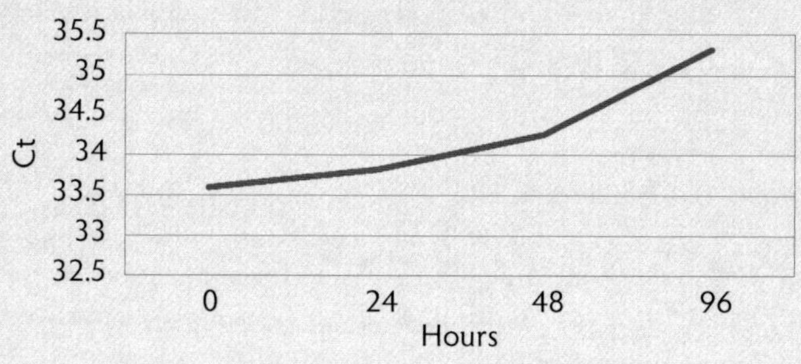

Discussion

Shipping and handling conditions do have the ability to affect PCR results and IAV-S detection. While 98% of all expected positive samples tested positive, the data showed that storage time and temperature affect degradation of viral nucleic acids causing some samples to test negative. Veterinarians should dilute OF samples 1:1 with PBS and ship samples frozen or near frozen as quickly as possible to the lab to help optimize PCR results based on the results in this study. Further investigation is warranted to identify the optimal dilution ratio of PBS to oral fluids.

References

1. Van Cuong et al. Serological and virological surveillance for porcine reproductive and respiratory syndrome virus, porcine circovirus type 2, and influenza A viruses among smallholder swine farms of the Mekong Delta, Vietnam. *Journal of Swine Health and Production,* 2014 Sep-Oct, Vol. 22(5), pp. 224-231.

Use of topical spray containing lidocaine to mitigate pain in pigs undergoing castration

Sara Davenport; Meghann K. Pierdon, VMD; Thomas D. Parsons VMD, PhD, DACAW

Introduction

Piglets routinely undergo pain-inducing procedures like tail docking and castration. Many countries now require some pain mitigation for these procedures, but in the United States, castration commonly is performed without any form of anesthesia or analgesia. Animal welfare concerns have motivated some pork buyers to request pain mitigation for these routine procedures despite a dearth of scientifically validated options available to producers. The objective of this study was to determine whether the use of wound care spray with lidocaine alleviated pain post-castration based on behavioral observation.

Materials and methods

At the Penn Vet Swine Teaching and Research Center, boars were castrated at 7-9 days of age. Within a litter, half of the males were randomly assigned to receive the lidocaine spray (L), and the other half was castrated without spray (C). Gilts from the same litter were used as non-castrated controls (NC). Castration was performed by making two vertical incisions on the scrotum. The testes were then freed by tearing the spermatic cord. 4 × 4 pads were soaked prior to the procedure in lidocaine spray and dabbed on the incision sites for boars receiving treatment for three seconds. 231 boars were castrated; of these, 129 received treatment. The piglets were then observed using scan sampling at ten minute intervals following return to the pen for two hours. Additionally, video recording of the first hour back in the pen post-castration was continuously coded using Noldus Observer V. 11.5 for 157 pigs (56 L, 51 C, and 50 NC pigs). Behaviors chosen for scan sampling and video coding included lying down alone, lying down in a group, walking, standing, playing, nosing the pen or sow, and suckling. Statistics were performed using Stata v. 13.1. All data was not normal, and a Kruskal Wallis non-parametric test was used. Bonferroni adjustment was used for multiple comparisons. Medians are reported unless otherwise noted.

Results

Table 1 summarizes the results of the video data. C animals laid alone more often and suckled, played, walked, and nosed less than NC pigs. L piglets did not isolate themselves like C pigs. However, compared to NC piglets, L piglets walked and nosed less. Play and suckling by L pigs was not different than C or NC piglets. Scan sampling data encompassed a larger number of animals over a longer period of time, and repeated the video findings on lying alone and suckling.

Discussion

Previous studies have shown that time spent suckling and social cohesion are decreased in pigs that are castrated relative to controls.[1] The behavior changes we found between C and NC piglets extend the list of behaviors changed by castration pain. The behavior of L pigs compared to the C and NC piglets suggests that the lidocaine spray does mitigate some, but not all, of these pain related behaviors. This study highlights that further research is needed to identify better methods of pain control in neonatal piglets undergoing painful procedures.

References

1. Hay, M. et al. Assessment of pain induced by castration in piglets: behavioral and physiological responses over the subsequent 5 days. *Applied Animal Behaviour Science*. 2003;82.

Acknowledgements

This work was funded by the Pennsylvania Pork Producers Council, NY Farmers, and the ASPCA.

Table 1: Video results

Treatment	Lying alone (median)	Suckling (median)	Playing (mean rank)	Walking (mean rank)	Nosing (mean rank)
C	418.1[a]	432.1[c]	72[e]	69[g]	65[j]
L	155.1[b]	599.1[c,d]	74[e,f]	75[g]	71[j]
NC	45.9[b]	661.2[d]	95[f]	97[h,i]	105[k]

Different superscripts in the same column indicate a significant result with a,b, g,h, j,k representing a difference at $P < .001$ and c,d, e,f, g,i at $P < .01$

Residue depletion profile of ampicillin trihydrate in cull sows

C. Mainquist-Whigham[1], BS; L. A. Karriker[1], DVM, MS, DACVPM; J. Bates[1], DVM;
R. Gehring[2], BVSc, MMedVet (Pharm), MRCVS, DACVCP; J. Coetzee[3], BVSc, Cert CHP, PhD, DACVCP
[1]Swine Medicine Education Center, VDPAM, Iowa State University, College of Veterinary Medicine, Ames, Iowa; [2]Department of Anatomy and Physiology, Institute of Computational Comparative Medicine, Kansas State University, Manhattan, Kansas; [3]Pharmacology Analytical Support team, VDPAM, Iowa State University, College of Veterinary Medicine, Ames, Iowa

Introduction

Penicillin has historically been used for the treatment of sows after farrowing. Due to changes in testing methods by the Food Safety and Inspection Service at slaughter plants, an increase in penicillin residues in edible tissues has occurred. This has led to a search for alternative treatment methods in sows, with an observable increase in use of ampicillin trihydrate (AT). Ampicillin, an aminopenicillin, has a similar structure and spectrum as penicillin, but the tissue residue depletion profile in cull sows is unknown, particularly at extralabel doses. Ampicillin is not labelled for use swine and there is no tolerance for residues in edible tissues at slaughter. The purpose of this study was to describe the tissue depletion profile of AT in cull sows following intramuscular administration of AT at either 6 mg/kg or 12 mg/kg bodyweight once daily for 3 days. It also correlates plasma and urine concentrations of AT with the drug residue depletion profile in tissues of cull sows over 40 days post-administration.

Materials and methods

Sixty cull sows were divided into two different treatment groups of thirty. Treatment Group 1 (TG1) was administered a dose of 6 mg/kg AT, and Treatment Group 2 (TG2) was administered a dose of 12 mg/kg AT once a day for three days. Each treatment group was divided into ten groups of three sows, each group corresponding to a necropsy time point 1, 3, 5, 8, 10, 12, 15, 20, 30, and 40 days post-administration of AT. Blood samples were collected on all sows before first administration of AT and before euthanasia on sows to be necropsied. Urine and tissue samples (liver, kidney, injection site, semimembranosus/semitendinosus muscle) were collected at necropsy. All samples were analyzed for AT residues using high-performance liquid chromatography (HPLC) with a level of quantification (LOQ) of 5 ppb.

Results and discussion

Ampicillin trihydrate was detected in the kidney samples of all sows in TG1 and TG2 at one day after the final treatment of AT. At least one sow at every necropsy timepoint in both TG1 and TG2 tested positive for AT residues at the injection site. No other tissue samples tested positive for AT residues. AT was detected in the plasma in all sows in TG1 and TG2 at one day post-administration. One sow in TG2 had AT residues detected in the plasma at three days post-administration. The urine in one sow in both TG1 and TG2 at day 10 post-administration tested positive for low levels of AT, and all but two sows had detectable urine AT concentrations at all time points prior to day 10.

Plasma concentrations of AT closely mirrored the concentration in the kidney tissues and could potentially be used as an indicator of kidney tissue residue levels ante mortem. There appears to be no dose related differences in residue depletion. However, AT was detected in injection site samples out to 40 days, which may raise some concerns with meat residues and the use of AT. The detection of AT in urine samples out to 10 days indicates that some AT is present in the sow and still being metabolized, despite the fact it is not consistently detected in tissues. Producers and practitioners should exercise caution when administering AT in place of penicillin. Further research is needed to better define the pharmacokinetics of AT in cull sows.

Characterization of genomic and pathogenic changes of porcine epidemic diarrhea virus after serial passages in cell culture

Holly Salzbrenner, BS; Qi Chen, DVM, MS; Phillip Gauger, DVM, PhD; Ganwu Li, DVM, PhD; Ying Zheng, MS; Michael Welch; Drew Magstadt, DVM, MS; James Carlson, BS; Rodger Main, DVM, PhD; Jianqiang Zhang MD, PhD
Department of Veterinary Diagnostic and Production Animal Medicine, Iowa State University, Ames, Iowa

Introduction

Since its emergence in the US in April 2013, porcine epidemic diarrhea virus (PEDV) has spread rapidly across the country and resulted in the estimated death of 8 million pigs in the first year, causing economic losses of $900 million to $1.8 billion.[1] In addition, PEDV has recently emerged or re-emerged in a number of countries including China, Japan, South Korea, Philippines, Thailand, Vietnam, Canada, Mexico, Germany, Belgium, France and Portugal.[2-10] So PEDV still remains a significant challenge to the global swine industries. Currently in US, there are two commercial killed PEDV vaccines (Zoetis and Harrisvaccines). However, some studies showed that these inactivated PEDV vaccines induced good IgG and IgA immune responses in herds previously exposed to PEDV but did not induce good IgA response in naïve pigs after vaccination.[11-12] Modified live virus (MLV) PEDV vaccine is more efficacious than the killed or subunit vaccines for inducing mucosal immunity. However, such a safe and efficacious MLV PEDV vaccine against the emerging US strains does not exist currently. The objectives of this study were to characterize the genomic and pathogenic changes of a US virulent PEDV prototype isolate after serial passages in cell culture and determine if any resultant viruses could be potential MLV vaccine candidate against PED.

Materials and methods

A US PEDV prototype strain cell culture isolate USA/IN19338/2013, previously isolated in our laboratory,[13] was serially passed in Vero cells for 100 passages. Pathogenic changes of selected viruses were evaluated in a neonatal pig model. Sixty piglets (5-day-old), free of PED virus and antibody, were randomly divided into 6 groups with 10 pigs per group. Pigs were fed with milk replacer. Groups 1 through 5 (G1-5) were orally inoculated with PEDV USA/IN19338/2013 at the passages 7, 25, 50, 75 and 100, respectively (10^4 $TCID_{50}$/ml, 10 ml/pig) and Group 6 (G6) received the same volume of virus-free culture media as negative control (Table 1). Clinical observations were recorded. Rectal swabs were collected at arrival and daily and tested by a PEDV nucleocapsid gene-based quantitative real-time RT-PCR.[14] Five pigs from each group were necropsied at 3 and 7 days post inoculation (DPI), respectively. Gross and microscopic lesions of small intestine, cecum and colon were examined and immunohistochemistry (IHC) staining was performed. Microscopic lesion severity was categorized as: 0 = no lesion, 1 = minimum villous atrophy, 2 = mild villous atrophy, 3 = moderate villous atrophy and 4 = severe villous atrophy. IHC staining was scored as: 0 = no signal, 1 = minimal staining, 2 = mild staining, 3 = moderate staining and 4 = severe staining. Whole genome sequences of the viruses P7, P25, P50, P75 and P100 were determined using next-generation sequencing technology.

Results and discussion

G6 (negative control) piglets remained negative throughout the study period. Piglets in G1-5 all had watery diarrhea at 1-4 DPI and mild to moderate diarrhea at 5-7 DPI. Piglets in G1-5 all

Table 1: Animal study design. Experimental infection of 5-day old piglets with PEDV USA/IN19338/2013 at different cell culture passages.

Group	Inoculation (0 DPI)	Necropsy (3 DPI)	Necropsy (7 DPI)
G1 (n = 10)	P7 (10^4 $TCID_{50}$/ml, 10ml/pig)	n = 5	n = 5
G2 (n = 10)	P25 (10^4 $TCID_{50}$/ml, 10ml/pig)	n = 5	n = 5
G3 (n = 10)	P50 (10^4 $TCID_{50}$/ml, 10ml/pig)	n = 5	n = 5
G4 (n = 10)	P75 (10^4 $TCID_{50}$/ml, 10ml/pig)	n = 5	n = 5
G5 (n = 10)	P100 (10^4 $TCID_{50}$/ml, 10ml/pig)	n = 5	n = 5
G6 (n = 10)	Virus-negative control	n = 5	n = 5

DPI: Days Post Inoculation

shed virus in rectal swabs with a trend that piglets inoculated with higher passage virus shed lower amount of viruses (Figure 1). Specifically, virus shedding levels were: G1 > G2 > G3, G4, G5 > G6 (> means significantly higher; but no significant differences among G3, G4 and G5). There was a trend that piglets inoculated with lower passage virus had more severe villous atrophy. At 3 DPI, microscopic lesions were G1, G2, G3 > G4, G5 > G6 in jejunum and G1, G2, G3 > G4, G5, G6 in ileum. The villus height were G1, G2, G3, G4, G5 < G6 in jejunum (< means significantly lower) and G1, G2, G3 < G4, G5, G6 in ileum at 3 DPI. The villus-height-to-crypt-depth ratios were: G1, G2, G3 < G4, G5 < G6 in jejunum and G1, G2, G3 < G4, G5, G6 in ileum at 3 DPI. The IHC staining in jejunum was: G1, G2, G3, G4 > G5 > G6; the IHC staining in ileum was: G1, G2, G3 > G4, G5 > G6 at 3 DPI. Differences between groups at 7 DPI were not very apparent.

Comparisons of the whole genome sequences revealed that nucleotide and deduced amino acid changes from P7 to P100 are mainly located in replicase non-structural protein (nsp) 2, nsp3, nsp4, nsp5, nsp6, nsp15, spike (S), ORF3, envelope (E), membrane (M), and nucleocaspid (N) proteins.

In summary, the virulent US PEDV prototype isolate USA/IN19338/2013 became gradually less pathogenic during serial passages in cell culture. Clinical observations, virus shedding in rectal swabs, histopathologic lesions, and IHC staining data suggest that the P75 and P100 viruses have been more attenuated than the P7, P25 and P50 viruses. However, the P75 and P100 viruses still can cause diarrhea and may not have been attenuated enough to be an MLV vaccine candidate. Continuous passage of the virus in cell culture is needed to obtain a fully attenuated vaccine candidate virus. Sixteen amino acid changes appear to be associated with virus attenuation under the conditions of this study. This study provides a strong basis for developing a MLV PEDV vaccine and for understanding the molecular mechanism of virus attenuation.

Figure 1: Virus shedding in rectal swabs of various inoculation groups during 0-7 days post inoculation (DPI).

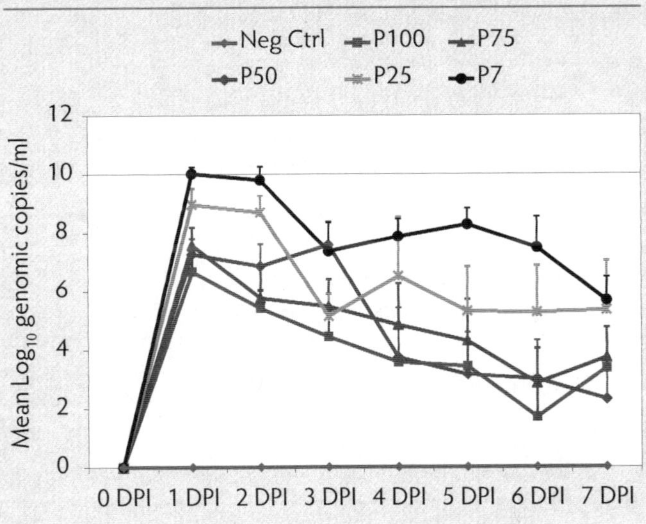

Acknowledgement

This study was supported by the Iowa State University College of Veterinary Medicine Seed Grant, USDA Formula Grant, and Zoetis. We are grateful to the Iowa State University Veterinary Diagnostic Laboratory faculty and staff for assistance with some testing. We also thank the Iowa State University LAR staff for animal care.

References

1. Paarlberg, P.L. 2014. Updated estimated economic welfare impacts of porcine epidemic diarrhea virus (PEDV). Available at: http://ageconsearch.umn.edu/bitstream/174517/2/14-4.Updated%20Estimated%20Economic%20Welfare%20Impacts%20of%20PEDV.pdf. Accessed 12 November 2015.

2. Grasland, B., Bigault, L., Bernard, C., Quenault, H., Toulouse, O., Fablet, C., Rose, N., Touzain, F. and Blanchard, Y., 2015. Complete genome sequence of a porcine epidemic diarrhea s gene indel strain isolated in france in december 2014. Genome Announc 3.

3. Mesquita, J.R., Hakze-van der Honing, R., Almeida, A., Lourenco, M., van der Poel, W.H. and Nascimento, M.S., 2015. Outbreak of Porcine Epidemic Diarrhea Virus in Portugal, 2015. Transbound Emerg Dis 62, 586-8.

4. Pasick, J., Berhane, Y., Ojkic, D., Maxie, G., Embury-Hyatt, C., Swekla, K., Handel, K., Fairles, J. and Alexandersen, S., 2014. Investigation into the role of potentially contaminated feed as a source of the first-detected outbreaks of porcine epidemic diarrhea in Canada. Transbound Emerg Dis 61, 397-410.

5. Puranaveja, S., Poolperm, P., Lertwatcharasarakul, P., Kesdaengsakonwut, S., Boonsoongnern, A., Urairong, K., Kitikoon, P., Choojai, P., Kedkovid, R., Teankum, K. and Thanawongnuwech, R., 2009. Chinese-like strain of porcine epidemic diarrhea virus, Thailand. Emerg Infect Dis 15, 1112-5.

6. Song, D. and Park, B., 2012. Porcine epidemic diarrhoea virus: a comprehensive review of molecular epidemiology, diagnosis, and vaccines. Virus Genes 44, 167-75.

7. Vui, D.T., Thanh, T.L., Tung, N., Srijangwad, A., Tripipat, T., Chuanasa, T. and Nilubol, D., 2015. Complete genome characterization of porcine epidemic diarrhea virus in Vietnam. Arch Virol 160, 1931-8.

8. Stadler, J., Zoels, S., Fux, R., Hanke, D., Pohlmann, A., Blome, S., Weissenbock, H., Weissenbacher-Lang, C., Ritzmann, M. and Ladinig, A., 2015. Emergence of porcine epidemic diarrhea virus in southern Germany. BMC Vet Res 11, 142.

9. Theuns, S., Conceicao-Neto, N., Christiaens, I., Zeller, M., Desmarets, L.M., Roukaerts, I.D., Acar, D.D., Heylen, E., Matthijnssens, J. and Nauwynck, H.J., 2015. Complete genome sequence of a porcine epidemic diarrhea virus from a novel outbreak in belgium, january 2015. Genome Announc 3.

10. Vlasova, A.N., Marthaler, D., Wang, Q., Culhane, M.R., Rossow, K.D., Rovira, A., Collins, J. and Saif, L.J., 2014. Distinct Characteristics and Complex Evolution of PEDV Strains, North America, May 2013-February 2014. Emerg Infect Dis 20, 1620-1628.

11. Thomas, P. (2014). Field experiences using porcine epidemic diarrhea virus (PEDV) vaccine in herds experiencing endemic disease. *In* "2014 Iowa State University Swine Disease Conferencee for Swine Practitioners", pp. 38-42, Ames, Iowa.

12. Schwartz, T.J. and Rademacher, C.J. (2015). Evaluation of the effects of PEDV vaccine on PEDv naïve and previously PEDv-exposed sows in a challenge model comparing immune response and preweaning mortality. *In* "2015 ISU James D. McKean Swine Disease Conferencee ", pp. 36-40, Ames, Iowa.

13. Chen, Q., Li, G., Stasko, J., Thomas, J.T., Stensland, W.R., Pillatzki, A.E., Gauger, P.C., Schwartz, K.J., Madson, D., Yoon, K.J., Stevenson, G.W., Burrough, E.R., Harmon, K.M., Main, R.G. and Zhang, J., 2014. Isolation and characterization of porcine epidemic diarrhea viruses associated with the 2013 disease outbreak among swine in the United States. J Clin Microbiol 52, 234-43.

14. Thomas, J.T., Chen, Q., Gauger, P.C., Gimenez-Lirola, L.G., Sinha, A., Harmon, K.M., Madson, D.M., Burrough, E.R., Magstadt, D.R., Salzbrenner, H.M., Welch, M.W., Yoon, K.J., Zimmerman, J.J. and Zhang, J., 2015. Effect of Porcine Epidemic Diarrhea Virus Infectious Doses on Infection Outcomes in Naive Conventional Neonatal and Weaned Pigs. PLoS One 10, e0139266.

Comparison of the detection of PRRSV antibodies from oral fluids using a commercial PRRS oral fluids assay and the modified PRRS X3 overnight protocol

C. Goodell[1], DVM, MS, PhD; S. Lizano[1], PhD; J. Fent[2], DVM; S. Stewart[2]; D. Baum[3], DVM, PhD; T. Gard[3]
[1]IDEXX, Westbrook, Maine; [2]Smithfield, Rose Hill, North Carolina; [3]Iowa State University, Ames, Iowa

Introduction

With the introduction of swine oral fluids as a valuable diagnostic tool for detecting both the presence of genetic material from and exposure to pathogens, veterinarians worldwide are adopting the use of oral fluid testing for routine monitoring programs. Commonly, monitoring for PRRSV includes rt-PCR testing of pen based oral fluids at nursery placement during PRRS control and elimination programs. However, the true understanding of whether PRRSV exists at very low prevalences in pigs from PRRS positive but stable sow flows, is through late nursery antibody monitoring, once maternal antibodies have waned (~7-10wks of age). Monitoring antibodies in young growing pigs using oral fluids is a highly sensitive method of determining early seroconversion in these large populations, and can be used to better manage (i.e. move) truly PRRSV negative groups of piglets particularly in 3 site systems.

Materials and methods

A field study was performed to compare test results from oral fluid field samples using the USDA licensed PRRS OF Test (OF) and the published overnight (ON) protocol.[1,2] Two hundred and ninety four expected antibody negative oral fluid samples from 34 field locations of a single production system were tested in two laboratories (A and C), and a subset (264) were tested at a third laboratory (B). Samples were tested within 1-5 days of collection in laboratories A and C, and after 1 freeze-thaw cycle at laboratory B. Laboratories A and B performed the ON protocol, while laboratory C performed testing using the commercial IDEXX PRRS OF Test. All three laboratories used the published cut off of $S/P \geq 0.40$, and all three centrifuged samples prior to testing. For Laboratory A, S/P values between 0.20 and 0.40 were considered suspect and therefore additionally defined.

Results

Negative ($S/P < 0.4$) results were obtained in 292, 264 and 272 of the oral fluid samples from Laboratories A, B and C, respectively (Table 1). 19 of Laboratory C S/P results ranged between 0.4 and 0.8.

From subsequent sampling and re-testing, 3 of the 34 expected negative field locations were determined to be seroconverting. These 3 locations represented 48 samples of which the PRRS OF Test detected 10 positive, whereas the ON protocol in Laboratory A detected only 2 positive and 2 suspects, at S/P ~0.2. Outside of the one confirmed seroconverting nursery, there were 9 other nursery sites represented. Age at sampling and transition diet make up could not be confirmed, therefore detection of residual maternal antibodies (several nursery sources were PRRS positive stable (2B)[3] sow farms) and/or the impact of sprayed dried plasma cannot be ruled out.

Conclusions

The commercial PRRS OF Test is a highly sensitive test which appeared to detect early seroconversion that was not detected by the modified ON protocol. The two ON protocols although providing equivalent negative categorical results, exhibited some difference in analytical S/P values. These differences may be due to the non-commercial components of the ON protocol adaptation to the commercial X3 kit.

References

1. Kittawornrat et al., Detection of Porcine reproductive and respiratory syndrome virus (PRRSV) antibodies in oral fluid specimens using a commercial PRRSV serum antibody enzyme-linked immunosorbent assay. *J Vet Diag Inv* 2012; 24(2): 262-269.

2. Kittawornrat et al., Ring test evaluation of the repeatability and reproducibility of a Porcine reproductive and respiratory syndrome virus oral fluid antibody enzyme-linked immunosorbent assay. *J Vet Diag Inv.* 2012; 24(6): 1057-1063.

3. Holtcamp et al., *J Swine Health and Prod* 2011; 19 (1): 44-56.

Table 1: Agreement between laboratories

		Laboratory A (ON protocol)			Laboratory B (ON protocol)	
		+	Sus	-	+	-
Laboratory C (PRRS OF)	+	2	3	17	0	12
	-	0	1	271	0	252

Descriptive study of swine enteric coronavirus disease (SECD) elimination and prevention in a genetic multiplication system in North America

J. P. Cano, DVM, PhD; T. Riek, DVM; R. Thompson, DVM, MS; J. Geiger, DVM, MS; W. Lyons, DVM; T. Snider, DVM, MPH
PIC, Hendersonville, Tennessee

Introduction

Porcine epidemic diarrhea virus (PEDV), porcine deltacoronavirus (PDCoV) and transmissible gastroenteritis virus (TGEV) are causal agents of SECD.[1] Diarrhea, vomiting and up to 100% mortality for 3-5 weeks are observed in suckling pigs infected with PEDV in North America (NA).[2] It was estimated that about 60% of the USA breeding herd was impacted by PEDV between May of 2013 and May of 2014 causing a shortfall of 12,500,000 market hogs (11%).[3] PDCoV was also recently reported in NA[4] with a similar but milder clinical presentation. Genetic analyses indicate relationship with isolates previously detected in Asia,[5] potentially introduced to the USA via contaminated tote bags in the Spring of 2013.[6] Because of the devastating impact experienced with PED in completely susceptible populations, the discussion in NA has focused around the feasibility and sustainability of control or elimination approaches. This abstract describes the elimination experience and the results of a prevention program for a 92.000 sow genetic multiplication system in NA.

Materials and methods

The PIC multiplication system in NA has accumulated a record of minimal impact from PRRS or *M. hyopneumoniae* in the past several years, but between November 2013 and April 2014, 47.4% of the breeding herds in the system became affected by SECD. Specifically, 26.4% were infected with PEDV only, 10.5% with PDCoV only and 10.5% with both viruses. Likely because their remote location and intensive biosecurity none of the genetic nucleus or sire line nucleus farms were affected.

The goal was immediately set to eliminate PEDV and/or PDCoV from the multiplication flow as soon as possible. When the presence of PEDV was confirmed, piglets were weaned early and moved to an offsite nursery for better survival, the gilt development units were loaded and controlled oral live virus exposure was initiated to all sows and gilts aiming to homogenize immunity to minimize the duration of shedding. To facilitate communication, a SEC herd infection status classification was structured for breeding herds (Table 1) and growing pig sites (Table 2) following the same logic of the PRRSV infection status classification.[7] In order to interrupt the introduction of susceptible individuals, every piglet was weaned off site and the herds were closed to replacement pigs until the third consecutive PCR-negative biweekly test. Negative sentinels were brought in at this time, and if no clinical or diagnostic evidence of infection was detected after five weeks, negative replacements entered the herd. Keeping litter integrity, minimizing piglet handling and intensifying cleaning and disinfection in the farrowing rooms was emphasized in all projects.

In the summer of 2014, a SECD prevention initiative was launched across the system focused on mitigating risk in five main areas: (1) feed ingredients, reception, manufacturing and delivery, (2) transportation decontamination and inspections, (3) people training and engagement, (4) manure management and (5) mortality disposal. PIC risk assessments for farms, feed mills and transport decontamination facilities were updated between September and December of 2014 (results not included) and corrective action plans were initiated when needed. In order to increase awareness about risk and biosecurity across the system, educational materials were developed and made available (`www.pic.com/biosecurity`). Also a biosecurity officer was designated in each breeding herd or flow and bi-weekly interactive webinars were carried out. The intensive surveillance program initiated early in 2014 was maintained to minimize the risk of moving infected breeding stock. The program included (a) daily evaluation of clinical signs by trained personnel, (b) routine testing based on oral fluids PCR and (c) contingency diagnostic investigations and herd health-holds any time clinical signs or unexpected diagnostic results were observed.

Results and discussion

By August of 2014 all breeding herds were consistently weaning negative pigs. By the end of October 2014 the lack of shedding had been verified through the use of sentinel pigs in all herds (Figure 1). It took 20,1 (7,1-28,1) weeks for the herds infected with PEDV and 14,6 (11,9-17) weeks for the herds infected with PDCoV to consistently wean PCR-negative pigs. A year later, no new or "reactivated" infections have been detected and 100% of the growing sites have stayed negative for SECD. Likely as the result of the surveillance program, the transportation biosecurity protocols and the quarantine at destination, no SECD was transmitted downstream. The successful elimination of PEDV and PDCoV from all herds as well as the lack of infections in the last year and a half are encouraging; however, the increasing proportion of pigs susceptible to SECD in the US represents a concern. The economic analysis of SECD elimination in commercial operations and the evaluation of the efficacy of current prevention strategies in a higher prevalence scenario still need to be conducted.

Table 1: SEC herd infection status classification in breeding herds.

Herd category	Mapping color	Code/stage	Definition	Diagnostic requirement
Positive unstable (I)	▲ Red	1	Herd with clinical signs consistent with SECD, infected and actively shedding the virus	Detection of PEDV, PDCoV or TGEV in swine tissues or feces by PCR, VI or IHC
Transitionally negative (II)	▲ Orange	2	Herd that has been infected where no clinical signs suggestive of SECD are observed, and virus is no longer detected. *USDA Reporting Negative*	Three consecutive PCR-negative tests at least two weeks apart from each other. Each time 30 rectal swabs from piglets of different litters are collected and tested in pools of five samples
Provisionally negative (III)	▲ Yellow	3	Herd that has been infected where no clinical signs suggestive of SECD are observed, and shedding of virus is not longer occurring	(a) Sentinels: lack of clinical signs or evidence of infection in a minimum of one naïve pig per 500 sows introduced to a transitionally negative herd. Sentinels should be allowed to roam the alley-ways for at least four hours per day. Fecal and serum samples are collected one, three and five weeks post introduction. PCR is performed in rectal swabs pooled in groups of five samples or less and individual serum samples are tested by ELISA or IFA **OR** (b) Extended testing: if naïve sentinels are not available, repeat testing for "Transitionally Negative" status, three consecutive PCR-negative tests at least two weeks apart from each other
Negative (IV)	▲ Green	4	Herd that has never been infected or has been provisionally negative with no clinical signs suggestive of SECD or detection of virus for at least six months	(a) Sentinels: negative serology test (ELISA or IFA) in at least 30 pigs six months after the introduction to the herd **OR** (b) Extended testing: continue the testing protocol for "Transitionally Negative" status, for three additional monthly tests (30 rectal swabs from piglets of different litters in pools of five) after the herd is "Provisionally Negative"

Note: polymerase chain reaction (PCR), virus isolation (VI), immunohistochemistry (IHC). Map code: ▲ *breeding herd.* With the lack of recent diagnostic evidence, the herd is classified as Positive (I) by default, unless clinical signs suggestive of SECD disease have never been observed in the breeding herd. Vet-to-vet communications are required prior to the first sale of pigs from the site, unless the herd is classified as negative (IV).

References

1. USDA APHIS VS. Novel swine enteric coronavirus disease case definition. 2014. www.aphis.usda.gov/animal_health/animal_dis_spec/swine/downloads/secd_case_definition.pdf.

2. Stevenson G., *et al.,* Emergence of Porcine epidemic diarrhea virus in the United States: clinical signs, lesions, and viral genomic sequences. 2013. Journal of Veterinary Diagnostic Investigation 25(5) 649–654.

3. Sawyer W. Rabobank Report: PEDv set to cause significant shortfall in North American Hog Market. 2014. http://rabobank-food-agribusiness-research.pr.co/73098-rabobank-report-pedv-set-to-cause-significant-shortfall-in-north-american-hog-market.

4. Marthaler D., *et al.,* Rapid detection, complete genome sequencing, and phylogenetic analysis of porcine deltacoronavirus. 2014. Journal of Emerging Infectious Diseases 20(8) [Internet] http://dx.doi.org/10.3201/eid2008.140526.

5. Vlasova A., *et al.,* Distinct Characteristics and Complex Evolution of PEDV Strains, North America, May 2013–February 2014. 2014. Emerging Infectious Diseases 20(10), 1620-8.

6. USDA APHIS VS. Swine Enteric Coronavirus Introduction to the United States: Root Cause Investigation Report. 2014. https://www.aphis.usda.gov/animal_health/animal_dis_spec/swine/downloads/secd_final_report.pdf

7. Holtkamp D., et al., Terminology for classifying swine herds by porcine reproductive and respiratory syndrome virus status. 2011. JSHAP 19(1) 44-56.

Table 2: SEC herd infection status classification in growing pig sites.

Herd category	Code/stage	Mapping color	Site type	Definition	Diagnostic requirement
Positive (I)	1	■ ● Red	AIAO	Current growing population or originating herd infected with SEC	Detection of PEDV, PDCoV or TGEV in swine tissues or feces by PCR, VI or IHC
			CF	Current growing population, previous groups or originating herd infected with SEC	
Negative (IV)	4	■ ● Green	AIAO	Site is depopulated, cleaned, disinfected and repopulated with pigs originated from a negative (IV) herd. OR, lack of clinical signs and virus detection in pigs originated from transitionally (II) or provisionally negative (III) herds	No testing is needed if the site is repopulated from a Negative (IV) herd and no clinical signs are observed. In pigs from transitionally (II) or provisionally (III) negative herds, three consecutive PCR-negative tests at least two weeks apart from each other are needed. Each time 30 rectal swabs (pools of five samples) collected from all air spaces OR two ropes per 1000 pigs (no pooling) need to be tested
			CF	No clinical signs or evidence of the virus in the population. Incoming pigs are originated from a Negative (IV) or provisionally negative (III) herd.	Three consecutive PCR-negative tests at least two weeks apart from each other are needed. Each time 30 rectal swabs (pools of five samples) collected from all air spaces OR two ropes per 1000 pigs (no pooling) need to be tested

Note: all-in/all-out (AIAO), continuous flow (CF). Polymerase chain reaction (PCR), virus isolation (VI), immunohistochemistry (IHC). Map code: ▪ nursery, ● finishing. In the lack of recent diagnostic evidence, the herd is classified as Positive (I) by default. Vet-to-vet communications are required prior to the first sale of pigs from the site, unless the herd is classified as negative (IV).

Acknowledgements

PIC's multiplication partners, farm crews, health team veterinarians (independent herd veterinarian), supply chain and technical service teams have been instrumental in the execution of the initiative. Drs Wayne Cast, Joe Connor, Steve Dritz, Laura Greiner, Tim Loula, Montse Torremorell and Noel Williams actively participated in the development of the prevention plan.

Figure 1: PEDV and PDCoV infection status by month in the genetic multiplication system.

PEDV	PDCoV	Both	
4	4	4	Negative
3	3	3	Provisionally negative
2	2	2	Stable
1	1	1	Positive

Site	Jun 13	Jul 13	Aug 13	Sep 13	Oct 13	Nov 13	Dec 13	Jan 14	Feb 14	Mar 14	Apr 14	May 14	Jun 14	Jul 14	Aug 14	Sep 14	Oct 14	Nov 14	Dec 14	Jan 15	Feb 15	Mar 15	Apr 15	May 15	Jun 15	Jul 15	Aug 15	Sep 15	Oct 15
1	4	4	4	4	4	4	4	4	4	4	4	4	4	4	4	4	4	4	4	4	4	4	4	4	4	4	4	4	4
2	4	4	4	4	4	1	0	0	1	1	1	2	2	2	2	2	2	3	3	4	4	4	4	4	4	4	4	4	4
3	4	4	4	4	4	4	4	4	4	4	4	4	4	4	4	4	4	4	4	4	4	4	4	4	4	4	4	4	4
4	4	4	4	4	4	4	4	1	1	1	2	2	2	2	2	3	3	4	4	4	4	4	4	4	4	4	4	4	4
5	4	4	4	4	4	4	4	4	4	4	4	4	4	4	4	4	4	4	4	4	4	4	4	4	4	4	4	4	4
6	4	4	4	4	4	4	4	4	4	4	4	4	4	4	4	4	4	4	4	4	4	4	4	4	4	4	4	4	4
7	4	4	4	4	4	4	1	1	1	1	1	2	3	3	3	3	3	4	4	4	4	4	4	4	4	4	4	4	4
8	4	4	4	4	4	4	4	4	4	4	1	1	1																
9	4	4	4	4	4	4	4	4	4	4	4	4	4	4	4	4	4	4	4	4	4	4	4	4	4	4	4	4	4
10	4	4	4	4	4	4	4	4	4	4	4	4	4	4	4	4	4	4	4	4	4	4	4	4	4	4	4	4	4
11	4	4	4	4	4	4	4	4	4	4	4	4	4	4	4	4	4	4	4	4	4	4	4	4	4	4	4	4	4
12	4	4	4	4	4	4	4	4	4	4	4	4	4	4	4	4	4	4	4	4	4	4	4	4	4	4	4	4	4
13	4	4	4	4	4	4	4	4	4	4	4	4	4	4	4	4	4	4	4	4	4	4	4	4	4	4	4	4	4
14	4	4	4	4	4	4	4	4	4	4	4	4	4	4	4	4	4	4	4	4	4	4	4	4	4	4	4	4	4
15	4	4	4	4	4	4	4	4	4	4	4	4	4	4	4	4	4	4	4	4	4	4	4	4	4	4	4	4	4
16	4	4	4	4	4	4	4	4	4	4	4	4	4	4	4	4	4	4	4	4	4	4	4	4	4	4	4	4	4
17	4	4	4	4	4	1	1	1	1	1	1	1	2	2	2	3	3	3	4	4	4	4	4	4	4	4	4	4	4
18	4	4	4	4	4	4	4	4	4	4	4	4	4	4	4	4	4	4	4	4	4	4	4	4	4	4	4	4	4
19	4	4	4	4	4	4	4	1	1	1	1	2	2	2	2	2	2	3	3	4	4	4	4	4	4	4	4	4	4
20	4	4	4	4	4	4	4	4	4	1	1	2	2	2	2	2	2	3	3	4	4	4	4	4	4	4	4	4	4
21	4	4	4	4	4	4	4	4	4	4	4	4	4	4	4	4	4	4	4	4	4	4	4	4	4	4	4	4	4
22	4	4	4	4	4	4	4	4	1	1	1	1	2	3	3	3	3	3	4	4	4	4	4	4	4	4	4	4	4
23	4	4	4	4	4	4	4	4	4	4	4	4	4	4	4	4	4	4	4	4	4	4	4	4	4	4	4	4	4
24	4	4	4	4	4	4	4	4	4	1	1	1	2	3	3	3	3	3	4	4	4	4	4	4	4	4	4	4	4
25	4	4	4	4	4	4	4	4	4	4	4	4	4	4	4	4	4	4	4	4	4	4	4	4	4	4	4	4	4
26	4	4	4	4	4	4	4	1	1	1	1	1	1	1	2	3	3	3	4	4	4	4	4	4	4	4	4	4	4
27	4	4	4	4	4	1	1	1	1	1	1	1	2	2	3	3	3	3	4	4	4	4	4	4	4	4	4	4	4
28	4	4	4	4	4	4	4	4	4	4	4	4	4	4	4	4	4	4	4	4	4	4	4	4	4	4	4	4	4
29	4	4	4	4	4	4	4	4	4	4	4	4	4	4	4	4	4	4	4	4	4	4	4	4	4	4	4	4	4
30	4	4	4	4	4	4	4	4	4	1	1	1	1	1	2	2	2	2	4	4	4	4	4	4	4	4	4	4	4
31	4	4	4	4	4	4	4	4	4	4	4	4	4	4	4	4	4	4	4	4	4	4	4	4	4	4	4	4	4
32	4	4	4	4	4	4	4	4	4	4	4	4	4	4	4	4	4	4	4	4	4	4	4	4	4	4	4	4	4
33	4	4	4	4	4	1	1	1	1	1	1	1	1	2	2	3	3	3	4	4	4	4	4	4	4	4	4	4	4
34	4	4	4	4	4	1	1	1	1	1	1	1	2	2	2	3	3	3	4	4	4	4	4	4	4	4	4	4	4
35	4	4	4	4	4	1	1	1	1	1	1	2	2	2	2	3	3	3	4	4	4	4	4	4	4	4	4	4	4
36	4	4	4	4	4	1	1	1	1	1	1	1	2	2	2	3	3	3	4	4	4	4	4	4	4	4	4	4	4
37	4	4	4	4	4	1	1	1	1	1	1	2	2	2	2	3	3	3	4	4	4	4	4	4	4	4	4	4	4
38	4	4	4	4	4	4	4	4	1	1	1	1	1	2	3	3	3	4	4	4	4	4	4	4	4	4	4	4	4

Ambitine™ Feed Technology: A late finishing intervention for pork production

D. McManus[1], DVM; M. Lachmann[1], PhD; B. de Rodas[1], PhD; T. Weeden[1], PhD; S. Crowder[1], MS; Suzanne Petersen[2], PhD
[1]Purina Animal Nutrition LLC; [2] PMI Nutritional Additives

Introduction

Modern swine production is characterized by housing large numbers of pigs in environmentally controlled buildings. These buildings are designed to house pigs transferred from sow units right after weaning, or from nursery facilities, where they will stay until they reach market weight. As the pigs approach the finishing phase the magnitude of the stressors they have to overcome also increases. These stressors include increased feed intake and competition for feeder space, greater heat production and limited pig space. This becomes even more relevant in conditions where the market incentivizes pork producers to finish pigs at heavier weights (290-300 lbs of body weight (BW)). Under these stressful conditions, the inclusion of functional components, such as plant extracts and organic acids, in the diets appear to be effective in improving the productive performance of pigs.[1,2] Ambitine™ Feed Technology (Ambitine™ FT) is a blend of plant extracts and acidifiers designed to help overcome the negative impact of the late finisher stressors on pig performance.[3] Two experiments were conducted in order to evaluate the effect of feeding Ambitine™ FT on the productive performance of pigs fed non-medicated diets during late finishing.

Material and methods

Experiment 1 (Exp.1): The first experiment was conducted at the Purina Animal Nutrition Center in Missouri, using a total of 168 mixed sex pigs, with an initial BW of 213.5 lbs, randomly allotted to 1 of 2 dietary treatments: Control or Ambitine™ FT. The dietary treatments were evaluated during the last 31 days in the finisher. The control treatment consisted of a non-medicated fortified corn-soybean meal-DDGS diet formulated to provide all required nutrients to support optimal growth. The Ambitine™ FT diet was similar to the control diet with the exception of replacing 2 lbs of corn with 2 lbs of Ambitine™ FT per imperial ton of complete feed. Both diets were formulated at 0.65% total ileal digestible lysine. The pigs were housed in an environmentally-controlled building, using 12 pens/treatment with 7 pigs/pen, with an available pig space of 8.5 sq ft/pig and with free access to feed and water.

Experiment 2 (Exp.2): The second experiment was conducted in a commercial setting at the Purina Applied Research Center, with the objective to evaluate the effect of feeding Ambitine™ FT and/or 150 ppm of Copper (Cu) in a non-medicated diet during late finishing. A total of 880 mixed sex pigs with an average initial BW of 209.7 lbs were housed in wean to finish facility, randomly assigned to 1 of 4 dietary treatments, with an available pig space for the finishing phase of 7.6 sq ft/pig. The experiment was designed as a 2 × 2 factorial. The 2 factors evaluated were the addition of Ambitine™ FT (0 and 2 lbs/imperial ton of complete feed) and Copper (0 and 150 ppm, from Copper Chloride); to derive 4 dietary treatments, fed over the last 42 days in the finisher (days 0 to 42), in 2 feeding phases (days 0 to 20 and days 20 to 42; 8 diets total, 2 diets/trt). The diets fed days 0 to 20 were formulated at 0.70% total ileal digestible lysine and diets fed days 20 to 42 were formulated at 0.65%. The 4 dietary treatments are described in Table 1.

In Exp.1 and Exp. 2, pigs had free access to feed and water. Feed samples were collected and analyzed for moisture, crude protein, fat, calcium, phosphorus, copper, sodium, zinc, chloride, magnesium and manganese. The analyzed nutrient levels were in accordance with the formulated nutrient levels. Production performance parameters including initial and final BW, feed consumption, weight gain and feed to gain ratio were recorded and analyzed using the mixed procedure of SAS.[4]

Results and discussion

In Exp.1: Pigs fed non-medicated diets with Ambitine™ FT during the last 31 days of the finisher period, had 5.8% greater ($P < 0.1$) average daily gain (ADG) and 4.2% improved ($P < 0.1$) feed conversion (F:G) compared to control fed pigs. Dietary treatments, however, did not significantly ($P > 0.2$) affect average daily feed intake (ADFI). By the end of the experimental period, pigs fed Ambitine™ FT were 4.4 lbs heavier than pigs fed the control diet (Table 2).

In Exp.2: Feeding Ambitine™ FT with or without copper significantly improved ADG ($P = 0.04$) and tended to improve F:G ($P = 0.07$) during the last 22 days of feeding and, the overall feed efficiency over the entire 42 days period prior to market, when compared with feeding the control diet with or without 150 ppm of copper (Table 3). However, this effect was not demonstrated during the initial 20 days of feeding, indicating the need of a longer exposure time to the compounds, to capture the benefits of feeding Ambitine™ FT. Overall, all treatments were able to numerically improve total gain. Pigs fed Ambitine™ FT during the last 42 days in the finisher exhibited 5.23 lbs more total gain (8.5% improvement), pigs fed Ambitine™ FT + copper resulted in the highest numerical improvement in total gain at 7.97 lbs (12.96%), and copper alone was able to improve total gain by

Table 1: Dietary treatments in a 2 × 2 factorial arrangement to evaluate the addition of Ambitine™ FT and Copper over the last 42 days in the finisher (days 0 to 42, Purina Applied Research Center – FT 140F-15).

		Dietary treatment			
	Days on feed	**Control (Non-Med)**	**Ambitine™ FT 2 lbs/ton**	**Ambitine™ FT 2 lbs/ton + 150 ppm Cu**	**150 ppm Cu**
Dietary Lys, %	0 - 20	0.70 TID	0.70 TID	0.70 TID	0.70 TID
	20- 42	0.65 TID	0.65 TID	0.65 TID	0.65 TID

Table 2: Exp.1: Productive performance of pigs fed non-medicated diets with and without Ambitine™ FT over the last 31 days in the finisher (Purina Animal Nutrition Center – HF739).

	Treatment				
Performance variable	**Control**	**Ambitine™ FT**	**SEM**	**Difference**	***P*-value**
Initial BW, lbs	213.00	214.10	1.29		
Final BW, lbs	271.90	276.30	2.1	4.4 lbs	
Total gain, lb	58.90	62.20		3.3 lbs (5.6%)	
ADG, lb	1.89	2.00	0.04	5.8%	< 0.1
ADFI, lb	6.35	6.42	0.10	1.1%	> 0.2
F:G	3.36	3.22	0.04	4.2%	< 0.1

4.57 lbs (7.43%). Feed intakes were not significantly affected by any treatment when compared to the control. Feed efficiency was significantly ($P < 0.05$) improved by feeding Ambitine™ FT and Ambitine™ FT + Copper by 4.5% and 4.4% respectively, when compared to the control. Copper alone did not significantly improvement feed efficiency. None of the treatments affected pig mortality and morbidity (Table 4).

In summary, these results indicate that Ambitine™ FT improve performance of late finishing pigs. Additionally, 150 ppm of copper can be added with Ambitine™ FT to increase total weight gain. However, there is no added benefit to the feed conversion versus feeding Ambitine™ FT alone.

References

1. J. H. Eisemann and E. van Heugten. Response of pigs to dietary inclusion of formic acid and ammonium formate. J. Anim. Sci. 2007. 85:1530–1539.

2. Tung, C. M. and J. E. Pettigrew, NA. Critical review of acidifiers. Pork Checkoff. Available at: `http://old.pork.org/filelibrary/animalscience/reviewofacidifiers.pdf`. Accessed: 09 September 2015.

3. ™ Trade marked by PMI Nutritional Additives.

4. SAS Institute Inc., Cary, NC.

Table 3: Exp.2: Productive performance of pigs fed non-medicated diets with and without Ambitine™ FT and 150 ppm of Copper over the last 42 days in the finisher (Purina Applied Research Center – FT 140F-15).

Performance variable	Dietary treatment: Control	Ambitine™ FT	Ambitine™ FT + 150 ppm Cu	150 ppm Cu	SEM	Main effects P =	Ambitine™ * Cu interaction P <
Initial BW, lbs	**210.46**	**209.04**	**209.44**	**209.83**	**3.84**	**0.987**	
Period 1, 0-20							
Final BW, lbs	252.77	251.42	252.54	252.95	3.38	0.976	
Total gain, lbs	42.31	42.38	43.10	43.12			
ADG, lbs	2.116	2.118	2.156	2.156	0.05	0.860	
ADFI, lbs	6.086	5.942	6.045	6.094	0.07	0.310	
F:G	2.881	2.816	2.819	2.835	0.05	0.681	
Period 2, 20-42							
Initial BW, lbs	231.43	231.74	235.74	233.52	2.58	0.568	
Final BW, lbs	271.96	275.77	278.91	275.90	2.73	0.312	
Total gain, lbs	40.53	44.03	43.17	42.38			
ADG, lbs	1.845b	2.002[a]	1.964[a]	1.899[ab]	0.05	0.040	0.01
ADFI, lbs	6.563	6.506	6.415	6.376	0.12	0.628	
F:G	3.581[b]	3.268[a]	3.276[a]	3.380[a]	0.08	0.007	0.01
Cumulative, 0-42							
Total Gain	61.5	66.73	69.47	66.07			
ADG, lbs	2.012	2.074	2.077	2.047	0.04	0.451	
ADFI, lbs	6.256	6.145	6.171	6.179	0.07	0.589	
F:G	3.114[b]	2.974[a]	2.977[a]	3.026[ab]	0.04	0.036	0.02

Table 4: Exp.2: Mortality and morbidity of pigs fed non-medicated diets with and without Ambitine™ FT and 150 ppm of Copper over the last 42 days in the finisher (Purina Applied Research Center – FT 140F-15).

Performance variable	Dietary treatment: Control	Ambitine™ FT	Ambitine™ FT + 150 ppm Cu	150 ppm Cu	SEM	P =
Mortality/morbidity, #	0	0	0	1		
Mortality/morbidity, %	0.00	0.00	0.00	0.54	0.27	0.41

Biometrics – Intelligent sensors for swine production

Guido Klement, Dr. agr.; Simon Lague; Patrick van Dijk, B.Eng., B.ICT
Fancom BV, Panningen, Netherlands

Introduction

As the world struggles with the dilemma between 'acceptable' and the 'practicality' of growing 40 billion animals for meat, Fancom sees an opportunity to offer a realistic alternative. Many consider iFarming to be a future philosophy for meat production in the intensive farming industry where the interests of the farmer, consumer and society are balanced. Market developments convey a clear message: "in 2050 the world population is expected to reach 9 billion." Together with ongoing economic growth in developing regions, this will lead to a massive demand for animal protein. This implies that meat production has to intensify alongside a responsibility with regard to animal welfare. At the same time the move towards metropolitan farming will introduce additional challenges such as the acceptance of meat production in our back yards and the need to reduce emissions to zero footprints.

Continuous and automatic monitoring

iFarming or intelligent farming is characterized by the automatic and continuous monitoring of ambient conditions in the house which is also combined with animal-based measurements that register welfare. Developments in sensor technologies enable an increasing number of facets of animal behaviour to be monitored. Fancom works closely with Leuven University in Belgium to develop innovative technologies that allow the farmer to better understand the behaviour of his animals and to improve the health and welfare that leads to better performance.

Camera technology

Fancom's eYeNamic™ System monitors pigs and poultry in the house when the farmer is not present and cannot determine whether his animals are showing signs of healthy activity. The system uses three cameras mounted in the ridge of the house to continuously monitor the floor below. Analysis software translates these camera images into an index (Figure 1) for animal distribution and activity. Both measures are valuable indicators of animal welfare. The system allows the behaviour of a group of animals to be followed by the minute. Abnormal behaviour is immediately visible, enabling the farmer to respond in time before any abnormalities can affect the performance of the animals. From this system, he has the certainty of no unpleasant surprises at the end of a round!

Figure 1: Fancom eYenamic cameras measuring animal activity and distribution.

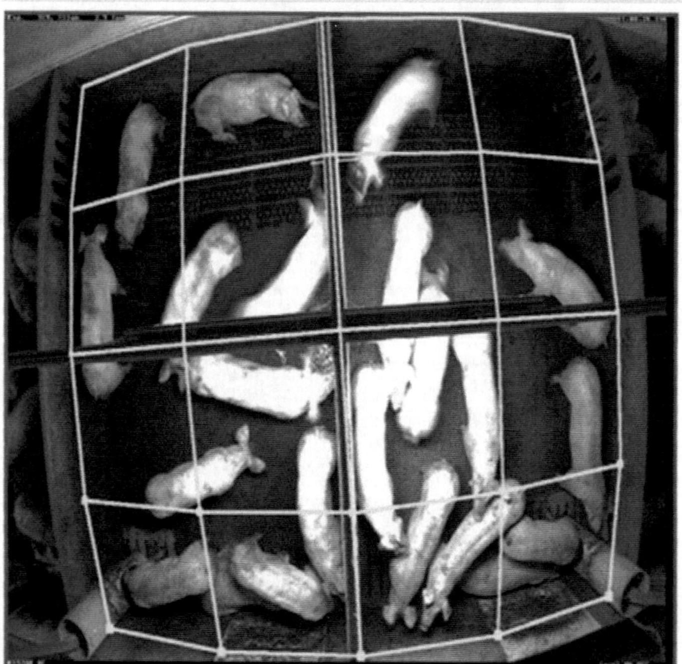

Activity			
1	5	48	6
21	51	26	22
26	31	13	10
2	30	21	9

Figure 2: Fancom eYeScan weighing pigs by camera.

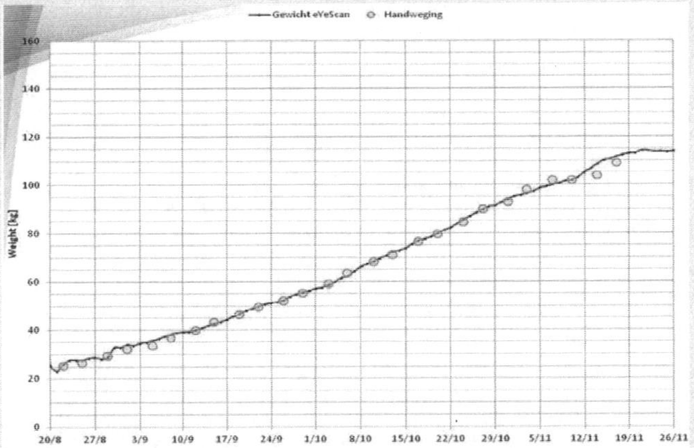

Aggression in pigs

Using the same technology, Fancom is contributing to research into the automatic monitoring of aggressive behaviour in pigs. The ultimate objective is to develop a system that manages behaviour using negative conditioning techniques to help prevent unwanted aggression, another influencing factor in the health and welfare of the group.

Monitoring growth with cameras

Fancom's eYeScan™ System is an automatic growth monitor for fatteners, again using camera technology. An image is made each time a pig starts eating. This enables the computer to accurately calculate the animal's weight through algorithms. This process is entirely automatic, labour saving and stress-free. The system allows for the growth behaviour of a group of fatteners to be followed from day to day. Graphs and tables (Figure 2) clearly illustrate the increasing line in the average growth of the animals.

Sound

The Fancom Pig Cough Monitor continuously monitors the occurrence of coughing in a compartment of pigs. The monitoring is accomplished by the automated analysis of the recorded sounds. With this innovation, a continuous observation of the respiratory health status of pigs becomes possible.

The system consists of a control with analysing software and two microphones installed at several locations to record sound. The recorded sound is then filtered initially to distinguish general pig sounds from coughing sounds. This information is graphically presented (Figure 3) to the end-user (farmer, veterinarian, etc.) and alarms are raised when thresholds are exceeded. From there the farmer can determine the acceptable level of coughing against a pre-set norm.

This Fancom product aims to help farmers reduce the use of antibiotics with the goal of a healthier outcome for both pigs and humans. Moreover current antibiotic policies present an enormous financial risk for the farmer. Fancom's pig cough monitor provides a viable answer to the problem by using the benefits of sound to identify respiratory disease early. Healthy animals will perform better, leading to more consistent growth, end weight and greater profit. At the same time, sick pigs still need to be treated with antibiotics. However improved and faster observation of the animals reduces the need for antibiotics which can then be reserved for use over shorter periods to fewer animals.

Apart from the obvious financial benefit, the system also gives veterinarians the ability to meet the challenge of continuously providing more detailed information about current threats. Not only can they consult in a quick and economic way, they can also compare the results of different farms and manage the threat of a wider spread infection within their region.

Water monitoring

Measurement of the water intake of pigs is one of the simplest and most effective tools to monitor the status of the animals. Volume and dynamics of water intake is related to diseases and health problems, feed intake, light intensity and equipment malfunctions. Water usage provides a very quick insight into the bio response of the animals. By reacting quickly to deviations, it is possible to manage growth rates as a consequence of addressing production problems quickly and properly, to reduce the use of antibiotics and prevent diseases. Through detection of characteristic drinking patterns it is possible to detect disease of animals in an early stage. By doing this, it is possible to reduce the use of antibiotics which is healthier for both the animals and humans. Today's levels can create serious problems in consumer health with the potential effect of serious disease. Combining evidence the animals not only had access to water but also through their bio responses to how they made use of the water supply in an appropriate way helps with the certification of food labeling. Improved and objectively-measured welfare of the animals can lead to higher prices of sold meat. The system evaluates the dynamic water intake patterns and generates an early alarm in case unexpected deviations occur. On the basis of measured animal-related variables over a period, a predicted value is calculated (Figure 4). If the deviation between the predicted value and the measured value on that moment is above the allowed level, this is noted as an incident and an alarm. The alarms are divided into soft and hard alarms depending on the severity of detected problem.

Turning data into useful information

The results of this biometry is then relayed to central systems where it is integrated with the output from climate and feed computers. The resulting data can be employed to generate a lot of useful information about the environment the animals are raised in, their feed and water intake and how they respond to this in terms of growth and health. This then leads to better insight into animal performance, operational processes and financial consequences. Planning and business intelligence software then allows

Figure 3: Fancom pig cough monitor.

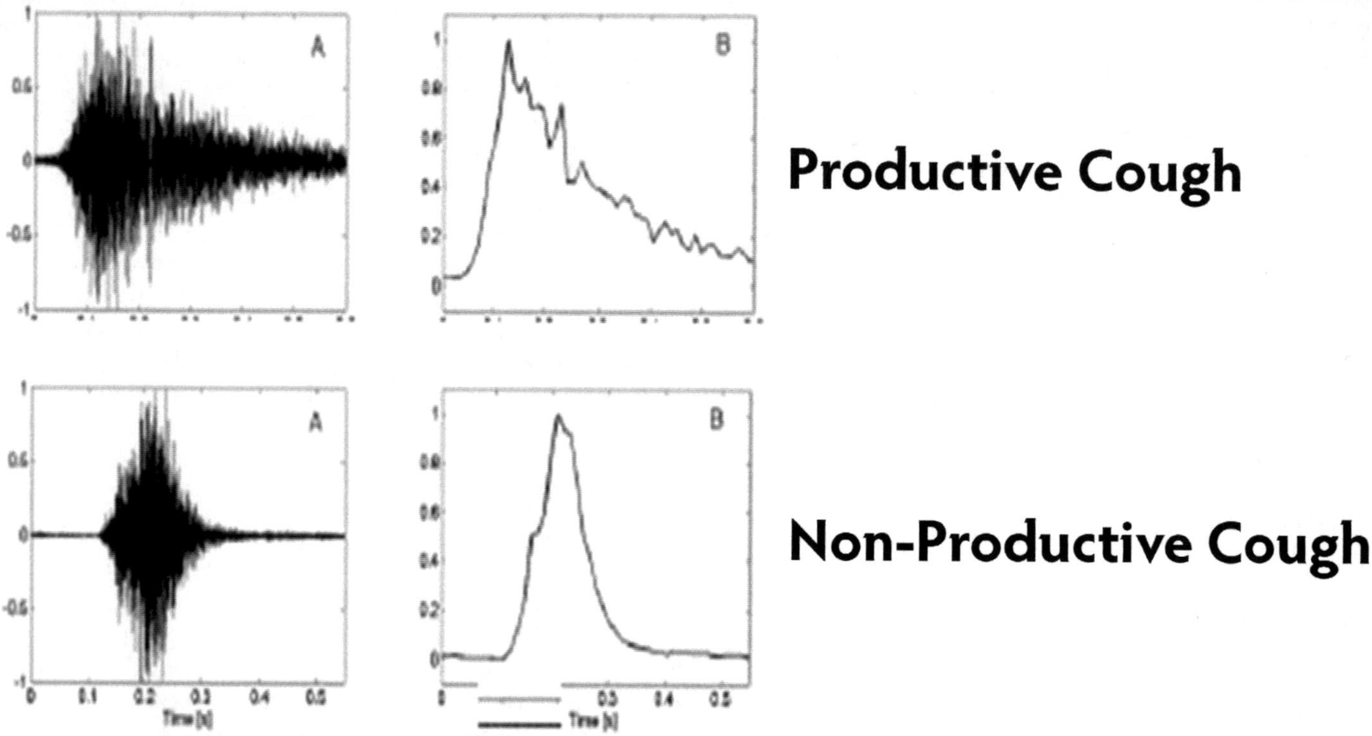

Figure 4: Water monitor.

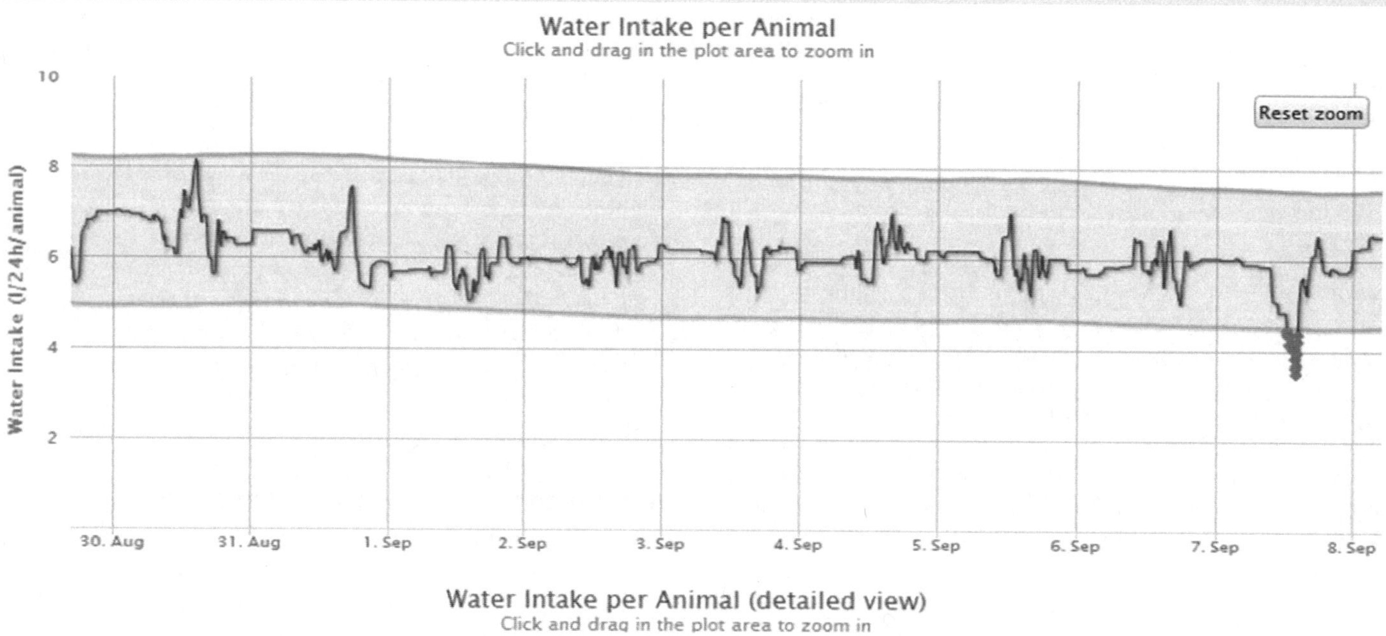

the farmer to compare, adjust where necessary and hopefully improve results in future rounds. Another benefit of this approach is that producers can document for the supply chain how animals are raised in order to safeguard their "license to produce."

Automated welfare monitor

There is a growing need to monitor the welfare of animals grown for meat production. Besides the increasing public demand to produce meat in a more friendly way, good animal care is the key for good productivity and health and thus for an economically viable business. Animal welfare has many different aspects and is hard to measure. Scientists within Europe have invested a lot in developing methodologies for assessing animal welfare at the farm level. To perform this welfare assessment, an expert is needed to carry out a time consuming and complicated process that only delivers a snap shot on the welfare situation in the house. In short this approach is impractical.

Fancom, using its integrated systems, developed a simplified automated welfare dashboard for conventional rearing systems The dashboard is based on a selection of the welfare criteria of the welfare quality assessment protocol. Only the criteria that have a direct link to current available sensors and measuring systems are used in this first version of Fancom's Automated Welfare Monitor. In this way animal welfare is assessed 24/7 without the need for a human observer in the house.

Early adoption

As with any period of early adoption, it quickly becomes apparent the data produced is valuable to a broad spectrum of stakeholders. As already mentioned, central to this is the farmer or producer who not only focuses on the financial aspects of his business but also on the need to satisfy consumers and legislators that the meat has been produced safely and in an acceptable way. The list is, however, extensive with feed and nutrition consultants helping the farmer get best return on investment (ROI), veterinaries focusing on the health of the animals, geneticists looking to get the best out of their breeds and slaughter houses with better planning.

Others that demonstrate a keen interest are banks looking for best performing entrepreneurs and, in turn, new business opportunities as well as insurance companies trying to reduce risk.

Conclusion

Fancom is convinced the industry will require these methods for healthy profits and in order to continue complying with the growing demands for animal proteins and the growing influence of today's critical consumers. This year, the company will be delivering its first automatic welfare monitor. Could this be the start of a new standard for an industry that has been searching acceptability for some time?

If you believe you are a stakeholder in the future of intensive farming, Fancom would be more than happy to discuss with you its vision for the future.

Xylanase reduces wean to finish mortality; potential mechanisms of action

D. Nolan, DVM; R. Cabrera, PhD; C. Sparks, PhD
Huvepharma, Inc., Peachtree City, Georgia

Introduction

Xylanases have successfully been used in animal production and industrial applications for over 2 decades. When xylanases were first used to improve digestibility of feedstuffs in animal agriculture the target group was poultry and most often wheat and rye diets that were highest in arabinoxylans, which is the primary substrate for the enzyme. Poultry studies with Hostazym® and other xylanases have shown an improvement in body weight gain and feed efficiency. The likely mode of action can be attributed to reduced digesta viscosity which permits greater access of digesta enzymes such as bile salts and amylase to come in contact with feed particles and stimulates a positive shift in the gut microbiota from potentially pathogenic to beneficial bacteria. Because of economic advantages realized in the production of poultry, xylanases are currently used in the majority of poultry diets world-wide. Whereas the use of a xylanase in swine diets has been much slower to be implemented and only recently gained very significant momentum. The slow implementation of enzyme technology in the swine industry has been at least partially due to the enzyme companies recommending swine nutritionists use a higher than actual energy uplift when xylanase is supplemented in swine diets, yielding inconsistent results. Xylanase feeding trials in swine typically demonstrate a feed conversion and/or gain advantage to xylanase in approximately 50% of the trials. However, it has now been demonstrated that feeding of xylanase in wean-to-finish or grow-finish trials consistently provides an economic advantage over negative control diets due in large part to the reduction in mortality and morbidity. Morbidity, although measured differently in each system, is those pigs which fail to thrive and almost certainly will be culled and sold to a lower value alternative market or bring less value at harvest. Differences in mortality, morbidity and culls are very difficult to detect in the typical research trial due to high coefficient of variation (CV), low number of experimental units and animals, and over all low mortality compared to normal industry basis due to superb animal care given in a trial situation. Further complicating the detection of a mortality effect, deads and removals are almost always removed or standardized across treatments in the final analysis.

Hostazym and other xylanases reduce mortality

Recently, presentations at multiple swine conferences[1,2] have featured the work of a cross section of individuals demonstrating improvements in the viability of pigs or reductions in mortality, and when measured a reduction in animals removed from the study, for all 3 competing xylanase products currently available in the US. The reduction in mortality appears to be extremely repeatable and observed in high or low health studies. These studies have been held in both controlled research environments and more typical field settings. Most of the trials were not originally designed to investigate pig viability, however the reduction of mortality and morbidity (when measured) is repeatable. A review of nine unpublished feeding trials with xylanase vs. a negative control have all demonstrated an improvement in mortality. The reduction in mortality ranges from 14.5-47.6% with an average of 32.8%. Three unpublished trails reported pigs that were either removed for death and all other reasons. Of those trials the morbidity and mortality combined (viability) was improved by an average 32% when pigs were fed diets containing a xylanase compared to pigs fed the control diets.

Proposed xylanase mechanisms

There are many different mechanisms of action described in the literature related to xylanase and NSP enzymes. Although it is unlikely that all of the mechanisms that lead to digestive or uptake improvements are directly related to the reductions in mortality and morbidity, many may have an indirect impact. For instance, the modification or shift in the intestinal microbiome that leads to more lactic acid producing microbiome is considered a positive attribute related to gut health, there is no direct science that would indicate this shift in microbiome would generate a reduced chance of mortality in what is otherwise a healthy group of growing pigs. Recent studies continue to describe new measures associated with and in support of what has been commonly referred to as maintenance of a healthy gut. A recent NCSU report demonstrated that Hostazym reduced circulating tumor necrosis factor (TNF-α), one of many cytokines that are considered an inflammatory marker. TNF-α was reduced in pigs consuming either a corn-soy or corn-soy-DDGs diet.[3]

- Shift of the intestinal microbiome towards beneficial bacteria.[4,-7,9]
- Decreased viscosity of digesta.[3,6,9]
- Improved fat digestion.[7,8,10,11]
- Improved starch digestion.[10]
- Increased villi length and crypt depth.[3,11]

Hostazym® is a registered trademark of Huvepharma, AD

- Slower proliferation of gram positive cocci and harmful organisms, enterotoxigenic *E.coli* and *Salmonella.*[4]
- Improved gut health by reduction in inflammatory response.[3]

Conclusion

It may seem obvious to point out that death is experienced at an individual level and establishment of the threshold or the trigger that leads an animal to lose the constant battle between health and disease, or life and death, also occurs at the individual animal level, influenced by genetics, environment, nutrition, immune system insults and response, disease and a number of other factors. Through a variety of mechanisms related to ingredient digestion and improved gut health, xylanase has been demonstrated to improve the pigs balance in favor of health and life, thereby increasing the number of top market hogs by reducing mortality and morbidity and giving the producer an economic advantage.

References

1. Boyd D. Improving Finish Pig Viability by using Xylanase. Allen D Leman Swine Conference, presentation. 2015. Available at: http://traffic.libsyn.com/animalcast/150921_Leman_Enzymes_Boyd.mp4. Accessed 02 November 2015.

2. Boyd D. Improving livability with xylanase inclusion. *Pig Progress*, World Pork Expo, Video Presentation. 2015. Available at: http://www.pigprogress.net/partners/maximize-swine-production-efficiency/video-1/. Accessed 02 November 2015.

3. Chen H et al. Effects of supplemental xylanase on digesta viscosity, gut health and growth performance of nursery pigs. *Proceedings, Joint Annual Meeting, ADSA-ASAS.* 2015.

4. Zhang G et al. Effects of dietary supplementation of multi-enzyme on growth performance, nutrient digestibility small intestinal digestive enzyme activities, and large intestinal selected microbiota in weanling pigs. *Journal of Animal Science*. 2014. 92:2063-2069.

5. Hubener K et al. Bacterial responses to different dietary cereal types and xylanase supplementation in the intestine of the broiler chicken. *Archives of Animal Nutrition*. 2002. 36:167-187.

6. Vahjen W et al. Influence of xylanase-supplemented feed on the development of selected bacterial groups in the intestinal tract of broiler chicks. *Journal of Agriculture Science, Cambridge*. 1998. 130:489-500.

7. The effect of Hostazym X alone or in combination with phytase on digestibility of nutrients in grow-finish diets for pigs. *Huvepharma Technical Bulletin TB-30*. 2012.

8. Mathlouthi N et al. Xylanase and β-glucanase supplementation improve conjugated bile acid fraction in intestinal contents and increase villus size of small intestine wall in broiler chicks fed a rye based diet. *Journal of Animal Science*. 2002. 80:2773-2779.

9. Choct M et al. Anti-nutritive effect of wheat pentosans in broiler chickens: Roles of viscosity and gut microflora. *British Poultry Science*. 1992. 33(4):821-834.

10. Adeola O et al. Exogenous dietary xylanase ameliorates viscosity-induces anti-nutritional effect in wheat based diets for White Pekin Ducks. *British Journal of Nutrition*. 2004. Jul 92(1):87-94.

11. Mathlouthi N et al. Xylanase and beta-glucanase supplementation improve conjugated bile acid fraction in intestinal contents and increase villus size of small intestine wall in broiler chickens fed a rye-based diet. *Journal of Animal Science*. 2002 Nov 80(11):2773-2779.

A novel PCV2 qPCR and PCV2 ELISA. Powerful tools to Monitor PCVAD

Maartje Wilhelm[1], DVM; Eric van Esch[1], DVM; Tim Goode[2]; Alex Eggen[1], DVM
[1]BioChek, the Netherlands; [2]BioChek, USA

Introduction

Detecting and quantifying Porcine Circo Virus type 2 (PCV2) in both clinically and sub-clinically (PCV2 virus-) infected pigs is important. PCV2 virus, even in low quantities, has a detrimental effect on their performance. A lower Average Daily Weight Gain (ADWG) and a less efficient Feed Conversion Ratio (FCR) are mainly held responsible for the economic damage.[1] To monitor PCV2 Associated Disease (PCVAD) both PCV2 qPCR and ELISA tests have been developed. Virus neutralizing (VN) antibodies are detected by both IPMA and ELISA and they control PCV2 viremia.[2] PCV2 viremia is monitored by qPCR. Both test kits are developed and produced by BioChek and tested for their respective suitability.

Materials and methods

The PCV2 ELISA test is typically used to determine the presence and titer of maternally derived antibodies (MDA), sero-conversion after vaccination or field virus infection, to check the uniformity of immunity and to determine the right vaccination moment. The PCV2 qPCR is used for checking for presence and quantity (viral load) of PCV2 virus. The BioChek PCV2 ELISA test was compared to other tests present in the market and to the IPMA test, which is the gold standard for detecting PCV2 antibodies, and use was made of well-defined serum samples.[3]

The BioChek PCV2 qPCR participated in a Proficiency Testing Scheme (PTS) or ring test, organized by the Animal Health Service (AHS) (GD) organization, Deventer, the Netherlands. In this PTS, 8 different freeze dried samples were included and divided in 3 groups. The group with sample numbers 1, 5 and 7 contained PCV2 field virus strains in different dilutions. The group with sample numbers 2 and 4 did not contain any PCV2 virus (negative or PCV type 1 virus). The group with sample numbers 3, 6 and 8 contained a different PCV2 virus strain in several dilutions. The 8 different samples were tested at the BioChek BV laboratories in Reeuwijk, the Netherlands; using the BioChek PCV2 qPCR test kit from a routinely produced and released batch.

As part of the development of the BioChek PCV2 qPCR, serial dilutions of PCV2 virus were tested. The BioChek PCV2 qPCR contains a positive standard to help diagnostic laboratories check their internal procedures.

Results

The BioChek PCV2 ELISA showed an excellent correlation with the IPMA test. This was tested and reported by Pileri.[3]

The PTS testing for the BioChek qPCR revealed that the result of every individual test was in line with the result of the PTS leader. Sample numbers 2 and 4 (no PCV2 virus) scored negative in the BioChek PCV2 qPCR. Confirming the declared specificity of > 99%. PCV2 virus positive samples (1, 5 and 7) scored positive in the BioChek PCV2 qPCR. Number 5 was highly diluted; sample 7 contained slightly less PCV2 virus than number 1. This was reflected in the Cq values, 5 had the highest Cq value, while 7 was just higher than 1. For the PCV2 virus positive sample numbers 3, 6 and 8 a similar pattern was found: High dilution - high Cq value. Proving high sensitivity (high dilution and still positive) and good linearity.

The PCV2 virus serial dilutions testing revealed the same level of linearity as was seen in PTS testing and a sensitivity of as low as 2.8 on a $\log_{10}$ scale in viral copies per ml serum,[4] which is even more sensitive than a regular PCV2 PCR. Also the good linearity between the samples in different dilutions can be observed (Figure 1).

In Figure 2 the results of testing different batches of the PCV2 standard included in the BioChek qPCR test kit are shown. Excellent similarity between the samples originating from different batches of positive controls was found.

Discussion

PCVAD can effectively be controlled by vaccination. The vaccination efficiency depends on the immunity induced by the vaccine. VN antibodies are the 1st line of defense. When no sero-conversion after vaccination is observed and PCV2 viremia is detected, the vaccination scheme should be reconsidered. The goal of PCV2 vaccination is to raise pigs that do not suffer from PCVAD. For monitoring the efficiency of PCV vaccination the combined information on seroconversion obtained by ELISA and PCV2 viremia by qPCR is needed. The financial gains can be significant.[5]

References

1. Segales J, Porcine Circovirus type 2 (PCV2) infections: Clinical signs, pathology and laboratory diagnosis. Virus Research 164 (2012) 10 -19.

2. Fort et all, One dose of a porcine circovirus 2 (PCV2) sub-unit vaccine administered to 3-week-old conventional piglets elicits cell-mediated immunity and significantly reduces PCV2 viremia in an experimental model, Vaccine 27 (2009) 4031-4037.

3. Pileri et all, Comparison of the immune-peroxidase monolayer assay and three commercial ELISAs for detection of antibodies against porcine circovirus type 2. Vet Journal 2014.

4. Data on file BioChek laboratories.

5. Atlagich et all, AASV 2014, ESPHM 2014.

Figure 1: Serial dilutions of a spiked PCV2 virus sample. Note the excellent linearity. First line left contains 105 copies DNA per reaction and then 104, 103 and 102 copies per reaction.

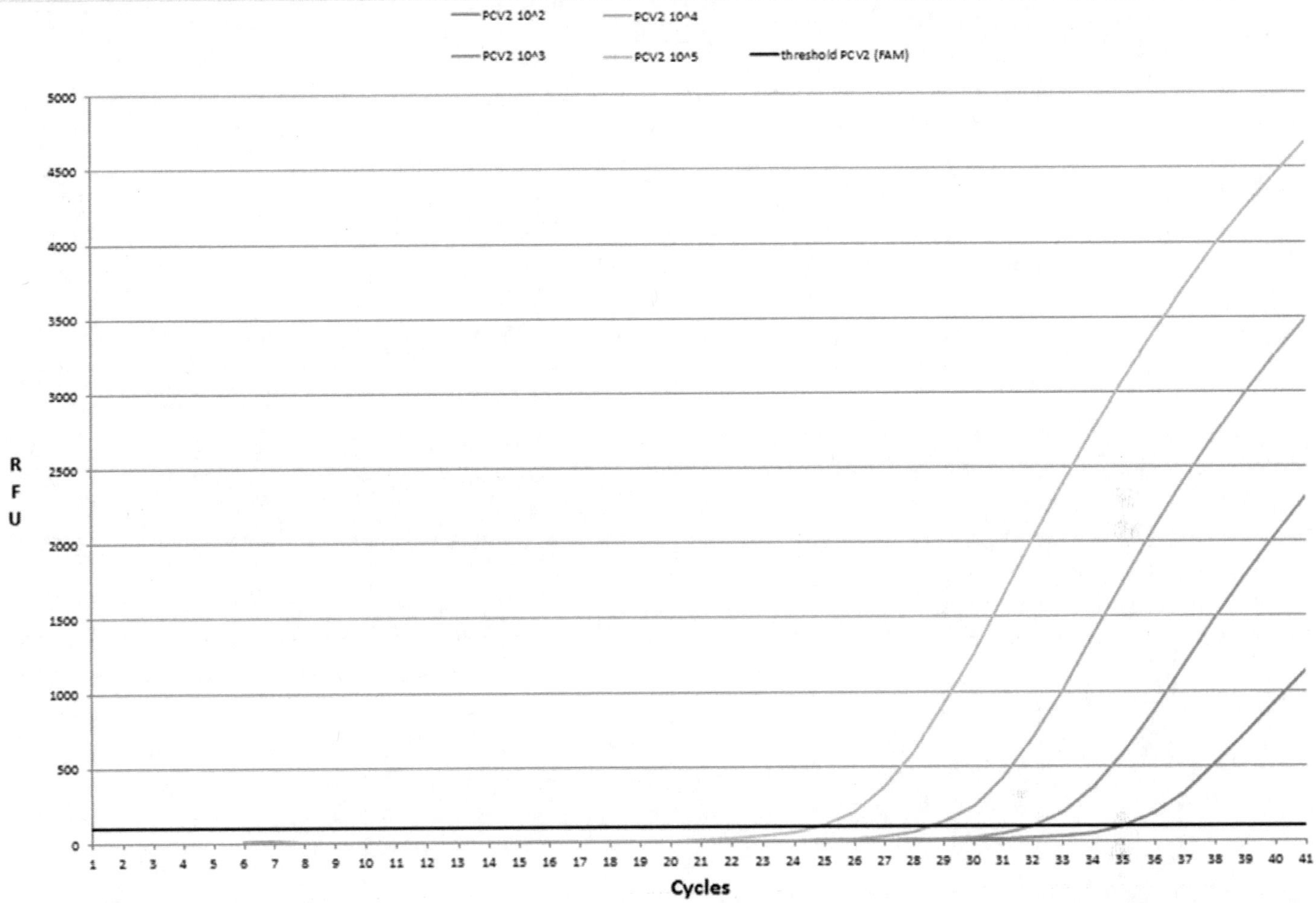

Figure 2: BioChek qPCR results of different batches of positive controls included in the PCV2 qPCR test kit. Note the excellent similarity in detecting the quantity of PCV2 virus.

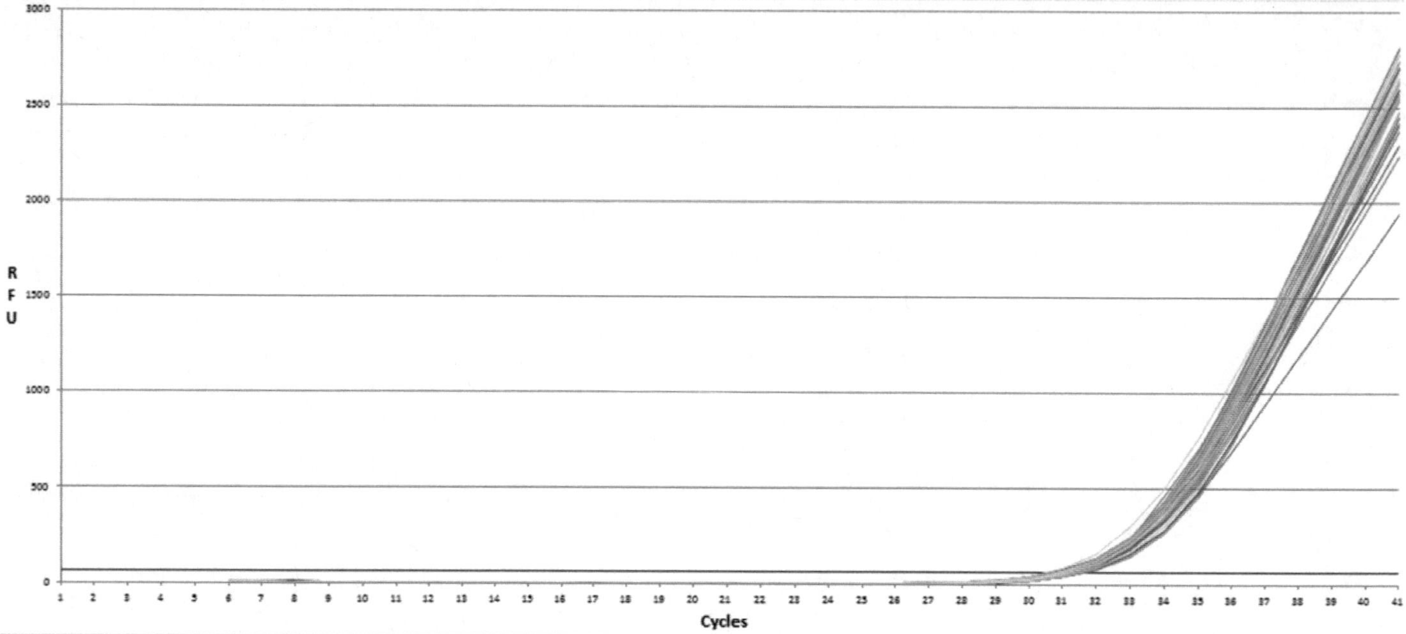

Advances in PRRS vaccination strategy: Vaccination schedules alternating MLVs and KVs increase efficiency

François Joisel[1], DVM, MSc, Univ. Doc.; Olivier Merdy[1], Agronomist; Tom Meyns[2], DVM, PhD, Dipl. ECPHM; Valérie Dekens[3], DVM; Sjouke Van Poucke[2], DVM, PhD; Guillaume Perreul[4], DVM; Benoit Boivent[4], DVM; Han Smits[5], DVM, MSc; Alberto Aguilera[1], DVM; Laurent Fischer[1], DVM, PhD; Catherine Charreyre[1], DVM, MSc

[1]Merial SAS, Lyon, France; [2]Merial Belgium, Diegem, Belgium; [3]Merial, Velserboek, the Netherlands; [4]Merial, SAS, Ancenis, France; [5]Merial North Europe and EMEA CoE

Introduction

Porcine reproductive and respiratory syndrome (PRRS) remains over the world, one of the most devastating scourges that plagues the swine industry. Control of PRRS virus (PRRSV) includes a large array of tools as internal and external biosecurity, set up of management and hygiene rules and also vaccination. Stabilization of the sow herd, in other words limitation of the virus circulation in the sows has a pivotal importance in the epidemiology of the virus in the farm. It has been clearly demonstrated that PRRSV virus circulation in the post-weaning phase was linked to PRRS related reproduction problems in sows and that even if the post-weaning phase was managed under continuous flow conditions, when sow herds were stable, no or minimal virus circulation was likely to occur up to 9-10 weeks of age.[1]

Two main types of commercial vaccines are currently available: modified live vaccines (MLV) or attenuated virus vaccines and inactivated vaccines (INV) or killed virus vaccines (KV). Vaccines manufactured using other technologies of antigen production are still experimental. Most of the vaccines contain only antigens of one of the genotype, i. e. type1 (previously European type) or type2 (previously American type). Killed vaccines used in the United States of America are mainly autogenous vaccines or customized made vaccines (CMV). The antigen fraction of these vaccines contain the PRRSV isolated from a farm, from an adjacent farm or, if it exists an epidemiological link, from a non-adjacent farm or even an as less heterogeneous as possible isolate.

In laboratory trials, viremia after homologous and heterologous challenge is significantly reduced in MLV vaccinated animals.[2] When the vaccine and the challenge strains belong to the same genotype, the level of protection seem to decline when genetic homology is decreasing.[3] However, after MLV vaccination, Prieto *et al.* (2008)[4] showed variable protection against PRRSV challenge strains of the same genotype even if they are genetically close indicating that cross protection is not only related to the type but is also to be considered isolate to isolate.

A type1 based KV (PROGRESSIS®, Merial, France) was shown to induce PRRSV-specific INF-γ secreting cells upon *ex vivo* stimulation with both type1 PRRSV strains[5,6] and a type2 one.[7] The same vaccine was shown to reduce viremia after homologous challenge[8] and to reduce viremia prevalence in piglet at weaning[9] in a large scale field study involving 17 herds monitored for two full reproduction cycles. PROGRESSIS repeated injections have also been shown to elicit sero-neutralizing antibodies[10] and to boost sero-neutralizing antibodies against heterologous strains naturally circulating in filed farms.[11]

In the fields, many different vaccination protocols against PRRSV have been used with various satisfaction. Regular vaccination with modified life vaccines[2] have also been shown to reduce PRRSV-related clinical signs. Viremia, virus shedding, lesions and clinical signs appear to be controlled to various extent. However, a second contact with the live virus usually has minimal immunological booster effect, i. e. no increase of sero-neutralizing antibody titres and no increase of specific cellular mediated immunity. Vaccination protocols including only KV have shown to be efficient in certain conditions: long term vaccination with inactivated vaccine at mid gestation,[9,12,13] mass vaccination repeated every 3 months.[14] However, it is a consensus that none of the existing vaccination protocol using a single vaccine does give long term full satisfying protection.

Priming the immune system with a vaccine produced with a defined technology and boostering the immune response with another vaccine based on a different technology has been already used in various species to fight against various pathogens, for instance BVD[15] or BRSV[16] in cattle, or FMDV in pigs.[17]

The purpose of this paper is to review several recent research trials and studies showing a benefit of MLV/KV mixed vaccination programs, either in terms of immune response or in the control of PRRS virus field circulation and subsequent improvement of pig flow performance.

Laboratory studies

Around a decade ago, using a type2-based KV (Promise®, not anymore currently commercially available) researchers showed in laboratory studies[18] as well as in the field[19] a significant increase of the immune response against PRRSV induced by vaccination schedules where a type2 MLV and this KV were included.

® PROGRESSIS is a registered trade mark of MERIAL SAS, a Sanofi Company, Lyon France

Recently using type1-based vaccines, a comparison between the immune responses induced by either a modified live vaccine (MLV) vaccination program or a combination of a MLV and a killed vaccine was made.[20] The experimental design was as follows: 32 piglets were randomly distributed in four groups (Table 1). A commercial type1 PRRS MLV was injected to the pigs of A, B and C at 1.5 months old, while group D pigs were kept as non-vaccinated controls. At 4.5 months of age, groups A and C received an injection of PROGRESSIS (2mL, IM) while group B received a second MLV. At 5.5 month of age, group C pigs received an additional PROGRESSIS injection at the same dose by the same route. At 6.5 months of age, all pigs were intranasally challenged using 2 mL of suspension of 10^6 tissue culture infective dose 50 ($TCID_{50}$) of a type1 PRRS virus, strain 2749 (99% similar to Lelystad virus based on ORF5 sequences, in saline). ELISA antibodies, neutralizing antibodies (NA) and cell-mediated immunity (CMI) were monitored, and viremia was evaluated after challenge.

Re-vaccination with PROGRESSIS induced ELISA responses similar to re-vaccination with MLV with no apparent booster effect (data not shown). However (Table 2), a significant increase in NA titres after challenge was detected in group C pigs, i.e. with the double KV booster when compared to the non-vaccinated control group and to the group having only received a MLV booster. The pigs that received one KV booster showed intermediary NA titres. The use of combined protocols (MLV+PROGRESSIS) was superior to the use of MLV alone in inducing cell mediated immunity, both after vaccination as well as when immunity was recalled by the challenge (Figure 1). After challenge, the virological protection was the same in the three different groups (data not shown here).

In conclusion, this study confirms that PRRS modified live vaccination induces as well as the natural challenge a weak immune response both in PRRSV-specific NA and in INF-γ producing cells. It was also confirmed that a second injection of the same MLV at a 3-month interval did not significantly boost the humoral immunity either quantified via ELISA or seroneutralization test or the cellular mediated immunity evaluated through PRRS-specific ING-γ secreting cell quantification. Conversely, a re-immunization with one KV injection induced a significantly higher CMI and a numerically higher NA responses. With two KV injections one month apart a significantly higher CMI and NA after vaccination as well as after the recall by the challenge was obtained. More practically, the combination of a primer inoculation for gilts with a MLV vaccine and KV revaccinations

Table 1: Experimental group design for the study (Diaz *et al.*)

		Experimental design (n = 32)						
		Month of age and treatment						
Group	Number of pigs	1.5 Treatment 1	2.5	3.5	4.5 Treatment 2	5.5 Treatment 3	6.5	7.5
A	8	MLV			MLV		Challenge	End
B	8	MLV			INV		Challenge	End
C	8	MLV			INV	INV	Challenge	End
D	8	Placebo			Placebo	Placebo	Challenge	End

Table 2: Neutralizing antibody evaluation. Results are expressed in proportion of positive pigs and mean titre (log_2) ± SD for positive pigs. Different superscripts in the same column indicate statistical differences ($P < 0.05$) between groups for proportions of positive pigs or for mean titres at a given time point (Diaz *et al.*)

		Months of age		Days post-inoculation (dpi)	
Group		4.5	5.5	6.5 (0 dpi)	14 dpi
A	Positive pigs	0/8	$4/8^{a}$	$5/8^{a}$	$7/8^{ab}$
	Mean titre (log_2) ± SD	–	1.5 ± 1	1.6 ± 0.9	2.1 ± 1.3^{b}
B	Positive pigs	1/8	$2/8^{ab}$	$5/8^{a}$	$8/8^{a}$
	Mean titre (log_2) ± SD	1 ± 0	2 ± 1.4	2.8 ± 0.8	3.2 ± 0.7^{ab}
C	Positive pigs	3/8	$5/8^{a}$	$6/8^{a}$	$8/8^{a}$
	Mean titre (log_2) ± SD	1 ± 0	1.4 ± 0.5	2.2 ± 1.6	3.6 ± 1.2^{a}
D	Positive pigs	0/8	$0/8^{b}$	$0/8^{b}$	$4/8^{b}$
	Mean titre (log_2) ± SD	–	–	–	2.2 ± 0.5^{b}

Figure 1: PRRS specific INF-γ secreting cells at each of the time points. Different letters regarded vertically mean a significant difference at $P < 0.05$ (Diaz *et al.*).

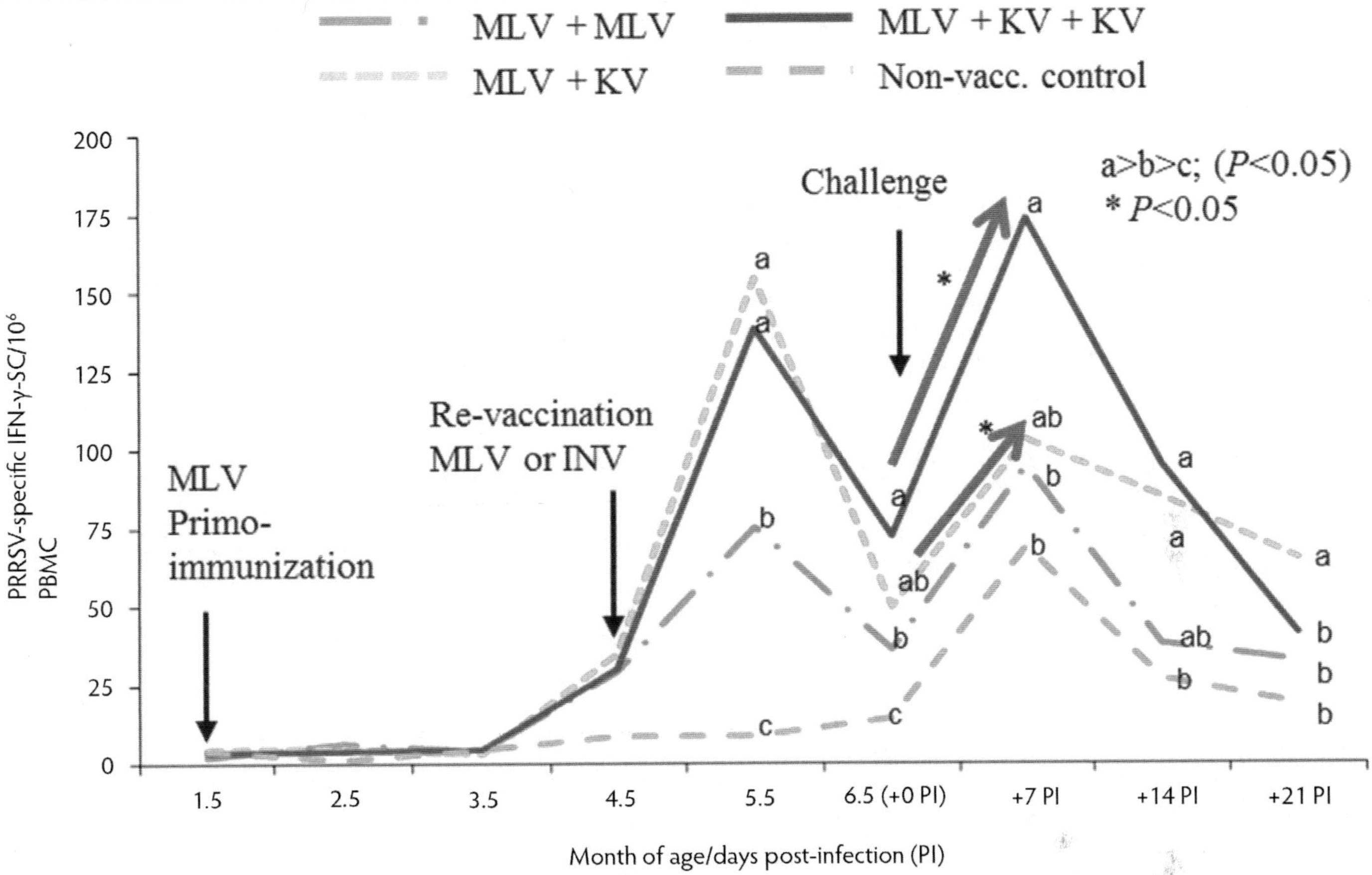

could reach similar or even superior levels of NA and CMI and with similar virological protection as repeated vaccination protocols with an attenuated strain.

Field case studies

The objective of the first study[21] was to assess the increase in PRRS antibody titre in piglets and the benefit on the PRRSV infection pattern in the same piglets during the nursery period with an additional PROGRESSIS vaccination of sows at day 90 of gestation (D90).

Data was collected in a well-managed 300-sow farrow-to-finish farm located in one of the most densely populated swine regions in The Netherlands. In this farm, long-lasting problems ligated with PRSSV in nursery used to be observed despite a sow vaccination with a commercial type1 MLV vaccine every reproduction cycle, 6 day after farrowing. The vaccination schedule was changed: the MLV vaccination in lactation was kept but a primary mass immunization with two injections 3 weeks apart, followed by a booster at D90 with PROGRESSIS. Serum samples were collected from piglets of 3 batches (2/3 batches using PROGRESSIS) at 4, 7 and 10 weeks of age for anti-PRRSV antibodies titration using an ELISA kit (Idexx) and with an IPMA test. Oral fluids were collected from the oldest piglets (7 and 10 weeks of age) of the two last batches for Multiplex PCR test against respiratory pathogens.

Before implementation of the additional vaccination with the killed vaccine, a circulation of PRRSV between 5 and 10 weeks of age was serologically confirmed in the batch. Following vaccination with PROGRESSIS, anti-PRRSV maternally-derived antibodies appeared to be higher and to last longer and no seroconversion against PRSSV was observed in piglet up to at least 8 weeks of age (Table 3). With this additional vaccination, PRRSV was not found in oral fluids during nursery except in one batch at 10 weeks of age.

Absence of the virus in oral fluid, although these result has nevertheless to be taken with caution regarding low specificity of the Multiplex PCR test, added to the absence of seroconversion in the course of the nursery phase in 3 consecutive batches indicate that the use of PROGRESSIS just before farrowing can be an efficient tool to induce a prolonged passive immunity and to reduce early infection in piglets.

A second case study[22] involved a 1200-sow farrow-to-finish farm, with 5 production lines. At the time of the problems, sows and gilts used to be blanket-vaccinated at with a type1 MLV PRRSV vaccine 3 times a year, on a regular basis. The nursery piglets were housed in 2 large buildings. In January 2012, severe clinical signs associated with PRRSV were observed. The breeding herd suffered from early farrowings and high number of stillborn and weak piglets. In nursery, performance was low and associated with high mortality rate (4.5%) and an abnormally increased use of antimicrobial drugs. Before implementation of the tested new vaccination schedule, ELISA (Idexx) serological data confirmed circulation of PRRSV in piglets (Table 4) and in sows and gilts (Table 5). PRRSV was also detected in blood by PCR in piglets of 3 and 7 weeks of age.

The vaccination program in the sows was changed to vaccination with a type1 MLV vaccine at day 60 combined with PROGRESSIS at day 90 of gestation.

In November 2012, around 11 months after the start of the new vaccination program, a clinical and serological stabilization of PRRSV circulation in the nursery piglets was observed (Table 4). At that time, the serological profile in sows and gilts was still indicative for recent circulation of PRRSV (titres > 3). In October 2013, after 22 months of vaccinating all sows and gilts repeatedly at day 90 of gestation with PROGRESSIS, also the titres of the sows and the gilts became uniform (Table 5), while S/P ratios in piglets at the end of nursery remained low. At the end of 2013, the mortality in the nursery was reduced to ≈ 2%. After the start of the vaccination with PROGRESSIS at day 90 of pregnancy, there was a relative quick recovery of the production results in the nursery. In addition after 22 months, there was a remarkable stabilization of the PRRSV circulation in sows and gilts.

These results suggest that the use of PROGRESSIS at D90 of gestation can be an efficient tool to induce an increased maternally derived immunity in piglets and to control virus circulation, in order to stabilize a whole herd.

A third case study[23] was carried out in a 400-sow PRRSV-positive farrow-to-finish farm operating in 5 farrowing batches of 80 sows in Brittany, France. Sows used to be vaccinated during lactation with a type1PRRSV MLV. No clinical signs were noticed in the suckling phase. Piglets used to be vaccinated against *M. hyopneumoniae* at weaning and were not vaccinated against PRRSV. Respiratory disorders used to occur on a recurrent basis in nursery and finishing barns needing frequent antimicrobial treatments. In December 2013, PRRSV infection was confirmed in two batches in the nursery, based on positive PRRSV PCR and clear ELISA (Idexx) seroconversion on both sera and oral fluids. In January 2014, one batch of breeders received an additional PRRSV vaccination at day 90 of gestation with PROGRESSIS. A cohort of 10

Table 3: Average PRRSV ELISA S/P results in piglets before and after implementation of an additional vaccination with PROGRESSIS® at D90 of gestation (Deckens *et al.*)

Piglet age (weeks)	4	7	10
Before – 1st batch	1.04	0.41	1.24
After – 2nd batch	1.79	1.07	0.57
After – 3rd batch	2.5	1.8	1

Table 4: PRRSV ELISA S/P ratios in sera collected from piglets of 6 and 10 weeks of age between Jan 2012 and Oct 2013 (Defoort *et al.*)

Month	01/2012	11/2012	10/2013
	2.14	2.96	1.04
	1.11	1.54	1.60
Piglets of 6 weeks of age	2.35	1.15	1.88
	2.34	0.69	2.57
	2.19	2.8	2.30
	2.55	0.28	0.67
	2.46	0.78	0.17
Piglets of 10 weeks of age	3.06	0.08	0.08
	2.25	0.34	0.55
	2.87	0.58	0.56

piglets born from these sows were monitored for the detection of PRRSV infection (PCR on oral fluids) and serological profiling (ELISA) at 8, 12 and 16 weeks of age.

The offspring of the KV vaccinated sows did not show any respiratory symptoms until slaughter. All except one (S/P ratio 0.66 at 8 weeks of age) ELISA results at 8 and 12 weeks of age were negative. Positive ELISA results on both sera and oral fluid samples at 16 weeks of age showed that PRRSV circulation had been delayed far beyond the end of the post weaning. In February 2014, while sows were vaccinated according to the historical vaccination protocol (only at day 6 in lactation with the MLV vaccine), respiratory symptoms re-occured in the nursery, suggesting an early PRRSV infection. In June 2014, the KV vaccination at 90 days of gestation was re-introduced as a routine practice leading to a reduction of the respiratory clinical signs in the pig flow. In October 2014, a transversal oral fluid sampling showed the persistence of maternal antibodies until 10 weeks of age, and the absence of PRRSV circulation until 12 weeks of age.

In conclusion, under the conditions of this case study, the implementation of mixed PRRSV vaccination program in sows combining MLV vaccination at day 6 of lactation and PROGRESSIS vaccination at 90 days of gestation resulted in a delayed PRRSV circulation, as well as a clear reduction of associated respiratory clinical signs. These results suggest that the use of PROGRESSIS around 3 weeks before farrowing can be an efficient tool to control early virus circulation in piglets and to lead to a total herd stabilization.

A fourth study[24] was run in a 1000-sow farrow-to-finish Dutch farm where piglets showed respiratory clinical signs linked with PRRSV between 4 and 10 weeks of age and confirmed by PCR analysis on broncho-alveolar lavages (BAL) fluid from 4 weeks of age onwards. Piglets were not vaccinated against PRRSV. The sow vaccination program consisted of the use of an type1 MLV at 6 days after farrowing and at 60 days of gestation. It was replaced by a vaccination with the MLV at 60 days of gestation and a booster with PROGRESSIS at 90 days of gestation. During 1 year after the implementation of the new program, cross-sectional collections for BAL fluid samples were performed in 4 to 10 WOA piglets and analyzed for PRRS RNA by PCR. Blood samples of 6 sows were taken at 4w, 3w and 1w *ante partum* for Idexx ELISA evaluation. For each sow, 3 piglets were randomly selected for blood sampling at 4, 7 and 10 weeks of age (WOA). PRRSV antibodies and viral DNA were assessed by Idexx ELISA and by PCR. Clinical signs and technical farm data were monitored before and after the change in vaccination protocol.

PCR analysis in weaned piglets showed PRRSV detection in BAL fluids and in sera shifting from 4 towards 10 WOA after implementing the MLV/KV mix vaccination program. The clinical signs in the weaned piglets clearly improved. Sows showed an anti-PRRSV Ab titre decrease between 4 and 3 weeks *ante partum* and a strong increase after PROGRESSIS vaccination. A high correlation between antibody levels in sows and 1week *ante partum* and 4 WoA piglets was observed, showing good colostrum management. MDA in piglets steadily decreased from 4 to 10 WoA, indicative for the absence of field infection with PRRSV.

After the change of the sow vaccination program, the technical performance of the herd improved for several parameters: number of weaned piglets per year increased from 28.2 to 31, the number of live born piglets increased from 32.2 to 34.7. In the post weaning age group, average daily weight gain increased from 302 to 320 g/day and the post weaning mortality decreased from 4.7 to 0.8 %.

Under the conditions of the study, the implementation of PROGRESSIS vaccination at the end of gestation reduced and delayed PRRSV circulation within the weaned piglets, while technical performance of the farm and clinical signs of PRRSV infection clearly improved.

Table 5: PRRSV ELISA S/P ratios in sera collected from sows and gilts between Jan 2012 and Oct 2013 (Defoort *et al.*)

Month	01/2012	11/2012	10/2013
	1.78	3.18	1.63
	2.26	1.52	1.44
Gilts	2.15	1.12	0.96
	0.87	2.14	1.69
	2.51	3.84	1.92
	0.9	4.23	2.29
	0.53	1.58	2.22
Sows	1.6	1.87	2.02
	3.16	3.25	2.14
	1.26	2.49	2.4

A PRRSV circulation survey[25] was made in the piglet flows of farrow-to-finish farms located in the most-dense pig farming region of France and where a PRRSV circulation was known to be occurring. In these farms, before the implementation of the additional KV vaccination of breeding females, the routine PRRSV vaccination scheme included a primo-immunization with a type1 PRRSV MLV in quarantine followed by booster injections at D6 of lactation and at D60 of gestation or blanket vaccination every 4 month. PROGRESSIS sow vaccination at D90 of gestation was added to the routine of the herd. PRRSV infection and antibody profiles were assessed by oral fluid (OF) sample collection in pigs. Samples were assayed for PRRSV antibody level using a commercial kit (Idexx PRRS OF Ab Test, OD value > 0.4 for positivity) and viral RNA presence using a RT-PCR technique (LSI VetMAX™, CT value < 40 for positivity). Eight transversal and fourteen piglet batch longidutinal follow-ups of 3 to 5 OF samples per batch from 5-7 weeks of age onward (overall 81 samples) were performed prior to and following the implementation of the KV additional injections in sows.

Figure 2 shows that no PRRSV RNA was detected in OF samples of piglets born from KV-vaccinated sows until 13 weeks of age. Virus circulation appeared to be delayed to late fattening (11/23 samples were positives from week 14 of life and later). Piglets born from the KV vaccinated dams showed a significantly higher anti-PRRSV ELISA Ab level after weaning, which steadily decreased up to 11 to 16 weeks of age depending on the batch. The seroconversion was thus delayed to a few weeks following the entry of the pigs in the fattening unit (data not shown).

Discussion and conclusion

The field studies run in many different locations and conditions in farms where PRRSV circulation occurred early in the pig flows confirm the research studies on MLV/KV mix vaccination programs showing a better stimulation of the immune mechanisms presumably involved in the protection against PRRSV, i.e. PRRSV-specific NA and INF-γ secreting cells. They also show that only an KV injection at the end of gestation in addition to various MLV sow vaccination schedules is able to delay the virus circulation in the pigs of the vaccinated dams and in many cases to postpone it after pigs are turned into the finishing barns. The reduction of PRRSV infection prevalence in the early ages of piglet is due to the addition of two effects of sow immunity reinforcement: i) a better transfer of maternally derived protective immunity and ii) a reduction of fetuses/piglet contamination

Figure 2: PRRSV RNA positive oral fluid samples in different age groups in a survey gathering results in 7 farms. Comparison between a MLV only and MLV + KV mix vaccination programs (Boivent *et al.*)

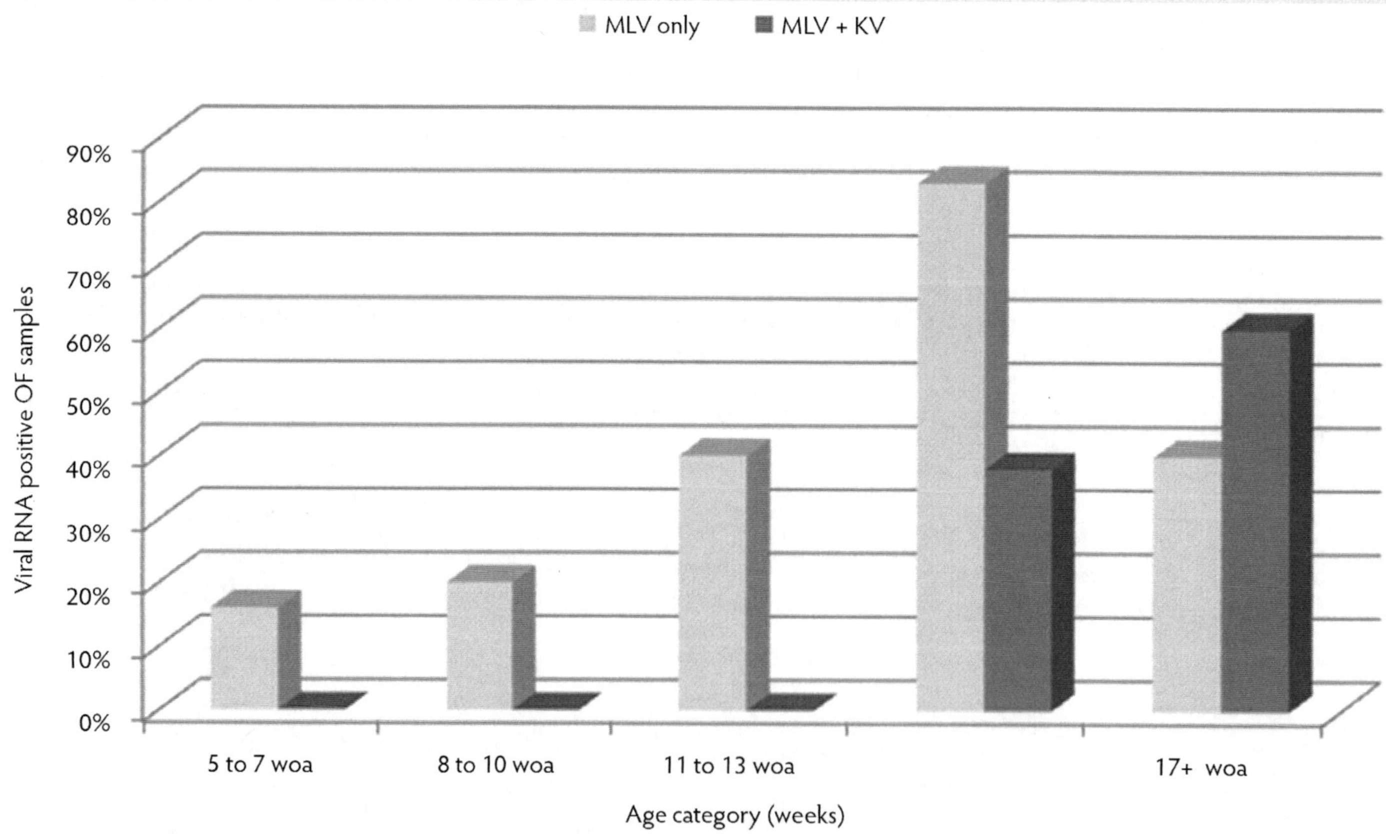

during the pre- and post-farrowing weeks of life of the baby pigs. This appears to be highly beneficial for the health and performance of the pigs, resulting in better growth, lower mortality and consequently in a lower consumption of antimicrobial drugs. In addition, on an immunological standpoint post-weaning is the time piglets are completing the construction of their immune system. Post weaning is also the time of immunization against a number of other pathogens (PCV2, M. hyo, APP, FMD, CSF, etc.). It thus appears as a very favorable effect to protect pigs against a powerful immune-modulating virus.

References

1. Candotti, P., Rota Nodari, S., Leotti, G., Longo, S., Joisel, F. 2006. Relationship between piglet PRRS seroprevalence at the end of nursery and reproduction disorders due to PRRS in sows. In Proceedings of the 19th IPVS Congress, København, Denmark. P O.02-01. Vol1: 54.

2. Martelli, P., Gozio, S., Ferrari, L., Rosina, S., De Angelis, E., Quintavalla, C., Bottarelli, E. and Borghetti, P. 2009. Efficacy of a modified live porcine reproductive and respiratory syndrome virus (PRRSV) vaccine in pigs naturally exposed to a heterologous european (Italian cluster) field strain: clinical protection and cell-mediated immunity. Vaccine. 27 (28): 3788-3789.

3. Mateu, E., and Diaz, I. 2008. The challenge of PRRS immunology. Vet J. 177(3):345-51.

4. Prieto, C., Alvarez, E., Martinez Lobo, J., Simarro, I. and Castro, J. M. 2008. Similarity of European porcine reproductive and respiratory syndrome virus strains to vaccine strain is not necessarily predictive of the degree of protective immunity conferred. The Veterinary Journal. 175 (3): 356-363.

5. Piras, F., Bollard, S., Laval, F., Joisel, F., Reynaud, G., Charreyre, C., Andreoni, C., Juillard V. 2005. Porcine reproductive and respiratory syndrome (PRRS) virus–specific interferon-γ^+ T-cell responses after PRRS virus infection or vaccination with an inactivated PRRS vaccine. Viral Immunology, 18(2): 381-389.

6. Zuckermann, F., Alvarez, E., Diaz, I., Christopher-Hennings, J., Doster, A., Brito, M., Osorio, F. 2007. Assessment of the efficacy of commercial porcine reproductive and respiratory syndrome virus (PRRSV) vaccines based on measurement of serologic response, frequency of gamma-IFN-γ producing cells and virological parameters of protection upon challenge. Vet Microbiol 123 (2007) 69-85.

7. Charreyre, C., Juillard, V., Piras, F., Richard, S., Joisel, F., Brun, A., Andreoni, C. 2007. Ex-vivo stimulation with different PRRSV strains for cellular mediated immunity monitoring of PROGRESSIS vaccinated pigs. In Proceedings of the 5th International Symposium on Emerging and Re-emerging Pig Diseases, Krakow, Poland, p144.

8. Reynaud, G., Charreyre, C., Tedeschi, L., Brun, A. , Joisel, F., Jeannin, P. 2004. Homologous PRRSV challenge after vaccination with an inactivated PRRS vaccine: PROGRESSIS® Merial. 18th IPVS Congress, Hamburg, Germany. 1: 429.

9. Reynaud, G., Chevallier, S., Brun, A., Herin, J. B., Albina, E. and Charreyre, C. 1999. Evaluation in the field of serological and virological efficacy of an inactivated PRRS vaccine in sows. In: Proceedings 3rd International Symposium on PRRS and Aujeszky disease. p281-282.

10. Joisel, F., Ferrari, M., Leotti, G., Dotti, S., Alborali, L., 2007. Evaluation of neutralizing antibodies in pigs repeatedly vaccinated with PROGRESSIS (Merial) an inactivated vaccine against PRRSV. 5th International Symposium on Emerging and Re-emerging Pig Diseases, Krakow, Poland, p 219.

11. Geldhof , M. F., Van Breedam, W., De Jong, E., Lopez Rodriguez, A., Karniychuk, U.U., Vanhee, M., Van Doorsselaere, J., Maes, D., Nauwynck, H.J. 2013 Antibody response and maternal immunity upon boosting PRRSV-immune sows with experimental farm-specific and commercial PRRSV vaccines. Vet Microbiol. 167(3-4):260-271.

12. Papatsiros, V. G., Alexopoulos, C., Kritas, S. K., Koptopoulos, G., Nauwynck, H. J., Pensaert, M. B. and Kyriakis, S. C. 2006. Long-term administration of a commercial porcine reproductive and respiratory syndrome virus (PRRSV)-inactivated vaccine in PRRSV-endemically infected sows. Journal of Veterinary Medicine. 53: 266-272.

13. Joisel, F., Reynaud, G., Charreyre, C., and Hérin, J. B. 2001. PRRS: Vaccination with a killed vaccine ; Field experience. The Pig Journal. 48: 120-137.

14. Bardini, R. 2014. Benefits of PROGRESSIS® vaccination in PRRS control, a field experience. In: Proceedings 6th ESPHM. p19.

15. Reber, A. J., Tanner, M., Okinaga, T., Woolums, A. R., Williams, S., Ensley, D. T. and Hurley, D. J. 2006. Evaluation of multiple immune parameters after vaccination with modified live or killed bovine viral diarrhea virus vaccines. Comparative Immunology, Microbiology & Infectious Diseases 29: 61-77.

16. Letellier, C., Boxus, M., Rosar, L., Toussaint, J. F., Walravens, K., Roels, S., Meyer, G., Letesson, J. J. and Kerkhofs, P. 2008. Vaccination of calves using the BRSV nucleocapsid protein in a DNA prime-protein boost strategy stimulates cell-mediated immunity and protects the lungs against BRSV replication and pathology. Vaccine. 26 (37): 4840-4848.

17. Li, Y., Stirling, C. M., Denyer, M. S., Hamblin, P., Hutchings, G., Takamatsu, H. H. and Barnett, P. V. 2008. Dramatic improvement in FMD DNA vaccine efficacy and cross-serotype antibody induction in pigs following a protein boost. Vaccine. 26 (21): 2647-2656.

18. Nilubol, D., Platt, K. B., Halbur, P. G., Torremorell, M. and Harris, D. L. 2004. The effect of a killed porcine reproductive and respiratory syndrome virus (PRRSV) vaccine treatment on virus shedding in previously PRRSV infected pigs. Veterinary Microbiology. 102 (1-2): 11-18.

19. Bassaganya-Riera, J., Thacker, B. J., Yu, S., Strait, E., Wannemuehler, M. J. and Thacker, E. L. 2005. Impact of immunization with porcine reproductive and respiratory syndrome virus on lymphoproliferative recall response of CD8+ T cells. Viral Immunology. 17 (1): 25-37.

20. Díaz, I., Gimeno, M., Callén, A., Pujols, J., López. S., Charreyre, C., Joisel, F., Mateu, E. 2013. Comparison of different vaccination schedules for sustaining the immune response againstporcine reproductive and respiratory syndrome virus. Vet J. 197(2):438-44.

21. Dekens, V., Bossers, M., Den Hartog, P., Seesing, E. and Meyns, T. 2013. Induction of maternal immunity by vaccination of sows at end of gestation with PROGRESSIS®. In: Proceedings 5th ESPHM. p190.

22. Defoort, P., Dekens, V., Meyns, T., Van Poucke, S. and Joisel, F. 2014. Stabilization of PRRSV circulation in a farm using a vaccination program with PROGRESSIS® at the end of gestation. In: Proceedings 23rd IPVS Congress. p565.

23. Willems, L., Boivent, B., Perreul, G., Merdy, O., Smits, H., Joisel, F. and Meyns, T. 2015. Stabilization of PRRSV circulation in a nursery and fattening unit following the implementation of mixed PRRSV vaccine program in breeders. In: Proceedings 7th ESPHM. p165.

24. Spaans, J., Verhaegen, A., Dekens, V., Smits, H., Meyns, T., Joisel, F. 2016. Effect of Dual Technology Prime Boost (DTPB) vaccination in sows on circulation of PRRSV in post-weaning piglets. Submitted at the 24th IPVS/8th ESPHM congress, Dublin, Rep. of Ireland.

25. Boivent, B., Perreul, G., Merdy, O., Joisel, F. *et al.* 2016. PRRS Dual Technology mix vaccination program (MLV/KV): virus circulation monitoring in pig flows using oral fluids. Submitted at the 24th IPVS/8th ESPHM congress, Dublin, Rep. of Ireland.

Improved swine influenza A virus (IAV-S) strain selection method affords cross-protection against heterologous challenge in the face of maternal antibodies

Paulraj Lawrence, PhD; Eric Bumgardner, MS; Brittanny Wiener, BS
Biological Development and Research Department, Newport Laboratories Inc., Worthington, Minnesota

Introduction

Swine origin influenza viruses are among the most important viral pathogens affecting the health of the US swine herd. The major subtypes that affect swine can be classified into several sub-clusters based on the sequence of the HA segment.[1] Genetic drift in the virus population has permitted the virus to evolve a significant degree of antigenic diversity, allowing the virus to evade immunity to many of the strains included in commercially available vaccines. Autogenous vaccines can be used to provide protection against viruses that evade immunity provided by commercial vaccines. However, recent work has shown that inactivated vaccines can cause vaccine associated enhancement of respiratory disease (VAERD).[2] Furthermore, maternal antibodies have been shown to interfere with inactivated vaccines when they are delivered early after weaning.[3] A recent study has demonstrated that intranasal delivery in combination with poly I:C adjuvant, can circumvent maternal antibody interference and afford broad protection.[4] The study presented here was designed to test whether a vaccine formulated using a proprietary IAV-S strain selection method based on serum neutralization and comparative genomic analysis would afford cross protection against a virulent challenge when delivered in the face of maternal antibodies.

Figure 1: Example of strain selection using phylogenetic analysis of cross-reactive isolates taken from within the H1 γ-cluster. The asterisk indicates the strain selected for use in vaccine formulation.

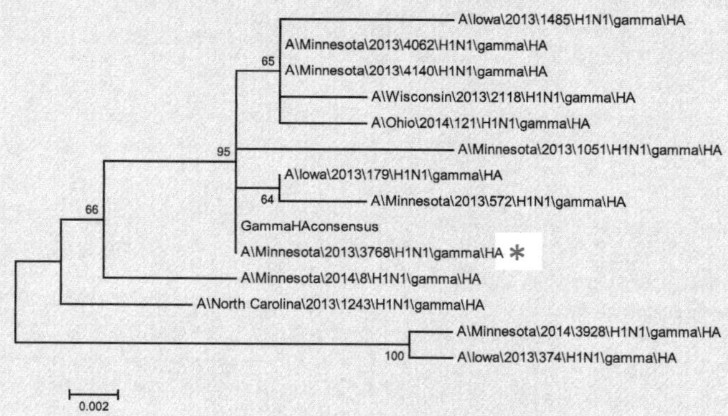

Materials and methods

An inactivated trivalent vaccine was formulated with a proprietary oil-in water adjuvant using strains selected from within the H1 δ and γ clusters and from H3 cluster IV. Candidate vaccine strains were selected for use in the experiment based on their susceptibility to neutralization with a panel of antisera described previously.[5] Hemagglutinin sequences from a panel of strains that exhibited significant cross-neutralization by antisera raised against different antigenic clusters within each subtype were used to develop a consensus sequence for each cluster. The strains were then compared against the consensus using a maximum likelihood method in MEGA 6.0 to determine which cross-reactive strain was most similar to the consensus (Figure 1). The most cross-reactive strain that bore the greatest homology to the consensus was used to formulate vaccine for use in the trial. The vaccine was formulated to contain equal quantities of inactivated antigen from each strain as well as a proprietary oil-in-water adjuvant. Sows were vaccinated two weeks prior to farrowing with MaxiVac Excell 5.0® (Merck). Piglets derived from vaccinated dams were brought to our research farm shortly after weaning at approximately four weeks of age and were bled on the first day of the study. Sera were tested using a competitive ELISA assay (SIV Ab Test, IDEXX Laboratories Inc.) (Figure 2) to verify that they were sero-positive for antibodies against SIV. Pigs were divided into three treatment groups. A group of eight animals received the trivalent inactivated vaccine in a proprietary adjuvant. A second group of eight animals received a mock vaccination containing phosphate buffered saline formulated with the same amount of adjuvant used in the trivalent vaccine. A third group of six animals did not receive any treatment or challenge and served as baseline controls. Treatments were delivered on days 0 and 21 of the study. Animals were challenged 14 days after their last vaccination with an H1N2 δ cluster strain, that was antigenically distinct from the strains included in the vaccine. Serum samples were taken immediately prior to challenge and were evaluatedfor neutralizing antibody titers against the strains included in the vaccine and the challenge isolate using a hemagglutination inhibition assay (HAI) (Figure 3). Following challenge, animals

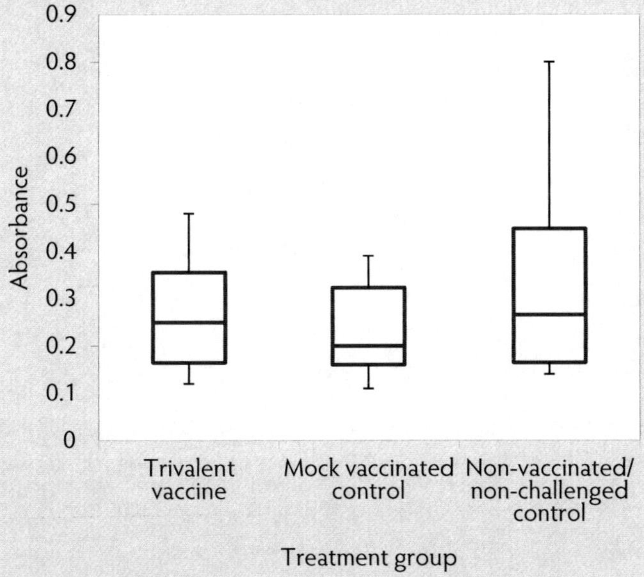

Figure 2: Summary of the results from the competitive ELISA using sera drawn on day one of the study.

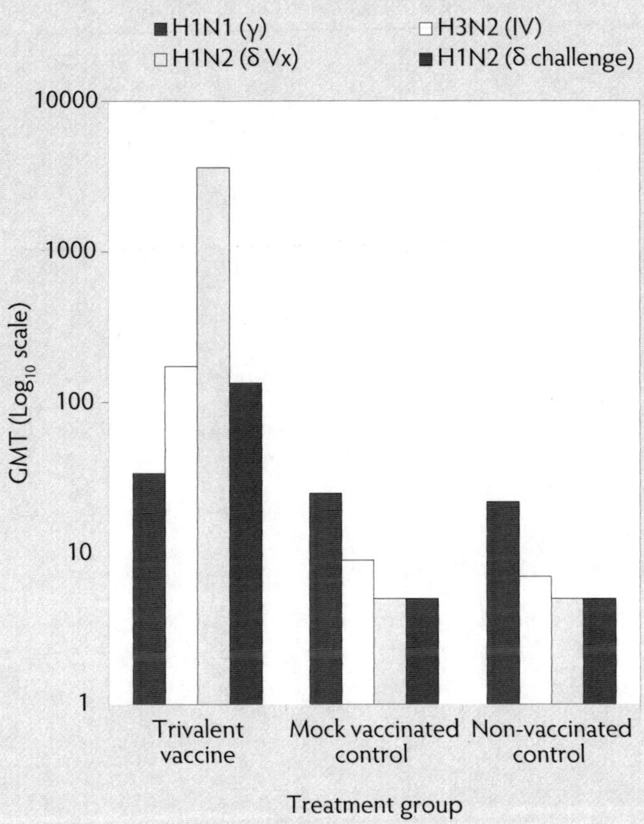

Figure 3: HAI data obtained from sera drawn on day 35 of the study, immediately before challenge.

were humanely euthanized and necropsied. The lung tissue was immediately assessed for gross pathology (Figure 4) and sections were fixed for microscopic histopathology scoring. (Figure 5).

Results and discussion

The results obtained from both gross lung lesion scoring and microscopic histopathology scoring indicated that the trivalent inactivated vaccine formulated with a proprietary adjuvant could induce cross-protection in weanling aged pigs. Despite the presence of maternal antibodies in pigs at the beginning of the study, significant HAI titers were obtained against both of the H1N2 (δ) strains used in the trial as well as the H3N2 (IV) strain. The HAI results also demonstrated that our proprietary strain selection method was able to successfully identify a H1N2 (δ) strain that could elicit cross-neutralizing antibodies against the heterologous challenge strain. Overall, the results of the study demonstrated that a trivalent inactivated vaccine formulated using our selection method was able to confer protection against a virulent heterologous challenge. This method could be applied on a regional basis to produce inactivated vaccines that would be expected to provide protection against modern strains circulating in the United States. Additionally, no evidence of VAERD was observed during the course of this study.

References

1. Vincent AL et al. Evaluation of hemagglutinin subtype 1 swine influenza viruses from the United States. *Vet Microbiol.* 2006;118:212.

2. Vincent AL et al. Failure of protection and enhanced pneumonia with a US H1N2 swine influenza virus in pigs vaccinated with an inactivated classical swine H1N1 vaccine. *Vet Microbiol.* 2008;126:310.

3. Kitikoon P et al. The immune response and maternal antibody interference to a heterologous H1N1 swine influenza virus infection following vaccination. *Vet Immunol Immunopathol.* 2006;112:117.

4. Thomas M et al. Poly I:C adjuvanted inactivated swine influenza vaccine induces heterologous protective immunity in pigs. *Vaccine.* 2015;33:542.

5. Hause BM et al. Genetic and antigenic characterization of recent human-like H1 (δ-cluster) swine influenza virus isolates. *J. Swine Health Prod.* 2011;19:268.

Figure 4: Results of gross lung lesion scoring.

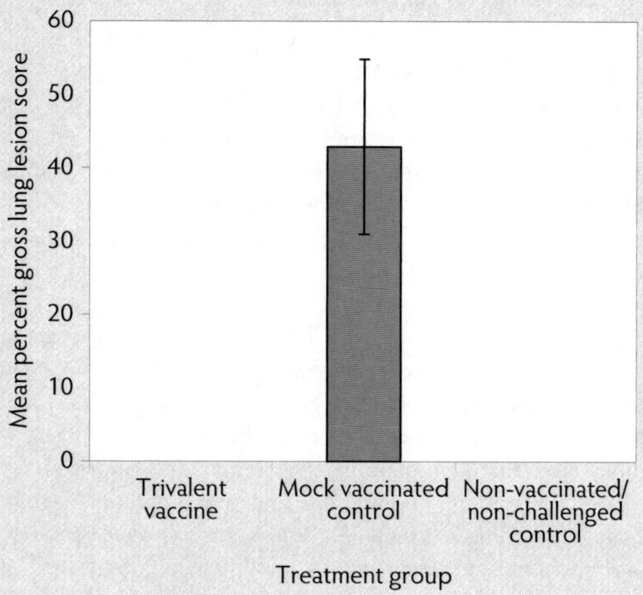

Figure 5: Results of microscopic histopathology scoring

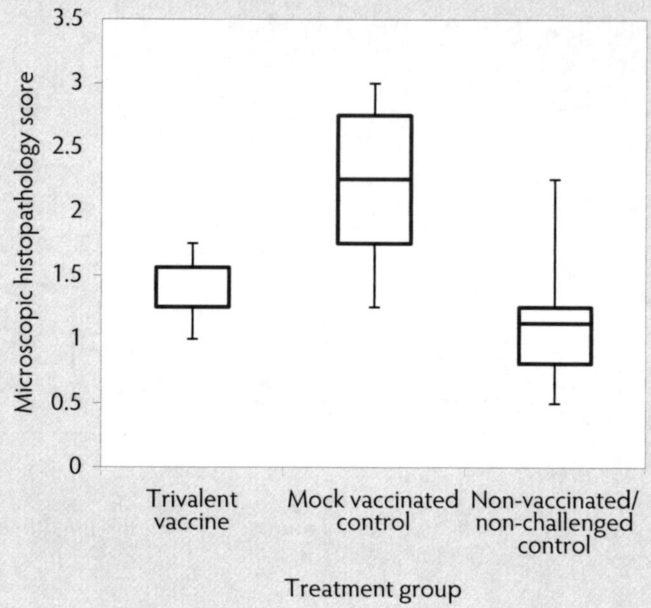

Utilizing a killed PRRSV vaccine in combination with *M. hyopneumoniae* and PCV2 vaccines; impact on weight gain in vaccinated animals

J. Kuecker[2], DVM; E.A. Bumgardner[1], MS; P.K. Lawrence[1], PhD; K. Wilson[1], DVM
[1]Newport Laboratories Inc., Worthington, Minnesota; [2]East Fork Swine, Algona, Iowa

Introduction

Porcine reproductive and respiratory syndrome virus (PRRSV) and *Mycoplasma hyopneumoniae* are two of the most commonly isolated pathogens of the porcine respiratory disease complex. In pig flows where both of these pathogens are present infection with *M. hyopneumoniae* often predisposes pigs to infection with PRRSV.[1] In an *in vitro* model, increased levels of inflammatory cytokines are induced by coinfection, which may continuously draw more macrophages to the site of infection in the lung and prolong the infection.[2] If this translates to the field, pneumonia induced by coinfection may result in a longer duration of PRRSV induced pneumonia than is observed when the infection is potentiated by the virus by itself. PRRSV strains currently circulating in the field possess a high level of genetic heterogeneity such that commercial vaccines are often unable to provide protection against heterologous strains of virus.[3] Recently, a new strain of PRRSV exhibiting a 1-7-4 RFLP pattern has been increasing in prevalence within the United States.[4] Preliminary reports from the field have described a lack of protection obtained with commercial vaccines against this virulent new strain. When *M. hyopneumoniae* is present on a farm, the outcome of infection with this virus may be worsened. New vaccination strategies are needed that can combat the combined pathologies associated with these two pathogens, particularly after weaning when pigs become most susceptible to infection. In the work presented here we performed a field study in which we tested four different vaccination strategies to determine whether they would be able to enhance weight gain from weaning to 12 weeks of age. The results demonstrate a significant weight advantage for animals that received an *M. hyopneumoniae* vaccination.

Materials and methods

A farm experiencing significant disease caused by PRRSV and *M. hyopneumoniae* was selected as the site for the study. Three week old weaned pigs were randomly divided into four treatment groups as they arrived at the barn and vaccinated as described in Table 1. Several days following group assignment it was noted that the start weight of the animals in groups A and B was significantly lower than the start weight for animals in groups C and D. The groups were assigned by different set of workers. However, since the treatments had already been administered it was not possible to reassign the animals to different groups in order to adjust for this variable. Therefore, the study was continued as planned. Animals were followed for 61 days and weighed on day 61. Serum samples were also drawn at both time points and were assessed for the presence of PRRSV by qPCR (Figure 1) and for neutralizing antibody titer using a Fluorescent Focus Neutralization assay (FFN) at South Dakota State University. Neutralizing antibody titers were measured against both the strains used in vaccination treatments as well as a heterologous strain, NADC20 (Figure 2). Weight gain was assessed between all groups using a two-way ANOVA ($\alpha = 0.05$) to determine whether the factors, start weight and treatment group, had a statistically significant impact on the outcome variable, finish weight. Groups were then ranked using Student's t-test ($\alpha = 0.05$) in order to visualize statistically significant differences between groups (Figure 3). Pigs were similarly assessed to determine whether vaccination with *M. hyopneumoniae* had a statistically significant impact on finish weight irrespective of the other components of the treatment given (Figure 4). We also assessed the possibility of interaction between start weight and Mycoplasma vaccination status on final weight by crossing these variables in our statistical model.

Table 1: Treatment groups included in study

Group	Treatment	n
A	Commercial PRRSV MLV Vaccine, circovac PCV2 vaccine, inactivated autogenous PRRSV vaccine	40
B	Commercial PRRSV MLV and Circovac PCV2 vaccine	40
C	Commercial PRRSV MLV, Circovac PCV2 vaccine, MailSail *Mycoplasma hyopneumoniae* vaccine, inactivated autogenous PRRSV vaccine	40
D	Commercial PRRSV MLV, circovac PCV2 vaccine, and MailSail *Mycoplasma hyopneumoniae* vaccine	40

Figure 1: qPCR performed on serum drawn on the first day of the study

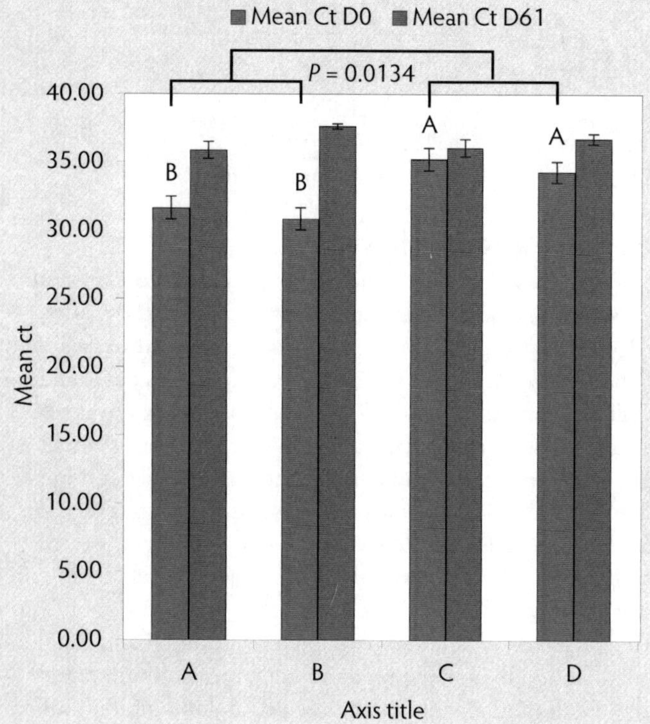

Results and discussion

The results of the study were complicated by the fact that animals in treatment groups A and B were significantly lighter than animals in groups C and D at the beginning of the study. When assessing the significance of the contribution of each of the two variables assessed, start weight was the most significant contributor toward the finish weight of the animals. However, vaccination against *M. hyopneumoniae* also contributed to finish weight and no interaction was observed between the two variables, demonstrating that both contributed to finish weight independently. No statistically significant interactions were present between start weights and Mycoplasma vaccination. Based on the qPCR results from serum drawn at the beginning of the study it seems that the difference in start weight may have been due to a more severe PRRSV infection in groups A and B. This evidence is further supported by the results of our FFN testing which demonstrated a statistically significant difference in neutralizing antibody titer against the 1-7-4 RFLP virus for pigs in group A at the end of the study. It seems likely that these pigs experienced a more severe challenge with this virus prior to vaccination. The inactivated vaccine delivered at day 0 might have served as a booster vaccination following this challenge. This may have resulted in enhanced neutralizing antibody titers at the end of the study in comparison to the other groups that either did not receive the vaccine, or were not as severely infected at the beginning of the study. When comparing treatments that resulted in a significant increase in weight gained by the end of the study the *M. hyopneumoniae* vaccination stood out as a factor that increased the ability of the pigs to gain weight. Pigs that received an *M. hyopneumoniae* vaccination finished the study an average of 11 pounds heavier than pigs that did not receive this treatment. Thus, a significant economic advantage can be assigned to *M. hyopneumoniae* vaccination. However, it is difficult to determine exactly how much of the weight gained might have been

Figure 2: FFN Titers from serum drawn on days 0 and 21. (*) indicates group that had significantly higher titers based on the results of a Student's t-test ($\alpha = 0.05$)

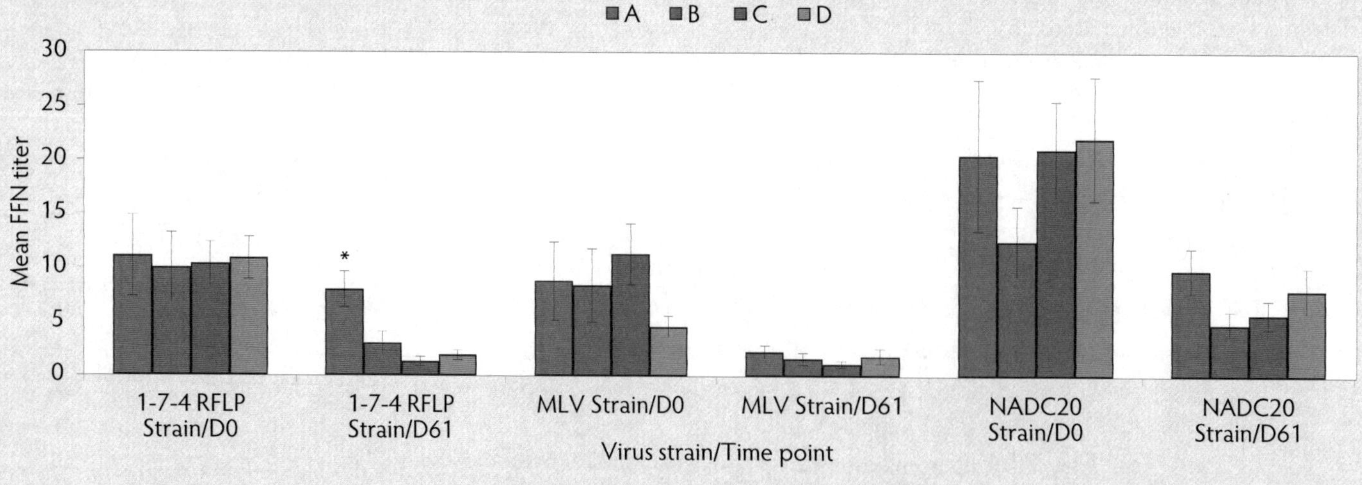

Figure 3: Mean weight of groups at days 0 and 61 of the study. Connecting letter report describes the results of at Student's t-test ($\alpha = 0.05$)

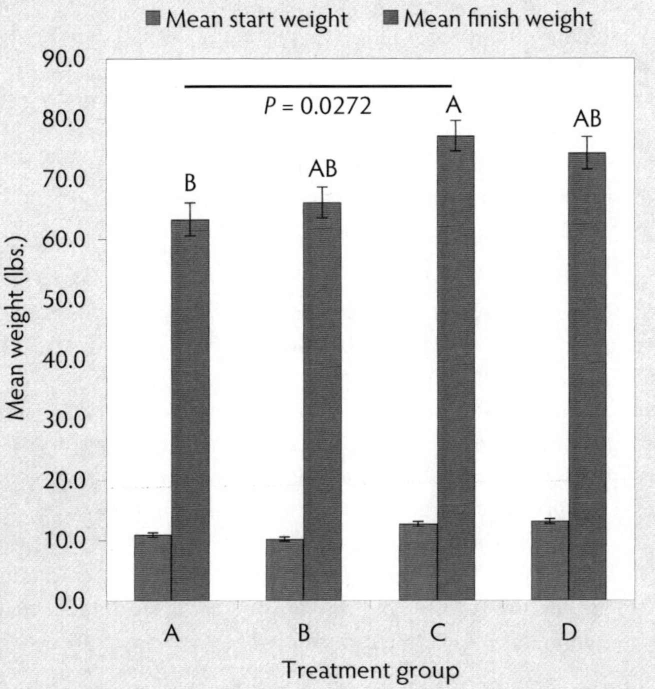

Figure 4: Comparison of finishing weights based on Mycoplasma vaccination status. *P* - value indicates the result of a two-way ANOVA

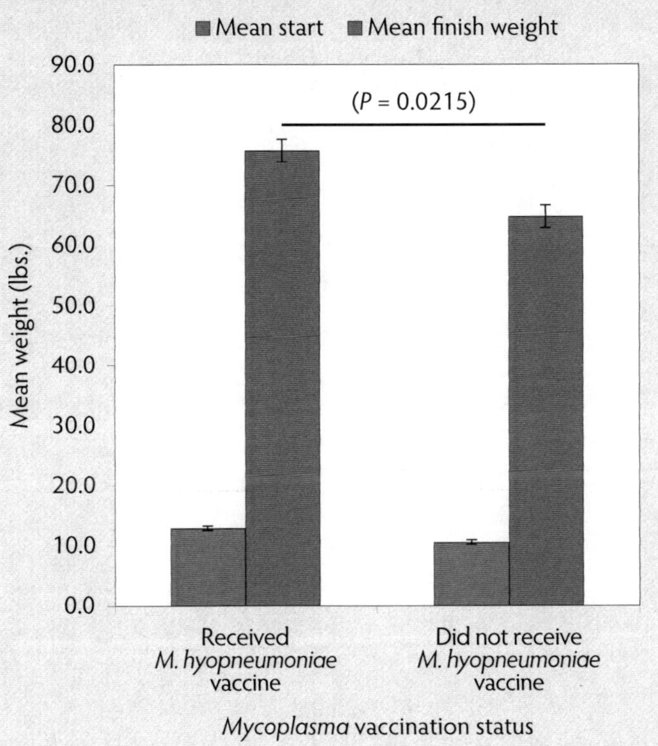

associated with vaccination versus the initial weight of the pigs in each group. These data imply that the benefit of *M. hyopneumoniae* vaccination is currently underestimated in the field. However, the results from this study should be interpreted with caution and should be repeated with treatment groups in which the animals start at a statistically equivalent weight.

References

1. Thacker EL et al. Mycoplasma hyopneumoniae potentiation of porcine reproductive and respiratory syndrome virus-induced pneumonia. *J. Clin. Microbiol.* 1999;37:620.

2. Thanawongnuwech R et al. Differential production of proinflammatory cytokines: in vitro PRRSV and Mycoplasma hyopneumoniae coinfection model. *Vet. Immunol. Immunopathol.* 2001;79:115.

3. Kimman TG et al. Challenges for porcine reproductive and respiratory syndrome virus (PRRSV) vaccinology. *Vaccine.* 2009; 27:3704.

4. Murtaugh M. PRRS '174' What are we learning about PRRSV evolution and characterization of new emerging strains of PRRSV? North American PRRS Symposium 2015;9.

The future of swine for research and xenotransplantation is here

Rexanne Struve, DVM

Struve Labs International, Inc. has been raising caesarian derived, colostrum deprived (CDCD) pigs since 1962. For over 5 decades, these pigs have been used; originally to start clean herds allowing producers to salvage their best genetics while breaking the disease cycle that robbed them of profits in their original herd, then for researchers to study new diseases as they emerged, since CDCD pigs can show the effects of a virus or bacterium without the confounding factor of other pathogens to influence the study pathogen. The future includes the use of swine products, be they blood components, tissues, or whole organs, to benefit the human health field. The possibilities are endless.

During the past 25 years, diseases such as parvo (PPV), pseudorabies (PRV), PRRSV, circo virus type 2 (PCV2), and torque tino virus (TTV) have been studied in CDCD pigs. These pigs are the gold standard for the study of single pathogens in swine. Specifically, Struve Labs pigs are the gold standard for PCV2 negative pigs. Work done previously shows that PRRSV, PPV, and swine influenza (SIV) all affect PCV2 research, so the question is, when studying PCV2, "is being negative to PCV2 enough?" or do we need to prove negativity to the extra 3; PRRSV, PPV, and SIV?

Because we produce thousands of CDCD pigs each year, Struve Labs has the ability to exhibit sow to piglet transmission (or lack thereof) of disease entities. Struve Labs has been very successful in procuring sows that produce PCV2 negative pigs. Because of our selective choice of sows, results of testing thousands of cord blood samples have shown 99.5% negativity to both PCV2 and PRRS viruses. Historical cord testing has shown approximately 95% negativity to PPV in well established sow herds. In 2 different incidences, when a low number of pigs were born with low PPV titers, retesting at 2 weeks of age showed almost all going negative. To date, hundreds of cord blood samples have been tested for SIV H1N1 and H3N2; none have been positive.

Traditionally, acceptable livability of CDCD pigs through the lab is 75%. For the past decade, Struve Labs has maintained a survivability rate of more than 95%, while rearing over 2000 pigs per year to 6 or 8 weeks of age. The reason Struve Labs has a 95+% livability of pigs is our longevity and experience in the field. We have systems in place; our staff has over 100 combined years of experience in C-sectioning and raising these special animals, and strict adherence to protocols is paramount to this success. Through years of procuring these pigs, anesthetic protocols have been developed that assure pig viability and quick sow recovery, which enables us to populate an entire set of 162 pigs within a 24-hour window.

If your research needs call for pigs of known status, highly negative to common pathogens for vaccine development, assay development, safety testing, or antigen/antibody production, Struve Labs is positioned to meet your needs at a higher standard and quantity than anyone else in the industry.

Struve Labs has been helping researchers in their quest for clean pigs to produce enzymes, pig tissues, and organ products for human health and treatments. In the fields of genetic modification and Xenotransplantation in human medicine, current practice includes the use of porcine heart valves, skin, and ligaments, with the ultimate goal being the transplanting of whole organs from pigs to humans. Struve Labs is in the forefront of verification of CDCD swine cleanliness and safety for this process, so our CDCD model has a profound impact on moving the industry forward. Because of the shortage of human organs for transplant, the ability to prolong life and provide organs for sick, rather than dying, humans is unfathomably exciting and appears to be within reach.

The CDCD standard is the top of the pyramid for starting specific pathogen free (SPF) and designated pathogen free (DPF) herds. Struve Labs is the premier provider of CDCD pigs in the US, and the National SPF program is a viable avenue for health certification. Most biomedical companies want some sort of verification process for SPF or DPF pigs. So if you have clients who are selling, or want to sell pigs for biomedical research, you have resources in the form of Struve Labs and the National SPF Association to help in this endeavor.

The assessment of the effectiveness of the Bock Industries Zephyr EXL non-penetrating captive bolt system for euthanasia of new-born and weaned piglets up to 10Kg

Andrew Grist and Stephen Wotton, MBE
School of Veterinary Sciences, University of Bristol, United Kingdom

Introduction

Modern pig production has an inherent requirement for the euthanasia of neonate piglets for various reasons including herd productivity, disease and underperformance. In the United Kingdom pre-weaning mortality averages 14.18% indicating one in twelve piglets in a litter that may require dispatch before the average weaning age of 26 days (average piglet weight 7 Kg). The traditional method of dispatch is blunt force trauma (BFT), either through holding the piglet by the hind-legs and hitting the head against a hard object or using some form of percussive device such as a 'priest' or hammer. Blunt force trauma as a method has several issues, firstly it relies on the ability of the operator to successfully perform the action, secondly the effects may not be reproducible and thirdly stockmen do not like performing the operation unless the animal appears ill and the method of euthanasia was perceived as being less painful to the animal.[1,2,3]

An alternative method that is applied in commercial units is the exposure of the piglets to carbon dioxide in groups in a container with up to 80% CO_2. CO_2 is known to produce an aversive response in the induction phase, which on average lasts for approximately 15 seconds before the onset of anaesthesia due to metabolic and respiratory acidosis, reducing the pH of the cerebrospinal fluid leading to neuronal dysfunction. Experiments have demonstrated that the aversion to CO_2 is such that pigs will avoid a food reward even when withheld food for up to 24 hours, if the CO_2 concentration is above 80%.[4] Pigs are also known to show aversion (escape attempts) when the rate of induction of unconsciousness is slow (e.g. during exposure to 40 to 70% by volume of carbon dioxide in air) and will demonstrate signs of respiratory distress at 40%.[5]

Mechanical killing via blunt force trauma has the advantage of speed of induction, reproducibility, less reliance on operator ability and with training, including the identification of post mortem movement that indicates an effective stun/kill, enhanced operator acceptability.

Mechanical stunning relies on imparting kinetic energy to the cranium to produce concussive effects within the brain, based on the velocity of the impact rather than the mass of the object.[6] The concussion produced by this impact is often associated with both haemorrhaging at the impact site ('coup') and further haemorrhaging opposite the impact site ('contra-coup')[7] due to the rotational and differential acceleration of the brain within the cranium.[8] Shearing forces are produced within the brain by the pressure waves producing vacuolation,[9] disruption of synaptic transmission[10] and depolarisation of neurons away from the impact site.[11]

The most common cause of death following brain injury is subdural haemorrhage due to direct injury to the cortical arteries and veins by the object, contusion and pulping of the cerebrum, or tearing veins that bridge the subdural space between the brain surface and the dural sinuses.[12]

There were initial concerns that the incomplete sutures in the newborn piglet that provide for cranial deformation during parturition may provide a form of elastic protection from the effect of BFT, in effect absorbing the blow. Previous studies (DEFRA MH0116) found that with non-penetrating captive bolt (NPCB) the skull development of the neonate piglet is sufficient that the mechanical BFT applied does produce enough kinetic energy to produce a stun-kill and work by Armstead[13] found that newborn piglets were particularly sensitive to brain injury.

A study to assess the effectiveness of a NPCB for euthanasia of suckling and weaned piglets using the Bock Industries Zephyr E pneumatic non-penetrating captive bolt was undertaken by the University of Guelph. This study used visual signs of sensibility and behavioural indicators including loss of rhythmic breathing, corneal reflex and response to painful stimuli to assess the effectiveness of the stun combined with post mortem of the head to assess skull and brain damage.[14] Following this research Bock Industries upgraded the Zephyr E to the Zephyr EXL, which has a higher velocity when operated at 120 psi than the Zephyr E (26 m/s *c.f.* 20 m/s) and hence develops a higher kinetic energy (26 J *c.f.* 20 J) (personal communication, Lines 2015).

Laboratory assessment

Piglets (n = 60) within 5 weight ranges - Birth to 3 kg, 3 to 5 kg, 7 to 9 kg and 9 to 11 kg, with an average liveweight 6.1 kg were anaesthetized and EEG recording electrodes were inserted to allow the recording of Visual Evoked Potentials (VEPs). The abolition of Flash VEPs is accepted in the EU as a method of assessing brain dysfunction as it measures the patency of a simple nerve pathway which when lost indicates an inability of the brain to receive and process external stimuli i.e. loss of residual consciousness. A strobe light flashed in the eye elicits a response in the visual cortex that can be measured.[15,16] Following recording of baseline VEPs in the anaesthetized state the piglet was shot once in the parietal position with a Bock Industries Zephyr EXL non-penetrating captive bolt powered by 120 psi air pressure. Movement scoring, behavioural indices of loss of brain function and VEPs were monitored throughout and for 3 minutes post shot. VEPs and brain stem reflexes (rhythmic breathing, corneal reflex, response to nose prick) were lost immediately in all piglets shot when the head was resting on a hard surface demonstrating the effectiveness of the Zephyr EXL to produce immediate brain death with a single shot.

Field study

The Zephyr EXL non-penetrating captive bolt was field trialled on-farm in a commercial setting. Over the course of two days 200 piglets selected by farm staff following their normal protocols were euthanized with the device and demonstrated no behavioural signs of recovery. Post mortem examination of the heads was undertaken confirming massive traumatic damage to the cerebrum with associated haemorrhage and bone plate shards forced down to the level of the corpus callosum in the majority of cases. Given a statistical confidence level of 95% the very maximum percentage of animals that would not immediately be stunned/killed by this system would be never more than 1.9%.

A second trial of 100 piglets demonstrated that under commercial production conditions it took less than seven seconds to select, place and shoot a piglet that compares favourably to the CO_2 system currently employed in terms of time and animal welfare.

References

1. Mort, M., I. Convery, J. Baxter, C. Bailey. 2008. Animal disease and human trauma: The psychosocial implications of the 2001 UK foot and mouth disease disaster. *J. App. Anim. Welf.* 11(2):133-148.

2. Whiting, T. L. and C. R. Marion. 2011a. Perpetration-induced traumatic stress-A risk for veterinarians involved in the destruction of healthy animals. *Can. Vet. J.* 52:794-796.

3. Matthis, J. S. 2004. Selected employee attributes and perceptions regarding methods and animal welfare concerns associated with swine euthanasia. PhD Diss. North Carolina State Univ., Raleigh, NC.

4. Raj, A.B.M., and Gregory, N.G., 1995. Welfare implications of the gas stunning of pigs 1. Determination of aversion to the initial inhalation of carbon dioxide or argon. *Animal Welfare*, 4: 273-280.

5. Raj, A.B.M., and Gregory, N.G., 1996. Welfare implications of the gas stunning of pigs 2. Stress of induction of anaesthesia. *Animal Welfare*, 5: 71-78.

6. Daly, C. C., Gregory N . G. and Wotton S . B . 1987 Captive bolt stunning of cattle: Effects on brain function and role of bolt velocity *Br. vet . J.* (1987) . 143, 574.

7. Ommaya, A.K., Grub, R.L. and Naumann, R.A., 1971. Coup and contre-coup injury: observations on the mechanics of visible brain injuries in the rhesus monkey. *Journal of Neurosurgery*, 35: 503-516.

8. Ommaya, A.K, Gennareli, T.A., 1974. Cerebral concussion and traumatic unconsciousness. *Brain* (1974) 97, 633-654.

9. Finnie, J.W., 1995. Neuropathological changes produced by non-penetrating percussive captive bolt stunning of cattle. *New Zealand Veterinary Journal*, 43: 183-185.

10. Gregory, N.G., 1998. Stunning and slaughter. In: *Animal Welfare and Meat Science*. Cabi. Publishing.

11. Somjen, G., 2001. Mechanisms of spreading depression and hypoxic spreading depression-like depolarization. *Physiological Reviews*, 81 (3): 1065-1096.

12. Millman, S.T. (2010) Mechanical euthanasia methods – process and physiology. 2010 AASV Annual Meeting: Implementing Knowledge.

13. Armstead, W.M. 1999 Cerebral hemodynamics after traumatic injury of immature brain. *Exp Toxic Patho* 1999; 51: 137-142.

14. Casey-Trott, T.M., Millman, S.T., Turner, P.V., Nykamp, S.G., Widowski, T.M. 2013. Effectiveness of a nonpenetrating captive bolt for euthanasia of piglets less than 3 d of age *J ANIM SCI* 2013, 91:5477-5484.

15. Gregory, N.G. and Wotton, S.B. 1984. Sheep slaughtering procedures. 2. Time to loss of brain responsiveness after exsanguination or cardiac arrest. *British Veterinary Journal.* 140, 354-360.

16. Guerit, M.J., 1999 Medical technology assessment: EEG and evoked potentials in the intensive care unit. *Clinical Neurophysiology*, 29: 301-317.

Safety and efficacy of a porcine circovirus type 2 (PCV2, killed) and *Mycoplasma hyopneumoniae* (M hyo, bacterin) vaccine

J. Dory, BS; L. Fischer, BS; J. Herczeg, DVM, PhD; A. Hicks, BS; H. Jin, PhD; J. Lopez-Virella, MS; S. Mahesh, MA; J. Newberry, DVM, MBA; T. Ng, PhD; W. Lu*, BVSc, PhD
Ceva Biomune R&D, Lenexa, Kansas

This report describes the safety and efficacy studies of a porcine circovirus type 2 (PCV2, killed) and *Mycoplasma hyopneumoniae* (M hyo, bacterin) vaccine (Code 41K5.20, unlicensed). Code 41K5.20 will be used to vaccinate healthy pigs at 3 weeks of age or older, as an aid of control of PCV2-associated viremia, virus shedding, lymphoid infection and lesion, and as an aid of control of M hyo-associated pneumonia.

Vaccine safety

PCV2 inactivation

Thirty (30) PCV2-viremic negative, CDCD pigs were assigned 6 groups (5/each) to confirm PCV2 inactivation of 5 PCV2 antigen lots (virus titer @ $10^{5.2\text{-}7.5}$ TCID50/ml). Each group was housed in a separated BLS-2 isolation room with individual air handling. At 4 wks of age, pigs in Groups 1-5 were each injected (IM) with 4 ml assigned PCV2 lot at both sides of neck (2 ml/side). Pigs in Group 6 served as controls. All pigs were bled and euthanized on 29 days post injection, lymphoid tissues (tonsil, bronchial and inguinal lymph nodes) were collected.

Laboratory test confirmed that all euthanized pigs were PCV2 viremia negative (by qPCR) and negative for PCV2 in lymphoid tissues (by IHC). These results confirmed complete PCV2 inactivation.

Adjuvant safety

Thirty (36) conventional 3 wk old, weaned pigs were assigned to 2 groups (18/each). In each group, 9 pigs were injected (IM) in the left neck with 2 ml "Adjuvant" (Code41K5.20 w/o antigens) and 2 ml in the right neck with a "Control" vaccine (a commercial swine vaccine with 21 day withdrawal approval); other 9 pigs in the group were injected with "Adjuvant" in the right neck and "Control" in the left neck.

Post-vac observation (n = 36 pigs): No clinical (general and local) abnormalities were recorded in the 1st week.

Palpations: On 21 day post-vac (n = 36 pigs), 2 swelling and no nodule were recorded in the "adjuvant" sites, compared to 7 swelling and 1 nodule were in the "Vaccine" sites; On 28 day post-vac (n = 18 pigs), no swellings and nodules were recorded in the "Adjuvant" sites, compared to 1 swelling and no nodule in the "Control" sites.

Gross pathology: On 21 day post-vac (n = 18 pigs) revealed no vaccine residues and 1 small Granuloma (< 2 mm) in both "adjuvant" and "Control" sites; On 28 day post-vac (n = 18 pigs), no vaccine residues and Granuloma were found in the "Adjuvant" sites, no vaccine residues and 3 granulomas (< 2 mm) were found in the "Control" sites.

Histopathology: No moderate and severe abnormal pathology findings in the two different vaccination sites in all pigs. This study confirmed that the safety performance of "Adjuvant" in Code 41K5.20 is as safe as the "Control" commercial swine vaccine with approved 21 day withdrawal claim.

Field safety

Field safety of two "Commercial dose" vaccines of Code 41K5.20 was evaluated at 3 sites (herds) in Iowa, Missouri, Illinois and Tennessee with a total of 2431 healthy commercial pigs (951 vaccinates and 1480 controls). Vaccination was given intramuscularly with 2 ml dose in the left neck at 3 weeks of age. The pigs were maintained under routine husbandry conditions and observed daily for 7 consecutive days post vaccination.

Among the 951 vaccinated pigs, only 2 pigs had observable, transient nodule in the injection sites on post vaccination days 1-3; no pigs showed any observable general adverse reactions. These results demonstrated the safety of Code 41K5.20 for use in commercial pigs when given IM at 3 weeks of age or older.

Vaccine efficacy

Two hundred (200) clinically healthy, weaned pigs were confirmed PCV2 viremia-free (by qPCR), and serology negative for PCV2, M hyo and PRRSV were used in two studies (M hyo fraction and PCV2 fraction, 100 /each). The 100 pigs in each study were assigned to 5 groups (20/each). The pigs in each group were vaccinated with one of 5 assigned/blind-coded "vaccines" including a Code 41K5.20, 3 commercial PCV2-M hyo vaccines (C1, C2 and C2) and a Placebo vaccine (CC). All vaccines were given IM in the left neck with a 2 ml dose or for label indications for commercial vaccines.

In the M hyo fraction study, pigs were intra-tracheally challenged, at 7 weeks post-vac, with a predetermined dose of a live M hyo material provided by ISU M Hyo laboratory. Necropsy was conducted 4 weeks post challenge and individual lung was removed and examined and scored for M hyo associated lung lesion (consolidation).

For PCV2 fraction study, the pigs were challenged (Intranasal and IM) with a predetermined dose of a live PCV2b material provided by KSU Dr. Dick Hesse Lab. Individual blood was withdrawn prior to challenge and weekly after challenge (7, 14, 21 and 28) for PCV2 viremia quantitation by qPCR. Lymphoid tissues (tonsil, bronchial and inguinal lymphoid nodes) were taken in 10% formalin solution, and tested for lymphoid infection (by IHC) and lesions at a Veterinary Diagnostic Laboratory.

The results are summarized in Table 1.

These results confirmed the clinical efficacy of Code 41.K5.20 "commercial dose" vaccine.

Table 1: Comparitive data for vaccine efficacy trial

Study	Parameter	41K5.20	C1	C2	C3	CC
M hyo	Lung lesion % (Gp avg)	2.2	2.1	2.7	7.0*	**4.9**
	Viremia titer (log_{10})	1.1	0.6	1.2	1.4	**3.8**
	Nasal shedding (log_{10})	3.2	3.4*	3.3	3.3	**4.0**
PCV2	Fecal shedding (log_{10})	2.5	2.0	2.6	2.3	**3.9**
	IHC scores (Gp avg)	1.6	2.6	1.8	2.3	**4.8**
	Histopath score (Gp avg)	0.2	0.6	0.4	0.3	**5.2**

* No statistical difference (reduction) compared to the Control group.

Utilization of commercial female data in maternal genetic improvement programs

D. W. Newcom[1], BS, MS, PhD; V. Duttlinger[2], AS; C. Witte[2], BS; M. Brubaker[2], BS, MS; S. E. Lawrence[2], BS, MS; C. Dematawewa[3], BS, PhD; K. J. Stalder[3], BS, MS, PhD

[1]National Swine Registry, West Lafayette, Indiana; [2]PureTek Genetics, Albion, Indiana; [3]Iowa State University, Ames, Iowa

Introduction

Maternal performance is encompassed in several of the key profitability drivers in commercial swine units. Purebred breeders and commercial breeding stock companies strive to improve reproductive performance in their maternal lines through genetic selection programs implemented in their nucleus herds. Maternal breeds or lines are crossed at the multiplication level to maximize heterosis in the resulting parent-stock female. The genes of genetically superior animals are passed to the multiplication level and ultimately the commercial level as quickly as possible to reduce genetic lag.

However, genetic selection programs are implemented at the nucleus level in high-health, intensely-managed herds of typically purebred/pureline animals. There may exist a genotype × environment interaction that limits the expression of purebred genetic improvement at the commercial level. The genetic improvement at the nucleus level does not always translate into improved performance at the commercial level.[1] The genetic correlation between reproduction at the nucleus level and the same traits at the commercial (purebreds and crossbred) level has been estimated to be less than 1 (in broilers), indicating selection for traits at the nucleus level will not maximize genetic improvement for the same traits at the commercial level.[2]

Genetic improvement programs in swine may indeed be missing an opportunity for additional genetic improvement by not including commercial crossbred data.[3] Additional records, in the form of commercial F1 daughters, will lead to improved accuracy of Estimated Breeding Values (EBV); one of the key components of genetic improvement programs. Ehlers et al. reported an increase in accuracy of EBV for reproductive traits on the order of 8% and significant reranking of both sires and dams when commercial crossbred data was included in the evaluation.[4] These results indicate a need to make nucleus selections utilizing a genetic evaluation system which includes reproductive records from pedigreed females at the commercial level.

Materials and methods

A data set including 30,355 purebred Yorkshire and Landrace litter records from 2010-2015 from the National Swine Registry's (NSR) Swine Testing and Genetic Evaluation System (STAGES™), 20,304 litter records from purebred Yorkshire and Landrace females producing F1 litters, and 31,915 litter records from pedigreed commercial parent-stock females was utilized to estimate variance components for a combined purebred and crossbred maternal genetic evaluation.[5] Purebred records included were Number Born Alive (P_NBA), Number Weaned (P_NW), and Litter Weaning Weight (P_LWT). Each of these traits were pre-adjusted according to NSR breed specific guidelines (P_NBA for parity and age at breeding; P_NW for parity, age at breeding, and number after transfer; P_LWT for age at weaning, parity, and number after transfer). The same traits were utilized from purebred sows farrowing pedigreed F1 litters; these traits were pre-adjusted as well. Commercial records utilized were Number Born Alive (C_NBA), Number Weaned (C_NW), and Litter Weaning Weight (C_LWT) and were unadjusted.

All traits were evaluated in a six-trait evaluation using REMLF90. Breed of sow (Yorkshire, Landrace, or F1) and breed of litter (Yorkshire, Landrace, F1, or Commercial) were included as fixed effects in the model. Contemporary group was defined as litters born in the same month in the same herd and included as a fixed effect for all traits. Parity, age at breeding, number after transfer and age at weaning were included in the six-trait evaluation as needed for the unadjusted commercial records. Service sire was included as a random effect for number born alive. Table 1 shows the summary statistics for the dataset used in this analysis.

Genetic parameters estimated from the study data were implemented into a daily genetic evaluation program for maternal line genetic improvement using BLUPF90. A Bio-economic selection index was calculated for maternal lines using EBV from the combined six-trait evaluation and economic values for traits from the NSR genetic evaluation system (STAGES™). Results from a sub-set of animals (one NSR firm with the largest number

Table 1: Summary statistics

Litter breed	Litter records	NBA[1]	NW[1]	LWT, kg[1]
Landrace	11,199	11.28	11.37	66.3
Yorkshire	19,156	11.25	11.80	64.4
F1	20,304	11.59	11.66	69.2
Commercial	31,915	11.92	10.77	71.5

[1] Number born alive, number weaned, litter weaning weight raw means - pre-adjusted for pure and F1 litters, unadjusted for commercial litters

of records in the commercial crossbred dataset; 2,363 Yorkshire boars and 8,479 gilts; 1,797 Landrace boars and 5,951 gilts) were further examined for changes in sire and dam ranking and differences in mean commercial index value when ranked on either the purebred or combined purebred and crossbred index.

Results and discussion

Table 2 shows heritability and genetic correlations from the six-trait analysis for purebred and crossbred reproductive traits. Results indicate heritability is low for each trait, being consistent with previous results.[6] The genetic correlation between purebred and crossbred traits is less than 1, indicating selection based on an index that includes only purebred data will not maximize the rate of genetic improvement in number born alive, number weaned, and litter weaning weight at the commercial level. Heritability for C_NW and C_LWT were lower than their purebred counterparts, possibly due to differences in on-farm management between nucleus and commercial farms. Number born alive, whether purebred or crossbred showed similar heritability; indicating number born alive may be a straightforward measure and counted the same way across herds. Interestingly, the genetic correlation of C_NBA with C_NW and C_LWT were negative. This, combined with the heritability near zero (0.03) for C_NWN, may be an indication of management differences between nucleus and commercial farms, where litter size is generally standardized to allow each female to nurse an approximately equal number of pigs. Cross-fostering, or the (lack of) record of cross-fostering at the commercial level, may also contribute to these observed differences.

Table 3 shows the number of sires represented and daughter records per sire utilized in the daily genetic evaluation process. Sires with daughters producing purebred and F1 litters average approximately 25 and 30 records, respectively. For sires with commercial F1 daughters in production, the number increases substantially to ~175. Additional daughter records in the form of commercial females add significant accuracy to Breeding Value estimation, further improving genetic gain.[7] Increasing the number of half-sib progeny from 20 to 100 improves accuracy of a sires' EBV from 0.45 to 0.75 for lowly heritable traits (0.05), such as number born alive, number weaned, and litter weaning weight.[8] In broilers, it has been shown that genetic gain is greater when more crossbred progeny are tested, regardless of varying family structures tested, due to the improved accuracy of selection.[2]

Of the 9,311 sires with purebred daughter records, 1,311 have records from purebred daughters farrowing F1 litters, and 194 have records from F1 daughters. Fewer boars are represented at the commercial level, due to the lack of concern over inbreeding typically accounted for in nucleus boar selection and use and the fact not all NSR members have customers with pedigreed F1 females.

In the subset of data from one NSR firm that accounted for > 90% of the commercial crossbred records, significant reranking of both sows and boars occurred when commercial data was included. Significant reranking did occur across breeds and genders, with Spearman rank correlation coefficients ranging from 0.77 to 0.85. Results are consistent with previous work utilizing purebred and crossbred swine reproductive data.[4] We also see a marked difference in selection index value between the top 5% boars or 25% gilts when ranked using the purebred versus combined purebred crossbred index. Differences in selection differential between the two ranking methods for boars are approximately 3 units for Yorkshires and 2 units for Landrace. With a value of $1.25

Table 2: Heritability and genetic correlations

		Pure			Cross		
	Trait	Number born alive	Number weaned	Litter weaning wt.	Number born alive	Number weaned	Litter weaning wt.
Pure	Number born alive	0.10	0.29	0.18	0.85	0.04	0.08
	Number weaned		0.06	0.53	0.15	0.56	0.38
	Litter weaning wt.			0.12	-0.11	0.47	0.86
Cross	Number born alive				0.10	-0.14	-0.25
	Number weaned					0.03	0.59
	Litter weaning wt.						0.06

[1] Heritability on diagonal; Genetic correlation above diagonal

Table 3: Summary statistics from daily genetic evaluation

Litter breed	Sires	Average[1]	Minimum[1]	Maximum[1]	Total records
Purebred	9,311	23.6	1	1,690	219,740
F1	1,486	30.0	1	1,759	44,580
Commercial	327	175.6	1	4,201	57,421

[1] Number of daughter litter records per sire

Table 4: Differences in maternal selection index value and rank correlations when selecting purebred boars and gilts incorporating either purebred or purebred and crossbred combined data.

Breed	Gender	Number[a]	Percentage[b]	Index1[c]	Index2[d]	Rank Correlation[e]
Yorkshire	Boars	2,363	5%	121.00	117.96	0.77
Yorkshire	Gilts	8,479	25%	117.33	116.10	0.83
Landrace	Boars	1,797	5%	121.85	119.95	0.81
Landrace	Gilts	5,951	25%	116.61	115.60	0.85

[a] Number of animals in the selection candidate pool
[b] Percentage selected
[c] Mean commercial index value of selected animals when ranked by commercial index
[d] Mean commercial index value of selected animals when ranked by purebred index
[e] Rank correlation coefficient between purebred and crossbred selection index value

per commercial daughter litter, per index unit, the value of this difference may not appear to be significant. With an average of 175 F1 daughter records per boar in this data set, the value would be $656.25 per Yorkshire sire and $437.50 per Landrace sire. However, assuming a normal working life in a boar stud, a purebred maternal boar could impact closer to 20,000 F1 commercial litters, indicating the value of the selection differential would be $75,000 for a Yorkshire sire and $50,000 for a Landrace sire. Differences in selection differential for the sows is roughly half the difference seen in the boars, more than likely due to the selection of five times as many sows compared to boars.

These analyses demonstrate maternal genetic evaluation systems should be updated to include pedigreed commercial female records to enhance selection programs for reproductive traits. The added expense and workload to manage and track multiplication and commercial performance should be off-set by the added improvement in reproduction at the commercial level.

References

1. McClaren, D.G., D.S. Buchanan, and R.L. Hintz. 1985. Sire ranking based upon purebred versus crossbred progeny performance in swine. J. Anim. Sci. 60:902.

2. Jiang, X., and A.F.Groen. 1999. Combined crossbred and purebred selection for reproduction traits in a broiler dam line. Journal of Anim. Breed. & Genetics. 116:2, pages 111-125.

3. Spilke, J.E., E. Groeneveld, and N. Mielenz. 1998. Joint purebred and crossbred (co)variance component estimation with pseudo multiple trait model: Loss in efficiency. J. Anim. Breed. Genet. 115:341.

4. Ehlers, M.J., J.W. Mabry, J.K. Bertrand, and K.J. Stalder. 2006. Breeding values, accuracies, and rank changes using purebred and combined purebred and crossbred information in swine. Prof. Anim. Sci. 22:183.

5. National Swine Registry. Swine Testing and Genetic Evaluation System. Available at: `http://nationalswine.com/pedigree_services/stages.php`. Accessed 1 November 2015.

6. Rothschild, M.F., and J.P. Bidanel. 1998. Biology and genetics of reproduction. Pages 313-344 in The Genetics of the Pig. M.F. Rothschild and A. Ruvinsky, ed. CAB International, Wallingford, Oxon, UK.

7. Bijma, P., and J.A.M Van Arendonk. 1998. Maximizing genetic gain for the sire line of a crossbreeding scheme utilizing both purebred and crossbred information. Anim. Sci. 66:529.

8. Bourdon, R.M. 1997. Genetic prediction. Pages 212-241 in Understanding Animal Breeding. Prentice-Hall, Inc. Upper Saddle River, New Jersey.

Porcilis® Ileitis: 20-week duration of immunity against *Lawsonia intracellularis* challenge

Frank Roerink, MS; Chandra Morgan, BS; Susan Knetter, MS, PhD;
Brad Thacker, DVM, PhD, DABVP-SHM; Erin Strait, DVM, PhD
Merck Animal Health, Desoto, Kansas

Introduction and objectives

Lawsonia intracellularis is the causative agent of porcine proliferative enteropathy, also referred to as ileitis. Despite currently available therapies, it remains an important enteric pathogen of growing pigs in today's swine industry. *Lawsonia intracellularis* is an obligate intracellular bacterium that infects epithelial cells of the intestine. The primary pathological change caused by this infection is an increased proliferation of immature epithelial cells of the intestinal crypts. Both acute and chronic forms of the disease exist and clinical signs can be mild to severe, and may include hemorrhage. While the cost of the disease varies widely based on factors such as concurrent diseases, and sanitation and drying between group of pigs, it is estimated to be between $2.73 and $19.79 per pig in the US annually.[1] The objective of this study was to demonstrate the efficacy of Porcilis Ileitis, a one-dose, ready to use, injectable vaccine, against virulent challenge five months after vaccination of three-week-old pigs.

Materials and methods

Eighty-five mixed-breed pigs, with low (≤ 160) *L. intracellularis* immunofluorescent antibody (IFA) titers at the time of vaccination, were utilized. The pigs were allowed to acclimate to their surroundings for three days prior to vaccination. Pigs were allocated to equalized treatments within each pen and were distributed so both vaccinates and placebo-treated controls were represented within each litter. At 22-26 days of age (24 days median), 40 pigs were given a single 2 mL Porcilis Ileitis vaccination intramuscularly in the neck. Another 40 pigs were injected with 2 mL of a placebo containing all the same components as the vaccine, but without *L. intracellularis* antigen. Four additional pigs served as unvaccinated sentinels. Pigs were individually observed and palpated for seven days following vaccination for adverse systemic events or injection site reactions.

Pigs were housed such that each pen contained approximately the same number of vaccinates and controls throughout the entire study. At the time of *L. intracellularis* challenge, which was at 23 weeks of age, the sentinel pigs were necropsied and evaluated for any evidence of *L. intracellularis* exposure prior to challenge. The remaining pigs were individually orally challenged with gut homogenate from pigs infected with a virulent field strain of *L. intracellularis*. Each animal was observed daily for clinical signs of ileitis following challenge.

A subset of 24 vaccinates and 25 placebo control pigs were necropsied at 21 days post challenge (dpc). Their ileal tissues were evaluated for gross ileitis lesion severity (scored 0-4) and length of lesion. Ileal mucosal scrapings were performed to determine the level of *L. intracellularis* in intestinal epithelial cells as measured by quantitative PCR (qPCR). Formalin-fixed tissues were submitted to an accredited veterinary diagnostic laboratory for evaluation of *L. intracellularis* antigen level by immunohistochemistry (IHC, scored 0-3) and for histopathological examination to evaluate microscopic lesions, primarily enterocyte proliferation, consistent with *L. intracellularis* infection (scored 0-2).

Another subset of 14 vaccinates and 15 placebo control pigs were fecal monitored for *L. intracellularis* shedding by qPCR at 0, 3, 6, 9, 12 and 15 weeks after vaccination, along with three days prior to challenge and then three times weekly up to 49 days post challenge (dpc) and at 56 dpc with the study concluded.

Serum samples were collected at vaccination, every 2-3 weeks thereafter and 3 days prior to challenge from all pigs, at 21 dpc (necropsied pigs only) and 56 dpc (fecal monitored pigs only) to measure antibody responses by IFA following vaccination and *L. intracellularis* challenge. Two-fold dilutions ranging from 1:40 to 1:5120 were performed.

Results and discussion

No systemic or injection site adverse reactions were observed in any pigs. Three pigs were removed from the study due to non-vaccine related reasons including one sentinel pig, and two vaccinates. No evidence of *L. intracellularis* was noted in sentinel pigs, up to the time of challenge, as evidenced by a lack of gross and microscopic lesions and the absence of *L. intracellularis* based on IHC and qPCR. The following results show that Porcilis Ileitis was found to be effective for 20 weeks following a single 2 mL vaccination.

By three weeks post vaccination, maternal antibody titers from pigs in the placebo group declined to less than 40 and remained there until challenge. By three weeks post vaccination, the average IFA titers in the vaccinate group demonstrated a significant increase from 43 to 1872. At challenge, the average IFA titers in the vaccinated group was 1250 while all controls remained seronegative (< 40). Following challenge, vaccinates demonstrated significantly higher IFA titers versus the control group, apparently due to an anamnestic response.

A pig was defined as having ileitis if both the gross lesion score was greater than one and the histopathological score was greater than zero. Three of the 24 vaccinates met both criteria (12.5%), while 16 of 25 control pigs (64.0%) met both criteria for ileitis diagnosis. This represents an 80.5% reduction of disease.

Pigs that presented with an IHC score greater than zero or a positive qPCR on ileal mucosal scrapings were classified as colonized by *L. intracellularis*. For this combined variable, 13 of 24 vaccinates (54.2%) were found to be colonized with *L. intracellularis*, while 23 of 25 control pigs (92%) were colonized, representing a 41.1% reduction. All pigs that were classified as colonized were qPCR positive. However, the rate of IHC-positive vaccinated pigs was only 8.3%, while the rate in control pigs was 76.0%, representing an 89.0% reduction. In addition, vaccination reduced the amount of *L. intracellularis* in mucosal scrapings by 68.4%. Overall, the vaccine significantly reduced the ability of *L. intracellularis* to colonize the small intestine.

Shedding of *L. intracellularis* after challenge was evaluated by qPCR on fecal samples, which were collected starting two days after challenge and continued three times per week until seven weeks post challenge. All fecal-monitored pigs were *L. intracellularis* negative by qPCR until after challenge. Both groups began shedding by four days post challenge. The average duration of shedding was 7.9 days for vaccinated pigs and 18.7 days for control pigs. Vaccination reduced the duration of shedding by 57.8%. In the control group, shedding of *L. intracellularis* continued up to 42 days post challenge, while vaccinates stopped shedding after 28 days post challenge. The average daily amount of *L. intracellularis* in the feces, as measured by qPCR, was reduced by 68.7% in vaccinates versus controls during the seven week post-challenge period.

This study demonstrates that Porcilis Ileitis was effective at reducing the ability of *L. intracellularis* to cause gross and microscopic lesions in the small intestine as well as decreasing replication of the organism at the site of infection. Serum antibody titers were detectable within three weeks of vaccination and a pronounced anamnestic response after challenge was observed, demonstrating a strong systemic immune response. More importantly, the level of immunity in the gut also appears to be strong based on the ability of the vaccine to reduce lesions, colonization and shedding following a virulent, experimental challenge. This strongly protective immunity was shown to last for at least 5 months after vaccination, clearly demonstrating the efficacy of Porcilis Ileitis against disease caused by *L. intracellularis.*

References

1.Veenhuizen MF et al. The potential economic impact of porcine proliferative enteropathy on the US swine industry. IPVS. 1998; 64.

Estimation of the economic value of the productivity differences following a change in PCV2 and *Mycoplasma hyopneumoniae* vaccines and PRRSV control efforts using historical production data

Derald Holtkamp[1], MS, DVM; Brad Thacker[2], DVM, PhD; Jack Creel[2], DVM
[1]Department of Veterinary Diagnostic and Production Animal Medicine, Iowa State University, Ames, Iowa;
[2]Merck Animal Health, DeSoto, Kansas

Introduction

When an animal health intervention, such as the addition of a vaccine or a vaccine change, is executed for a farm or production system, producers and veterinarians frequently desire to determine if the intervention was economically beneficial. Answering this question involves a two-step process. First, the difference in the values of key productivity indicators (KPIs) that are attributable to the intervention must be estimated. Second, the differences in the KPIs attributable to the intervention must be converted into economic terms. The most difficult step is the first. Historical production data may be used as a source of the estimated differences in the KPIs. The date the intervention was initiated separates the historical data into pre and post-intervention periods. Of interest are changes in the KPIs after the intervention was made. The goal is to estimate causality between the intervention and the KPI outcomes. This can be challenging because there are always confounding factors that may provide alternative explanations for any apparent causal association between the intervention and the KPI outcomes. For example, an apparent improvement in productivity associated with a change in vaccine in growing pigs might have been due, in whole or part, to the change in status of diseases other than the one targeted by the vaccine, seasonality, labor or management changes, changes in market weights, days on feed, diets or genetics. Furthermore, many diseases present as complexes, such as the porcine respiratory disease complex, where control of one pathogen indirectly improves control of other pathogens.

When historical data are used retrospectively to estimate causality between the intervention and the KPI outcomes a quasi-experimental design, sometimes called the pre-post–intervention design, is used. The quasi-experimental design is a non-randomized study design which makes it prone to confounding factors. For example, consider the weight of pigs at placement. In a randomized controlled study, if weight at placement may be associated with the outcome, every effort will be made to make sure there was not any systematic allocation of pigs to study groups that results in one treatment group of pigs that are heavier than another at placement. In non-randomized studies the opportunity to deal with confounding variables in the study design is absent but confounding variables may be included in the statistical analysis.

The goal is to explain as much of the differences in KPIs between the pre and post intervention periods that is not attributable to the intervention as possible. If the confounding effects are adequately represented in the statistical analysis, the remaining differences in the KPIs, adjusted for the other confounding effects, between the pre and post intervention periods may be more confidently attributed to the intervention.

Estimated differences in the KPI attributable to the intervention may then be converted into economic terms using an enterprise budgeting model with standard market pig and input prices and costs for before and after the change in vaccine.

The objective of this analysis was to use a retrospective quasi-experimental study design with historical production data, and an enterprise budgeting model to estimate the economic value of the productivity differences attributed to a change in porcine circovirus type 2 (PCV2) and *Mycoplasma hyopneumoniae* (M hyo) vaccines in growing pigs following efforts to control porcine reproductive and respiratory syndrome virus (PRRSV) in the breeding and nursery herds.

Materials and methods

The analysis was performed for growing pigs from an 8,000 sow production system in the Midwest USA. The production system is a 3-site system with breeding, nursery and finishing on separate sites. The labor and management for the breeding herds, nurseries and finishers were very stable for the study period with employee turnover less than 10 percent. The three breeding herd sites were in American Association of Swine Veterinarians (AASV) PRRSV category[1] I (Positive Unstable) or II (Positive Stable) for the duration of this study. All of the breeding herds were Positive Unstable (I) prior to the spring of 2013 when efforts to stabilize the herds for PRRSV were initiated. By June of 2013 all of the herds were Positive Stable (II). In late summer to early fall of 2013, following stabilization of the breeding herds, all of the continuous flow nurseries were depopulated and repopulated with pigs that were negative for PRRSV. In late fall of 2013, weaned pigs from one of the breeding herds were again positive for PRRSV by PCR. The continuous flow nurseries receiving pigs from that breeding herd subsequently were positive for PRRSV as well.

All breeding herds also experienced outbreaks of porcine epidemic diarrhea virus (PEDV) in November and December of 2013.

Prior to February of 2014 pigs were vaccinated for PCV2 and M hyo with a single, 2 ml dose of a combination vaccine (Vaccine A) at weaning, approximately 3 weeks of age. For a short period of time, a second dose of M hyo vaccine was administered 3 weeks later. In February of 2014, in response to higher than expected finishing mortality rates that persisted following the PRRSV control efforts, the vaccine was changed to a 2 dose regimen, 1 ml per dose, of Circumvent® PCV-M G2 (PCVM-G2) (Merck Animal Health, Summit, NJ), administered at 3 and 6 weeks of age. Finishing groups started in April of 2014, after a 7 to 8 week nursery phase, were the first to be vaccinated with PCVM-G2. The last finishing groups vaccinated with PCVM-G2 that closed by December 31, 2014, the last day for which data was available, were started in August of 2014. Therefore, data for finishing groups vaccinated with PCVM-G2 were only available for finishing groups started between April and August of 2014. Graphing the KPIs by month started, suggested that seasonality was a feature for most of the KPIs. To deal with the potentially confounding effect of seasonality, only finishing groups started between April and August of each year were included in the analysis. Because the production system began using MetaFarms® (MetaFarms Inc., Burnsville, Minnesota) in August of 2010, very few groups of finishing pigs were eligible in 2010. Therefore, only groups started in 2011 through 2014 were included. Data on groups of finishing pigs were extracted from the Closeout Summary Report in the Finishing Manager® module of MetaFarms®. Nursery closeout data was also extracted from the same report.

To explore potential confounders and to identify outliers, frequency distributions and box and whisker plots were generated by year for the average weight of pigs at placement, average weight at market and days on feed. Frequency distributions and box and whisker plots were also generated for the KPIs evaluated including average daily gain (ADG), feed conversion (FCR), mortality rate and cull rate. Two treatment groups (TrtGrp) were defined based on when each group of pigs was started. The first treatment group included all groups of pigs that were started from 2011 to 2013 and vaccinated with Vaccine A. The second study group included groups of pigs started in 2014 and vaccinated with PCVM-G2.

The KPIs were analyzed as response variables in linear regression models using standard statistical software (JMP® version Pro 12.0.1, SAS Institute Inc, Cary, North Carolina). Explanatory variables included TrtGrp and potential confounding variables. The resulting least squares means by TrtGrp represent means for each treatment group adjusted for the other explanatory variables included in the statistical model.

The economic simulator in the ResPig® Management System (**www.respig.com**; accessed October 30, 2015) is an enterprise budgeting model that estimates costs, revenues and profit for a period of time or group(s) of pigs. For this analysis, the model was set up to evaluate two scenarios, one for the period of time Vaccine A was used and the other for the period PCVM-G2 was used. Standard parameter values for facilities and pig flow; input prices and costs; and market pig prices were used so that the only differences in profit between the two scenarios was due to differences in the cost of the vaccines and in the KPI outcomes. For each scenario, the least square means of each KPI from the regression analysis were used in the economic simulator.

Results

There were 240 groups of finishing pigs that were started in April through August from 2011 through 2014. Box and whisker plots along with qualitative judgments about valid ranges of the variables for the production system were used to identify outlier finishing groups. Fourteen groups of pigs were removed because they had one or more variable with a KPI value outside the ranges shown in Table 1 with the remaining 226 included in the analysis. A summary of the number of groups, weight at placement, days on feed and market weight by year is included in Table 2. Compared to 2011 through 2013, the weight at placement, days on feed in the finisher and market weights were all higher in 2014.

A summary of nursery closeout data is included in Table 3. The values reported in Table 3 are for nursery groups that started in February through October to align with groups that received PCVM-G2 in 2014. While the days on feed in the nursery was higher in 2014, it was only partially responsible for the heavier weights at the end of the nursery phase. Average daily gain in the nursery was also. higher and mortality lower in 2014 compared to the 2011 through 2013 time period. The weights at the end of the nursery phase were a function of both the increased days on feed in the nursery and the improved productivity. Although the PCV2 vaccine was administered to pigs in the nursery phase it likely had very little impact on productivity in the nursery phase. Therefore, the results in the nursery suggest that other factors, including the PRRSV control efforts, likely impacted the productivity of pigs in the finisher independent of the vaccine change that occurred earlier in the year. Based on the preliminary summary of the closeout data in the nursery and finisher, the weight of pigs at placement in the finisher and days on feed in the finisher were both considered potentially important confounding variables for the analysis. Results from the regression analysis of finishing group KPI data are presented in Table 4. Both raw means and least square means are reported. The least square means are the within-group means adjusted for the weight of pigs at placement and days on feed in the finisher. Compared to the differences in the raw means between the treatment groups, the differences in the least square means are less for the cull rate and mortality but greater for ADG and FCR. All of the differences between Vaccine A compared to PCVM-G2 were significant at $P < 0.05$ except for the cull rate ($P = 0.087$). The weight of pigs at placement in the finisher was significant ($P < 0.05$) for all of the KPIs except for ADG ($P = 0.52$). Days on feed in the finisher were significant ($P < 0.05$) for all of the KPIs.

Table 1: Cutoff values for excluding finishing groups as outliers.

Finishing	Weight at placement (lbs)	Mortality rate	ADG (lb/d)	FCR	Market weight (lb)	Days on feed
Less than	30	0%	1.4	2.2	205	100
Greater than	80	25%	2.4	3.8	999	155

Table 2: Summary of finishing closeouts by year.

Year	Number of groups	Weight at placement (lbs)	Days on feed	Market weight (lbs)
2011	63	47.29	124	277.07
2012	62	45.81	128	257.62
2013	52	46.72	132	268.64
2014	49	53.36	131	295.74

Table 3: Summary of nursery closeouts by year.

Year	Number of groups	Weight at placement	Nursery mortality	Days on feed	Weight at end of nursery
2011	85	12.01	7.1%	47	47.21
2012	112	12.00	10.6%	47	46.76
2013	85	11.95	11.2%	49	49.97
2014	104	12.03	6.5%	50	54.38

Table 4; Summary of raw means and least squares means (LSM) from regression analysis of key productivity indicators in finishing.

	Mortality rate		Cull rate		ADG		FCR	
	Raw mean	LSM*	Raw mean	LSM	Raw mean	LSM*	Raw mean	LSM*
Vaccine A (2011-2013)	8.06%	8.03%	5.13%	4.80%	1.69	1.68	2.962	2.987
Circumvent PCV-M G2 (2014)	5.39%	5.49%	3.71%	4.90%	1.83	1.86	2.887	2.795

* Values are significantly different at $P < 0.05$

To determine the relative value of the productivity differences in Table 4, the economic simulator in the ResPig® Management System was used with standard input prices and costs; and market pig prices. The model was set up to evaluate two scenarios, one for Vaccine A and the other for PCVM-G2. The model evaluated profitability on an annual basis assuming 3,500 pigs placed per week, since this was approximately the average number that were placed in finishing during the study period. A fixed number of days on feed was modeled and the average market weight varied according to the weight of pigs at placement and ADG. Premiums and discounts for carcass weight were not considered. Therefore, the average market price for pigs sold in the primary market did not vary as the average market weight changed. Standard parameter values for facilities and pig flow; input prices and costs; and market pig prices (Table 5) were used so that the only difference in profit between the two scenarios was due to differences in the cost of the vaccines and in the cull rate, mortality rate, ADG and FCR. The least squares means of each of the KPIs in Table 4 for each scenario were used in the economic simulator. The difference in the least squares means for the cull rate were not statistically significant ($P > 0.05$), but were included nevertheless. *P*-values greater than 0.05 do not necessarily mean that there is not a true difference between the scenarios, only that there is a greater chance that there is not a true difference. In other words,

Table 5: Standard parameter values used in the economic simulator in ResPig®.

Parameter	Standard value in model	Source
Target days on feed, finisher	127	Closeout data; average for all closeouts included in analysis
Pigs placed per week	3,500	Closeout data; average for all closeouts included in analysis
Finisher spaces available	70,000	Calculated as function of pigs placed per week and target days on feed
Average weight at placement (pounds/pig)	49.21	Closeout data; average for all closeouts included in analysis
Carcass yield (%)	75%	Industry standard
Price of feeder pig ($/pig placed)	$69.08	United Stated Department of Agriculture (USDA). Weekly average, 2011 to 2014, 40 pound pig
Average finisher diet price (US$/ton of feed)	$242.72	Industry standard finishing diets. Corn, soybean meal (SBM) and distillers dried grains (DDG) from USDA. Monthly average prices from 2011 to 2014. Corn - Yellow #2, North Central Iowa; SBM - 46.5 to 48% protein, Decatur-Central Illinois, IL; DDG - 10% protein, Central Illinois, IL
Market pig price ($/pound carcass wt.)	$0.93	United Stated Department of Agriculture (USDA). Monthly average, 2011 to 2014, Iowa/Southern Minnesota barrow and gilt negotiated purchases on carcass basis
Average price of cull pigs ($/pig)	$50.00	Rough estimate based on market pig prices from 2011 to 2014

we were less confident that there was a true difference and whether to include the difference in the economic analysis was a judgment call. The direct cost of the vaccines was also included. For Vaccine A the cost was $1.50 per pig placed in the finisher and $1.65 for PCVM-G2. Labor costs of $0.04 per pig for Vaccine A and $0.08 per pig for PCVM-G2 were also included.

The individual contribution of the change in each of the key productivity indicators after the switch to PCVM-G2 is summarized in Table 6. The largest contribution to the increase in profitability was $5.54 per pig marketed from the 0.18 increase in the least squares means estimate of ADG. Results from the economic analysis of the changes in all of the key productivity indictors combined are in Table 7. The value of the improved productivity metrics, measured by the increase in profit, after the switch to PCVM-G2 was $14.96 per pig marketed.

Conclusions

This study demonstrates the use of historical production data, and an enterprise budgeting model to retrospectively estimate the economic value of the productivity differences following a change in porcine circovirus type 2 (PCV2) and *Mycoplasma hyopneumoniae* (M hyo) vaccines in growing pigs from a 1 dose regimen of Vaccine A at 3 weeks of age to a 2 dose regimen of Circumvent® PCV-M G2 at 3 and 6 weeks of age following efforts to control porcine reproductive and respiratory syndrome virus (PRRSV). The approach used addresses the challenge of isolating changes in KPI outcomes attributed to the vaccine change from those attributed to other causes like the improvements in the overall health of the herd due to efforts to control PRRSV made prior to the vaccine change. The weight of pigs at placement was intended to serve as a proxy for the improvements in the health of the herd following the PRRSV control efforts, however, because of the short duration of time, less than one year, for which data were available following the vaccine change the value of productivity improvements in 2014 should conservatively be attributed to the combined effect of the vaccine change and PRRSV control efforts. The largest contribution to the increase in profitability was the 0.18 increase in ADG which was valued at $5.54 per pig marketed. However, this value depended on two conditions. The first was that there were a fixed number of days on feed so that the average market weight varied according to the ADG. The average live market weight increased from 262 before the vaccine change to 285 pounds after. The second was that the average market price for pigs sold did not vary as the average market weight changed. The value of increasing average daily gain, however, can vary greatly and depends on how facilities are owned, leased or contracted as well as the amount of space available. The value of increasing average daily gain is greatest when pig space, which dictates how long groups of pigs can remain on feed before the

Table 6: Results of economic analysis with the economic simulator in ResPig®, contribution of each key productivity indicator (KPI). Values reported are changes from Vaccine A for each KPI.

	Mortality rate	Cull rate	ADG	FCR
Finisher mortality rate (% of pigs placed)	-2.5%			
Finisher cull rate (% of pigs placed)		0.1%		
Finisher average daily gain (pounds/day)			0.18	
Finisher feed to gain ratio (pounds of feed/pound of gain)				-0.192
Number of standard weight pigs marketed per year (pigs/year)	4,623	-182	0	0
Average live weight at market (pounds/pig)	0.00	0.00	22.81	0.00
Profit ($/pig marketed)	**$3.65**	**-$0.32**	**$5.54**	**$4.83**

Table 7: Results of economic analysis with the economic simulator in ResPig®, all key productivity indicators.

	Vaccine A	Circumvent PCV-M G2	Difference	Percentage difference
Number of standard weight pigs marketed per year (pigs/year)	158,649	163,090	4,441	2.8%
Average live weight at market (pounds/pig)	262.12	284.93	22.81	8.7%
Total annual revenue ($/pig marketed)	$176.14	$191.29	$15.15	8.6%
Total annual feeder pig cost ($/pig marketed)	$75.11	$73.09	-$2.02	-2.7%
Total annual feed cost ($/pig marketed)	$78.53	$80.21	$1.67	2.1%
Total annual non-feed variable cost ($/pig marketed)	$10.02	$10.78	$0.76	7.5%
Total annual fixed cost ($/pig marketed)	$14.14	$13.76	-$0.38	-2.7%
Total annual cost of production $/pig marketed)	$177.81	$177.84	$0.03	0.0%
Profit ($/pig marketed)	-$3.44	$11.52	**$14.96**	-434.4%

next group forces them out the door, is very limited. When space is limited, a larger percentage of pigs in every group are marketed at less than profit maximizing weights. In the production system evaluated, all of the facilities were owned, space was limited and pigs were frequently marketed at less than profit maximizing weights when ADG was low. The discounts for heavier weight carcasses from the packer to which pigs were shipped were relatively low limiting the impact of heavier market weights on the average market price. Therefore, the conditions under which the value of the ADG was valued for this analysis closely represented those of the production system during the study period.

Limitations of the analysis include the short duration of time for which data were available following the vaccine change which made it challenging to isolate the impact of the vaccine change from the improvements in the overall health of the herd attributed to the PRRSV control efforts made just prior to the vaccine change. In addition, the possibility of other confounding factors may have explained, in part, some of the changes in the KPIs. For example, all breeding herd sources of the pigs in this analysis experienced outbreaks of porcine epidemic diarrhea virus (PEDV) in November and December of 2013, just before the change in vaccine was made. While it seems unlikely that this would have contributed to improved productivity in finishing, the timing of this event creates some uncertainty.

Acknowledgements

The authors acknowledge the assistance of Mr. Brian Doyle of MetaFarms for retrieving the data used for this analysis.

References

1. Holtkamp D.J., Polson D.D., Torremorell M. and committee members Morrison B. (chair), Classen D.M., Becton L., Henry S., Rodibaugh M.T., Rowland R.R., Snelson H., Straw B., Yeske P., Zimmerman J. 2011. Terminology for classifying swine herds by PRRS virus status. J. Swine Health Prod. 19:1 44-56.

Circumvent® PCV-M G2: PCV2d challenge study and field production data update

Brad Thacker[1], DVM, PhD, DABVP-SHM; Richard Schlueter[1], DVM; Jack Creel[1], DVM; Darin Madson[2], DVM, PhD, DACVP; Richard Hesse[3], PhD
[1]Merck Animal Health, Desoto, Kansas; [2]Iowa State University Veterinary Diagnostic Laboratory, Ames, Iowa; [3]Kansas State Veterinary Diagnostic Laboratory, Manhattan, Kansas

Introduction and objectives

In early 2014, Merck Animal Health (MAH) introduced a new generation PCV2/*Mycoplasma hyopneumoniae* (M hyo) combination vaccine, Circumvent® PCV M G2 (PCV-M G2) (Merck Animal Health, Summit, NJ) and a monovalent PCV2 only vaccine, Circumvent® PCV G2 (PCV-G2). These vaccines differ from previous Circumvent products with regard to the recommended age at vaccination and dose volume. While previous Circumvent products were recommended for use at 3 and 6 weeks of age (2 mL per injection), PCV-M and PCV-G2 are recommended for as early as 3 days of age followed by a second vaccination 3 weeks later (1 mL per injection) or as a single vaccination (2 mL) at 3 weeks of age or greater. In this paper, we report on a laboratory vaccination/challenge study that demonstrates the effectiveness of PCV-M G2 against the new PCV2d variant, and an ongoing field production data analysis that found similar production performance between PCV-G2 and the previous product Circumvent® PCV (PCV-G1).

Challenge study- Although vaccination for PCV2 has been successful in most swine operations, PCV2 still remains an economically important disease, at least in part due to the high rate of disease in pigs that are not successfully immunized for various reasons. One potential cause for reduction in vaccine efficacy is the emergence of new PCV2 variants. One such variant was first reported in China and apparently spread to the US where it now appears to be the dominant variant found in pigs with clinical PCV2 infection.[1] This "Chinese mutant" variant is now designated as "PCV2d." As new variants are found in clinical cases, the question arises regarding the efficacy of current vaccines for protecting against these emerging variants.[2] In one report, commercial vaccines marketed in the US were found to be effective against experimental PCV2d challenge.[3] However, the experiment was contaminated by another PCV2 variant and one of the commercial vaccines used was the pioneer Circumvent PCV product (PCV-G1), not the newer PCV-G2. Accordingly, the objective of the experimental study reported here was to evaluate the efficacy of PCV-M G2 against a mutant PCV2 infection under laboratory conditions, using the same PRRSV co-infection challenge model used in the past with the exception of a PCV2d rather than PCV2b variant as the challenge virus.

Field production data update- Previously, we reported on the production record analysis of a swine operation located in the Midwest USA.[4,5] Weaned pigs are purchased from two sow farms and nearly all pigs are kept in separate flows from the time of arrival until marketing. One flow consists of terminal line animals where all progeny are sold for slaughter. The other flow consists of maternal line animals where a large portion of the gilts are sold as replacements. Both flows are considered to be of high health status. Nursery and finishing sites contain one to several barns and are managed all-in, all-out by site. The previous reports addressed differences in production performance between the flows, which over time used different PCV2 vaccination programs. Originally, a one dose PCV2 vaccine, CircoFLEX® (Boehringer Ingelheim Vetmedica, Inc., St. Joseph, MO) was used in both flows. The maternal flow was switched to PCV-G1 in response to concerns about viremia in gilts that were sold for breeding purposes. After a period of time, the producer judged that the maternal flow was outperforming the terminal flow and decided to switch the terminal flow to Fostera™ PCV (Zoetis, Florham, NJ) in order to continue using a one dose product in this flow. Over time, the performance of the terminal flow still lagged behind the maternal flow. In mid-2013, the producer switched the terminal flow to PCV-G1. In early 2014, both flows were switched to PCV-G2 soon after the product arrived on the market. The objective of the production data analysis reported here was to compare the performance of PCV-G1 to PCV0-G2 within each flow.

Materials and methods

Challenge study- The source herd was a 3,200 sow inventory, breed-to-wean farm located in the Midwestern USA that was PRRS virus naïve, M hyo free and PCV2 antibody positive. Three to four weeks prior to farrowing, approximately 50 sows were bled and their serum tested by IFA for PCV2 antibody, for PCV2 by PCR, and by ELISA for detecting M hyo antibodies in order to determine PCV2 maternal antibody levels, avoid using pigs from viremic sows and verify the M hyo free status of the herd, respectively. Pigs in eligible litters were tagged, weighed, their gender recorded, allotted to treatment group and vaccinated at 2-5 days of age. Assignment to treatment groups was done within litter such that each treatment group was equally represented within a litter. All pigs were bled at 13-16 days of age and their serum tested for PCV2 viremia by PCR (pools of 5) to verify that the pigs were PCV2 viremia negative prior to shipment to a contract research facility at 17-20 days of age.

With regard to experimental design, four treatment groups were evaluated; the groups are shown in Table 1. Pigs that were injected twice received 1 ml of vaccine each time. The 1 × 2 mL, 3 weeks of age (WOA) group received a single 2 ml dose. After movement, the pigs were bled at approximately 3, 6, 9, 11, 12 and 13 WOA. The pigs were challenged at 9 weeks of age. The PCV2 challenge inoculum, containing a PCV2d tissue homogenate and PRRSV, was provided and administered intra-nasally by Dr. Richard

Table 1: PCV2 IFA results

Treatment group (PCV-M G2)	No. pigs	Group reciprocal geomean IFA titers by week of age			
		3	6	9	13
Control	8	50	47[a]	< 20[a]	5120
2 × 1 mL at 3 DOA & 3 WOA	11	85	1202[b]	1754[b]	5120
1 × 2 mL at 3 WOA	10	70	640[b]	2079[b]	5120
2 × 1 mL at 3 & 6 WOA	10	70	320[b]	4777[c]	5120

[a,b,c] Values with different superscripts within a column were significantly different ($P < 0.05$) by ANOVA.

Hesse of the Kansas State Veterinary Diagnostic Laboratory (KS-VDL), Manhattan, Kansas. Necropsies were performed at 13 weeks of age. At necropsy, fresh and fixed tonsil, mediastinal and bronchial lymph, nodes and lung were collected for assessment of virus levels and microscopic lesions.

For laboratory testing, quantitative PCR testing of serum (individual pigs) and tissue (composite of the four tissues collected from each pig) was done at the Iowa State University Veterinary Diagnostic Laboratory (ISU-VDL). Results are reported as copies per mL and cycle times (CT). Immunochemistry (IHC) testing was done to detect PCV2 antigen at the ISU-VDL and an intensity score (0-3) was applied by Dr. Madson based on previously reported studies.[6] Similarly, microscopic lesion evaluation of tissues was performed by Dr. Madson and lesion severity scores for lymphoid tissues were based on previously reported studies.[6] For this study, the lesion score was calculated by adding together the scores (0-3 each) for lymphoid depletion and granulomatous inflammation. For serologic testing, PCV2 antibodies were measured by IFA at the KS-VDL. IFA titers are reported as the reciprocal of the group geometric mean. The highest dilution tested was 1:5120.

Statistix™ 10 (Analytic Software, Tallahassee, FL) was used to perform all statistical analyses. Continuous variable data are presented as group means and were analyzed by ANOVA . Pair wise comparisons were done by the Least Significant Difference. Binary variable (positive/negative) data are presented as the number positive/number tested.

Field production data update- Vaccination with either Circumvent product was done per label at 3 and 6 weeks of age using 2 mL for each injection with PCV-G1 and at 3 and 6 weeks of age using 1 mL for each injection with PCV-G2. Production data was collected by site as each group was closed-out and the site was emptied prior to the arrival of the next group. The range of group start dates by flow and vaccine used is shown in Table 3. Only finisher data was evaluated as nursery performance was the same in both flows. *Statistix*™ 10 was used to perform all statistical analyses. The data was analyzed on a close-out group basis or an individual pig basis. For the close-out based analyses, the group average daily gain (ADG) and feed conversion rate (FCR), mortality rate and culled rate were tested by ANOVA to identify significant differences between the vaccines PCV-G1 vs. PCV-G2 within each flow.

Analyses based on the individual pig were performed by chi-square to detect differences in the mortality and culled rates between the vaccines/flows.

Results and discussion

Challenge study- PCR testing at the time of challenge revealed PCV2 infection was occurring prior to challenge in 2 out of the 12 litters in the study. The litters were housed by pen with solid pen dividers between pens. Sequencing indicated that this contaminant was a PCV2a variant. Further sequencing of positive samples from all other control pigs at 14 days post challenge indicated that the contamination occurred in 2 additional pens. Accordingly, data generated from these 4 pens (out of 12 total) were excluded from the analysis. The original design was to have 14 or 15 pigs per treatment group. The final number of pigs used is presented in Tables 1 and 2.

IFA results are summarized in Table 1. Titers at 3 WOA were low in all groups. At 6 WOA, all vaccinated groups had higher geomean titers than the control group and were similar to each other although numerically, the group vaccinated twice at 3 days of age (DOA) and 3 WOA had the highest geomean titer. At 9 WOA, the group vaccinated at 3 and 6 WOA had the highest titer and the control pigs were all seronegative. The other two vaccinated groups had intermediate titers. At necropsy (13 WOA), all pigs had titers ≥ 5120.

Viremia, tissue infection and microscopic lesion data are summarized in Table 2. For all parameters, the level or severity of infection and lesions was significantly greater in the control group and there were no differences among the three vaccinated groups. The level of viremia overall in the control group was nearly 1,000 times higher compared to a previous study[7] but similar to another, more recent study.[8] In both previous studies, a PCV2b variant was used for the challenge. The studies differed by the age at inoculation, which was 7 WOA in the study where the viremia level was similar (high) to the current study.

Overall, PCV-M G2 was able to protect pigs against a PCV2d challenge. The percent reduction in viremia and tissue infection levels compared to controls ranged from 98.4% to 99.9%. Likewise, both microscopic lesion and IHC scores were significantly lower in the vaccinated groups compared to the controls.

Table 2: PCV2 viremia levels following challenge and tissue lesion and virus levels at necropsy

Treatment group (PCV-M G2)	No. pigs	Lymph node scores: Lesion (0-6)	Lymph node scores: IHC (0-3)	Tissue virus levels: Copies/ reaction (X 10^6)	Tissue virus levels: % reduced	Total viremia levels: Copies/ reaction (X10^6) (No. pos.)	Total viremia levels: % reduced
Control*	8	2.22	1.67	50,064.10		15,252.77 (8/8)	
2 × 1 mL, 3 DOA & 3 WOA	11	0.36	0.36	792.90	98.4%	11.11 (6/11)	99.9%
1 × 2 mL, 3 WOA	10	0	0	61.47	99.9%	0.26 (4/10)	100.0%
2 × 1 mL, 3 & 6 WOA	10	0.20	0.20	2.15	100.0%	0.03 (3/10)	100.0%

* For all variables, control group values were significantly greater than the three vaccinated groups ($P < 0.05$) by ANOVA.

Table 3: Close-out group summary data by flow and vaccine

Flow	Terminal		Maternal	
1st group start date	7/3/2013	5/1/2014	7/13/2013	5/5/2014
Last group start date	3/14/2014	5/14/2015	3/4/2014	6/12/2015
Vaccine	PCV-G1	PCV-G2	PCV-G1	PCV-G2
No. groups	33	60	25	37
No. head started	68,074	131,463	71,371	103,819
Mortality	2.32%[a]	1.58%[b]	1.77%[a]	1.54%[b]
Culled	1.14%	1.03%	0.68%[a]	0.37%[b]
Daily gain (lb)	2.028	2.072	1.956	1.941
Feed conv. ratio	2.664	2.612	2.682	2.622

[a,b] Values with different superscripts within a row and within each flow were significantly different ($P < 0.05$) by chi-square test.

Field production data update- Production performance data by flow and vaccine used are presented in Table 3. For daily gain and feed conversion ratio, there were no differences between the vaccines used within each flow. In the terminal flow, mortality rate was significantly lower in the PCV-G2 vaccinated pigs. In the maternal flow, the mortality and culled rates were significantly lower in the PCV-G2 vaccinated groups. However, these differences were relatively small and become statistically significant due to the large sample size. Overall, the performance of PCV-G2 appears to be at least equal to, if not slightly better than, PCV-G1.

References

1. Guo, L. J., Lu, Y. H., Wei, Y. W., Huang, L. P., Liu, C. M. 2010. Porcine circovirus type 2 (PCV2): genetic variation and newly emerging genotypes in China. Viro J 7: 273.
2. Xiao, C., Halbur, P. G., Opriessnig, T. 2012. Complete genome sequence of a novel porcine circovirus type 2b variant present in cases of vaccine failures in the United States. J Virol 86 (22): 12469.
3. Opriessnig, T., Gerber, P. F., Xiao, C., Halbur, P. G., Matzinger, S. R., Meng, X. 2014.Commercial PCV2a-based vaccines are effective in protecting naturally PCV2b-infected finisher pigs against experimental challenge with a 2012 mutant PCV2. Vaccine 32: 4342-4348.
4. Thacker, B., Blomme, R., Holtkamp, D., Creel, J. 2014. Field comparison of PCV2 vaccines: A retrospective production data analysis. In: Proc. 45th Annual Meeting of the American Association of Swine Veterinarians, Dallas, Texas, pp. 139-143.
5. Thacker, B., Creel, J., Blomme, R. 2015. PCV2 vaccination update: Field experiences and serological test comparisons. In: Proc. 46th Annual Meeting of the American Association of Swine Veterinarians, Orlando, Florida, pp. 205-208.
6. Opriessnig, T., Thacker, E.L., Yu, S., Fenaux, M., Meng, X.J., Halbur, P.G. 2004. Experimental reproduction of postweaning multisystemic wasting syndrome in pigs by dual infection with *Mycoplasma hyopneumoniae* and porcine circovirus type 2. Vet Pathol 41:624-640.
7. Thacker, B., Kniffen, T., Hesse, R., Madson, D, Strait,, E. 2012 Laboratory studies on Circumvent® PCV M: Combination *Mycoplasma hyopneumoniae* and porcine circovirus type 2 vaccine. In: Proc. 43th Annual Meeting of the American Association of Swine Veterinarians, Denver, Colorado, pp. 123-126.
8. Thacker, B., Schlueter, R. Engle, M., Creel, J., Wilson, W., Madson, D. 2015. Circumvent® PCV M G2: Impact of vaccination timing and interval on PCV2 efficacy and serological responses. In: Proc. 46th Annual Meeting of the American Association of Swine Veterinarians, Orlando, Florida, pp. 210-214.

Effectiveness of avilamycin (Kavault™) for reducing the incidence and severity of colibacillosis in post-weaned pigs

Thomas Marsteller, DVM; Brandon Carter, DVM, MS; Kevin Eggers, DVM; Eric Christianson, DVM; Mark Hammer, DVM; Robert Evelsizer, DVM; Marcos Rostagno, DVM, PhD; Jason Hammelman, DVM, MS
Elanco Animal Health, Greenfield, Indiana

Introduction

Post-weaning colibacillosis is a common enteric nursery disease in swine.[1,2,3] The disease is caused by enterotoxigenic *Escherichia coli* (ETEC). ETEC colonizes the intestine, and upon attachment to enterocytes, produces heat labile (LT) and heat-stable (STa and STb) enterotoxins. These enterotoxins trigger intracellular signaling pathways inducing massive fluid secretion leading to a secretory diarrhea. There are numerous strains of *E. coli* that may cause post-weaning colibacillosis. The most common post-weaning ETEC strains containing fimbrial (pili) adhesins K88 (F4) and F18 are listed in Table 1 below.[3]

Clinical signs of post-weaning colibacillosis are watery diarrhea, dehydration, swollen eyelids, and death loss. The histopathological lesions are typically unrewarding. Weaned pigs will usually experience the onset of disease within the first seven days after weaning; however the disease may last up to 22 days without intervention.[4] Weaned pigs undergo a very stressful period in their life as dietary, environmental and social changes occur very abruptly. The disruption from sow's milk to cereal based diets and nursery environments that lack proper temperature controls and flooring hygiene are common risk factors for young pigs to develop post-weaning colibacillosis. Weaning age is also documented to be an important risk factor. McLamb et. al. demonstrated that early weaning (16 days) versus later weaning (20 days) has a profound impact on subsequent clinical severity and intestinal injury in response to an ETEC challenge.[5] Viral pathogens, such as rotavirus and porcine epidemic diarrhea virus, may predispose nursery pigs to development of post-weaning colibacillosis.[6] Accurate diagnosis by the herd veterinarian is important to identify the primary and secondary causes of any nursery enteric disease prior to interventions.

Avilamycin (Kavault™) is an oligosaccharide in the orthosomycin class that has unique activity on *E. coli*. The mode of action of avilamycin consists of decreasing fimbriae development on *E. coli* bacteria, thus preventing attachment to the enterocyte, and prevention of disease in post-weaned pigs.[7]

A confirmatory study conducted in South Dakota evaluated the effectiveness of avilamycin in newly weaned pigs following oral *E. coli* challenge in an induced post-weaning diarrhea model.[8,9] Pigs were assigned to one of four treatments: non- medicated feed (negative control), a positive control group (carbadox 50 g/ton), or one of two avilamycin feed concentrations (36 or 73 g/ton). Treatment feed administration was initiated on study day 0, and pigs were challenged with hemolytic *E. coli* on study days 4, 5, and 6. Treatment feeds were administered for the duration of the 32-day treatment phase. Each animal was observed daily from study day 0 through study day 32 for diarrhea, perianal inflammation, depression, and appetite scoring. Fecal samples were collected throughout the study for quantitative assessment of hemolytic *E. coli* shedding. Effectiveness was assessed based on relative differences between treatment groups for diarrhea scores to define prevalence, incidence, and severity of disease during the treatment phase. Pigs administered either avilamycin 34 g/ton or avilamycin 73 g/ton had statistically significant reductions in diarrhea

Table 1: Characteristics of *E. coli* strains associated with post-weaning colibacillosis in pigs.

Serogroup	Fimbriae	Toxin(s)	Hemolysins
O8	K88 (F4)	LT, STb ± STa	Yes
O149	K88 (F4)	LT, STb ± STa	Yes
O157	K88 (F4)	LT, STb ± STa	Yes
O138	F18ab, F18ac	STa, STb ± Stx2e	Yes
O139	F18ab	STa, STb ± Stx2e	Yes
O141	F18ac	STa, STb ± Stx2e	Yes
O157	F18ac	STa, STb ± Stx2e	Yes

incidence and severity relative to negative controls ($P < 0.05$). Diarrhea incidence in avilamycin 73 g/ton and positive control treatments were significantly improved relative to avilamycin 34 g/ton ($P < 0.05$). Based on the results of this study, an inclusion rate of 73 g/ton avilamycin in the feed was chosen to evaluate in the clinical field effectiveness study described in this abstract.

Materials and methods

A four site clinical study was completed in 2011 to confirm the following objective: To demonstrate effectiveness of avilamycin premix administered in feed at 73 g/ton for reduction in incidence and overall severity of diarrhea in the presence of pathogenic *Escherichia coli* in groups of newly weaned pigs.[9] The four study sites were completed in Iowa (2), Nebraska, and California locations.

At each study site, eligible study animals had a history of exposure to one or more risk factors associated with post-weaning colibacillosis (e.g., weaning, transport, relocation, commingling, diet change, etc.) and they were determined to be healthy based on individual pig physical examination conducted by the study investigator prior to study enrollment. The 200 eligible pigs (barrows and gilts) at each location were blocked by body weight, and randomly allocated to pens (20 pens/site; 10 pigs/pen/site) on day 0. Pigs were approximately 3 weeks of age at receipt, and weighed approximately 12 lbs. Pigs were sourced from commercial US swine production facilities and were uniquely identified with bilateral, identical ear tags. The pigs were housed in indoor, concrete pens with either solid or slatted flooring representative of industry standards, and mechanical and/or curtain-sided ventilation facilities.

A commercial pelleted feed ration appropriate for the pigs' age and stage of production was provided *ad libitum* throughout the study. Medicated (avilamycin 73 g/ton) and non-medicated treatment feeds were administered on days 0-21. Medicated or non-medicated phase 1 post-weaning rations were issued *ad libitum* on study days 0 through 7. Remaining feed was removed from feeders and weighed on study day 7. Feeders were refilled with medicated or non- medicated phase 2 post-weaning rations. Phase 2 nursery rations were issued *ad libitum* for the remainder of treatment phase (study days 7 through 21). Non-medicated post-treatment feed was administered on days 22-28 to all pigs enrolled in the study. Dietary zinc was included in the diet at nutritional levels, not therapeutic levels. Non-medicated water was provided *ad libitum* for the duration of the study. The animals did not receive any injectable, water soluble medications, or *E. coli* vaccinations prior to or during the study.

All study animals were assigned clinical observation scores on days 0-28 for diarrhea, depression, and gauntness. Diarrhea incidence was calculated as the percentage of animals in the population (number animals enrolled less non-colibacillosis removals) with diarrhea score ≥ 2 (0-3 scale) on one or more study days during the treatment phase (days 0-21). Mean clinical scores for diarrhea severity, depression, and gauntness were calculated as the sum of each score divided by total head days in each treatment group. Diarrhea scores collected were: 0 = Normal feces; no diarrhea present; 1 = Semi-solid (cow-pie consistency); 2 = Loose with some solid material (oatmeal consistency); 3 = Watery feces with little or no solid material. Depression scores collected were: 0 = bright, alert, and responsive; 1 = may stand isolated but will quickly respond to stimulation; 2 = may stand isolated with head down and possible signs of muscle weakness; delayed response to stimulation; 3 = severely depressed; recumbent and reluctant to rise. Gauntness scores collected were: 0 = normal abdominal fill; flank is full and round; 1 = decreased gut fill; flank is flat; 2 = severely gaunt; flank is hollow. For these efficacy variables, heterogeneity of variance among sites was tested using Levene's test (Levene, 1960). Levene's test detected significant heterogeneity of variance among sites for residual component in the analysis of mean depression score ($P < 0.005$). A weighted analysis was conducted for final analysis of mean depression score only.

In order to substantiate the presence of pathogenic *E. coli*, fecal bacterial samples were collected during the study from both treated and untreated groups. Samples were collected from each pig, daily for a seven day period. Since the goal was to confirm the presence of β-hemolytic *E. coli* from each enrolled pig, swabs were streaked only if β-hemolytic *E. coli* had not already been identified from any given pig. Microbiological data was analyzed for prevalence of β-hemolytic *E. coli* at each study site.

Other data of interest collected were: bodyweight (Days 0, 21), average daily feed intake, average daily gain, and gain efficiency. These variables were analyzed using the same linear mixed model as primary efficacy variables. These variables were calculated on a pen basis and summarized in Table 2.

Results and discussion

Diarrhea incidence and mean clinical scores for diarrhea severity, depression, and gauntness were the primary outcome variables to evaluate avilamycin treatment effect on nursery pig colibacillosis for 21 days. Diarrhea incidence rates were 62.44% and 90.78% in avilamycin and negative control treatment groups, respectively ($P < 0.01$). Animals in the avilamycin treatment group also had significantly lower ($P < 0.01$) mean diarrhea severity score (0.43) relative to negative control (0.87). Mean depression scores were 0.12 and 0.18 in avilamycin and negative control treatment groups, respectively ($P < 0.04$). No difference in gauntness score was observed between treatments.

Fecal swab samples were collected from 200 pigs at each site by the four investigators. A total of 3,589 samples were received and 3,569 samples were processed for the isolation of β-hemolytic *E. coli*. The prevalence of pigs positive for β-hemolytic *E. coli* at each study site was: 91%, 69%, 98% and 85%. The

Table 2: Least squares means for the effects of avilamycin inclusion level on incidence and overall severity of diarrhea caused by pathogenic *Escherichia coli* in pens of post-weaned pigs.

	Avilamycin premix inclusion level, g/ton			
Item	**0**	**73**	**SEM**	***P*-value**
No. of pens	40	40	-	-
No. of pigs starting study, day 0	400	400	-	-
No. pig removals, non-*E. coli* reasons[1]	18	19	-	-
No. pigs in efficacy analysis	382	381	-	-
No. pig removals, *E. coli* diagnosis[2]	72	56	-	-
Bodyweight, lb				
Start of 21 day treatment period	12.04	12.05	0.52	-
End of 21 day treatment period	17.88[c]	19.67[d]	2.35	0.03
Diarrhea incidence, %	90.78[a]	62.44[b]	4.00 and 10.73[3]	< 0.01
Mean diarrhea severity score	0.87[a]	0.43[b]	0.01	< 0.01
Mean depression score	0.18[c]	0.12[d]	0.01	0.04
Mean gauntness score	0.25	0.16	0.01	0.10
ADG, lb/d	0.29[c]	0.38[d]	0.08	0.05
ADFI, lb/d	0.51	0.55	0.10	0.09
G:F, lb:lb	0.51	0.57	0.08	0.47

[a,b] Least squares means are significantly different ($P \leq 0.01$).
[c,d] Least squares means are significantly different ($P \leq 0.05$).
[1] Non-*E. coli* No. pig removals and deviations were not included in the efficacy analysis.
[2] *E. coli* No. pig removals, based on *E. coli* diagnosis, were included in the efficacy analysis.
[3] Diarrhea Incidence %, SEM of Avilamycin 0 and 73 g/ton respectively.

study population at all four sites exhibited a diarrhea outbreak typical of post-weaning colibacillosis.

Other variables of interest included body weight on day 21, average daily gain, average daily feed intake, gain efficiency, and feed efficiency. Animals in the avilamycin treatment group had heavier bodyweight on study day 21 ($P < 0.03$) and average daily gain ($P < 0.05$) during the treatment phase relative to negative control animals.

During the treatment phase, 56 animals in the avilamycin treatment group and 72 animals in the control group were removed for humane reasons and the animals were diagnosed with colibacillosis disease. These animals were included in the efficacy analysis. During the treatment phase, 18 animals in the avilamycin treatment group and 19 animals in the control group were removed for non-colibacillosis and protocol deviation reasons. These animals were not included in the colibacillosis efficacy analysis. Clinical scores associated with colibacillosis did not increase when treatment feed was removed for 7 days, during study days 21-28.

These data demonstrate effectiveness of avilamycin administered in feed at 73 g/ton to reduce diarrhea incidence and severity associated with *Escherichia coli* in groups of newly weaned pigs during the post-weaned phase of production. Effective May 2015, avilamycin was approved for use in pigs that are at risk of developing, but not yet showing clinical signs of, diarrhea in the presence of pathogenic *Escherichia coli* and is approved for use in pigs up to 14 weeks of age.[9]

A valid veterinary feed directive (VFD) is required to dispense Kavault™. The feeding directions are to feed at 73 grams avilamycin per ton of Type C medicated feed (80 ppm) as the sole ration for 21 consecutive days. The veterinarian may direct feeding for up to a total of 42 consecutive days, based on the clinical assessment.[9] Any animal feed bearing or containing this drug will be fed to animals only by or on a lawful veterinary feed directive issued by a licensed veterinarian in the course of their professional practice.

References

1. Fairbrother JM, Gyles CL. In: Zimmerman JJ, Karriker LA, Ramirez A, Schwartz KJ, Stevenson GW. Diseases of Swine. 10th Edition, Wiley-Blackwell Publishing, 2012 Colibacillosis. pg 723-749.

2. Gyles CL, Fairbrother JM. In: Gyles CL, Prescott JF, Songer JG, Thoen CO. Pathogenesis of Bacterial Infections in Animals. 4th Edition, Wiley-Blackwell Publishing, 2010. *Escherichia coli*. pg 267-308.

3. Francis DH. Enterotoxigenic *Escherichia coli* infection in pigs and its diagnosis. *J Swine Health Prod.* 2002; 10(4):171:175.

4. Erlandson K. et al. Characterization of nursery pig colibacillosis in a natural infection study. Allen D Leman Swine Conference, Recent Research Reports, Volume 42, 2015 pg 32.

5. McLamb BL, Gibson AJ, Overman EL, Stahl C, Moeser AJ (2013) Early Weaning Stress in Pigs Impairs Innate Mucosal Immune Responses to Enterotoxigenic *E. coli* Challenge and Exacerbates Intestinal Injury and Clinical Disease. PLoS ONE 8(4): e59838. doi:10.1371/journal.pone.0059838.

6. Guggenbiller DJ, et al. Impact of Neo-Terramycin®, Mecadox®, Denagard® plus CTC and Denagard® feeding programs on nursery pigs with PEDV, *Salmonella* and *E. coli*. AASV 2015, pg 135-138.

7. Rostagno M, Pelger G. Exposure to the antibiotic avilamycin inhibits *E. coli* fimbriae and attachment. IPVS 2014 P246, pg 238.

8. Marsteller T. et al. Efficacy of Kavault™ (avilamycin) in pigs challenged with *Escherichia coli* in an induced challenge model. Allen D Leman Swine Conference, Recent Research Reports, Volume 42, 2015 pg 30.

9. Freedom of information summary. Original new animal drug application. Nada 141-438. Kavault. Avilamycin Type A Medicated Article, May 8, 2015.

Effectiveness of tilmicosin phosphate (Pulmotil® AC) for controlling respiratory disease associated with *Mycoplasma hyopneumoniae* in swine in the presence of porcine reproductive and respiratory syndrome virus (PRRSV)

Daniel Ivers[1], MS; Chris Puls[1], PhD; Tom Weber[1], PhD; Daniel Mowrey[1], PhD; Matt Ritter[1], PhD; Keith Baker[1], MS; Lyle Kesl[2], DVM, PhD; Tom Marsteller[1], DVM

[1]Elanco Animal Health, Greenfield, Indiana; [2]Veterinary Resources, Inc., Ames, Iowa

Introduction

Swine respiratory disease (SRD) is associated with multiple bacterial and viral respiratory pathogens. Clinical disease is characterized by decreased rate of growth, decreased feed efficiency, anorexia, fever, dyspnea, and mortality. In the field, co-infections of multiple bacteria (e.g., *Mycoplasma hyopneumoniae*) with or without viral pathogens (e.g., PRRSV) are common and a leading cause of SRD. Experimental models have demonstrated that a single infection with either *Mycoplasma hyopneumoniae* (M hyo) or a viral pathogen usually results in mild or unapparent and transient disease.[1] However, it has also been demonstrated that M hyo infections potentiate the severity and the duration of disease in animals co-infected with porcine reproductive and respiratory syndrome virus (PRRSV) supporting the criticality of controlling M hyo in herds affected with PRRSV.[1] Tilmicosin phosphate (Pulmotil® AC) was approved in the US in Spring 2014 for the control of swine respiratory disease associated with *Pasteurella multocida* and *Haemophilus parasuis* in groups of swine in buildings where a respiratory disease outbreak is diagnosed.[2] Tilmicosin has also been shown to have antimicrobial effects against M hyo in *in vitro* settings.[3] Therefore, this study evaluated the effectiveness of tilmicosin phosphate administered via the drinking water for controlling respiratory disease associated with M hyo in swine in the presence of PRRSV.

Materials and methods

This study was carried out using a randomized complete block design (blocking factor = location within the facility) with 2 tilmicosin inclusion levels (0 vs. 200 ppm) in the drinking water within each of 2 infection groups [Co-Infection (M hyo + PRRSV) and Single-Infection (M hyo only)]. Different facilities were used for each infection group. A total of 362 crossbred serologically M hyo and PRRSV naïve pigs (barrows and gilts) were used on the study. The study included a total of 20 pens for the Co-Infection group and 30 pens for the Single-Infection group with 6 pigs per pen. In each infection group, there were 20 pens that each had 1 additional sentinel pig that was used for determining tilmicosin treatment initiation. Pen was the experimental unit for all measures.

On Day 0, animals allocated to the Co-Infection group were experimentally challenged with 10 mL of *Mycoplasma hyopneumoniae* (10^4 CCU/mL) inoculum and 2 mL of PRRSV inoculum (7.58×10^4 $TCID_{50}$/mL of strain VDL 2385; Iowa State University, Ames, IA). On Day 1, animals allocated to the Single-Infection group were experimentally challenged with 10 mL of M hyo inoculum only (10^4 CCU/mL). Within infection group, five pre-designated sentinel pigs were allocated to each necropsy day, which were scheduled for Days 3, 5, 7 and 10 of the study for the Co-Infection group and for Days 12, 14, 16 and 22 of the study for the Single-Infection group.

For each necropsy day, total lung lesion percentage was calculated for each pig using the following equation:

Total lung lesion, % = Left cranial lobe × 20.0% + Left caudal lobe × 27.5% + Right cranial lobe × 10.0% + Right middle lobe × 10.0% + Right caudal lobe × 27.5% + Accessory lobe × 5.0%.

Tilmicosin treatment for an infection group was initiated if at least 3 of the 5 sentinel pigs had lung lesions on greater than 5% of the lung. The Co-Infection group met criteria on Day 5 necropsy, and the Single-Infection group met criteria on Day 12 necropsy. Remaining sentinel pigs were used as replacements for animals that had died prior to treatment initiation or were euthanized. Each pen had 6 pigs in the pen at the start of tilmicosin treatment initiation.

Tilmicosin phosphate was added at either 0 ppm or 200 ppm of the drinking water for 5 days. Pigs were euthanized at the end of a 4-day post-treatment period at which time lung lesions were collected on all pigs. No animals entered the food chain.

Animals had ad libitum access to water and non-medicated feed for the duration of the study. Animals were weighed at the start of tilmicosin treatment initiation and again immediately prior to euthanasia. Feed delivered to the feeders and feed remaining in the feeder at the time of pig weighing was recorded. Water delivered to the pens for the 5 day treatment period and water remaining at the end of each day was recorded.

Lung lesion data for each infection group was analyzed separately. Within each group, data were analyzed using a linear mixed model analysis using the Proc Mixed procedure of SAS® v9.2 (SAS Institute, Cary, NC). The model included the fixed effects of tilmicosin inclusion level and random effect of block. A P-value ≤ 0.05 was considered significant. Within each Infection group, water usage, ADG, ADFI, G:F, and F:G were analyzed using a linear mixed model analysis using the Proc Mixed procedure of SAS®. The model included the fixed effects of tilmicosin inclusion level and random effect of block. Tukey's multiple comparison procedure was conducted for mean comparison. A P-value ≤ 0.10 was considered significant. Body weight was calculated using arithmetic means.

Results and discussion

Results for the analysis of the Co-Infection group are presented in Table 1. Pigs administered 200 ppm tilmicosin phosphate in the drinking water for 5 days had a lower percentage of lung lesions than the control (0 ppm) (31.74% vs. 43.04%, respectively; $P = 0.0004$). Water usage (+19.3%), ADG (+74.7%), ADFI (+24.5%), and G:F (+42.3%) were greater ($P \leq 0.10$) and F:G (-33.6%) was lower in the 200 ppm group when compared to the control group.

Results for the analysis of the Single-Infection group are presented in Table 2. Pigs administered 200 ppm tilmicosin phosphate in the drinking water for 5 days had a lower percentage of lung lesions than the control (28.26% vs. 21.01%, respectively; $P = 0.0050$). Water usage (+17.4%), ADG (+18.4%), ADFI (+7.0%), and G:F (+10.3%) were greater ($P \leq 0.10$) and F:G (-9.6%) was lower in the 200 ppm group when compared to the control group.

Similar to previous research, the Co-Infection model (M hyo + PRRSV) resulted in greater lung lesion percentage than the Single-Infection model with M hyo alone.[1] This is the first published study carried out in the US that has evaluated the effects of tilmicosin phosphate delivered via the drinking water on controlling swine respiratory disease associated with M hyo alone or in the presence of PRRSV. However, feeding tilmicosin has been shown to be effective in lowering the percentage of lung lesions associated with M hyo in previous research.[4,5] The results of this study suggest that tilmicosin phosphate, when added to the drinking water for 5 days, is effective at controlling respiratory disease caused by *Mycoplasma hyopneumoniae*, as measured by a reduction in lung lesions.

References

1. Thacker, E. L., P. G. Halbur, R. F. Ross, R. Thanawongnuwech, and B. J. Thacker. 1999. *Mycoplasma hyopneumoniae* potentiation of porcine reproductive and respiratory syndrome virus-induced pneumonia. Journal of Clinical Microbiology. 37:620-627.

2. FDA. 2014. Freedom of Information Summary. NADA 141-361 Pulmotil AC. Available at: `http://www.fda.gov/downloads/animalveterinary/products/approvedanimaldrugproducts/foiadrugsummaries/ucm396900.pdf`. Accessed 18 August 2015.

3. Wu, C. C., T. R. Shryock, T. L. Lin, and M. F. Veenhuizen. 1997. Testing antimicrobial susceptibility against Mycoplasma hyopneumoniae in vitro. Swine Health and Production. 5:227-230.

4. Reinbold, J. B., T. E. Weber, K. K. Keffaber, and T. A. Marsteller. 2013. Rationale for the use of tilmicosin for the control of bacterial pneumonia complicated by coinfection with *Mycoplasma hyopneumoniae*. Page 234 of the 2013 Allen D. Leman Swine Conference Proceedings.

5. Stipkovits, L., D. Miller, R. Glavits, L. Fodor, and D. Burch. 2001. Treatment of pigs experimentally infected with Mycoplasma hyopneumoniae, Pasteurella multocida, and Actinobacillus pleuropneumoniae with various antibiotics. The Canadian Journal of Veterinary Research. 65:213-222.

Table 1: Least squares means for the effects of tilmicosin phosphate inclusion level on lung lesion percentage, water usage, and growth performance of nursery pigs infected with *Mycoplasma hyopneumoniae* and PRRSV[1] (Co-Infection).

	Tilmicosin inclusion level, ppm[2]			
Item	**0**	**200**	**SEM**	***P*-value**
No. of pens	10	10	-	-
BW, lb				
Start of 5 day treatment period	22.1 ± 1.17	22.5 ± 0.77	-	-
End of 4 day post-treatment period	24.9 ± 1.81	27.6 ± 1.53	-	-
Lung lesions, %[3]	43.04	31.74	1.935	0.0004
Water usage, lb/d[4]	1.64	1.95	0.079	0.0159
ADG, lb/d[5]	0.32	0.56	0.035	0.0003
ADFI, lb/d[5]	0.63	0.78	0.031	0.0009
G:F, lb:lb[5]	0.50	0.71	0.031	0.0009
F:G, lb:lb[5]	2.13	1.41	0.141	0.0056

[1] PRRSV: porcine reproductive and respiratory syndrome virus.
[2] Tilmicosin phosphate administered in drinking water for 5 day period and was followed by 4 day post-treatment period.
[3] Lung lesions = Left cranial lobe × 20.0% + Left caudal lobe × 27.5% + Right cranial lobe × 10.0% + Right middle lobe × 10.0% + Right caudal lobe × 27.5% + Accessory lobe × 5.0%.
[4] Water usage measured during the 5 day treatment period.
[5] Measured from start of 5 day treatment period to end of 4 day post-treatment period.

Table 2: Least squares means for the effects of tilmicosin phosphate inclusion level on lung lesion percentage, water usage, and growth performance of nursery pigs infected with *Mycoplasma hyopneumoniae* (Single-Infection).

	Tilmicosin inclusion level, ppm[1]			
Item	**0**	**200**	**SEM**	***P*-value**
No. of pens	15	15	-	-
BW, lb				
Start of 5 day treatment period	29.6 ± 1.08	29.5 ± 1.10	-	-
End of 4 day post-treatment period	37.8 ± 1.69	39.1 ± 1.52	-	-
Lung lesions, %[2]	28.26	21.01	1.564	0.0050
Water usage, lb/d[3]	3.11	3.65	0.137	0.0036
ADG, lb/d[4]	0.91	1.08	0.031	0.0008
ADFI, lb/d[4]	1.29	1.38	0.033	0.0476
G:F, lb:lb[4]	0.71	0.78	0.010	< 0.0001
F:G, lb:lb[4]	1.42	1.29	0.020	0.0001

1 Tilmicosin phosphate administered in drinking water for 5 day period and was followed by 4 day post-treatment period.
2 Lung lesions = Left cranial lobe × 20.0% + Left caudal lobe × 27.5% + Right cranial lobe × 10.0% + Right middle lobe × 10.0% + Right caudal lobe × 27.5% + Accessory lobe × 5.0%.
3 Water usage measured during the 5 day treatment period.
4 Measured from start of 5 day treatment period to end of 4 day post-treatment period.

Relationship between pig birth weight and survival in a swine nucleus herd

C. Abell, PhD; J. Sonderman, PhD; T. Rathje, PhD
DNA Genetics, Columbus, Nebraska

Introduction

Swine maternal line selection programs place large emphasis on litter size. This has been evidenced by marked improvements in litter size over the past several years. Industry-wide total born increased by 0.14 pigs per year from 2009 to 2014 with an average of 13.5 total born per litter in 2014.[1] While larger litters result in more revenue for producers by lowering sow farm costs per pig, there are diminishing returns in profit with additional pigs.[2] Improving sow production efficiency through increasing litter size can adversely impact pig quality. A negative correlation between litter size and birth weight has been reported.[3] Optimizing profitability for a commercial swine operation would require pig quality to be maintained with increased litter sizes. The objective of this study was to evaluate the relationship between pig birth weight and survivability as a pig quality measure.

Materials and methods

From September 2014 to November 2015, birth weights were recorded on 3 genetic lines at 6 sow farms. Every live born pig was weighed individually. A total of 91905 birth weight records were collected. At all sow farms, pigs are weaned once a week at 3 weeks of age. Three survival traits were defined for the analysis. Pre-weaning survival was 0 for all pigs that were not alive at 21 days of age given that they were born alive and 1 for all other pigs. Nursery survival was defined as 0 for all pigs that were not alive at 11 weeks of age given that they were alive at 21 days of age and 1 for all other pigs that were alive at 21 days of age. Finisher survival was defined as 0 for all pigs that were not alive at 23 weeks of age given that they were alive at 11 weeks of age and 1 for all other pigs that were alive at 11 weeks of age. A generalized linear mixed model for a Bernoulli random variable was used to analyze all three survivability traits. Genetic line and sex were fitted as fixed effects and birth weight was fitted as a covariate for all three traits. Total born was also included as a covariate for pre-weaning survival. The random effects for pre-weaning and nursery survival were litter and Year/Herd/Month of birth. Litter and pen were fitted as random effects for finisher survival.

Results and discussion

The relationship between birth weight and pre-weaning survival, nursery survival, and finisher survival are shown in Figures 1, 2, and 3, respectively. The effect of birthweight was significant ($P < 0.05$) for pre-weaning and nursery survival. There was a tendency for an effect of birth weight on finisher survival ($P < 0.10$). For the population used in this study, raising the average birth weight by 0.1 kg would improve pre-weaning survival by 4.3% and nursery survival by 0.5%. Finisher survival would not be impacted by a 0.1 kg increase in the average birth weight. The results of this study indicate that pig birth weight is related to survival and should be considered as a trait to improve pig quality. It is important for swine genetic companies to understand the relationship between birth weight and survival if the goal is to maintain pig quality while increasing litter size. Before incorporating birth weight into a genetic selection program, the genetic parameters between birth weight and other traits in the index would need to be evaluated.

References

1. Stalder. 2009-2014 analysis of US pork industry productivity. National Pork Board, Des Moines, IA. 2014.

2. Widmar et al. Integrated on-farm decision making: economic implications of increased variation in litter size. Southern Agricultural Economics Association Annual Meeting, 2011.

3. Abell et al. Effect of litter size on piglet birth weight and birth weight as an indicator of pig quality. ADSA-ASAS Midwest Meeting. 2015;36.

Figure 1: Relationship between pig birth weight and pre-weaning survivability.[1]

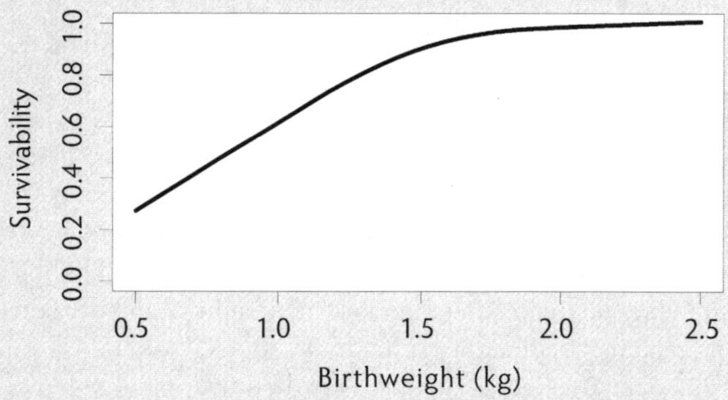

[1] Pre-weaning survival was defined at day 21 post-farrowing.

Figure 2: Relationship between pig birth weight and nursery survivability given the pig was alive at weaning.[1]

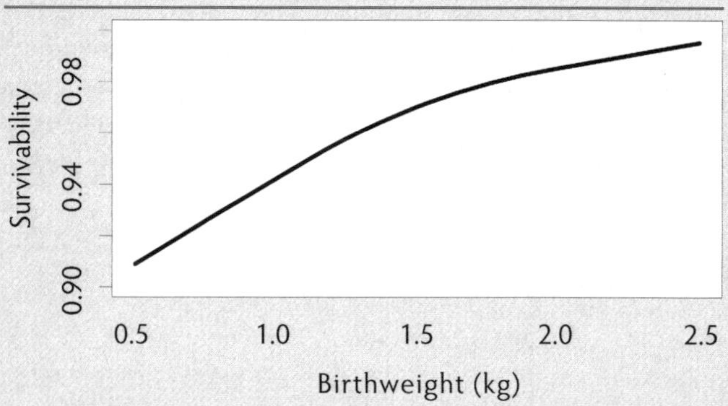

[1] Pigs were weaned at 3 weeks of age and in the nursery for 8 weeks.

Figure 3: Relationship between pig birth weight and finisher survivability given the pig was alive at the end of the nursery period.[1]

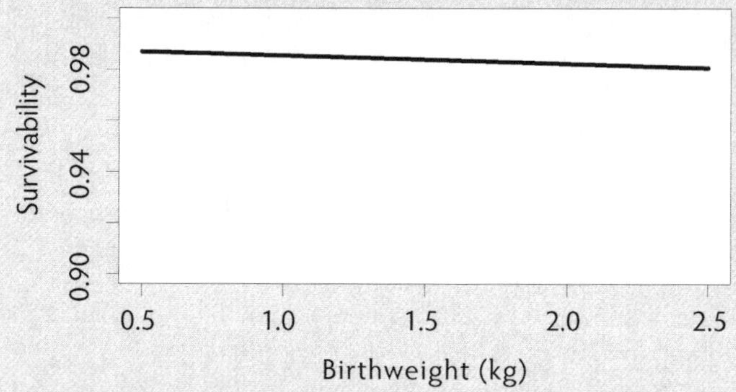

[1] Pigs entered the finisher at 11 weeks of age and were there for 12 weeks.

Evaluation of efficacy, safety and stability of a novel adjuvant system in conventional and advanced vaccine platform technologies for use with various antigens in swine vaccines

T. Miller, PhD; M. Pfannenstiel, PhD; W. Swafford, DVM
Vaxliant Corporation, Lincoln, Nebraska

Introduction

Vaccine platform technologies for next generation biologic medicinal products have evolved quickly both in the elements used to construct the vaccine platforms but also in the strategies for delivery to the immune system of both animals and humans. To meet the demands of efficacy, safety, and potency needs of herd management and next generation biologics, adjuvant systems will also need to evolve in both research and product development. VaxLiant®, Inc. has a novel adjuvant system that satisfies the versatility of the next generation of biologics, as well as, for optimizing response to conventional vaccines such as conventional killed swine influenza virus (SIV). SIV vaccines that have been prepared for use in pigs were shown to increase in HAI response when formulations were optimized using VaxLiant® adjuvant components. This adjuvant system has been used with a variety of vaccine platforms and infectious disease models for use in autogenous, conditional, or fully licensed swine vaccines. The adjuvant system does not use any oil and water format, aluminum hydryoxide, surfactant, or other harsh detergent that can irritate the site of inoculation. Instead, the vaccine uses an amphipatic lipid that is emulsified with a bioadhesive polymer using microfluidics technology; the formulations can be used *in vitro* to analyze stimulation of immune signals in cultured cells, as well as polarizing the immune response in animal models. Features of the adjuvant system, determined during the optimization of its use with SIV vaccines, include the potential for great breadth and utilization in the vaccine industry.

Materials and methods

The adjuvants used for this study included three formulations of BioMize™ manufactured by the VaxLiant corporation located in Lincoln, NE. The three adjuvants from VaxLiant® are indicated as BioMize™ 0220, BioMize™ 0219, and BioMize™ 0218. These adjuvants differ in the concentration of a proprietary lipid and polymer that make up the base emulsion of the adjuvant. The adjuvants are supplied as sterile solutions in 5X concentrations and are diluted one part adjuvant to 4 parts antigen for administration.

To combine adjuvant with antigen for a 2 ml dose vaccine, 0.4 ml of the adjuvant was added to 1.6 ml of inactivated SIV and mixed by gentle inversion until the solution was homogenous. No mechanical mixing or homogenization step is required.

The SIV antigen used in the study was a commercially available freeze-dried killed SIV vaccine containing the H1N1 (H1N2-like H1N1) and H3N cluster I Texas-like. The freeze dried plug was resuspended in sterile phosphate buffered saline (PBS) using 80% of the recommended diluent for resuspension. Once the vaccine plug was in suspension, the adjuvant was added and mixed as described above.

Transmission electron microscopy was performed by the Central Facility for Electron Microscopy at the University of Nebraska (Lincoln, NE) and the scanning electron microscopy at Mount Allison University Digital Microscopy Facility (Sackville, New Brunswick, Canada). Adjuvant particles for scanning electron microscopy were prepared by placing droplets of osmium tetroxide (2% solution from Electron Microscopy Sciences) next to droplets of adjuvant sample on dental wax filter within a Petri dish (Fisher). The Petri dish was sealed and placed in a vacuum chamber overnight. Portions of the filters containing the adjuvant sample droplets were then mounted on a 10 mm diameter aluminum support stub and coated with collodial graphite. The samples were then coated with 5nm of gold using a Hummer 6.2 sputtering unit and examined by JEOLJSM SEM equipped with an Oxford Inca Energy 200 EDS system and Deben XYR stage system. The transmission electron microscopy was performed by drying the adjuvant onto a copper grid by placing a drop of diluted adjuvant onto the grid and absorbing the excess water from the droplet using whatman filter paper from the bottom side of the grid. The copper grid was then air dried in a closed glass Petri dish until used for microscopy.

For vaccination, 25 weaning age commercial pigs 2-3 weeks were obtained from Amore Pork, Inc (Battle Lake, MN) and held in standard biosafety level 2 containment facilities for 12 days prior to inoculation. The 25 pigs were separated into 4 groups. Three of the treatment groups had 7 pigs each and the sentinel control group had 4 animals. The pigs were ear tagged and randomly distributed in pens prior to inoculation. Each pig was inoculated intramuscularly (IM) with two ml of inactivated SIV vaccine containing adjuvant as described above, except the sentinel control group, which remained untreated. At 21 days after the primary inoculation, blood was collected from all pigs. At 28 days after the primary inoculation, three pigs from each treatment group receiving the inactivated SIV + adjuvant were given a second inoculation IM; all remaining pigs remained untreated for the second inoculation. At 21 days post secondary inoculation (42 days), the study was terminated. Blood was drawn for serum isolation at day one prior to inoculation, at day 21 and at day 42 at the end of the study.

The serum from each pig was analyzed for hemagglutination inhibition (HAI) antibody titers using 96 well U-bottom plates (Falcon). The virus used for hemagglutination was the resuspended and inactivated virus from the commercial vaccine. The HAI titers of each virus H1 and H3 serotypes were first determined and then each virus preparation was diluted to 4-8 HA units per 25 ul. Serum from each pig was serial diluted two-fold in PBS with a final volume of 25 ul per well. The SIV virus (25ul/well diluted to 4-8HA units) was added to the plate and the virus and serum samples were allowed to incubate at room temperature for 30 minutes to provide time for the serum to neutralize binding by the virus. Chicken red blood cells (cRBC) diluted to 1.0% were added to the plate (50ul/well) and the plate was allowed to incubate for 1hour at 2-7°C. The reciprocal of the last dilution of antisera that no longer inhibited the RBC agglutination was considered the titer of the sample.

Results and discussion

Vaxliant® adjuvants are unique in that they are emulsions that do not contain aluminum hydroxide, oil, detergents, or surfactants. The core emulsion of VaxLiant® adjuvant formulations is comprised of a bioadhesive lipid and polymer that are mixed using microfluidics technology to develop a consistent size and viscosity. Figure 1a shows a scanning electron microscopy picture of an emulsion particle of ENABL™ S3 formulation, which consists of the same core emulsion as the BioMize™ formulations used in this study. The mean particle size shown in this figure is approximately 1 micron (µ). However, the adjuvant system provides for a formulation capability to combine different amounts of lipid and polymer together and the microfluidics technology allows different sized molecules to be produced. This is an important feature when considering modes of delivery such as transdermal and mucosal vaccination regimens. Figure 1b shows a transmission electron microscopy picture of the ENABL™ S3 adjuvant particle dried onto an electron microscopy copper grid and magnified 50,000 fold. The information provided in these figures suggests the lipid polymer structure is comprised of a lattice or network both on the surface but also for the internal intra-particle structure. Because the lipid can associate with the lattice structure, it provides the ability to have a larger surface area for antigen binding, which is a feature of this adjuvant system. The bioadhesive lipid used in the vaccine formulation in the study described below in Figures 3 and 4, can form a bilaminar lipid layer that is similar to that of native cell membranes. The adjuvant particle can thus bind to and orient proteins and other molecules that are typically delivered or act at the cell surface. This is depicted in Figure 2 using computer graphics to aid in visualizing the model for the adjuvant particle.

Figure 1: (A) Scanning electron micrograph of an osmified ENABL™ S3 adjuvant emulsion particle. Average mean size of the particle is 1 micron. Magnification is 10,000 X. **(B)**Transmission Electron Micrograph of an air dried sample of ENABL™ S3 adjuvant. Magnification is 50,000X.

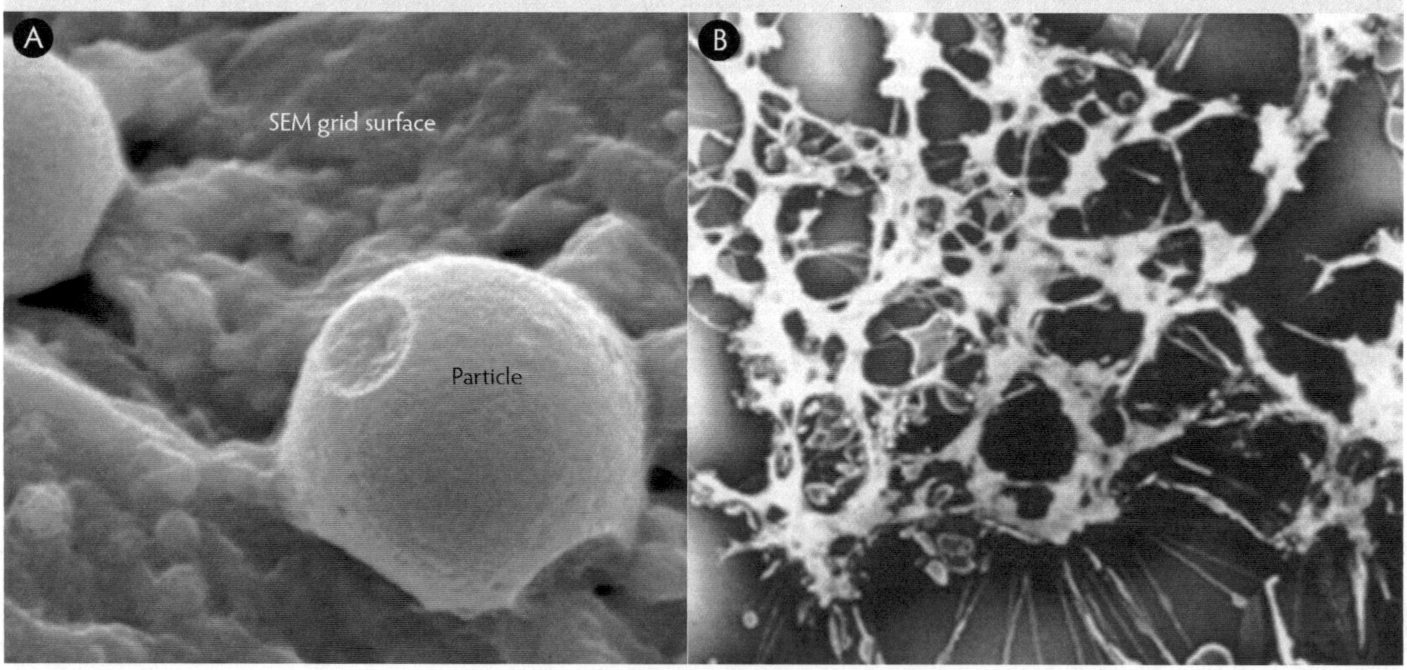

Figure 2: (A) Computer graphic model of VaxLiant adjuvant ENABL™. The polymer forms a lattice structure particle in aqueous solution. **(B)**The lipid is represented by the rough and rounded edged (green) coated particle. Proteins or other macromolecules that can typically associate with a lipid membrane can then bind to the lipid polymer structure (darker, rounded particles irregulary scattered on the intra-particle matrix).

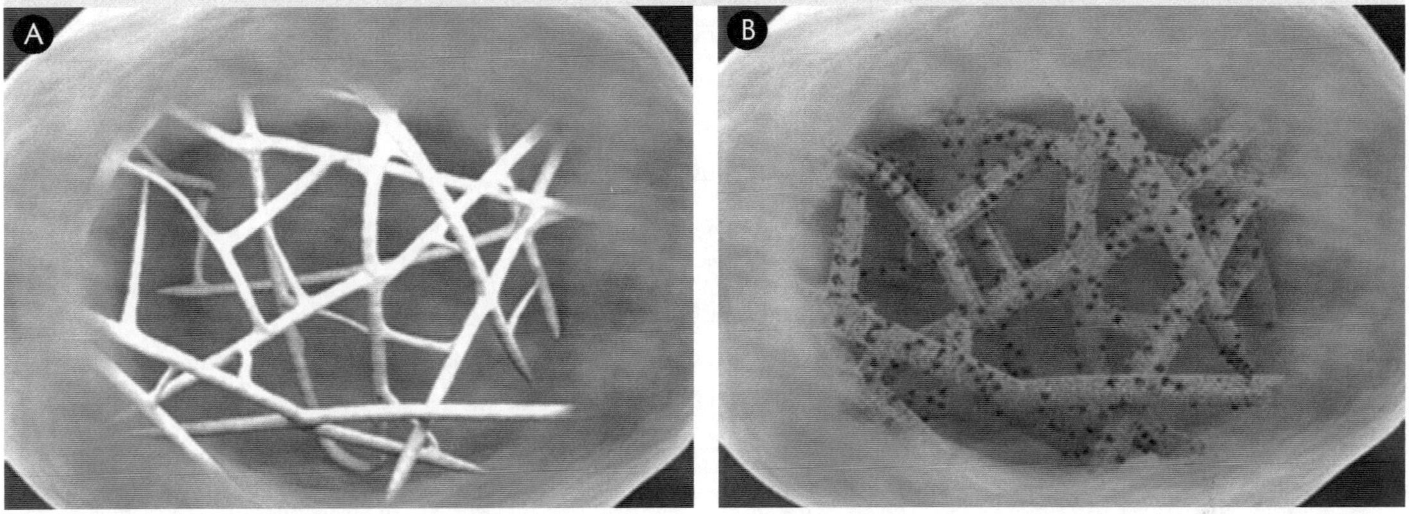

The adjuvants are safe for use in that ENABL™ S1, S2 and S3 have met the safety requirements to satisfy the minimal withdrawal time of 21 days by VS Memorandum 800.51[1] (data not shown). In addition, all excipients and additives used in animal vaccines, which include adjuvants, must also show that there is no unwanted pathology associated with the immune response at the inoculation site. For ENABL™ both SQ and IM routes of administration have been approved for the minimum 21 days withdrawal (minimum allowed) by Center of Veterinary Biologics (CVB) for cattle and pigs.

The adjuvants are stable both at cold storage temperature 2-7°C and room temperature 18-27°C. Table 1 shows the results of data compiled over a 2 year period on several biophysical features of the adjuvant including particle size, pH, and viscosity. The particle size adjuvant consistency and other general features have remained stable for more than two years. In addition, adjuvant formulations that have undergone freeze drying and resuspension to their original volume will regain their biophysical features as released from manufacturing (data not shown). When these adjuvants are used in animals to determine whether they retain their potency the adjuvants perform the same or better when using bovine serum albumin as the test antigen in a mouse immunization regimen (data not shown).

Figure 3 shows the hemagglutination inhibition (HAI) results of nursery age piglets (4-5 weeks of age) vaccinated with the H1N2 isolate of inactivated SIV vaccine containing either BioMize™ 0220, BioMize™ 0219, or BioMize™ 0218. These adjvuants have progressively increasing amounts of bioadhesive lipid and polymer from the base formulation ENABL™ S3. As the amount of lipid and polymer are increased in the base formulation the amount of antibody response increases, especially after two doses. Likewise Figure 4 indicates the results of HAI response to the H3N2 serotype increases as the base formulation of ENABL™ S3 changes to progressively higher concentrations of lipid and polymer. Collectively these data suggest that antibody response to both important serotypes of SIV can be enhanced in a dose related manner to adjuvant formulations in the VaxLiant adjuvant system. The titers achieved after one dose and evaluated at 21 days post primary inoculation were considered a good titer in that they are high enough to provide protection against challenge to homologous seroyptes. Furthermore, the titers could be maintained through 42 days after one dose. However, when boosted at 21 days with a second identical dose of inactivated H1N2/H3N2 inactivated SIV vaccine, the titers increased to a very high level indicating that the animals would probably have been protected against heterologous H1 or H3 serotypes (Figure 3 and 4).

References

1. VS memorandum 800.51. 2015. Additives in Administered Animal Biological Products.

Figure 3: Hemagglutination inhibition titer for H1N2 serotype SIV using three separate formulations of type 3 for swine. ENABL™ S3 is base formulation for ENABL™ type 3 formulation for ENABL™. BioMize™ 0218 thru -0220 are progressively different in concentration from the core lipid and polymer than ENAB™ S3.

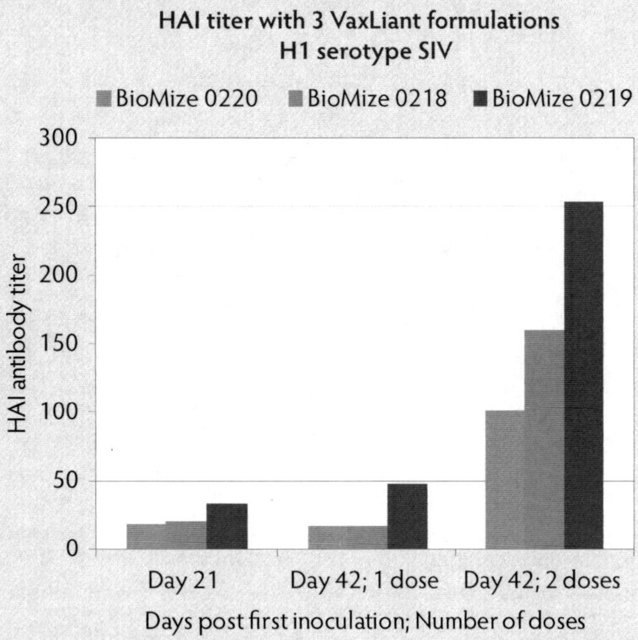

Figure 4: Hemagglutination inhibition titer for H3N2 serotype SIV using three separate formulations of type 3 for swine. ENABL™ S3 is base formulation for ENABL™ type 3 formulation for ENABL™. BioMize™ 0218 thru -0220 are progressively different in concentration from the core lipid and polymer than ENABL™ S3.

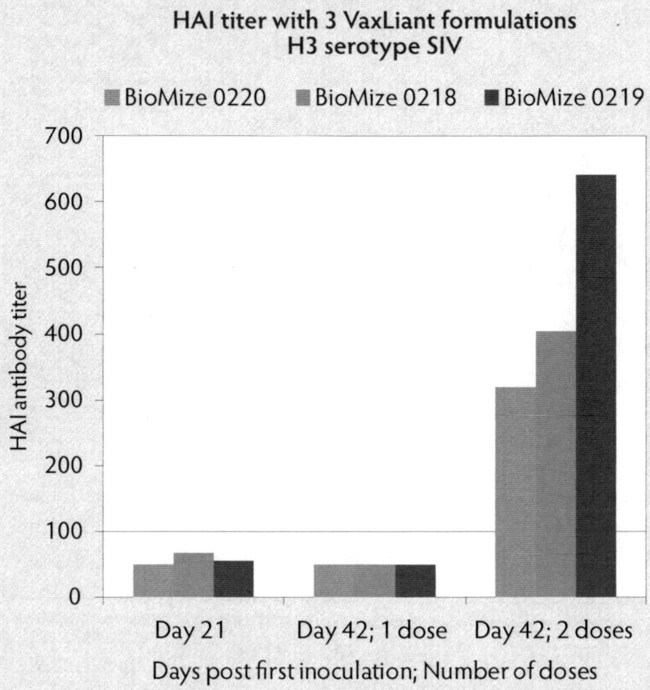

Table 1: Comparisons of biophysical characteristics for 2 year dating in liquid form.

Test	Test date	Adjuvant formula type 01	Adjuvant formula type 01	Adjuvant formula type 06	Adjuvant formula type 06
		2-7°C	**RT**	**2.7°C**	**RT**
pH	Original	7.0	7.0	7.1	7.1
	@ 2 yr dating	7.0	6.8	7.2	7.1
Viscosity (Cst)	Original	16	16	21	21
	@ 2 yr dating	17	17	25	26
Mean particle size (µ)	Original	0.33	0.33	0.64	.64
	@ 2 yr dating	0.66	.61	.58	.59

Establishing *Mycoplasma hyopneumoniae* herd status classification criteria for breeding herds

L. Galina Pantoja[1], DVM, PhD; M.J. Clavijo[2], DVM, PhD
[1]Zoetis, Florham Park, New Jersey; [2]Pig Improvement Company, Hendersonville, Tennessee

Introduction

Currently, a standardized terminology describing the *M. hyopneumoniae* (Mhp) status of breeding herds has not been established. The benefits of a clear and concise classification system are manifold; facilitating communication between swine producers, veterinarians, diagnosticians and breeding stock companies, evaluating strategies for disease control and supporting regional control and elimination efforts. The objective of this document is to propose a Mhp herd status classification system that incorporates comprehensible diagnostic criteria based on biologically relevant features of Mhp. This herd status classification was created utilizing the input from the Zoetis Veterinary Advisory Board, a group of twenty US swine veterinarians who provide veterinary service to swine operations in the United States. Furthermore, the authors followed the terminology for classifying herds by PRRSV infection status defined by Holtkamp and collaborators.[1]

Herd status classification

The proposed herd-status classification focusses on the sow herd, a population of animals which includes the breeding animals and their offspring. If a gilt development unit (GDU) is located within the premises, the GDU will be considered part of the breeding herd and the group of animals with the lower Mph health status in the premises will determine the status of the entire site.

Four major herd-status categories are proposed for breeding herds: positive unstable (I), positive stable (II), provisionally negative (III), and negative (IV) (Table 1).

Positive unstable (I): breeding herds experiencing Mhp- associated clinical signs and Mhp is detected within the respiratory tract of weaning-age pigs. These populations may be serologically positive. Herds going through a Mhp outbreak in breeding animals and those that have not performed the necessary testing are considered category I.

Positive stable (II): breeding herds that do not show Mhp-associated clinical signs and pulmonary lesions. These herds show a reduced prevalence of Mhp in the upper respiratory tract of weaning-age pigs (< 10% of positive laryngeal swabs) for at least 90 days. To fall into this category 4 consecutive samplings of 45 laryngeal swabs from weaning –age pigs every 30 days should reveal no more than 10% of pigs positive in each sampling. The breeding herd population is serologically positive. Herds undergoing elimination efforts (i.e. herd closure) fall into this category.

Provisionally negative (III): breeding herds that do not show Mhp-associated clinical signs and Mhp is not detected, but the population remains serologically positive. Category III herds should have at least 90 days of no positive results from weaning-age pigs in samples from the respiratory tract (4 consecutive samplings of 30 laryngeal swabs every 30 days). Negative but vaccinated herds are considered category III. These commercial farms are not infected but implement continued use of Mhp vaccine in replacement gilts that entered the breeding herd due to the high cost of an acute outbreak in a naive farm.

Negative (IV): breeding herds that do not show Mhp-associated clinical signs, Mhp is not detected in any type of sample and the population is serologically negative. Herds undergoing elimination efforts should have been category III for at least one year and the herd would have rolled over to fall into category IV. Newly established herds and those that went through complete depopulation and repopulation efforts fall within category IV, after no positive serology results at least 60 days after placement of animals. A minimum of 30 sow samples (estimated prevalence of 10%) and ideally 57 samples (estimated prevalence of 5%) have to be collected to detect one positive.

Considerations

Diagnostic criteria

Mhp-associated clinical signs with disease are defined as a dry non-productive cough, exacerbated by physical exertion. Furthermore other clinical signs that may be present; fever, decreased appetite and labored breathing.[2] Clinical signs are not pathognomonic for Mhp and other respiratory pathogens must be ruled out. Microscopic lesions consist of lobular distribution of peribronchiolar and perivascular lymphocytic cuffing.[2] Alveoli and airways may contain serous fluid with a few macrophages and neutrophils. Airway epithelium is intact, and sometimes slightly hyperplastic.[2] Microscopic lesions are nonspecific and can be similar to those observed with viral agents.

Detection of the agent within lung lesions or the upper respiratory tract can be achieved using several tests. Immunohistochemistry (IHC), fluorescent antibody (FA), in situ hybridization (ISH) and polymerase chain reaction (PCR) testing are routinely utilized by diagnostic laboratories. However, PCR is recommended as the ideal testing method. Although culturing is confirmatory, it is less practical than the other methods and not frequently used.

Table 1: *Mycoplasma hyopneumoniae* breeding herd-status classification criteria.

Herd category	Criteria: Clinical signs	Criteria: Antigen detection in respiratory tract	Criteria: Serology	Description and diagnostic requirements
Positive unstable (I)	Yes	Yes	Maybe	Clinical signs consistent with Mhp are present and Mhp is detected within lesions and/or in the respiratory tract. Recently infected populations may be serologically negative. Untested herds are category I by default.
Positive stable (II)	No	Yes < 10% of weaning-age pigs positive for antigen within respiratory tract	Yes	Mhp characteristic clinical signs or lesions are not detected. At least 90 days of less than 10% of due to be weaned pigs positive by PCR in 4 consecutive samplings of 45 laryngeal swabs at days 0, 30, 60 and 90.*
Provisionally negative (III)	No	No	Yes	At least 90 days of negative results from weaning-age pigs by PCR in 4 consecutive samplings of 30 laryngeal swabs at days 0, 30, 60 and 90.* Herds undergoing Mhp elimination, no clinical signs or positive tests from respiratory tract of replacement animals, at least 60 days after the initial introduction. Negative but vaccinated herds are considered category III
Negative (IV)	No	No	No	At least one year as category III and have gone through a complete sow rollover. At completion of rollover testing must take place. Newly established herds and herds that underwent complete depopulation and repopulation. A minimum of 30 serum samples from sows should test negative by ELISA.*

* http:epitools.ausvet.com/au/content.php?page=PrevalenceSS. Accessed April 27, 2015.

The selection of the optimal sampling site in live animals will likely be affected by timing of infection. For example, it is possible that recently infected pigs harbor Mhp in the nasal cavity while non-acute pigs are more likely to harbor Mhp from the larynx, trachea or deeper sections of the respiratory tract. Current literature suggests laryngeal swabs as the preferred sample type for Mhp over nasal swabs and tracheo-bronchial lavage.[3] However, these methods were evaluated under experimental conditions where pigs were inoculated intratracheally, bypassing the nasal cavity. Finally, seroconversion within a population can take several weeks to be detected by ELISA[2,4] and therefore timing should be considered.

For sample size calculations several factors must be considered, the assumed true prevalence, the level of precision, the sensitivity and specificity of the diagnostic test and the population size. For the proposed classification, to detect a 10% prevalence of Mhp infection in weaning-age pigs (i.e., for a positive stable herd), at least 45 samples should be collected, from a population of 1000 pigs.[5] On the other hand, when the objective is to detect at least one positive, at least 30 samples should be collected, assuming the prevalence of Mhp is 10% (i.e., provisionally negative) or 57 samples if 5% (i.e., negative) prevalence is assumed.[5] Both of these calculations assume nearly perfect tests. But diagnostic test available today for Mhp are not perfect. Therefore, veterinarians should recognize these limitations when considering sample sizes and interpreting diagnostic results.

Definition of stability

A positive correlation has been reported between the presence of Mhp in the upper respiratory tract at weaning and the extension of pulmonary lesions at slaughter.[6] Based on those findings, reasonable goals for control should be to produce litters that are either negative or with low Mhp prevalence at weaning. This classification has taken the suggested cutoff point of 10% of infected weaning-age pigs to determine if a population has reached stability. This herd-status classification can be further refined as more information about prevalence at weaning and disease in the finisher stages becomes available.

Introduction of replacement gilts in the breeding herd plays a critical role in the Mhp stability a herd. Negative breeding herds that introduced positive replacement gilts are likely to become positive unstable (I) or stable (II). Positive breeding herds that introduced positive or negative replacements are likely to be positive unstable (I) or stable (II). Provisionally negative or negative herds that introduced negative replacement gilts are likely to maintain their provisionally negative (III) or negative (IV) status.

Growing-pig herd classification

A herd-status classification in the growing-finishing herd may be useful in the future. There is no clear definition of what an unstable growing-finishing site is. Recent reports suggest that a specific coughing index and serology thresholds have been associated with Mhp disease in finishing pigs. A coughing index (calculated as the percentage of pigs coughing per minute of observation) ≥ 2.5% and a Mhp-ELISA seroprevalence > 50% was reported to be an indicator of enzootic pneumonia, and therefore, could be an indicator of lack of stability, in finishing pigs.[7]

Conclusion

This Mhp herd-status classification is a first proposal to be utilized as a foundation for discussions in the swine industry. It is the authors desire to initiate collaboration efforts, to identify areas of improvement, and to recognize areas that warrant further research.

References

1. Holtkamp DJ, Polson DD, Torremorell M. Terminology for classifying swine herds by porcine reproductive and respiratory syndrome virus status. J Swine Health and Prod. 2011;19(1):44-56.

2. Thacker E. Mycoplasmal diseases. In: Straw BE, Zimmerman JJ, D'Allaire S, et al, eds. Diseases of Swine. 9th ed. Oxford, UK: Blackwell Publishing Ltd; 2004:701-717.

3. Pieters M, and Rovira A. Comparison of various samples types for detection of Mycoplasma hyopneumoniae in recently infected pigs. Allen D. Leman Swine Conference, St. Paul, MN. 2013;75-76.

4. Sibila M et al. Chronological study of *Mycoplasma hyopneumoniae* infection, seroconversion and associated lung lesions in vaccinated and non-vaccinated pigs. Vet Microbiol. 2007;122:97-107.

5. EpiTools epidemiological calculators. Accessed April 27, 2015. `http:epitools.ausvet.com/au/content.php?page=PrevalenceSS`

6. Fano E, et al. Effect of *Mycoplasma hyopneumoniae* colonization at weaning on disease severity in growing pigs. Can J Vet Res. 2007;71:195-200 .

7. Nathues H et al. Value of the clinical examination in diagnosing enzootic pneumonia in fattening pigs. Vet J. 2012;193:443-447.

Evaluation of Fostera® PRRS cross-protection against a contemporary Linage 1 (RFLP 1-7-4) PRRS virus

Jose Angulo[1], DVM; Jay G. Calvert,[2] PhD; Gene Nemechek,[1] DVM; Michael K. Senn,[1] DVM, MS; Marcia L. Keith[2], BS; Lucas P. Taylor,[2] MS; Douglas S. Pearce,[2] BS; Chadwick Brice,[2] BS; M. Corinne Lenz,[2] BS, MS
[1]Zoetis, Florham Park, New Jersey; [2]Zoetis, Kalamazoo, Michigan

Introduction

Porcine reproductive and respiratory syndrome virus (PRRSV) is a highly infectious RNA virus causing an important economic impact on the US swine industry. The disease shows a clear seasonal pattern of increased reproductive outbreaks, beginning in the fall, followed by an increase of respiratory clinical signs during winter.[1] PRRSV genetic diversity is extensive and every year new variants, many causing severe clinical impact, are detected. Emerging PRRS viruses represent a common threat for pig producers and veterinarians because field strains differ in virulence, transmission and persistence. Restriction Fragment Length Polymorphism (RFLP) is commonly used in the US to identify PRRSv strains, yet this diagnostic technique has its limitations. However, it is commonly reported along with ORF5 sequencing as a practical way for swine practitioners to track PRRS viruses in farms, large systems and even geographical areas as part of regional control projects.[2]

A new PRRSV (Genotype 2, Lineage 1) with an RFLP pattern of 1-7-4 emerged recently in North Carolina and subsequently appeared in the upper Midwest. This PRRSV strain has been reported to cause up to 20% mortality in growing pigs, 0.4 to 16.5% mortality in sows, and a high proportion of abortions.[3] This virus has also been reported to negatively impact pig growth, with a reduction in profit of $2.87/head.[4]

The objective of this study was to evaluate the protective efficacy of Fostera® PRRS in pigs vaccinated at three weeks of age and challenged with a virulent Lineage 1 PRRSV (RFLP 1-7-4) at seven weeks of age.

Material and methods

The challenge strain (Type 2, Lineage 1, RFLP pattern 1-7-4) used in this study was collected and isolated in 2015 from a farm located in North Carolina that suffered severe reproductive and respiratory clinical signs. It was tested for 9CFR sterility, 9CFR mycoplasma and porcine extraneous viruses by Zoetis biological control labs in Lincoln NE, having a final challenge titer of 1.0 x 10^4 $TCID_{50}$/4 ml dose. The challenge strain showed an ORF5 nucleotide genetic distance of 13.4% from Fostera® PRRS.

The study was conducted in a BSL-2 containment facility at Midwest Veterinary Services Inc. located in Oakland Nebraska. The study design was reviewed and approved by the Site Animal Care and Use Committed (IACUC). Forty-six healthy, PRRS-negative (by IDEXX ELISA) pigs were obtained from a commercial sow farm at ~3 weeks of age. Forty pigs were randomly allocated to two treatment groups: T01 Non-vaccinated (N = 20) and T02 Vaccinated (N = 20). A strict control group (NTX) also was included in the study design as an environmental control that received neither vaccination nor challenge (N = 6), this group was housed in separate pens of three pigs each, one pen with each treatment group. NTX pigs were removed and humanely euthanized prior to challenge and examined for lung lesions to monitor for confounding disease factors.

T01 and T02 groups were housed in separate rooms by treatment during the vaccination phase. Pigs were vaccinated at ~3 weeks of age then comingled four weeks post-vaccination then challenged with virulent PRRS 1-7-4 intranasally (1 ml in each nostril) and intramuscularly (2 ml in right neck) giving a total dose of 1.0×10^4 $TCID_{50}$/dose. After challenge, T01 and T02 groups were observed for 10 days and humanely euthanized for lung lesion evaluation. Figure 1 shows a summary of the study design and a timeline for clinical observations, weights and sampling.

The primary variable was macroscopic lung lesions, while secondary variables included: incidence of clinical signs, level and duration of viremia, innate immune response (IFN-α) and Average Daily Gain (ADG) post-challenge. The data were summarized and analyzed using a centralized data management system (SAS/STAT Vo. 9.3, SAS institute, Cary, NC), and only post-comingling data from T01 and T02 were statistically analyzed.

Transformed lung lesions were analyzed with a general linear mixed model and pairwise comparisons were made between treatment groups. Frequency distributions of clinical signs were described as the proportion of animals ever observed with an abnormal condition in the following categories: general condition; depression; respiratory distress and coughing. RT-qPCR and IFN-α results were transformed and analyzed by linear combinations of the parameter estimates used in *a priori* contrast after testing for a significant ($P \leq 0.05$) treatment effect or treatment by time point interaction. Average Daily Gain (ADG) post-challenge was analyzed using the linear functions of the least squares means for body weights for calculating ADG estimates.

Results and discussion

The T01 (Non-vaccinated) group remained PRRSV negative throughout the vaccination period and no confounding disease factors were detected in the strict control group (NTX).

Figure 2 shows mean percentage of lung with lesions at 10 days post-challenge, where T02 (vaccinated) had significantly fewer lung lesions ($P = 0.0066$) as compared to the T01 (non-vaccinated) group. Post-challenge, T02 (vaccinated) groups showed less incidence of clinical signs. The most prevalent abnormal clinical signs observed after challenge were depression, general condition and respiratory distress (Figure 3). For the depression category, 100% of non-vaccinated pigs were affected as opposed to only 65% of vaccinated pigs. For general condition and respiratory distress, incidences of abnormal clinical signs in the T01 (non-vaccinated) group (60%) were double those in the T02 (vaccinated) group (30%).

Viremia levels measured by RT-qPCR post-challenge indicated all pigs in group T01 and T02 were positive for PRRSV by three days post-challenge (Day 31) and remained positive through the final bleeding before end of the study (Day 37). T02 (Vaccinated) pigs had significantly less virus in serum at all post-challenge sampling points ($P \le 0.05$) (Figure 4). Interferon-alpha (IFN-α) titers were analyzed on days 27, 31, 33, 35 and 37 in both treatment groups. Prior to challenge, both groups were

Figure 1: Study design & timeline

TX	N	Vaccination	Challenge	Necropsy
NTX	6	NA	NA	Day 27
T01	20	Non-vaccinated	1-7-4 Isolate (Day 28)	10 days post-challenge (Day 38)
T02	20	Fostera® PRRS		

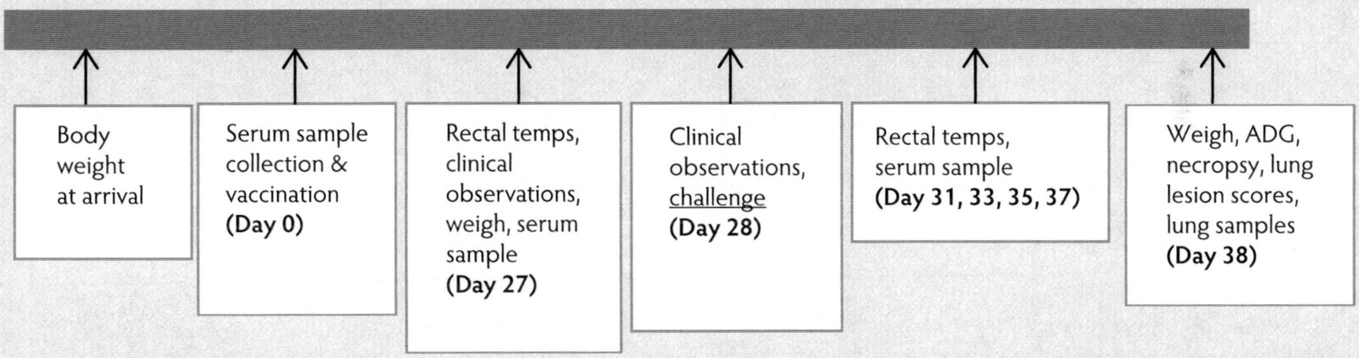

Figure 2: Back transformed least squares means percent lung with lesions.

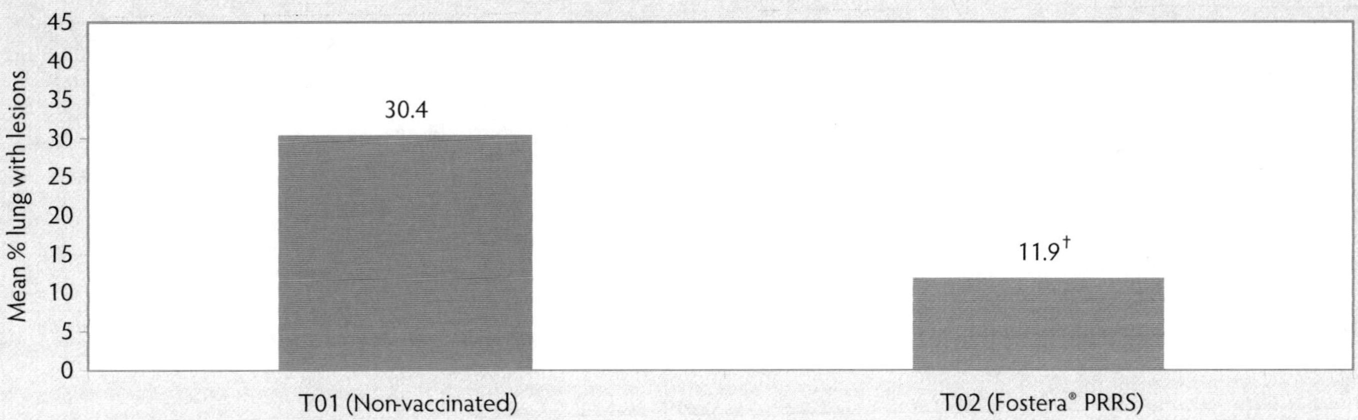

TX	N	% lung with lesions	Standard error	Lower 95% confidence interval	Upper 95% confidence interval	Range
T01	20	30.4	4.37	21	40.7	5.9-58.0
T02	20	11.9*	3.08	5.9	19.7	0.9-62.5

† Significant difference $P \le 0.05$
* $P = 0.0066$

very low (T01 ≤ 1.184 pg/ml, T02 ≤ 0.028 pg/ml). After challenge, titers increased in all pigs, however, titers were significantly lower in T02 compared to T01 on days 31, 33 and 35. Figure 5 shows a reduction of viremia and IFN-α post-challenge in the T02 (vaccinated) group compared to the T01 (non-vaccinated) group. IFN-α induced by vaccination (not sampled in this study) stimulates the production of pro-inflammatory cytokines and the establishment of acquired immunity. The strength of the interferon response is largely determined by the magnitude of viral replication. Thus high levels of viremia following challenge of naïve pigs with virulent PRRS can result in high levels of IFN-α, which in turn can directly contribute to clinical signs (malaise, depression, inappetence, fever) and poor growth performance. In contrast, reduced post-challenge viremia in Fostera® PRRS vaccinated pigs induces only low levels of IFN-α and thereby mitigates clinical signs, providing better conditions for continued growth in the face of a virulent PRRSV challenge.

Statistical comparisons of body weight and average daily gain (ADG) were made between treatment groups during the

Figure 3: Incidence of clinical signs post challenge. Proportion of animals ever observed with an abnormal condition.

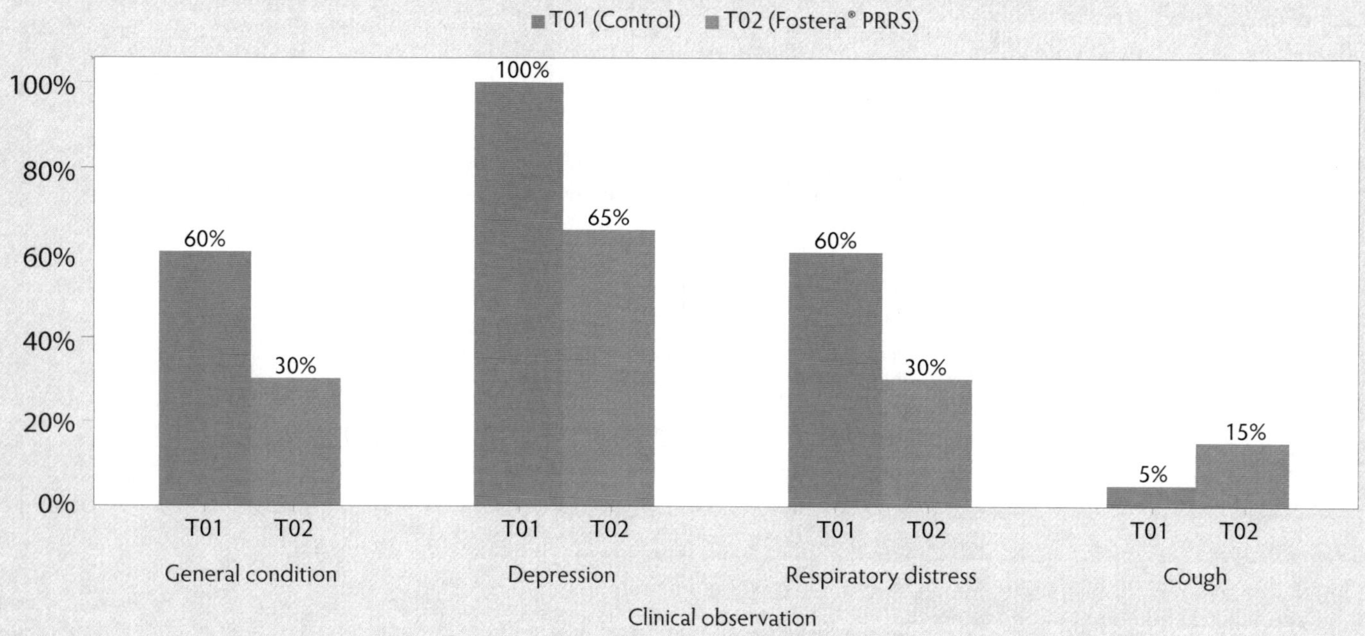

Figure 4: Level and duration of viremia (RT-qPCR)

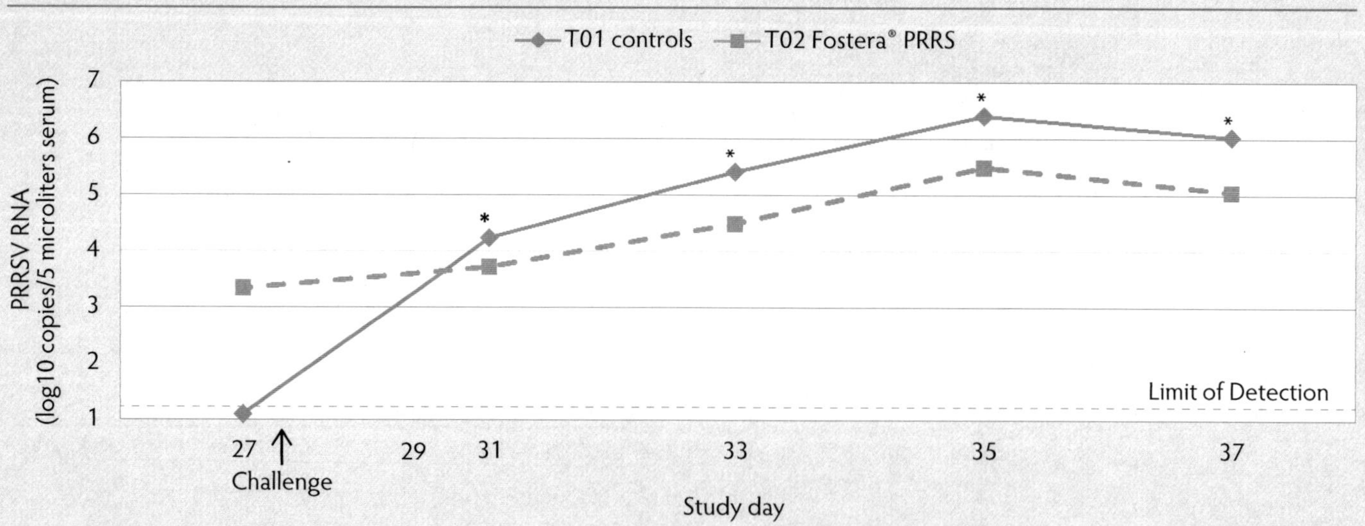

* Significant difference $P \leq 0.05$

post-challenge phase. Prior to challenge (day 27) and prior to day of necropsy (day 37), weights were not significantly different between groups. However, average daily gain during the 10 days post-challenge period was significantly higher in T02 (Vaccinated) than in T01 (Non-vaccinated) ($P = 0.0001$) (Figure 6).

Conclusions

This study demonstrates a protective vaccine effect in pigs vaccinated with Fostera® PRRS, showing a significant reduction of lung lesions caused by a virulent Lineage 1 (RFLP 1-7-4) PRRSV. In addition, pigs vaccinated with Fostera® PRRS had a 50% reduction in the clinical signs "general condition" and "respiratory distress" and a 35% reduction in "depression" compared to non-vaccinated pigs. Vaccinated pigs had significantly reduced viremia which translates to a decreased risk for spread of field PRRSV, and a reduced IFN-α spike. In addition, this study showed significantly higher ADG in vaccinated pigs during the 10 days post-challenge phase.

Figure 5: Innate immune response (IFN-alpha)

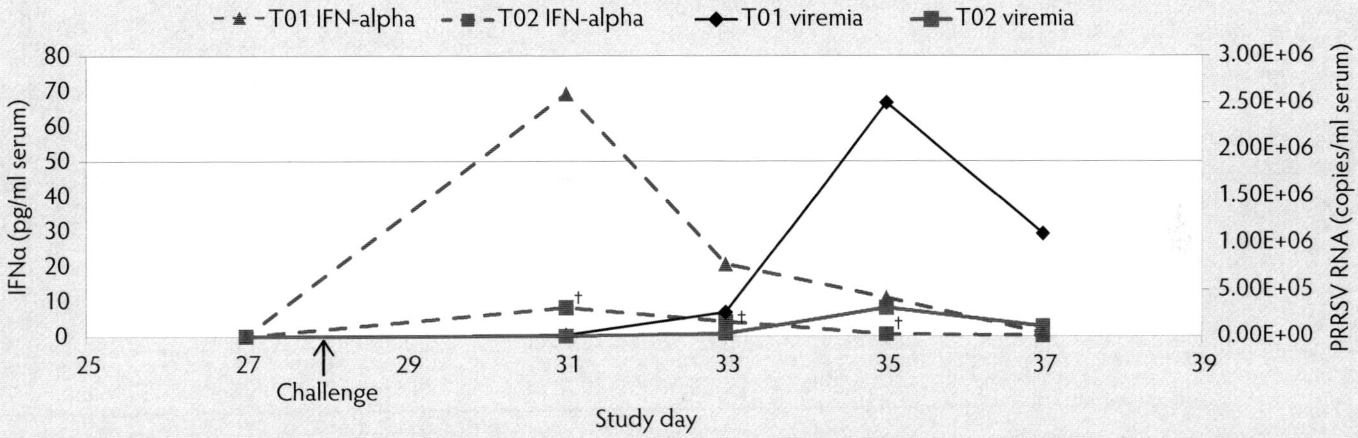

† Significant difference $P \leq 0.05$ for mean Serum IFN-Alpha titers.

Figure 6: Average Daily Gain (ADG) 10 days post-challenge

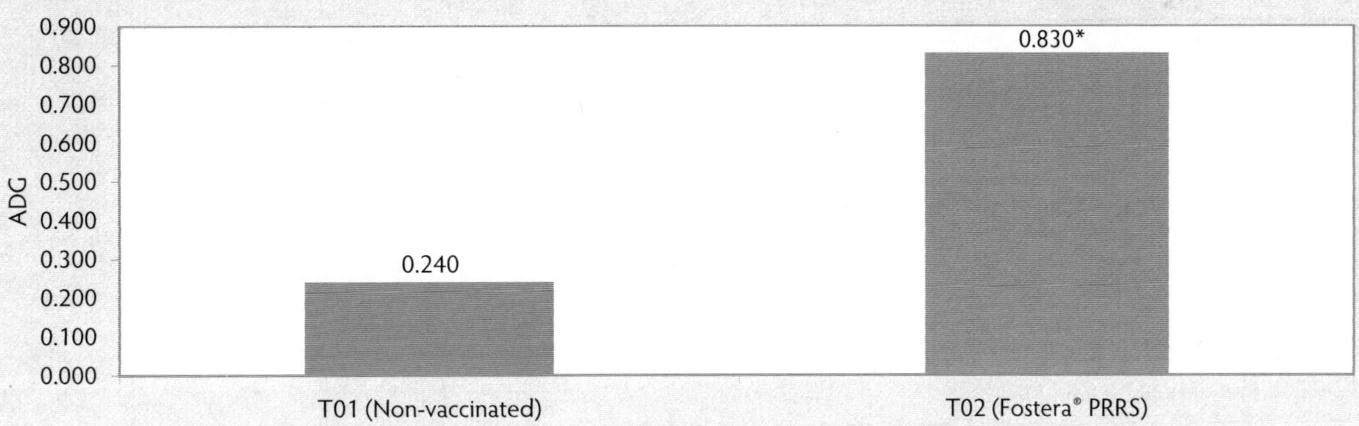

	Day 5	Day 27	Day 37	ADG D27 – D37
T01 (Non-vaccinated)	13.1	56.5	**58.9**	**0.24**
T02 (Fostera® PRRS)	12.9	54.1	**62.3**	**0.83***

* Significant difference $P \leq 0.05$
* $P = 0.0001$

References

1. Rovira A. PRRS diagnostic trends: What changed with PRRS behavior? AASV Proceedings. 2015. Pp. 509-511.

2. Murtaugh M. Use and interpretation of sequencing in PRRSV control programs: inside the black box. AASV Seminar #5 Full Sequencing Circle Proceedings. 2013. Pp 17-23.

3. Johnson A. A., Smith C., Almond G. Update of PRRSV 1-7-4 strain in North Carolina. AASV Proceedings. 2015. Pp 513-515.

4. Linhares D., Levis I., Wedel K., Betlach C., Morrison B. Assessing impact of 1-7-4 on wean to finish pig health. Leman Conference proceedings. 2015.

Comparison of immunological response, shedding profiles, and production performance between Fostera® PRRS and Ingelvac PRRS®MLV in the face of a heterologous PRRSV challenge

S. Dee[1]; B. O'Brien[2]; J. Nerem[1]; B. Morrison[3]; D. Hanson[1]; D. Amodie[2]
[1]Pipestone Veterinary Services, Pipestone, Minnesota; [2]Zoetis Inc., Florham Park, New Jersey;
[3]University of Minnesota, St. Paul, Minnesota

Introduction

Porcine reproductive and respiratory syndrome (PRRS) is widely recognized as one of the most costly diseases impacting pork production in the US. [1]Modified live virus (MLV) vaccines like Fostera®PRRS and Ingelvac PRRS®MLV are widely used to help mitigate the impact of field PRRSV introductions. Performance data for both of these MLV vaccines have demonstrated efficacy against genetically divergent PRRSV strains but have never been compared directly against one another in the US.[2,3] Therefore, the objective of this study was to compare the immunological response, shedding profiles and production performance between Fostera®PRRS and Ingelvac PRRS®MLV in the face of a lineage 1 PRRSV challenge.

Materials and methods

Two thousand two hundred and eighty three-week-old mixed breed swine were sorted by weight and gender and randomly allocated into three experiment groups: non-vaccinated and challenged controls (T01, n = 732), vaccinated with Fostera® PRRS and challenged (T02, n = 775), and vaccinated with Ingelvac PRRS® MLV and challenged (T03, n = 773). Pigs were sourced from a PRRSV negative, M hyo stable, and AIV-S positive farm. Following allocation, treatment groups were weaned into 3 separate locations. Vaccinated pigs (T02 and T03) were administered one 2 mL dose of either Fostera® PRRS or Ingelvac PRRS® MLV vaccine intramuscularly at 3 weeks of age (study day 0) while control pigs (T01) received no injection at this time. Four weeks post-vaccination, all treatment groups were commingled (27 pigs/pen) and challenged intramuscularly with 2 mls (2.8 logs of virus) of a lineage 1 PRRSV (R) with RFLP 1-26-2. A two hour cold stress was applied to pigs 7 day after PRRSV challenge. Pigs were then challenged intramuscularly with 2 mls (1.6 logs of virus) of a second lineage 1 PRRSV (W) with RFLP 1-26-2. Both field viruses were historically described as mildly virulent in commercial farms and were found to be 85-87% homologous to both vaccines and 5% heterologous to one another. Pigs that were moved to the sick pen were removed from the study at that time and all treatments and mortalities were recorded. Rectal temperatures (T02 = 38 and T03 = 39) and lethargy scores were obtained 6 and 18 hours post-vaccination. Nasal swabs were performed on 6 pigs at weaning for IAV-S pcr and ORF5 sequencing. Twenty three blocks of pigs were necropsied at ISU VDL 10 days after receiving PRRSV (R). The percentage of the surface area of each lung lobe affected with pneumonia was visually estimated by a single pathologist who was blinded to treatment group. Gross lesions consistent with bacterial consolidation were also scored for each lung set. Five sections of each right lung were collected, fixed, and scored for severity of interstitial pneumonia (0-6 scale) by a single histopathologist blinded to treatment group. Viral loads (qPCR) and PRRSV ORF5 sequencing was performed on the lungs. AIV-S PCR and sequencing and general bacterial and viral testing were performed on a subset of tissues to characterize secondary infections in the necropsied animals. Forty pigs per treatment group (paired) were bled at d0, 4, 7, 14, 21, 28, 35, 55, 74, 104, and 132 for PRRSV pcr (pooled by 5) and ORF5 sequencing and PRRSV ELISA was performed on T02 and T03 samples at d0, 7, 14, and 28. IFN α was analyzed using Meso Scale Discovery Assay (Kalamazoo, MI) on 10 individuals per vaccinated group at d0, 3, 5, and 7 and 10 pigs per treatment group at d28, 31, 33, and 35. Viremia was measured on serum collected from 10 pigs per treatment group at d35, 42, 55, and 74 by qPCR and the three samples with the highest viral load per treatment group was sequenced. PCR for PCV2, M hyo. and AIV-S was performed on oral fluids (6 ropes/room) obtained at d14, 27, 74, 104, and 132 and PRRSV PCR was performed on T01 oral fluid samples at d14 and 27. Aerosol samples were collected using two liquid cyclonic collectors (each dedicated to T02 and T03) using previously published techniques daily during the vaccination phase (d0-27).[4] Each collector was capable of collecting 400 L air/min. Four air samples were collected at 30 minute intervals between 6 and 8 am at 1 meter from each exhaust fan as well as inside the barn approximately 3 feet directly above the animals. Air samples were tested by PRRSV PCR and any positive samples had virus isolation on MARC cells and PRRSV ORF5 sequencing performed. All diagnostic testing was performed at ISU VDL. Pigs were weighed at allocation (d-1 to -3), d14, d27, d55 and d138 and average daily gain (ADG) was calculated for each time period. On the final weigh day (d138) the number of defect culls (physical deformities like umbilical hernia, tail bite, lames), light culls (animal back-calculated at < 230 lbs. on final marketing day), and full value pigs were assigned.

Descriptive statistics was used for d0-27 when treatment groups were housed separately. After d28, the individual pig was the experimental unit. This study was designed as a split-plot. The whole-plot factor was gender while the split plot factor was treatment. The whole-plot experimental design was a completely randomized design (CRD) and the split plot experimental design was a randomized complete block design (CRBD). Body weights, Average Daily Gain, Diagnostic (transformed by the Log_{10} or Log_2), lung lesions scores (transformed by arcsin(square-root)) and number of re-treatments were analyzed by a Generalized Linear Mixed Model (GLMM) approach. Using the Proc Mixed Procedure (SAS 9.3, Cary NC) these data were analyzed with a model that considered the fixed effects of treatment, gender and the interaction of treatment-by-gender and the random effects of commingling room, pen-within-room, block-within-pen-and-room and the residual error. Wean weight was included as a covariate in the analyses of all other weight days as well as ADG. Full value, defect culls, light pigs and removals/mortalities were analyzed by SAS Proc Glimmix. Treatment and treatment-by-gender were tested at the 5% level of significance. Treatment Least Squares Means (LSMeans) were calculated for each group. Comparison of LSMeans was performed by the two-sided Student's t-test (Fisher protected) if the overall analysis of treatment was significant at the 5% level.

Results

All weaned pigs were confirmed PRRSV negative on PCR and ELISA. Consistent with clinical signs observed in the weaned pigs, nasal swabs obtained from weaned pigs confirmed the presence of two strains of IAV-S (H1N1 and H1N2). Post-weaning diarrhea (historically *E. coli*) was evident during the first two weeks post-weaning in both vaccinated groups but not in the control group. Fostera®PRRS group was IAV-S positive (confirmed by pcr and clinical signs) at the time of first PRRSV challenge. Secondary infections characterized 10 days post-PRRSV challenge were hemolytic *E. coli*, *Rotavirus* type A, B, and C, AIV-S and *Strept. Suis*. Clinical signs were seen in pigs following the first PRRSV challenge (R) but were minimal after the second (PRRSV-W). Oral fluid results show evidence of late-finishing IAV-S and M hyo circulation in the barn at d132. There were numerical reductions in total lung lesion scores of vaccinates versus control animals (T01 = 18.7, T02 = 15.4, T03 = 15.8). There was a trend in pigs vaccinated with Fostera®PRRS for a reduction in total percent lung lesions without gross bacterial consolidation versus the control group (T01 = 17.7 vs T02 = 13.1; $P = 0.13$). There were no differences noted between treatment groups in mean histological scores (T01 = 3.14, T02 = 2.89, T03 = 2.58; $P \leq 0.05$) or tissue viral loads (T01 = 3.7×10^7, T02 = 3.0×10^7, T03 = 1.1×10^7; $P \leq 0.05$).

Both vaccines had significantly higher ADG during the wean to finish period (T02 = 1.56 and T03 = 1.57) than the control group (1.54) but were not different from one another ($P \leq 0.05$). Likewise, final weights for both pigs vaccinated were significantly higher at the end of the study than controls (T01 = 226.1, T02 = 228.4, T03 = 229.3, $P \leq 0.05$). All comparative weights and ADG are shown in Table 1. Pigs vaccinated with Fostera®PRRS weighed 0.4 lbs. more than pigs vaccinated with Ingelvac PRRS®MLV two weeks post-vaccination. There were no differences between any of the treatment groups in the additional production performance variables measured (% full value pigs, light culls, defect culls, mortality and removals). There was also no difference between groups on treatment rate.

Fostera®PRRS and Ingelvac PRRS®MLV both demonstrated minimal aerosol shed post-vaccination. There was only 1 positive PRRSV pcr sample for each vaccinated group isolated from sampling performed outside of the exhaust fan. Likewise, there was little aerosol shed during the vaccination phase on interior barn samples collected (4 positive PRRSV pcr for Fostera®PRRS and 1 positive for Ingelvac PRRS®MLV). PRRSV pcr CT values for the positive air samples ranged from 34-36.7 and there was insufficient virus present to either sequence or grow on MARC cells. Post-vaccination rectal temperatures were not different between groups at either 6 or 18 hours and averaged 103.4°F at 6 hours and 103.2°F at 18 hours. There were more lethargic pigs 6 hours post-vaccination in the group vaccinated with Ingelvac PRRS®MLV than those vaccinated with Fostera®PRRS (11 vs. 1 pig) but no difference at 18 hours.

As expected, both vaccination groups were 100% viremic 7 days following PRRSV vaccination. Figure 1 shows viremia profile of each treatment group over the course of the trial. Viremia began to drop off first in the Fostera®PRRS group at d55. All groups were no longer shedding PRRSV in serum by d104. Pigs vaccinated with Fostera®PRRS had a significant reduction of viremia 7 days post-challenge with PRRSV when compared with the other two treatment groups (see Figure 2). All vaccinated pigs were PRRSV ELISA positive 14 days post-vaccination. Both vaccinated groups showed a mild increase in IFNα following challenge with PRRSV-R but this response was significantly lower than that of controls at 3 and 5 days post challenge (See Figure 3).

PRRS sequencing confirmed that there was no cross-contamination of groups during the vaccination phase. After PRRSv challenge, only challenge viruses were sequenced out of serum samples. Only the PRRSV-R strain was sequenced out of Fostera®PRRS group (on d35,42,55,74) while the PRRSV-W strain was the predominate strain recovered from both the control and Ingelvac PRRS®MLV at d55 (18 days after PRRSV-W challenge).

Discussion

Consistent with results in the field, both modified-live PRRSV vaccines (Fostera®PRRS and Ingelvac PRRS®MLV) allowed pigs to keep growing in the face of a PRRSV challenge and more so than unvaccinated animals. As expected, there were no differences in production performance between Fostera®PRRS and Ingelvac PRRS®MLV. There was a trend for reduced lung lesions

Table 1: Comparison of weights and ADG (LSMeans)

Variable (lbs)	Control	Fostera®PRRS	Ingelvac PRRS®MLV
allocation weight	11.84	11.85	11.83
D14 weight	17.60	16.63	16.23
D27 weight	30.45[a]	28.40[b]	27.82[c]
D55 weight	61.89	62.06	62.80
D138 weight	226.1[b]	228.4[a]	229.3[a]
ADG d0-27	0.67	0.59	0.57
ADG d28-138	1.77[b]	1.81[a]	1.83[a]
ADG wtof	1.54[b]	1.56[a]	1.57[a]

[a,b,c] Means in rows with different superscripts are significantly different ($P \leq 0.05$).

in the Fostera®PRRS versus the controls. It is important to note that the lack of significance between treatment groups on lung lesions is most likely a consequence of the mildly virulent PRRSV strains used in this study.

While no conclusions about vaccine setback can be drawn between vaccinates and controls given the post-weaning diarrhea that affected the vaccine groups and not the control group and given they were housed separately, the 0.4 pound advantage in the Fostera®PRRS two weeks post-vaccination and reduced lethargy scores immediately following vaccination warrant further investigation into setback differences between these MLV vaccines.

Previous studies have demonstrated that there is minimal aerosol shed in pigs vaccinated with Fostera®PRRS and this study's findings are in agreement with the prior work done in this area.[5,6] This study like others previously published show a reduction in viremia post-PRRSV challenge in pigs vaccinated with Fostera®PRRS.[2] This reduction in viremia was only observed in the Fostera®PRRS group in this study.

When comparing immunological response of both vaccines, both vaccines elicited a transient controlled elicitation of IFNα production following vaccination. Likewise, because of the primed immune system both vaccines exhibited a more controlled response in the face of PRRSV challenge versus the unvaccinated pigs as evidenced by the significantly reduced IFNα immediately post-challenge. Interferons are induced by double-stranded RNA and other by-products of viral replication and the magnitude of the IFN response is largely determined by the magnitude of viral replication. In this study, we see vaccination with Fostera®PRRS was sufficient to reduce peak viremia and peak IFN levels following challenge with an antigenically diverse 1-26-2 PRRS virus.

This study demonstrated the ability of Fostera®PRRS to protect pigs in the face of a lineage 1 1-26-2 PRRSV challenge.

References

1. Holtkamp D. et al. Assessment of the economic impact of porcine reproductive and respiratory syndrome virus on United States pork producers. *Swine Health and Production*. 2013. 21:72-84.

2. Angulo J. et al. Efficacy of a PRRSV MLV vaccine against a genetically diverse range of PRRSV isolates. *Leman Proceedings*. 2015. 16.

3. Patterson A. et al. Modified live PRRSV vaccination is efficacious following challenge with eight genetically diverse PRRSV isolates. *AASV proceedings*. 2014. 425-426.

4. Pitkin A, Deen J, Dee S. Use of a production region model to assess the airborne spread of porcine reproductive and respiratory syndrome virus. *Vet Microbiol*.2009. 136:1-7.

5. Yeske P. Evaluation of airborne shedding of Fostera PRRS MLV vaccine from growing pigs raised under commercial conditions. *Leman proceedings*. 2013. 191.

6. Samson A. et al. Evaluation of aerosol shedding in growing pigs following administration of porcine reproductive and respiratory syndrome virus vaccine. *AASV proceedings*. 2013. 89-90.

Figure 1: PRRSV viremia in serum.

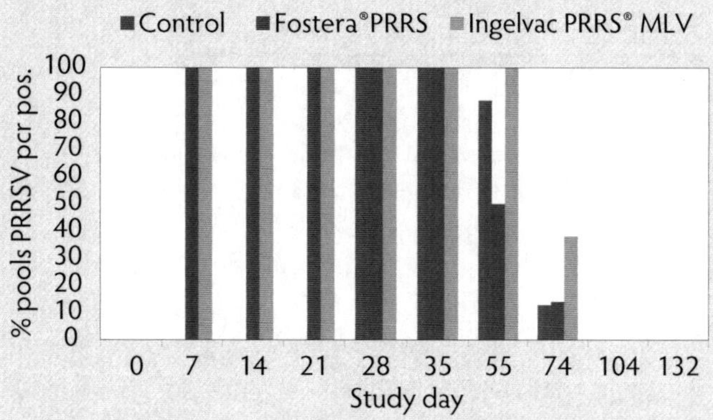

Figure 2: PRRSV viremia (qPCR) in serum post-challenge with PRRSV (Back transformed LS means).

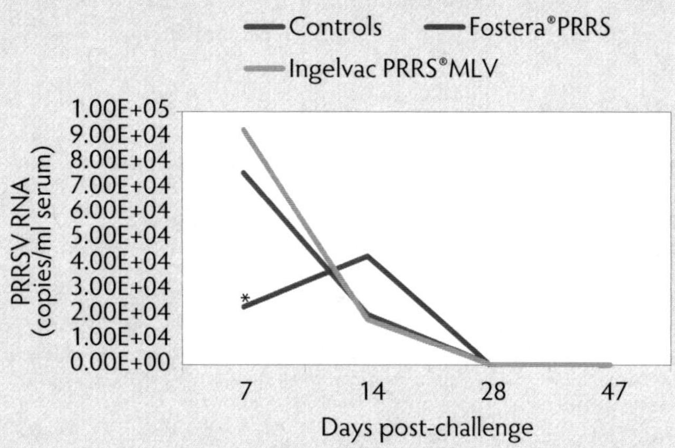

* Significant difference in viremia at $P \leq 0.05$.

Figure 3: IFNα levels post-challenge.

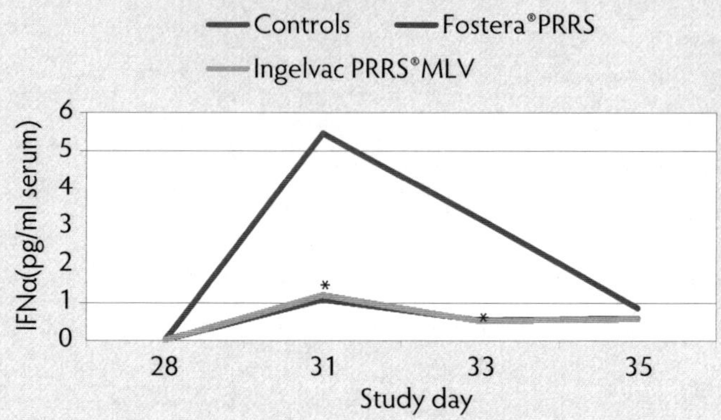

*Significant difference in IFN-alpha of $P \leq 0.05$.

Field evaluation of the intranasal route as an alternative for administering *Streptococcus suis* and *Haemophilus parasuis* bacterins

R. Jones [1],DVM; B. Heins[1], DVM; G. Weaver [2], DVM; J. Kula[2], DVM, MAM; B. Lin [2],DVM, PhD; W. Johnson [2], BS
[1]Livestock Veterinary Services, Kinston, North Carolina; [2]MVP Laboratories, Inc., Omaha, Nebraska

Introduction

Major economic losses in commercial swine nursery facilities caused by *Streptococcus suis*-related infections and by Glässer's Disease attributable to *Haemophilus parasuis* have been well documented.[1,2] The majority of the mortality observed in these young pigs occurs between three and five weeks post-weaning for *S. suis* infections and between four and six weeks post-weaning for *H. parasuis* infections. Control of clinical disease associated with *S. suis* and with *H. parasuis* has been attempted by means of antibiotic treatment, by controlled exposure, and by vaccination, using either inactivated commercial or autogenous bacterins. An accepted strategy for protecting weaned pigs entering this stage of production involves vaccination using appropriate bacterins at the standard time of piglet processing followed by revaccination at weaning. The present study investigated the subsequent nursery performance of pigs administered a federally-licensed autogenous *S. suis/H. parasuis* bacterin applying this vaccination regimen by the intranasal (IN) route as an alternative to the traditional intramuscular (IM) route of injection.

Materials and methods

Animals

A commercial sow farm of 2,000 sows in North Carolina experiencing significant nursery mortality from *S. suis* and/or *H. parasuis* infection was selected for evaluation of the different vaccination routes/regimens of piglets. Nursery pigs from this source were positive for porcine reproductive and respiratory syndrome virus (PRRSV), porcine circovirus type 2 (PCV2), influenza A (IAV-S), and *Mycoplasma hyopneumoniae.* Pigs were weaned at 21 days of age and transferred to a single-source, off-site nursery. The nursery contained four rooms which were filled with approximately 900 pigs each. When clinical disease occurred in the nursery, individual animals were given an antibiotic injection. No feed or mass water medications were administered.

Vaccine development

The two licensed autogenous vaccines used in the present study were prepared at MVP Laboratories using multiple isolates of *S. suis* (serovars 1/2, 2, and 3) and *H. parasuis* (serovars 2, 7, and non-typeable). These specific bacteria were identified from previous farm visits and by subsequent diagnostics to be the strains primarily responsible for causing clinical disease and mortality in the nursery. The bacterins were prepared using a formalin-inactivated inoculum adjuvanted with aluminum hydroxide. The vaccines differed in their production in that they were further adjuvanted with either an oil-in-water adjuvant (intramuscular injection) or with a proprietary adjuvant (intranasal administration).

Experimental design

Piglets were enrolled at the sow farm into the field investigation over five consecutive weeks of production. Enrollment occurred at the time of processing at 3-5 days of age when animals were identified to the respective treatment groups by colored ear tag. Animals were randomly assigned to one of four vaccination regimens (Table 1): 1) IN administration at processing and repeated at weaning; 2) IN administration at processing and IM injection at weaning; 3) IM injection at processing and repeated at weaning; 4) IN administration at processing only. Piglets within litters were vaccinated so as to equally represent each of the four treatment groups as much as possible. At weaning, treatment groups of pigs were not sorted to separate pens. The animals were randomly mixed together in order to promote uniform pathogen exposure. However, for the purpose of data collection the test groups were maintained separately by week of production (week 1 through week 5).

Laboratory testing

From pigs in the week 3 production group, 10 serum samples were collected from each of the four treatment groups at approximately three weeks after the weaning vaccinations. The samples were analyzed by homologous ELISA using soluble proteins obtained from *H. parasuis* vaccine strains as coated antigens in an assay method described previously.[3]

Parameters measured

Over the five weeks of study observation, the parameters recorded were pre-weaning mortality, percent nursery mortality, percent treated animals (antibiotic injected upon presence of clinical disease), and the percent poor quality pigs at end of nursery phase. Due to a miscommunication with the farm staff, week 5 data were recorded in a manner not valid for use in the study database.

Statistical analysis

Data collected by treatment group for production weeks 1 through 4 were analyzed. The week of production was used as the

Table 1: Study design

Treatment	N	IN vacc.† at processing	IN vacc.‡ at weaning	IM vacc.† at processing	IM vacc.‡ at weaning
T01	911	√	√	---	---
T02	911	√	---	---	√
T03	929	---	---	√	√
T04	848	√	---	---	---

† 1.0 mL dose of autogenous *S. suis/H. parasuis* bacterin
‡ 2.0 mL dose of autogenous *S. suis/H. parasuis* bacterin

experimental unit. Main effects ANOVA was performed with percent nursery mortality, total percent mortality (pre-weaning mortality plus nursery mortality), percent treated animals, and percent poor quality pigs as the dependent variables and vaccination treatment as the independent variable.

Results and discussion

During the nursery phase (Figure 1), the IN/IN and IN/IM vaccinated groups had average mortality of 3.70% (sd 2.77%) and 9.13% (sd 6.67%), respectively. The average mortality observed in the IM/IM vaccinated group was 5.03% (sd 3.49%) whereas the average mortality observed in the IN vaccinated (only at processing) group was 1.42% (sd 1.79%). Across treatments, the type of vaccine administered did not significantly affect the mortality rate in the nursery ($P > 0.05$).

For total mortality (Figure 2) which included both the pre-weaning and nursery mortality observed, the IN/IN and IN/IM groups had average mortality of 12.7% (sd 2.61%) and 14.5% (sd 7.68%), respectively. The IM/IM group was 12% (sd 3.67%) whereas the average total mortality observed in the IN (only at processing) group was 11.3% (sd 1.58%). Across treatments, the type of vaccine administered did not significantly affect overall mortality ($P > 0.05$).

For animals treated (Figure 3), the IN/IN and IN/IM groups had an average treatment rate of 8.78% (sd 3.84%) and 9.53% (sd 6.17%), respectively. The IM/IM group was 7.63% (sd 3.53%) while the average treatment rate observed in the IN (only at processing) group was 6.88% (sd 4.82%). Across treatments, the type of vaccine administered did not significantly affect the rate of antibiotic treatment of pigs during the nursery phase ($P > 0.05$).

For nursery pigs grading as poor quality (Figure 4), the IN/IN and IN/IM groups had average poor quality of 22.60% (sd 4.91%) and 24.53% (sd 8.39%), respectively. The IM/IM group was 22.65% (sd 6.02%) whereas the rate of poor quality pigs observed in the IN (only at processing) group was 22.23% (sd 11.63%). Across treatments, the type of vaccine administered did not significantly affect the percentage of poor quality pigs observed at the end of the nursery phase ($P > 0.05$).

Of the animals initially enrolled at processing, the nursery performance for the parameters measured showed no differences when comparing vaccination treatment groups either by route of administration or by regimen.

It is well accepted that killed vaccines stimulate humoral immunity resulting in the production of both IgG and IgM antibodies. Interestingly, the majority of animals in each of the treatment groups in this investigation also responded with measurable ELISA IgA titers to *H. parasuis* (vaccine strain) indicating that the pigs had largely (> 80%) seroconverted (Table 2). The lower rate of IgA titers elicited by the IN (only at processing) group was the exception to this trend. In addition, the IN/IM and IM/IM groups were the only treatments that resulted in > 80% rate of seroconversion as measured by IgG titer. Pigs in this study appeared to respond very similarly between the IN and IM routes at the time of processing when boosted at weaning by the IM route.

In the commercial setting, the intranasal administration technique has several distinct advantages when compared with traditional parenteral vaccination including ease of administration at processing, lessened risk of needle loss, and reduction of workload. Furthermore, the intranasal route may be less susceptible to maternal immunity interference. The overall comparable results when using intranasal administration in this study invite further investigation into this novel approach as an alternative to parenteral injection.

Acknowledgement

The authors would like to thank Dr. Jeff Zimmerman, Iowa State University College of Veterinary Medicine, for his assistance in analyzing the data.

References

1. Gottschalk, M. Streptococcosis. In *Diseases of Swine*. 10th Edition. Iowa State University Press, pp. 841-855.

2. Oliveira, S. et al. (2002). The role of *Haemophilus parasuis* in nursery mortality. In Proceedings Allen D Leman Swine Conference, pp. 111–113. University of Minnesota, USA: Veterinary Outreach Programs.

3. Lin, B. et al. Antibody response of young pigs to autogenous *Haemophilus parasuis* vaccine. AASV 2007:207.

Figure 1: % Nursery mortality.

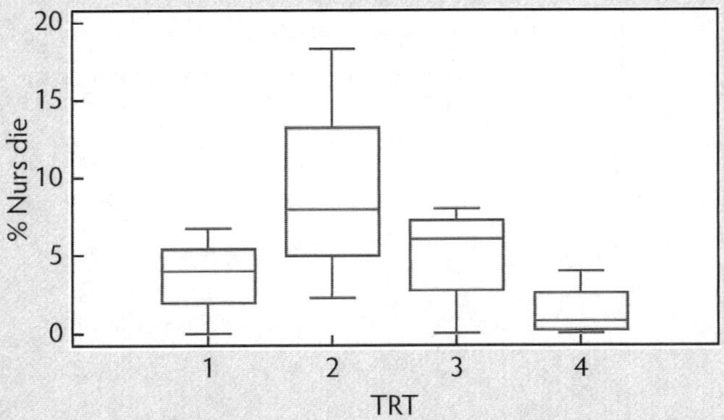

Figure 2: % Total mortality (PWM + Nursery mortality).

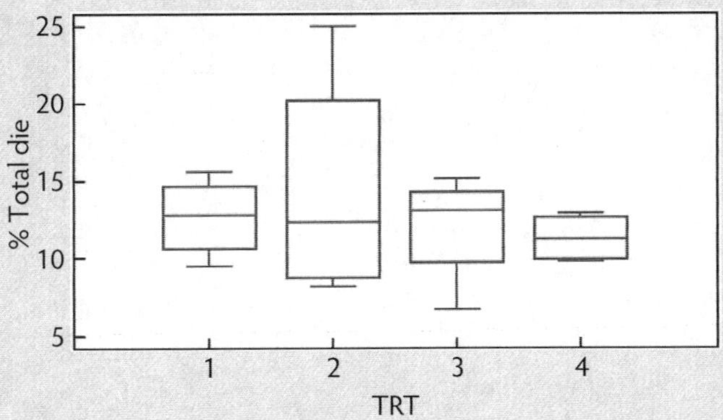

Figure 3: % Treated animals during nursery phase.

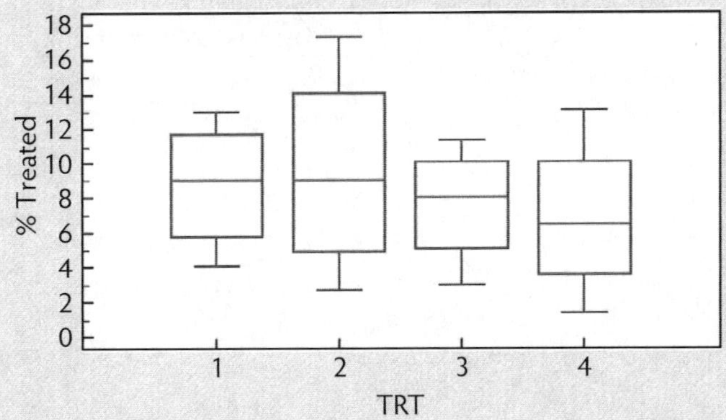

Figure 4: % Poor quality pigs at end of nursery phase.

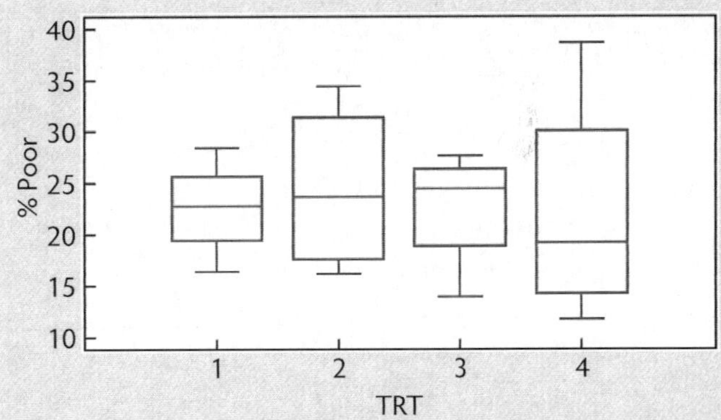

Table 2: *H. parasuis*-specific ELISA results.

Treatment	IgA response			IgG response		
	OD[1,2]	Seroconversion[3]	End-point titer	OD[1,2]	Seroconversion[3]	End-point titer
	0.358	+	200	0.245	-	0
	0.548	+	400	0.315	+	200
	0.604	+	800	0.407	+	200
	0.447	+	200	0.306	+	200
T01	0.351	+	200	0.220	-	0
(IN/IN)	0.263	-	0	0.275	-	0
	0.400	+	200	0.211	-	0
	0.400	+	200	0.264	-	0
	0.947	+	800	0.262	-	0
	0.354	+	200	0.258	-	0
Average	**0.467**		**320**	**0.276**		**60**
	0.272	-	0	0.325	+	200
	0.467	+	200	0.740	+	800
	0.552	+	400	0.415	+	200
	0.381	+	200	0.657	+	800
T02	0.340	+	200	0.617	+	800
(IN/IM)	0.378	+	200	0.633	+	800
	0.661	+	800	0.546	+	400
	0.365	+	200	0.369	+	200
	0.392	+	200	0.600	+	800
	0.553	+	400	0.491	+	200
Average	**0.436**		**280**	**0.542**		**540**
	0.453	+	200	0.639	+	800
	0.250	-	0	0.286	-	0
	0.591	+	400	1.076	+	1600
	1.393	+	3200	0.789	+	800
T03	0.358	+	200	0.498	+	200
(IM/IM)	0.578	+	400	0.463	+	200
	0.798	+	800	0.847	+	800
	0.415	+	200	0.639	+	800
	0.431	+	200	0.918	+	800
	0.432	+	200	0.227	-	0
Average	**0.570**		**580**	**0.638**		**560**

1 All samples diluted to 1:200.
2 Mean OD reading of positive control and negative control approximated 1.500 and 0.200, respectively (at 1:400).
3 Cut-off point: S/N >2.00.

Table continued on next page

Table 2 Continued:

Treatment	IgA response			IgG response		
	OD[1,2]	Seroconversion[3]	End-point titer	OD[1,2]	Seroconversion[3]	End-point titer
	0.268	-	0	0.226	-	0
	0.297	-	0	0.311	+	200
	0.354	+	200	0.515	+	400
	0.350	+	200	0.501	+	400
T04	0.341	+	200	0.533	+	400
(IN 1X)	0.270	-	0	0.423	+	200
	0.537	+	400	0.289	-	0
	0.214	-	0	0.260	-	0
	0.577	+	400	0.636	+	800
	0.277	-	0	0.318	+	200
Average	**0.348**		**140**	**0.401**		**260**

1 All samples diluted to 1:200.
2 Mean OD reading of positive control and negative control approximated 1.500 and 0.200, respectively (at 1:400).
3 Cut-off point: S/N > 2.00.

Quantitative assessment of prevalence and severity to understand impact of *Lactobacillus acidophilus* fermentation product (SynGenX®) in herd with endemic K88 and Salmonella post-wean

S. Probst Miller[1], DVM; A. Ramirez[2], DVM, MPH, PhD, Diplomate ACVPM; B. E. Bass[3]; J. W. Frank[3]
[1]AgCreate Solutions; Inc., [2]Iowa State University; [3]Diamond V

Introduction

Pathogenic *E. coli*, such as K88 and F18, can cause disease in pigs post weaning.[2] Salmonella may also be problematic as a primary or secondary infection.[3] This case study examines a herd with a history of endemic K88 *E. coli* and Salmonella diarrhea post wean (Figures 2). Signs at this farm included acutely: diarrhea and progressive dehydration; subacutely: acute signs plus gradual fat loss; and chronically: muscle loss and rough hair coat (Figure 1). In this paper we refer to the degree of progression of these signs as acute, subacute, or chronic fallouts (Figure 1). This diarrhea and pathogen combination has been present and repeatedly diagnosed since the flow was established 5+ years ago. Histopathology shows coliforms adhered to microvilli of intestinal epithelial cells and some necrosis of villi.[4] Whether the Salmonella is primary or secondary is often hard to discern.[3] While control has been sporadic with both vaccination and medication attempts, the most consistent control has occurred with various oral antibiotic water medications. Currently Neomycin therapy is used; however in spite of treatment, recovery has been slow and significant fallouts occur resulting in the need for follow-up therapy or extended therapy. The farm desired less dependence on antibiotics to control Colibacillosis and Salmonellosis in this herd. In an effort to increase treatment efficacy and potentially reduce dependence on antibiotics and support health and performance of nursery pigs, a *Lactobacillus acidophilus* fermentation product (SynGenX®; Diamond V, Cedar Rapids, IA) was examined as support to current control methods. This product has been shown to alter enteric bacteria concentrations and the immune response and therefore may assist pigs in coping with both *E. coli* and Salmonella before they cause disease.[1] SynGenX® was evaluated as a part of therapy (implemented after diarrhea started in herd) and as a part of prevention (implemented immediately post entry).

Materials and methods

SynGenX® as a part of therapeutic application

In the therapy group, subpopulations of 150 out of 326 pigs on the South (control) and 160 out of 315 pigs on the North (treatment); SynGenX®, 2 pounds per ton fed continuously) were assessed each week post entry for prevalence and severity of clinical signs. These groups were across the aisle from each other in the same air space. Pigs were not treated with Neomycin or SynGenX® for 2 weeks post entry until diarrhea was established in the groups (Figure 2). Quantitative assessments of severity and prevalence were done weekly. At 2 weeks post entry, a 36% diarrhea prevalence was observed on the treatment side and a 40% diarrhea prevalence was observed on the control side. Diagnostics (live pigs submitted to lab) were taken from both sides of the barn; and both K88 *E. coli* and Salmonella were isolated from a pig in the treatment group and from a pig in the control group in the same barn.[4] With diarrhea and diagnosis established, the typical treatment for diarrhea in this herd (Neomycin) was administered to both groups. SynGenX® was included in the feed on the treatment side of the barn. Quantitative assessments on severity and prevalence were done in both subpopulations once a week. The clinical signs of Fallout, Respiratory, Diarrhea, Lameness,

Figure 1: Examples of fallout severity levels

Figure 2: Example of relatively equal diarrhea and fallouts prior to implementation of therapy.

Neurologic and Other were measured each week for both prevalence and severity of respective signs in the barn's subpopulation. Severity was divided into the following categories: acute, subacute, chronic, or euthanize/cull (Figures 1). Death loss, stools per pen, and injectable treatments were measured on both sides of the barn as well.

SynGenX® as a part of a prevention plan

Investigators monitored the impact of SynGenX® (2 pounds per ton fed continuously through study) and Neomycin in the water (1 week) on prevalence and severity of clinical signs compared to pigs receiving therapy with 1 week of Neomycin alone. Five barn sections of approximately 600 pigs per 4 sections and 1200 pigs per 1 section were enrolled in this study. Sections were defined by equivalent air space. In equivalent air spaces, the treatment side was the north side of each barn and the control was the south side of each barn. Severity was divided into the following categories: acute, subacute, chronic, or euthanize/cull. The clinical signs of Fallout, Respiratory, Diarrhea, Lameness, Neurologic and Other were measured each week for both prevalence and severity of respective signs in the barn's measured population. Death loss, stools per pen, and injectable treatments were measured on both sides of the barn as well. Statistical analysis was performed using Fisher's Exact test for proportions using MedCalc 15.11 (MedCalc Software, Ostend, Belgium). Pairwise comparisons with a P value < 0.05 were considered statistically significant.

Results

SynGenX® as a part of therapeutic application

In this treatment barn, 2 weeks after SynGenX® was implemented, this barn saw a fallout reduction trend as early as 1 week post administration with 33% fallout on the control side and 21% fallouts on the treatment side and saw significant reduction ($P < 0.001$) in total fallouts at 3 weeks post administration with 11% fallouts on the control side and 1% fallouts on the treatment side (Figure 4, Table 1). One week post implementation of SynGenX® we saw a significant reduction in the amount of chronic diarrhea ($P = 0.012$) (Figure 5) and a significant reduction in the amount of diarrhea stools per pen ($P = 0.009$) (Figure 6).

SynGenX® as a part of a prevention plan

The difference in death loss was statistically significant ($P = 0.003$) (Figure 7). Total death loss on the treatment side was 0.84%. Total death loss on the control was 1.91%. In this study, death loss on the treatment side was 56% less than the control. The difference in number of necessary individual treatments was statistically significant as well. Total injectable treatments in the treatment group were 5.4% of the herd. Total injectable treatments on the control side of the herd were 9.1%. (Figure 8) In this study the number of necessary individual treatments on the treatment side was 40% less than the control. In all prevention barns, total fallouts were significantly reduced at two weeks post entry when the diarrhea is typically most severe (Table 1). The severity level that was reduced varied depending on the barn section. At 2 weeks post entry, acute fallouts were significantly reduced post entry in 3/5 barns (Barn 3 NE $P = 0.016$, Barn 3 NW $P < 0.0001$, Barn 4 NE $P = 0.031$); subacute fallouts were significantly reduced in 3/5 barns (Barn 1 N $P = 0.011$, Barn 3 NW $P < 0.001$, Barn 4 NW $P = 0.041$); chronic fallouts were significantly reduced at 2 weeks post entry in 1/5 barns (Barn 3 NW $P = 0.002$). Of interest, in our therapeutic application barn, where no Neomycin or SynGenX was instigated yet in our treatment group, there was no reduction in fallouts on the treatment vs the control side at 2 weeks post entry. However 3 weeks after SynGenX® was implemented this barn saw a significant reduction in total fallouts ($P < 0.001$) (Table 1). Stools per pen were significantly reduced in barns experiencing diarrhea at 2 weeks post entry (Barn 3 NE $P < 0.001$, Barn 3 NW $P < 0.001$, and Barn 4 SW $P = 0.004$) and at 5 weeks post entry (Barn 1 N $P < 0.001$, Barn 2 NE $P = 0.009$, Barn 3 NW $P = 0.035$).

Figure 3: Comparison of treatment (left) to control (right) two weeks post initiation of therapy.

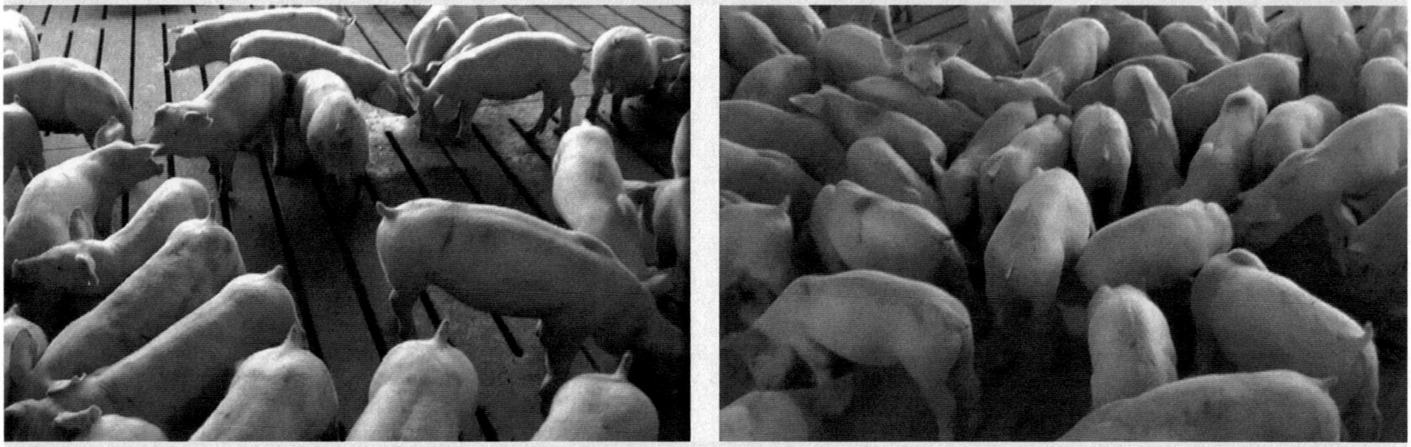

Figure 4: Therapeutic application group: Total fallout comparison--Treatment instigated was one week of Neomycin in both the NE (treatment) and SE (control) sides and implementation of SynGenX® fed continuously on NE (treatment) side.

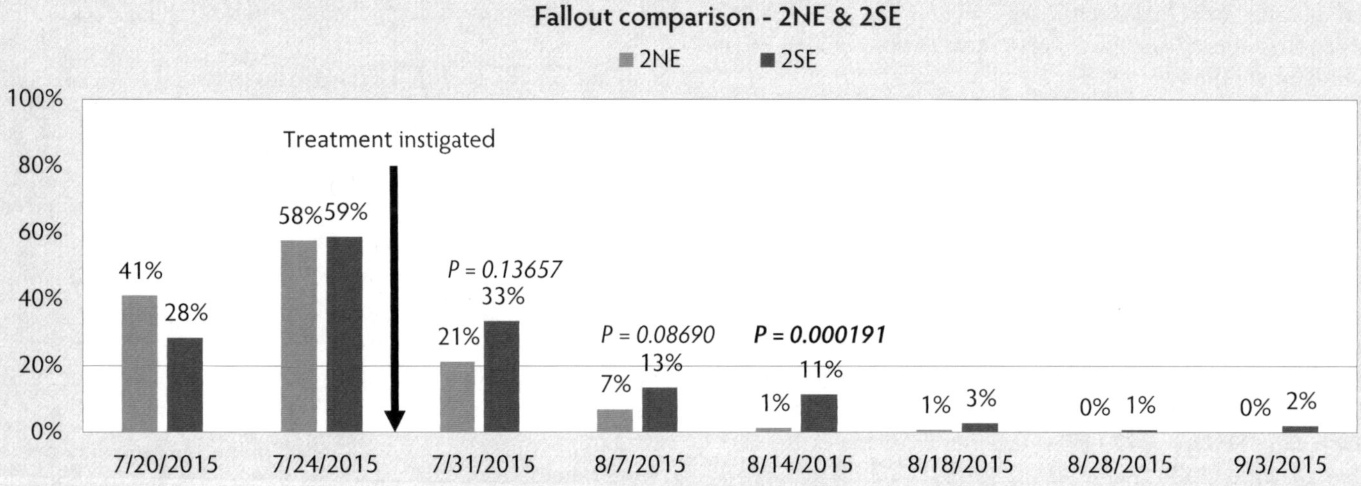

Conclusions and discussion

Regular quantitative assessment of prevalence and severity of symptoms is essential to understanding impact of therapeutic and preventative intervention. Therapy with SynGenX® reduces death loss and the number of pigs that needed individual treatment. Use showed reduced fallouts and reduced diarrhea in pigs who received SynGenX® with their antibiotic therapy alone. Clinically, in the therapy groups, the reduction of signs was visually apparent on the treatment side vs the control (Figure 3). Additional studies for replacing antibiotic therapy completely are warranted. In this case, the producer was able to meet his goal of using less antibiotic because what he used was more effective when used in conjunction with SynGenX® so second treatments or extended treatments were not necessary in treatment groups.

Acknowledgements

Iowa State for statistical analysis of final results. Diamond V for product used in this study. Case study farm owners and staff for their time and cooperation.

References

1. Bass, B.E. et al, Effects of all-natural fermentation products or an antibiotic on weaned pig health and performance. Leman Conference Proceedings 2014.

2. Francis, D, Post-weaning E. coli – diagnosis, treatment, control and its effect on subsequent growth performance. American Association of Swine Vets: 2004: 495-499.

3. Schwartz, K, Salmonella diagnostics and interpretation: What does it all mean?. College of Veterinary Medicine, Iowa State University, Ames, Iowa: 55-68.

4. University of Illinois College of Veterinary Medicine, Veterinary diagnostics, 2015.

Table 1: Preventative treatment groups all had significantly less fallouts than controls at 2 weeks post entry. In the therapeutic application barn, there was no SynGenX® present at this time and there was no statistically significant difference. However 3 weeks after SynGenX® was in the feed there was a statistically significant difference in the number of fallouts in the treatment group vs the control.

	Week 1	Week 2	Week 3	Week 4	Week 5
Barn 1 Preventative	*P* = 0.518072	***P* = 0.018543**	*P* = 0.082262	*P* = 0.000002	*P* = 0.896421
Barn 2 E Therapy	**P* = 0.202587	**P* = 1	*P* = 0.136578	*P* = 0.086909	***P* = 0.000191**
Barn 3 E Preventative	*P* = 0.652865	***P* = 0.021376**	*P* = 1	*P* = 0.502984	
Barn 3 W Preventative	*P* = 0.005057	***P* = < 0.000001**	*P* = 0.003055	*P* = 0.000005	*P* = 0.746332
Barn 4 E Preventative	*P* = 0.002471	***P* = 0.001622**			
Barn 4 W Preventative	*P* = 0.011632	***P* = 0.08801**			

* no SynGenX® present

Figure 5: Therapeutic application group: Chronic Diarrhea Comparison--Treatment instigated was one week of Neomycin in both the NE (treatment) and SE (control) sides and implementation of SynGenX® fed continuously on NE (treatment) side.

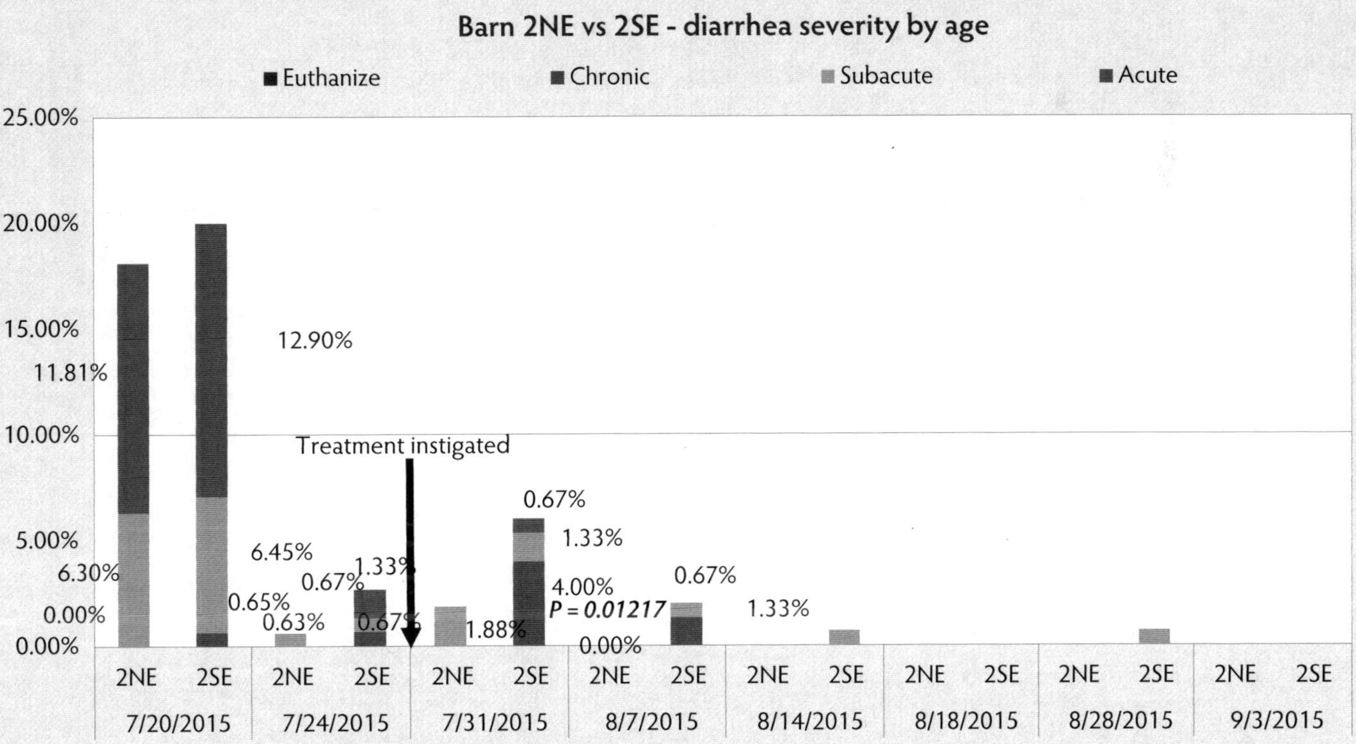

Figure 6: Therapeutic application group: Total stool in pen comparison--Treatment instigated was one week of Neomycin in both the NE (treatment) and SE (control) sides and implementation of SynGenX® fed continuously on NE (treatment) side.

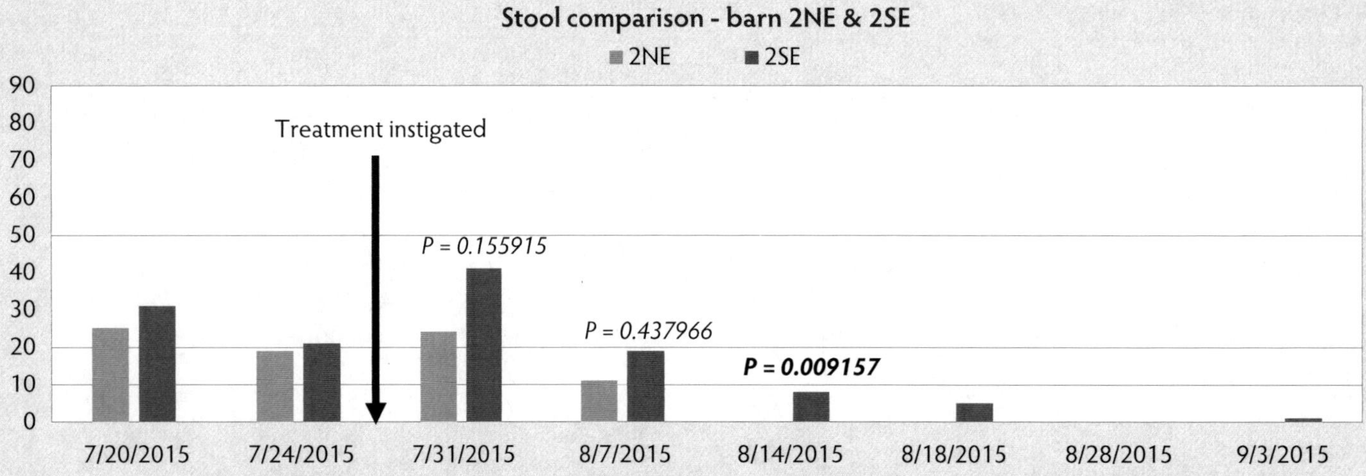

Figure 7: Preventative application group: Death Comparison--Both groups received Neomycin for one week in water. Treatment group (North) received SynGenX® continuously at 2 pounds per ton. Total death count over time across all barns. North is treatment. South is control.

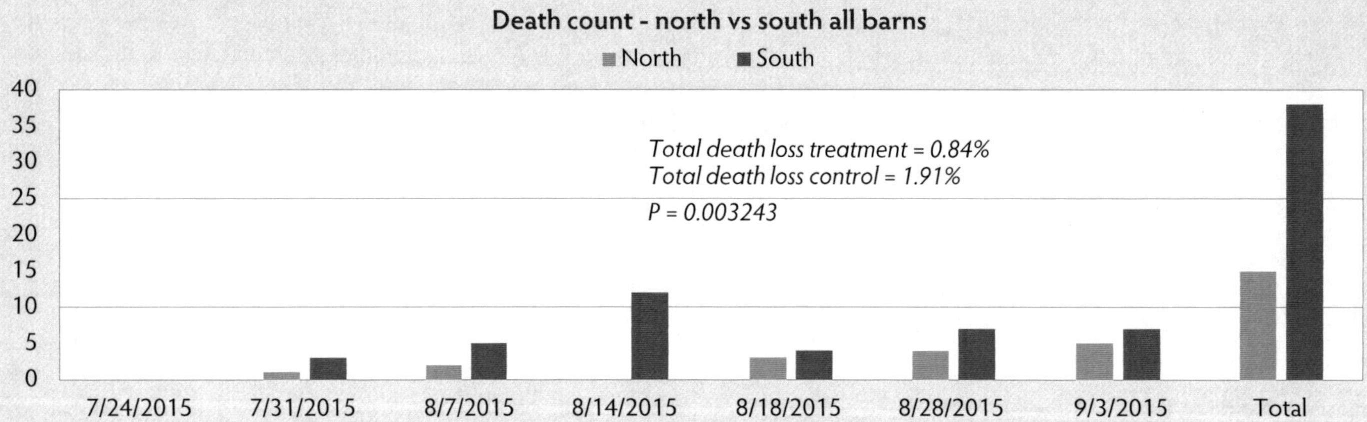

Figure 8: Preventative application group: Injectable Treatment Comparison-- Both groups received Neomycin for one week in water. Treatment group (North) received SynGenX® continuously at 2 pounds per ton.

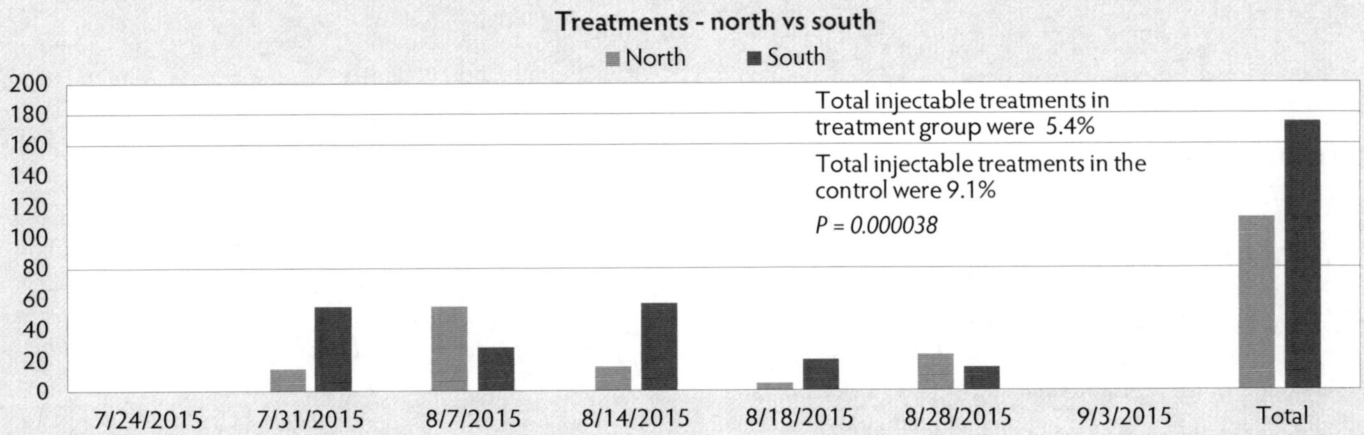

New discoveries for injectable fat-soluble vitamins for sows and piglets

Robert L. Stuart, PhD
Stuart Products, Inc., Bedford, Texas

Introduction

Fat-soluble vitamins A, D and E are vital nutrients for all swine, although precise requirements for all segments of production have not been clearly elucidated. Furthermore, the method of administration can dramatically alter vitamin status of swine. Injecting fat-soluble vitamins has been shown to be more efficiently utilized when compared to oral gavage or feed supplementation during critical times of production. Of the many metabolic functions of fat-soluble vitamins, immunity and muscle function are two that can be dramatically improved by injecting fat-soluble vitamins. Sows and gilts benefit from injections of fat-soluble vitamins pre-partum and pre-breeding. Pigs are born deficient in fat-soluble vitamins A, D and E and largely depend on colostrum and milk for adequate vitamin A and E needs provided the gestating gilt or sow has received optimum vitamin quantities in the gestating diet. Injection of fat-soluble vitamins at processing dramatically improves status of nursing pigs; especially vitamin D since it is poorly transferred in colostrum and milk. After weaning, vitamin E status drops dramatically due to the inability of pigs to utilize vitamin E-acetate regardless of level of vitamin E-acetate in nursery diets. Supplementing water with micellized, non-esterified vitamin E (d-alpha-tocopherol) assists in maintaining adequate vitamin E status of the post-weaned pig.[1] Lindemann et al. 2012 and Stuart et al., 2015, reviewed surveys of vitamin status during stages of production.[2,3]

The purpose of this paper is to review studies when injectable fat-soluble vitamin products were evaluated in gilts, pre-farrowing sows, and newborn pigs to evaluate the effectiveness of injecting fat-soluble vitamins.

Vitamins E, A and D injection dramatically increases serum and ovary fat-soluble vitamin status in gilts

Ovarian follicular fluids contain approximately four times more polyunsaturated fatty acids compared to serum.[4] Since ovaries have such high levels of polyunsaturated fatty acids, enhancing fat-soluble vitamins may improve antioxidant protection in ovaries and possibly improve ovarian function.[5] Eight gilts were injected with 5 mL VITAL E-Repro and eight gilts were not injected. Gilts were bled daily and ovaries were harvested 72 hours post injection to determine the effects of injection on serum levels of vitamins A, D and E, and ovarian levels of vitamins A and E (Lindemann, unpublished). Serum levels of vitamin A were increased by 39% (0.22 vs 0.31 μg/mL), serum vitamin D status increased by 302% (29 ng/mL vs 116.6 ng/mL) and vitamin E status was increased by 821% (2.52 vs 23.21 μg/mL). Vitamin E and vitamin A levels in ovaries were dramatically increased. Vitamin E level were increased by 75.8% (4.4 vs 7.7 μg/g) ($P < 0.05$), and vitamin A level were increased by 27.3% (0.49 vs 0.63 μg/g) ($P < 0.05$). These data imply that ovaries from injected gilts may have improved antioxidant capacity and possibly improved ovarian function.

Pre-partum injection of vitamins E, A and D reduces delivery time

Originally, the purpose of injecting sows pre-partum with vitamins E, A and D was to improve fat-soluble vitamin status of the newborn pig. Producers were reporting that a pre-partum injection of vitamins E, A and D resulted in sows having a shorter delivery time compared to sows not injected with fat-soluble vitamins. To verify those field reports, a study was conducted at Michigan State University to determine the impact of injecting multiparious sows on d 107 of gestation on length of delivery time.[6] Following confirmation of pregnancy, PEDV-exposed sows and gilts were given injection of either saline or 1500 I.U. vitamin E, 1,000,000 I.U. vitamin A-palmitate, and 500,000 I.U. vitamin D (5 mL VITAL E-Repro, Stuart Products, Inc.). Sow parity, body condition score were not different between treatments. Average farrowing time for saline or vitamin injected sows was 287 min and 190 min, respectively ($P = 0.05$). Number of pigs born and number of pigs weaned were not significantly different. Mean litter birth and weaning weights were not different (Table 1). Field studies with larger number of sows are being conducted to confirm the university study findings.

The recommended program for gilts and sows is to inject 5 mL VITAL E-Repro one week pre-partum and again at weaning. The pre-partum injection is to reduce farrowing time and improve vitamin status of the nursing pigs. The injection at weaning is to improve embryo quality and uniformity.

Injecting newborn pigs with vitamins A, D and E at processing is superior to a single oral dose

Pigs are born with low vitamins E, A, and D status. In non-supplemented pigs, Figure 1 shows that colostrum intake improved vitamin A and vitamin E status, but not vitamin D (Jang et al. 2015).[7] Since previous research on effectiveness of oral vitamin D was with 40,000 I.U. vitamin D given in a single oral dose,[8]

Table 1: Effects of pre-partum injection of vitamins E, A and D on sow farrowing time (Becerra et al., 2015).

Parameter	Saline Injection	VITAL E-Repro Injection (5 mL)	Probability
Sows (n)	22	22	
Pigs born live (n)	12.0	10.5	*P* = 0.10
Farrowing time (min)	**287**	**190**	***P* = 0.05**
Farrowing time/pig (min)	**23.92**	**18.09 (-32%)**	
Litter birth weight (kg)	15.1	13.5	*P* = 0.32
Pigs weaned (n)	11.6	9.7	*P* = 0.06
Litter wean weight (kg)	59.1	57.5	*P* = 0.81

Figure 1: Time course of plasma vitamin status of non-supplemented pigs from day-1 through weaning (day 21).

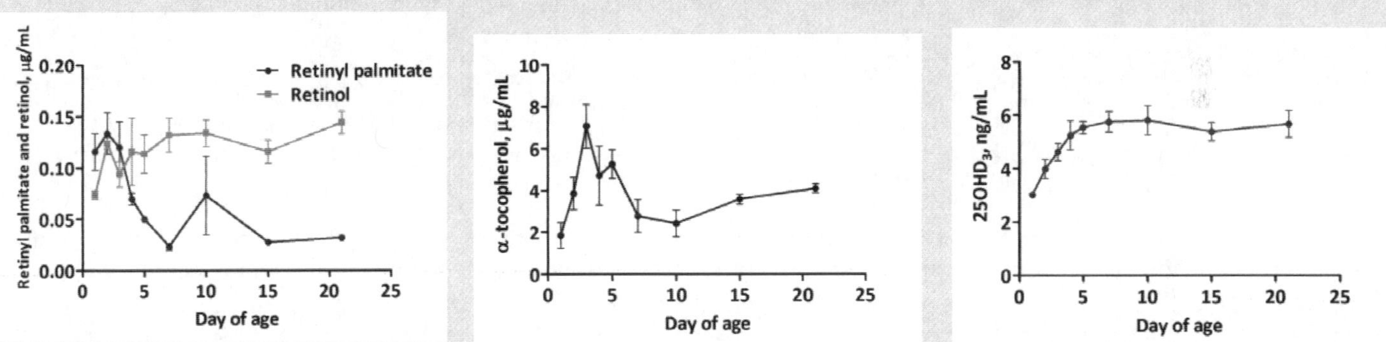

that same concentration of vitamin D was chosen to be utilized by Jang et al., 2014, 2015. In the first experiment, pigs were administered either saline, oral fat-soluble vitamins, or injected fat-soluble vitamins.[9] The concentrations of vitamins were the same for the two routes of administration (400 I.U. vitamin E (d-alpha-tocopherol), 40,000 I.U. vitamin A (retinyl-palmitate), and 40,000 I.U. vitamin D (cholecalciferol). Pigs were bled initially, d 10 and at weaning and plasma analyzed for alpha-tocopherol, retinol, and 25-OH-D3. The study demonstrated that injecting fat-soluble vitamins was approximately twice as effective as a single oral dose with less variation among pigs. The vitamin D injection response persisted through weaning (d 21) compared to either control or oral administration. The second experiment was done to more precisely quantify plasma vitamin concentrations prior to d 10 post-administration.[7] Pigs were bled on d 0, 1, 2, 3, 4, 6, 9, 14, and 20 and plasma analyzed for vitamins E, A and 25-OH-D3. Maximum plasma concentrations of 25 OH D3 increased to peaks at d 3 post-administration (Table 2). Plasma alpha-tocopherol concentrations for both oral ($P < 0.10$) and injection ($P < 0.05$) treatments were greater than in the control treatment until day 6 post-administration, the decline following the peak was faster in the oral treatment than in the injection treatment (Table 3) Only the injection treatment had an increase in plasma retinyl palmitate concentration with the peak at d 1 post-administration and did not increase with the oral treatment (Table 4). Plasma retinol was not different among the three treatments (Data not shown). The relative bioavailability estimate of the oral administration was determined to be 55% of the injection route of administration.

From these results, the recommended program for newborn pigs is to inject 0.5 mL VITAL E-Newborn at processing to provide 250 I.U. vitamin E, 25,000 I.U. vitamin A, and 25,000 I.U. vitamin D. The variation among serum concentrations was much less with the injection compared to the oral dose suggesting that some of the oral dose may not have been ingested. Due to enhanced utilization, An injection quantity of 40,000 I.U. vitamin D is too great.

Conclusion

Of the three fat-soluble vitamins, vitamin E appears to be the most deficient in all post-weaning phases of swine production, while vitamin D supplementation may be most critical in the nursing pig due to low milk content. Reduced farrowing time due to pre-partum injection of fat-soluble vitamins, reduction in mulberry heart, and reduction in transit losses are areas of continued research.

Table 2: Effect of administration routes of fat-soluble vitamins on 25-hydroxycholecalciferol concentrations (ng/mL) for suckling pigs (Jang et al, 2015).[1-3]

	Treatment			
Item	Control	Oral	Injection	SEM
Day				
0	3.0	2.9	3.1	0.20
1	4.0[c]	138.8[b]	212.5[a]	18.33
2	4.6[c]	155.4[b]	235.4[a]	10.37
3	5.3[c]	145.9[b]	222.8[a]	8.55
4	5.5[c]	145.2[b]	203.6[a]	9.60
6	5.8[c]	108.2[b]	174.9[a]	7.68
9	5.8[c]	64.3[b]	115.4[a]	8.73
14	5.4[c]	33.2[b]	67.4[a]	5.40
20	5.7[c]	16.3[b]	39.2[a]	3.18

[a-c] Means within the same row without a common superscript differ ($P < 0.05$).

[1] Values are least squares means that represent data from 5 observations per treatment for each day with each observation a value for a pooled sample from 3 pigs. There was 1 value below the detection limit (on d 0) and 1 outlier (high, on d 20) in the control treatment.

[2] Control, no vitamin administration; Oral, oral administration with 40,000 IU of vitamin D_3, 40,000 IU of vitamin A as retinyl palmitate, 400 IU of vitamin E as d-α-tocopherol; Injection, i.m. injection with 40,000 IU of vitamin D_3, 40,000 IU of vitamin A as retinyl palmitate, 400 IU of vitamin E as d-α-tocopherol.[5]Day after vitamin administration except d 0 is before administration.

[3] Day after vitamin administration except d 0 is before administration.

Table 3. Effect of administration routes of fat-soluble vitamins on α-tocopherol concentrations (μg/mL) for suckling pigs (Jang et al. 2015).[1,2]

	Treatment			
Item	Control	Oral	Injection	SEM
Day				
0	1.8	2.1	1.6	0.48
1	3.9[c]	14.3[b]	26.1[a]	2.30
2	6.7[b,y]	12.4[b,x]	23.6[a,z]	1.95
3	4.7[b,y]	10.0[b,x]	21.3[a,z]	2.05
4	4.7[b]	6.2[b]	17.9[a]	1.36
6	2.8[b,y]	7.4[ab,x]	10.4[a,x]	1.68
9	2.4	2.5	1.5	0.88
14	3.5[b]	3.3[b]	4.9[a]	0.39
20	4.1[b]	3.8[b]	5.7[a]	0.41

[a-c] Means within the same row without a common superscript differ ($P < 0.05$).

[x-z] Means within the same row without a common superscript differ ($P < 0.10$).

[1] Values are least squares means that represent data from 5 observations per treatment for each day with each observation a value for a pooled sample from 3 pigs. There are 6 (1 on d 0, 2 on d 2, 2 on d 4, and 1 on d 20), 2 (2 on d 6), and 3 (1 on d 0 and 2 on d 9) values below the detection limit in the control, oral and injection treatments, respectively, and 2 outliers (1 on d 1, high; and 1 on d 4, low) in the injection treatment.

[2] See table 2 for treatment identification.

Table 4. Effect of administration routes of fat-soluble vitamins on retinyl palmitate concentrations (μg/mL) for suckling pigs (Jang et al., 2015).[1,2]

Item	Treatment			SEM
	Control	Oral	Injection	
Day				
0	0.12	0.14	0.15	0.015
1	0.13[b]	0.25[b]	2.17[a]	0.208
2	0.12[b]	0.12[b]	0.60[a]	0.049
3	0.07[b]	0.08[b]	0.37[a]	0.043
4	0.05[b]	0.06[b]	0.14[a]	0.023
6	0.05	0.04	0.12	0.042
9	0.08	0.04	0.02	0.020
14	0.03	0.03	0.03	0.002
20	0.03	0.03	0.03	0.003

[a,b] Means within the same row without a common superscript differ ($P < 0.05$).

[1] Values are least squares means that represent data from 5 observations per treatment for each day with each observation a value for a pooled sample from 3 pigs. There are 9 (2 on d 3, 2 on d 4, 2 on d 6, 2 on d 9, and 1 on d 14), and 6 (1 on d 2, 1 on d 4, 3 on d 9, and 1 on d 14) values below the detection limit in the control and oral treatments, respectively, and 1 (on d 1, high) and 3 (1 on d 0, low; 1 on d 1, high; and 1 on d 2, high) outliers in the oral and injection treatments, respectively.

[2] See table 2 for treatment identification.

References

1. Specht, T.A., D.C. Mahan, N.D. Fastinger and R.L. Stuart. 2003. Effects of d-alpha-tocopherol alcohol or acetate in water soluble or emulsified form to the drinking water of weanling pig. J. Anim. Sci. 81 (Suppl. 2): p 88 (Abstract No. 356).

2. Lindemann, M. D., R. L. Stuart and J. Goihl. 2012. Oral and injectable fat-soluble vitamin programs for sows, newborn and weaned pigs. Proc. AASV Annual Meeting. p 168-170.

3. Stuart, R. L., P. G. Brinkman, W. D. Safley, and J. J. Stuart. 2015. New findings for injectable and oral fat-soluble vitamins. Proc. AASV Annual Meeting.

4. Yao, J. K., R. J. Ryan, and P. J. Dyck. 1980. The porcine ovarian follicle. VI. Comparison of fatty acid composition of serum and follicular fluid at different developmental stages. Biol. of Reprod. 22:141-147.

5. Agarwal A., A. Aponte, M. Beena, J Premkumar, A. Shaman and S. Gupta. 2012. The effects of oxidative stress on female reproduction: a review. Reproductive Biology and Endocrinology. 10:1-31.

6. Becerra, R., J., T. Gebhardt, J.E. Link, K. Turner, R. L. Stuart and G. M. Hill. 2015. Are porcine epidemic diarrhea virus (PEDv) exposed gilts and sows farrowing problems improved by vitamin E? J. Anim. Sci. Midwest Meeting. Mar. 15-18, 2015.

7. Jang, Y. D., J. Y. Ma, J. S. Monegue, H. J. Monegue, R. L. Stuart, and M. D. Lindemann. 2015. Temporal plasma vitamin concentrations are altered by fat-soluble vitamin administration in suckling pigs. J. Anim. Sci. 93:5273-5282.

7. Tousignant, S. J. P., S. C. Henry, A. Rovira and R. B. Morrison. 2013. Effect of oral vitamin D3 supplementation on growth and serum 25-hydroxy vitamin D levels of pigs up to 7 weeks of age. J Swine Health Prod. 21:94–98.

8. Jang, Y. D., M. D. Lindemann, H. J. Monegue and R. L. Stuart. 2014. The effects of fat-soluble vitamin administration on plasma vitamin status of nursing pigs differ when provided by oral administration or injection. Asian-Australas J Anim. Sci. 27: 674–682.

Comparative efficacy evaluation of two modified-live PRRS vaccines

G. Haiwick, BS; A. Neubauer, Dr. med vet, DACVM; J. Hermann, PhD; M. Roof, PhD; B. Fergen, MS, PhD; R. Philips, DVM
Boehringer Ingelheim Vetmedica, Inc., Saint Joseph, Missouri

Introduction

The swine industry continues to experience significant losses due to PRRS virus infections of sow farms and growing pigs.[1] Modified-live vaccine (MLV) can significantly reduce PRRSV induced lung lesions following a heterologous challenge.[2] In addition, it has been demonstrated that the implementation of a systematic methodology for PRRS control that utilizes modified-live vaccine for the control of wild type-PRRSV infections can mitigate the consequences of infection on health and performance.[3,4] It is necessary for modified-live vaccines to be effective against heterologous challenge with current PRRSV field isolates. The objective of this study was to directly compare the efficacy of Ingelvac PRRS® MLV (BIVI-Boehringer Ingelheim Vetmedica, Inc., St. Joseph, MO) and Prime Pac™ PRRS + (Merck Animal Health, Omaha, NE) against a heterologous PRRSV challenge in a standard respiratory challenge model in growing pigs.

Materials and methods

A randomized, blinded vaccination-challenge study was performed in seventy, three week old pigs from a PRRS naïve source, and confirmed PCR negative for PRRS. The study consisted of four treatment groups of pigs. Group 1; n = 20, was vaccinated with Ingelvac PRRS® MLV (2 ml I.M. per label recommendations) and Group 2; n = 20, was vaccinated with Prime Pac PRRS (1 ml I.M. per label recommendations). Group 3; n = 20, was a non-vaccinated challenge control treatment group. Group 4; n = 10, was an Ingelvac PRRS® MLV vaccinated and non-challenged treatment group. Groups 1, 2 and 4 were vaccinated on day 0. Groups 1, 2 and 3 were challenged on day 28 intranasally with 2.0 ml (1.0 ml/nostril) of inoculum containing 4.1 log $TCID_{50}$/ml of virulent PRRSV SDSU-73. Ingelvac PRRS® MLV and Prime Pac PRRS are 89.7% and 91.3% homologous to SDSU-73 respectively based on ORF 5 sequence. On day 42, 10 vaccinated pigs were randomly selected from Groups 1 and 2, along with 10 challenge control pigs from group 3. The pigs were necropsied and lungs were collected and scored for percent pneumonia associated with PRRSV exposure. The study was terminated on day 70 (42 days following challenge), and the remaining pigs were necropsied and lungs were scored for percent pneumonia. Blood was collected from all pigs on days 0, 7, 14, 21, 28, 31, 33, 35, 38, 42, and weekly thereafter until day 70 to assess the serologic response and viremia. Viremia testing for PRRSV was done by BIVI-Health Management Center (HMC Ames, Iowa) using Quantitative PCR and results were reported as "positive" or "negative". Rectal temperatures were collected on the day prior to challenge (day 27), and daily for 14 days until day 42. Pigs were weighed at day 0, day of challenge (day 28), and termination of the study (day 70) to assess daily weight gain (ADWG). The study design is summarized in Table 1.

Results

Lung lesions: The day 42 (14 days post-challenge) lung lesion assessment found both vaccinated treatment groups to have reduced gross lung lesion scores compared to the non-vaccinated challenged treatment group. The median percent lung lesion scores for Ingelvac PRRS® MLV and Prime Pac PRRS were 0.5% and 1.3% respectively, compared to a median lung lesion score of 28.6% for

Table 1: Study design

Group	No. Inglevac PRRS® MLV vaccinated pigs	No. Prime Pac PRRS vaccinated pigs	No. challenge control pigs	PRRSV SDSU-73 challenge dosage ($Log_{10}TCID_{50}$/ml)	Necropsy subset of groups 1,2 & 3 (lung lesion evaluation)	Necropsy Remaining pigs
	Day 0	Day 0	Day 0	Day 28	Day 42	Day 70
1	20	-	-	4.1	10	10
2	-	20	-	4.1	10	10
3	-	-	20	4.1	10	10
4	10	-	-	None		10

the non-vaccinated challenged treatment group. At day 70 (termination of the study); as expected, lung lesions had mostly resolved, and the overall group median lung lesion scores were low and similar among treatment groups. Post-challenge lung lesion results are summarized in Table 2.

Viremia: Following challenge, all pigs became viremic by day 31, and all pigs were still viremic at the time of necropsy on day 42. Following day 42 (14 days post-challenge) of the study, viremia begins decreasing in all groups of pigs until termination of the study on day 70. The decline in post-challenge viremia begins earlier in the Ingelvac PRRS® MLV vaccinated group. From day 42 to day 63 of the study, the Ingelvac PRRS® MLV vaccinated group demonstrates less percent PCR positive pigs than the Prime Pac PRRS vaccinated pigs and challenge control pigs. The Prime Pac PRRS vaccinated pigs viremia pattern following challenge was similar to the challenge control pigs. Post-challenge viremia results are summarized in Figure 1.

Temperature: The Ingelvac PRRS® MLV vaccinated pigs maintained lower average temperatures throughout the challenge phase of the study; (days 28-42), when compared to the challenge controls. Ingelvac PRRS® MLV vaccinated pigs also demonstrated lower average temperatures compared to the Prime Pac PRRS vaccinated pigs during the first week post-challenge. During the first week post-challenge, average temperatures in the Prime Pac PRRS vaccinated pigs were similar to the challenge control pigs. Post-challenge temperature results are summarized in Figure 2.

Average daily weight gain: ADWG was evaluated during the post-challenge period of the study (days 28-70). The Ingelvac PRRS® MLV vaccinated group demonstrated a 17% higher ADWG compared to both the Prime Pac PRRS vaccinated group and the non-vaccinated challenged controls. ADWG for the Ingelvac PRRS® MLV group was 1.41 lbs/day compared to 1.17 lbs/day and 1.18 lbs/day for the Prime Pac PRRS and challenge control groups respectively. ADWG for the Prime Pac PRRS and challenge control groups were similar. The vaccinated non-challenged group demonstrated the best ADWG across treatment groups at 1.67 lbs/day, indicating the virulent PRRSV challenge has a biologic impact as measured by ADWG. The ADWG results are summarized in Table 3.

Table 2: Day 42 lung lesions (median %) in vaccinated and non-vaccinated pigs challenged with PRRSV SDSU-73.

Group	Treatment	Median percent lung score
1	Ingelvac PRRS® MLV	0.5
2	Prime Pac PRRS	1.3
3	Challenge control (non-vaccinated)	28.6

Figure 1: Percentage of viremic pigs per treatment group following PRRSV SDSU-73 challenge

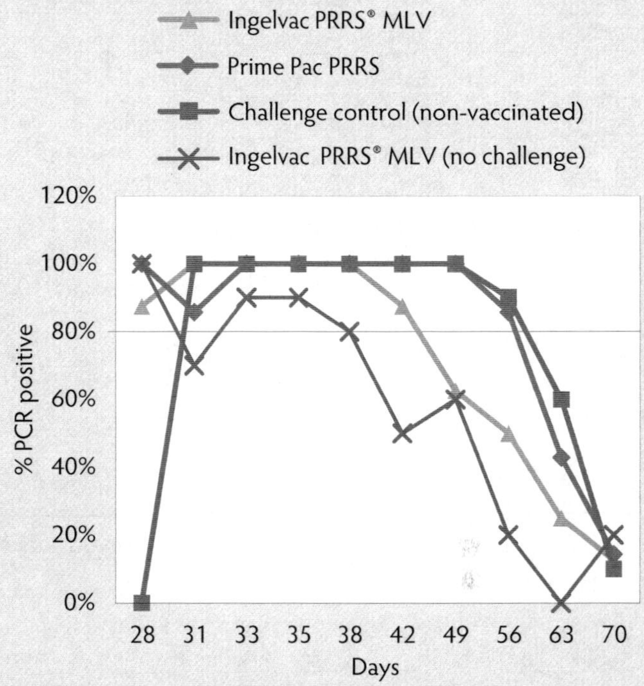

Figure 2: Average daily temperature by treatment group

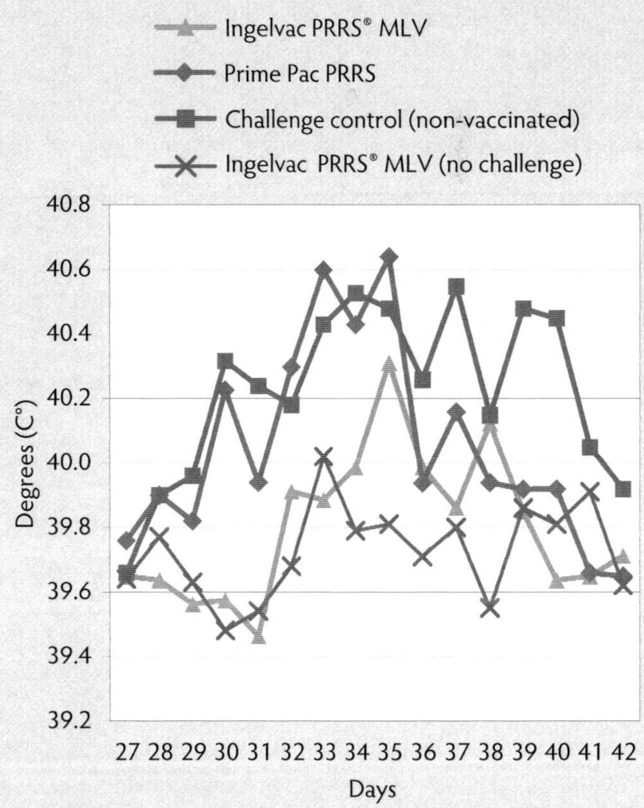

Table 3: Average daily weight gain (ADWG)

Group	Treatment group	ADWG in lbs.
1	Ingelvac PRRS® MLV	1.41
2	Prime Pac PRRS	1.17
3	Challenge control (non-vaccinated)	1.18
4	Vaccinated control (non-challenged)	1.67

Discussion and conclusions

The objective of this study was to compare the efficacy of Ingelvac PRRS® MLV and Prime Pac PRRS against a virulent heterologous PRRSV in a respiratory challenge model. The reduction of PRRSV induced lung lesions following challenge is one of the primary measures of efficacy. In this study, both vaccines demonstrated a large and similar reduction in PRRSV associated lung lesions compared to non-vaccinated controls 14 days after challenge.

Other measures of efficacy include evaluation of viremia, temperature and ADWG post-challenge. In this study the reduction of viremia following challenge occurred earlier in the Ingelvac PRRS® MLV vaccinated group compared to the Prime Pac PRRS vaccinated group, and the pattern of viremia reduction in the Prime Pac PRRS vaccinated group was similar to the non-vaccinated controls. In this study, the Ingelvac PRRS® MLV vaccinated group maintained lower average temperatures throughout the challenge phase of the study compared to the Prime Pac PRRS and challenge control groups. Reducing fever following challenge and infection is an important factor in mitigating the consequences of infection and helping pigs return to normal feed and water consumption. PRRSV infection can negatively impact pig performance as measured by ADWG. Mitigating the negative biological impact PRRSV infection can have on ADWG can have significant economic benefit. In this study, the Ingelvac PRRS® MLV group had a 17% higher ADWG when compared to both the Prime Pac PRRS group and the challenge control group. The ADWG in the Prime Pac PRRS group was not different than the non-vaccinated challenge controls. Potentially, the slower mitigation of viremia and average temperature following challenge in the Prime Pac PRRS group had an impact on overall ADWG.

Modified-live PRRS vaccines have been shown to be efficacious against heterologous PRRSV challenge.[2] This study is another example demonstrating the ability of modified-live PRRS vaccines to protect against a relevant PRRSV challenge. Implementing vaccine in a systematic methodology for PRRS control can mitigate the consequences of infection and subsequently improve the health and performance of pigs.

References

1. Holtkamp, D., et al. 2013. Assessment of the economic impact of porcine reproductive and respiratory syndrome virus on United States pork producers. *J Swine Health and Prod.* 21(2):72-84.

2. Patterson, A., et al. 2013. Modified live PRRSV vaccination is efficacious following challenge with eight genetically diverse PRRSV isolates. *A.D. Leman Swine Conference Proceedings.*

3. Oropeza, A., et al. 2012. A methodical approach to PRRS management, a success story of production improvements. *IPVS Proceedings.*

4. Kolb, J., et al. 2015. A summary of four large scale system-based PRRS control projects. *AASV Proceedings.*

Evaluation of porcine reproductive and respiratory syndrome virus (PRRSV) challenge dose in vaccinated pigs

G. Haiwick, BS; A. Neubauer, Dr. med. vet, DACVM; J. Hermann, PhD; M. Roof, PhD; B. Fergen, MS, PhD; R. Philips, DVM
Boehringer Ingelheim Vetmedica, Inc., Saint Joseph, Missouri

Introduction

The swine industry continues to experience health challenges and economic losses due to PRRSV infections of sow farms and growing pigs.[1] The infectious dose of PRRSV by intranasal, intramuscular and aerosol routes has been shown to be very low, confirming PRRSV to be highly infectious.[2,3,4] Ingelvac PRRS® MLV has demonstrated protection against heterologous PRRSV challenge and can significantly reduce PRRSV induced lung lesions and mitigate the consequences of infection following challenge.[5] The objective of this study was to evaluate the effect of PRRSV challenge dose in Ingelvac PRRS® MLV (BIVI-Boehringer Ingelheim Vetmedica, Inc., St. Joseph, MO) vaccinated pigs in a respiratory challenge model.

Materials and methods

Experimental design: A randomized, blinded vaccination-challenge study was performed in ninety, three week old pigs from a PRRS naïve source, and confirmed PCR negative for PRRS. The study consisted of five treatment groups of pigs. Groups 1-5; n = 10 per group, were vaccinated with Ingelvac PRRS® MLV (2 ml I.M. per label recommendations). Groups 1-5 were vaccinated on day 0. Forty pigs served as non-vaccinated challenge controls (Groups 1-4; n = 10 per group) and were housed separately from any vaccinates during the vaccination phase. The study design is summarized in Table 1.

Challenge: Groups 1-4 were challenged on day 28 intranasally with 2.0 ml (1.0 ml/nostril) of inoculum of virulent PRRSV SDSU-73. Housing during the challenge phase was by group and PRRSV challenge. Group 5 was an Ingelvac PRRS® MLV vaccinated and non-challenged treatment group.

Viremia: Blood was collected from all pigs on days 0, 7, 14, 21, 28, 31, 33, 35, 38, 42, and weekly thereafter until day 70 to assess the serologic response and viremia. Viremia testing for PRRSV was assessed by BIVI-Health Management Center (HMC Ames, Iowa) using Quantitative PCR and results were reported as "positive" or "negative".

Temperature: Rectal temperatures were collected on the day prior to challenge (day 27), and daily for 14 days until day 42. Pyrexia was defined as a rectal temperature > 40.0°C.

Weight gain: Pigs were weighed at day 0, day of challenge (day 28), and termination of the study (day 70) to assess daily weight gain (ADWG).

Statistics: Results were summarized via descriptive statistics by day, challenge dose and group. For number of days pyrexic and ADWG post-challenge, a linear regression model incorporating treatment and challenge dose was utilized, with model simplification based on stepwise removal of non-significant model terms. A *P*-value < 0.05 was used to indicate statistical significance.

Table 1: Study design

Group	No. Inglevac PRRS® MLV vaccinated pigs	No. non-vaccinated challenge control pigs	PRRSV SDSU-73 challenge dosage ($Log_{10}TCID_{50}$/ml)	Study termination
	Day 0	Day 0	Day 28	Day 70
1	10	10	4	
2	10	10	3	
3	10	10	2	
4	10	10	1	
5	10	-	None	

Results

Viremia: Following challenge, all pigs in the Ingelvac PRRS® MLV vaccinated treatment groups 1 and 2, (4 and 3 log challenge), became viremic by day 31. Following day 42 (14 days post-challenge) of the study, viremia begins decreasing in the vaccinated pigs until termination of the study on day 70. From day 42 to day 70 of the study, the vaccinated pigs in groups 1 and 2 demonstrate less percent PCR positive pigs than the non-vaccinated challenge control pigs in groups challenged with 4 and 3 logs of virus, indicating a reduction in post-challenge viremia in vaccinated pigs. As seen in groups 1 and 2, the Ingelvac PRRS® MLV vaccinated pigs in groups 3 and 4, (2 and 1 log challenge), demonstrate less percent PCR positive pigs than the non-vaccinated challenge control pigs. The pattern of viremia following challenge in vaccinated groups 3 and 4 was similar to the vaccinated non-challenged pigs in group 5. As the challenge dose decreases the percentage of viremic pigs in the vaccinated pigs in the vaccinated groups decreases. At all challenge doses, the non-vaccinated and challenged pigs show a similar post-challenge viremia profile.

Post-challenge viremia results are summarized in Figures 1 and 2.

Temperature: At each challenge dose, there was a statistically significant decrease in the number of days pyrexic post-challenge between vaccinated and non-vaccinated treatment groups. The Ingelvac PRRS® MLV vaccinated pigs maintained a lower average temperature throughout the challenge phase of the study; (days 28-42), when compared to the non-vaccinated challenge controls at all challenge doses. At PRRSV challenge doses of 2 logs or less, the post-challenge average temperatures of the Ingelvac PRRS® MLV vaccinated pigs are similar to the temperatures of the vaccinated non-challenged pigs. Post-challenge temperature results are summarized in Table 2 and Figures 3, 4 and 5.

Average daily weight gain: ADWG was evaluated for the post-challenge portion of the study (days 28-70) for all treatment groups. The Ingelvac PRRS® MLV vaccinated groups demonstrated higher ADWG compared to the non-vaccinated challenged control groups at all challenge doses. The difference in ADWG between vaccinates and non-vaccinates was statistically significant in the 3 log, 2 log and 1 log challenge dose groups. The ADWG for the Ingelvac PRRS® MLV vaccinated groups challenged with 4 logs, 3 logs, 2 logs and 1 log of PRRSV were 1.41, 1.29, 1.70 and 1.64 lbs/day respectively; compared to ADWG of the non-vaccinated challenged control groups challenged with 4 logs, 3 logs, 2 logs and 1 log of PRRSV at 1.18, 1.06, 1.15 and 1.23 lbs/day respectively. There was a significant increase ($P < 0.05$) in the ADWG based on the challenge dose in the Ingelvac PRRS® MLV vaccinated groups, as the ADWG increased 0.085 for each one log decrease in challenge dose. There was a measurable negative ADWG impact in the non-vaccinated challenged groups at all challenge doses, with no significant difference in ADWG across all challenge doses. The vaccinated non-challenged group demonstrated an ADWG of 1.67 lbs/day, indicating the virulent PRRSV

Figure 1: Percentage of viremic pigs per treatment group challenged with 4 logs of PRRSV SDSU-73.

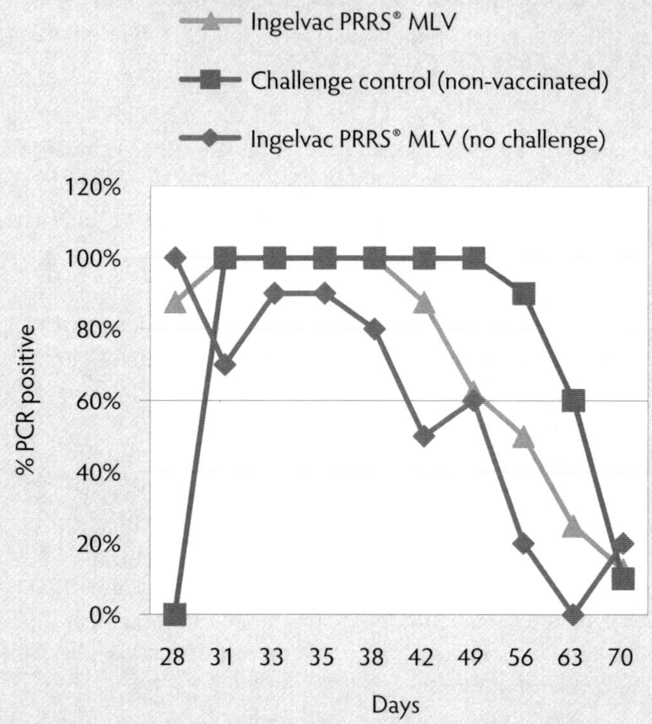

Figure 2: Percentage of viremic pigs per treatment group challenged with 2 logs of PRRSV SDSU-73

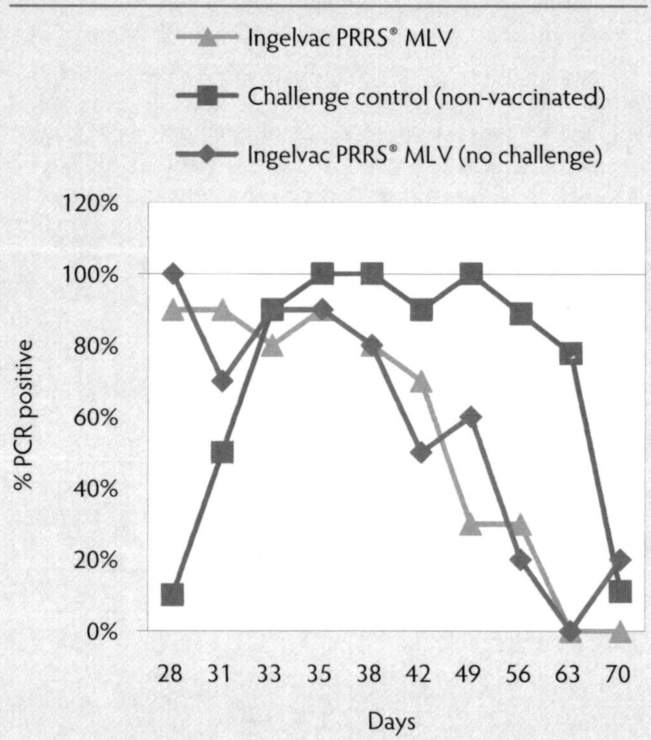

Table 2: Mean number days pyrexic post-challenge.

Treatment group	4 log challenge	3 log challenge	2 log challenge	1 log challenge	No-challenge
Ingelvac PRRS® MLV	4.4[1]	4.2[1]	1.0[1]	1.4[1]	1.8
Challenge control (non-vaccinated)	11.2	8.8	10.0	6.0	-

[1] Statistically significant difference ($P < 0.05$) in number days pyrexic between Ingelvac PRRS® MLV and Challenge control groups based on model prediction.

Figure 3: Average daily temperature per treatment group challenged with 4 logs of PRRSV

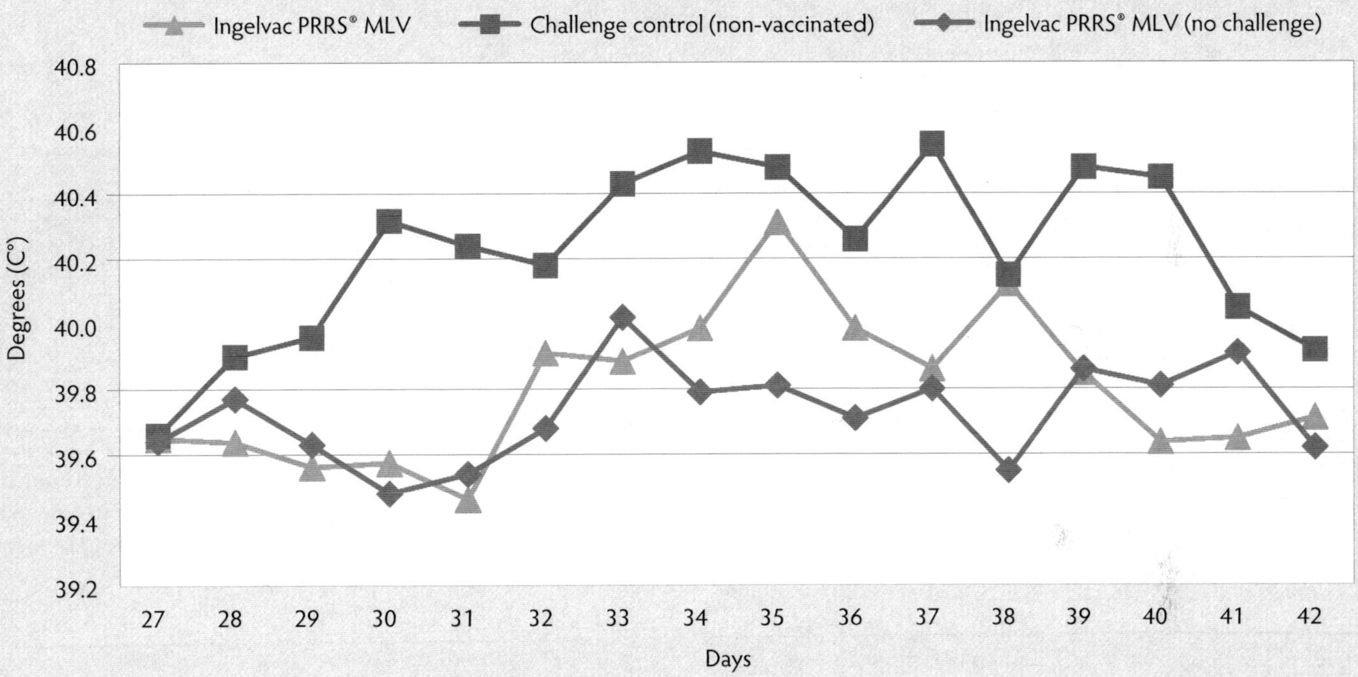

challenge has a biologic impact as measured by ADWG when compared to non-vaccinated challenge groups. The ADWG of the Ingelvac PRRS® MLV vaccinated groups challenged with 2 logs of PRRSV or less was numerically similar to the ADWG of the vaccinated non-challenged group, indicating limited post-challenge ADWG impact in vaccinated pigs. The ADWG results are summarized in Table 3 and Figure 6.

Discussion and conclusions

The objective of this study was to evaluate the effect of PRRSV challenge dose in vaccinated pigs. The PRRSV challenge isolate used in this study was SDSU-73, and is 89.7% homologous to Ingelvac PRRS® MLV based on ORF 5 sequence. In this study, Ingelvac PRRS® MLV vaccinated pigs demonstrated mitigation of the consequences of PRRSV infection compared to non-vaccinated challenge pigs at all challenge doses with a reduction in post-challenge viremia, temperature and increased ADWG. The Inglevac PRRS® MLV vaccinated groups had better ADWG compared to non-vaccinated challenge controls at all challenge doses demonstrating that vaccination mitigated the negative impact of the PRRSV challenge on ADWG. Compared to non-vaccinated challenged pigs, the increase in ADWG of the Ingelvac PRRS® MLV vaccinated pigs was statisticially significant ($P < 0.05$) in the 3, 2 and 1 log challenge groups, and at $P < 0.07$ in the 4 log challenge group. The number of days pyrexic in the non-vaccinated challenge control group seemed to be affected by challenge dose, whereas ADWG seemed not to be affected. For all endpoints, there was little indication of a difference between 0, 1 and 2 log challenge in vaccinated animals, indicating that there is a challenge dose effect in vaccinated pigs, and at a challenge of 2 logs or less, the consequences of challenge in vaccinated pigs were similar to non-challenged pigs. Identification of the PRRSV strain measured by post-challenge viremia is in process at the time of the writing of this paper. Conversely, in the non-vaccinated challenged pigs, the post-challenge viremia and impact on ADWG were similar across

Figure 4: Average daily temperature per treatment group challenged with 2 logs of PRRSV

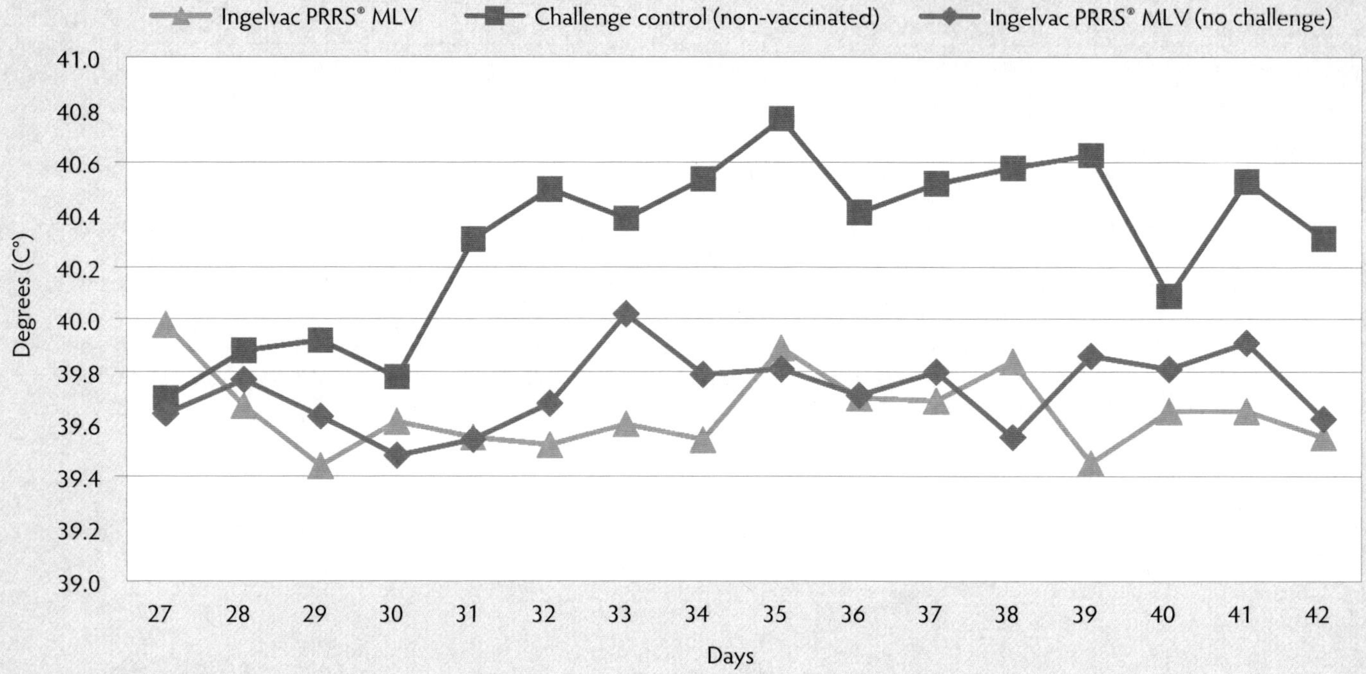

all challenge doses, indicating no challenge dose effect and a measurable impact at all challenge doses in non-vaccinated pigs.

This study is another example demonstrating the ability of Ingelvac PRRS® MLV vaccine to protect against a relevant PRRSV challenge and mitigate the biologic consequences of infection. The study also demonstrates a challenge dose effect in vaccinated pigs; as challenge dose decreases, the mitigation of the consequences of infection improves. Based on challenge dose, the

Figure 5: Number of days pyrexic by group and challenge dose

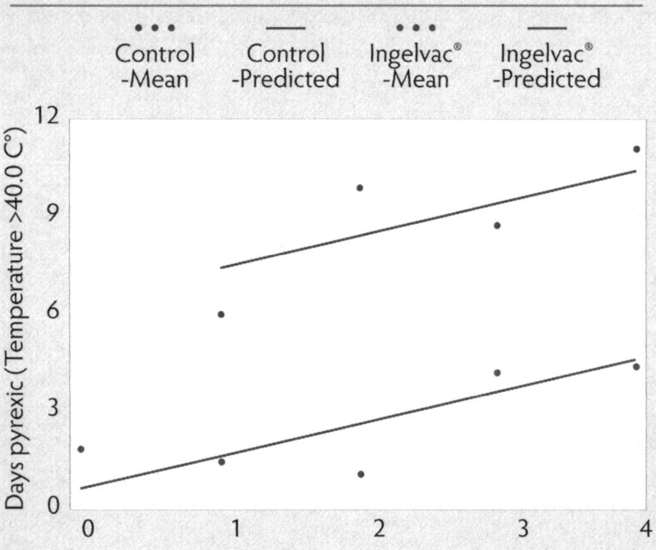

consequences of challenge in vaccinated pigs can be similar to non-challenged pigs. The results of this study have relevance in the field as implementation of vaccine in a systematic and coordinated methodology along with the other complimentary components for PRRS control in system-based and area-regional control programs can mitigate the consequences of PRRSV infection subsequently improving health and performance.

References

1. Holtkamp, D., et al. 2013. Assessment of the economic impact of porcine reproductive and respiratory syndrome virus on United States pork producers. *J Swine Health and Prod.* 21(2):72-84.

2. Yoon, K. J., et al. 1999. Effect of challenge dose and route on porcine reproductive and respiratory syndrome virus (PRRSV) infection in young swine. *Vet Res.* 30:629-638.

3. Hermann, J., et al. 2007. Susceptibility of young pigs to porcine reproductive and respiratory syndrome virus (PRRSV) infection by aerosol route of exposure. *AASV Proceedings.*

4. Cutler, T., et al. 2010. Median infection dose (ID50) of PRRSV isolate MN-184 for young pigs via aerosol exposure. *IPVS Proceedings.*

5. Patterson, A., et al. 2013. Modified live PRRSV vaccination is efficacious following challenge with eight genetically diverse PRRSV isolates. *A.D. Leman Swine Conference Proceedings.*

Table 3: Average daily weight gain in lbs. (ADWG).

Treatment group	4 log challenge	3 log challenge	2 log challenge	1 log challenge	No-challenge
Ingelvac PRRS® MLV	1.41	1.29[1]	1.70[1]	1.64[1]	1.67
Challenge control (non-vaccinated)	1.18	1.06	1.15	1.23	-

[1] Statistically significant difference ($P < 0.05$) in ADWG between Ingelvac PRRS® MLV and Challenge Control groups based on model prediction.

Figure 6: Average daily weight gain over days 28 to 70 by group and challenge dose

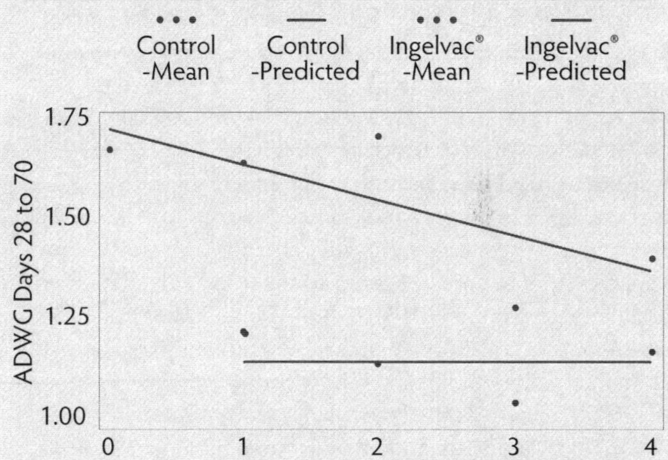

PCV2 vaccine cross-protection: Identification of sequences in successfully vaccinated field cases

Brian Payne[1], DVM; Amy Jacobs[2], BS; Cheryl Dvorak[3], PhD; Michael Murtaugh[3], PhD
[1]Boehringer Ingelheim Vetmedica, Inc., Saint Joseph, Missouri; [2]Health Management Center, Boehringer Ingelheim Vetmedica, Inc., Ames, Iowa; [3]Department of Veterinary and Biomedical Sciences, University of Minnesota, Saint Paul, Minnesota

Introduction

Primary control for porcine circovirus type 2 (PCV2) is vaccination of pigs around the age of weaning with PCV2 vaccines which are efficacious at controlling the disease when administered to the pigs according to label. Vaccination does not eliminate PCV2 exposure in the barns; therefore, PCV2 can still be found in successfully vaccinated herds even with optimal protection. In 2012, it was reported that a novel mutation of PCV2 could be a possible cause of disease in pigs but vaccination non-compliance could not be ruled out.[1] Since then, current commercial PCV2 vaccines have been shown to be efficacious against this described novel strain.[2,3] Schwartz reported that there had been a shift from PCV2b to PCV2d (new designation for the PCV2 mutation reported by Opriessnig in 2013) in the industry, at least from clinical cases submitted to the Iowa State University Veterinary Diagnostic Laboratory.[4] It is necessary to evaluate emerging isolates in control cases (PCV2 exposed pigs, vaccinated with PCV2 vaccines with no PCVAD signs and with satisfied owners and veterinarians on control) as well as clinical cases in order to determine the importance of novel mutations in the industry. The primary objective of this study was to identify PCV2 isolates exposing herds where nursery and finisher production performance was meeting the systems' expectations, no clinical PCVAD signs were apparent and both the ownership and veterinary staff were satisfied with their use of Ingelvac CircoFLEX® at weaning age.

Materials and methods

Farms (n = 48, 10 unique systems) in Iowa, North Carolina, Minnesota, Illinois, and Ohio (Figure 1) were selected for sampling based on the criteria that, historically and currently, 1) pigs are vaccinated with Ingelvac CircoFLEX® at weaning age, 2) both the pig owner and veterinarian were satisfied with the vaccine, 3) pig performance was meeting the systems' expectations and 4) no clinical health issues suggestive of PCVAD were present. Follow-up with the veterinarians through finishing of these particular groups was completed to confirm the absence of PCVAD. Serum (n = 432 samples; 23 farms), oral fluids (n = 168 samples; 24 farms) and lung homogenate (n = 1 sample) from the farms was collected. Only one attempt to collect at each farm was made. In some cases, pigs in the treatment pens with non-related ailments were sampled. Individual pig serum and oral fluid samples were PCV2 tested at the Health Management Center (HMC, Ames, IA) using TaqMan real-time PCR reagents (Life Technologies Corp., NY, USA) with a detection limit of below 3.5 genomic equivalents/reaction. Lung tissues were sent to Iowa State University Veterinary Diagnostic Lab for PCV2 PCR. Farms' (n = 16) samples were targeted for lower Cq values (i.e. higher virus content) with a PCR cycle quantity (Cq) ≤ 32 for serum (n = 16), Cq ≤ 30 for oral fluids (n = 11) and for tissues (n = 1; Cq=19.3) to be sent to the Dept. of Vet. and Biomed. Sci. (U of MN, St. Paul, MN) for PCV2 capsid gene sequencing with quality analysis performed within DNAstar (Madison, WI). Each isolate was further characterized as PCV2a, PCV2b or PCV2d based on standards and dendrograms were created with both Mega software (Tempe, AZ) and Disease Bioportal (UC Davis, California).

Results

Twenty-seven of 48 farms had PCV2 PCR positives (12, 15 and 5 from serum, oral fluids and lung homogenate, respectively) of which 44 samples (24, 17 and 3 from serum, oral fluids and lung homogenate, respectively) were sent for sequencing. There was an overall 91% sequencing success rate on the submitted samples (Table 1, 87% for serum, 94% for oral fluids, 100% for tissue homogenate). Forty total samples were successfully sequenced and were put into radial and traditional dendrograms for evaluation. Sequences were characterized as PCV2a (n = 3), PCV2b (n = 21) or PCV2d (n = 16) and placed into linear and radial dendrograms for visual examination (Figures 2 and 3).

Discussion

PCV2 is ubiquitous and although there is evidence that vaccination strategies have decreased the presence of PCV2 in populations of pigs, it remains in the barn environment and in vaccinated populations. Past experiences of the authors indicated that selecting pigs in treatment pens to sample had a better chance of detecting PCV2 in vaccinated populations, regardless of vaccination protocol. There are numerous reasons why this is the case in individual pigs such as co-infections, immune suppression, chronicity of non-PCVAD pathogens, or the individual pig was not properly PCV2 vaccinated (failed compliance). To answer the question of which strains of PCV2 are in a barn, this method of sampling was deemed appropriate for this project. The PCR percentages do not indicate overall population status as this was an intentionally biased sampling protocol to answer a particular question, which PCV2 strains are circulating in the population.

Figure 1: Bioportal (University of California, Davis) map displaying states where sequencing of select samples were successful

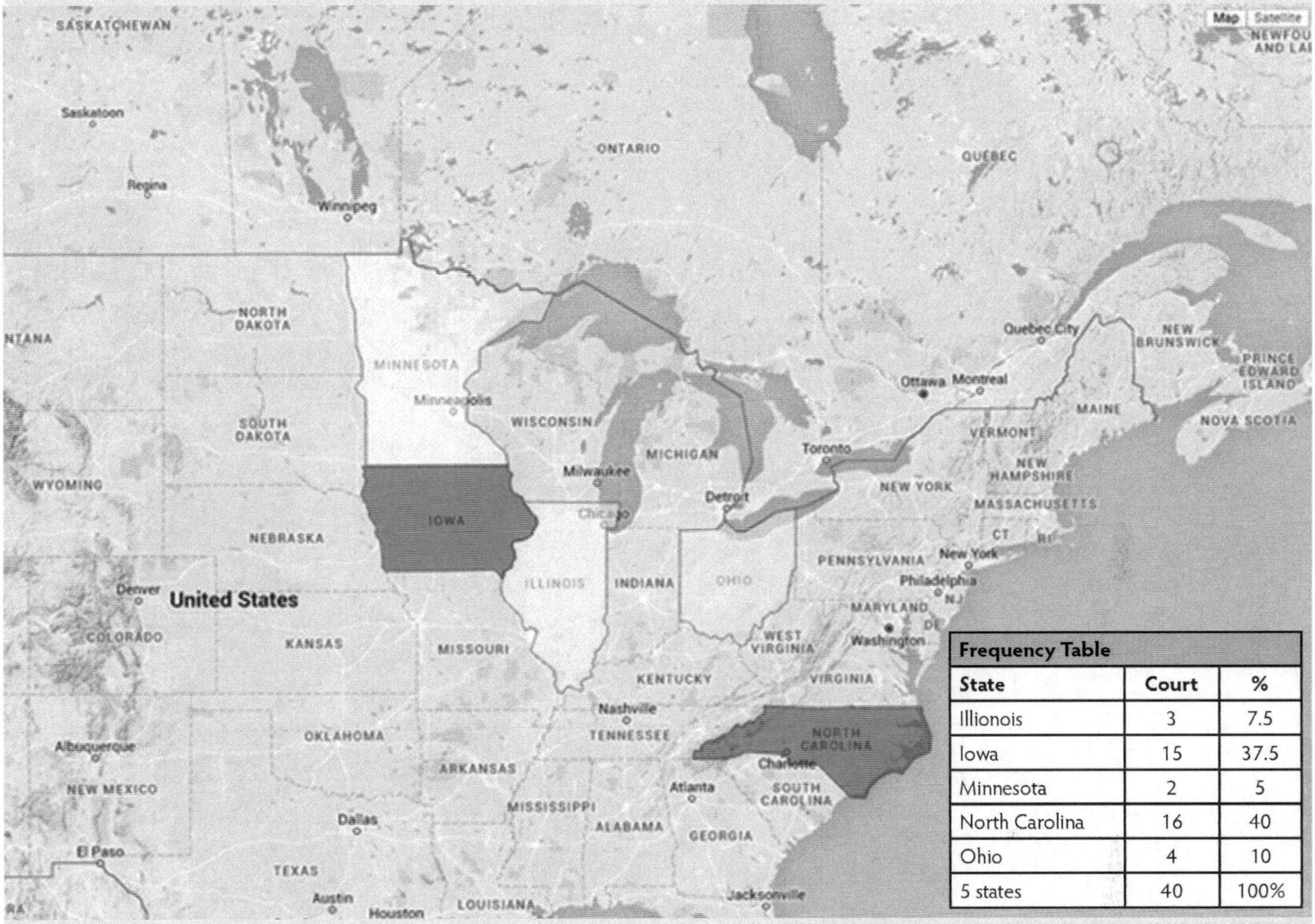

Frequency Table		
State	**Court**	**%**
Illionois	3	7.5
Iowa	15	37.5
Minnesota	2	5
North Carolina	16	40
Ohio	4	10
5 states	40	100%

Table 1: PCR positivity (HMC, Ames, IA) and sequencing success (U of MN) of select serum and oral fluid samples.

		Serum	Oral fluid
No. samples tested, total		432	168
No. and percent positive PCR, %		50/432 (11.6%)	67/168 (39.9%)
Sequencing success			
	Cq < 30	14/14 (100.0%)	13/13 (100.0%)
	Cq 30-32	7/7 (100.0%)	3/4 (75.0%)
	Cq > 32	0/3 (0.0%)	0/0

Figure 2: Bioportal (University of California, Davis) PCV2 traditional linear dendrogram rooted to PCV2 Reference sequence JF290418. Notations of "A", "B", and "D" indicate the designation of PCV2a, PCV2b and PCV2d, respectively. Only those 40 cases that were successfully sequenced in this project are displayed.

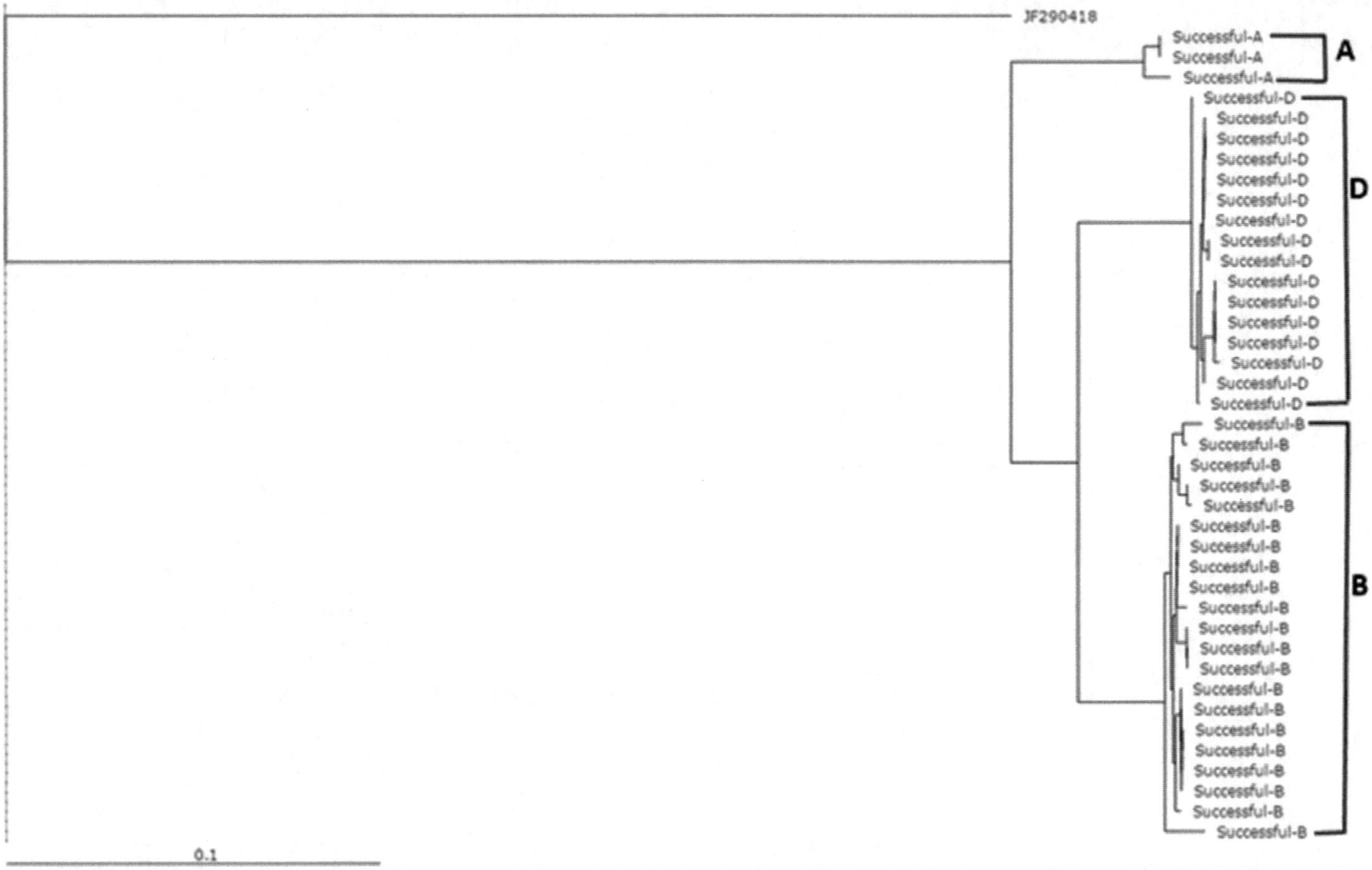

Recommendations from Iowa State University Veterinary Diagnostic Lab (Gauger, 2014, personal communication) to bias the PCV2 PCR positive samples based on Cq level (see Materials and Methods) to submit for sequencing allowed for a more economically beneficial success rate on sequencing. Some samples, as indicated in Figure 2, fell outside of this recommendation. Some higher Cq samples may have been successfully sequenced; however, they were not tested. This would have accounted for samples from 11 farms which may have under represented sequences presented here. No farms with multiple sequences on samples had homologies less than 100% (amino acid comparison), indicating that the same strain was found multiple times in the same farm in multiple pigs. Traditional sequencing tends to bias towards a predominate virus in a sample if multiple viruses are present; therefore, the use of traditional sequencing in this evaluation could have masked strains if dual infection was present.

Although there were forty sequences and they are classified as PCV2a, PCV2b or PCV2d, via capsid sequencing there was only a maximum 3% heterology (via nucleotide comparison) amongst these samples. As indicated in the radial dendrograms (Figure 3) there is very little heterology amongst all the samples collected by the U of MN veterinary diagnostic laboratory along with this sample set. This is a reasonable and expected finding for PCV2 which is a DNA virus and very little mutation is expected, compared to RNA viruses like IAV-S (influenza A virus in swine) and PRRSV (porcine reproductive and respiratory syndrome virus).

Pigs vaccinated with Ingelvac CircoFLEX®, in this study, showed no clinical signs of PCVAD and met the systems' performance expectations, per the attending veterinarian, while still being exposed to a wide-range of modern PCV2 strains including PCV2a, PCV2b and PCV2d. Numerous studies evaluating the efficacy of Ingelvac CircoFLEX® in clinical and field situations have shown it has been efficacious in at least 23 additional PCV2 challenges.[2-3,5-10] Although mutations may occur the commercial PCV2 vaccines appear to remain efficacious.

Figure 3: Megasoftware (Tempe, AZ) PCV2 radial dendrogram. The 40 individual cases from this project are noted by colored shapes with case ID labels next to the sequence and nucleotide sequences were compared amongst the U of MN Veterinary Diagnostic Lab submissions since 2005 (n >700 sequences). PCV2a, PCV2b and PCV2d groups are indicated in boldface.

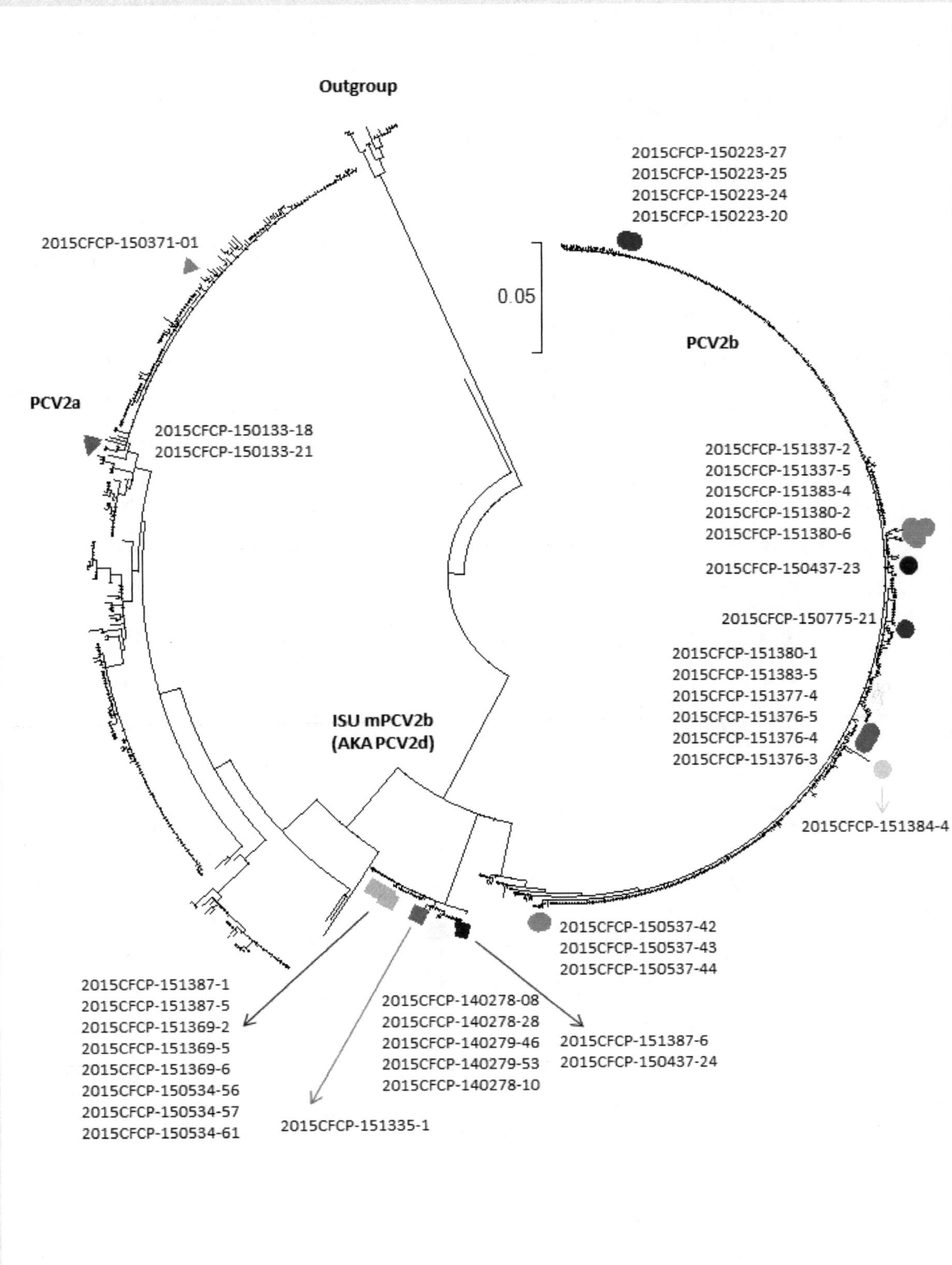

References

1. Opriessnig et al, Field investigations of perceived PCV2 vaccine failures and experimental challenge trials that support transitioning to a PCV2b-based vaccine, AASV. 2013;37-8.

2. Opriessnig et al, Commercial PCV2a-based vaccines are effective in protecting naturally PCV2b-infected finisher pigs against experimental challenge with a 2012 mutant PCV2, *Vaccine*. 2014; 32.4342–4348

3. Jeong et al, Comparison of three commercial one-dose porcine circovirus type 2 (PCV2) vaccines in a herd with concurrent circulation of PCV2b and mutant PCV2b cross protection paper, *Vet Micro*. 2015. 177;43–52.

4. Schwartz et al, PCV2: Tools to assess the subclinical impact, AASV. 2015;487-91.

5. Desrosiers et al, Use of a one-dose subunit vaccine to prevent losses associated with porcine circovirus type 2 *JSHAP*. 2009;148-54.

6. Cline et al, Efficacy of immunising pigs against porcine circovirus type 2 at three or six weeks of age. *Vet Record*, 2008;1-4.

7. Haiwick et al, A single dose, PCV2a based ORF2 subunit vaccine cross-protects against challenge with a recent PCV2b strain. AASV. 2014;129-30.

8. Rodier et al, A PCV2a based ORF2 subunit vaccine cross protects against challenge with a recent PCV2b strain AASV. 2014.

9. Licensing data, on file.

10. Takahagi et al, Differential effects of porcine circovirus type 2 (PCV2) vaccination on PCV2 genotypes at Japanese pig farms. *J. Vet. Med. Sci.* 72(1): 35–41, 2010.

Innovative polymeric adjuvant for PCV2 vaccination

Sophia Xu[1], BVM PhD; Juliette Ben Arous[2], PhD; Jérome Gaucheron[2], PhD; Laurent Dupuis[2], PhD
[1]SEPPIC Inc., Fairfield, New Jersey; [2]SEPPIC, Paris La Défense, France

Introduction

Porcine circovirus associated diseases (PCVADs) are economically important diseases of domestic pigs caused by porcine circovirus type 2 (PCV2). In order to control porcine circovirus associated diseases and reduce body weight losses in pig farming, adjuvanted inactivated or subunit PCV2 vaccines are applied to swine on a routine basis in all parts of the world.[1] Pigs are sensitive animals which need specific vaccine formulations to prevent local reactions and preserve meat quality at the injection site. Therefore, water continuous phase adjuvants such as oil-in-water (O/W) adjuvants, water-in-oil-in-water (W/O/W) adjuvants or polymeric adjuvants are usually preferred for injectable swine vaccination.[2]

Vaccine adjuvants are a key parameter in modern vaccination in order to increase the level of the induced immune response, the duration of the vaccinal protection and the orientation of the immune response. Most adjuvant formulations are composed of synthetic components combined to create a specific galenic antigen presentation. Moreover, most of the time, adjuvant components have intrinsic immunomodulator properties. The vehicle effect can be influenced by the evolution during storage or by chemical degradations (oxidation, precipitations) of adjuvant components. More specifically, the stability of vaccines can be affected by the interaction of the adjuvant formulation with the antigenic media, and particularly during storage at room temperature. Some cell lines used to produce PCV2 antigen may also contain some components (e.g. enzyme) which disturb the vaccine's stability. One solution consists in decreasing antigen enzymatic activity by high temperature treatment[3] but this method is not compatible with sensitive immunogenic epitopes.

The Montanide™ range of adjuvants has been shown to be safe and efficient in diverse swine vaccines.[4-7] We have developed a new robust polyacrylic polymer based liquid adjuvant formulation which is ready to disperse. Here we show that Montanide™ Gel 02 (Gel 02) allows the formulation of stable, safe and efficient PCV2 vaccines.

Materials and methods

First, new resisting adjuvant Montanide™ Gel 02 (polymer) was compared to reference adjuvants for galenic properties. To assess the stabilizing properties of the adjuvant, PCV2 vaccines and placebo formulations were formulated with adjuvants and stored at 4°C (normal conditions) or 20°C (destabilizing conditions) for up to 1 year. Stability of formulations was assessed after storage by visual observation and default intensity quotation.

Efficacy and safety profile of Montanide™ Gel 02 was then assessed in swine. In a first swine trial, groups of 10 3-weeks old pigs were vaccinated at D0 with a commercial PCV2 vaccine or a PCV2 vaccine based on Gel 02 adjuvant (used at 10% in the vaccine). Safety was assessed by temperature and behavioral measurement during the trial and macroscopic and histological examination of the injection site after slaughter at 18 weeks post injection. Antibody titers in blood were also measured by ELISA titration every 2 weeks all along the trial.

In a second swine field trial, 300 3-weeks old pigs (mixed Duroc, Landrace and Large white) of a farm susceptible to PCV2 infections were randomly separated in 3 groups of 100 animals. All pigs were seronegative against circovirus at the beginning of the trial. 2 groups of pigs were vaccinated at D0 with a commercial PCV2 vaccine or a recombinant PCV2 vaccine based on Gel 02. A non-vaccinated group was also analyzed as a control for natural infection. Behavior of the animals was controlled on 3 days after vaccination. For 10 pigs in each group, body temperature and local reaction at the injection site was measured at 0, 4h, 24h and 48h post vaccination. At the end of the test, 10 pigs per group were slaughtered and local reactions at the injection sites were assessed after dissection. At D0 (before vaccination), D30 and D120, blood samples were taken from 10 pigs in each group. Specific antibody titers in blood serum were assessed by specific ELISA titration. To assess the protection conferred by the vaccine in field settings submitted to PCV2 infection, average daily body weight gain was assessed in each group to measure protection against subclinical PCV2 infection.

Results

Formulations based on Gel 02 were perfectly stable over time, in presence of all tested buffers, and even in destabilizing conditions.

In both swine trials, body temperature increase after injection was < 0.5°C in all groups. Transient local reactions (small edemas) at the injection site and transient depressive states of the animals were observed at a low frequency in vaccinated groups (Table 1). All disappeared after 1 to 3 days. No lesions were observed at the slaughterhouse at the injection site. These results show that vaccine based on Montanide™ Gel 02 has an acceptable safety profile in swine which is similar to commercial formulations.

During the field trial, titers at D0 showed that all pigs were PCV2 seronegative at the beginning of the trial. At D120, there was no difference in antibody titers between all 4 groups, including the non-vaccinated control group. This result shows that the animals have been as expected in contact with circovirus during the trial, and have therefore been submitted to a natural challenge. At D30, a lower variability among animals was observed in the commercial group compared to other vaccinated groups.

As the animals have been submitted to PCV2 infection during the trial, average daily body weight gain of the animals is the best marker of performance of PCV2 vaccination. Non-vaccinated control animals had reduced average body weight gain and lower probability of survival compared to vaccinated groups. Vaccine based on Gel 02 induced a similar protection (measured as the average body weight gain during the trial) compared to a commercial PCV2 vaccine (Table 2). The results show that resisting adjuvants induce a similar protection compared to commercial vaccine.

Discussion and conclusions

Vaccine formulations can show stability defects over the long term, due to chemical degradations of the adjuvant (emulsions or polymer solutions) by enzymatic degradation or oxidation linked to the chemical properties of the antigenic media. These defects are usually enhanced and quickened by destabilizing storage conditions (especially in uncontrolled temperature conditions). We have shown that new resisting polymeric and emulsion Montanide™ adjuvants increase the stability of swine vaccine formulations in presence of destabilizing antigenic media or in stressing storage conditions.

The goal of vaccine formulation is to produce vaccines that are robust and stable in a large range of conditions, have an acceptable safety profile, and induce a strong and specific disease protection. These parameters are interdependent and depend also strongly on the vaccine adjuvant. Adjuvant design should thus aim at reaching the best possible balance between stability, efficacy and safety of the final vaccine.

In this study we have shown that new resisting Montanide™ adjuvant have safety and efficacy profiles similar to reference standard adjuvant formulations in a PCV2 vaccine model. Resisting adjuvant Montanide™ Gel 02 indeed showed an acceptable safety level in swine, and allowed the formulation of highly protective vaccines against PCV2. This new line of adjuvants will help to improve long term stability and thus long-term efficacy of pig vaccines which are based on destabilizing antigens or stored in stressing conditions. These results show that the polymeric adjuvant Montanide™ Gel 02 can allow the development of safe and efficient PCV2 vaccines.

References

1. Beach NM, Meng XJ. Efficacy and future prospects of commercially available and experimental vaccines against porcine circovirus type 2 (PCV2). *Virus Research.* 2012; 164:33- 42.

2. Heegaard PMH, Dedieu L, Johnson N, Le Potier MF, Mockey M, Mutinelli F, Vahlenkamp T, Vascellari M, Sørensen NS. Adjuvants and delivery systems in veterinary vaccinology:current state and future developments. *Arch Virol.* 2011 ;156:183–202.

3. Goodyear MD, Huether MJ, Mannan, RM, Oien, NL. Heat treated bacterins, and emulsion vaccines prepared from such heat treated bacterins. US Patent Application 2008/0112970. May 15, 2008.

4. Aucouturier J, Dupuis L, Ganne V. Adjuvants designed for veterinary and human vaccines. *Vaccine.* 2001;19: 2666–2672.

5. Cox SJ, Aggarwal N, Statham RJ, Barnett PV. Longevity of antibody and cytokine responses following vaccination with high potency emergency FMD vaccines. *Vaccine.* 2003;21:1336–1347.

6. Crisci E, Fraile L, Moreno N, Blanco E, Cabezón R, Costa C, Mussá T, Baratelli M, Martinez-Orellana P, Ganges L, Martínez J, Bárcena J, Montoya M. Chimeric calicivirus-like particles elicit specific immune responses in pigs.*Vaccine.* 2012;30: 2427–2439.

7. Aucouturier J, Ganne V, Laval A. Efficacy and safety of new adjuvants. *Ann NY Acad Sci* 2000;916:600-604.

Table 1: Frequency of pigs showing depressive state and of local reactions at the injection site after vaccination.

Groups	Transient edema at injection: Frequency	Transient depressive state: Frequency
1- Montanide™ Gel 02	1/10	6/100
2- Commercial	2/10	7/100

Table 2: Average daily body weight gain (± s.e.m) and survival rate at D120

Groups	Average daily weight gain
1- Montanide™ Gel 02	697 ± 7 g
2- Commercial	698 ± 7 g
3- Not vaccinated	**664 ± 6 g**

Herd health management solutions: The future is now

J.T. Holck[1], DVM, MS, MBA; A. Yarnell[2], BS; B. Cowan[2], BS, MS; L. Scott[2], BS
[1]Feed His People, LLC, Gilbert, Iowa; [2]GlobalVetLINK, Ames, Iowa

Introduction

The ability to align real-time diagnostic data with Veterinary Feed Directives (VFDs) can support a veterinarian-client-patient relationship (VCPR), an expectation of food animal practitioners with the implementation of Guidance Document 213.[1]

GlobalVetLINK, LLC (GVL®) created FeedLINK® electronic Veterinary Feed Directive (eVFD) system in 2005 and has ten years of experience providing the solution to swine practitioners. In 2014, GVL acquired LabLINK HIMS™ Diagnostic System, a diagnostic database developed by Carthage Veterinary Services in 2008. Currently, GVL is developing ScriptLINK® electronic prescription system, which will assist food animal practitioners in compliance with prescribing antimicrobials in water.

Together, these solutions offer a software suite which provides food animal practitioners a Herd Health Management Solution to aid in compliance with the new antimicrobial regulations. This system also provides tools to support better decision making when prescribing interventions for clients, while helping practitioners save time.

FeedLINK®

FeedLINK eVFD system is a web-based software solution, developed to assist in the streamlined fulfillment and tracking of VFDs. GVL launched FeedLINK in 2005 and continues to invest in further developments and enhancements to ensure full regulatory compliance. Another major objective of the system is streamlining the communication responsibility of veterinarians by automatically transmitting VFDs to both the feed distributor and producer. By subscribing to GVL, veterinarians, producers, and feed distributors gain access to their VFDs in a secure, CFR Part 11 compliant database, thus fulfilling the two year storage requirements in Guidance 120.[2]

Two minutes equals 90% reduction in time

The time required for generating a VFD by hand was recently estimated at approximately 7.5 minutes.[3] This estimate does not include communication or storage obligations required by clients and Guidance 120, which more than double the estimate to 20 minutes per VFD. The process for generating an electronic VFD with FeedLINK is estimated at 2 minutes; a 90% reduction in time. Practitioners can save client premises information and auto-populate up to 45 separate animal sites to further expedite the VFD process.

Built-in GVL SmartEngine™ Technology

The simplified eVFD process provided by FeedLINK is driven by the patent pending GVL SmartEngine™ technology. SmartEngine guides the user through the labelled selections for a specific product to provide assurance to veterinarians, producers and feed distributors that each VFD is consistent with label instructions. Veterinary feed directive drugs can be searched by company, product, or indication of use by species. Once a product is selected, SmartEngine only allows legal options for 1) indication of use, 2) dosage, 3) stage of production, 4) combination products, and 5) expiration date (Table 1).

With the addition of more VFD products to manage in January 2017, SmartEngine technology is designed to help simplify the creation and management of electronic VFDs. By providing VFD expiration date reminders to all stakeholders as well as other features, the system provides compliance assurances to the veterinarian, client, and feed distributor.

Integration with current feed and production software

GVL is also investing in the integration of FeedLINK into already existing feed and production software systems to further streamline the VFD process. An example is the integration of

Table 1: SmartEngine™ parameters

VFD parameter	Examples
Indications of use	Control of porcine proliferative enteropathy (PPE) associated with *Lawsonia intracellularis* infection in groups of swine in buildings experiencing an outbreak of PPE.
Combination feeding with other drugs	Tylosin and Ractopamine
Dosage and duration	38 g/ton for 7 days
Production class	Grow-finish
Expiration date	90 days

FeedLINK into Prairie Systems, a Spencer, Iowa based company offering a proprietary web-based animal management platform. The platform, known as Feed Allocation System (FAS), ensures animals receive the right feed at the right time. FAS also considers production constraints including an eVFD. By integrating with Prairie Systems using the industry approved standard the actual feed order process will not change, but simply be enhanced as FAS will automatically check for a valid eVFD in the FeedLINK database. If no VFD exists to enable the order to be completed, a request for an eVFD can be automatically generated on the behalf of the producer and sent to the veterinarian identified with a VCPR through FeedLINK. This request may contain all the necessary information required, such as but not limited to requested medication, animal group details, farm location, and mill information, to initiate a request of the eVFD. The veterinarian will then review the request and either revise the VFD request, electronically sign, or decline the request. The integration of two third party systems working collaboratively to ensure each and every feed order is valid prior to reaching the manufacturing partner is an industry first.

Mobile friendly

The FeedLINK eVFD system is mobile friendly. Whether on a smartphone or a tablet device, the FeedLINK system is ready to go anywhere that has an internet connection. This mobile connectivity allows veterinarian users to create and send an eVFD while on a farm visit, reducing the time it takes to start treatment.

LabLINK HIMS™

LabLINK HIMS is a real-time web-based software system that provides digital veterinary laboratory diagnostics for disease monitoring, analysis, and pathogen response. GVL acquired LabLINK HIMS in 2014 and continues to build on its initial strength of rapidly aggregating diagnostic information from multiple veterinary diagnostic laboratories, seven of which are currently connected with LabLINK HIMS (Table 2). With a focus on fulfilling client's evolving needs, GVL's investment in development to date has included enhanced reporting and analysis features, auto-population of Disease BioPortal, and support for diagnostic justification toward VFDs and prescriptions. As a trusted and independent third party company, GVL ensures that all client data is safe and secure.

Built-in reports for trend analysis

As practitioners, we are accustomed to receiving diagnostic results on a case-by-case basis, often having to manually enter these results into spreadsheets or other databases to assess trends over time. LabLINK HIMS is the first veterinary diagnostic database which provides practitioners with built-in reports to group cases from one or more labs over time for analysis. The twelve built-in reports display diagnostic data in tabular and/or graphical format and are available on both a case level and sample level basis. Cases may be grouped by a variety of parameters, such as but not limited to farm, site, flow, project, disease, and practitioner. Recently GVL launched the "chart builder" customized reporting tool which provides practitioners the flexibility to build their own reports, including data content, comparisons, and the ability to save the chart settings for on-going reporting.

Upload historical data and internal laboratory results

All clients have the option to have historical data electronically transferred into LabLINK HIMS, immediately providing a searchable database of cases from multiple laboratories over multiple years. While any amount of historical data may be imported, the recommendation is up to three years to maximize the amount of submission data consistency. To help ensure data consistency for reporting purposes, GVL provides a historical data clean-up service upon direction from the client.

In addition to historical data, diagnostic results from internal labs or other private laboratories can also be imported into LabLINK HIMS. A standard import feature allows clinics and production companies to import diagnostic results immediately upon completion to ensure LabLINK HIMS contains all of a company's diagnostic information. Customized import templates can also be developed to further streamline the upload process.

Save time – auto-populate Disease BioPortal!

Disease BioPortal is a geospatial visualization tool that is utilized in PRRSV area regional control (ARC) projects as well as

Table 2: Laboratories connected with LabLINK HIMS™

Laboratories connected with LabLINK HIMS™	Location
Boehringer Inglheim Vetmedica Inc. Health Management Center	Ames, Iowa
GeneSeek	Lincoln, Nebraska
Iowa State University Veterinary Diagnostic Laboratory	Ames, Iowa
Kansas State Veterinary Diagnostic Laboratory	Manhattan, Kansas
South Dakota State Animal Disease Research and Diagnostic Laboratory	Brookings, South Dakota
University of Illinois Veterinary Diagnostic Laboratory	Urbana, Illinois
University of Minnesota Veterinary Diagnostic Laboratory	St. Paul, Minnesota

by individual production companies and producers. Manually importing diagnostic information into Disease BioPortal can be a time-consuming process. Now through a partnership between GVL, Iowa State University Veterinary Diagnostic Laboratory (ISU VDL), and the University of California Davis, data aggregated in LabLINK HIMS can be automatically and electronically transferred with seamless connectivity (Figure 1) to Disease BioPortal via the ISU VDL Secure Animal Health Diagnostic Database (SAHDD). SAHDD provides site classification of PRRSV, MHYO, SIV, and PEDV prior to the data being auto-populated in Disease BioPortal.[4] Alternatively, LabLINK HIMS clients can also automatically populate Disease BioPortal directly to use the geospatial visualization and dendogram assessment tools.

Data integrity

GlobalVetLINK's objective is to be the world's most trusted, secured and independent repository of animal health data, providing clients the confidence in the security of their data. LabLINK HIMS clients, for example, provide permission for GVL to access their specific diagnostic data directly from laboratories. Each client's data is secure and segregated from other clients' data. The client is then provided with individual unique user accounts for each specified user within their company. As an independent third party company, GlobalVetLINK understands the high importance of a client's data security as well as the significance to their operation, and GVL ensures client data is secure.

ScriptLINK®

ScriptLINK™ Electronic Veterinary Prescription Certification application will provide veterinarians a secure digital Rx software system for multiple species and delivery methods. In the oversight of water soluble antibiotic usage, this system will provide online prescription certification solutions for veterinarians. ScriptLINK will optimize prescribing, fulfillment and tracking of all prescription medication supporting the judicious use of antibiotics in food animals.

Summary

GlobalVetLINK's software suite provides food animal practitioners a "one-stop-shop" of herd health management solutions to aid in compliance with the new antimicrobial regulations and support better decision making when prescribing interventions for clients all while helping practitioners save time.

References

1. Guidance #213 - Guidance for industry on new animal drugs and new animal drug combination products administered in or on medicated feed or drinking water of food producing animals. Available at www.fda.gov.
2. Guidance #120 – Veterinary feed directive regulation questions and answers. Available at www.fda.gov.
3. Thomas P. Countdown to Guidance 213 Practitioner View – Company. 2015 Allen D. Leman Swine Conference.
4. Main R et al. New Tools for Monitoring the Health Status of Swine Herds. 2015 ISU James McKean Swine Disease Conference.

Figure 1: Automated data flow - from your lab into BioPortal.

Protective efficacy of a replicon particle vaccine in both naïve sows and previously exposed gilts against porcine epidemic diarrhea virus

K. Crawford[1], DVM, MS; M. Mogler[1], PhD; J. Hicks[2], DVM; D.L. Harris[1], DVM, PhD
[1]Harrisvaccines, Ames, Iowa; [2]Leading Edge Advanced Research, Carroll, Iowa

Introduction

Alphavirus-based replicon RNA particle (RP) vaccines are capable of stimulating a robust and balanced immune response in vaccinated animals.[1] RP vaccines expressing the hemagglutinin gene of Swine Influenza Virus (SIV), delivered parenterally, have been shown to protect against disease after homologous challenge.[2,3,4] An RP vaccine for Porcine Epidemic Diarrhea Virus (PEDV) was formulated by incorporating the genetic sequence of the Spike glycoprotein of PEDV into an alphavirus replicon derived from Venezuelan equine encephalitis (VEE) virus strain TC-83. The USDA conditionally licensed this vaccine, PED RNA, in mid-2014. The purpose of this study was to evaluate the protective efficacy of the PED RNA vaccine delivered intramuscularly in both naïve and previously exposed dams, and to measure overall litter mortality associated with PEDV infection after a subsequent PEDV challenge in piglets.

Materials and methods

Two trial protocols were used to evaluate the PED RNA vaccine, utilizing either naive sows or PEDV-exposed gilts. In the naive dam trial pregnant sows (n = 16), with no prior exposure to PEDV were vaccinated with either PED RNA vaccine or placebo at various time points. The naïve sows were vaccinated approximately 8, 5, and 1 week pre-farrowing. In the previously exposed dam protocol, pregnant gilts (n = 7), previously exposed to PEDV approximately 10 months prior to farrowing were vaccinated approximately 4 and 1 week pre-farrowing. All animals were allowed to farrow naturally and no piglets were cross-fostered between dams. All piglets were challenged between 2-6 days of age depending on the dam's farrowing date.

On Day 0 (DPC 0), all piglets from the naive sow trial were orally challenged with 2 mL containing 10^3 $TCID_{50}$ of cell culture- derived porcine epidemic diarrhea virus (PEDV/CO/2013). In the previously exposed gilt protocol, piglets were orally challenged with 2 mL of filter-sterilized gut homogenate in PBS, containing approximately 10^8 genome copies by quantitative RTPCR. The gut homogenate was from a single passage of PEDV/CO/2013 in a litter of SPF neonatal piglets, and was determined to be PCR-negative for rotavirus A/B/C, PCV2, PRRSV, TGEV, and SDCoV.

Clinical observations and fecal scores were recorded for each dam and piglet through DPC14. Piglets that died prior to study termination were necropsied and probable cause of death was recorded. Moribund piglets were humanely euthanized at the discretion of the attending veterinarian. On DPC14, all remaining piglets were euthanized and necropsied. Litter mortality was calculated for each sow/gilt. Blood samples were collected from all dams at the following time points: vaccination days, farrowing, challenge, DPC7, and at study termination on DPC14. Blood samples were collected from all piglets on DPC0 and from all remaining piglets at DPC14. Colostrum was collected from all dams within 24 hours of farrowing and milk collected at DPC7. To assess immune response to vaccination and challenge, sera and milk samples were analyzed for PEDV-neutralizing antibodies using the fluorescent focus neutralization (FFN) test.

Results and discussion

The average litter mortality for both trials can be found in Table 1. The overall average mortality rates for the control litters and vaccinate litters in the naïve sow trial were 91% and 69%, respectively. Litter mortality rates for vaccinated dams were significantly lower than those of control dams by ANOVA ($P = 0.02$). This result is consistent with previously reported results from smaller trials using the same challenge methodology.

The PEDV FFN titers can be found in Table 2. As expected, the naïve control sows were negative at farrowing, at challenge and in the colostrum and milk samples. In contrast, all vaccinated sows had neutralizing antibody titers in sera at farrowing, challenge, post-challenge, and in colostrum and milk samples.

Table 1: Average litter mortality for both control and vaccinate groups from the naïve sow and previously exposed gilt studies.

Treatment	n =	Avg. litter mortality
Naive sow protocol		
Sham controls	6	91%
PED RNA vaccinates	10	69%
Previously-exposed gilt protocol		
Sham controls	3	59%
PED RNA vaccinates	4	45%

Table 2: Geometric mean inverse titers by sample type for both control and vaccinate groups from the PEDV-naïve sow and previously exposed gilt studies at various time points throughout the experiment. "nd" = no data at time of writing.

Group	Farrowing		Challenge	DPC7	
	Serum	Colostrum	Serum	Serum	Milk
Naive sow protocol					
Sham controls	< 20	< 20	< 20	20	< 20
PED RNA vaccinates	85	367	121	452	65
Previously-exposed gilt protocol					
Sham controls	160	1280	403	nd	nd
PED RNA vaccinates	1280	> 2560	1810	nd	nd

In the previously exposed gilt trial, the average mortality rates for the control litters and vaccinate litters were 59% and 45%, respectively. This difference was not found to be statistically significant, perhaps due in part to smaller group sizes. The PEDV FFN titers for vaccinated gilts were consistently of higher magnitude than control gilts, including in both sera and colostrum, consistent with previous field and clinical experience.

In contrast to some other reports, our experience with PEDV challenge of previously exposed gilts suggests that significant piglet mortality can occur despite prior dam infection. The specific factors responsible for herd-level resistance to so-called "re-breaks" will require more detailed study. The nature of PED makes studying this at-risk population in a laboratory setting quite difficult.

In summary, these data demonstrate the induction of robust anti-PEDV immune responses in two distinct target populations using PED RNA vaccine. Significant reductions in litter mortality were achieved with a naive dam, suckling piglet challenge model. Additional data and discussion will be presented at the meeting.

References

1. Rayner, J.O., Dryga, S.A., Kamrud, K.I. 2002. Alphavirus vectors and vaccination. *Rev Med Virol*;12:279-296

2. Vander Veen, R.L., Mogler, M.A., Russell, B.J., Loynachan, A.T., Harris, D.L., Kamrud, K.L. 2013. Haemagglutinin and nucleoprotein replicon particle vaccination of swine protects against the pandemic H1N1 2009 virus. *Vet Rec*;173:344 doi:10.1136/vr.101741.

3. Bosworth, B., Erdman, M.M., Stine, D.L., Harris, I., Irwin, C., Jens, M., Loynachan, A., Kamrud, K., Harris, D.L. 2010. Replicon particle vaccine protects against influenza. *Comp Immunol Microbiol Infect Dis*:33:e99-e103. doi:10.1016/j.cimid.2010.05.002.

4. Erdman, M.M., Kamrud, K.L., Harris, D.L., Smith, J. 2010. Alphavirus replicon particle vaccines developed for use in humans induce high levels of antibodies to influenza virus hemagglutinin in swine: proof of concept. *Vaccine*;28:594-596.

RNA particle vaccines: Bringing solutions to emerging and rapidly mutating pathogens

Channing Sebo, DVM; Joel Harris; Mark Mogler, DVM, PhD; Kimberly Crawford, DVM
Harrisvaccines, Ames, Iowa

Summary

RNA Particle vaccines provide solutions to rapidly emerging and mutating pathogens. New threats occur frequently and the RNA Particle technology can respond quickly, in only 6-10 weeks with a new vaccine. 100% farm specific vaccines can be customized to the strains currently circulating on farm for diseases caused by influenza A virus – swine (IAV-S), porcine respiratory and reproductive syndrome virus (PRRSV), and rotavirus. Rapid changes in strains can be difficult to track. To help veterinarians monitor their farms, Harrisvaccines offers herd sequence monitoring through the use of customizable dendrograms for these pathogens.

Technology overview

Harrisvaccines uses the same RNA particle (RP) platform technology to make all of the vaccines. The platform uses an alphavirus, Venezuelan equine encephalitis (VEE) virus strain TC-83 to make RPs. RPs are single-cycle, propagation-defective particles that are not able to spread beyond the initial infected cells. RPs are produced by removing the VEE structural genes from the replicon RNA vector and inserting a gene of interest (GOI). This platform, along with helper genes are electroporated into vero cells, then incubated and harvested. RP vaccines have been evaluated in many different species of animals as well as humans with a proven record of safety and efficacy. Multiple RPs can be combined into a single dose, decreasing the number of vaccinations that an animal receives. Other advantages of the RP technology include resistance to anti-vector immunity and DIVA (differentiating infected from vaccinated animals) compatibility. This technology can also be used to combat foreign animal diseases such as High Pathogenic Avian Influenza, Foot and Mouth Disease, and Classical Swine Fever.

What is the gene of interest?

Each disease that a vaccine targets uses a specific gene of interest (GOI) or sequence to induce a robust and balanced immune response. Examples of these GOI include the following: Herd specific GOI are IAV-S (HA gene), Rotavirus (VP7 gene) and PRRS (ORF 5). For the rest of the products a conserved gene is used to provide protection and a herd specific approach is not required. These diseases include: porcine epidemic diarrhea (PED) (Spike gene); swine dysentery, *Mycoplasma hyopneumonia, E. coli* (K88 pilus gene); and for *Clostridium difficile* (Toxins A & B) and *C. perfringens* type A & C (Toxins alpha, beta1 & beta2). Any of these products can be formulated into a multivalent combination vaccine.

Dendrograms

For diseases such as IAV-S, rotavirus and PRRS, Harrisvaccines recommends 100% herd specific vaccines strategies. Diagnostic samples are submitted to either a state or private lab and sequenced. The sequence is then electronically submitted to Harrisvaccines without samples ever entering the production facility. This electronic sequence is synthesized directly into vaccine. Managing the sequences and herd specific vaccines can be difficult for veterinarians and clinics to do. To assist our customers, Harrisvaccines vaccines makes customizable dendrograms to see relationships among the strains circulating within their system and most accurately make vaccine selections. Viruses mutate and emerge so if a new strain enters your system, our goal is not only for quick detection, but to change the vaccine just as rapidly too; switching out the genes that are specific to the new strain. Dendrograms allow veterinarians to look at both a farm and regional pathogen sequence prevalence to make the best vaccine decisions for each farm.

Influenza A Virus – Swine

Harrisvaccines currently has over 150 RPs made for IAV-S due to the large variation in the nucleotide sequences of the gene of interest (GOI) region, the HA gene. Within the H1 subtype, the Beta RPs have 8.3% homology variation. For the Gamma H1 subtype, there is 4.95% variation, for the Pandemic H1, the maximum variation is 16.5%. For the Delta H1 subtype the variation is even larger at 23.72%. Within the H3 subtype there is 15.55% variation. This database is growing daily as new RPs are made specific to each farm.

Rotavirus

The VP7 region of the rotavirus sequence is target for vaccine production. The VP7 region is highly variable with over 128 RPs currently in our database. With so much variation, again a herd specific approach is required. For rotavirus A, Harrisvaccines has a maximum variation of 26.99% within the current RPs. For rotavirus B there is 47.18% and for rotavirus C, 29.82%.

Porcine respiratory reproductive syndrome

For PRRSV the target sequence is the ORF 5 region for vaccine production with 71 RPs currently in our database. Within the North American strains there is a variation of 23.0% in the RPs and within the European strains there is a variation of 12.44%.

Conclusion

Harrisvaccines' RP vaccine technology can be very valuable asset in pork production. Unique advantages included a balance immune response, high safety profile and DIVA compatibility. With such a high degree of strain variation for several swine pathogens, a herd specific approach is required for diseases such as IAV-S, PRRS and Rotavirus. Managing herd specific vaccines can be challenging for veterinarians to manage so Harrisvaccines offers customizable dendrograms to assist in both herd monitoring and vaccine selection. Combining RPs for multiple pathogens for multivalent vaccines is another unique advantage of the RP technology.

References

1. Erdman, M.M., et al. 2010. Alphavirus replicon particle vaccines developed for use in humans induce high levels of antibodies to influenza virus hemagglutinin in swine: Proof of concept. *Vaccine.* 28: 594–596.

2. Vander Veen, RL et al. 2012 Alphavirus replicon vaccines. *Animal Health Research Reviews. 1-9.*

3. Vander Veen RL et al. 2009. Rapid development of an efficacious swine vaccine for novel H1N1. PLoS Currents Influenza 1: RRN1123.

Spermatotoxic potential of insemination lubricant

Chris Kuster[1], DVM, MS, PhD; Julie Robinson[1], BS, PhD; Wai-May Lim[1], BSc, MS; Peggy Anne Hawkins[2], DVM, MS
[1]Kuster Research and Consulting, Inc., Geneseo, Illinois; [2]Aurora Pharmaceutical, Northfield, Minnesota

Introduction

In a typical 2400 sow breeding barn approximately 100 sows are artificially inseminated each week. If every sow receives two inseminations, 200 times a week sperm may contact the lubricant used to assist insertion of the AI catheter. If the lubricant is toxic to the sperm there is a chance of reduced reproductive potential. Although insemination lubricants may claim they are "non-spermicidal", preliminary results indicate spermatotoxic effects of a popular brand of lubricant across a range of exposure levels and contact times.

The specific aim of this study was to evaluate the effect of two different catheter lubricants on boar sperm motility, acrosome integrity and membrane viability using computer-assisted semen analysis (CASA) and flow cytometry. Samples extended in both a short-term and a long-term extender were evaluated across three different lubricant concentrations (1%, 2.5% and 5% v/v) at multiple durations of exposure (0, 30, 60 min).

Materials and methods

Insemination lubricants

Kuster Research and Consulting, Inc. (KRC) performed replicate testing on extended semen samples using two different insemination lubricants: Clarity, supplied by Aurora Pharmaceutical (Northfield, MN) and Product B – a frequently used commercial catheter lubricant purchased from an industry vendor.

Semen extenders

Semen doses from two different extender classifications: short-term, (BTS: IMV USA, Maple Grove, MN) and long-term, proprietary non-BSA , (NUTRIXcell+, IMV USA, Maple Grove, MN) were analyzed across three lubricant concentrations (1%, 2.5% and 5% v/v) and three exposure times (0, 30 and 60 minutes). A third extender representing the category of extenders containing bovine serum albumin (BSA) will be added for the meeting presentation.

Boar semen doses

- Pooled semen doses were obtained from commercial boar studs during routine collection.
- Ejaculates were partially diluted, pooled and fully extended in either short-term (n = 4) or NutriXcell+ (n = 6) according to the stud targets of approximately 35-50 million sperm per ml.
- Insemination doses were cooled and maintained at 17°C during same day transportation or overnight shipping to KRC.
- Only pooled semen doses that met standard production criteria (75% total motility; 75% normal morphology) were included in this study.

Semen incubation with insemination lubricants

Treatment groups: Extended semen samples were mixed 1:1 with lubricant diluted in extender to a total volume of 1ml and incubated at room temperature (RT, 20-22°C) for 0, 30 or 60 minutes for CASA analysis, and 60 minutes prior to flow cytometry. Control samples: Semen doses were mixed 1:1 with the same extender to match the sperm concentration in the treatment groups. Undiluted control samples were also incubated at RT.

CASA motility analysis

A CASA system (IVOS, Hamilton Thorne Biosciences, Beverly, MA) was used to analyze gross and progressive motility in control and treatment groups following exposure to lubricant (1%, 2.5% or 5% v/v) at 0, 30 or 60min.

Flow cytometric analysis

Sperm plasma membrane viability and acrosome integrity were assessed at a single time point (60 min), using the fluorescent dyes propidium iodide (PI) and fluorescein-labelled *Arachis hypogaea* (peanut) agglutinin (FITC-PNA) to determine the percentage of viable sperm with intact acrosomes (VIA) in both control and treatment groups. The VIA assays were performed on a Guava EasyCyte Plus flow cytometer (Millipore Corp., Hayward, CA).

Results

Statistics

Further replicates are planned with statistics presented at the meeting pending attainment of full sample size.

Effect of insemination lubricant on boar sperm motility

Incubation of extended semen with Clarity gel induced only mild sperm agglutination in some samples, while sperm agglutination was routinely noted at higher levels when exposed to Product B. Motility remained comparable to the Controls for samples exposed to Clarity in the long-term extender (Table 1 and Figure 2), with relatively small differences at the longer time points in the short-term extender (Table 1 and Figure 1). In contrast, sperm motility reductions were apparent after exposing semen to Product B for all lubricant concentrations in both short-term and long-term extenders (Figure 1 and Figure 2). For Product B, sperm motility was reduced to ≤ 60% for all time points at the 2.5% and 5% levels in the short term extender (Figure 1), as well as for all time points at the 5% level in BSA-free long term extender (Figure 2).

Table 1: Total sperm motility (average % ± SD) after 30 minutes exposure to lubricant.

Treatment/extender type	Control	Lubricant concentration 1% (v/v)	2.5% (v/v)	5% (v/v)
Clarity – short term	78 ± 6%	78 ± 3%	76 ± 3%	72 ± 6%
Product B – short term	78 ± 6%	65 ± 3%	47 ± 7%	23 ± 14%
Clarity – long term	85 ± 4%	83 ± 5%	81 ± 5%	81 ± 6%
Product B – long term	85 ± 4%	67 ± 6%	65 ± 6%	33 ± 14%

Effect of insemination lubricant on sperm viability and acrosome integrity

There was no apparent effect of Clarity lubricant on sperm VIA results (percentage of sperm with viable plasma membranes and intact acrosomes) in either short-term or long-term extenders as determined by flow cytometry (Figures 3 and 4). However, a downward trend in VIA was observed as the exposure concentration increased for Product B. The average control sperm VIA was 76.4 ± 3.6% in short-term extender and 59.1% ± 8.5% following exposure to Product B (5% v/v) for 60 minutes (Figure 3). A lower magnitude trend was visible in VIA results for sperm exposed to Product B in the long-term extender.

Conclusions

Although insemination lubricants may claim non-spermicidal status, the current data demonstrates that spermatotoxicity varies by lubricant, semen extender, lubricant concentration and duration of exposure. These effects can be observed immediately after exposure, and tend to become more severe over time. In the current study, motility suppression was a more sensitive indicator than sperm membrane viability status for the spermatotoxic lubricant. Exposure to Product B for specific concentration*time combinations reduced motility to the extent that litter size could be compromised.[1,2] In contrast, exposure to the highest concentration of Clarity gel (5% v/v) for the longest duration (60 min) had virtually no effect on sperm viability or acrosome status, with a relatively small change in motility for only the short-term extender. The BSA-free long-term extender appeared partially protective against the membrane toxicity effects of Product B when compared to the short-term extender. Based on both the sperm motility and viability results obtained in this study, Product B expresses spermicidal effects, while Clarity lubricant does not.

References

1. Hirai M et al. Objectively Measured Sperm Motility and Sperm Head Morphometry in Boars (*Sus scrofa*): Relation to Fertility and Seminal Plasma Growth Factors. *J Androl*. 2001;22:104-110.

2. Broekhuijse MLWJ et al. Application of computer-assisted semen analysis to explain variations in pig fertility. *JAS*. 2015;90:779-789.

Figure 1: Total sperm motility in short-term extender following exposure to lubricant over increasing time.

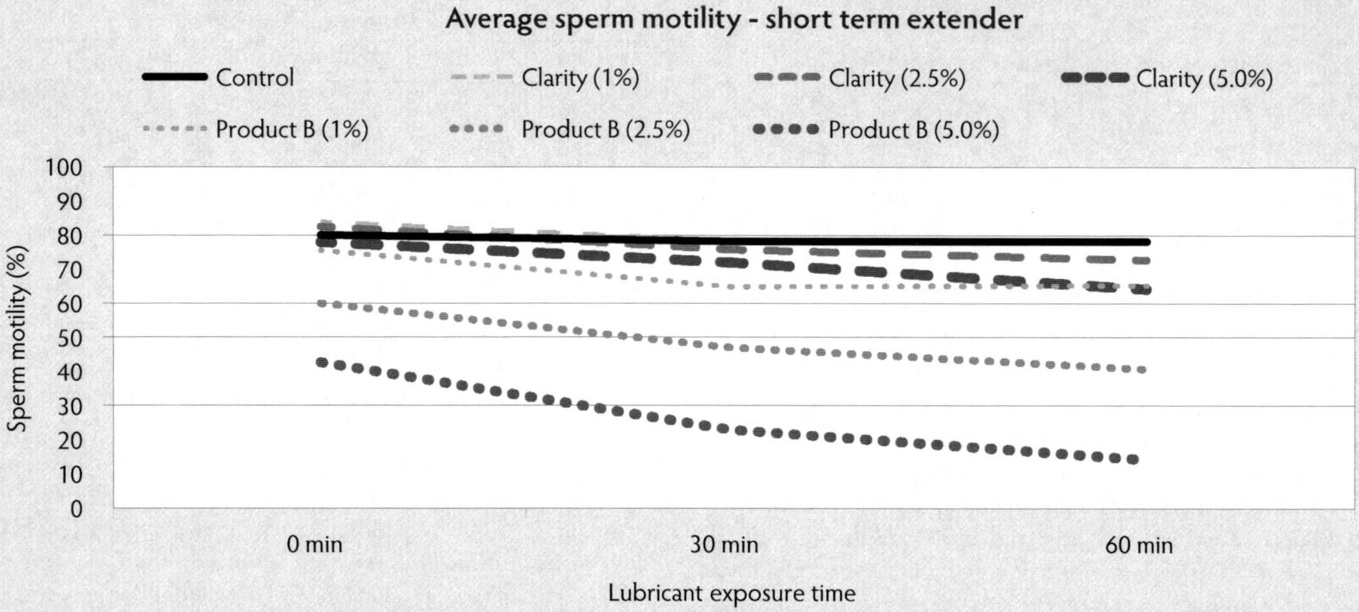

Figure 2: Total sperm motility in long-term (non-BSA) extender following exposure to lubricant over increasing time.

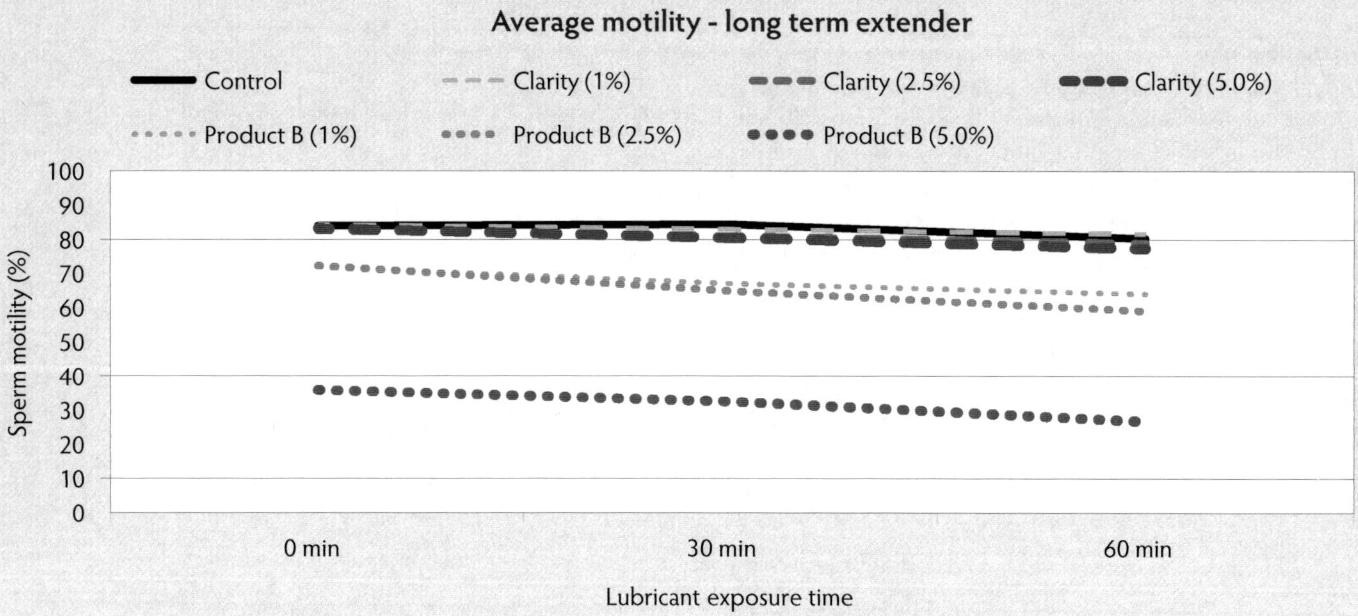

Figure 3: Sperm VIA (viable with intact acrosomes) in short-term extender.

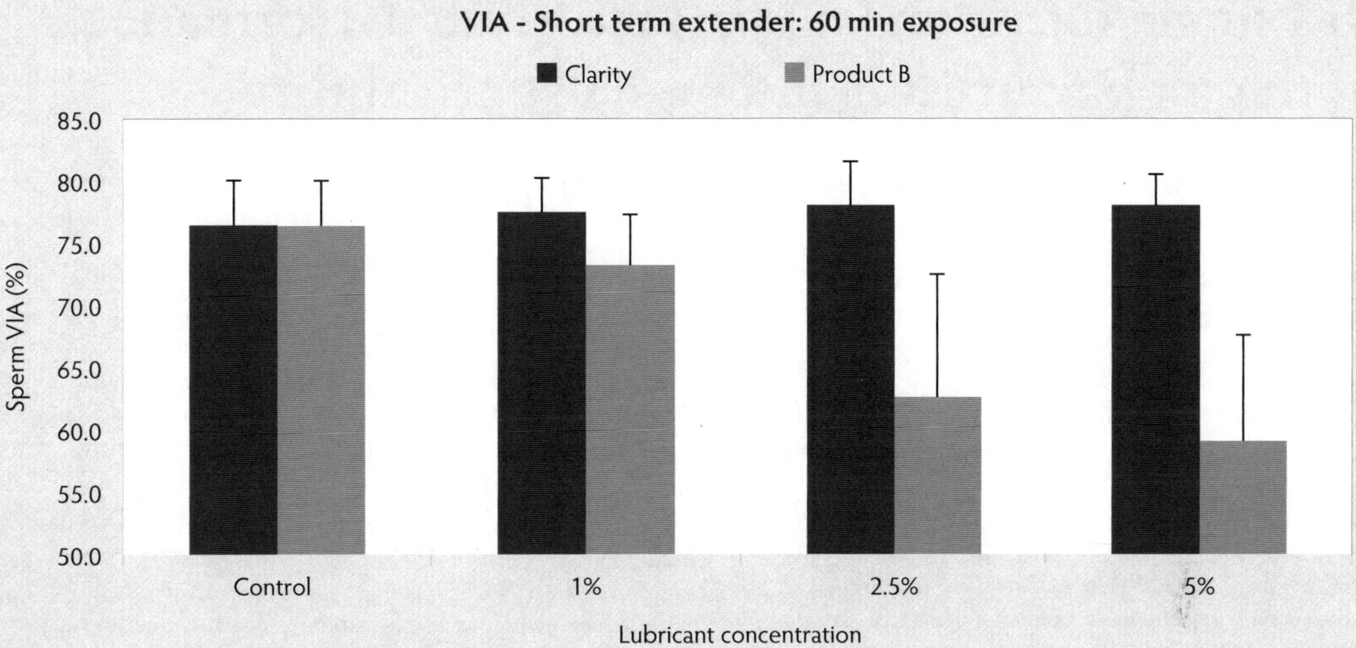

Figure 4: Sperm VIA (viable with intact acrosomes) in long-term extender after 60 minute lubricant exposure.

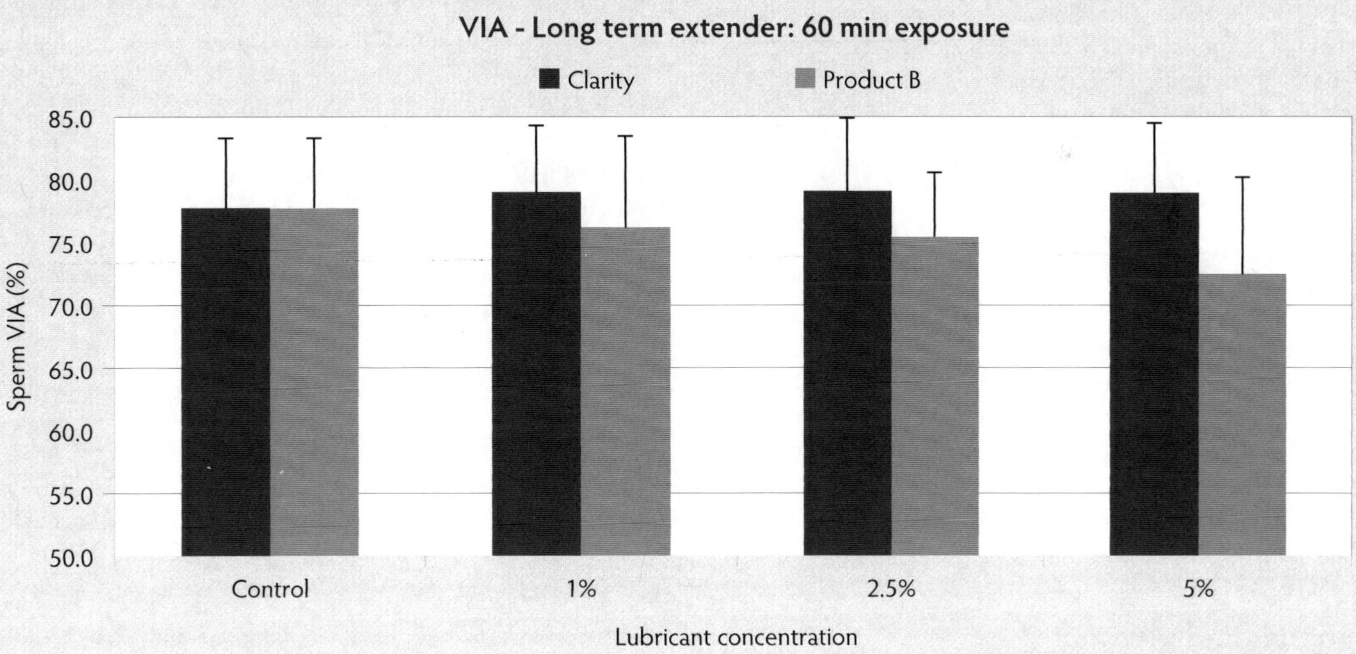

Effect of induction of farrowing following OvuGel® and single fixed time AI on farrowing performance, timing of parturition and piglet survivability

P. E. Yeske[1], DVM, MS; M. E. Johnston[2], MS; C. Betlach[3], BS; M. Sun[4], BA; A. R. Hanson[1], PhD, MS
[1]SVC Research, Saint Peter, Minnesota; [2]JBS-United Animal Health, Sheridan, Indiana; [3]University of Minnesota, Saint Paul, Minnesota; [4]Gustavus Adolphus College, Saint Peter, Minnesota

Introduction

The swine industry continues to improve in litter size[1]. However, larger litters may be accompanied by increased number of stillborn piglets and preweaning mortality. Some farms have gone to 24/7 farrowing care to reduce the numbers of stillborns and improve the preweaning mortality. Other farms without enough labor to do 24/7 assistance have used induction of farrowing to be able to predict the timing of farrowing event. This allows the farm crew to attend and assist in the farrowing process and work to reduce the stillborns and preweaning mortality. Stillborn pigs and preweaning mortality contribute to significant economic losses to the pork industry. The use of OvuGel® (JBS United Animal Health) concentrates farrowings into a narrower interval of time, thus facilitating the allocation of human resources. However, its use in combination with other methods of manipulating the timing of parturition via induction has not been explored.

Objectives

Evaluate the effect of induction of farrowing on performance, time of parturition and piglet survivability in sows treated with OvuGel® and inseminated at a fixed time.

We hypothesized that induced sows would:

- Farrow over fewer days in the week.
- Farrow more litters between the hours of 7 AM and 3 PM (when staff is available to assist sows farrowing).
- Have more live pigs born, reduce the percentage of stillborn piglets.
- Have reduced preweaning mortality by assuring piglets got a good start.
- Wean more piglets

Materials and methods

One hundred fifty one sows (parity 1 to 12, average 3.6) were stratified by parity and breeding group and randomly assigned to be either induced on d 113 post insemination with a 2 ml intramuscular injection of Lutalyse® or no induction. All sows were administered OvuGel® on d 4 post weaning and inseminated using single dose fixed time insemination on d 5 post weaning (with insemination occurring at 22-24 hours after OvuGel treatment). Routine farrowing and sow management procedures were practiced by farm personnel. Piglets were tagged and weighed individually within approximately 12 h of birth and before weaning (mean = 20.0 days). Pigs were cross-fostered within treatment, and cross-fostering was recorded. Numbers of pigs born live, still born, mummified, and pigs that died pre-weaning were recorded. Time of farrowing was recorded or estimated by visual assessment of pigs.

Statistical analysis was done using litter as the experimental unit. Data were analyzed to evaluate the effect of induction using ANOVA procedures in JMP (7.0.1, SAS Institute). Farrowing group and parity were included as random effects in the model. Only litters with complete weaning and farrowing data and all pigs accounted for were analyzed (n = 121).

Number of mortalities, reason for mortalities, number of stillborn, and number of mummified piglets were evaluated using a chi-squared test. The chi-squared test evaluated the association of treatment with the occurrence of at least one mortality, still born or mummified fetus.

Results and discussion

Due to this study being done on a commercial farm by commercial farm staff, a total of 30 litters were excluded because they lacked either birth data or wean data or because all born piglets were not accounted for as dead, transferred or weaned pigs (n = 61 or 60 for Control or Induced, respectively).

Gestation length was shorter for sows induced to farrow compared to controls (113.9 vs 114.8 d, $P < 0.01$; Table 1). A greater proportion of sows induced to farrow gave birth on d 114 (90% vs 29.5% for Induce vs Controls, $P < 0.01$; Table 2 and Figure 1). A greater proportion of induced sows farrowed during a standard 8 hour workday (70% vs 46% for Induced vs Controls, $P < 0.01$; Table 2).

The control sows had greater total pigs born compared to sows induced to farrow (14.2 vs 13.1 pigs, $P < 0.05$; Table 1 and Figure 2) but pigs born live was not different (13.1 vs 12.2 pigs, $P = 0.23$ Table 1 and Figure 2). This unintended difference in total born must have been related to chance since all areas that could be controlled were done in the design. Mean parity (3.8 vs 3.7, $P > 0.05$ Table 1) and distribution of parity were similar across the treatments (Figure 3). Birth weight was not different between induced and control (3.0 vs 3.0 lb., $P = 0.87$ Table 1 and Figure 2).

Table 1: Effects of sow treatment on farrowing performance for sows with complete data (n =121).

Item	CON	Induce	SEM	*P*-value
No sows	61	60		
Parity	3.8	3.7	0.4	0.856
Gestation length, d	114.8	113.9	0.1	**< 0.001**
Total born (Live + SB+Mum)	14.2	13.1	0.8	**0.038**
Born Live	13.1	12.2	0.7	0.228
Born live, % Total	91.5	95.9	1.5	**0.022**
Pigs started after cross foster	11.3	11.8	0.4	0.215
Stillborn	1.1	0.5	0.2	**0.031**
Stillborn, % of total born	8.5	4.1	1.5	**0.022**
Mummified	0.4	0.4	0.2	0.980
Total birth wt (SB + live+Mum), lb	41.6	37.1	2.0	**0.014**
Average Birth wt, lb	3.0	3.0	0.1	0.873
SD Birth wt, lb	0.63	0.56	0.02	**0.034**
CV of Birth wt, %	21.4	19.4	0.9	0.089
No weaned, pigs/litter	10.07	10.74	0.36	**0.034**
Wean wt, lb/litter	125.09	126.46	4.67	0.804
Average wean wt, lb/pig	12.47	11.88	0.25	0.099
SD of Wean wt	1.94	1.85	0.08	0.337
CV of Wean wt, %	15.78	15.96	0.68	0.831
ADG, lb/pig	0.52	0.50	0.01	0.172
SD of ADG	0.09	0.09	0.00	0.725
CV of ADG, %	17.93	18.65	0.80	0.450
lactation length, days	20.08	19.85	0.20	0.354
Pre-weaning mortality, % of pigs after cross fostering	9.88	6.87	2.04	0.143
Total mortality, no pigs	74	55	-	-
Total mortality, pigs/litter	1.20	0.91	0.28	0.309
Laid on, pigs/litter	0.40	0.39	0.09	0.936
Low viability, pigs/litter	0.33	0.24	0.22	0.655
Runt, pigs/litter	0.11	0.02	0.07	0.364
Starved out, pigs/litter	0.08	0.06	0.05	0.749
Dead other, pigs/litter	0.26	0.08	0.11	0.258

[1] Only includes 121 litters with both birth and weaning data and all born piglets accounted for as dead, transferred, or weaned

Table 2: Effect of sow treatment on pre-weaning mortality for litters with complete data (n = 121).

	Control		Induced		
	No[1]	Yes	No	Yes	*P*-value[2]
Number of litters					
Born between 7 AM & 3 PM	33	28	18	42	**0.001**
Born on d 114	43	18	6	54	**< 0.001**
Still born	28	33	38	22	0.054
Mummified	50	11	50	10	0.843
Total mortality	25	36	36	24	**0.037**
Laid on	40	21	45	15	0.257
Low viability	48	13	50	10	0.515
Runt	60	1	59	1	0.991
Starved out	57	4	56	4	0.981
Dead other	55	6	56	4	0.527
% of litters					
Born between 7 AM & 3 PM	54.1	45.9	30.0	70.0	**0.001**
Born on d 114	70.5	29.5	10.0	90.0	**< 0.001**
Still born	45.9	54.1	63.3	36.7	0.054
Mummified	82.0	18.0	83.3	16.7	0.843
Total mortality	41.0	59.0	60.0	40.0	**0.037**
Laid on	65.6	34.4	75.0	25.0	0.257
Low viability	78.7	21.3	83.3	16.7	0.515
Runt	98.4	1.6	98.3	1.7	0.991
Starved out	93.4	6.6	93.3	6.7	0.981
Dead other	90.2	9.8	93.3	6.7	0.527

[1] No = did not occur for the litter, yes = occurred for one or more pigs in the litter
[2] Chi squared test

Induced sows had fewer stillborn pigs relative to controls (4.1% vs 8.5% $P \leq 0.05$ Table 1 and Figure 4), and stillborn pigs were reduced for induced compared to control (0.5 vs 1.1 pigs/litter, $P = 0.03$ Table 1 and Figure 2). The incidence of stillborn pigs was less for induced sows relative to control sows (36.7 vs. 54.1% of litters, $P < 0.05$ Table 2).

The number of pigs started after cross fostering did not significantly differ across treatments, but was numerically greater for induced sows. Sows induced to farrow weaned more pigs per litter ($P = 0.013$), but their pigs weighed less than those from control sows at weaning ($P = 0.10$). Variation in weaning weight and ADG were not affected by treatment. Pre-weaning mortality was approximately 8% across all treatments. Preweaning mortality was less for litters of induced sows relative to control (6.7% vs 9.9%, $P = 0.14$; Table 2 and Figure 4). Incidence of mortality was less for litters of induced sows relative to control (40 vs 59%, $P = 0.04$; Table 2).

Using a simple partial budget to estimate the value of induction the margin over cost would be over $21.00 or an ROI of 2.7:1 (if both the cost of induction and OvuGel are included in the cost)

Figure 1: Length of gestation and interval of farrowing was longer for CON relative to induced sows

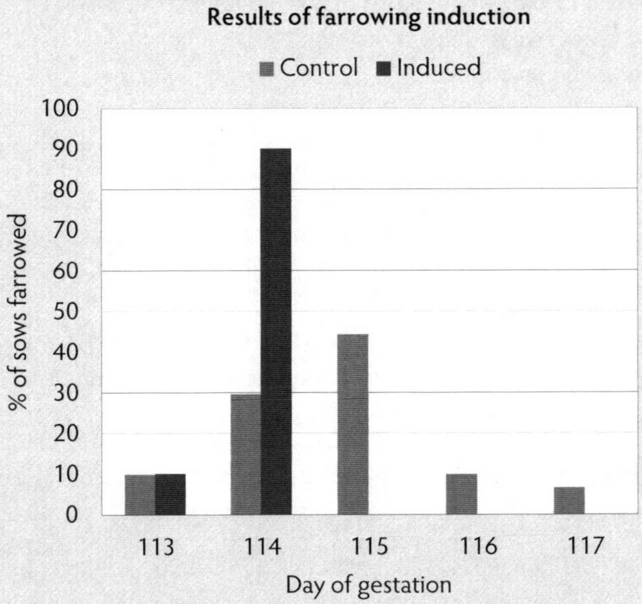

Figure 2: Number of total born, born alive, stillborns and birth weight.

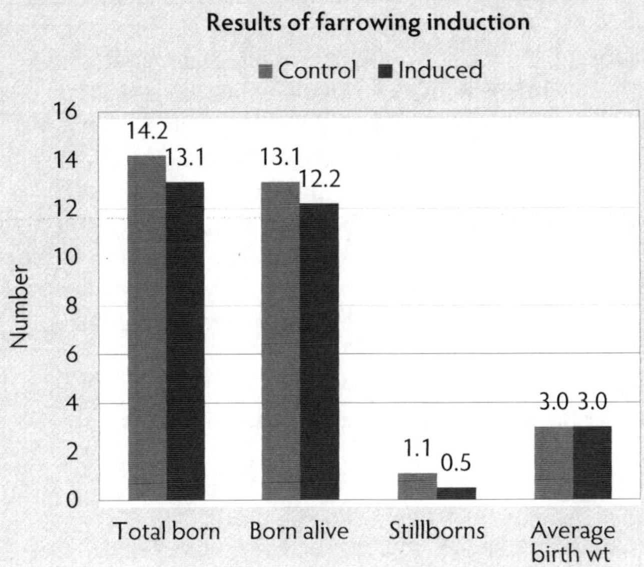

Figure 3: Distribution of parity was similar for control and induced sows

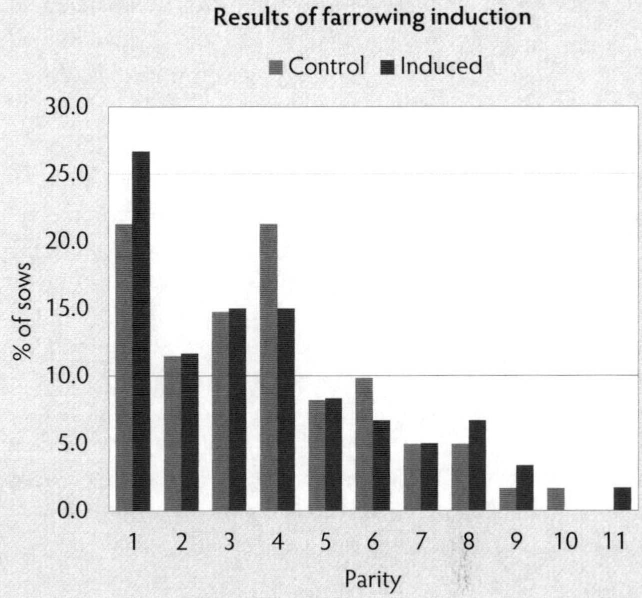

Figure 4: Percentage of stillborns and preweaning mortality.

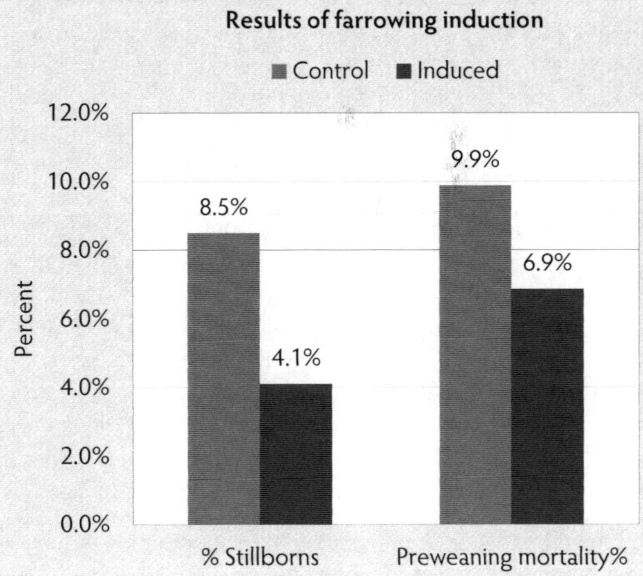

assuming all cost are for the savings in the farrowing stillborns and preweaning mortality.

Induction of sows to farrow may facilitate allocation of labor during intense periods of farrowing, thus optimizing pig survival and reducing the incidence of stillborn pigs which resulted in more pigs weaned.

References

1. USDA Quarterly Hogs and Pigs Report

New generation live intranasal PRRSV vaccine

Federico A. Zuckermann, DVM, PhD; Gabriela Calzada-Nova, DVM, PhD; Robert J. Husmann, DVM, MS; Steve Berger; Aaron Gilbertie
Aptimmune Biologics, Inc., Champaign, Illinois

Introduction

For the last 25 years porcine reproductive and respiratory syndrome virus (PRRSV) has caused significant economic losses to the pork industry. Although many strategies have been implemented to control the spread of this virus, none have been sufficiently effective. It is apparent that a PRRSV control program requires an effective vaccine. In this regard, intranasal immunization against viral respiratory infections is known to provide a superior level of protective immunity when compared to vaccines delivered by other routes. Aptimmune is completing the development of a novel live PRRSV vaccine, using a virus strain of North American-like genotype, termed G16X, to be administered to swine intranasally. According to the classification by Shi et al.,[1] the G16X virus belongs to lineage 5 (Dr. Fred Leung, personal communication). Because G16X was derived from a naturally non-virulent virus, serial passage of the virus in a non-porcine host cell was not necessary to attenuate the virus. The G16X virus efficiently replicates in porcine alveolar macrophages and the vaccine will be produced using a novel and proprietary porcine macrophage cell line, ZMAC. By virtue of G16X being grown in its natural host cell, this vaccine virus readily replicates in pigs, without selective pressure for the virus to mutate to gain replication fitness in a vaccinated host. This G16X virus vaccine should provide a significant safety advantage as a vaccine candidate. Inoculation of G16X directly into the respiratory tract, where it will readily replicate, will stimulate the nasopharynx-associated lymphoid tissue (NALT) and the bronchial-associated lymphoid tissue (BALT), eliciting a mucosal immune response in the upper and lower airways. A number of experiments have been conducted to characterize the biological properties of the G16X as an intranasal PRRS vaccine. Here, the most salient properties of the G16X virus in relation to the characteristics required for an intranasal vaccine are described. In addition, results obtained earlier demonstrating the level of cross-protective immunity afforded by this virus when given by the conventional intramuscular injection are also presented.

Replication of G16X in its host

Protective mucosal immune responses are most effectively induced by mucosal immunization. This can be achieved with live virus vaccines that replicate in the host while activating cells of the innate immune system. This activation causes the innate cytokine response that is necessary to promote the development of mucosal adaptive immunity. Following intranasal vaccination, G16X replicates abundantly in the respiratory tract, which is manifest by viremia that is detected by 4 days after virus exposure. The viremia peaks between day 7 and 10, then begins to gradually decline with no infectious virus being detectable by 31 to 33 days after exposure (Figure 1). The observation that the delivery of the G16X vaccine causes a viremia of limited duration bodes well for an expected significant safety advantage for this vaccine.

Stimulation of innate immunity by G16X

Interferon (IFN)-α constitutes the principal innate immune response of the host to combat viral infections. In addition, the production of IFN-α by plasmacytoid dendritic cells (pDCs) in response to viruses plays a critical role in promoting the development of adaptive immunity.[2] The effectiveness of live attenuated viruses as mucosal vaccines is due in part to their ability to activate multiple innate responses and the importance of innate immunity in the development of adaptive immune responses is becoming increasingly clear.[3] However, PRRSV of type 2 genotype typically not only fails to stimulate the production of IFN-α by pDC, but also has the ability to inhibit the IFN-α-producing

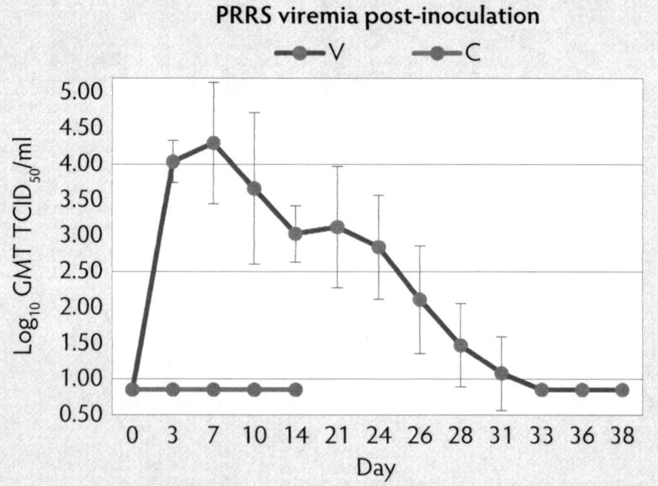

Figure 1: Extent of viremia in pigs after intranasal vaccination with G16X. Serum samples were collected from mock-vaccinated (n = 2) and G16X-vaccinated animals (n = 6) immediately prior to and at the indicated days after intranasal vaccination with G16X. The virus load in the serum was examined for the presence of infectious virus ($TCID_{50}$) in ZMAC cells. Results are presented as the geometric mean for each group. V = G16X-vaccinated; C=mock-vaccinated. Assay level of detection is ≤ 0.85 log_{10}.

function of this important cell[4]. Notably, the G16X virus has the unique ability to allow IFN-α-producing function of pDCs. An additional well-acknowledged and troubling property of PRRSV is that it fails to stimulate the production of IFN-α by its natural host cell, namely alveolar macrophages.[5] Again, G16X has the exceptional property of stimulating the production of IFN-α of alveolar macrophages. These biological properties exhibited by G16X are consistent with the observed ability of this vaccine strain to stimulate a vigorous IFN-α response that can be detected in the serum of animals within 4 days after vaccination and subsides by day 7. Given all of the concepts described above, it seems likely that the ability of the G16X vaccine to elicit an IFN-α response contributes to the cross-protective immunity that has been demonstrated with this vaccine. An example of this property of the experimental G16X vaccine is described below.

Stimulation of adaptive immunity by G16X

Ideally, vaccination at a single site would provide both humoral and cell-mediated protection, not only at the relevant mucosal surface, but also throughout the body. In this regard, nasal vaccination has shown particular potential. Nasal administration of vaccines has induced specific mucosal IgA antibody responses in the salivary glands, upper and lower respiratory tracts of several species including swine.[6] In addition, nasal immunization studies in humans and mice produced greater systemic antibody responses than other mucosal immunization routes.[3] Accordingly, our studies have shown that nasal immunization with the G16X vaccine elicits a systemic antibody response of similar magnitude as the intramuscular administration of the same virus (Figure 2). The presence of systemic immunity following mucosal immunization is considered an indication that a strong mucosal immune response has developed.

Stimulation of cross-protective immunity by G16X

As described above, upon administration to swine, the G16X vaccine stimulates an early systemic IFN-α response, which should contribute to the development of a robust adaptive immune response. Previous studies have shown that when the G16X vaccine was administered intramuscularly, it was capable of eliciting the development of broadly cross-protective immunity against PRRSV of North American-like genotype, which according to the phylogenetic classification by Shi et al.,[1] belong to lineages 1 and 8. The stimulation of cross-protective vaccine efficacy against a representative of the famous 1-22-2 family of PRRSV that has caused numerous outbreaks in South West Ontario in Q4 2010/Q1 2011 and belongs to the highly virulent PRRSV North American-like genotype lineage 1 is shown in Figure 3. The protective efficacy of the G16X vaccine against this genetically divergent (heterologous) virus was demonstrated by the curtailment of the viremia that ensued in vaccinated pigs following their challenge with virulent virus and the subsequent complete elimination of infectious virus as compared to the

Figure 2: PRRSV antibody response in G16X intranasally vaccinated pigs. Results are presented as ELISA S/P ratios at day of serum sampling post-vaccination (Day 0).

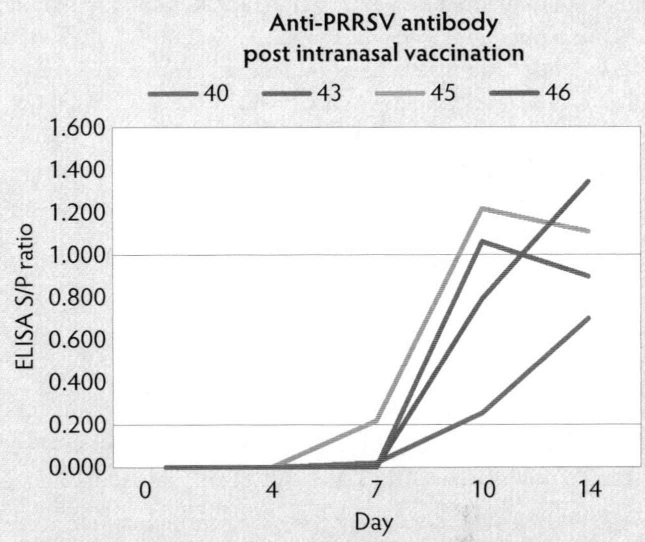

Figure 3: Frequency of viremia in G16X and mock-vaccinated pigs after challenge with LTX1. Serum samples were collected from mock-vaccinated (n = 10) and G16X-vaccinated (n = 20) animals immediately prior to and at the indicated days after challenge with the wild-type PRRSV LTX1. The virus load in the serum was examined for the presence of infectious virus ($TCID_{50}$) in ZMAC cells. Results are presented as % of viremic pigs for each group.

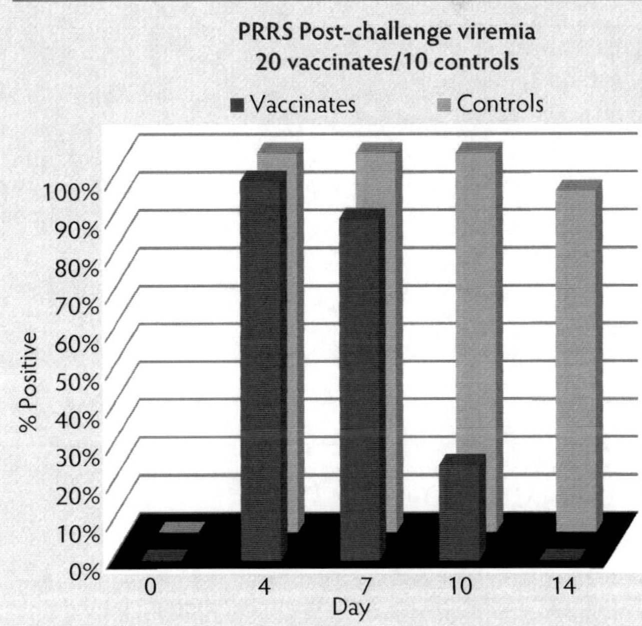

unabated viremia observed in unvaccinated and identically challenged animals. Additional parameters indicating the presence of protective immunity included: reduction of morbidity, elimination of the challenge virus from the lung, reduction of lung pathology. An improvement of the overall health of the vaccinated animals was clearly indicated by the rate of weight gain exhibited by the vaccinated and challenged animals, which at 14 days after challenge, exhibited a much improved weight gain as compared to the unvaccinated and challenged animals. The growth rate of the vaccinated pigs was 80% of that attained by the strict (unchallenged) control pigs. Due to the severity of the challenge, the unvaccinated and challenged animals only realized 25% of the growth attained by the strict controls.

Conclusion

The availability* of a highly efficacious, safe and intranasally delivered G16X PRRSV vaccine capable of providing both mucosal and broad cross-protective immunity, should prove to be an effective tool to diminish the heavy economic losses due to PRRS.

*The experimental G16X vaccine has not been approved for use in the United States.

References

1. Shi, M., Lemy, P., Singh B., Suchard, M.A., Murtaugh, M.P., Carman, S., D'Allaire, S., Deslisle, B., Lambert, M.E., Gagnon, C.A., Ge, L., Qu, Y., Yoo, D., Homes, E.C., Leung, F.C. 2010. The spread of type 2 Porcine Reproductive and Respiratory Syndrome Virus (PRRSV) in North America: a phylogeographic approach. Virology. 447, 146-154.

2. Colonna, M., Trinchieri, G., Liu, Y.J. 2004. Plasmacytoid dendritic cells in immunity. Nat. Immunol. 5:1219-1226.

3. Neutra, M.R., Kozlowski, P.A. 2006. Mucosal vaccines: the promise and the challenge. Nature Reviews Immunol. 6:148-158.

4. Calzada-Nova G, Schnitzlein W, Husmann R, Zuckermann FA. 2011. North American porcine reproductive and respiratory viruses inhibit type I interferon production by plasmacytoid dendritic cells. J. Virol. doi:10.1128/JVI.01616-10. Vol. 85 (6) 2703-2713.

5. Albina, E., Carrat, C., Charley, B., 1998. Interferon-alpha response to swine arterivirus (PoAV), the porcine reproductive and respiratory syndrome virus. J. Interferon Cytokine Res. 18, 485-490.

6. Hughes HR, Vincent AL, Brockmeier SL, Gauger PC, Pena L, Santos J, Braucher DR, Perez DR, Loving CL. 2015. Oral Fluids as a Live-Animal Sample Source for Evaluating Cross-Reactivity and Cross-Protection following Intranasal Influenza A Virus Vaccination in Pigs. Clin Vaccine Immunol. 22:1109-20.

Reviewing a novel approach to classifying porcine reproductive and respiratory syndrome virus – *MJPRRS®* Grouping Technology

Keith Kinsley[1], DVM; Dwain Guggenbiller[1], DVM, MBA; Douglas Weiss[1], DVM; Ronald Nimmo[1], PhD; B. K. Kim[2], PhD
[1]Phibro Animal Health Corporation, Teaneck, New Jersey; [2] MJ Biologics, Mankato, Minnesota

Introduction

Porcine reproductive and respiratory syndrome virus (PRRSv) has remained highly elusive. Breakthroughs in the science of PRRSv immunization often do not hold up for the long-term when applied to populations and ever-changing biological systems. This is apparent when evaluating diagnostic descriptors and methods of virus identification and comparison used today. Since the late 1980's producers and veterinarians alike have battled PRRSv and struggled as to how to best catalogue and compare these viruses within and between farms and systems. The methods available to this point, Restriction Fragment Length Polymorphisms (RFLP), dendrograms, and Nucleotide (Nt) % Homology comparisons fall short when attempting to explain large clinical presentations when only small genetic differences have been identified in the PRRSv ORF 5 sequence.

There exists today a technology that is not only capable of describing differences in clinical presentations (population's response to the disease regardless of sequence homology), but also of more clearly identifying new introductions versus internal mutations. *MJPRRS®* Grouping Technology, developed by MJ Biologics, is a unique means of comparing and classifying PRRSv based on ORF 5 amino acid sequences and on-farm clinical presentations. This technology, which is the subject of this article, was invented by Dr. BK Kim of MJ Biologics and licensed to Phibro Animal Health Corporation in January 2015. This method of "grouping" PRRS viruses has helped many producers and systems better understand their key biosecurity intervention points as well as to better control pig flow health and maximize system throughput. In scenarios where virus elimination has proven to be both difficult and costly, the understanding of common viral mutations through application of *MJPRRS* Grouping Technology affords an option for addressing these challenging farms.

PRRS diagnostics

The most common strategies for identifying PRRSv outbreaks and the specific viral characteristics involved have been dominated by Polymerase Chain Reaction (PCR), virus sequencing, RFLP, dendrograms and Nt homology tables. This is not to infer that any of these are poor methodologies for comparing PRRSv strains, but, most provide insufficient information because they often fail to distinguish between PRRSv strains, particularly when small changes in the viruses equate to large clinical changes in the infected population.

MJPRRS Grouping Technology utilizes clinical presentations and Open Reading Frame (ORF 5) nucleotide and amino acid sequences to categorize different physical viral presentations. This identification, while not as visually impressive as a list of RFLP or nucleotide homology tables, can be highly accurate when describing a clinical presentation on a farm. This is especially the case when the farm has a history of prior PRRSv exposure and highly genetically similar ORF 5 viral sequences.

MJPRRS vaccine is unique

There are at least two unique parts to this technology. The first is a patented method for producing an immune response in a subject (US Patent No. 8,142,788). The immune response is produced by evaluating and choosing viral vaccine candidates based on differences in physical presentations ("how the pig sees PRRSv" vs. "how we as scientists see PRRSv"). The second is a method of preparing a vaccine containing PRRS virus proteins and antigens from PRRS virus infected cells (US Patent No. 7,776,537), which was patented by the University of Minnesota and has been sublicensed by MJ Biologics to Phibro. The vaccine composition is prepared by virus selection, preparation and antigen harvest. We will not focus on vaccine preparation here, but rather on the selection of the correct virus for inclusion in the vaccine.

Unlike RFLP, dendrograms, or Nt homology comparisons, highly similar viruses as measured by these methods are not always similar enough when viewed through the lens of clinical signs in a population (the pig's immune response to specific antigens). From our basic understanding of molecular biology, we recall that DNA is transcribed into RNA, and RNA is translated into an amino acid sequence (or protein structure). A portion of these protein structures form exposed envelope proteins that serve the virus by seeking out host cell receptors. Although there is a high level of conservation in DNA-based organisms, RNA viruses are not as highly conserved or proofread. So, errors are common and often advantageous to strain survival. PRRS is a RNA virus. A single Nt change, if in the right position(s), can change the resulting protein structure and the virus' physical presentation. See the example in Table 1. Table 1 demonstrates the effect of changing one Nt in a neucleotide sequence. In some cases, a change in one Nt makes no

Table 1: Wobble Base Effect

1st Nt base	2nd Nt base	3rd Nt base	Resulting amino acid
A	T	T	Isoleucine
A	T	C	Isoleucine
A	T	A	Isoleucine
A	T	G	Methionine
A	A	T	Asparagine
A	A	C	Asparagine
A	A	A	Lysine
A	A	G	Lysine

difference in the resulting amino acid. In other cases, a single Nt change can result in production of a totally different amino acid, thus modifying the resulting protein. This Nt change is called the Wobble Base Effect.

The Wobble Base Effect can result in a highly variable amino acid with the simple change identified in Table 1. If the Nts in the first or second positions of this codon are changed, then the possibilities are even more dramatic. A change in the wobble base occurring in a critical area, could make a significant change in the resulting protein, thus changing the physical presentation of PRRSv to this population of animals, even if this is the only change to occur. This change is small numerically, 1 Nt out of 603 Nts or 0.17% difference in nucleotide homology. Although likely to not have changed this virus' RFLP or position on a production system's dendrogram sufficiently so as to catch the viewer's attention, the possibility still exists that the clinical effects are catching the producer's attention in this situation.

The analysis of these physical changes to the protein structure is known as *MJPRRS* Virus Grouping. Through evaluation of 10,000-plus viral sequences by MJ Biologics, all viruses have fit neatly into one of 25 physical viral presentations: 3 – major groups D, S, and European (E), and 8 or 9 individual subgroupings of these (D-1 thru D-8, including D4b, S-1 thru S-8, and E-1 thru E-8). These group relationships are shown in Figure 1.

These *MJPRRS* virus groupings are independent of both virus Nt homology and RFLP. This evaluation occurs at the amino acid (protein structure) level. For example, the same RFLP will often represent 3-5 different *MJPRRS* virus groups while a single *MJPRRS* viral group may contain viruses with 5-8 (or more) different RFLP patterns. Viruses immediately neighboring each other on a system's dendrogram may represent different *MJPRRS* viral groups. At the same time, those on one end or the other of the dendrogram may be more physically alike in appearance to the pig. *MJPRRS* virus grouping helps to explain how small, seemingly irrelevant, changes in Nt homology, or in sequence proximity in a NT homology table or dendrogram, can appear clinically different in a population of animals.

How to utilize *MJPRRS* grouping technology

Producers and their veterinarians who are using *MJPRRS* vaccine have embraced the technology of virus group identification. The main reason behind this acceptance is that the previously unexplainable is now explained. Historical comparisons, RFLP analysis, and dendrograms failed to describe why the virus clinically appeared different when it was highly genetically related to the last disease break. Matching a clinical picture with historical PRRSv challenges based off no more than *MJPRRS* grouping analysis often helps to obtain acceptance of the technology. Evaluating the previous viral challenges and associated clinical pictures from a farm with a difficult PRRSv situation is often a good place to start for those unfamiliar with *MJPRRS* Grouping Technology. A farm, or system, with a rocky history of PRRSv stability, will often allow for the linkage of epidemiologic description and clinical signs.

Characteristics of a farm (system) with PRRSv stability issues are as follows:

1. Weaned piglet numbers vary frequently;
2. Tested pools of piglet serum oscillate between PRRSv PCR positive and negative;
3. Piglet populations fail to test PRRSv PCR negative despite lengthy farm closures; and
4. Clinical disease in sows/gilts despite exposure to 98% + similar PRRSv by Nt homology.

All of the above factors are further complicated by: 1) the need for replacement animals, 2) sustained poor production and 3) increasing costs of production.

By gathering of 18-24 months of the most recent PRRSv ORF 5 sequence information, much of this story can be elucidated. There is nothing magical about the 18-24 month period of time except that this will often catch at least one PRRS break with a new introduction and an internal virus mutation or two. ORF 5 sequences can be obtained and grouped by MJ Biologics typically within 48-72 hours and discussion can begin as to why this population of animals may not be responding clinically as would be expected.

Figure 1: *MJPRRS®* Immunological Groupings

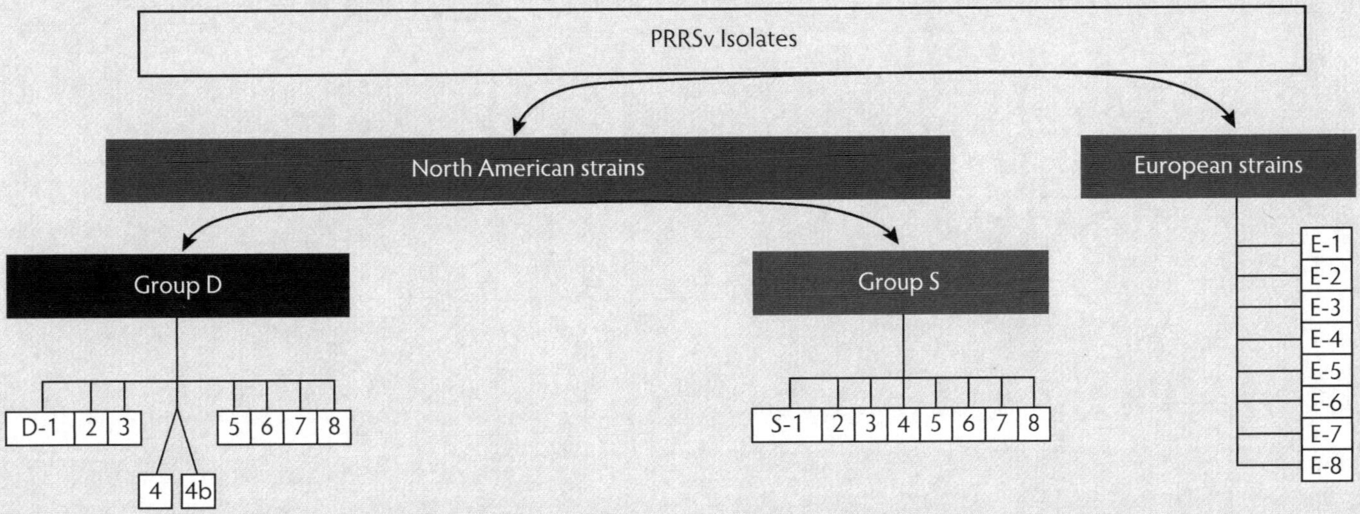

Differentiation between new viral introductions and internal viral mutations is critical to maintaining a PRRSv control program. New viral introductions usually fuel a full-blown investigation as to how it happened: "How did the virus get here?" Mutations, in spite of being a natural phenomenon, add the element of unpredictability to an already problematic situation. To date, controlling these mutations has not been possible. If common mutations could be identified, PRRSv control programs could be tailored to reduce the clinical severity of these mutations as well as to control the current virus challenge within the farm.

Comparisons of amino acid sequences using *MJPRRS* Grouping Technology have identified mutations that are likely to have occurred within the farm population versus those that occurred with a new viral introduction (different origin ORF 5 amino acid sequence). This differentiation is possible even within the same physical groupings of a virus. For example, two D-4 viruses can be of varied origin. The alternative is also true, that a small mutation can give rise to a different *MJPRRS* group from the original virus – for example a D-4 virus becoming a D-5 virus with a small Nt change (refer to Table 1).

MJPRRS Grouping Technology can also help identify patterns (viral group partnerships), or mutations that are likely to occur from one specific virus group to another. One instance of this is the relationship of D-4, D-5, and D-6 viral groups. These viral groups can be introduced independently of one another. But, they are equally or more likely to be found as mutations in the same population over time (See Figure 2).

Figure 2 shows the progression of PRRS outbreaks in one farm over a period of time from December 2010 to December 2012. The actual dendrogram is attached without scale on the left and corresponds with the date of clinical signs and RFLP rows across the top. Some variant groups are positioned immediately adjacent to one another. This dendrogram was produced by the supporting diagnostic laboratory in this case with all the historical farm information attached. The only farm contributing to this comparison of viruses was the farm in question.

The farm was experiencing clinical signs related to PRRSv circulation every 5-8 months. Neither RFLP nor Nt homology could explain this clinical phenomenon.

The only way to explain each occurrence of these clinical presentations was the *MJPRRS* Grouping Technology. As is demonstrated in Figure 2, these viral mutations are highly similar and non-distinguishable by all other comparison methods, but each date identified represented a new clinical presentation. The *MJPRRS* Grouping Technology was able to differentiate the breaks as caused by D4b, D4, D5, D6 and D4.

The most powerful, and likely most beneficial, capability of *MJPRRS* Grouping Technology is the ability to track viral movements

epidemiologically throughout a large geographic region. Many of the most devastating PRRS breaks in the last 18-24 months may have been connected to 1-7-4 RFLP patterns. However, there were many of these breaks that were identified as involving PRRS viruses with the RFLP patterns of 1-7-3, 1-21-4, 1-10-4, or even 1-4-4. These cases were similar in clinical appearance, but did not match exactly by RFLP. These viruses were dominant in many regions of the United States. Using the *MJPRRS* Grouping Technology, approximately 60-65% were identified as D-7. Despite the high percentage of 1-7-4 viruses also being D-7s, not all 1-7-4 RFLPs are D-7 MJ grouped viruses. *MJPRRS* Grouping Technology provides a method for identifying and monitoring virus movement in a more defined manner than a combination of RFLP, dendrograms, and Nt Homology. When Porcine Epidemic Diarrhea virus (PEDv) went from a low (zero) prevalence situation to a high prevalence situation in a short period of time,

Figure 2: Internal mutations over time

Date	RFLP	Jan 2012	Dec 2012	Dec 2011	Dec 2010	May 2010	MJ Group
Dec 2012	1-2-2	93.2%	100.0%	99.7%	98.7%	99.3%	D-4
Dec 2011	1-2-2	93.0%	99.7%	100%	98.7%	99.3%	D-5
Dec 2010	1-18-2	92.9%	98.7%	98.7%	100%	99.3%	D-4b
May 2011	1-2-2	92.5%	99.3%	99.3%	99.3%	100%	D-4
Jan 2012	1-18-2	100%	93.2%	93.0%	92.9%	92.5%	D-6

MJ PRRS Grouping Technology explains occurrence of clinical signs:
D4b --> D4 --> D5 --> D6 --> D4

it was considered an epidemic. The H1N1 "Swine Flu" of 2009 was very similar. Both instances were classified as epidemics. If the identification of a specific PRRS virus presentation (D-7 in this case) can be correlated by similar monitoring, shouldn't the same epidemic status be applied?

Figure 3 shows a historical picture of PRRSv groups from 2005 to 2014. From back to front, D-7 prevalence as a total of all D-groups identified in the US from 2005-2013 was very low. By the end of 2014, D-7 was the second most frequently identified D-group virus and rising. Current monitoring would suggest that this trend has continued at an average of 35-40% monthly. In nearly all cases identified, D-7 has been associated with a new clinical break. Projects tracking RFLP as a method for viral classification and comparison may miss this subtlety.

Presently the D-7 virus group has spread quickly throughout the United States. Only the *MJPRRS* Grouping Technology has been able to demonstrate the similarity of the virus causing these outbreaks. The strength of *MJPRRS* Grouping Technology is the ability to identify similarities where traditional methods of comparisons have failed to classify PRRS viruses as being related. This can have a meaningful impact on population (as well as regional) virus control programs through recognition of common virus movement and potentially source of origin.

Conclusions

MJPRRS Grouping Technology is a tool for classifying and differentiating PRRSv based on their different physical properties – as identified by the ORF 5 amino acid sequence. To date, all PRRSv that have been analyzed have fit reasonably well into one of the 25 established MJ groupings. This grouping technology has allowed for some degree of predictability with regards to potential mutations and has facilitated tracking of viral epidemics even where Nt % Homology comparisons and RFLP analysis would suggest that different viruses may be involved. *MJPRRS* Grouping Technology also offers an explanation in situations where small changes in Nt % Homology equate to clinical presentations on a farm.

Taken alone, any one of these aspects helps to better grasp PRRSv control in a region, system, or on an individual farm. All facets being used concurrently can not only aid in the selection of viruses for future autogenous vaccine production, but can also predict which mutations are likely, and can improve monitoring of new introductions to a given region or across the country. This capability allows producers and veterinarians alike to aggressively

Figure 3: Historical grouping percentages 2005-2014 - % D group

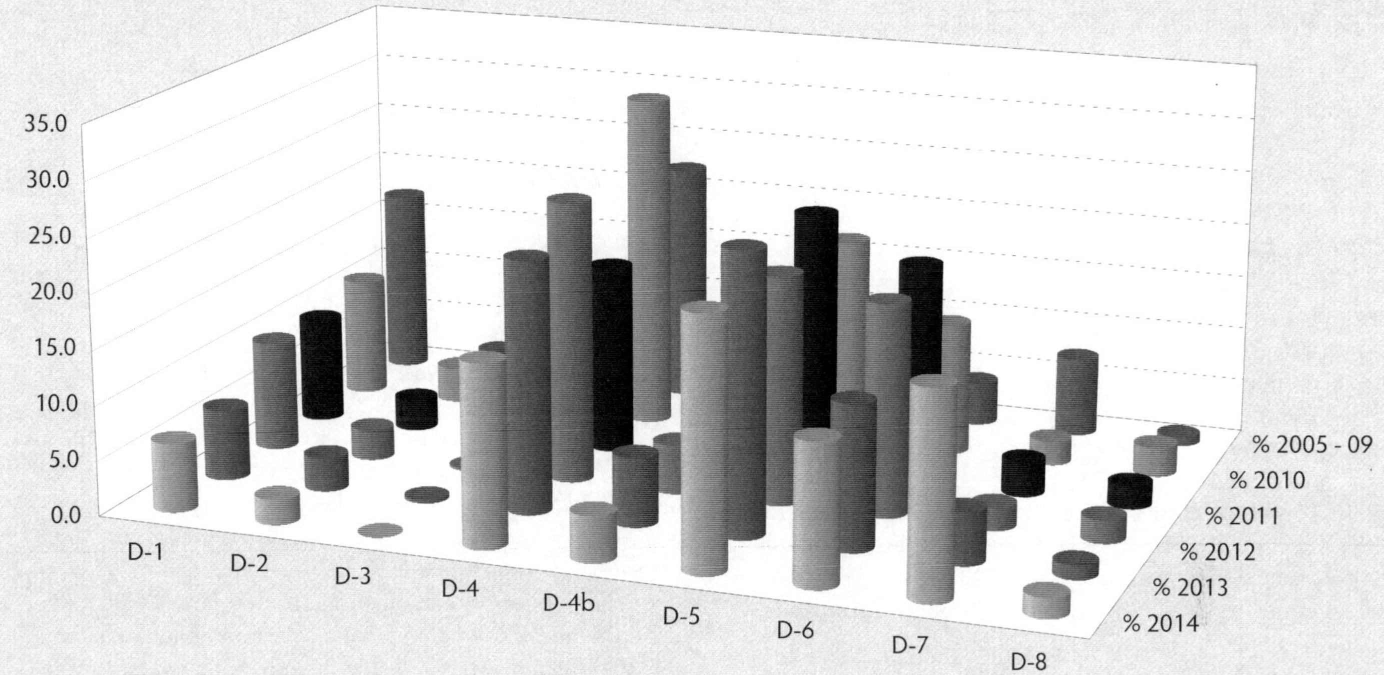

and precisely address the PRRS virus on their farms. As well, appropriate interventions can be implemented with respect to their internal and external biosecurity measures. *MJPRRS* Grouping Technology best prepares for the next PRRSv epidemic wave.

Acknowledgments

The authors would like to thank MVP Laboratories, especially Drs. Karen Brown and Jeff Kula for their multiple reviews. The authors would also like to thank all employees of MJ Biologics; without their tireless work none of this would be possible.

Experimental infection of US swine with China origin PRV (HeN1 variant)

S. Swenson, DVM, MS, PhD; T. Sturgill, DVM, PhD; R. Clayton, MS; K. Mozingo; L. Koster; M. Jenkins-Moore; T. Beach; A. Predgen, DVM, MPS, MS; DVM; D. Toms
National Veterinary Services Laboratories, USDA, APHIS, VS, Ames, Iowa

Introduction

Pseudorabies (PR), also known as Aujeszky's disease, is a viral disease primarily of pigs but can also affect other farmed species such as cattle and companion animals such as dogs. Infection in pigs can result in reproductive disorders including stillbirths and abortions; respiratory disease including sneezing, coughing, and pneumonia; neurologic disease including tremors and seizures; and death. Clinical presentation is age-dependent with younger pigs demonstrating more severe clinical signs. Pseudorabies virus (PRV) was eradicated from the US domestic swine population in 2004 through the use of gene-deleted vaccines and companion serologic assays to identify infected herds for depopulation. In late 2011/early 2012 severe PR was reported in China in PRV vaccinated herds.[1,2] Clinical signs included reproductive, respiratory, and neurologic disease, with mortality rates up to 50% in affected piglets. In 2013 and 2014, porcine epidemic diarrhea virus and porcine deltacoronavirus, respectively, were detected for the first time in US swine.[3,4] Genetic characteristics of the viruses indicated close homology to viruses circulating in China. With introduction of these new viruses into the US, there is a subsequent interest in developing a better understanding of the recently detected PRV variant in China. This understanding needs to include both the ability of US laboratories to detect the presence of the virus, should it enter the US, and the protective capability of US licensed, commercial PRV vaccines. The HeN1 strain of PRV was generously made available to the National Veterinary Services Laboratories (NVSL) by Professor Zhijun Tian and Dr. Jinmei Peng at the Harbin Veterinary Research Institute, Harbin, China. Two trials were designed to address diagnostic capabilities and vaccine efficacy. The first study was performed to evaluate the best challenge dose and to collect samples for diagnostic test evaluation. The second study was to evaluate vaccine efficacy. At the time of submission for these proceedings, testing for the first study was not yet completed and the vaccine study was underway.

Materials and methods

The HeN1 strain of PRV was imported into the US under USDA permit and shipped to the Foreign Animal Disease Diagnostic Laboratory (FADDL), Plum Island, NY. The virus underwent safety testing prior to shipping to NVSL, Ames, IA. Virus was propagated and titrated in PK15 cells to generate sufficient volume for experimental studies. Virus was diluted to approximately $10^4 TCID_{50}$/ml, $10^5 TCID_{50}$/ml, $10^6 TCID_{50}$/ml, and $10^7 TCID_{50}$/ml, frozen in small aliquots to allow for use in both studies, and individual vials thawed and titrated to determine actual titer. For the titration study, pigs were approximately 5-6 weeks of age at the time of challenge with five pigs in each of the four challenge groups. Pigs were intranasally challenged with one ml of virus split between the two nostrils. Each challenge group was housed individually. Personnel worked from lowest viral dose to highest viral dose and were required to wear full face respirators and gloves, complete a full body shower, and change clothes between each group. Pigs were monitored in person daily and periodically monitored throughout the day and night via security cameras for 23 days. Individual nasal and oral swabs were collected just prior to challenge and on days 2, 4, 7, and 9 post-challenge. Blood samples were collected via orbital sinus just prior to challenge, days 7, 10, 15, 18 post-challenge, and at the time of euthanasia. Brain, lung, tonsil, and spleen were collected at necropsy either at the time of euthanasia or death. A real-time PCR targeting the PRV gB gene[5] was used to test tissues, nasal, and oral swabs. Virus isolation was attempted only on the nasal swabs and tissues.

Results and discussion

Onset of clinical signs began on day two post-challenge with more pigs from the highest viral dose group demonstrating the earliest clinical signs. Initial clinical presentation included depression and respiratory signs, which started with sneezing, with subsequent development of coughing and serous, purulent, or caseous nasal discharge. Tremors were observed as early as day three post-challenge, typically located in the rear quarters, but eventually progressed to the whole body in some pigs. Additional neurologic signs included saw-horse stance, circling, head tilt, head shaking, aimless wandering, and seizures. One pig in the highest dose group died acutely on day five post-challenge. Eight pigs in the two highest virus dose groups were euthanized on days five to eight, primarily due to neurologic signs. One pig with tremors was euthanized due to severe respiratory distress. The most consistent gross and histopathological finding was interstitial pneumonia, often presenting in a nodular pattern. Lymphoplasmacytic perivascular cuffing, meningoencephalitis, and gliosis were most marked in the pigs examined at days 6 and 8 post-challenge. Viral shedding was detected in nasal swabs at first collection on day two post-challenge in the two highest viral dose groups. PCR detection of virus occurred from oral swabs and all four tissue types. Seroconversion was detected by the three US licensed, commercially available serologic assays. This titration study demonstrated that viral dose impacts clinical presentation, with common observation of neurologic and respiratory clinical signs. Severe disease occurred when 5-6 week old pigs were challenged with the higher doses of the HeN1 PRV strain. Disease severity was such that pigs had to be euthanized for humane reasons. Use of security cameras allowed for frequent monitoring of the animals and demonstrated the importance of frequent evaluation of pigs for clinical signs. Diagnostic assays currently available in the US include virus isolation, PCR, and serologic assays. All were capable of detecting infection with the HeN1 strain of PRV.

References

1. An T-Q et al. Pseudorabies virus variant in Bartha-K61-vaccinated pigs, China, 2012. Emerg Infect Dis 2013;19:1749.

2. Yu X et al. Pathogenic pseudorabies virus, China, 2012. Emerg Infect Dis 2014;20:102.

3. Stevenson G et al. Emergence of porcine epidemic diarrhea virus in the United States: clinical signs, lesions, and viral genome sequence. J Vet Diagn Invest 2013;25:649.

4. Wang L et al. Detection and genetic characterization of deltacoronavirus virus in pigs, Ohio, USA, 2014. Emerg Infect Dis 2014;20:1227.

5. Ma W et al. Development of realtime polymerase chain reaction assays for rapid detection and differentiation of wild-type pseudorabies and gene-deleted vaccine viruses. J Vet Diagn Invest 2008;20:440.

Identification of *Mycoplasma hyopneumoniae* shedding gilts in the field

Chris Deegan[1]; Laura Dalquist[2]; Maria Pieters[1]
[1]College of Veterinary Medicine, University of Minnesota; [2]Swine Vet Center

Introduction

Mycoplasma hyopneumoniae (M hyo) affects the respiratory tract of swine and leads to large economic losses in growing pigs each year. To help reduce these losses, producers will attempt to eliminate dam-to-piglet transmission by exposing their naïve gilts to shedding pigs and then allowing a "cool down" period before their first farrowing. However, accurately identifying shedding gilts in the field, to be placed in direct contact with naïve gilts, remains a difficult task for swine veterinarians. In this investigation we tested the hypothesis that dry cough, serology in unvaccinated gilts, and time point since initial exposure to M hyo, can be used to correctly identify shedding gilts in the field, and that laryngeal swabs are a sensitive sample type to confirm M hyo shedding status.

Materials and methods

Three gilt development units (GDUs) managed in continuous flow continuous where older, shedding gilts were placed in the pens with naïve gilts in order to expose them, were selected for this investigation. A total of 84 gilts were sampled based on the following criteria: 1. Selection of at least 5 gilts from various lots exposed to M hyo at different time points, 2. Gilts in the GDU were not vaccinated against M hyo at any time before and after arrival to GDU, and 3. Presence of a variable number of coughers and non-coughers within each lot. Coughers and non-cougher gilts were identified by listening for and marking coughing pigs with livestock marker for 15 minutes. Laryngeal swabs and blood samples were collected from all selected gilts, a coughing status (yes or no) was recorded, as well as time since exposure to M hyo infected gilts. Time since exposure to M hyo was determined by farm records and discussion with the attending veterinarian. Laryngeal swabs were processed using a species-specific real time PCR to detect the gilt's shedding status and was then compared to the coughing status and time since exposure. Blood samples were processed using Idexx® ELISA in order to detect antibodies to M hyo to help confirm the gilt's exposure to M hyo (as gilts were unvaccinated).

Results and discussion

The % of gilts positive to M hyo by real-time PCR in laryngeal swabs, as well as antibody detection in serum, and the percentage of coughing gilts are graphed in Figure 1. At 75 and 100 days after exposure all tested gilts were determined positive for M hyo antibodies. However, vaccination of gilts is a common practice in the field and for that reason serology alone should not be used to identify infected gilts. It is important to note that a significant proportion of gilts in the +125 days after exposure group were negative for M hyo antibodies, which could have been the result of non-uniform exposure for that specific group. Also, serology and laryngeal swab real-time PCR results were only correlated during the acute phase of infection only, suggesting that serology is not a good indicator of M hyo shedding, as already described by other researchers. Under the conditions of this investigation, a high number of non-coughing gilts were detected positive to M hyo via laryngeal swab PCR, thus suggesting that cough alone was a poor predictor of M hyo shedding status, especially at +125 days since exposure (chronic phase of infection). It has to be noted that cough was observed in a 15-min time period, and while this should be enough time to observe pigs with a moderate dry cough, it may have led to unidentified coughing pigs being incorrectly classified as non-coughing pigs. However, a 15-min observation period was used in this investigation in order to mimic the time that producers or veterinarians would dedicate to such activity when in the barns. Time after exposure to M hyo infected pigs suggested that shedding of the bacterium occurs mostly during the acute phase of infection and that it declines during the chronic phase, even though an important proportion of gilts are still potentially shedding the pathogen. Nevertheless, the 100% shedding detected in the group +125 days exposed to M hyo suggests that shedding may increase at different times

Figure 1

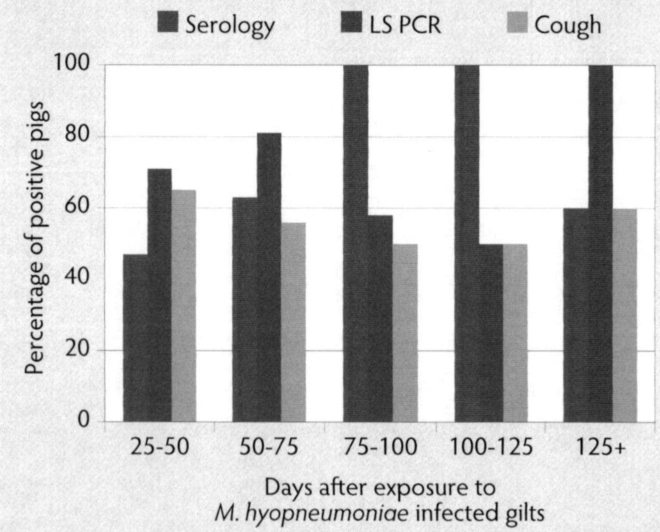

after infection, or that infection in that group did not take place at the same rate as in the other groups. Our results point out that consistency of infection with *M. hyopneumoniae* by exposing young gilts to older infected ones may not be easily achieved. In summary, results from this investigation indicate that cough, serology, and time after exposure, alone or in combination, were poor indicators of *Mycoplasma hyopneumoniae* shedding in gilts, as determined by real-time PCR in laryngeal swabs.

Risk assessment for the introduction of African swine fever virus (ASFV) into the United States

Jessica Rotschafer[1], PhD; Solenne Costard[2], DVM, MS, PhD; Andres Perez[1], DVM, PhD; Timothy Boyer[3], PhD
[1]Veterinary Population Medicine, College of Veterinary Medicine, University of Minnesota, Saint Paul, Minnesota; [2]EpiX Analytics, Boulder, Colorado; [3]Food Protection and Defense Institute, University of Minnesota, Saint Paul, Minnesota

Introduction

African swine fever (ASF) is a highly contagious viral disease of swine that is transmitted by direct contact between infectious and susceptible animals or indirect contact via *Ornithodoros* spp. ticks. Some strains can cause extremely high mortality and there is no available vaccine. As such, introduction of ASFV into domestic swine production systems in the United States has the potential to be economically devastating to swine producers. Recent eastward spread of ASF in wild and domestic pig populations from Russia suggests that the range of this virus is increasing and that the US may be at increased risk of introduction. The objective of this study was to assess five different pathways for the virus to enter individual states and Puerto Rico by adapting a framework that had been previously developed for the European Union.

Materials and methods

The five pathways were legal pig imports, legal and illegal importation of pork products, transport-associated fomites, and feral swine. The outcome was a ranking of each of the 50 states and Puerto Rico according to their risk of introduction of ASFV via each of the 5 pathways. For the legal import of live pigs and pork products pathways, quantitative assessments were carried out as detailed data on imports were available. Because detailed data on imports of illegal pork products, transport-associated fomites, and feral swine were not available, semi-quantitative assessments for those three pathways were performed using a number of factors as proxy indicators, such as international travelers passing through the states from ASF-affected countries and cargo ships arriving to US ports from ASF affected countries. Feral swine were assessed as a means of introducing ASF based on their potential to become infected with the virus as well as their potential interaction with domestic swine.

Results

Interestingly, the different pathways identified different states as having the highest risk of ASF introduction. While Canada is officially free of ASF, the legal pig import pathway indicated that Iowa, Minnesota, South Dakota, Michigan, and Nebraska were the five highest ranked states at risk of introduction of ASF via the legal pig import pathway. The overall monthly risk of introduction of ASF to a state via live pig import from Canada was 0 to 0.0032. Canada was the source country with the largest volume of legal pork product importation and assessment of the legal pork product import pathway revealed that Pennsylvania, New Jersey, Michigan, New York, and Florida had the highest risk of importing ASF-infected pork products (range of monthly probabilities of introduction: 0-0.065). Importation of legal processed pork products through these states may prove particularly important as the ASF virus is stable in processed pork products. Illegal importation of pork products identified states with high levels of international passenger traffic as potential entry points. Assessment of the transportation-associated fomite pathway is limited to a ranking of states according to the factors that were chosen as proxies for risk of introduction of the virus and is primarily aimed at identifying those states most susceptible to viral entry. Similarly, the feral swine route of introduction produces a relative ranking of the states, however the geographical separation between current ASF outbreaks and US feral swine suggests that introduction via infected feral swine is currently unlikely.

Conclusions

The strength of this model lies in the in-depth assessment of different pathways to introduce not only ASF, but other emerging swine diseases while identifying the pathway of greatest risk for each individual state. This framework demonstrates that different states may be susceptible to disease introduction via different routes of introduction and that ASF-surveillance measures should be developed on a state by state basis.

Comparing biopsy instruments and techniques for antemortem joint tissue collection

Emily Mahan-Riggs[1], BS; Paisley Canning[2], DVM; Bailey Arruda[3], DVM, PhD; Locke Karriker[2], DVM, MS, DACVPM
[1]North Carolina State University, Raleigh, North Carolina; [2]Swine Medicine Education Center, Iowa State University, Ames, Iowa; [3]Iowa State University Veterinary Diagnostic Laboratory, Ames, Iowa

Introduction

Infectious arthritis in late finishing hogs is reported by practitioners to be a cause of lost revenue. One veterinary diagnostic laboratory reports that *Mycoplasma hyosynoviae* arthritis cases are increasing in number.[1] Currently, implementation of effective intervention measures requires a postmortem diagnosis. Postmortem sampling provides a one-time snap shot of the disease process and does not allow for the observation of treatment outcomes. In addition, postmortem samples often correlate poorly to clinical outcomes and predicted treatment efficacy. Therefore, an antemortem biopsy technique would benefit producers by diagnosing causes of lameness without adding to mortalities. In order to develop this type of sampling, the best biopsy instrument to use must be determined.

Materials and methods

Four biopsy instruments made by Argon Medical Devices, Inc: the SuperCore™ (biopsy gun, semi-automatic, half core), Biopince™ (biopsy gun, automatic, full core), Jawz™ (forcep, manual), and TruCore™ II (biopsy gun, automatic, half core) were assessed. The assessment consisted of multiple parameters: Anatomical approach to the joint, leg position, usability (time required to collect sample, durability, personnel required), utility of a sheath or guide needle, successful attainment of target tissue (synovial or joint capsule) and biopsy quality. These tools were tested repeatedly using 12 rear and four front legs from euthanized finisher pigs. Usability of each instrument was recorded on a scoring rubric. Elbow, stifle, and hock joints were biopsied then opened for gross examination to assess the impact of the technique and instrument on the native tissue. Samples from each biopsy instrument were pooled in formalin and submitted to a state veterinary diagnostic laboratory to evaluate tissue type and specimen quality.

Results and discussion

Three of the four instruments returned samples containing synovium/joint capsule. The SuperCore™ produced 13 of 21 samples (62%) confirmed as synovium/joint capsule with minimal crush artifact. The BioPince™ only had 4 of 9 samples (44%) confirmed as synovium/joint capsule, and it also produced good quality samples with minimal crush artifact. The Jawz™ produced 3 of 7 samples (43%) from the synovium/joint capsule, but the quality of samples was fair due to moderate crush artifact. The TruCore™ II did not produce any synovium/joint capsule samples, and only one sample collected was adequate for histopathology evaluation. Stifle and hock joints were the easiest joints from which to obtain a biopsy sample based on the scoring rubric criteria.

The biopsy instruments are disposable but were used multiple times in this project. The SuperCore™ was easy to position for synovial tissue collection but could bend if maneuvered aggressively in the joint without ensuring the protective sheath was fully extended beyond the sample needle. The Biopince™ was extremely sturdy during multiple uses, but hitting bone resulted in destruction of the biopsy mechanism of this gun. The Jawz™ was easily controlled but required two people to operate and was the largest (7 French). The sample collection portion bent easily on the TruCore™ II.

Conclusion

The SuperCore™ was the best biopsy instrument for collecting samples for histopathology. Applied practically to antemortem sampling, the tissue collected via biopsy can be used for PCR or histopathology. Joint fluid could also be collected at the time of biopsy to be used for PCR and culture. Together, these results can aid in the diagnosis of infectious causes of lameness via clinical pathology, bacterial culture, PCR and histopathology while avoiding euthanasia and allowing the observation of treatment efficacy in the affected animal.

References

1. Gomes Neto JC, Gauger PC, Strait EL, et al. *Mycoplasma*-associated arthritis: Critical points for diagnosis. *J Swine Health Prod.* 2012;20(2):82-86.

Serum inoculation of a finishing swine operation to aid the elimination of porcine reproductive and respiratory syndrome

R. Tenbergen[1], MS; Allister Scorgie[2], DVM
[1]Ontario Veterinary College, University of Guelph; Guelph, Ontario, Canada
[2]Demeter Swine Veterinarians; Levis, Quebec, Canada

A 4000-head finishing operation in south-western Ontario detected antibodies to porcine reproductive and respiratory syndrome virus (PRRSV) during routine sampling in market pigs. The producer did not observe any clinical effects of disease other than a mild cough 2-3 weeks prior to the positive testing. The finishing barn has two wings with a total of 17 rooms and is managed as continuous flow receiving 1000 11-week-old grower pigs from a single source every 4 weeks.

Immediately following the positive results, testing was conducted on the site, which revealed a mixed level of infection. The last group of pigs accepted (16 weeks old at testing) was both PCR- and ELISA-negative, whereas the previously accepted group of pigs (20 weeks old at testing) was both PCR- and ELISA-positive. Testing of market pigs (24 weeks old at testing) revealed that pigs were both PCR- and ELISA-negative. These results raised the concern that the virus was spreading slowly throughout the barn.

After much discussion, serum inoculation of the herd was planned. Sampling was performed for the production of a serum injection and results confirmed that a mixed level of infection existed throughout the barn. Quantification of the amount of virus was performed on a pool of all of the samples collected (34 ml of serum) and showed a low level of viral particles per ml (20 viral particles/ml). Prior to inoculation, the barn was loaded with 1000 7-week-old pigs, in addition to the normal 1000 11-week-old pigs, after which the herd was closed with no naïve incoming pigs planned for a 7-week period. Due to the low amount of virus obtained, a total of 100 doses were prepared to inoculate approximately 5% of the pigs in each age group. In addition, confirmed positive pigs (PCR- and ELISA-positive) that were marked during testing were mixed into the presumed negative rooms in order to increase the chances of viral transmission.

One week following serum inoculation, all of the rooms tested PCR-positive with the exception one (Table 1). Again, pigs that were confirmed positive (marked during testing) were mixed into the negative room to increase the chances of viral transmission. Testing prior to introduction of naïve pigs, 5 weeks after serum inoculation, revealed PCR-positive results (Table 1). However, a financial decision was made to bring the next group of 1000 naïve grower pigs into the finishing unit. The new entrants were isolated in a segregated wing of the barn for a total of 4 weeks before requiring more room. This adjustment allowed an 11-week period for the pigs that underwent serum inoculation to stop shedding the virus before having to introduce any naive pigs into the group. Testing of the isolated group of pigs was carried out after a period of 5 weeks in the barn and revealed PCR- and ELISA-negative results (Table 1). In addition, following the introduction of further naïve pigs, testing has revealed PCR- and ELISA-negative results.

Even with such a low viral inoculum in a low proportion of the population, serum inoculation aided in the apparent stabilization of this herd. Pigs that have entered the site PRRSV-negative have remained negative, presumably due to an absence of viral source. However, PRRSV can persist for a long period of time after initial infection, and therefore, more testing is scheduled before the next group of naïve grower pigs is introduced into the barn.

Table 1: Results from blood testing following serum exposure.

Date	Test	Room ID (N)	Result (PCR Ct; ELISA S/P)	Age (weeks)	Time after serum inoculation (weeks)
June 5, 2015	PCR[a]	10 (N = 5)	Negative (> 40)	8	1
		9 (N = 5)	Positive (31.5)	12	
		11 (N = 5)	Positive (28.5)	12	
		13 (N = 5)	Positive (27.4)	16	
		14 (N = 5)	Positive (32.3)	16	
July 3, 2015	PCR[a]	10 (N = 5)	Positive (28.4)	12	5
		12 (N = 5)	Positive (30.4)	12	
		13 (N = 5)	Positive (30.2)	12	
		14 (N = 5)	Positive (27.6)	12	
		15 (N = 5)	Positive (26.0)	12	
		5 (N = 5)	Negative	16	
		9 (N = 5)	Positive (30.7)	16	
		10 (N = 5)	Positive (30.9)	16	
		11 (N = 5)	Positive (35.5)	16	
		7 (N = 5)	Negative	20	
		5 (N = 5)	Negative	24	
August 18, 2015	PCR[a]	10 (N = 5)	Negative	12	12
		9 (N = 5)	Negative	16	
		13 (N = 5)	Negative	16	
		14 (N = 5)	Negative	16	
		15 (N = 5)	Negative	16	
		16 (N = 5)	Negative	16	
		17 (N = 5)	Negative	16	
	ELISA[b]	10 (N = 5)	Negative (0.000)	12	
		9 (N = 5)	Negative (0.000-0.002)	16	
		13 (N = 10)	Negative (0.000-0.033)	16	
		14 (N = 10)	Negative (0.000-0.029)	16	
		15 (N = 5)	Negative (0.000)	16	
		16 (N = 5)	Negative (0.000)	16	
		17 (N = 5)	Negative (0.000)	16	

[a] Tetratcore PRRSV real-time RT-PCR.
[b] IDEXX PRRSV X3 antibody ELISA.

Differentiation of *Pasteurella multocida* field strains isolated from growing-finishing pigs that died of bronchopneumonia using PCR-RFLP analysis of the *omph* gene

Kara Telfer[1], BS; Boh Chang Lin[2], DVM, MS, PhD
[1]Iowa State University, Ames, Iowa; [2]MVP Laboratories, Inc. Omaha, Nebraska

Introduction

Recent studies have indicated that *Pasteurella multocida* is more of a primary contributor to pneumonia in swine than previously thought.[1,2] The problem with vaccinating for *P. multocida* is that modified live vaccines have been known to produce disease in animals, and killed bacterins have only been able to provide homologous immunity.[3] For this reason, we chose to look at the *omph* gene for outer membrane protein H (OmpH), a porin found in the outer membrane. OmpH is a major component of the cell envelope, is highly conserved, and has been shown to provide cross-protective immunity for *Pasteurella* strains.[4] By determining differences between isolates based on OmpH, we will be able to choose *P. multocida* strains for a more effective autogenous vaccine.

Figure 1: Percentage of *omph*-RLFP patterns in *P. multocida* isolated from swine with bronchopneumonia

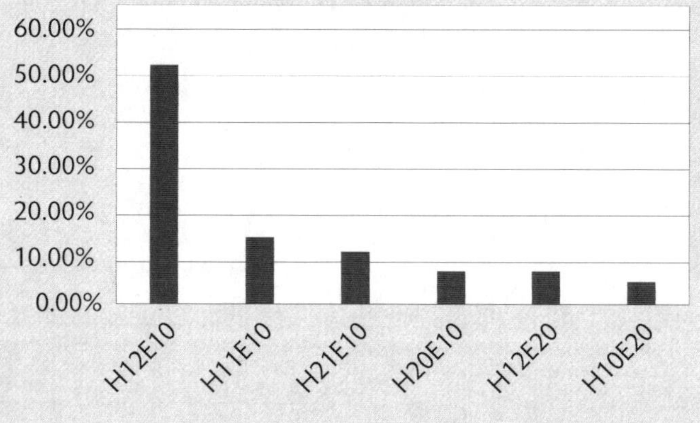

Methods

Fifty-two samples of *Pasteurella multocida* isolated from pigs that died of bronchopneumonia were characterized using DNA fingerprinting and Restriction Fragment Length Polymorphism (RFLP). Samples were chosen from strains isolated at MVP Laboratories, Inc. between 10/03/2013 and 6/26/2015. DNA was extracted from each sample and PCR for the *omph* gene was run. We then used HindIII and EcoR1 restriction endonucleases to digest the PCR product. This product was run through 2% agarose gel and bands were evaluated to characterize the isolate. We also used bacterial neuraminidase testing to evaluate virulence of the isolates and used a chi-square test to see if each RFLP pattern corresponded to a certain neuraminidase titer.

Results

Of the 52 isolates we saw 6 different *omph*-RFLP patterns with the H12E10 pattern comprising 52% of the isolates (Figure 1). We also saw variability of *omph*-RFLP patterns between isolates from the same producer. Low numbers of some RFLP patterns inhibited chi-square testing for significant neuraminidase values. However, H12E10 significantly correlated to neuraminidase endpoint titers of 1:10 and 1:100.

Discussion

We were able to use a method that can easily be done in a lab to identify differences between *Pasteurella multocida* isolates of the same serogroup. Swine farms where *Pasteurella multocida* is contributing to outbreaks of pneumonia could benefit from this sort of typing because it ensures that all relevant *omph*-RFLP types are included in the isolates chosen for autogenous vaccine production. Further studies with larger sample numbers will need to be done for a more thorough understanding of whether or not an isolate's *omph*-RFLP pattern corresponds to a certain virulence.

References

1. Brockmeier SL, Halbur PG, Thacker EL. Porcine Respiratory Disease Complex. In: Brogden KA, Guthmiller JM, editors. Polymicrobial Diseases. Washington (DC): ASM Press; 2002. Chapter 13.Available from: http://www.ncbi.nlm.nih.gov/books/NBK2481/

2. Mores, N, Mores, MA, Oliveira Filho, JX, Klein, CS, Caron, L, Coldebella, A. Rebellato, R. Sluszz, T, Guedes, RMC, Alberton, GC, Dutra, V, Morena, AH, Sobestiansky J, Matos, MPC, Barcellos, DSN, Study of the pathogenesis and development of diagnostic methods and control of pneumonic pasteurellosis in pigs. Symposium EMBRAPA (Brazilian Agricultural Research Corporation) Labex of Animal Health. Aug. 28-30, 2012. Embrapa project #: 02.10.06.016-00.

3. Lee, Jeongmin, Seoyoung Kang, Shin in Park, Hee-Jong Woo, and Moosik Kwon. Molecular cloning and characterization of the gene for outer membrane protein H in *Pasteurella multocida* (D:4) Isolate from pigs with atrophic rhinitis symptoms in Korea. J. Microbiol Biotechnol. (2004);14(6):1343-1349.

4. Luo Y, Glisson JR, Jackwood MW, Hancock REW, Bains M, Cheng IN, Wang C, Cloning and characterization of the major outer membrane protein gene (ompH) of *Pasteurella multocida* X-73. J of Bacteriology. 1997;179(24):7856-7864.

Cognitive bias testing of commercial and heritage-breed gilts

J. Piergiovanni; K. M. Horback, PhD; T. D. Parsons, VMD, PhD, DACAW
University of Pennsylvania School of Veterinary Medicine, Kennett Square, Pennsylvania

Introduction

Swine producers are being asked to transition their sows out of individual gestation stalls to group housing. In group housing, the sow faces greater freedom of movement and a more complex social environment. After years of selection to perform in individual stalls, the appropriateness of today's commercial genetics for group housing is unclear. Heritage breeds have not been exposed to the same selective pressure and might harbor behavioral traits advantageous to the more complex environments of group housing. The goal of this project was to assess breed differences in the cognition of replacement gilts using operant conditioning and a judgment bias task. This type of assessment has been utilized in many different species to assess the affective state.[1]

Materials and methods

At the Penn Vet Swine Teaching and Research Center, gestating animals and replacement gilts are housed in dynamic groups and fed via electronic sow feeding stations (Compident 7, Schauer Agrotronic). DNA Line 241 (n = 11) and registered purebred Tamworth (n = 13) gilts were trained a "go/no-go" operant task in a 7 × 2.4m arena. The positive "go" cue (feed reward of 200g of corn-soy feed) was placed in the far left or right corner of the field, while the negative "no-go" cue was placed in the opposite corner and punished with a red flag wave and no access to feed. During testing, gilts were given ambiguous trials with a bowl placed equidistant from the positive and negative cues. A cognitive bias index (CBI) was calculated for these sows based on their latency to approach the first ambiguous cue that was scaled such that 0 would reflect a failure to approach the ambiguous cue whereas 1 represents an approach comparable to a positive cue. Values are expressed as mean +/- SEM. Mann-Whitney U tests ($\alpha = 0.05$) were conducted to determine significant differences in learning and bias between the breeds.

Results and discussion

The DNA gilts were able to begin training after 3 to 5 minutes of habituation to the arena whereas Tamworth gilts required 15 minutes of habituation for two days prior to training. Despite this difference in habituation, there was no significant difference (Mann-Whitney U, $P > 0.05$) in learning ability between the breeds; defined as the average number of flags (punishments) needed for the gilt to learn the "no-go" task. While not statistically significant ($P = 0.2$), the DNA gilts approached the negative cue on the test days (32 ± 7.2%) more often than the Tamworths (19.2 ± 6.4%). There was not a significant difference ($P > 0.05$) between the breeds in the average percent of approach (DNA: 66.8 ± 9.8%, Tamworth: 73.8 ± 9.6%) or speed of approach (DNA: 26.1 ± 5.2s, Tamworth: 23.6 ± 4.8s) to the ambiguous cue as seen in Figure 1. And there was not a significant difference ($P > 0.05$) in the average CBI between the breeds (DNA: 0.58 ± 0.1, Tamworth: 0.56 ± 0.1). Seven DNA gilts had positive biases (CBI > 0.8) and four DNA gilts had negative biases (CBI = 0), while nine Tamworths gilts had positive biases (CBI > 0.6) and four Tamworth gilts had negative biases (CBI ≤ 0). The lack of difference in the cognitive abilities between DNA Line 241 and registered purebred Tamworths is striking given their markedly different selection programs over that last several decades. The Tamworth gilts did require longer habituation times to the training arena and may reflect differences in other behavioral attributes. Once habituated, no differences were observed between the breeds in learning ability or judgment bias. Also, both breeds willingly participated in the trials and were handled with minimal effort as the days of training went on. This indicated that both breeds were able to adapt and perform the tasks without additional stress contributing to our cognitive bias study. Our work suggests that different breeds or genetic lines may not be as important for optimization of group housing as managing trait differences between individual animals.

Acknowledgements

Supported by Pennsylvania Pork Producers Council, New York Farmers, and Swiss Village Foundation.

Figure 1: Mean proportion of approach (in %) for each trial type in Tamworth and Danbred (DNA 241) breeds.

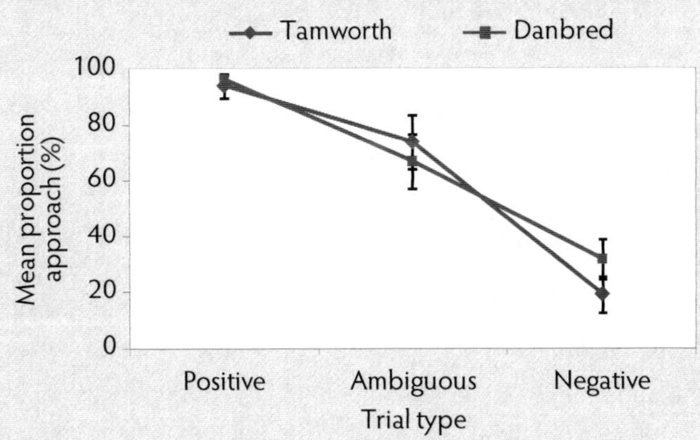

References

1. Mendl M et al. Cognitive bias as an indicator of animal emotion and welfare: Emerging evidence and underlying mechanisms *Appl Anim Behav Sci*. 2009; 118:161-181.

Determining the effects of pooling placenta umbilical cord serum (PUCS) on polymerase chain reaction (PCR) sensitivity

Donna Drebes[1], BS; Jana Morgan[2], DVM; Thomas Wetzell[2], DVM; Dale Polson[2], DVM; Allison Spitz[2], MS; Wayne Chittick[2], PhD; Brad Leuwerke[3], DVM

[1]University of Minnesota CVM; [2]Boehringer Ingelheim Vetmedica Inc.; [3]Swine Vet Center

Introduction

Even with widespread vaccination to protect swine against porcine circovirus type 2 (PCV2), cases of porcine circovirus associated disease (PCVAD) are still diagnosed. In those circumstances, suggested follow up is to access vertical transmission of PCV2. Previous studies demonstrate PCV2 prevalence can be determined by collecting presuckle piglet serum[1] or colostrum.[2] In addition, recent work has shown PUCS are a more efficient alternative allowing prevalence determination without compromising sensitivity.[3] However, the sensitivity of pooling PUCS samples to gain a broader perspective of prevalence while decreasing costs is unknown. This study was designed to determine how many PUCS samples could be pooled before PCR testing no longer accurately reflects PCV2 vertical transmission prevalence.

Materials and methods

For this trial, the experimental unit was the individual sow PUCS. The current reference standard for PUCS is a pool of 3 to 5 umbilical cords from the same placenta. To complete the needed replicates of each pool, approximately thirty placentas were collected from a variety of farms. Collections were taken at least seven days apart to ensure placentas were from different farrowing groups. Placentas were collected by staff as soon as possible after sow cleaned and refrigerated. Umbilical cord contents were collected within twenty four hours of collection and sent to the Boehringer Ingelheim Health Management Center in Ames, Iowa. PCR testing was performed to determine the quantity of viral DNA present. After obtaining an individual cycle threshold (CT), the PUCS samples were pooled. Pools were created by combining a PCR positive PUCS sample (CT value of 34 to 38); with 1, 2, or 4 PCR negative PUCS samples.

Results and discussion

Ninety three replicates were obtained. A power calculation was performed using 0.80 for beta, 0.05 for alpha, and 0.50 for sigma. This resulted in the ninety three replicates detecting a 20% difference (delta). The results of pooling were 81.7% positive in when combining one negative and one positive, 76.3% positive in when combining one positive and two negatives, and 63.4% positive in when combining one positive and four negatives. Overall, the higher the initial CT value the less able the PCR was to detect a positive when pooled with negative samples. In addition, the effect of higher initial CT on pooling CT is displayed in figure 1. Once again, it can be seen that increasing initial CT will decrease sensitivity of pooling. In regards to application of the information, PUCS can become a tool which allows veterinarians to evaluate PCV2 viral load at the sow herd.

References

1. Baumert, Fangman, and Rustvold. Comparing placenta and presuckle piglet PCV2 status between two breeding sites. Allen D Leman Swine Conference. 2013.

2. Gerber PF, Garrocho FM, Lana AM, Lobato ZI, Serum antibodies and shedding of infectious porcine circovirus 2 into colostrum and milk of vaccinated and unvaccinated naturally infected sows. Vet J 188:240–242, 2011.

3. Grotenhuis, Fangman, and Baumert. Using placental umbilical cord serum to determine porcine circovirus type 2 status of breeding herd. AASV. 2014.

Figure 1: Effect of higher initial CT on pooling CT.

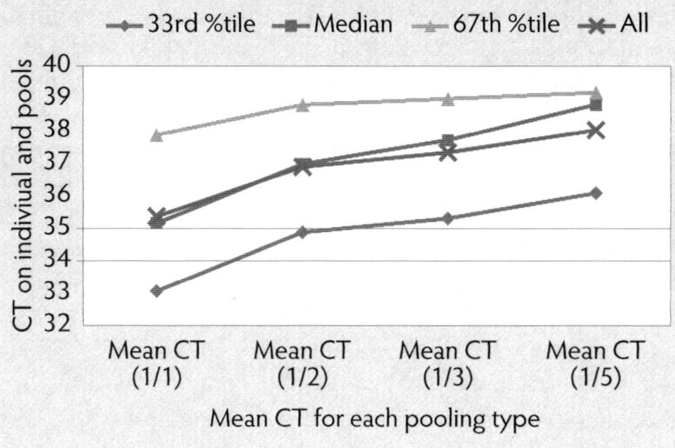

Evaluating the use of the EnviroBootie™ to detect *Salmonella* at five breed-to-wean sites

Kayla Blake[1]; Thomas Fangman[2], DVM, MS, DABV; Greg Cline[2], DVM; Jessica Seate[2], DVM
[1]Auburn University, CVM; [2]Boehringer Ingelheim Vetmedica, Inc.

Introduction

The EnviroBootie™ (Hardy Industries) has been used to detect *Salmonella enteritidis* in poultry[1] and *Salmonella choleraesuis* and *typhimurium* in swine.[2] The objective of this study was to utilize the EnviroBootie™ to identify Salmonella groups at 5 breed to wean sites suspected as having elevated Salmonella prevalence.

Materials and methods

The 5 suspect sites chosen for evaluation had previously demonstrated > 30% prevalence of Salmonella ELISA antibodies in 5 week old piglets originating from these sites.[3] The suspect sites evaluated demonstrated Salmonella risk factors to include; poor sanitation practices, visible rodents and/or feces and open gutter flush systems. The control site was a suspected low prevalence farm following the same general layout and management as the other sites visited. The control site was selected as the 4 week old seroconversion was < 10% of pigs tested. Biosecurity practices upon entry at each site included; Tyvek coveralls, clean boots and plastic boot covers. Sanitized collection kits (12) utilized at each site included: 2 gloves and 2 EnviroBooties™ contained in individual whirl-pack bags (24 total booties/site). After securing the EnviroBooties™ (1 on each foot) the technician walked behind each row of sows (2 rows/alley) taking slow and deliberate steps. The technician also walked in pens, hallways and chutes. After walking to the end of each area to be sampled, the environmental booties were taken off and put back into its individual whirl-pack bag and sealed. This walk was repeated 12 times in each barn. Prior to leaving the site the 24 bootie samples were immediately placed on ice and overnight shipped to Iowa State University Veterinary Diagnostic Lab for Salmonella culture.

Results

Table 1 shows the different groupings of positive isolates found at the 5 suspect sites and 1 control site. The Group B Salmonella isolate was found at every site, and Group C1, was found at all but one site. Positive results were identified in places that appeared fecal material free, such as pens, hallways, and chutes.

Conclusions and discussion

The EnviroBooties™ utilized to culture and identify Salmonella species picked up many positive samples at the 5 sites that demonstrated ≥ 30% positive findings for Salmonella ELISA titers in 5-week-old pigs.[3] Multiple Salmonella groups were identified and were widely diverse among the different sites. Salmonella species were evident in all environments but clinically relevant in weaned pigs where the breed-to-wean environments where rodents were observed. Observable signs of rodents served as a visual risk factor suggestive of Salmonella challenge. The EnviroBootie™ is a sensitive method for picking up Salmonella isolates. Five representative samples from each site have been forwarded to the NVSL in Ames, IA for further serotyping (30 samples total). Results will be reported as they become available.

References

1. Federal Register: http://www.gpo.gov/fdsys/pkg/FR-2009-07-09/pdf/E9-16119.pdf.

2. Fangman T, Cline G. 2015. Evaluating The Use of Envirobootie™ to Detect Lawsonia and Salmonella From Known Positive Concrete Surfaces. Allen D Leman Swine Conference.

3. Fangman T, Baumert D, et al. 2015 Salmonella Elisa Serology from Weaned Pigs as an Indicator of Salmonella Circulation in Breed to Wean Sites. Allen D Leman Swine Conference.

Table 1: *Salmonella* groups

	Grouping of positive samples				
Farm	B	C1	E	Poly	Untype
A	8	0	0	2	1
B	5	3	0	1	2
C	1	11	0	0	0
D	1	5	1	0	0
E	11	1	0	2	1
F/cntrl	2	5	0	6	0

Case study: Investigation of porcine reproductive and respiratory syndrome positive PCR results from a gilt development unit with a negative status

Nicole Thomas; Erin Lowe, DVM; Abbey Harding, DVM; James Lowe, DVM, MS, Dip ABVP

Introduction

Porcine reproductive and respiratory syndrome (PRRS) costs the US swine industry an estimated $664 million each year. A key part of managing PRRS virus (PRRSV) in a pork production system is routine verification of the PRRS negative status of multiplication and gilt development units (GDU). This limits the potential losses by preventing the unintentional spread of virus to downstream sites. This case study describes a diagnostic investigation into positive oral fluid polymerase chain reaction (PCR) results from an expected PRRSV negative GDU, with subsequent PRRSV negative status confirmation of the population.

Background

A 3900 head GDU was located within a PRRSV area-regional control (ARC) project that was experiencing new PRRSV outbreaks. The production system was also experiencing recent PRRSV outbreaks in other sites, including the GDU's source herd. Following the outbreak at the source, pig movement was terminated and additional testing was completed at the GDU; The GDU remained PRRS negative. On May 13th (Day 0) the GDU tested positive for PRRSV by PCR during routine monthly monitoring of oral fluids in the absence of clinical signs consistent with a PRRSV infection.

Investigation

An ORF5 sequence, requested upon receipt of the positive results, was compared to sequences within the ARC project as well as in a statewide sequence database. Although PRRSV was prevalent in the area, the GDU sequence was 6% and 4.5% different than the most closely related sequence in the databases, respectively. A discussion on Day 7 with the diagnostic lab about these findings led to simultaneous retesting of the original Day 0 oral fluid samples and resampling of oral fluids. The Day 7 samples tested negative by PRRSV PCR and negative by PRRSV Enzyme-linked immunosorbent assay (ELISA). On Day 15 the diagnostic lab retested the original samples and re-extracted samples from the original Day 0 template for retesting. All retests indicated positive PRRSV PCR results with similar cycle times as had been initially reported. Sequencing from each retest and re-extraction was performed on Day 16 and resulted in sequences very similar to the originally reported Day 0 sequence. The matching results from the Day 0 testing and the subsequent retesting indicate that the sample did contain PRRSV RNA, however the negative PCR and ELISA results collected Day 7 demonstrate that the herd did not seroconvert to PRRSV. As laboratory contamination was a concern, the diagnostic laboratory reviewed and assured that the AAVDL procedure guidelines were met with all samples. Additionally, they reviewed and sequenced all PRRSV positive samples in the lab for the 2 weeks prior to Day 0 testing and found no similar PRRSV results. As lab contamination was highly unlikely, alternative sources of contamination were investigated. Records indicate that the samples were packaged alone and were unlikely to be contaminated in transit to the lab. Sampling kits were sent directly to the site, and unlikely to be contaminated in transport or storage in company sites or vehicles prior to use.

Conclusion

Veterinarians use routine diagnostic sampling frequently in the swine industry. Thus it is imperative when interpreting results of tests that the whole clinical picture of the site be assessed and that a veterinarian communicates with the diagnostic laboratory, so that interpretations of results and follow up recommendations can be made. Although the exact source of contamination of this case was not determined, this case study demonstrates a systematic approach to ruling out some of the many potential sources of possible contamination that may lead to a PCR positive result.

Effect of antimicrobial administration on the microbiome of the pig

Katharine Kancer, BS; Brian M. Aldridge, DVM, MS, PhD; James F. Lowe, DVM, MS

Problem

The use of antimicrobials in livestock production has been widely criticized with the increase of drug resistant microbes. Negative public perception on the use of antimicrobials in animal agriculture, as well as recent legislative changes in select European countries, have led US veterinarians and producers to foresee a future where antimicrobial use will be more limited. In order to overcome these challenges, production systems must adapt to reduce the overall need for antimicrobials. A key step is this process will be to shift from pathogen focused to host focused approaches in controlling disease. This entails optimizing host defenses against infectious disease, primarily through improving local innate and adaptive immune defenses. The microbiome of the mucosal surfaces plays an important role in regulating local and systemic immune defenses in a complex cross talk between the neurologic system, immune system and the microbiome. We know that antibiotic administration upsets the delicate microbial ecosystem in the gut and on mucosal surfaces. This has an unknown but significant impact on local immune defenses. Here we report our investigation into the changes in the resident microflora of the upper respiratory and gastrointestinal tracts over time following the administration of five antimicrobials commonly used in swine production. Understanding these changes are a critical step in designing comprehensive health management programs that optimize local immunity to minimize disease and the need for antibiotics.

Objective

To measure and compare the effects of different classes of antibiotics on upper respiratory and fecal microflora over time.

Materials and methods

Five antimicrobial treatment groups each consisting of four, eight week old piglets were treated with one of the antimicrobials on day 0 at label dose and route: Tulathromycin, Ceftiofur Crystalline free acid, Ceftiofur hydrochloride, Oxytetracycline, and Procaine Penicillin G. Deep nasal and fecal swabs were taken from each pig before administration (day 0) and again on days 1, 3, 7, and 14. Pigs had not received any antibiotics in the 4 weeks prior to enrollment. Once collected, all samples were placed on dry ice and held at -20°C until analysis. To assess microbial diversity high throughput sequencing was used. Briefly, samples were processed for total DNA using MoBio DNA Isolation kits following manufacturer's instructions. We used PCR to amplify and enrich both the V1-V3 and V4 regions of the 16S ribosomal RNA and sequenced the samples using the current generation MiSeq (Illumina). We assembled the sequences using FastX Toolkit and UCHIME, and performed taxonomic-assessment using the RDP Classifier. Total microbial richness was assessed through rarefaction analyses and diversity was determined by Shannon's diversity indices.

Results

Preliminary analysis shows a significant shift in respiratory and fecal microbiomes after administration. Final results including duration and antibiotic class effects will be presented.

Discussion

We know microorganisms are transferred from dam to offspring at parturition and during the neonatal period. We also know the composition of the microbiome has been shown to impact the severity of respiratory disease due to influenza. This suggests that management practices in the neonatal period could markedly impact the microbiome during the growing period. Thus affecting a pig's response to infectious disease challenges by altering mucosal immunity. These data suggest that antibiotics might be one of the factors that alter the long-term composition of the microbiome if they are given during key periods. This could have long-term impacts on overall pig health and the need for future antibiotic interventions.

Effect of pre-farrow ceftiofur sodium (Naxcel®) administration on *Streptococcus suis* colonization of peri-parturient females and their litters

Daniel Gascho[1], BS; Duane Long[2], DVM; Lynette Holman[3], DVM, Todd Wolff[4], DVM, MPH
[1]Purdue University College of Veterinary Medicine, West Lafayette, Indiana;
[2]Four-Star Veterinary Services, Mexico, Indiana; [3]Kalmbach Swine Management, Ohio; [4]Zoetis, Florham Park, New Jersey

Introduction

Streptococcus suis (*S. suis*) is one of the most important bacterial pathogens encountered in swine production. Characterized by acute meningitis, polyarthritis, polyserositis, bronchopneumonia, or sudden death, it often results in economic losses through increased mortality and medication costs.[1] Research suggests that sows shed multiple *S. suis* serotypes in their vaginal secretions that transfer to piglets' dorsal surface and oral cavity during birth.[2] The primary objective of this study was to evaluate if a pre-farrow injection of ceftiofur sodium will decrease *S. suis* colonization of sow vaginal secretions and result in decreased colonization of piglets during parturition. The secondary objective was to evaluate the effect on nursery mortality and the number of individual pig treatments.

Materials and methods

The study was conducted on a 2500-sow, PRRSV-positive unstable farrow-to-wean unit. 200 sows were blocked by due date, then by parity, and randomly assigned one of two treatment groups. Sows in Treatment 1 (T1) were used as a control group. Sows in Treatment 2 (T2) were given an injection of ceftiofur sodium (Naxcel®, Florham Park, NJ) at a dosage of 5 mg/kg on day 114 of gestation. T1 sows that did not farrow on or between days 114 and 117 of gestation, and T2 sows that did not farrow within the window of 1-60 hours post injection were excluded from the study. Research has demonstrated that Naxcel® reaches plasma levels above minimum inhibitory concentration for *S. suis* within 1 hour, and lasts for at least 60 hours (Pfizer, 2008). Post-farrowing, 149 sows from seven rooms remained in the trial (T1 = 78, T2 = 71). Guarded swabs (Starswab®, Starplex®, Cleveland, TN) of vaginal secretions were taken from 20 randomly selected sows (T1 = 10, T2 = 10) pre-treatment, 24 hours post-treatment, and post-farrow. From each of their litters, oropharyngeal swabs were taken from four piglets within 24 hours of birth. All swabs were sent to the University of Minnesota VDL for *S. suis* culture and isolation. Cross-fostering was allowed only within a treatment group and room, post-swabbing. Farm protocol of an injection of ceftiofur crystalline free acid (Excede®, Florham Park, NJ) at processing and tulathromycin (Draxxin®, Florham Park, NJ) at weaning was followed for all pigs. 102 litters (1575 piglets) were placed in an off-site, four-room, 1600-head nursery. (T1 = 620, T2 = 595). Approximately 200 piglets from each treatment group were placed in rooms 1-3. Separate treatment groups were not allowed physical contact. Piglets were monitored in the nursery for five weeks, and mortalities and treatments were recorded. Statistical analysis was completed using the chi-squared test by Carthage Innovative Swine Solutions, LLC.

Results/implications

Pre-farrow use of ceftiofur sodium resulted in a significant ($P < 0.05$) decrease in number of positive sows at farrowing, and a significant ($P < 0.05$) reduction in the number of positive piglets farrowed, nursery mortality, and number of individual piglet treatments (Table 1). In conclusion, Pre-farrow use of ceftiofur sodium may serve as an effective and economical method of *S. suis* control.

References

1. Gottschalk, M., Higgins, R., & Quessy, S. (1999). Dilemma of the Virulence of *Streptococcus suis* Strains. Journal of Clinical Microbiology, 37(12), 4202–4203.

2. Amass SF, Clark LK, et al. *Streptococcus suis* colonization of piglets during parturition. J Swine Health Prod 1996;4(6):269-272.

Table 1: Effect of Ceftiofur sodium application on *S. suis* presence in swabs collected from sows vaginal secretions and piglets throats, and on the incidence of mortality during the nursery period[1]

Item	Treatment		
	No Ceftiofur	Ceftiofur	*P*-value[2]
Sow vaginal swabs[3]			
Number of swabs	10	10	-
Percentage of positive swabs, %			
Before Ceftiofur treatment	60%	70%	0.64
24 h post-Ceftiofur treatment	70%	10%	0.006
Post-Farrowing	80%	10%	0.002
Piglet throat swabs[4]			
Number of swabs	40	40	-
Percentage of positive swabs, %	50%	25%	0.021
Nursery[5]			
Number of pigs	620	595	-
Number of injections given	31	14	-
Mortality, %[6]	5.00	1.85	0.003

1 Intramuscular injection of Ceftiofur sodium was given to the Ceftiofur treatment sows on day 114 of gestation at a dose of 5 mg/kg of BW.
2 Comparison of proportions of positive swabs and mortality rates with and without the Ceftiofur were carried out using the chi-square (χ^2) test for proportions
3 Swabs of vaginal secretions were taken from 20 sows selected randomly.
4 From each litter of sows where vaginal swabs were collected, oropharyngeal swabs were taken from four piglets within 24 hours of birth.
5 Nursery period = 5 weeks.
6 Odds ratio: pigs from sows not treated with Ceftiofur died 2.8 times more than pigs from treated sows.

Determining the prevalence of *Mycoplasma hyopneumoniae* colonization of weaned piglets from gilt litters

Thomas Wurtz[1], BS; Laura Greiner[2], PhD; John Waddell[3], DVM; Eduardo Fano[3], DVM MSc, PhD; Roy Edler[3], MS
[1]Washington State University, Pullman, Washington; [2]Carthage Innovative Swine Solutions, Carthage, Illinois;
[3]Boehringer Ingelheim Vetmedica Inc., Saint Joseph, Missouri

Objective

To determine the colonization rate of *Mycoplasma hyopneumoniae* in weaned piglets from gilt litters using laryngeal swabs.

Introduction

Mycoplasma hyopneumoniae (*M. hyopneumoniae*) initiates respiratory health challenges that result in economic losses during the finishing period. Replacement gilts have been suspected to disseminate the disease resulting in variable percentages of piglets being colonized at weaning. Therefore, this study was designed to measure the prevalence of *M. hyopneumoniae* colonization in weaned piglets from positive versus negative gilts.

Methods

A 2,500 head farrow to wean sow farm was studied, where 26 week-old negative *M. hyopneumoniae* breeding age gilts had entered a positive breeding herd. The gilts were vaccinated for *M. hyopneumoniae* at 2 weeks after entry into herd. Animals requiring treatment with antimicrobials known to target *M. hyopneumoniae* were removed from the study.

Eighty-four randomly selected pregnant gilts were sampled one week prior to farrowing to identify 33 dams for the study. Dams were sampled again one day prior to weaning. Oropharyngeal mucous was collected and the mucous samples were transferred to COPAN eSwabs, cooled and shipped overnight for real time polymerase chain reaction testing.

Dams were assigned to one of five farrowing rooms, each with 56 crates, according to due date and *M. hyopneumoniae* status. Fostering of piglets was limited to removing average sized piglets in litters whose piglet number exceeded the dam's teat number. Piglets were processed and vaccinated with 1mL of a mycoplasma vaccine at three days of age. One day prior to weaning, oropharyngeal swabs were collected from the piglets using the same type of swab used in the sows. Entire litters were considered colonized if at least one piglet tested positive. The dam was considered the experimental unit for calculations. The data was organized in a two by two contingency table and odds ratio was calculated.

Results

The prevalence of piglet *M. hyopneumoniae* colonization was 32%, coming from 11 out of 33 weaned litters. Piglets from colonized litters were 5.4 times ($P = 0.067$) more likely to come from the 19 positive dams than from the 14 negative dams.

Discussion

M. hyopneumoniae's tropism for ciliated epithelium, sampling type, and piglet age could account for the lower than expected prevalence of colonization. Although there is a 5 fold increase in the odds of a positive litter, the increase is not statistically significant at the 0.05 level. This could be due to sample size limitations. The prevalence varied considerably between farrowing rooms, suggesting there are other intra-farm factors acting as epidemiologic agents of this disease. In order to decrease the financial impact of *M. hyopneumoniae*, efforts should be made to reduce the number of positive dams in the farrowing house.

Evaluation of diagnostic tests for Circumvent PCV-M G2 compliance monitoring

T. Girard[1], BS; P. Thomas[2], DVM
[1]Iowa State University College, Ames, Iowa; [2]Smithfield Hog Production Division, Midwest Division

Introduction

Effective immunization against porcine circovirus type 2 (PCV2) and *Mycoplasma hyopneumoniae* (Mhp) is dependent on strict compliance with vaccination protocols. Circumvent® PCV-M G2 (PCV-M G2) is a polyantigen vaccine composed of PCV2, Mhp and baculovirus antigenic components and results in development of measurable antibody responses after vaccination. Accordingly, serodiagnostic tools are potentially useful for evaluating vaccination compliance.

Objective

The objective of this study was to evaluate several serological assays for potential use for PCV-M G2 compliance monitoring. The polyantigenic nature of the vaccine allows for the possibility of multiple serological evaluations of compliance whether the vaccine is administered by label with either one or two dose regimens.

Materials and methods

Three week old pigs were allotted to the following PCV-M G2 vaccinated groups: Group A pigs (n = 30) were vaccinated with a single 2 ml dose at 3 weeks of age (WOA); Group B pigs (n = 30) were vaccinated with a 1 ml dose at 3 and 6 WOA; Group C (n = 15) pigs served as non-vaccinated controls; and Group D pigs (n = 30) were vaccinated with 1 ml at 3 and 6 WOA in a blinded fashion by the operation's vaccination crew. Pigs were bled at three and nine weeks of age. Serum samples were evaluated for PCV2 antibodies by full range dilution immunofluorescence assay (IFA), four dilution IFA, Ingenasa PCV2 IgG ELISA and ISU-VDL PCV ELISA; Mhp antibodies by the IDEXX ELISA; and baculovius antibodies by the ISU-VDL BV ELISA.

Results and discussion

The results provide the following findings: 1) consistent PCV2 immune response was observed at 9 WOA for groups A, B and D with the full range dilution IFA and the Ingenasa ELISA; 2) consistent PCV2 immune response was observed in groups B and D with the four dilution IFA; 3) consistent Mhp immune response was observed in groups B and D with the IDEXX ELISA; and 4) consistent baculovirus immune response was observed with the BV ELISA in group B only (Table 1). Group C pigs remained consistently negative on all assays at 9 weeks of age.

All of the assays produced a consistent immune response when the vaccine was used in a two dose regimen. With the one dose regimen, only PCV2 antibodies were measurable by the full range dilution IFA and the Ingenasa ELISA. In the field, positive results with all of the PCV2 and Mhp assays could be due to natural infection so interpretation of the results needs to be considered in light of potential field exposure prior to or during the vaccination period. However, this study provides guidance for interpreting negative results. Accordingly, these assays can be used as indicators of compliance failure.

Table 1: Results of serological testing before and after vaccination with Circumvent® PCV-M G2

Test	Variable	WOA	2 mL, 3 WOA	1 mL, 3 & 6 WOA Student	1 mL, 3 & 6 WOA crew	Control
ISU BV-ELISA	S/P ratio	3	0.137	0.111	0.134	0.159
		9	0.132	0.236	0.171	0.085
	No. pos.	3	10/30	8/30	11/30	10/30
		9	9/30	27/29	21/30	1/15
	No. responded	----	9/30	27/29	21/30	1/15
Ingenasa PCV2 IgG ELISA	S/P ratio	3	190.6	167.7	156.4	147.0
		9	677.0	41,852.7	27,779.9	106.4
	No. pos.	3	16/30	12/30	15/30	4/15
		9	30/30	29/29	30/30	0/15
	No. responded	----	30/30	29/29	30/30	0/15
IDEXX M hyo ELISA	S/P ratio	3	0.184	0.142	0.112	0.125
		9	0.086	1.131	1.126	-0.035
	No. pos.	3	2/30	2/30	2/30	0/15
		9	2/30	27/29	26/30	0/15
	No. responded	----	2/30	27/29	26/30	0/15
ISU 4-dilution IFA	Geomean titer	3	83.8	113.1	145.9	145.9
		9	412.6	852.6	806.3	175.5
	No. responded	----	15/30	24/29	25/30	2/15
KSU full range IFA	Geomean titer	3	55.3	59.2	69.6	76.4
		9	1,539.9	4,436.0	3,537.7	41.9
	No. responded	----	27/30	29/29	29/30	1/15

Management practices implemented following an outbreak of PRRS in commercial swine breeding herds in North America

Kathryn N. Duda, BS; James F. Lowe, DVM, MS

Introduction and problem

PRRS continues to infect a high proportion of sow farms in the US annually. It is well accepted today that to minimize the impact of PRRS following a new infection a suite of tools need to be applied. These tools can be grouped into two basic groups, those that maximize host immunity and those that minimize viral load. Expansive research has been done under both controlled and field settings on best approaches to maximizing immunity. The research about minimizing viral load has not reached the same level of completeness. There have been many routes of intra-herd PRRS transmission identified over the last 20 years. The data from these multiple studies were used by the American Association Swine Veterinarians PRRS Task Force to develop a set of recommended management practices based on the level of PRRS virus transmission in a herd. To date there have been no formal studies examining application and relative value of each of these recommended management practices. In 2012 we informally collected data from six large production systems/veterinary clinics about the recommendation of specific management practices in herds following PRRS outbreaks which demonstrated a great deal of variation in the practices that veterinarians recommend. This lack of evidence is preventing swine production herds and their veterinarians from developing evidence based effective management practices when a PRRS outbreak occurs.

Methods

Veterinarians practicing in NA who were members of AASV were asked to report which management practices that they had suggested for the last 5 PRRS outbreaks in herds under their direct management using an electronic survey format. Questions about management practices were based on the AASV PRRS Task Force recommendations.

Results and discussion

Preliminary data (N = 50 herds, of a target of 250) were analyzed. The goal of the PRRS management plan was evenly distributed between elimination and control of the virus at the time of the outbreak. Conditions and location of the herd (area pig density, flexibility in production numbers, and previous success in stabilizing PRRS transmission and preventing new introductions) were reported as primary reasons to choose the management plan goal. There was a wide variation in practice implementation across herds and veterinarians. "All-in, All-out" management, washing, disinfecting and drying of all farrowing crates/pens between litters was common across herds in the immediate post break period regardless of herd goal. Methods for monitoring and sampling the herd to determine herd status showed consistency with respect to numbers and sampling of due to wean pigs with common use of PRRS ORF5 rtPCR testing. In phase 2 of the project we will compare production outcomes (total piglet losses from baseline) and the sub metrics that determine total herd productivity between herds with high and low levels of PRRS specific management practices that were implemented. Management practice data from all herds in the survey will be presented. The project will begin to fill the gap of knowledge about the value of specific management practices in lowering viral load following novel PRRS introductions in breeding herds. Frist it will provide clinicians an objective measure of how their recommendations compare to professional norms and common practices. Secondly it will provide the first object assessment under field conditions of the effectiveness of a suite of practices in improving production following a novel PRRS introduction which will be an important step in reaching an evidence based approach to PRRS management. Applying an evidence based approach will lead to an improved predictability in outcomes which overtime will increase the confidence of producers in the profession's recommendations.

Determining effects of transport factors on transport loss in a commercial production system

Megan Pieters[1]; James Mark Hammer[2], DVM; Thomas Marsteller[2], DVM; Bruce Ervin[2]; Jessica Risser[2], DVM; Christy Goldhawk[2], PhD
[1]Iowa State University; [2]Elanco Animal Health

Extensive studies have shown that transportation loss is multi-factorial, however, few have evaluated the utility of pre-transport pig evaluations in determining the post-transport welfare of pigs. The objective of this study was to determine if transport factors, as well as the pre-transport condition of the pigs, in a commercial swine operation played a role in transport losses. In this study, 69 Load Site Assessments were completed at 26 different barns, with different 24 truck drivers, representing 11,482 pigs. Pigs were located in the eastern corn-belt within 300 miles from the harvesting facility. Load Site Assessment forms were created to collect objective data on pig, facility design, and trailers, based on results from a previous project efficiency study.[1] Stressed pigs were identified in the barn during load out before entering the trailer as red (open mouth breathing, muscle tremors, and or skin discoloration), or only green (only high pitch vocalization for greater than two seconds), pigs were marked exclusively in one of the categories. Marked pigs were observed through the packing plant and the number of red or green dead at arrival (DOA) and non-ambulatory at arrival or in the plant (NA) were recorded. Facility differences included an observation of the angle of the loading chute (greater or less than 20 degrees), as well as the number of times a group of pigs stopped when encouraged forward motion during load out, and the average size of loading groups. Standard least squared regression models were used to evaluate the effect of pig and pre-transport factors on the number of pigs marked as red prior to transport, as well as post-transport DOA and NA, using JMP v9.0. A total of 317 pigs were marked as red (2.8%) and 20 as green (0.2%) prior to transport (Table 1). More pigs were marked red prior to transport when loading ramps were greater than 20° (2.9 ± 0.83 vs 5.7 ± 0.68, $P = 0.02$), and the number of pigs marked red increased by 0.14 ± 0.06 for every degree increase in ambient temperature during loading, within the range of 43 to 88 °F that occurred in this study ($P = 0.02$). Overall, 0.2% pigs were DOA, and 0.3% were NA. Of the 317 pigs marked red prior to transport, 7 (2.2%) were DOA and 1 (0.3%) were NA. However, 25% of DOA pigs were pigs that had been marked red prior to transport and pigs marked red prior to transport had 11.7 times the risk of being DOA compared to pigs that were not marked prior to transport (95 % CI = 5.01 to 27.3; $P < 0.01$). Zero of the 20 pigs marked green prior to transport were DOA or NA following transport. There was no statistical association between the number of red or green pigs in a load and the number of DOA or NA pigs after transport. There was a tendency for DOA pigs to increase when the loading ramp was greater than 20° (0.13 ± 0.17 vs 0.58 ± 0.15, $P = 0.07$). The number of NA pigs tended to increase by 0.09 ± 0.05 for every increase in lbs/ft^2 within the range of 44 to 62 lbs/ft^2 in the current study ($P = 0.08$). The findings of this study are consistent with previous transport data on angle of the loading chute.[2] The study was designed to be applicable to provide support for data to make adjustments to loading procedures; such as ramp design, space allowance, and decreasing the opportunities for physical and vocal signs of stress during load out.

References

1. Thomas, P. 2015. Proficiency analysis. Elanco Animal Health

2. Fraser, D., Phillips, P.A. and Thompson, B.K., 1986. A test of a free-access two level pen for fattening pigs. Anim. Prod., 42: 269-274.

Table 1: Post-transport outcome based on pre-transport signs of stress on pigs identified during load out.

		Post-transport outcome		
		DOA	NA	None
Pre transport visual signs of stress category	Red[a]	7	1	309
	Green[b]	0	0	20
	None	21	44	11080

[a] Pigs identified during transport showing signs of open mouth breathing, muscle tremors, and or skin discoloration.
[b] Pigs identified during transport with a high pitch vocalization for greater than two seconds.

Use of oral fluids and fecal samples in sow herds for detection of *Erysipelothrix rhusiopathiae*

Claire Behnke[1]; Amanda Sponheim[2], DVM; Deb Murray[3], DVM
[1]University of Illinois CVM; [2]Boehringer Ingelheim Vetmedica, Inc.; [3]New Fashion Pork

Introduction

Erysipelothrix rhusiopathiae (*E. rhusiopathiae*) can cause abortions and sudden deaths in sows. Additionally, piglet exposure can result in chronic cases of swine erysipelas (SE) downstream, which is often displayed as lameness in market age pigs. Increases in the withdrawal times for penicillin have resulted in a reduction in its use, which may increase the incidence of SE in swine. In order to determine appropriate checkpoints for surveillance and understand pathways of transmission, increased awareness of when swine are most likely to shed *E. rhusiopathiae* in the sow herd is needed. Polymerase chain reaction (PCR) has been used as a reliable method to detect *E. rhusiopathiae* in tissues of animals, but limited work has been done to demonstrate shedding of the bacteria in oral fluid or feces. *Erysipelothrix* DNA can be detected in oral fluids for up to 7 days post infection in a controlled setting, but the ability to detect intermittent shedding using oral fluids is still unknown.[1] This pilot study aims to determine the appropriateness of oral fluids and fecal samples for surveillance of shedding of *E. rhusiopathiae*.

Material and methods

In this pilot study, three 6,000-head sow farms were sampled. All 3 sow herds historically have exhibited clinical signs of SE, and chronic signs of SE also are seen downstream. In order to monitor for shedding of the bacteria, oral fluids and fecal samples were collected from 10 individual females each in early gestation (6-7 weeks bred), mid gestation (12-13 weeks bred), late gestation (15-16 weeks bred), open sows, and lactating sows. Two oral fluids and pooled fecal samples were collected from group-housed gilts over 25 weeks of age for a total of 52 oral fluid and 52 fecal samples from each site. Samples were analyzed via PCR for the presence of *E. rhusiopathiae, E. tonsillarum* and *E. sp2.*

Results and discussion

One hundred forty-seven oral fluid samples, and 150 fecal samples were collected and tested. One oral fluid sample collected from 30-week-old group housed gilts was positive for *Erysipelothrix* sp. by PCR. All other samples were negative. The high number of negative results suggests that the prevalence of *E. rhusiopathiae* shedding at any particular time within each herd sampled may have been below the level of detection for the sample size collected. Shedding may also be intermittent and/or short lived. Based on these results, conclusions cannot be made about which reproductive stage experiences a higher prevalence of shedding. Additionally, while not conclusive, this study suggests that individual oral fluids and fecal samples may not be reliable methods for *E. rhusiopathiae* surveillance.

Results

1. Gimenez-Lirola et al. Improving ante mortem diagnosis of *Erysipelothrix rhusiopathiae* infection by use of oral fluids for bacterial, nucleic acid, and antibody detection. *J Microbio Meth.* 2013;92:113-121.

Effect of injectable fat-soluble vitamins pre-farrow on piglet number born alive

Daniel Carreño[1], BS; Glen Almond[2], DVM, PhD
[1]North Carolina State University College of Veterinary Medicine, Raleigh, North Carolina;
[2]Murphy-Brown, LLC – South Central Division, Tar Heel, North Carolina

Introduction

Fat-soluble vitamins A, D and E are required nutrients through all phases of swine production. Sows and gilts benefit from injections of fat-soluble vitamins pre-farrow. Pigs are born deficient in fat-soluble vitamins, they largely depend on colostrum and milk for adequate needs provided the gestating gilt or sow. Vitamin E impacts sow productivity, more specifically number born alive (NBA), colostrum and milk yield, number weaned and wean weight. Current industry levels of Vitamin E range from 66-150 IU/kg. Using injectable fat soluble vitamins may be an alternative way to provide piglets with the appropriate vitamin levels and could lead to an increase in number of piglets born alive.

Materials and methods

495 multiparous sows were allotted into two Vital E treatments prior to placement into farrowing. Sows were blocked by breed group and by parity to ensure that the Vital E Repro treatments were allotted evenly among parity and breed group. This project was conducted on two commercial sow farms. 235 multiparous sows were injected with fat soluble vitamins (IM) 7 days pre-farrow and 260 multiparous sows served as the control. Sows were distributed into parity categories: gilt, P1-3, and P4+ to standardize the treatments amongst parity distribution.

Discussion and implications

Results were possibly influenced by the developmental stage of the fetus at the time Vital E was administered. Injecting 5cc of fat soluble vitamins (Vital E Repro) 7 days pre-farrow did not increase number of piglets born alive.

Results

Table 1: Effect of Stuart Products' Vital-E fat soluble injectable administered 7 days pre-farrow on litter traits*.

	Treatments			*P*-value		
	Vital E	Control	SE	TRT	Parity	Farm
No. litters on trial	235	260				
Teat count	13.5	13.7	0.113	0.324	0.088	0.014
Total born alive per litter	13.2	13.3	0.307	0.873	0.003	0.867
No. born alive per litter	11.3	11.4	0.323	0.930	0.022	0.155
No. mummies per litter	1.1	0.9	0.244	0.683	0.876	0.107
No. stillborns per litter	0.8	1.0	0.118	0.309	0.314	0.904

Statistical differences are considered $P < 0.05$ and trends $P < 0.10$.

The effect of sampling strategy on porcine epidemic diarrhea virus (PEDV) associated gut morphology and immunohistochemistry scores

Jessica Davenport, BS; Paisley Canning, DVM; Jessica Bates, DVM; Chelsea Ruston, BS; Kristin Skoland, BS; Abbey Canon, MPH, DVM; Darin Madson, DVM, PhD, DACVP; Chong Wang, PhD, MS; Locke Karriker, DVM, MS Diplomate, DACVPM

Introduction

Porcine epidemic diarrhea virus (PEDV) lesions are confined to the small intestine and consist of villous atrophy and fusion. Histological examination, immunohistochemistry (IHC), and villous-to-crypt height ratio (VCR) indicate the presence of PEDV and provide estimates of the disease severity, but rely heavily on high quality fixed tissues and the selection of affected gut regions for sampling. The objective of this trial was to compare lesions, IHC scores and VCRs between PEDV positive and negative pigs and between two different intestinal sampling strategies, the W and L methods, to generate recommendations on intestinal sampling strategies to optimize diagnostic testing results.

Materials and methods

Twenty, 15-day-old pigs from a PEDV naïve herd were randomly allocated to treatment groups, PEDV negative (PN) and PEDV positive (PP), and housed in a research facility. Fecal swabs from all pigs tested PEDV negative by PCR on arrival and at challenge. At 36 days of age, the PN pigs received 10mL of PEDV negative cell culture and PP received 10mL of 10^4TCID50/ml PEDV positive cell culture via oral gavage. Four days post-inoculation, each pig was necropsied and nine, 1-cm small intestinal samples were collected and fixed. Figure 1 displays the sample collection points for the W method (red) and L method (green). A VCR, atrophic enteritis score (AE), and IHC score was determined for each sample by a diagnostician blinded to treatment allocation.

Results and discussion

SAS (v9.4) was used to create three generalized linear mixed models for AE, IHC, and VCR as response variables and with groups (PN and PP) and intestinal segment as explanatory variables. With the exception of W1 (duodenum) ($P = 0.0995$), IHC scores for PP were moderate to high and significantly greater ($P < 0.0012$) than scores for the PN group whose mean IHC score was 0.

For PP, IHC values significantly decreased between the proximal jejunum and distal segments within both the L and W methods ($P = 0.0309$). VCR values were significantly lower ($P < 0.05$) for PP than PN for jejunal segments only, but there was no significant difference in VCR values between those segments (all $P > 0.05$). For all 9 segments, the AE scores were significantly higher in the

Figure 1: Sampling locations within the small intestine for the W and L collection methods.

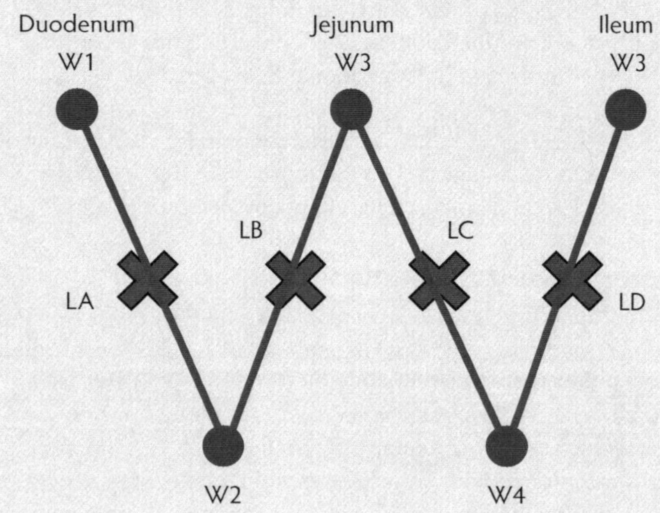

PP group versus the PN group (all $P \leq 0.0181$). Within both L and W methods, the AE values decreased from proximal jejunum to ileum; this change was statistically significant ($P = 0.0053$ for L and $P = 0.0164$ for W).

This study demonstrates that IHC, AE and VCR scores do differ significantly between PEDV positive and PEDV negative nursery pigs at 4 days post-infection. With immediate tissue fixation, samples from the proximal to distal jejunum yielded the highest IHC and AE scores and lowest VCR ratios, suggesting that sampling be concentrated to this area. Increasing the number of samples did not improve detection in this model likely due to uniform exposure. In the field when infection is unknown and sampling conditions are suboptimal, sampling several jejunum pieces would increase the likelihood of identifying lesions and pathogens in acute exposure.

Effects of multi-dose iron dextran and vitamin E on hemoglobin concentration and weight gain in neonatal pigs

S. Foster[1], BS; R. Berghaus[2], DVM, MS, PhD, ACVPM; R. Jones[3], DVM; B. Heins[3], DVM, MFAM
[1]University of Georgia CVM, DVM, Athens, Georgia; [2]Department of Population Health, University of Georgia CVM, Athens, Georgia; [3]Livestock Veterinary Services, Kinston, North Carolina

Introduction

Iron is commonly administered to piglets shortly after farrowing to prevent iron deficiency and consequent anemia. Despite decades of research regarding the use of iron dextran in confinement-reared swine, there remains quite a bit of variability with regard to the absorption and utilization of iron dextran as measured by hemoglobin concentration. An additional iron dextran injection before weaning has been shown to improve the performance of nursery pigs.[1,2] Vitamin E deficient pigs have a lower hemoglobin concentrations, resulting in decreased iron utilization.[3] This study investigated the use of vitamin E as a means to improve the absorption and utilization of iron in neonatal swine. The objectives of the study were to 1) determine if the industry standard iron administration is sufficient to meet the needs of today's neonatal pigs, 2) determine if vitamin E supplementation at processing increased hemoglobin concentration and average daily gain (ADG), and 3) determine if an additional iron injection at weaning increased hemoglobin concentrations and ADG three weeks post-weaning.

Materials and methods

Pigs (n = 388) were enrolled into one of four treatment groups (A-D). All groups received 1cc (200 mg/ml) iron dextran (Uniferon®, Pharmacosmos) at processing (< 24 hrs. post-farrow). Group A was the control group. Groups B and D were supplemented with 250 IU vitamin E (VITAL E®–Newborn, Stuart Products) at processing. Group C and D received an additional 1cc iron dextran at weaning (15-21 days post-farrow). All pigs were weighed at processing, weaning, and 22 days post-weaning. Blood was collected at weaning and 22 days post-weaning from all the pigs for hemoglobin concentrations. Vitamin E analysis was performed on ten randomly chosen blood samples from each treatment collected at weaning. Post-hoc pairwise comparisons were performed using the Boferroni procedure.

Results and discussion

There were no differences between the groups at weaning with respect to either ADG ($P = 0.190$) or hemoglobin concentration ($P = 0.917$). At 22 days post-weaning, groups A and B had hemoglobin concentrations significantly lower than both groups receiving additional injections of iron dextran (C&D) ($P < 0.001$). However, there was no significant difference between treatments with respect to ADG at any time ($P = 0.386, 0.186$) (Table 1). No difference was found from serum vitamin E concentrations between treatments ($P = 0.227$). From these results, it appears that 200 mg iron dextran is sufficient to meet iron requirements of neonatal pigs. It is possible that the difference in length of time of weaning and the timing of the second iron injection may have an impact on performance. It is likely that nursing piglets received adequate levels of vitamin E from the sow's milk and the additional supplementation did not improve performance. More research should be performed to investigate if pigs could benefit from vitamin E supplementation at weaning. A larger sample size may be needed to show significance.

References

1. Haugegaard J et al. Effect of supplementing fast-growing, late weaned piglets twice with 200 mg iron dextran intramuscularly. *The Pig Journal*. 2008; 61: 69-73.

2. Kamphues J et al. Effects of a 2nd iron injection in suckling piglets on iron retention and performance before and after weaning. In: Proc. 12th IPVS Cong., August 1992, The Hague, The Netherlands, p. 601.

3. Nafstad I. Studies of hematology and bone marrow morphology in vitamin E-deficient pigs. *Pathologia veterinaria*. 1965; 2: 277-287.

Table 1: Mean ADG and hemoglobin (Hb) concentrations for weanling pigs.

Group	†Mean ADG processing to weaning (lbs)	†Mean ADG processing to 22 days post-weaning (lbs)	†Mean [Hb] at weaning (g/dl)	†Mean [Hb] at 22 days post-weaning (g/dl)
A	0.546[a]	0.508[a]	10.93[a]	9.53[a]
B	0.530[a]	0.511[a]	10.96[a]	9.61[a]
C	0.569[a]	0.524[a]	10.94[a]	10.10[b]
D	0.548[a]	0.496[a]	10.86[a]	10.08[b]

† Within columns, estimated marginal means with a superscript in common do not differ with a level of significance of 5% over all comparisons.

Impact of grain particle size reduction technologies on pig performance and economics

Jordan Gebhardt, BS; Steve Dritz, DVM, PhD; Mike Tokach, PhD; Joel DeRouchey, PhD; Robert Goodband, PhD; Jason Woodworth, PhD; Charles Stark, PhD; Cassandra Jones, PhD
Kansas State University, Manhattan, Kansas

Introduction

Grinding grain is a critical factor in swine production profitability due to its impact on nutrient digestibility, directly impacting feed cost per pig which accounts for 2/3 of the cost of production. A 2-high roller mill or hammer mill has been the standard to grind corn; however, it is difficult to achieve high throughput and a consistent particle size below 600 µm. Therefore, many feed mills in the US have installed 3-high roller mills to more consistently achieve target particle sizes. A limited number of feed mills are using 4-high roller mills to achieve even finer and more consistent particle size targets. Unfortunately, there is limited evidence that 3- or 4-high roller mills will improve income over feed costs (IOFC) . In addition, an increased prevalence of gastric ulcers has been observed in field and university-based studies when particle size is reduced below 600 µm, further complicating this issue. Therefore, the objective of this study was to evaluate the effects of different roller mill configurations on nursery and finishing pig performance, feed mill throughput and electricity costs, and IOFC.

Materials and methods

The same diets were fed in each experiment but with the corn fraction processed using: 2 sets of rolls (2-high, approximately 667 µm), 3 sets of rolls (3-high, approximately 536 µm), 4 sets of rolls in a fine grind configuration (4-high fine, approximately 350 µm), or 4 sets of rolls in a coarse grind configuration (4-high coarse, approximately 478 µm). All corn was processed in a commercial feed mill using the same 4-high roller mill, but opening the various pairs of rollers to create the 4 experimental treatments. In Exp. 1, 320 nursery pigs were randomly allotted to 16 pens per treatment for a 21-d growth trial. In Exp. 2, 90 nursery pigs were used in a feed preference trial. In Exp. 3, 922 finishing pigs were used in a 97-d experiment, with pens allotted to 1 of the 4 treatments with 11 pens per treatment in a randomized complete block design. Data were analyzed using a general linear mixed model.

Results and discussion

In Exp. 1, there were no differences ($P > 0.05$) in nursery pig ADG, ADFI, F/G, or IOFC between roller mill configurations (Table 1). In Exp. 2, when given a choice, nursery pigs preferred ($P < 0.05$) diets manufactured with corn ground to a coarse particle size (667 vs 350 µm, and 478 vs 350 µm). In Exp. 3, finishing pigs fed diets containing corn ground with the 2-high configuration had the greatest ($P < 0.05$) ADFI and ADG, and pigs fed diets using the 4-high fine configuration had the lowest ADFI and ADG. There were no differences ($P > 0.05$) in F/G, caloric efficiency, or carcass characteristics. There was no change in ulcer prevalence, with only 3.4% removals and only 1 pig removed with an ulcer. Feed cost/lb of gain was lowest ($P < 0.05$) for 4-high coarse and revenue/pig was greatest ($P < 0.05$) for the 2-high and 4-high coarse configurations. Income over feed cost was lowest ($P < 0.05$) for the 4-high fine configuration ($68.23); however, there were no differences ($P > 0.05$) in IOFC among the other milling configurations ($70.99, $70.06, $71.70; 2-high, 3-high, 4-high coarse, respectively). The 2-high configuration had the lowest ($P < 0.05$) grinding electricity cost. Mill throughput (ton/hr) was greatest ($P < 0.05$) for the 4-high coarse and 2-high configurations. Consistent with other recent studies, fine ground corn (< 600 µm) in meal-based nursery and finishing diets reduced feed intake and growth rate without improving feed efficiency. Diets with corn ground to fine particle sizes generally have reduced flowability, which can lead to bridging in bulk bins and out of feed events. Therefore, in this study there was no advantage to installing a 4-high roller mill for reducing corn particle size below 600 µm.

Table 1: Impact of grain particle size reduction technologies on pig performance and economics

Item:	Roller mill configuration[1]				SEM	Probability, $P <$
	2-high	3-high	4-high fine	4-high coarse		
Mill economics						
Grinding rate, ton/hr[2]	13.39[a]	12.07[b]	8.24c	14.46[a]	0.449	0.001
Electricity cost, $/ton[3]	0.366[d]	0.541[c]	1.011[a]	0.612[b]	0.0321	0.001
Exp. 1, Nursery performance[4]						
ADG, lb	1.33	1.33	1.28	1.32	0.023	0.479
ADFI, lb	2.06	2.04	1.98	1.98	0.038	0.350
F/G	1.55	1.54	1.54	1.50	0.015	0.122
IOFC, $[5]	11.98	12.01	11.61	12.10	0.219	0.412
Exp. 2, Nursery preference, ADFI, lb[6]						
2-high vs. 4-high fine	1.15	---	0.57	---	0.069	0.001
2-high vs. 4-high coarse	0.89	---	---	0.87	0.058	0.778
4-high fine vs. 4-high coarse	---	---	0.63	1.10	0.084	0.003
Exp. 3, Grow/finish performance[7]						
ADG, lb	2.18[a]	2.13[b]	2.07[c]	2.14[a,b]	0.014	0.001
ADFI, lb	6.20[a]	6.03[b]	5.83[c]	6.02[b]	0.047	0.001
F/G	2.85	2.83	2.82	2.81	0.013	0.147
IOFC, $[5]	70.99[a]	70.06[a]	68.23[b]	71.70[a]	0.571	0.001

1 Corn was ground using 2 sets of rolls (approximately 667 µm), 3 sets of rolls (approximately 536 µm), 4 sets of rolls in a fine-grind configuration (approximately 350 µm), or 4 sets of rolls in a coarse-grind configuration (approximately 478 µm).
2 Grinding rate = tons corn ground/hour.
3 Electricity cost = $/ton ground corn. Cost of electricity was fixed at $0.13/kWh.
4 A total of 320 nursery pigs (DNA 400 × 200, initial BW=24 lb) were used in a 21-d study with 5 pigs per pen and 16 replications per treatment.
5 Income over feed cost (IOFC) = total revenue/pig – feed cost/pig.
6 A total of 90 pigs (PIC 327 × 1050, initial BW=27 lb) were used in a 7-d preference trial with 5 pigs per pen and 6 replications per comparison.
7 A total of 922 finisher pigs [PIC TR4 × (FAST Large white × PIC Landrace), initial BW = 88.3 lb] were used in a 97-d study with 21 pigs per pen and 11 replications per treatment.
a,b,c Means with unlike superscripts differ ($P < 0.05$).

Salmonella enterica: A case study of enterocolitis in Oklahoma and Iowa swine herds

Kylie Glisson; Amelia Naher, DVM; Kate Dion, DVM; Nathan Schaefer, DVM; Jessica Seate, DVM

Salmonella enterica spp. is a common etiologic agent of clinical disease in swine and potential source of contamination of pork products. The two most common serovar in swine include *Salmonella enterica* serovar Cholerasuis (serogroup C1) or *Salmonella enterica* serovar Typhimurium (serogroup B), both of which can cause enteritis or enterocolitis in growing pigs. Sources of *Salmonella enterica* are spread via the oral-fecal route, with clinical *Salmonella enterica* outbreaks often occurring after periods of increased stress. However, clinical signs may vary depending on virulence, dose, and resistance, with some exposed populations of swine appearing asymptomatic.

Case history & statement of problem

This investigation followed a flow of pigs from a 10,000 head sow farm in OK. The pigs are weaned into facilities both in IA and OK. Farms in OK are company-owned, shower in, 10,000 head, hotel style nurseries. In comparison, farms in IA are contract, 2200-8800 head wean-to-finish farms of various barn styles. Although all farms follow a similar vaccination and treatment plan, the pigs that are sent to IA routinely exhibit clinical enterocolitis due to *Salmonella* spp. serogroup B, while OK farms remain asymptomatic. The objective of this study was to investigate and describe epidemiologic characteristics of *Salmonella enterica* and evaluate associations between management and disease occurrence in downstream flows.

Materials and methods

Groups of pigs were tested in IA and OK at 3, 6, 9, 12, 16, 20, and 24 weeks of age. From each age group, six fresh free-catch fecals, and fifteen individual serum samples total were collected randomly throughout each enrolled barn to determine the exposure, prevalence and serogroup of *Salmonella enterica* at the site. Serum samples were submitted to the Boehringer Ingelheim Health Management Center (HMC; Ames, IA) and tested via the VETSIGN™ *Salmonella*-Ab ELISA (Svanova, Uppsala, Sweden). Fecal samples were submitted to the Iowa State Veterinary Diagnostic Laboratory (Ames, IA), for double enrichment culture, with positive cultures selected for *Salmonella enterica* sero-grouping. In addition, a questionnaire to evaluate differences in management was developed and administered to veterinarians and production managers in both IA and OK.

Results

The serological response to *Salmonella* in both the IA and OK flows demonstrated maternal antibody decay, (detectable peak at 3 weeks of age) with the majority of pigs (86.7% and 93.3%) ELISA negative by 9 weeks of age (Figure 1). Following field *Salmonella* exposure, the IA flow had peak serological response at 16 weeks of age (93.3% ELISA positive), indicating peak environmental exposure at 11 weeks of age. In contrast the OK flow had peak serological response at 20 weeks of age (93.3% ELISA positive), indicating peak environmental exposure at 15 weeks of age. Fecal cultures isolated *Salmonella* spp. serogroup B in IA pigs at 3, 6, and 12 weeks of age. *Salmonella* spp. serogroup B was isolated in OK pigs at 9 weeks of age only, despite expected positives at 12 and 16 weeks of age. These age groups are being resampled and *Salmonella enterica* sero-grouping results are pending.

The management questionnaire highlighted contract vs. owned facilities, flooring type, field managers, geography, pit depth, and manure handling as potential management practices that increase *Salmonella* environmental dose and timing of exposure.

With both flows demonstrating *Salmonella* spp. serogroup B field exposure, a combination of management practices aimed at reducing environmental exposure and early pig vaccination with a *Salmonella* vaccine containing *Salmonella enterica* serovar Typhimurium prior to fecal antigen exposure is recommended.

Figure 1: Summary of results from the VETSIGN™ *Salmonella*-Ab ELISA in both the Iowa and Oklahoma flows (HMC).

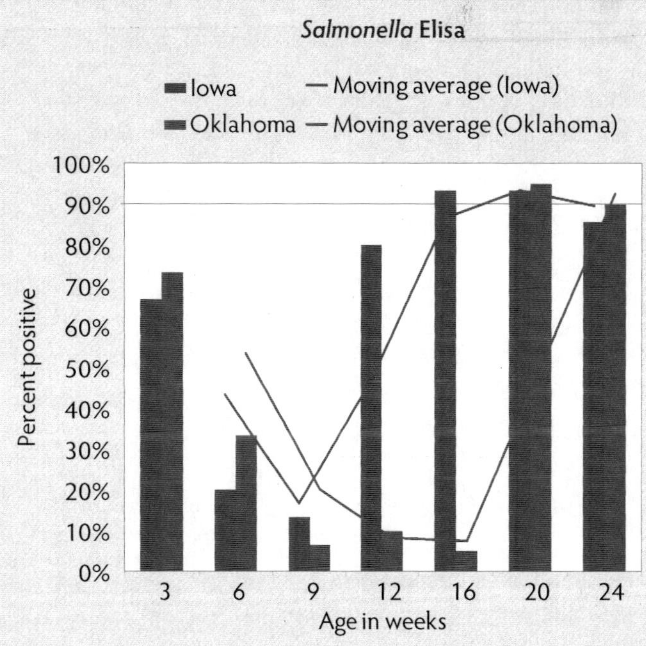

Evaluation of iron injection protocols in piglets and their effect on wean weights and hemoglobin levels

Tiffany Haegele[1], BS; Ashley DeDecker[2], PhD; Glen Almond[1], DVM, PhD
[1]North Carolina State University College of Veterinary Medicine, Raleigh, North Carolina;
[2]Smithfield Foods, Inc.

Introduction

The US market is reaching higher production goal as pork is becoming more popular. Producers are looking to find ways to increase market weights and production efficiency. Anemia in piglets is a well-known issue for producers due to sows housed on slates and inadequate milk transfer of iron for piglets. Current operating procedure on most farms is a single injection of iron dextran 100-200mg prior to day 7 post partum. The goal of this study was to challenge this SOP by administering iron dosages varying from 150mg-300mg at different time points post partum and tracking weights of the piglets post weaning. The goal of the study is to determine optimum dosages, frequency, and day of administration to improve iron stores and subsequently weaning weights through a stronger immune system and increased feed efficiency.

Materials and methods

650 sows were selected at random into the study and were closely monitored as they were moved from gestation to farrowing. Farrowing date and time was closely monitored so piglets could be enrolled within 24hrs of parturition. Piglets in each litter were assigned to a treatment group based on weight and then rotated for each subsequent litter to get equal representation. The heaviest piglet in the first group was assigned to treatment group 1, the second heaviest to group 2, the third heaviest to group 3, and the fourth heaviest to group 4. After the fourth piglet the 5th heaviest piglet was assigned to group one, etc. until all the piglets were enrolled. Enrollment consisted of placing a colored ID tag in the left ear that correlated with the treatment group, an RFID tag in the right ear, an injection of penicillin IM (as per the SOP of the farm) and a dose of iron dextran IM based on the treatment group. Treatment group 1 received 150mg of iron within 24 hours post partum and another dose on day 3. Group 2 received 200mg of iron on day 3. Group 3 received 300mg of iron on day 3. Group 4 received 200mg on day 3 and at weaning. At weaning the piglets were weighed and those in treatment group 4 received an addition iron dose injection.

Blood samples were taken from a subsample of piglets at weaning then at 3 weeks and 5 weeks post placement in the nursery. A total of 48 samples per breeding group per wean week were collected. Hemoglobin levels were determined using a HemoCue analyzer.

Results and discussion

Over 4000 pigs were enrolled in into the four treatment groups. There was no significant difference between the hemoglobin levels among the pigs of different groups. Analysis showed that there were no significant differences between the wean weights among the pigs in different groups. The data concludes that administration of iron at varying dosages, frequencies, and timing post partum showed equivalent results to the current SOP administration of iron. It cannot be determined that changes to these protocols will effectively improve nursery growth and performance. At this time farms may choose to use any of the four methods described that suit their SOP and yield similar results.

Improved influenza A virus isolation using non-woven fabric for nasal wipe collection

C. T. Hammons, BS; N. Bliss, MS; J. M. Nolting; A. S. Bowman, DVM, PhD
Department of Veterinary Preventive Medicine, College of Veterinary Medicine, The Ohio State University, Columbus, Ohio

Introduction

Influenza A virus (IAV) is a burden on swine production, as infections result in respiratory illness, high fevers, abortions, and deaths. Control of IAV is difficult due in part to the rapid, ongoing evolution of these viruses in swine populations. Genomic reassortment between IAV strains is common in pigs, which allows for novel strains to rapidly emerge and cause disease.[1] With the ability to be transmitted through droplets and direct contact, IAV is able to rapidly spread through immunologically naive swine populations. Additionally, the virus has been known to become endemic in some herds. Viral surveillance is used to monitor for new IAV introductions, estimate vaccine effectiveness, and assess overall herd health. Current antemortem sample collection methods used for IAV surveillance among swine include oral fluids, nasal swabs, and nasal wipes.[2,3] Nasal swabs are the well-established gold standard IAV collection method.[4] Although oral fluids allow for sampling groups of pigs without restraint, oral fluids are less sensitive for viral culture and virus sequencing than nasal swabs. While not as sensitive as the gold standard nasal swab, nasal wipes allow veterinarians to test pigs without restraint and produce a sample suitable for viral isolation and viral genome sequencing.[5] Although cotton gauze is commonly used as a nasal wipe substrate, preliminary data suggested that other wipe materials could improve IAV recovery. The objective of this study was to investigate the potential for improvement of IAV isolation from nasal wipe specimens by testing commercially available fabrics.

Materials and methods

Six non-woven polyester fabrics, blinded as A through F, were compared to cotton gauze by testing for molecular IAV detection and viable IAV recovery. Three 25.08 cm^2 swatches of each fabric were inoculated with 0.20 mL contemporary swine-origin H3N2 (2.01×10^8 $TCID_{50}$/swatch), placed into vials containing 5 ml viral transport media, and frozen at -80°C until tested with quantitative reverse transcription polymerase chain reaction (qRT-PCR) and inoculated into Madin Darby canine kidney cells for tissue culture infective dose ($TCID_{50}$). Pairwise comparison of viral culture ($TCID_{50}$) and qRT-PCR (copy number) results between fabrics was performed using Mann-Whitney rank sum.[6,7]

Results and discussion

While there were no significant differences in IAV detection with qRT-PCR between cotton gauze and the non-woven fabrics, Fabrics A, B, and C yielded significantly more viable IAV than cotton gauze ($P \leq 0.05$). Fabric A also yielded significantly more viable IAV than both Fabrics B and C ($P = 0.046$). Our results demonstrated that Fabric A was significantly better than cotton gauze for recovery of infectious IAV *in vitro.* The findings from this study facilitate improved nasal wipe sampling methods for IAV detection in pigs. Practitioners collecting nasal wipes for IAV testing can expect improved viral isolation rates and better sequencing success by using Fabric A instead of cotton gauze.

References

1. Ma W et al. The pig as a mixing vessel for influenza viruses: human and veterinary implications. *Journal of Molecular and Genetic Medicine.* 2008; 3 (158-166).

2. Detmer SE et al. Detection of influenza A virus in porcine oral fluid samples. *Journal of Veterinary Diagnostic Investigation.* 2011; 23: 241-247.

3. Dobesh K et al. Utilizing snout wiping for PCR detection and virus isolation of SIV in nursery pigs. 44th Annual Meeting of the American Association of Swine Veterinarians. 2013; 69-70.

4. Van Reeth K et al. Influenza virus; in Zimmerman JJ (ed.): *Diseases of Swine, 10th edn.* Wiley-Blackwell, 2012; 557-571.

5. Edwards JL et al. Utility of snout wipe samples for influenza A virus surveillance in exhibition swine populations. Influenza and Other Respiratory Viruses. 2014; 8(5), 574-579.

6. Reed LJ and Muench H. A simple method of estimating fifty per cent end points. *American Journal of Epidemiology.* 1938; 27:493-497.

7. Mann HB and Whitney DR. On a test of whether one or two random variables are stochastically larger than the other. *The Annals of Mathematical Statistics.* 1947; 18(1): 50-60.

Determination of a maternal antibody decay curve for influenza A virus in swine

Henry Johnson[1], BS; Joel Nerem[2], DVM; Tom Wetzell[3], DVM; Scott Vander Poel[2], DVM; Adam Schelkopf[2], DVM
[1]University of Illinois, Urbana, Illinois; [2]Pipestone Veterinary Services, Pipestone, Minnesota;
[3]Boehringer Ingelheim Vetmedica, Saint Joseph, Missouri

Introduction

The ability of Influenza A Virus in swine (IAV-S) to infect and undergo reassortment within all ages has caused significant respiratory disease problems and economic loses. IAV-S vaccine implementation in piglets could serve to greatly reduce clinical signs, shedding duration, and overall viral load during a disease outbreak. An accurate and reliable maternal antibody decay curve will provide valuable information as vaccination timing strategies are implemented. The objective of the study was to build on previous work to compare IAV-S maternal antibody rate of decay across multiple farms of similar IAV-S status. (Fleck, 2002).

Materials and methods

Four separate groups of animals from four sow farms were selected for the study. Four Oral Fluid (OF) samples were collected by each farm at the beginning of each month and submitted for routine IAV-S PCR testing. Each farm had seven consecutive months of OF wean age piglet samples that were IAV-S Polymerase Chain Reaction (PCR) negative. The four groups of animals were also housed in growing barns that were single sourced from the identified IAV-S stable sow farm and follow an all-in, all-out movement to minimize lateral IAV-S exposure. Thirty-six animals consisting of 1 male and 1 female per litter, from each group were selected randomly across parity 1, 2-3, and 4 or greater. Each animal was individually identified with an ear tag. Serum and nasal swab (Copan nylon flocked swab transported in Amies Media, Copan USA) samples were collected from each identified animal at 3-5 days of age and the day prior to weaning. Serum was collected from the identified animals at 6, 9, 12, and 15 weeks of age. OF samples were collected from pens at time of serum collection. Sample testing was done at Iowa State University Veterinary Diagnostic Lab and the Boehringer Ingelheim Health Management Center Diagnostic Laboratory in Ames, IA. Nasal swab and OF samples were tested for the presence of IAV-S via PCR to identify potential IAV-S circulation that could occur around sampling events. Serum titers were individually tested for IAV-S nucleoprotein (NP) ELISA and Hemagglutination Inhibition assay for H1N1 and H3N2.

Results

Sample results were not available yet at the time of this writing but will be by publication time. Results will be evaluated with a logistic regression analysis at the end of the study.

Discussion

Many production systems across the industry are exposed to IAV-S periodically throughout the year, especially during the fall months. Providing an accurate IAV-S maternal antibody titer decay curve will help veterinarians further understand passive immunity for IAV-S. The decay curve will also serve as a baseline for potential vaccine interference if an IAV-S vaccine becomes a viable strategy in the future to increase immunity against IAV-S.

Sow lameness scoring: Reliability and correlations with other indicators of lameness

Jane Karpowicz; Meghann K. Pierdon, VMD; Thomas D. Parsons, VMD, PhD, DACAW

Statement of the problem

Lameness is of great importance to the swine industry because it contributes to economic losses associated with decreased production and involuntary culling as well as negatively impacting sow welfare. The reliability of one popular sow lameness scoring tool was only fair when quantified.[1] Thus, the development of a simple yet effective sow lameness scoring tool is need for both swine producers and researchers.

Objective

To establish a locomotion scoring system that is reliable between observers and test whether that scoring method correlates with other indicators of lameness.

Materials and methods

At the Penn Vet Swine Teaching and Research Center sows were housed in a large pre-implantation dynamic pen fed by 2 electronic sow feeders. Over 200 sows were motion scored at the walk on a solid concrete area by 3 trained scorers using a newly devised locomotion scoring system. The new scale ranged from 1 to 3 where a score of 1 was normal, a score of 2 indicated a moderate lameness, and 3 indicated a severe lameness. After gestation, sows farrowed in a 7 × 7 ft hinged style farrowing crate that is open until the sows are at day 113 of gestation. This time in the open hinged crate prior to farrowing was used to video record 26 sows on their 112th day of gestation. These sows in farrowing were motion scored on the new scale and their foot lesions were scored using Zinpro's Feet First® scale. Video footage was coded using Noldus Observer XT 11.5 and included the time spent lying down, sitting, standing, walking, eating, drinking and nosing. Farrowing production data was collected 2-4 days post farrowing. Stata v. 13.1 was used to calculate Cohen's Kappa (K), to compare non-normally distributed behavioral data with the Kruskal Wallis non-parametric test, as well as for point bi-serial correlations.

Significant results

Overall there was moderate reliability between observers using a new 1-3 scale for all three locomotion scores (K = 0.50). Substantial reliability for a locomotion scores of 3 was observed (K = 0.66), while moderate reliability was found for scores 1 (K = 0.54) and 2 (K = 0.44)(Table 1).[2] The percent agreement between observers, across all scores, was 80.5%. Reducing the lameness scoring tool to a 2 point scale - lame sows (score > 1) and non-lame (score ≤1) did not appreciably improve the overall Kappa as it remained at moderate reliability (K = 0.54)(Table 1). The video data showed that the time spent walking by lame sows was significantly less than non-lame sows ($H(1)=3.97$, $P<.05$). There was also a significant correlation between being motion scored as lame and having higher foot lesion scores ($R_{pb}=0.389$, $P < 0.01$). There was no correlation found between farrowing productivity and lameness scores.

Discussion of how results can be applied by practitioners

A reliable locomotion scoring system that is both universal and easy to implement is important in order to treat sows effectively and to conduct useful research. Our results indicate that our 3 point scoring scale provides an overall moderate agreement between observers and is shown to be a simple and fairly effective method of assessing lameness. The correlation between lame sows and higher total claw lesions and the fact that the lame sows were shown to spend less time walking, demonstrates that our motion scoring system correlates with both anatomical and behavioral indicators of lameness. The scale however still yielded only moderate agreement between observers and work should continue to improve the reliability of subjective measures of lameness. The use of a reliable locomotion scoring system in sows promises a better understanding of how lameness is affecting productivity and improves our ability to diagnose and treat lameness.

References

1. Pierdon, M., Parsons, T. Reliability of a Four Point Sow Lameness Score Across Multiple Observers, *Leman Conference Proc.* St.Paul, MN; 2015.

2. Landi JR, Koch GG. The measurement of observer agreement for categorical data, *Biometrics.* 1977; 33:159-74.

Acknowledgements

This work was funded by the Pennsylvania Pork Producers Council, NY farmers and Zinpro Performance Minerals.

Table 1: Kappa levels for all lameness categories

Lameness category	Kappa	Kappa level	*P*
1	.54	Moderate	< .001
2	.44	Moderate	< .001
3	.66	Substantial	< .001
Lame/not lame	.54	Moderate	< .001

Porcine epidemic diarrhea: The value of repeat testing

C. Mesquita[1]; K. Kuruppu[2]; S. Burlatschenko[3], DVM
[1]Veterinary Science and Policy, OMAFRA, Elora, Ontario, Canada; [2]Ontario Veterinary College, Ontario, Canada;
[3]Goshen Ridge Veterinary Services, Tillsonburg, Ontario, Canada

Porcine epidemic diarrhea (PED) is in decline in Ontario. Over half of the herds infected with PED have eliminated the virus by creating herd immunity and preventing re-infection through enhanced biosecurity protocols, cleaning and disinfection.

Herds that are unsuccessful at eliminating PED experience a chronic form of the disease, where clinical signs sporadically reappear in suckling and weaned pigs. Monitoring these farms using PCR testing is used to identify locations where virus is persisting. Funding to Ontario Pork from the Ontario Ministry of Agriculture, Food offsets the cost of laboratory fees for herds working to eliminate PED. This enhanced monitoring for PED has occasionally resulted in some unexpected findings.

For example, Farm A is a single site farrow to finish operation that continued to experience chronic PED symptoms 10 months after the initial infection. The entire sow herd was re-exposed to virus prior to a second attempt at virus elimination. Weekly sampling was performed in all production areas following the second virus exposure. Four weeks post-exposure, the farm began to wean PCR negative pigs. These pigs were placed in clean and disinfected nursery rooms. Weekly monitoring indicated these pigs remained PCR-negative. Throughout the monitoring period, the finishing area, which consisted of 12 separate rooms, consistently tested PCR-positive in half or more of the rooms. The results for three weeks of testing in the finishing rooms are shown in Table1. The large decline in positive test results between May 22 and June 2 was unexpected. Retesting on June 9 was consistent with a significant decline in PCR-positive findings. The owners reported no difference in clinical signs between the test-positive and test negative weeks.

PCR test results for PED from one week to the next varied more than expected on this farm. This case shows that it may be difficult to predict virus survival in a production area based on a single set of samples. Had only the samples been taken on May 22, five rooms would have been assessed as positive, yet they had discordant results on subsequent testing.

Serial testing provides a more representative sample of the PED status in a herd for veterinarians and producers during an elimination process.

Table 1: PED-PCR results from samples taken in 12 finishing rooms on Farm A over a three-week period.

	Ct value*		
Room #	**22-May**	**02-Jun**	**09-Jun**
1	> 40.00	> 40.00	> 40.00
2	> 40.00	> 40.00	> 40.00
3	> 40.00	> 40.00	> 40.00
4	**32.67**	> 40.00	> 40.00
5	**35.38**	> 40.00	> 40.00
6	**35.41**	> 40.00	> 40.00
7	**26.93**	> 40.00	**26.75**
8	**33.03**	> 40.00	> 40.00
9	> 40.00	> 40.00	> 40.00
10	> 40.00	> 40.00	> 40.00
11	**37.09**	> 40.00	> 40.00
12	> 40.00	> 40.00	> 40.00

* A Ct value < 37 is a positive test, 37-39.99 is suspicious and > 40 is not detected.

See one, do one, teach one: Increasing swine disease response and diagnostic capacity by using a pyramid based training technique to get boots on the ground

N. Lauterbach, BS; A. Canon, DVM, MPH; J. Bates, DVM; K. Skoland, BS; C. Ruston, BS; J. Ellingso, DVM, MS; P. Thomas, DVM, MS; N. Hershberger, BS; T. Te Grotenhuis, BS; K. Hayman, BS; M. Breuer; A. Buckley, DVM; A. Koethe; K. Gerardy, BS; D. Holtkamp, DVM, MS; L. Karriker, DVM, MS, DACVPM
Swine Medicine Education Center, Iowa State University, Ames, Iowa

Problem and objective

Few swine-focused veterinarians would be available to respond during a transboundary, emerging, or zoonotic disease outbreak. In the United States, 65.4 million pigs on > 63,000 farms creates a farms-to-veterinarian ratio of 75:1. Deploying non-swine veterinarians to collect diagnostic samples during an outbreak could be necessary to prevent disastrous economic loss. The objective of this study was to address the critical need for rapidly deployable diagnostic collection and disease response by developing and evaluating a train-the-trainer teaching mechanism that could be used to efficiently train non-swine focused veterinarians and production personnel on collection of swine diagnostic samples.

Materials and methods

A one day veterinary student wetlab was divided into three phases: pre-instruction, instruction, and post-instruction. Pre-instruction: participants did not receive training, but were given necessary and distractor supplies to collect blood, oral fluids, nasal swabs, and feces. They were evaluated clinically and by written assessment, resulting in a pre-instruction score. Instruction: students viewed instructional videos and participated in hands-on, pyramid based clinical instruction, initiated by a single, competent veterinarian. At each pyramid level, each teacher demonstrated and each student practiced sample collection. Post-instruction: analogous to pre-instruction, students were given the same supplies and asked to collect the same samples. Clinical evaluation was conducted by the same evaluators; two days after the wetlab, participants took a similar written quiz assessing retention and resulting in an overall post-instruction score.

Results

Approximately half of the participants (22/45) had no previous swine experience. During pre-instruction, only two participants were able to demonstrate proper blood collection technique; during post-instruction, 27 participants demonstrated proper blood collection, and all 45 selected appropriate supplies and indicated correct needle placement. During pre-instruction, few could collect adequate fecal (12/45) or nasal (9/45) samples; after instruction, all 45 participants were able to successfully collect those samples. During pre-instruction, only two groups (three participants per group) successfully collected oral fluids; after instruction, all 15 groups were able to successfully collect oral fluids. No participant recognized the correct techniques through photos and written assessments pre-instruction, but participants saw an average improvement of 48% in their written assessments post-instruction, indicating improvement in knowledge and retention.

Application to practitioners and the US swine industry

In the case of a transboundary, emerging, or zoonotic disease, this training method could alleviate the strain put on swine-focused veterinarians and conserve resources, while quickly and efficiently getting boots on the ground to collect necessary diagnostic samples. While tested with swine, this template could be applied to any food animal species for any disease outbreak response. This template was capable of starting with a single competent trainer and rapidly expanding the skill pool. The ability to get trained boots on the ground quickly is a critical step in being prepared for the next threat to the swine industry and to US animal agriculture.

The effects of different vaccination dosing and timing protocols on porcine circovirus type 2 viremia and post vaccination serologic responses

Claire LeFevre[1], BS; Emily Byers[2], DVM
[1]University of Wisconsin School of Veterinary Medicine, Madison, Wisconsin;
[2]Smithfield Hog Production Division, Warsaw, North Carolina

Introduction

Porcine circovirus type 2 and *Mycoplasma hyopneumoniae* (M hyo) continue to cause significant economic losses to the US swine industry. Circumvent® PCV-M G2 (Merck Animal Health, Summit, NJ) provides protection against both of these pathogens while allowing producers to choose from two labeled dosing options. Producers may choose to administer a single, 2 mL intramuscular dose at 3 weeks of age or older or administer two doses allowing a lower-volume initial dose of 1 mL intramuscularly as early as 3 days of age followed by a second 1 mL dose 3 weeks later. Several studies have been performed that evaluate the effect of porcine circovirus type 2 (PCV2) vaccination dose and timing on viremia following vaccination. In 2015, Crozier et al found that decreasing the interval between PCV2 vaccinations in a two dose regimen to 8 or 11 days resulted in higher viremia levels following field exposure.[1] The objective of this study was to determine the impact of different vaccination dosing and timing protocols on PCV2 viremia and post vaccination serologic responses.

Materials and methods

This study was conducted in four separate flows within a large production system. One flow was a parity 0 to 1 sow farm that provided sows for the three other sow farms. Pigs (n = 521) were randomly assigned by weight within litter to treatment groups just prior to weaning at 21-28 days of age. All pigs received the total 2 mL labeled dose of PCV-M G2 but with different dosing and timing protocols as shown in Table 1. Upon placement in the finisher, pigs were evenly dispersed throughout the barn by treatment group and mixed with general population non-study pigs. Pigs were weighed at allocation, at 9 weeks of age (WOA), and just prior to marketing at 24 WOA. Blood samples were collected at allocation and again at approximately 7, 9, 16, and 24 WOA. Serum was tested for PCV2 viremia by PCR in pools of five at the Iowa State University Veterinary Diagnostic Lab. Environmental challenge was assessed by monitoring oral fluids for PCV2 by PCR at each blood sampling. Serum was tested for PCV2 antibodies by IFA, reported as the reciprocal of the geometric mean titer and M hyo antibodies by ELISA performed at 3, 7, and 9 WOA.

Results

No pigs were viremic at 3, 7, 9, 16, and 24 WOA. Oral fluid rope samples at 7, 9, 16, and 24 WOA were PCV2 PCR positive but at a relatively low level. Serological results are shown in Table 1. Pigs vaccinated at 3 and 7 WOA seroconverted to M hyo at a high rate in the presence of M hyo maternal antibody at 3 WOA. PCV2 maternal antibody at 3 WOA reduced IFA titers at 9 weeks of age. In the 1 mL, 2X, 3&7 WOA vaccine group, pigs with titers ≤ 640 at 3 WOA had a geometric mean titer of 3,044 at 9 WOA while pigs with titers ≥ 1,280 had 1,428. Average daily gain (ADG) for groups A, B, and C were 0.827, 0.842, and 0.834 pounds in the nursery and 1.991, 2.012, and 2.029 pounds in the finisher, respectively. Groups did not differ significantly in ADG by ANOVA. Mortality rates by group were 4.6%, 5.7%, and 5.2%, respectively and did not differ significantly by chi-square test.

Discussion

No viremia was detected in the pigs throughout the duration of the study, possibly due to the low level of challenge as assessed by oral fluid samples. The reason for the low level of challenge is unknown. Previous testing of these flows indicated that active PCV2 infection was occurring. However, this finding points out the limitation of field studies with regard to the potential variability in the endogenous challenge level. Serologic testing shows the 1 mL, 2X, 3 & 7 WOA group pigs are able to mount an immune response to both M hyo and PCV2 in the presence of maternal antibody. Overall, the pigs gained well and gain in groups vaccinated twice was not affected by the additional handling required by a second injection of PCV-M G2.

References

1. Crozier et al. 2015. The effects of vaccination timing on porcine circovirus type 2 viremia. *In:* Proc. 46th Annual Meeting of the American Association of Swine Veterinarians, Orlando, Florida, pp. 91-94.

Table 1: Serum results at 3 and 9 weeks of age

Vaccination group	No. of pigs	M hyo ELISA positive (SP ratio)		PCV2 IFA - Geomean titer	
		3 WOA	9 WOA	3 WOA	9 WOA
A: 2 mL, 1X, 3 only WOA	174	12/40 (0.316)	4/38 (0.101)	597	640
B: 1 mL, 2X, 3&7 WOA	174	12/38 (0.333)	36/37 (1.205)	727	2123
C: 1 mL, 2X, 3&9 WOA	173	13/39 (0.355)	0/36 (0.015)	545	395

Salmonella shedding in nursery pigs fed with a high or low complexity diet

Jane Newman[1], BSc; Brandon Lillie[1], DVM, PhD, DipACVP; Kees de Lange[2], MSc, PhD; Robert Friendship[3], DVM, MSc, DipABVP; Vahab Farzan[3], DVM, MSc, PhD
[1]Department of Pathobiology, [2]Department of Animal and Poultry Science,
[3]Department of Population Medicine, University of Guelph, Guelph, Ontario, Canada

Background

In addition to being a public health issue, *Salmonella* have the potential to cause clinical disease in pigs, resulting in significant economic loss. Pigs are often asymptomatic carriers and during periods of stress, such as weaning, may shed *Salmonella*. The *Salmonella* shedding rate may be higher in nursery pigs compared to finisher pigs, due to multiple factors, including the stress of weaning. Nursery management, health status, biosecurity, and diet composition may also influence *Salmonella* shedding. There is some concern that feeding low complexity nursery diets, which may reduce feeding costs by up to $2.50 per pig, can leave pigs vulnerable to disease challenges.[1] In order to enhance food safety, reduce clinical salmonellosis and improve growth performance, *Salmonella* shedding in nursery pigs needs to be considered when contemplating a switch to a cheaper ration. This study aims to examine the impact of a low complexity nursery diet on *Salmonella* shedding during the nursery phase.

Materials and methods

Four farrow-to-feeder sow barns were visited between May and September 2015. On each farm, 60 piglets were selected from 10 sows and ear tagged within 24-96 hours after farrowing. In total, 228 piglets were weaned at 20-26 days of age and transported to 4 different nursery barns. In the nursery, piglets were housed in two different pens and randomly assigned to either a high protein complexity (HC, containing blood plasma and high levels of whey) or a low protein complexity (LC, largely corn and soybean meal based) nursery diet.[1] Fecal and/or rectal swabs were collected at weaning and at the end of the nursery phase and cultured for *Salmonella*. Presumptive *Salmonella* colonies were tested with polyvalent *Salmonella* antisera. A logistic regression method was used to compare the prevalence of *Salmonella* shedding at weaning and at the end of the nursery period and to determine the impact of diet complexity on *Salmonella* shedding.

Results and discussion

In total, 228 and 218 piglets were tested for *Salmonella* shedding at weaning and the end of the nursery phase, respectively. Prevalence of *Salmonella* shedding at weaning and the end of the nursery stage on each farm is shown in Table 1. Overall, *Salmonella* were recovered from 33 (14.5%) and 23 (10.5%) piglets at weaning and the end of nursery period, respectively. There was no significant difference in *Salmonella* shedding among piglets at placement into the HC (16.7%) and LC group (12.3%) at weaning ($P = 0.3$). However, at the end of nursery period, pigs fed with the LC diet had a higher shedding rate (14.8%) as compared to pigs fed with HC diet (6.4%) (Table 1). The odds of *Salmonella* shedding in LC diet pigs was 2.6 times higher than in pigs receiving the HC diet ($P < 0.05$).

This study suggests utilization of a low complexity nursery diet may have an adverse impact on *Salmonella* shedding in nursery pigs. The pigs fed a low complexity diet may have a less robust immune response, giving them a low resistance to infectious agents. Also, it is possible that the low complexity diet has lower digestibility, resulting in feed passing into the caecum and colon where it may alter the microbiome by encouraging the proliferation of different micro-organisms in favor of *Salmonella*. It may be necessary, if using an inexpensive, reduced animal protein diet, to consider the implementation of management control measures or altered genetics to reduce *Salmonella* shedding.

References

1. Skinner, L.D. et al. 2014. Impact of nursery feeding program on subsequent growth performance, carcass quality, meat quality, physical and chemical body composition of growing-finishing pigs. J Anim Sci. 92: 1044-1054.

Table 1: *Salmonella* shedding at weaning and at the end of the nursery phase in pigs fed a low complexity (LC) or high complexity (HC) diet on four farms (A, B, C, D).

Farm	Number (%) of *Salmonella* positive pigs					
	At weaning			End of nursery		
	LC	HC	Total	LC	HC	Total
A	1 (3.6)	2 (7.4)	3 (5.5)	12 (48)	4 (16.0)	16 (32.0)
B	0 (0.0)	5 (17.9)	5 (8.8)	0 (0.0)	0 (0.0)	0 (0.0)
C	10 (34.5)	8 (27.6)	18 (31.0)	1 (3.4)	2 (6.9)	3 (5.2)
D	3 (10.7)	4 (13.3)	7 (12.0)	3 (10.7)	1 (3.3)	4 (6.9)
Total	**14 (12.3)**	**19 (16.7)**	**33 (14.5)**	**16 (14.5)***	**7 (6.2)***	**23 (10.3)**

* Different $P < 0.05$

A comparison of NP ELISA and HI tests identifying influenza A virus maternal antibody levels in processing-aged piglets

M. O'Loughlin[1]; K. Bretey[2], DVM; R. Edler[2], MS; B. Payne[2], DVM; M. Sprague[3], DVM; B. Bishop[4], DVM
[1]Iowa State University, Ames, Iowa; [2]Boehringer Ingelheim Vetmedica, Inc., Saint Joseph, Missouri;
[3]AMVC, Audubon, Iowa; [4]Veterinary Medical Center, Worthington, Minnesota

Introduction

A better understanding of maternal antibody (MAb) levels that may contribute to vaccine response interference will aid the animal health industry in its attempt to develop the most effective vaccines for swine production. In this surveillance study, we utilized two distinct diagnostic tests, hemagglutination inhibition (HI) and nucleoprotein-blocking ELISA (NP ELISA), to determine the levels of MAb to influenza A virus in swine (IAV-S) in processing-aged pigs and to identify correlation and determine strength between the two tests.

Materials and methods

Sow farms (n = 5) were sampled three times with a minimum of two weeks between samplings at any one farm. Eight sows were selected from each parity group: 1, 2-3, and 4+, and a different pool of 24 farrowed sows was selected at each collection. Piglets 2-6 days of age, one barrow and one gilt were tested from each litter for a total of 48 serum samples per collection. Serum samples (n = 720; 144 per site) were tested for the presence of IAV-S antibodies by NP ELISA, IAV H1N1, and H3N2 HI. The mean values for the collected data were used, and results were analyzed using linear regression analysis models evaluating HI results versus NP ELISA S/N results.

Results

Percent positives for each test were 99.7%, 79.4% and 76.2% for NP ELISA (< 0.67 S/N), H1N1 HI (≥ 40 titer), and H3N3 HI (≥ 40 titer), respectively. A significant regression equation was identified for the H1N1 HI ($P = 0.002$, $R^2 = 0.65$) compared to the NP ELISA, while a weak association was identified for the H3N2 HI ($P = 0.17$, $R^2 = 0.34$).

Discussion

With a blocking-ELISA, lower S/N ratios correspond to increased positivity. As shown on the graph, in this survey, high HI titers corresponded to low NP ELISA S/N ratios. HI results are dependent on the IAV-S strain in which the pig is exposed, and the different strains in a sow vaccine or field-strain exposure can impact HI titers. This could explain the correlation between the H1N1 HI and NP ELISA tests, but not the H3N2 HI. Ongoing work with larger sample sets, multiple farms, and different geographical locations is underway to further assess the industry MAb levels and the best method of testing for them.

Figure 1: Graphical representation of H1N1 (blue) and H2N2 (red) MAbHI titers plotted against NP ELISA S/N titers.

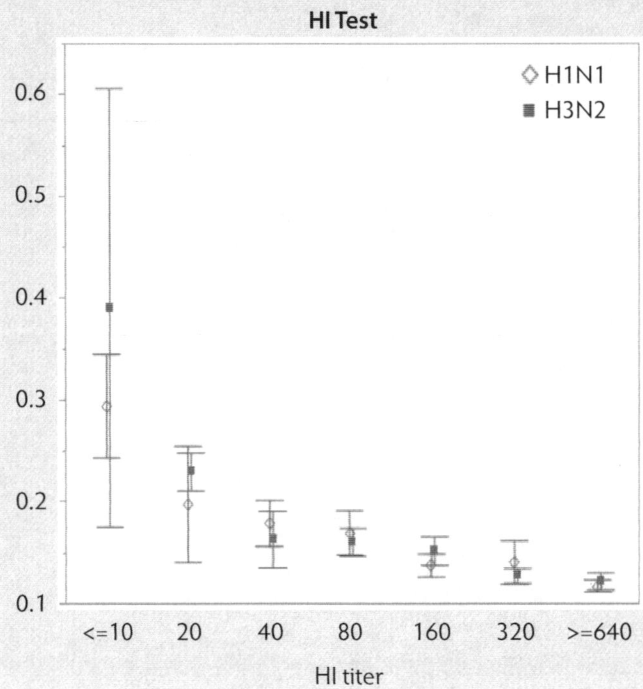

Comparison of macro-gut health parameters in PEDV positive and negative nursery pigs

Chelsea Ruston, BS; Paisley Canning, DVM; Jessica Davenport, BS; Jessica Bates, DVM; Kristin Skoland, BS; Nicholas Lauterbach BS; Emily Mahan-Riggs BS; Locke Karriker, DVM, MS, ACVPM

Introduction

Macroscopic parameters of intestinal health such as gut length, gut weight, percent dry matter of colon and cecal contents, fecal scoring and average daily gain (ADG) might be used in addition to mortality and morbidity to assess the efficacy of supportive care products for viral enteric diseases such as porcine epidemic diarrhea virus (PEDV). It is unclear whether PEDV has an effect on macroscopic gut health parameters in nursery pigs. The objective of study is to compare macroscopic parameters of intestinal health between PEDV positive and negative nursery pigs.

Material and methods

Twenty pigs, processed at 15 days of age were randomly allocated into a control group (C) and PEDV positive group (PP). Weights of all pigs were taken at age 15, 18, 36 and 40 days of age. At thirty-six days of age, C was inoculated via oral gavage with 10mL of PEDV negative culture inoculum and PP was inoculated with 10 ml of PEDV cell culture inoculum (1 × 104 TCID50/ml). Four days post-inoculation, pigs were euthanized. The small intestine was separated from the large intestine and the small intestine was weighed (g) and its length measured (cm). Cecal and colonic contents were also collected. Colon contents were scored on a 1 to 4 scale by a veterinarian blinded to the treatment groups (0 = normal, 1 = mild diarrhea 2 = moderate diarrhea, 4 = severe diarrhea) (Figure 1). The cecal and colonic contents were weighed (g) and dried in a convection oven at 69°C for 48 hours and percent dry matter was calculated.

Results and discussion

There was a significant reduction in cecal dry matter content between PP and C. There was an estimated mean difference of 3.37% (SE = 0.67%, p value 0.0378). For average daily gain between the two groups, there was an estimated difference of 0.142 kg (SE = 0.2382, p-value 0.2565). There was not a significant difference in intestinal length (*P*- value 0.9457), weight (*P*-value 0.4291), colon dry matter content (*P*-value 0.1312) and fecal score (*P*-value 0.2567) between C and PP. At necropsy, all pigs in C were PEDV PCR negative while CT values for PP ranged from 14.00 to 28.60.

One parameter that may be helpful is cecal dry matter content for insight on disease severity and to provide an objective and quantitative assessment of fecal consistency. Since there was no difference in fecal score or colonic dry matter between C and PP, it may be clinically challenging to detect PEDV infection in nursery pigs based on the presence of diarrhea alone. Thus, it is worth testing pigs for PEDV via PCR on oral fluids or fecal swabs to characterize the disease status of a nursery group even if diarrhea is not present. Since there were no major differences between the two treatment groups for these parameters, other means of assessing the effect of supportive care products such as mortality and morbidity are warranted.

Figure 1: Displays the colon content score rubric used for each pig's colon contents in the study.

1 = Normal | **2 = Mild diarrhea** | **3 = Moderate diarrhea** | **4 = Severe diarrhea**

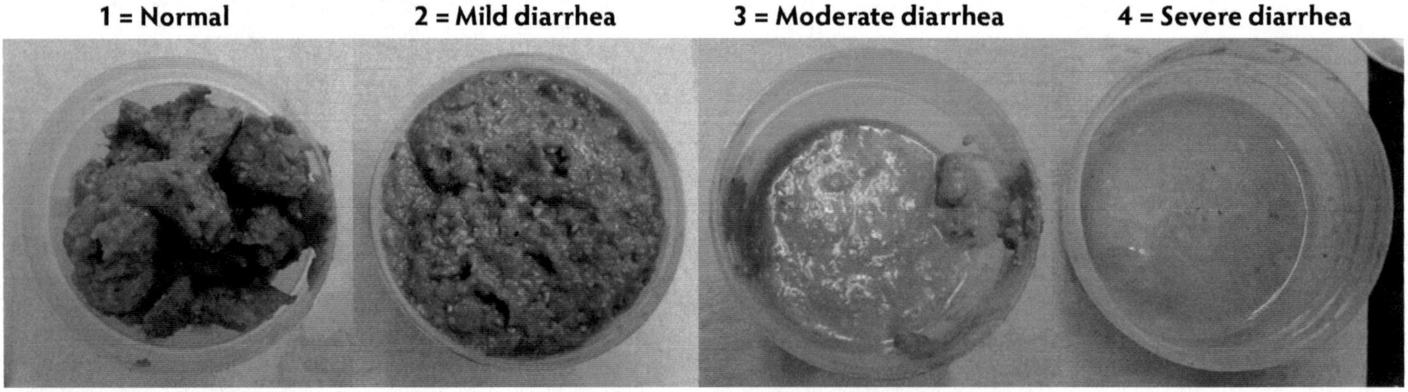

1 = Normal: The contents have some form within the container and appear thick.

2 = Mild diarrhea: The contents have little to no form, but appear thick and grainy.

3 = Moderate diarrhea: The contents appear to have a pudding consistency; not formed.

4 = Severe diarrhea: The contents have a watery consistency; not formed.

Incidence of porcine coronaviruses and porcine rotaviruses in an Indiana swine operation following clinical onset of acute diarrhea

C. Schnur[1], BS; D. Ragland[3], DVM, PhD; R. Pogranichniy[1,2], DVM, MS, PhD
[1]Indiana Animal Disease Diagnostic Laboratory; [2]Department of Comparative Pathobiology;
[3]Department of Veterinary Clinical Science, College of Veterinary Medicine, Purdue University, West Lafayette, Indiana

Introduction

Infections with porcine coronaviruses — including porcine epidemic diarrhea virus (PEDV), transmissible gastroenteritis (TGE), and porcine deltacoronavirus (PDCoV) — and porcine rotaviruses A, B, and C (RVA, RVB RVC) have caused devastating losses to the US swine industry.[1] These RNA viruses are highly contagious diseases of swine and inflict similar acute clinical symptoms, including severe diarrhea and subsequent dehydration, emaciation, and death.[2-4] Current research has focused on characterizing PEDV and PDCoV, which were identified as novel, emerging diseases in the United States.[3] By examining an Indiana farrow-to-finish hog farm four days following onset of diarrhea, this study attempts to elucidate any significant correlations between swine age, viral incidence, and viral co-infection of diarrheal viruses.

Materials and methods

On 30 January 2014, Fecal swab samples from 139 pigs were collected from an Indiana farrow-to-finish hog operation and categorized into 15 age groups with 6-10 samples each ranging from suckling piglets to breeding sows. Samples were extracted for total RNA, assayed by multiplex real-time reverse transcriptase polymerase chain reaction (RT-PCR), evaluated for presence of PEDV, TGE PDCoV, RVA, RVB and RVC, and analyzed for significant variance in viral load among groups.

Results and discussion

Notably, PEDV was present in all samples of all age groups (100% incidence), and variance among groups was significant ($P < .05$); highest viral loads were present in 36-60 day-old, post-weaned pigs and lowest loads from suckling piglets younger than 7 days-old (Figure 1). TGE and PDCoV were detected in very low prevalence and in concurrence with PEDV.

Rotavirus A and C were detected exclusively in pigs 11-60 days old, while Rotavirus B was detected in pigs 11-22 days old. A significant viral load variance in groups infected with Rotavirus A ($P = 0.0049$), but not with Rotavirus B or C, was detected.

The overall results of this study support that PEDV infection and viral load may depend on age of pigs. Porcine rotavirus was observed to infect post-weaned piglets exclusively and in concurrence with PEDV, while PEDV infected all ages, but with varying viral load among age groups. Identifying incidence of viral diseases within swine herds exhibiting early-stage diarrhea is necessary for understanding viral spread and susceptibility. This data could potentially lead to improved virus prevention and control measures on this particular swine farrow-to-finish operation.

References

1. Paarlberg, P.L. Updated Estimated Economic Welfare Impacts of PEDV. Department of Agricultural Economics, Purdue University. 2014.

2. Zimmerman, J, Karriker, L, Ramirez, A, Schwartz, K, & Stevenson, G (Eds.). *Diseases of Swine* (10th ed.). Wiley-Blackwell. 2012.

3. Wang, L, Byrum, B, Zhang, Y. Detection and Genetic Characterization of Deltacoronavirus in Pigs, Ohio, USA, 2014. *Emerging Infectious Diseases*. July 2014. 20(7): 1227–1230.

4. AASV Quick Facts: Porcine Epidemic Diarrhea. American Association of Swine Veterinarians. Available at: `https://www.aasv.org/pedv/AASVPEDQF.pdf`. Accessed 18 November 201.

Figure 1: Average CT values for PEDV, TGE, and PDCoV across age groups during the study. Not shown are CT's of positive virus controls: PEDV + (CT : 21.06), TGE + (CT : 23.03), and PDCoV + (CT : 19.40).

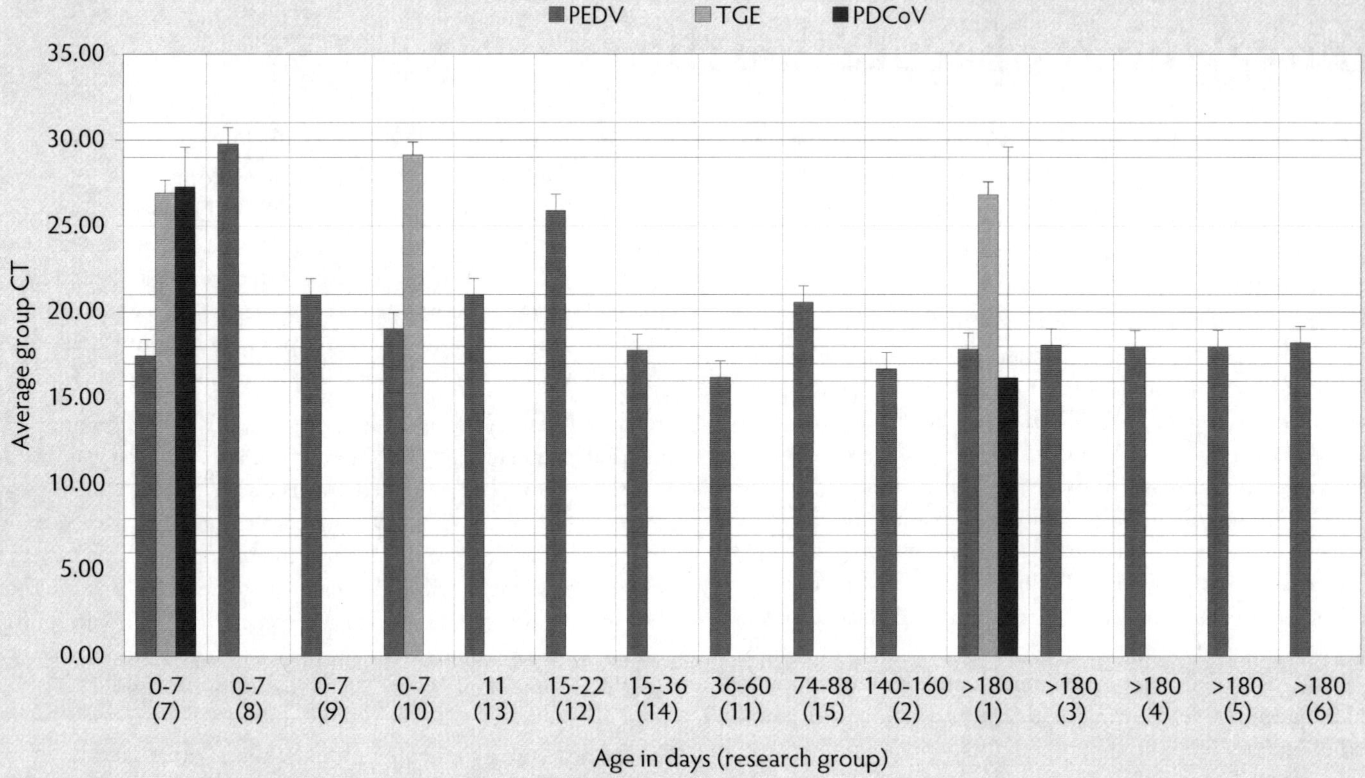

Do sows and gilts previously infected with PEDV respond to PEDV vaccination?

B. Setness[1], BS; J. Bjustrom-Kraft[1], BS; K. Woodard[1], DVM; L. Giménez-Lirola[1], PhD; Y. Sun[1], BS; P. Lasley[2], MS, DVM; Q. Chen[1], DVM, MS; J. Zhang[1], MD, MS, PhD; E. Nelson[3], PA, PhD; D. Baum[1], DVM, MS, PhD; P. Gauger[1], DVM; J. Zimmerman[1], DVM, MS, PhD; R. Main[1], DVM, PhD
[1]Iowa State University; [2]Murphy-Brown of Missouri LLC; [3]South Dakota State University

Introduction

The objective of this study was to characterize the immune response of sows and gilts to (previously infected with PEDV) alternative PEDV vaccines and regimes prior to farrowing.

Materials and methods

Multiparous sows (n = 36) and gilts (n = 36) were randomly assigned to one of five vaccine treatments; 33 gilts and 31 sows actually completed the study (Table 1). Samples were collected and tested using the PEDV whole virus (WV) IgG ELISA, the WV IgA ELISA, and an FFN assay. The sows in this study were infected with PEDV via natural exposure December 2013 and the gilts were intentionally exposed at 13 weeks of age on August 28th, 2014.

Results

ELISA S/P (means) and FFN (geometric mean titers) are reported. Comparisons of the means suggest that vaccines effectively increased IgG, IgA, and neutralizing antibody in gilts in all diagnostic specimens tested (statistical analyses in progress). The effect of vaccination was less dramatic in sows, but was still evident. These data suggest that PEDV vaccination of previously exposed gilts and sows may be a useful tool in PEDV control with either killed vaccine or replicon particle RNA vaccine.

Discussion

In the present study we characterized the immune response of sows and gilts to alternative vaccines and regimes prior to farrowing. Based off of the means of the samples collected, it was evident that IgG, IgA, and neutralizing antibody titers in gilts and sows can be increased with the use of specific vaccination regimes. However, there was a greater increase in antibody titers in the gilt vaccination groups compared to the sow vaccination group. It is also evident that the 2 week pre-farrowing vaccination protocols enhanced a greater antibody response compared to the 5 and 2 week pre-farrowing protocols. PEDV vaccine protocols continue to be a challenge for swine producers to date. Upon statistical analyses, the evidence of different immune responses of sows and gilts to alternative vaccines and regimes prior to farrowing will aid in preventing future outbreaks of PEDV.

Table 1: Experimental design

Trt*	Vaccination protocol	Animals	Samples
1	Unvaccinated (control)	5 sows, 5 gilts	1. Serum at farrowing
2	1 ml IM - 2 weeks pre-farrow	7 sows, 6 gilts	2. Colostrum ≤ 24 hr post farrowing.
3	1 ml IM - 5 and 2 weeks pre-farrow	6 sows, 8 gilts	
4	2 ml IM - 2 weeks pre-farrow	6 sows, 7 gilts	3. Milk at 3 days post farrowing
5	2 ml IM - 5 and 2 weeks pre-farrow	7 sows, 7 gilts	

* Treatment 2, 3 - Harrisvaccines™ PEDV Vaccine; 4, 5 - Zoetis PEDV vaccine, killed virus

Development of a snatch farrowed-colostrum deprived piglet challenge model for porcine rotavirus C

Brooke Smith; Ryan McCann; Patrick Roady; James Lowe

Problem

Porcine rotavirus (PRoV) is a ubiquitous, environmentally stable pathogen in the swine industry, responsible for causing severe enteritis in nursing and post-weaned piglets. There are three known phenotypes of PRoV: A, B and C. Reliable culture methods have not been successfully developed for PRoV-C impeding evaluation of vaccines and other interventions under controlled conditions. In this study we describe a novel snatch farrowed-colostrum deprived (SF-CD) neonatal piglet challenge model for PRoV-C to allow for the assessment of potential interventions under controlled conditions. The objective was to determine if animals infected with rotavirus could be differentiated from control animals by presence of diarrhea, weight gain and morphological changes in the gastrointestinal tract over a 72 hour period.

Methods

To validate that clinical observation of diarrhea was a symptom of rotavirus infection and not a nutritional factor, different sources of milk replacer were evaluated along with a the PRoV-C challenge. Sixteen total piglets were placed into four treatment groups: Control pigs that remained on the sow until completion of the experiment (CS), Non-challenged SF-CD pigs provided a milk replacer free of porcine products intended for use in pigs, Non-challenged SF-CD provided a milk replacer developed for humans, and rotavirus-challenged SF-CD pigs on porcine milk replacer. On experiment day -1 piglets in the CR, CH and RR groups were snatch-farrowed from four litters and randomly assigned a treatment group within litter. On Day 0, RR piglets were inoculated with an intestinal homogenate that contained PRoV-C. 1 pig from the CH and CS groups were euthanized prior to enrollment resulting in N = 3 for the CH and CR treatments and N = 4 for the RR treatment. Throughout the course of the study (Day -1 to Day 3), all piglets were housed in individual pens, their respective diets fed via bottle every two hours and observed for diarrhea by a blinded participant every 24 hours while following strict infection control protocols. At 72 hours post-challenge, all piglets remaining in the study were euthanized and necropsies were performed, to collect samples for real-time qPCR and histological analysis.

Results

5/10 SF-CD pigs that were enrolled were removed from the study prior to D3; CH = 3, CR = 0, RR = 2 due to sepsis or bronchopneumonia. None of the CR pigs exhibited diarrhea and 2/4 RR pigs had diarrhea on day 1 but not on other days. CR pigs had a significant increase in weight gain from D0-D3 (390g vs. 70 kg, $P = 0.02$). 4/4 pigs the RR group were PCR positive (Mean CT = 26.1) for PRoV-C but not PRoV-A or PRoV-B. CR or CS pigs were considered suspect PCR positive (Mean CT 31.3 and 34.2 respectively), which was most likely due potential cross contamination during necropsy processing.

Discussion

These data suggest both the challenge process and the infection control procedures were effective. To further refine the model we will conduct dose titration experiments to calculate the Pig Infectious Dose 50 (PID50). We conclude that this model will be suitable for assessing the efficacy of interventions against PRoV-C in a controlled manner which is a critical step for the development of evidence based interventions for swine practitioners.

Disease surveillance in an all natural system

C. Stewart, BS; T. Schwartz, DVM; J. Bates, DVM; A. Canon, DVM, MPH; P. Canning, DVM; L. Karriker, DVM, MS; K. Skoland, BS; A. Ramirez, DVM, MPH, PhD

Problem

An all-natural swine production company uses multiple small outdoor sow farms to breed and farrow piglets for their production system. Each farm has a different disease and health status which creates problems during later phases of production when commingling occurs. The goal of this disease surveillance project was to determine the health status of each farm and ultimately create more homogenous flows within the system.

Herd history

Individual farms (n = 37) were included in this case study. These farms are located within a 10 mile radius of each other in southern Missouri. Piglets from these farms flow into commingled nurseries and finishers in another state. Diagnostics are rarely performed at these farms so all were of unknown health status. Due to the nature of the all-natural system, antibiotics are not allowed for use in market animals. Sows receive a prefarrow vaccination to provide maternal antibodies for *Clostridium perfringens* type C and *Escherichia coli* (K99, K88, 987P and F41) as well as a prebreeding vaccination for parvovirus, leptospirosis and erysipelas. Piglets are vaccinated for *Lawsonia intracellularis, Streptococcus suis, Salmonella Choleraesuis,* and *Escherichia coli*.

Data collection and diagnostics

Six centrally located sites were set up for producers to bring pigs for sample collection. These pigs were approximately six to 10 weeks of age. Each farmer provided one to three pigs for humane euthanasia and necropsy. A complete herd history was taken from each farmer. A full set of fresh and fixed tissues was taken, beginning with the brain to limit contamination in the field necropsy setting. Fecal samples and any clinically relevant samples (e.g. joint swab, pleural, peritoneal swabs) based on history or gross lesions were taken.

Results

Gross lesions seen on necropsy were variable but included thin-walled intestines, white spots on the liver, thickened intestines and lung consolidation. None (0/37) of the tests were positive for porcine reproductive and respiratory syndrome virus (PRRSV) via PCR. Only one farm had a history of lameness, but 22% (8/37) of samples tested positive for *Mycoplasma hyorhinis* by PCR. One submission had bronchointerstitial pneumonia lesions suspicious of a previous influenza A infection, but no samples (0/37) tested positive for influenza A by PCR. At least one submission had lesions consistent with ileitis, and 46% (16/35) of submissions tested positive for *Lawsonia intracellularis* by PCR. Rotavirus PCR was performed on 30% (11/37) of the submissions. 73% (8/11) tested positive for rotavirus. 68% (25/37) of the submissions revealed at least one kind of intestinal parasite on fecal floatation, primarily *Strongyloides* and ascarids (Table 1).

Analysis and recommendations

Diagnostic results were interpreted in correlation to clinical signs and gross pathology. *Lawsonia intracellularis* and rotavirus were determined to be the primary pathogens of concern. It is best to create flows based on the positive status of these pathogens. The internal parasite burden was also significant and commercial parasiticides, which can be used in all natural production systems, should be implemented to decrease the parasitic burden.

Take home message

It is important to recognize the prevalence of pathogens in outdoor swine rearing facilities in order to manage a homogenous and healthy group of animals. Diagnostics are a critical tool that provides necessary information to better maintain pig health. It is also essential to use this recognition and create a system that works for the betterment of the groups and creates a healthy flow. Extensive swine production systems are challenged with some of the same health problems and flow variability as intensive production systems during commingling.

Table 1: Prevalence of disease based on diagnostic tests

Diagnostic test	Prevalence (out of 37 farms)	Percent prevalence
PRRSV X3 PCR	0/37	0%
Influenza A PCR	0/37	0%
Mycoplasma hyorhinis PCR	8/37	21.6%
Lawsonia intracellularis PCR	16/35	45.7%
Rotavirus (A, B or C) PCR	8/11*	21.6%
Fecal floats	25/37	67.5%
Hemolytic *E. coli* culture	19/36	51.3%

* Rotavirus PCR was only performed if gross lesions were evident; only 11 submissions fit the criteria for testing

Comparison of sample types for assessing porcine circovirus type 2 status of sow herds

Corrine Stoffel[1], BS; Trevor Schwartz[2], DVM; Amber Stricker[2], DVM, MS
[1]University of Illinois, Urbana, Illinois; [2]Suidae Health and Production, Algona, Iowa

Introduction

The typical timing of porcine circovirus type 2 (PCV2) clinical disease in pigs is during the late nursery and early finishing period of production. Clinical disease can occur early in the nursery phase, indicating the pigs are actively infected with PCV2 prior to weaning, decreasing effectiveness of vaccination. In order to explain the temporal progression of infection, many different sampling techniques have been used but verification of a gold standard has been difficult to define. Testing in the sow herd has utilized PCR for PCV2 antigen testing of samples including sow serum, pre-suckling piglet serum, post colostrum pig let serum up to wean age, placental umbilical cord serum (PUCS), stillborn/mummy tissues, and failure to thrive piglet tissues. In total, this testing attempts to determine the "stability" of the sow herd for PCV2. To date, there are few, if any, data sets that enable comparison of these techniques in an attempt to establish a gold standard for determining an unstable farm.

Objective

The objective of this study was to compare PUCS, colostrum, serum from post-suckle live pigs, and serum from sows as sample types for PCR for agent detection to assess stability of sow herds for PCV2.

Materials and methods

During July of 2015, samples were acquired from twenty-eight sows and litters from 3 unrelated sow units. Sows were selected on the basis of parity, (P1, P2-3, P4+). Within 12 hours of farrowing, colostrum from each sow and pooled PUCS from 5 placental units were collected. Samples were refrigerated and submitted to Iowa State University Veterinary Diagnostic Laboratory for immediate testing. Based on testing of those results, serum from selected sows and pooled serum from 5 suckling piglets within respective litters was collected.

Results

The following table presents the PCV2 PCR results; the level of positive samples was low (Table 1).

Conclusions

The low level of infection in these herds precluded any comparison of the diagnostic sampling protocols. We attribute this low level to the vaccination protocols that are used on replacement gilts that enter these herds. Experience gained will enable us to decide which method is easiest to implement on a herd-by-herd basis. Future studies should be performed in herds with a higher prevalence of infection.

Table 1: Sampling methods comparison for PCV2 at the sow farm

		No. positive by sample type			
Farm	**Parity group**	**PUCS**	**Colostrum**	**Sow serum**	**Pig serum**
A	P1	0/2	1/2	0/2	0/10
	P2-3	0/3	1/3	0/3	0/15
	P4+	0/3	0/3	0/3	0/15
B	P1	0/0	0/0	0/0	0/0
	P2-3	0/1	0/1	0/1	0/5
	P4+	0/6	1/3	0/6	0/30
C	P1	0/3	0/3	0/3	0/15
	P2-3	1/5	0/5	1/5	0/25

The interaction between swine health and low crude protein, amino acid supplemented diets

C. Vonderohe, MS; J. S. Radcliffe, PhD
Purdue University, West Lafayette, Indiana

Recent interest and actions to mitigate environmental impact of animal agriculture has resulted in critical examination of current animal production practices. It has also led to greater interest in technologies that mitigate production losses and nutrient excretion. Synthetic amino acids (SAA) are added to reduced crude protein (CP) swine diets to more accurately meet amino acid needs, and therefore improve nitrogen retention by the animal. The objective of this literature review is to examine effects of low protein, high SAA diets on swine health.

Nutritionists use SAA to formulate reduced CP diets where limiting amino acid needs are met by adding SAA. This improves nitrogen retention because SAA are more digestible, and the precision that they lend to diet formulation results in less total protein fed, and therefore excreted into the environment. This is of interest to swine veterinarians because feeding an extremely low CP, highly SAA supplemented diets can negatively affect growth performance but may positively affect health of the animal.

Use of SAA is consistently associated with reductions in N excretion.[1] However, use of ultra-high concentrations of SAA, achieved by using synthetic SAA to balance the diet up to the seventh limiting amino acid, is also associated with increased variability in growth performance and carcass composition.[2,3] In most of these cases, pigs fed the highest concentration of SAA and lowest crude protein (CP) concentration had poorer growth performance. However, these reductions in performance do not appear to be related to poor enteric or systemic health.

Some investigators[3] hypothesize that reductions in performance observed with low CP, high SAA diets is due to dietary amino acid imbalances. Others[1] hypothesize that increases in back fat and alterations of carcass characteristics in pigs fed high-SAA diets are due to changes in energy and N metabolism. This theory is supported by Jones et al.,[4] who reported that inclusion of high concentrations of SAA decreased energy retention. Other research indicates that high concentrations of SAA reduce digesta pH and therefore reduce pathogen survivability in the GI tract. Nyachoti et al.[5] indicated that SAA improve enteric health because their use is associated with fewer toxic metabolites, such as ammonia. SAA have also been reported to reduce cecal pH in nursery pigs, thereby reducing potentially pathogenic bacteria.[6] Low CP diets fed to E.coli challenged pigs after weaning are associated with a lower incidence of diarrhea and decreased colonic pH, indicating lower pathogenic bacterial survival throughout the GI tract.[7]

The pH reduction in the gastrointestinal tract, reduction of toxic metabolites and metabolic changes associated with low-CP, high-SAA diets indicates a potential to improve enteric health of swine. It is important for veterinarians to understand interactions between nutrition and animal health, particularly in relation to industry-significant topics such as the environmental footprint of swine production. In this case, more research is warranted to better understand the interaction between high concentrations of SAA and animal health and performance.

References

1. Kerr BJ, Easter RA. Effect of feeding reduced protein, amino-acid supplemented diets on nitrogen and energy balance in grower pigs. *J animal sci.* 1995;73:3000-3008.

2. Kerr BJ, McKeith FK, Easter RA. Effect on performance and carcass characteristics of nursery to finisher pigs fed reduced crude protein, animo acid-supplemented diets. *J animal sci.* 1995;73:433-440.

3. Figueroa JL, Lewis AJ, Miller PS, Fischer RL, Gomez RS, Diedrichsen RM. Nitrogen metabolism and growth performance of gilts fed standard corn-soybean meal diets or low-crude protein amino acid-supplemented diets. *J animal sci.* 2002;80:2911-2919.

4. Jones AM, Kelly DT, Richert BT, Maxwell CV, Radcliffe JS. Effect of feeding reduced-CP, amino acid supplemented diets on dietary nitrogen and energy utilization and volatile fatty acid excretion in wean-to-finish swine. *Proc. Midwest ASAS.* Des Moines, Iowa. 2014;16.

5. Nyachoti CM, Omogbenigum FO, Rademacher M, Blank G. Performance responses and indicators of gastrointestinal health in early-weaned pigs fed low-protein amino acid-supplemented diets. *J animal sci.* 2006;84:125-134.

6. Htoo JK, Araiza BA, Sauer WC, Rademacher M, Zhang Y, Cervantes M, Zijlstra RT. Effect of dietary protein content on ileal amino acid digestibility, growth performance and formation of microbial metabolites in ileal and cecal digesta of early weaned pigs. *J animal sci.* 2007;85:3303-3312.

7. Wellock IJ, Fortomaris PD, Houdijk GM, Kyriazakis I. Effects of dietary protein supply, weaning age and experimental enterotoxigenic Escherechia coli infection on newly weaned pigs: health. *Animal.* 2008;6:834-842.

Patterns of resting position in group housed gestating sows

H. Wensil-Strow; K. Horback, PhD; M. Pierdon, VMD; T. Parsons, VMD, PhD, DACAW
University of Pennsylvania School of Veterinary Medicine, Kennett Square, Pennsylvania

Introduction

Public concern over sow welfare is driving producers to seek alternatives to current husbandry practices. Group housing allows gestating sows to have increased freedom of movement and social contact with their conspecifics. Little is known, however, about the details of these social interactions of sows or their spatial use of large pens. Social network analysis (SNA) can be used to analyze relationships among individuals and the patterns of relationships in a large group of animals. To determine the consistency of sow resting location in a large pen over time, and in order to better understand individual movement within a pen, this study examined whether individual sows have preferred resting locations and whether sows have preferred resting partners (i.e., social cliques).

Materials and methods

At the Penn Vet Swine Teaching and Research Center, gestating sows and gilts are housed in dynamic groups and fed via electronic sow feeding stations (Compident 7, Schauer Agrotronic). The sow pen allows for 2.4m^2 per sow, includes 2 electronic sow feeding stations, and offers a variety of resting options including several 3m × 2.2m concrete lying beds and two recessed 3.7m × 8.2m deep bedded straw pits. The resting location of each sow was recorded once a day in the early evening for 22 days. The pen layout was divided into x and y coordinates and the position of each sow's shoulder was recorded on a map of the barn. The coordinates of each pig were then transcribed into a data sheet. Data was collected on over 200 sows during the test period. A subset of 29 sows identified as having 10 or more location data points and studied further. Individual sow location was visualized by plotting the Cartesian coordinates of each daily observation. A hierarchical clustering algorithm (R Statistical Package) was used to identify discrete, repeated resting positions or a so-called cluster. Non-parametric Kruskal Wallis and Wilcoxon signed rank test were used to assess factors impacting the spatial distribution of sow resting location. A social network analysis was conducted, using igraph[1] (R Statistical Package), to determine social cliques based on the number of days a sow was recorded to be within two body lengths (~4 m) of another sow.

Results and discussion

All sows studied exhibited at least one cluster of preferred resting locations. The distribution of number of location clusters per sow was: 1-13%, 2-74%, 3-13% and was significantly different from the null hypothesis of no clusters (V= 435, $P < .001$). Neither time in pen nor age of the animal impacted the number of location clusters per sow. A cliquish analysis describes the extent at which the social network contains subgroups that exclude others from them socially. There were a total of 6 large cliques of sows based on resting location within two body lengths of another sow. Each clique contained 22 individuals (Figure 1). Future research may investigate whether sows form long-term preferential associations or 'friendships' within group pens[2] as well as the welfare implications of these social preferences. These findings provide information on individual sow movement and use of a pen in a group-housing situation. Interestingly with time sow resting location is not random and defined locations for laying preferences can be discerned. It is also striking that nearly 90% of the sow had more than one preferred resting locations but utilization of those spaces was not correlated in time. Further research will be needed to determine how these preferred resting locations

Figure 1: A network diagram (or sociomatrix) representing the social relationships between the largest clique of 22 individuals in the dynamic sow pen. Each yellow circle, called a node, represents a sow (A-V) and the width of each line, called an edge, represents the number of times the two sows are seen resting within two body lengths of each other.

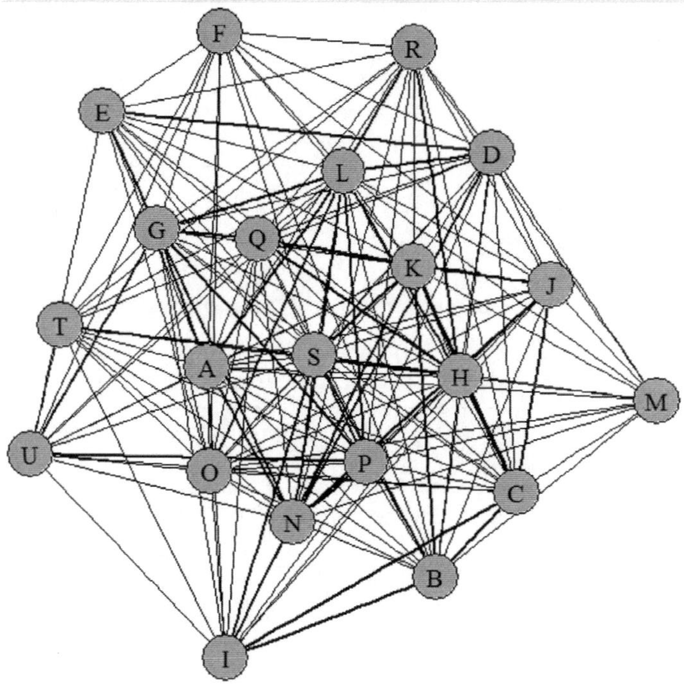

arise and when they are specifically utilized. This work promises a better understanding of how to both design and manage large pens of group housed gestating sows.

Acknowledgements
Supported by Pennsylvania Pork Producers Council, New York Farmers, and Swiss Village Foundation.

References

1. Csardi G, Nepusz T. The igraph software package for complex network research, *InterJournal, Complex Systems,* 2006: 1695.

2. Durrell, J. L., et al. Do pigs form preferential associations? *Appl. Anim. Behav. Sci.* 2004; 89: 41-52.

Conditions for development of cross neutralizing antibodies to porcine reproductive and respiratory syndrome virus

J. Young[1], BS; K. Martins[2], DVM, MS; S. Robinson[2], BVSc, MVSc, PhD; D. Ngo[2], BS; M. Murtaugh[2], PhD
[1]University of Minnesota, College of Veterinary Medicine, Saint Paul, Minnesota;
[2]University of Minnesota, Veterinary and Biomedical Sciences, Saint Paul, Minnesota

Introduction

Porcine reproductive and respiratory syndrome virus (PRRSV) is the major disease of swine worldwide, causing more than $660 million in damages annually in US herds alone. PRRSV spreads rapidly throughout farms resulting in decreased appetite and fever in sows, with abortions, stillbirths and weak-born piglets. In growing pigs, respiratory disease, secondary infections, mortality and reduced growth are typical. Symptomatic acute infection is followed by subclinical infection and transmissibility. Good vaccination strategies are important for herds already infected or in endemic areas, but efficacy varies due in part to high levels of genetic variation in virus and pigs. Improving vaccination efficacy and reproducibility is a challenge since mechanisms of immune protection are not yet established. Because neutralizing antibody titers are an established indicator of protection against viral infection, the aim of this study was to investigate virus exposure conditions that facilitate the development of cross-neutralizing antibodies. We hypothesized that high viral load drives titer and avidity maturation, while viral attenuation will reduce viral load and neutralizing antibody avidity. Avidity is defined as the strength of antibody/antigen interaction.

Materials and methods

Serum samples were obtained from a study performed by the University of Minnesota in collaboration with Boehringer Ingelheim Vetmedica.[1] Swine, from a PRRSV free facility, divided into three age groups, were challenged with either Ingelvac® PRRS ATP attenuated virus vaccine or its parental isolate, virulent JA142 PRRSV. Age groups included swine of ages three weeks (piglets, ATP n = 11, JA142 n = 5), 16 to 20 weeks (finishers, ATP n = 10, JA142 n = 8), and third parity (± 1) sows (ATP n = 9, JA142 n = 5). Serum collected at day 63 post challenge was analyzed using a guanidine hydrochloride (GuHCl) treatment ELISA assay. GuHCl was previously shown to disrupt weakly bound antibody/antigen interactions, while leaving stronger antibody/antigen interactions intact. OD 450nm values obtained from the ELISA were then used to calculate individual avidity indices, or strength of antibody/antigen binding, using the area under the curve avidity index equation.[2] Day 63 serum was also analyzed using an ELISA-based PRRSV neutralization assay to detect percent inhibition of infection.

Results and discussion

Viral load was shown to have a strong correlation with antibody avidity ($R^2 = 0.9325$). Higher viral titers may be leading to increased B-cell differentiation in the lymph node germinal center, resulting in expanded affinity maturation and thus higher avidity. Antibody response varied by age group (Figure 1). Within the sow and finisher age groups, virulent virus infection was associated with increased avidity, increased viral load and increased serum neutralizing titer. Piglets on average showed higher serum avidities and viral loads than sows and finishers when challenged with either attenuated or virulent virus. These findings support unpublished observations that antibody levels are positively associated with avidity and neutralizing activity. The results of this study show that avidity ELISA testing of antibody reactivity to a non-structural protein antigen retrodicts viral virulence as measured by peak viral titer. Since virulence is difficult to quantify in the laboratory and cannot be determined in an outbreak situation, the avidity ELISA may be useful for assessing the severity of specific viral outbreaks in retrospective studies.

References

1. Klinge, K.L. et al. Age-dependent Resistance to Porcine Reproductive and Respiratory Syndrome Virus Replication in Swine. *Virol J* 6 (2009): 177.

2. Perciani, C.T. et al. Improved Method to Calculate the Antibody Avidity Index. *J. Clin. Lab. Anal.* 21 (2007): 201-206.

Figure 1: Using the guanidine hydrochloride (GuHC1) ELISA. OD 450nm values were obtained from day 63 serum of each group over 2-fold series of serum dilutions with and without GuHC1 treatments. Robust antibody responses were evident in piglets to both the virulent and attenuated virus challenge. A dramatic increase in antibody response was evident against virulent JA142 compared with attenuated ATP in finishers. Sows mounted a similar antibody response to both JA142 and ATP.

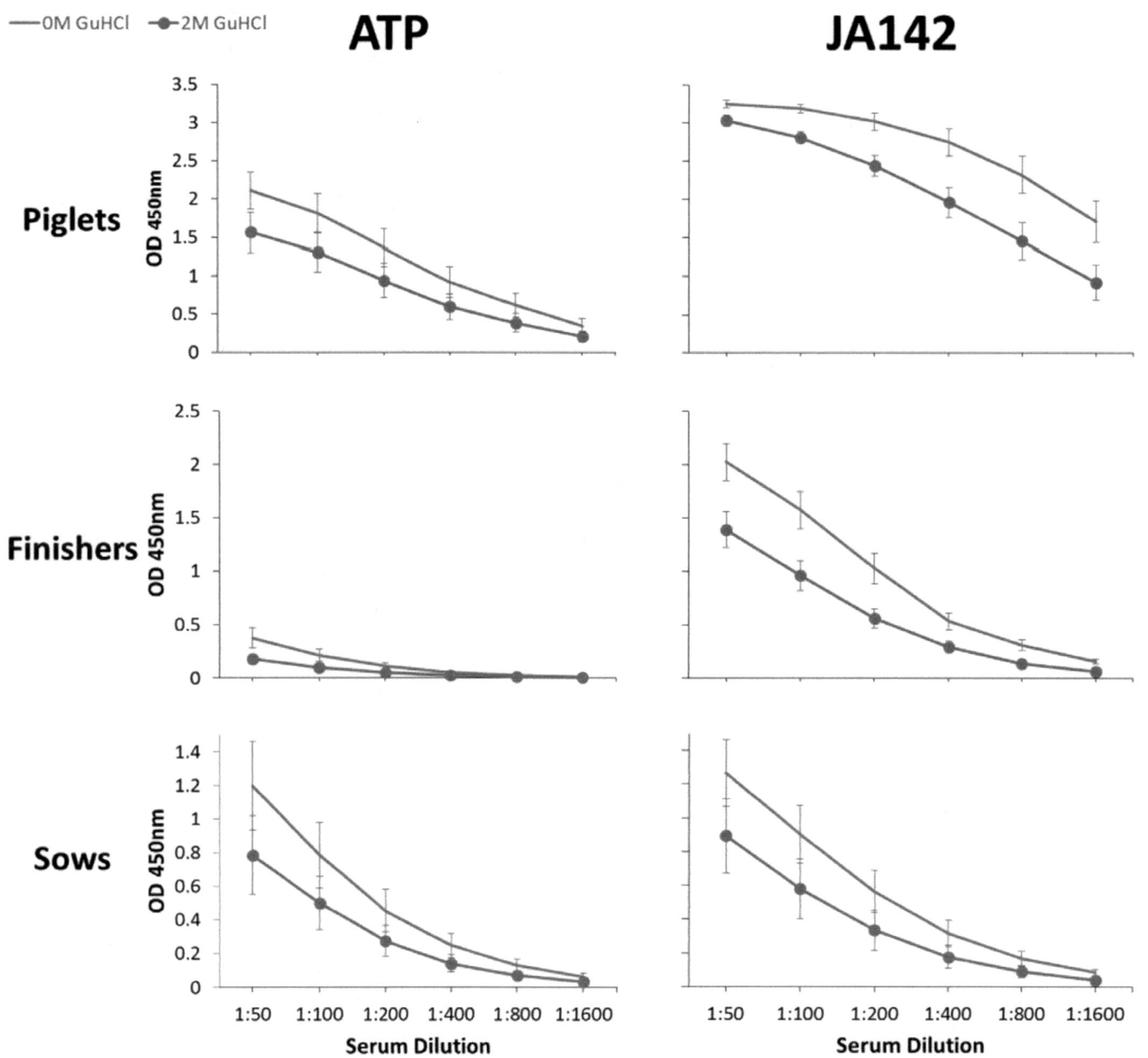

Financial impact of endemic and epidemic pathogens in growing pigs

J. Alvarez[1], DVM, PhD; J. Sarradell[2], DVM, PhD; M. Torremorell[1], DVM, PhD; R. Morrison[1], DVM, PhD; A. Perez[1], DVM, PhD

[1]Department of Veterinary Population Medicine, College of Veterinary Medicine, University of Minnesota, Saint Paul, Minnesota; [2]Department of General and Systemic Pathology, Veterinary Faculty, National University of Rosario, Casilda, Argentina

Introduction

Despite the efforts invested on the control of endemic diseases in the US swine industry, quantification of their financial impact throughout the production chain is often hampered by the lack of data or data analysis capabilities.[1] There are studies that have evaluated the effect of disease in sow farms, however there are important knowledge gaps regarding the full impact of endemic and epidemic diseases in growing pigs. We hypothesized that the analysis of production records routinely collected in growing pigs together with laboratory and clinical information on the presence of selected endemic pathogens in sow and growing pig farms can help to fully evaluate their production impact, thus helping to inform the decision-process with regards to the implementation of control/eradication strategies. Three diseases [porcine epidemic diarrhea (PED), influenza A (IA) and porcine reproductive and respiratory syndrome (PRRS)] were selected based on their economic importance, widespread distribution, and different epidemic/endemic nature in the US swine industry.[1-3]

Materials and methods

Close-out performance records from wean-to-finish (WF) sites from two large systems located in the Midwest were screened to identify groups of pigs that had tested positive at weaning for at least one of the three selected pathogens. In the case of PED infection, performance [average daily gain (ADG), average daily feed intake (ADFI), feed conversion ratio (FCR) and post-weaning mortality] of the last batch of pigs weaned before detection (clinical symptoms + laboratory confirmation) of PED virus (PEDV) in the sow farm (control batch) was compared with that of the first batch of pigs weaned after a PEDV outbreak (case batch). For IA virus (IAV) and PRRS virus (PRRSV), post-weaning mortality of positive and negative batches to either or both pathogens (determined by PCR in oral fluids at weaning) was compared using a Bayesian mixed effects negative binomial multivariable regression model that accounted for potential confounders such as flow, date at weaning, days on feed and batch size.

Results and discussion

Detection of any of the three pathogens in the sow farms in pigs at weaning resulted in poorer performance of the pigs throughout the growing stage. Specifically, an average increase of 11% post-weaning mortality, and decreased ADG and increased FCR was observed in the first batches weaned after PEDV was detected in a farm. This demonstrates that PEDV infection, in addition to causing increased mortality in suckling pigs, also results in a poorer performance of the surviving pigs.[2] Presence of IAV and PRRSV in groups of pigs at weaning was also associated with an increase in post-weaning mortality (median increase of 13% and 34% over the system-specific baseline mortality, what would translate into a raw increase of 0.65% and 1.7%, respectively), thus suggesting that early exposure to these pathogens could affect the pigs' survival throughout the growing period.[3] Overall, our results demonstrate that the combination of information from surveillance protocols in sow farms and production records in growing pigs can help to quantify the full impact of endemic and epidemic diseases beyond the breeding herd, and provide a methodological framework that could be extended to include other confounders and infectious agents.

References

1. Holtkamp DK et al. Assessment of the economic impact of porcine reproductive and respiratory syndrome virus on United States pork producers. *J Swine Health Prod* 2013; 21:72

2. Alvarez J et al. Impact of porcine epidemic diarrhea on performance of growing pigs. *PLoS One* 2015; 10: e0120532.

3. Alvarez J et al. Association of the presence of influenza A virus and porcine reproductive and respiratory syndrome virus in sow farms with post-weaning mortality. *Prev Vet Med* 2015; 121;240

Description and use of network data to inform disease control and prevention strategies

Andréia G. Arruda[1], DVM, MS; Robert Friendship[1], DVM, MS, DipABVP; Jane Carpenter[2], DVM, Karen Hand[3], PhD; Zvonimir Poljak[1], DVM, MS, PhD
[1]Department of Population Medicine, University of Guelph, Guelph, Ontario, Canada; [2]Woodstock, Ontario, Canada; [3]Strategic Solutions Group, Puslinch, Ontario, Canada

Introduction

The importance of networks in the spread and control of infectious diseases has been recognized in the last decade for different animal diseases. The current North American swine industry is highly connected, with multiple sites that belong to large production systems, and service providers that indirectly connect sites. The objectives of this study were to describe static relationships between swine sites and their service providers including transportation, feed, semen, gilt and boar companies; and to extract parameters from network analysis to be used in risk factor and cluster analyses for porcine reproductive and respiratory syndrome (PRRS).

Materials and methods

Data for this project was obtained from the Ontario area regional control and elimination (ARC&E) projects, which currently contains over 40% of all swine sites of the province. Swine producers enrolled in this voluntary project answered a standardized questionnaire with demographics, biosecurity and network information. Porcine reproductive and respiratory syndrome status was obtained through a combination of historical information and on-farm sampling by herd veterinarians. Data management and descriptive statistics were conducted using SAS 9.3, and network analysis was conducted using UCINET 6. All the above-mentioned networks were visualized as two-mode and converted to one-mode networks. Two-mode (i.e. affiliation) networks relate two different types of entities, in this case swine sites and companies. These networks, when converted to one-mode, provide information on how swine sites are indirectly connected (sites that share companies are linked and companies are eliminated). Network and site-level parameters were extracted using one-mode networks. A "network component" was defined as a group of at least two swine sites being connected. A larger network combining all networks was constructed and used to extract parameters for univariable logistic regression models. The main site-level network variables of interest were degree centrality (defined as the number of links for each site), and betweeness centrality (defined as the number of times a site falls along the shortest path between other sites, an important measure of how to best control the network flow). Univariable models were constructed separately for breeding and growing pig sites using SAS 9.3 to investigate network parameters as risk factors for being positive for PRRS virus.

Results and discussion

A total of 816 swine sites were enrolled in the ARC&E project. The majority of the sites (50.25%) were finisher sites, followed by farrow-to-wean (15.07%) and nurseries (12.13%). The average number of animals per site was 1,404. Forty-three percent of the sites reported having all-in/ all-out pig flow, and approximately 35% reported having a shower facility at the site entry. Approximately 29% of all sites were PRRS positive, 43% were negative, and 27% had unknown status at time of data gathering. The network that presented the larger number of components was the gilt network, followed by the truck and boar networks (48, 35 and 33 components, respectively). The gilt and boar networks were the ones that could be more easily fragmented. When all networks were combined only five components were present, which indicates that the vast majority of swine sites were indirectly connected.

Univariable analysis showed that, for growing pig sites (n = 557), both betweeness centrality and degree centrality (measures included a combination of all networks) were positively and statistically associated with being positive for PRRS ($P < 0.01$). For breeding sites (n = 259), degree centrality was positively and statistically associated with being positive for PRRS ($P = 0.02$), but betweeness centrality was not statistically associated ($P = 0.08$). Detailed multivariable analysis with inclusion of individual networks and cluster analysis is pending.

Biotherapeutics as alternatives to antibiotics: Effects of adenoviral delivered IFN-α and G-CSF on innate and adaptive immunity in swine

Susan L. Brockmeier, DVM, PhD; Crystal L. Loving, PhD; Marcus E. Kehrli, DVM, PhD
USDA, ARS, National Animal Disease Center, Ames, Iowa

Introduction

Acceptable alternatives to the use of antibiotics in swine practice need to be explored. The use of immunomodulators is a promising area for therapeutic, prophylactic, and metaphylactic use to prevent and combat infectious disease during periods of peak disease incidence. We developed a method to circumvent the need for production of a recombinant cytokine by using a replication-defective adenovirus vector to express cytokines of interest, including interferon-α (IFN-α) or porcine granulocyte colony-stimulating factor (G-CSF). Type I interferons, such as IFN-α, contribute to innate antiviral immunity by promoting production of antiviral mediators and also play a role in the adaptive immune response. G-CSF enhances neutrophil production and release from the bone marrow and is already licensed for use in humans for treatment of neutropenia and prevention of bacterial infections in those with compromised immunity.

Materials and methods

Several experiments were conducted in weaned pigs administered the vector expressing porcine IFN-α before, simultaneously, or after exposure to virulent or an attenuated vaccine strain of porcine reproductive and respiratory syndrome virus (PRRSV) and clinical disease, viremia, immune responses, and transmission were followed.

In other studies weaned pigs were administered the vector expressing porcine G-CSF to follow the physiologic response and a preliminary challenge study looking at protective effects against *Streptococcus suis* challenge was conducted.

Results and discussion

Pigs administered the vector expressing porcine IFN-α and challenged with a moderately virulent PRRSV had lower febrile responses and decreased percentage of lung involvement. Viremia was delayed and there was a decrease in viral load in the sera of pigs. In addition, there was an increase in the number of virus-specific IFN-γ secreting cells, as well as an altered cytokine profile in the lung 14 days post-infection, indicating that the presence of IFN-α at the time of infection can alter innate and adaptive immune responses to PRRSV. We further explored the use of IFN-α as an adjuvant given with attenuated PRRSV virus vaccine and found that it appeared to abolish replication of the vaccine virus as no virus was recovered and the pigs given IFN-α did not seroconvert. Current studies are examining its use as a metaphylactic treatment for an outbreak of PRRSV to both treat and prevent transmission of the virus.

Intramuscular administration of the vector expressing porcine G-CSF was found to elicit a sustained neutrophilia, lasting nearly 3 weeks. In a pilot study using G-CSF with a *Streptococcus suis* challenge, only 1/4 of the pigs given G-CSF developed disease, while 3/4 of the non-treated pigs developed Streptococcal disease. Thus, it is possible to deliver G-CSF to pigs for a sustained increase in circulating neutrophil numbers in pigs, which may be a useful alternative to antibiotics for prevention of infectious disease, especially during times of stress and pathogen exposure such as post-weaning.

Acknowledgements

The authors would like to acknowledge partial funding for this project was from the National Pork Board.

Poster 43

An emerging H1N2 sub-cluster within the alpha H1 cluster of influenza A viruses of swine

Lauren L. Redies[1]; Marie R. Culhane[2], DVM, PhD; Susan E. Detmer[1], DVM, PhD

[1]Dept. of Veterinary Pathology, University of Saskatchewan, Saskatoon, Saskatchewan, Canada;

[2]Veterinary Population Medicine, University of Minnesota, Saint Paul, Minnesota

Introduction

Over the last 15 years, genetically and antigenically distinct groups of influenza A viruses in swine (IAV-S) have emerged.[1] The examples of 2009 pandemic H1N1 virus and variant H3N2 viruses of swine-origin underscore the threat that IAV-S pose to public health.[2,3] Consequently, there is greater need for increased IAV-S surveillance globally. This is particularly important in regions where there is very little historical sequence data available. In this study, IAV-S surveillance was conducted to examine H1N1 and H1N2 sequences from Western Canada.

Materials and methods

Diagnostic samples of lung tissue and nasal swabs were collected from pigs in Alberta (AB), Saskatchewan (SK) and Manitoba (MB) by the herd veterinarian or trained farm staff to conduct both active and passive surveillance. Over a two-year period, active surveillance was conducted monthly on 11 farms in SK and bi-monthly on 6 farms in MB. Passive surveillance was conducted through routine submissions submitted to Prairie Diagnostic Services, Inc. (PDS) and the University of Minnesota Veterinary Diagnostic Lab (UMVDL). Samples were screened for IAV-S by CFIA or USDA certified Matrix PCR tests. Positive samples were subtyped and virus isolation and sequencing of the hemagglutinin (HA) gene were attempted at UMVDL. Phylogenetic analyses were conducted in MEGA 6 using Neighbor-Joining and Maximum Composite Likelihood methods with bootstrapping of 500 replicates.

Results and discussion

Phylogenetic analysis of the HA genes of 85 IAV-S revealed that the Western Canadian H1 viruses (2012-2015) were within the pH1N1 and H1α clusters (Figure 1).[1] Within the H1α cluster there were two distinct sub-clusters. These included the H1N1 alpha viruses known to previously circulate in this region and a new cluster of H1N2 alpha viruses first detected in MB in the fall of 2013. Most of the H1N2 alpha viruses contained the same 2 amino acid (AA) deletion (amino acid positions 129 and 130 in the H1 numbering system), which is similar to the 1AA human-like deletion observed in the H1 delta-cluster viruses. This H1N2 strain containing a unique deletion at amino acid positions 129 and 130 continues to be the dominant IAV-S detected in Western Canada 2015.

While 17/36 virus sequences are from the 6 farms in one MB system, 12 are from unrelated farms in MB, 1 unrelated farm in SK and 6 farms in the United States. Additionally, one strain from AB and two strains from MB are within the H1N2 alpha sub-cluster, but do not contain this deletion. The maintenance and spread of this virus in the region suggests that deletions at sites 129 and 130 may provide a selective advantage. AA Residue 130 is located in the receptor-binding pocket. Removal of 2AA at this site will potentially reduce the size and change the shape of the binding pocket, thereby changing the ability of the virus to bind to sialic acid residues and infect host cells. It would also change the shape of the surrounding antigenic sites where antibodies would bind.

These results highlight the importance of IAV-S surveillance within overlooked regions, and is thus crucial for understanding the ecology and spatial dissemination of swine diseases. In order to develop effective control strategies for emerging influenza strains, we need to know what types of viruses are circulating. We also need to know the genetic relationships of these viruses to each other, as well as their relationship to viruses in other regions.

References

1. Anderson TK, et al. Characterization of co-circulating swine influenza A viruses in North America and the identification of a novel H1 genetic clade with antigenic significance. *Virus Res* 2015;201:24-31.

2. Nelson MI, et al. Evolution of novel reassortant A/H3N2 influenza viruses in North American swine and humans, 2009-2011. *J Virol* 2012;86:8872-8.

3. Nelson MI, et al. Continual Reintroduction of Human Pandemic H1N1 Influenza A Viruses into Swine in the United States, 2009 to 2014. *J Virol* 2015;89:6218-6226.

Figure 1: Phylogenetic relationships of H1N1 and H1N2 influenza A viruses from North American swine. The evolutionary history was inferred using the Neighbor-Joining method. The percentage of replicate trees in which the associated taxa clustered together in the bootstrap test (500 replicates) are shown next to the branches. The evolutionary distances were computed using the Maximum Composite Likelihood method and are in the units of the number of base substitutions per site. The analysis involved 89 nucleotide sequences. There were a total of 1630 positions in the final dataset. Evolutionary analyses were conducted in MEGA 6 software.

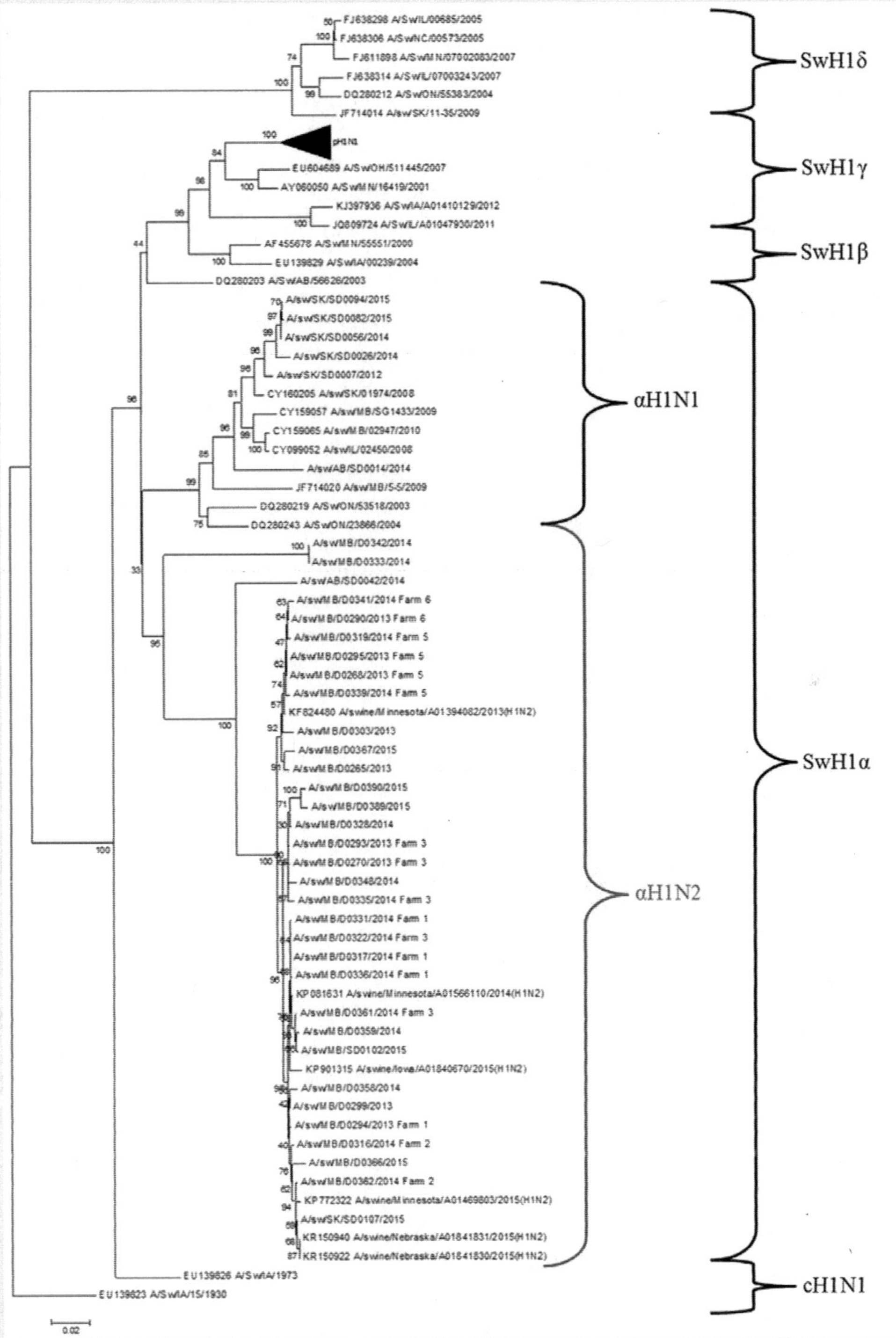

The effect of agitation and dose position during storage on semen quality

Kara Stewart[1], PhD; Douglas D. Groth[2], DVM; Hanneke Feitsma[3], DVM
[1]Purdue University, West Lafayette, Indiana; [2]Carthage Veterinary Service, Ltd., Carthage, Illinois;
[3] Minnesota Swine Reproduction Center, Northfield, Minnesota

Introduction

Over the last 30 years, the North American swine industry has adopted the technology of artificial insemination (AI). Current recommendations are to store the semen doses horizontally to increase the surface area of sperm having contact with the semen extender and to gently agitate the samples 1-2 times per day to prevent sedimentation.[1,2,3] The objective of this study was to evaluate different semen storage protocols with regard to storage position and agitation on semen quality over 10 days in storage.

Materials and method

Semen was collected from 6 boars and divided into 2 pools of 3 ejaculates each and then diluted in 4 different long-term commercial semen extenders within 15 minutes of collection. Doses, each containing 2.75 Billion sperm cells demonstrating 80% or greater motility, and 70% or greater normal morphology, were prepared for the 4 treatment groups: semen samples stored horizontally or vertically (H, V) and agitated 1×/day or no agitation (A, NA) for 10 days at 17° C. For each analysis day (day 1, 3, 7 and 10) separate doses was prepared. Semen quality analysis was performed analyzing the motility (Ceros II, IMV Technologies), morphology (bright-field microscopy evaluating, 200 cells), live-dead assessment (LIVE/DEAD® Sperm Viability Kit, Life Technologies), acrosome reaction (using calcium ionophore), and sperm chromatin structure assay (SCSA, acridine orange stain). The entire study was repeated with the same 6 boars 2 weeks after the first study. Data was analyzed using analysis of variance procedures for repeated measures in SAS® v. 9.3. Orthogonal contrasts were used to compare horizontal storage to vertical storage.

Result

There was no effect of the extender or pool on the treatment results. Motility and progressive motility were greater for horizontally stored samples compared to vertically stored samples ($P \le 0.001$ for both). No motility differences were observed between agitated and not agitated semen in storage. There were no differences in percentage of morphologically normal cells ($P = 0.1112$) and all treatment groups averaged > 80% normal morphology. Additionally, there were no effects of treatment on the percentage of normal acrosomes ($P = 0.4487$), or in the percentage of proximal droplets ($P = 0.8126$). There were statistically significant differences among treatments in percentage distal droplets ($P = 0.0354$) and tail abnormalities ($P = 0.0183$). There was a treatment by day interaction in the percent of cells with distal midpiece reflexes (DMR, $P = 0.0036$) with no agitation and vertical storage showing increased DMR at day 10 of storage. There were no differences among treatments in the percentage of live cells ($P = 0.5321$). There was a significant effect of treatment on percentage of cells undergoing the acrosome reaction ($P = 0.0205$), whereas Day significantly impacted this as well ($P < 0.0001$). However, there was not a treatment by day interaction ($P = 0.8744$). A significant treatment effect on the defragmentation index, following acridine orange staining was observed
($P = 0.0065$). (Table 1)

Conclusion

Overall conclusion was that there were no sperm motility differences observed between agitated and none agitated semen doses during 10 days of storage, whereas storage in horizontal position showed, although not always significant, but slightly better results. The significant treatment differences for other sperm quality parameters were, due to the low level, of no real effect and difficult to explain.

Acknowledgement

We would like to thank the Bear Creek Genetics, LLC boar stud for their kind collaboration.

References

1. Simmet, C., D. Rath, and S. Lorton. 1998. Does Sedimentation of Liquid Boar Semen Influence Semen Quality During Storage. Proceedings International Pig Veterinary Society Congress, Birmingham, England.

2. Flowers, W.L. 1996. Semen evaluation, extension, packaging and transportation methods. Proceedings of the 27th Meeting of the American Association of Swine Practitioners. Nashville, TN. Pp 469-479.

3. Rodriguez-Gil, J.E., and T. Rigau. 1995. Effects of slight agitation on the quality of refrigerated boar sperm. Animal Reproduction Science 39:141-49.

Table 1: Treatment effect on motility, progressive motility, morphologic normal cells, normal acrosomes, tail abnormalities, DMR, proximal and distal cytoplasmic droplets.

Variable	Treatment means ± SE				*P*-value	
	NAV	NAH	AV	AH	Treatment	Treatment *Day
% Motile cells	57.5 ± 2.0[a]	63.2 ± 1.5[b]	58.9 ± 2.3[a,c]	62.2 ± 1.6[b,c]	0.0004	0.0516
% Progressive motile cells	49.7 ± 1.9[a]	55.5 ± 1.4[b]	51.9 ± 2.1[c]	55.1 ± 1.5[d]	< 0.0001	0.0061
% Morphologic normal cells	90.0 ± 0.7	88.9 ± 0.6	90.4 ± 0.6	89.4 ± 0.5	0.1112	0.3151
% Normal acrosomes	83.9 ± 1.8	84.4 ± 1.7	83.7 ± 1.8	83.4 ± 1.8	0.454	0.7164
% Tail abnormalities	4.6 ± 0.5[a]	3.3 ± 0.3[b]	2.9 ± 0.4[b]	3.3 ± 0.4[b]	0.0183	0.5482
% DMR	3.3 ± 0.4	2.7 ± 0.3	2.5 ± 0.3	2.6 ± 0.4	0.2561	0.0036
% Proximal droplets	4.3 ± 0.5	4.8 ± 0.4	4.4 ± 0.6	4.8 ± 0.4	0.8126	0.6951
% Distal droplets	9.2 ± 0.8[a]	11.9 ± 0.8[b,c]	10.4 ± 0.8[a,c,d]	11.7 ± 1.0[b,d]	0.0354	0.7896

NAV: no agitation, vertical storage; NAH: no agitation, horizontal storage; AV: agitation, vertical storage; AH: agitation, horizontal storage; [a,b,c,d] Results with different superscripts differ significantly *($P < 0.05$)*.

Dynamics of influenza virus transmission in a swine herd and analysis of risk factors for recurrent infections

J.B. Ferreira[1], DVM, MSc; H. Grgić[1], DVM, MSc, PhD; R. Friendship[1], DVM, MSc; G. Wideman[2], DVM; È. Nagy[3], DVM, PhD DSc; Z. Poljak[1], DVM, MSc, PhD
[1]Department of Population Medicine, University of Guelph, Guelph, Ontario, Canada; [2]South-West Veterinary Services, Stratford, Ontario; [3]Department of Pathobiology, University of Guelph, Guelph, Ontario, Canada

Introduction

Influenza in swine is considered to be a relatively simple infection with short duration of infectiousness. At the population level, the disease is commonly recognized by the sudden appearance of respiratory signs and also by rapid recovery of the sick animals. Results from recent field studies, however, suggest that circulation of influenza A virus (IAV) in individual animals may not be as simple as previously thought. The epidemiology of influenza in large multi-site and multi-source herds, in particular, is not well described. Therefore, the objectives of this study were to describe patterns of IAV infection in pigs after weaning in multi-source nursery herds, and to determine factors that contribute to such patterns.

Materials and methods

The nursery included in the study was a ~2200-head barn operated on an all-in/all-out basis by site. Pigs from 5 different sow herds were mixed in 4 rooms (each with 24 pens). In the first 2 hours of arrival to the nursery barn, 400 pigs were selected for the initial virological testing. Additionally, 81 and 75 pigs were included in the longitudinal study for ongoing weekly testing for influenza virus, respectively for trial 1 and 2 in two different periods. Virus isolation and propagation were done in Madin-Darby canine kidney (MDCK) cells. Serology was performed by hemagglutination inhibition (HI) using 6 different influenza viruses. Risk factor analysis for virological positivity and likelihood of recurrent infection was conducted using logistic regression model and a Cox's regression model that accounts for recurrent events (PWP-CP). Full genome sequencing of the isolated viruses was performed by the Illumina Miseq platform.

Results and discussion

In trial 1 no pigs tested positive for influenza virus at the first sampling, but all pigs included in the longitudinal study were virus-positive at some point in time, with 53% being positive more than once. At one specific moment all pigs were positive and shedding virus. Overall, 47%, 10%, and 43% of pigs were positive on one sampling, subsequent samplings and recurrently; respectively. There was a significant difference in positivity of pigs among the sow sources ($P < 0.01$). In trial 2 a different pattern was observed with less than 50% of pigs being positive. Similar to trial 1, there was a significant difference among sources on having positive results in the virological test ($P < 0.01$). In trial 1, pigs with higher titers of heterologous maternal immunity were more likely to have recurrent infections ($P < 0.01$). Having maternal immunity for different pathogens, such as PRRSV and *Mycoplasma hyopneumonia*, did not play a role as a risk factor for recurrent infection ($P > 0.05$). Sequencing of the HA gene indicated that the viruses isolated in trial 1 (H3N2) and trial 2 (H1N1) had a high level of identity (> 99.5%).

Results from these studies indicate that influenza virus can circulate during the nursery phase in a cyclical pattern. Although such a pattern could be due to a variety of mechanisms, the results obtained in this study suggest that reinfection of pigs with the same virus likely play a role in the development of such a pattern.

Evaluation of the risk of a feed mill being contaminated with PEDV or PDCoV

L. Greiner, PhD
Carthage Innovative Swine Solutions, LLC, Carthage, Illinois

Introduction

With the outbreak of PEDV and PDCoV in the United States, many questions have arisen concerning the movement of the viruses throughout the United States. One area of interest was the involvement of transport vehicles onto farms. A common transport vehicle that visits multiple farms in a day is the feed truck. With both feed trucks and staff returning from farms and incoming ingredient trucks delivering feedstuffs, the question is what is the risk of various areas of the mill to be found positive for either virus on any given day. Therefore, the objective of this study was to investigate the risk of key areas in the mill of testing positive for either PEDV or PDCoV.

Materials and methods

Twenty-four feed mills from various regions (Eastern, Midwest, Western, and Northern Midwest) in the US were evaluated in this study to assess the risk of different areas within a mill to test positive for either porcine epidemic diarrhea virus (PEDV) or swine delta coronavirus (PDCoV). Samples were collected at each feed mill for up to 5 days. Sample areas included: both foot pedals of feed delivery truck, bulk ingredient unloading pit, inside mixer/pellet cooler, mill office floor, inside feed compartment on feed truck, and incoming bagged ingredient truck (inside of truck near site of off-loading). Samples were collected using a technique similar to the Swiffer® technique described by Dee et al. In brief, a wet gauze pad was used as the transfer material and was placed into 25 mL of phosphate buffered saline (PBS). The gauze was then squeezed multiple times in the solution and then the supernatant was poured into a sterile, transfer tube. The samples were submitted to diagnostic facilities for PCR testing for PEDV and PDCoV. The results of the feed mills for each collection site within the mill and for the mill itself were compiled. Feed mills were assigned a letter for anonymity during analysis and final report. Data were analyzed as probabilities and risk assessments for each area within the mill being tested. In addition, a break-down of the risk of contamination of the mills that already serviced positive herds was evaluated.

Results and discussion

Raw data percentages demonstrated that no samples tested positive for PEDV. However, 5% of the truck foot pedals and 1% of the bulk ingredient pit tested suspect for PEDV. Porcine delta coronavirus was found on 3.4% of the foot pedals of the trucks and 2.2% of the office floors tested suspect. Of the feed mills tested, 75% of the mills were providing feed to herds infected with PEDV and 21% were feeding herds infected with PDCoV. An additional 33% of the feed mills reported "unknown" of the PDCoV status of the herds that they were feeding. As the number of days increase, the probability of a positive/suspect result also increases for locations such as the foot pedals, office floor, bulk ingredient pit and the inside of the feed truck compartment. Based on the results by day and the percentage of mills that were feeding positive herds, after one day of negative test results on foot pedals, there is a 74% chance that the foot pedals could test positive for PEDV with additional days of testing. When evaluating the probability of a PDCoV positive/suspect result on the foot pedals, there is a 28% chance that the mill will have a positive result in the next four days after the first day of negative results regardless of the health status of the herds being fed from that mill. One feed mill that was servicing PDCoV negative herds tested PCR suspect on the office floor, which indicates that the health status of the herds being serviced by the feed mill may not be the only potential cause for a positive/suspect finding. In conclusion, this study demonstrates that both PEDV and PDCoV virus particles can be detected at different locations associated with a feed mill. This study does not determine if the virus particles are infectious and therefore, only demonstrates the probably of detecting virus particles at the various locations based on the number of days testing occurs and also based on the health status of the herds being fed from the mill.

Assessment of the immunocrit ratio assay for evaluation of colostrum quality in sows induced to farrow and inseminated using single dose fixed time insemination

A. R. Hanson[1], PhD, MS; J. L. Vallet[2], PhD; M. E. Johnston[3], MS; P. E. Yeske[1], DVM, MS
[1]SVC Research, Saint Peter, Minnesota; [2]US Department of Agriculture, Agriculture Research Service, Clay Center, Nebraska; [3]JBS-United Animal Health, Sheridan, Indiana

Introduction

Sufficient intake of quality colostrum is essential for the success of newborn piglets. Relatively high immunoglobulin (Ig) content generally indicates colostrum of high quality. IgG is the predominant Ig in colostrum. The immunocrit ratio assay has been developed as a simple method to assess concentrations of IgG in porcine serum,[1] but it is unknown if this assay has similar utility for colostrum or milk.

Objective

The objective of this study was to evaluate the relationship between concentrations of IgG and the results of the immunocrit ratio (ICR) assay in sow milk.

Materials and methods

One hundred fifty one sows (parity 2 to 13) were stratified by parity and breeding group and randomly assigned to be either induced on d 113 post insemination (2 ml intramuscular injection of Lutalyse®) or not induced. All sows were administered Ovugel® (JBS-United) on d 4 post weaning and inseminated using single dose fixed time insemination on d 5 post weaning. Milk samples were collected and serum obtained from 3 pigs/litter within 18.25 h of farrowing (median = 3 h). Milk was analyzed for IgG using a commercial ELISA (Immunology Consultants Laboratory, Inc.; Kit E-5G) and the ICR assay described elsewhere with duration of centrifugation extended to 15 min.[1] Serum samples were evaluated for ICR as described elsewhere.[1] Relationships between continuous variables were assessed using linear regression in JMP 7.0 (SAS Institute, Inc.). Evaluation of treatment effects was conducted using ANOVA procedures in JMP 7.0, and the model included effects of parity, sample collection time, and farrowing group.

Results and discussion

Three milk samples were omitted because a Cook's distance test indicated they were outliers. Treatment did not significantly affect pig serum or milk ICR or IgG levels in milk ($P > 0.20$). Litter mean serum ICR had weak negative associations with milk ICR and IgG ($R^2 = 0.03$ and 0.03, respectively; $P \leq 0.05$).

Both ICR and IgG declined as the time of sample collection post farrowing increased ($R^2 = 0.20$ and 0.16, respectively; $P < 0.01$). The ICR and IgG concentrations of milk were moderately correlated ($R^2 = 0.41$; $P < 0.01$; Figure 1). Lack of a strong relationship indicates that components in milk in addition to IgG react in the ICR assay. Samples with low IgG also had low ICR indicating that the assay could be used to screen for colostrum quality. Furthermore, the immunocrit is considerably less expensive and easier to conduct than alternative techniques to assess specific Ig.

References

1.Vallet, J. L. 2013. A simple novel measure of passive transfer of maternal immunoglobulin is indicative of preweaning mortality in piglets. Vet. J. 195. 91-97.

Figure 1: Immunocrit ratio related positively with IgG concentration of sow milk

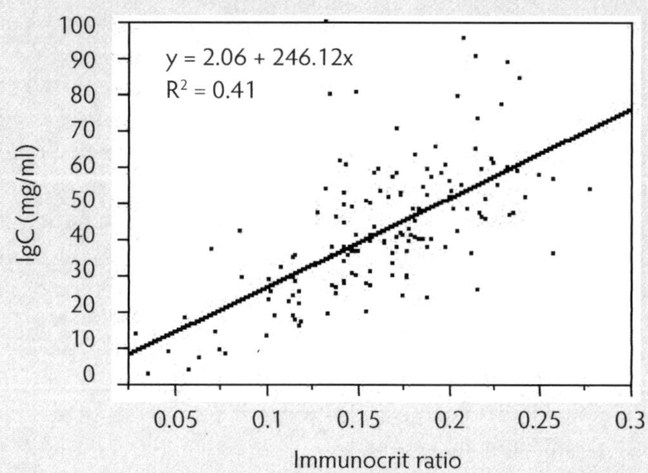

Absence of human innate immune evasion complex in LA-MRSA ST5 strains isolated from pigs, swine facilities, and humans with swine contact

Samantha Hau[1], DVM; Timothy Frana[1], DVM, PhD, MPH; Peter R Davies[2], PhD;
Jisun Sun[2], DVM, MS; Tracy L Nicholson[3], PhD
[1]Department of Veterinary Diagnostic and Production Animal Medicine, College of Veterinary Medicine, Iowa State University, Ames, Iowa; [2]Department of Veterinary Population Medicine, College of Veterinary Medicine, University of Minnesota, Saint Paul, Minnesota; [3]National Animal Disease Center, Agricultural Research Service, USDA, Ames, Iowa

Introduction

Since its first ties to swine, livestock associated methicillin-resistant *Staphylococcus aureus* (LA-MRSA) has raised public health concerns because livestock may serve as the largest reservoir of MRSA outside a hospital setting. In contrast to Europe and Asia, where the primary sequence types (ST) of swine associated LA-MRSA are ST398 and ST9 respectively, US isolates are more diverse with ST5, ST398, and ST9 being isolated.[1] The presence of ST5 MRSA in swine increases concern because, unlike ST398 and ST9, ST5 MRSA is among the most prevalent lineages causing human disease.[2] A primary factor contributing to the clinical impact of ST5 MRSA in the hospital (HA-MRSA) and community (CA-MRSA) is the capacity of ST5 isolates to acquire mobile genetic elements carrying virulence factors and antimicrobial resistance genes. Previous reports establishing the absence of most virulence genes from swine associated ST398 isolates[3] and the lack of disease reports in humans occupationally exposed to swine carrying ST5 MRSA, suggests virulence factors contributing to human disease may also be uncommon in swine associated ST5 strains.

Materials and methods

ST5 isolates collected from swine, swine facilities, and humans with short- and long-term swine contact (LA-MRSA) and humans with no swine contact (HA-/CA-MRSA) were PCR screened for the presence of human-specific virulence factors carried by β-hemolysin converting bacteriophages including: staphylococcal complement inhibitor, chemotaxis inhibitory protein, staphylokinase, staphylococcal enterotoxin A, and staphylococcal enterotoxin P. Southern blotting was used to confirm if the β-hemolysin gene was intact.

Results

None of the LA-MRSA ST5 isolates screened carried the virulence factors found in β-hemolysin converting bacteriophages (0/82). An intact β-hemolysin gene was also found in all isolates (82/82) indicating the absence of β-hemolysin converting bacteriophages in these isolates. The β-hemolysin converting bacteriophage was found to be present in 90.4% (66/73) of isolates from humans with no swine contact ($P < 0.0001$). The PCR results for the genes tested are broken down by isolate and can be found in Table 1.

Conclusion

A clear genetic difference was noted in MRSA ST5 isolates from swine associated sources compared to isolates from humans with no swine contact. The findings here indicate a potential genetic adaption to swine colonization and a loss of virulence in swine associated LA-MRSA ST5 isolates. This parallels previous reports examining LA-MRSA ST398 and suggests a generally reduced capacity of livestock adapted *S. aureus* to cause significant disease in immunocompetent humans.

References

1. Frana TS, et al. Isolation and characterization of methicillin-resistant Staphylococcus aureus from pork farms and visiting veterinary students. *PLoS One*. 2012; 8: e53738.

2. Monecke S, et al. A field guide to pandemic, epidemic and sporadic clones of methicillin-resistant Staphylococcus aureus. *PLoS One*. 2011; 6: e17936.

3. Price LB, et al. Staphylococcus aureus CC398: host adaptation and emergence of methicillin resistance in livestock. *MBio* 2012; 3: e00306.

Table 1: Immune-evasion complex and β-hemolysin converting bacteriophage screening results for all isolates.

Isolate source		Gene tested						
		scn	*chp*	*sak*	*sea*	*sep*	*int*	Intact *hlb*[a]
Agricultural	Human: short-term contact	0 (0/9)[b]	0 (0/9)	0 (0/9)	0 (0/9)	0 (0/9)	0 (0/9)	100 (9/9)
	Human: long-term contact	0 (0/9)	0 (0/9)	0 (0/9)	0 (0/9)	0 (0/9)	0 (0/9)	100 (9/9)
	Pig	0 (0/38)	0 (0/38)	0 (0/38)	0 (0/38)	0 (0/38)	0 (0/38)	100 (38/38)
	Environment	0 (0/26)	0 (0/26)	0 (0/26)	0 (0/26)	0 (0/26)	0 (0/26)	100 (26/26)
Clinical		90.4 (66/73)	87.7 (64/73)	90.4 (66/73)	1.4 (1/73)	37.0 (27/73)	90.4 (66/73)	9.6 (7/73)

a Data reported represent the results from Southern blotting.
b Data reported as percent of isolates positive for each gene tested. Number of positive isolates is noted in parenthesis

PEDV shedding patterns and antibody kinetics in commercial growing pigs

J. Bjustrom-Kraft[1], BS; K. Woodard[1], DVM; L. Giménez-Lirola[1], PhD; M. Rotolo[1], DVM; C. Wang[1], PhD; P. Lasley[2], DVM, MS; Q. Chen[1], DVM, MS; J. Zhang[1], MD, PhD; D. Baum[1], DVM, PhD; P. Gauger[1], DVM, PhD; K.J. Yoon[1], DVM, PhD; J. Zimmerman[1], DVM, PhD; R. Main[1], DVM, PhD
[1]Iowa State University Veterinary Diagnostic Laboratory, Ames, Iowa; [2]Murphy Brown of Missouri, Princeton, Missouri

Introduction

The aims of this prospective longitudinal field study were to (1) describe the pattern of PEDV shedding and comparative sensitivity of alternative sampling methods in growing pigs as revealed by RT-PCR testing of individual pig fecal swabs, pen fecal samples, and pen oral fluids (OF); (2) describe the kinetics of PEDV antibody by ELISA (IgA, IgG) testing of individual pig serum and pen OF samples; and (3) establish cutoffs and performance estimates for PEDV "whole virus" IgA and IgG ELISAs (WV IgA or IgG ELISA).

Materials and methods

Samples to achieve these aims were collected from two production sites (one PEDV positive; one PEDV negative). Farm 1 was a 52-pen WTF barn stocked with ~800 pigs. Pen samples (feces and OFs) and pig samples (fecal swabs and serum) were collected from the same 6 pens and a convenience sample of 5 pigs in each of the 6 pens at placement and at ~2-week intervals for 27 weeks. At ~13 weeks of age, this PEDV-negative population was exposed to PEDV using standard field exposure techniques. Farm 2 consisted of 3 identical 40-pen WTF barns, each stocked with ~900 pigs. Pen OF samples were collected from 36 pens (4 pens were not stocked) in each of the 3 barns and serum samples were collected from a convenience sample of 20 pigs in 2 pens (10 pigs per pen) in each barn. Sampling began at two weeks post-placement and was done weekly for a total of 9 samplings. Pen feces, OF specimens and fecal swabs were tested by PEDV RT-PCR; OF and sera were tested by PEDV WV ELISA (IgG, IgA) using procedures available for routine testing at the ISU Veterinary Diagnostic Laboratory.

Results

On Farm 2, all oral fluid samples collected during the monitoring period were PEDV RT-PCR-negative. On Farm 1, PEDV was detected by RT-PCR at the first sampling post exposure (6 DPE) in individual fecal swabs, pen fecal samples and pen OF. The last RT-PCR positives were detected in fecal swabs and OFs on 69 DPE. An analysis of the adjusted RT-PCR Cq values showed differences in the concentration of PEDV in the three sample types over time ($P = 0.0005$). At 6 DPE, the concentration of PEDV was higher in pen feces compared to fecal swabs ($P = 0.0001$) and oral fluids ($P = 0.0088$). At all sampling points through 69 DPE, no significant difference was detected in the concentration of virus in oral fluid and pen feces samples. Overall, the highest percentage of positive samples was observed in OF. At 27, 41 and 55 DPE (17, 19, 21 weeks of age) the data showed significant differences between sample types ($P < 0.03$). IgG and IgA responses were detected in the OF and serum samples collected at 13 DPE. The OF IgA response continued to increase through 97 DPE, while serum IgA responses peaked at 27 DPE.

Conclusions

Antibody (IgG and IgA) kinetics in oral fluids and comparison with the serum antibody responses has not previously been reported. PEDV IgG and IgA are readily detected in both serum and oral fluid by 10 - 14 days post exposure. The oral fluid IgA response is particularly noteworthy because of the level and duration of response. To the authors' knowledge, no one has looked at oral fluid IgG or IgA. Additional research is needed to evaluate the ability of the OF IgA response to predict neonatal protection against PEDV.

Dietary iron deficiency and excess affect intestinal permeability and inflammation in pigs

Y. Li[1], PhD; S. Hansen[2], PhD; L. Borst[3], DVM, PhD; J. Spears[4], PhD; A. Moeser[1]
[1]Gastrointestinal Stress Biology Laboratory, Department of Large Animal Clinical Sciences, College of Veterinary Medicine, Michigan State University, East Lansing, Michigan; [2]Department of Animal Science, Iowa State University, Ames, Iowa; [3]Department of Population Health and Pathobiology, College of Veterinary Medicine, North Carolina State University, Raleigh, North Carolina; [4]Department of Animal Science, North Carolina State University, Raleigh, North Carolina

Introduction

Iron (Fe) deficiency is one of the most common nutritional disorders worldwide. Poor Fe intake, impaired Fe absorption and chronic blood loss (inflammatory bowel disease) cause Fe deficient anemia.[1] However, dietary Fe fortification could result in gut dysfunction, diarrhea and infection related morbidity.[2] The conflict between Fe deficiency anemia and the health issue of overdose Fe supplementation has spurned greater interest in the role of dietary iron intake on human intestinal health. In agricultural pig production, Fe supplementation to young pigs to prevent Fe deficiency in the rapidly growing pig. However, commercial starter diets often exceed this requirement by up to five-fold, due to Fe being supplemented to the diet, and the fact that several swine ingredients contain excessive amounts of Fe, including blood meal, dicalcium phosphate and limestone. In contrast with the importance and prevalence of Fe deficiency and the common use of oral Fe supplementation, very little is known about the influence of deficient or excess dietary Fe on intestinal function. In the present study we investigated the effects of feeding newly weaned pigs a diet low, adequate, or high in Fe on the duodenal mucosal barrier function, morphology and inflammation. We hypothesized that feeding low or high Fe diets to pigs would be disruptive to duodenal barrier function and morphology relative to pigs fed adequate Fe diets and feeding high Fe diets would induce intestinal inflammation.

Materials and methods

Twenty-four weanling male pigs (21 d of age and 5.5 ± 0.4 kg BW) were assigned to one of 3 treatments: 1) Fe-deficient basal diet [low Fe; (L-Fe)], 2) Basal diet with 100 mg supplemental Fe/kg DM (FeSO4) [adequate Fe; (A-Fe)], 3) Basal diet with 500 mg supplemental Fe/kg DM [high Fe; (H-Fe)]. After 33 days of trial, all pigs were euthanized and samples of proximal duodenum were collected for Ussing chamber assay, histopathology, and gene and protein analysis. Transepithelial electrical resistance (TER; Ω•cm2), short-circuit current (Isc), and mucosal-to-serosal fluxes of [3H]mannitol of proximal duodenum was measured by Ussing chamber, as descripted previously.[16] Duodenal tissues were 10% formalin-fixed, Paraffin-embedded, and stained (5 µm section) with either Prussian blue (Fe staining) or HE for histological analysis. Villi were measured using the 20× objective. Neutrophil counts were performed at 40× magnification using a 0.043 mm2 grid and were expressed as number of neutrophils/mm2. Total 3 µg RNA samples were isolated from frozen duodenal scrapings, and used for cDNA synthesis. Semi-quantitative real-time PCR was used to determine the relative quantities of transcripts of the genes of interest: 60S ribosomal subunit (RPL4), beta-actin (ACTB), tumor necrosis factor alpha (TNFA), interleukin 1 beta (IL1B1), interleukin 6 (IL6), interleukin 8 (IL8), interleukin 10 (IL10) and interferon gamma (IFNG). The gene expression normalized to the geometric mean of RPL4 and ACTB. Data were analyzed by using the general linear model procedure of SAS (version 9.3; SAS Institute).

Results and discussion

The basal diet was analyzed to contain 20 mg/kg DM Fe, which fulfilled approximately 25% of iron requirement of young pigs. Animal growth rate was significantly diminished by L-Fe treatment. Anemia was noted in L-Fe pigs by the end of the study (d 32) as evidenced by hemoglobin concentrations of 62, 128, and 133 g/L for L-Fe, A-Fe, and H-Fe pigs, respectively. Intestinal TER was reduced ($P < 0.05$) in pigs fed L-Fe and H-Fe diets compared with A-Fe controls, while 3H-mannitol permeability was greater ($P < 0.05$) in H-Fe pigs. Also, Isc was greater in both L-Fe and H-Fe compared with the A-Fe group. Villus heights in the duodenum were reduced in H-Fe pigs compared with other dietary treatments ($P < 0.05$). Intestinal mucosal and liver malondialdehyde (MDA) concentrations, a marker of oxidative stress, were increased in a dose-dependent manner from low to high dietary Fe concentration ($P < 0.05$). Neutrophil counts were higher in both L-Fe and H-Fe groups compared with A-Fe group which was in line with increased ($P < 0.05$) relative gene expression of selected pro-inflammatory cytokines (TNFα, IL1β, IL6 and IL8). Both dietary Fe deficiency and excess have a significant deleterious impact on intestinal physiology and inflammation. These data suggest that adequate, but not excessive dietary Fe are required to maintain intestinal barrier health and function. Understanding the effect of dietary Fe on the intestinal integrity could alter the strategy of iron supplementation and anemia therapy.

References

1. Stein J et al. Diagnosis and management of iron deficiency anemia in patients with IBD. *Nat Rev Gastroenterol Hepatol.* 2010;7(11):599–610.

2. Jaeggi T et al. Iron fortification adversely affects the gut microbiome, increases pathogen abundance and induces intestinal inflammation in Kenyan infants. *Gut.* 2015;64(5):731-42.

3. Moeser AJ et al. Recovery of mucosal barrier function in ischemic porcine ileum and colon is stimulated by a novel agonist of the ClC-2 chloride channel, lubiprostone. *Am J Physiol Gastrointest Liver Physiol.* 2007;292(2):G647-56.

Routine manure removal from swine operations: A potential mechanism for pathogen dispersion

M. L. Millerick-May[1], MSc, PhD; G. A. May[2], E. Ferry[2], M. E. Gemus-Benjamin[3] MS, DVM
[1]College of Human Medicine, Michigan State University, East Lansing, Michigan; [2] Extension Agriculture and Agribusiness, Michigan State University, East Lansing, Michigan; [3]College of Veterinary Medicine, Michigan State University, East Lansing, Michigan

Background

In early 2013, the porcine epidemic diarrhea virus (PEDV) was identified in the United States, with infection spreading quickly to over 30 states and mortality reaching almost 100% in suckling pigs. In early 2014, quickly following the PEDV epidemic, a porcine deltacoronavirus (PDCoV) was identified in two states in the US and rapidly spread throughout the remainder of the country and into Canada, also resulting in significant losses. While nationwide surveillance data for both PEDV and PDCoV suggests that incidence and prevalence are on the decline, current reporting requirements likely result in a significant underestimate of the actual disease burden as only 'new' outbreaks within a naïve herd or in herds where there has been a demonstrated 'disease-free' period are required to be reported. At present, there is a gap in knowledge relative to the identification of routine management practices performed at livestock facilities that – even when in compliance with established biosecurity protocols - have the potential to disperse pathogens to uninfected herds, and re-infect herds at previously affected sites. As a result, and based on recent studies indicating extended survival of pathogens in manure we hypothesized that PEDV was likely to be present in bioaerosol(s) generated during routine manure removal activities.

Methods

Ten swine production facilities were recruited to participate in our pilot study from April-October 2015. Bulk samples were collected from storage pits/lagoons to verify the presence of PEDV prior to pump-out. Three sets of samples were collected during pump-out for each of three 'loads' (tankers) filled at each site. Swab samples were collected immediately inside the port/hatch into which the slurry was pumped as an indicator of concentration of PEDV in slurry as pumped into the tank. All samples collected were submitted to the Animal Disease Research and Diagnostic Laboratory at South Dakota State University for analysis.

Results

All (100%) bulk samples (storage pits/lagoons) tested positive for PEDV with cycle thresholds (Ct) ranging from 23.12-33.15. Ninety-percent of swab samples collected from the inside hatch of the manure hauler tested positive for PEDV with Ct's ranging from 29.07-36.20. Four of ten (40%) sets of bioaerosol samples were positive for PEDV (Ct's 31.85-37.33), with three of the four that tested negative not having pigs with clinical signs on site for > 4 months. At one site not testing positive for PEDV in bioaerosol, porcine reproductive and respiratory syndrome virus (PRRSV) was captured in one site not positive for PEDV, and more than half of samples testing positive for PEDV also tested positive for PDCoV. In all cases where the results of bioaerosol sampling were positive for pathogen (PEDV, PDCoV, PRRSV), aerosol (mist), splashing, and roadway puddling was observed.

Discussion

Results indicate that pathogen containing bioaerosol have the potential to be dispersed during routine manure removal practices. Cycle thresholds observed, while 'high',are consistent with what would be expected as a result of ambient dilution. Current biosecurity procedures that address direct contact of a surface with contaminated fecal matter are likely to be insufficient in preventing spread of pathogens. Management activities external to the unit and not in direct contact with pigs are also important factors that may cause spread not only to units within close proximity (reinfection), but possibly units at significant distances from the point of aerosol generation. Our results indicate that future studies are warranted with an emphasis on management practices resulting in little or no generation of pathogen laden aerosol (as well as splashing/puddling), results from which could provide the rationale for development of best practices aimed at pathogen control at the 'source', rather than trying to control exposures at the level of the 'receiver' (animal).

Support for our pilot study was provided by the Michigan Pork Producers Association and the Michigan Alliance for Animal Agriculture.

Association of growth performance with clinical signs following co-infection with porcine reproductive and respiratory syndrome virus (PRRSV) and porcine circovirus type 2 (PCV2) in nursery pigs

Megan C. Niederwerder, DVM, PhD; Steve S. Dritz, DVM, PhD; Raymond R. R. Rowland, MA, PhD
Department of Diagnostic Medicine/Pathobiology, College of Veterinary Medicine,
Kansas State University, Manhattan, Kansas

Introduction

Average daily gain (ADG) is used in swine production as an objective measurement of overall health and performance. On a global scale, co-infections involving porcine reproductive and respiratory syndrome virus (PRRSV) and porcine circovirus type 2 (PCV2) are common and contribute to several disease syndromes that reduce ADG. Clinical signs associated with PRRSV and PCV2 co-infections are diverse and include pallor or jaundice, lethargy or depression, muscle wasting, decreased body condition, dyspnea, open-mouth breathing, rhinorrhea, diarrhea, aural cyanosis, lameness, joint effusion and coughing. The purpose of this study was to standardize collection of clinical data in a co-infected nursery pig model, document the time course of clinical disease post-infection, and investigate the impact of clinical disease on animal welfare and ADG.

Materials and methods

Thirteen clinical phenotypes were documented and medical records were compiled for 476 nursery pigs (average age 53.3 ± 6.9 days) co-infected with PRRSV and PCV2. In addition to the clinical phenotypes listed above, the administration of veterinary treatment was also documented as an indicator of overall morbidity. Phenotypes were documented daily for each pig until 42 days post-infection (dpi) or until death. ADG was calculated from 0-42 dpi. ADG was compared between pigs with and without each clinical phenotype using the unpaired t-test. Linear regression was used to analyze the strength of the association between the duration of each clinical phenotype and ADG.

Results and discussion

Aural cyanosis was the only clinical phenotype that peaked in the first half of the post-infection period (0-20 dpi) and correlated with the peak in PRRSV viremia. In contrast, the majority of clinical phenotypes peaked in the second half of the post-infection period (21-42 dpi) and corresponded to increased PCV2 replication. The presence of muscle wasting was associated with the greatest reduction (62%) in ADG; 0.31 kg and 0.81 kg for pigs with and without muscle wasting, respectively ($P < 0.0001$). Whereas, the duration of dyspnea had the strongest negative association ($r = -0.51$) with ADG (Figure 1A). Large reductions in ADG were associated with a single day of open mouth breathing (0.227 kg, Figure 1B). Lameness and aural cyanosis were the only two clinical phenotypes where no significant association was detected with ADG for either clinical presence or duration. In this experimental model, respiratory distress and muscle wasting were the clinical phenotypes associated with the greatest impact on ADG. This study provides insight into which clinical signs may predict poor performance in nursery pigs and how detailed clinical data collection may provide an opportunity for early intervention and improved outcome. Overall, reducing causes of respiratory and wasting disease should be considered high priority for improving animal welfare and growth performance in swine production.

Acknowledgements

This work was supported by the USDA NIFA Award #2013-68004-20362.

Figure 1: Linear regression plots showing association between duration of dyspnea **(A)** and open-mouth breathing **(B)** with average daily gain (ADG). ADG is shown for each pig from 0-42 dpi (■) or until death (○). Linear regression line calculated for surviving pigs only (■).

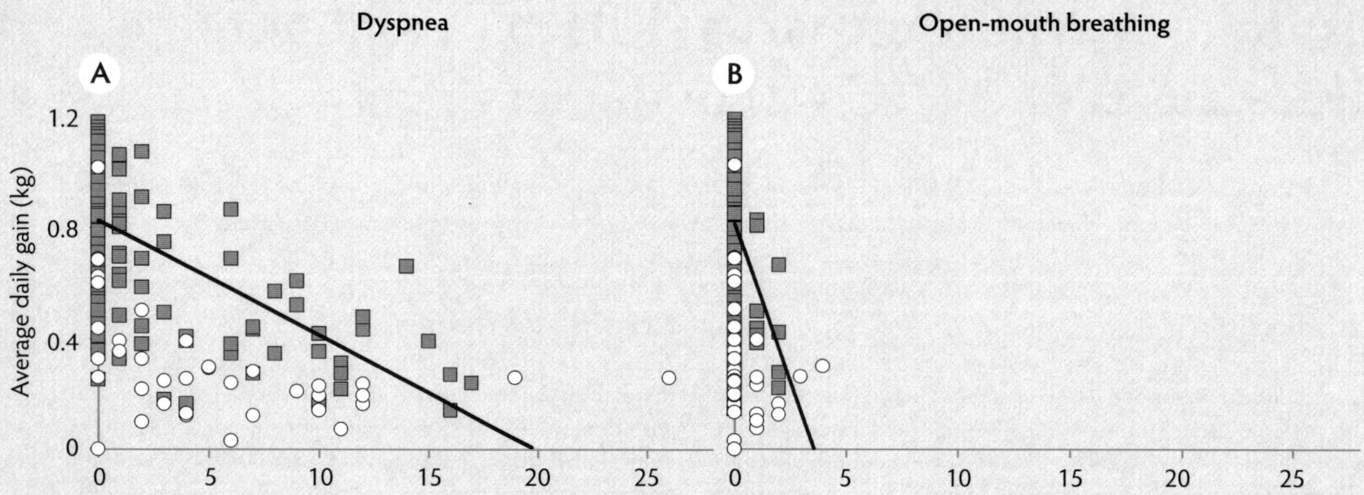

Early weaning induces long lasting effects on gastrointestinal health and susceptibility to subsequent stressors in grower and finisher pigs

Calvin S. Pohl[1], DVM; Julia E. Medland[2]; Laura L. Edwards[2]; Kristin Bagley[2]; Adam J. Moeser[1], DVM, PhD
[1]Gastrointestinal Stress Biology Laboratory, Department of Large Animal Clinical Sciences, College of Veterinary Medicine, Michigan State University, East Lansing, Michigan; [2]Department of Population Health and Pathobiology, College of Veterinary Medicine, North Carolina State University, Raleigh, North Carolina

Objective

Piglet weaning age can have profound outcome on health late in life. Early weaned pigs are susceptible to GI infection, and may benefit from immune stabilizing drugs.[1,2] In the current study, we sought to determine the impact of early weaning on both 1) long-term gastrointestinal barrier and immune function and 2) susceptibility to lose gastrointestinal barrier integrity when exposed to subsequent stressors.

Methods

Early life stress in pigs was induced by weaning piglets either early, at 16 days, or late, at 28 days, and comparing differences in adult fecal score, intestinal permeability, mast cell activation, and susceptibility to social mixing stress. Yorkshire-duroc cross females and castrated male pigs were used in this study and all co-housed in the same barn with 6 pigs per pen. For long-term barrier function and immunological studies, pig ileum was harvested from 23 week old pigs that had been early or late weaned. Barrier function and immunological responses were measured using Ussing chambers, immunohistochemistry, and mast cell mediator activity assays. For stress susceptibility studies, 8 week old early or late weaned pigs were subjected to mixing stress and 24 hours later ileum was harvested for measurement of intestinal permeability in Ussing chambers. There were 8 pigs per each weaning group, and each weaning group was equally split into 4 males and 4 females to study the effect of early weaning on sex.

Results

Compared with late-weaned pigs, early weaned pigs exhibited chronic intermittent diarrhea (measured as fecal scores and % of days exhibiting diarrhea) that was observed throughout the post-weaning period and as they reached sexually maturity. Underlying pathologies due to early weaning, occurring concurrently with the noted diarrhea, were increased intestinal permeability (Figures 1A and 1B), increased mast cell activation, and increased susceptibility to intestinal barrier defects following a subsequent psychosocial stress (Figures 1C and 1D). Despite the marked disturbances in gut function and mast cell activation observed in early weaned pigs, there was no evidence of histological inflammation observed. The clinical presentation of diarrhea, increased intestinal permeability, and mast cell activation due to early weaning predominated in adult females rather than adult males.

Implication

Early weaning causes lasting gastrointestinal barrier defects and innate immune changes, particularly in female pigs. Furthermore, early weaned pigs exhibit exacerbated intestinal responses to subsequent stressors later in life. Considering that early weaned pigs have baseline gastrointestinal defects throughout production, they might be considered a cohort of pigs in production that are susceptible to enteric infectious diseases, and may require specialized management.

References

1. McLamb BL, Gibson AJ, Overman EL, et al. Early weaning stress in pigs impairs innate mucosal immune responses to enterotoxigenic E. coli challenge and exacerbates intestinal injury and clinical disease. PloS one 2013;8(4):e59838.

2. Mereu A, Tedo G, Moeser AJ, et al. Cromolyn-mediated improvement of intestinal barrier function is associated with enhanced piglet performance after weaning. BMC Vet Res 2015;11(1):274.

Figure 1: Early weaning stress induces permanent intestinal epithelial barrier defects. Ileum intestinal epithelial barrier permeability to **A)** 180 kDa mannitol and **B)** 5,000 kDa inulin on Ussing Chambers of both late wean control (LWC) and early wean stressed (EWS) males and females. **C)** Increased intestinal permeability inEWS pigs compared to LWC following mixing stress, and **D)** mast cell stabilizing drug blocks elevated intestinal permeability in mixed early wean stressed pigs.

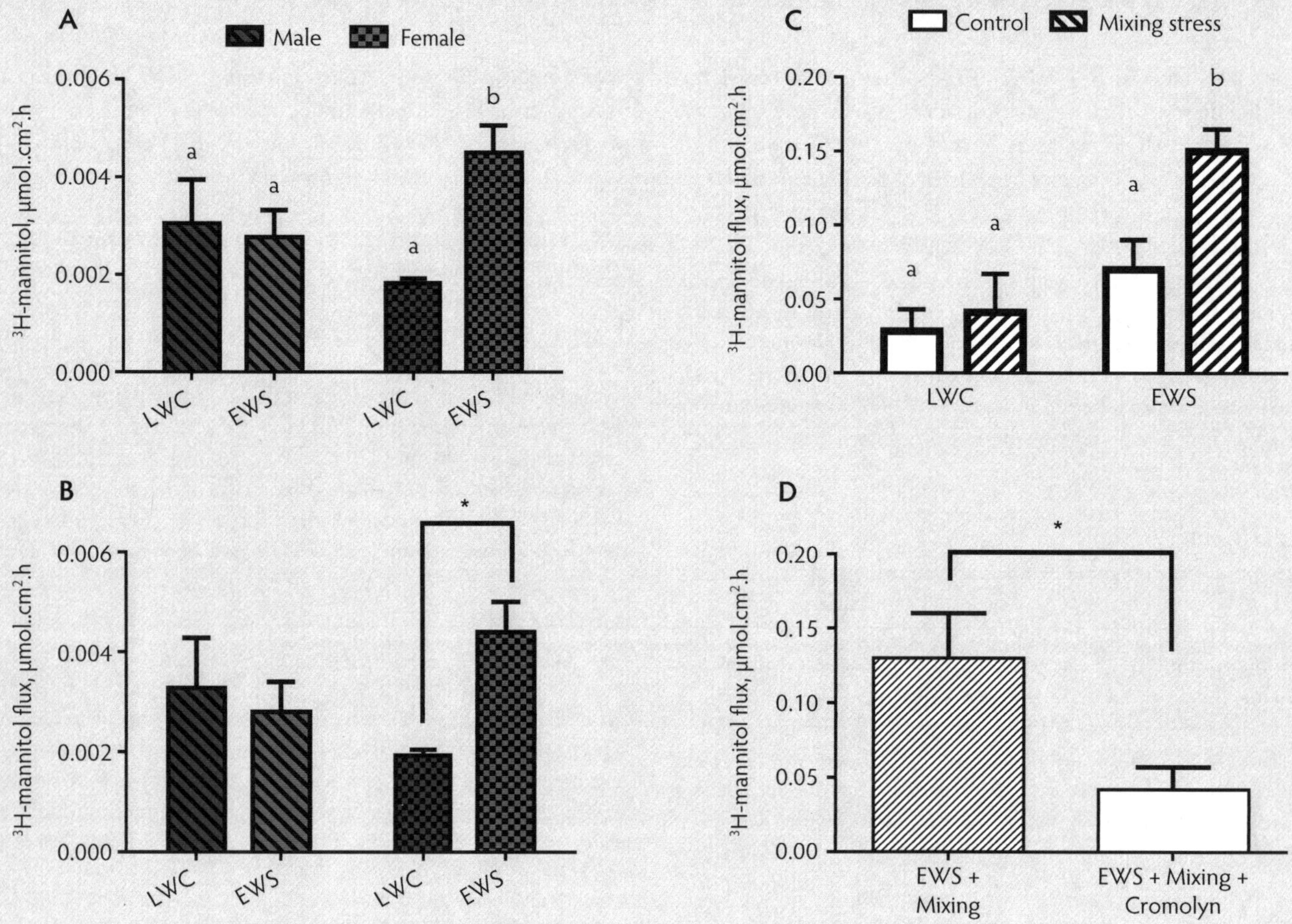

Does circulating antibody affect the course of PEDV infection in neonatal piglets?

K. Poonsuk, DVM, MS; L. G. Giménez-Lirola, PhD; J. Zhang, MD, PhD; P. Arruda, DVM, PhD; Q. Chen, MS; L. Correa da Silva Carrion; R. Magtoto, MS; P. Piñyero, DVM, PhD; L. Sarmento, DVM, PhD; C. Wang, PhD; K. J. Yoon, DVM, PhD; J. Zimmerman, DVM, PhD; R. Main, DVM, PhD
College of Veterinary Medicine, Iowa State University, Ames, Iowa

Introduction

The work described herein used a "passive transfer model" to explore the impact of maternal (colostral) antibody on the course of PEDV replication and neonatal health. The question asked was, "*Does circulating antibody affect the course of PEDV infection in piglets?*"

Methods

Piglets (n = 62; 2-to-3 days of age) from 6 PEDV IFA-negative sows were intraperitoneally (IP) administered 1 of 6 levels of concentrated PEDV antibody sufficient to achieve circulating FFN antibody titers of (< 1:8, 1:5.3, 1:6.1, 1:8, 1:17.1, and 1:32). 24 h later, all piglets were inoculated with PEDV and then observed through DPI 14 or until humane euthanasia was necessary. Piglets remained on the dam throughout the observation period.

Data and samples collected on a daily basis included sow milk, piglet fecal samples, piglet clinical signs, body weight, and body temperature. Serum samples were collected from piglets at DPI -1, 0, and 14, or at the time of humane euthanasia. Fecal samples were tested by PEDV rRT-PCR. Piglet serum samples were tested for PEDV IgG and IgA by a whole virus-based ELISA and for PEDV FFN antibody. The effects of treatment on the outcomes measured were analyzed using repeated measures ANOVA.

Results

- No difference among treated and non-treated pigs was detected in body weight, PEDV fecal shedding, and PEDV antibody responses
- Compared to the antibody-negative control group, a significant difference in time to death (treatment 5, $P < 0.05$), or a trend toward a difference (treatment 6, $P = 0.11$) was detected.
- Piglets that received concentrated antibody returned to normal body temperature sooner than control piglets, i.e., on DPIs 4, 5, 6, and 8 (P-value < 0.05).

Conclusions

Combining the results of this experiment with previous work in the literature leads to the conclusion that both colostral antibody (IgG) and maternal secretory IgA in milk contribute to the protection of the neonatal pig against PEDV infection.

Investigating the seeder-to-naïve ratio for successful exposure of gilts to *Mycoplasma hyopneumoniae*

Luiza R. Roos, DVM; Nitipong Homwong, DVM, MS; Maria Pieters, DVM, PhD
Department of Veterinary Population Medicine, College of Veterinary Medicine, University of Minnesota, Saint Paul, Minnesota

Introduction

Mycoplasma hyopneumoniae is a swine respiratory pathogen of major significance, associated with Enzootic Pneumonia and the Porcine Respiratory Disease Complex. Transmitted mainly by direct contact, infection with *M. hyopneumoniae* can last up to 7 months after experimental inoculation and a large percentage of infected animals can become chronic carriers of the pathogen capable of infecting susceptible pigs.[1] Dam shedding status at farrowing is considered the main risk factor for high prevalence of *M. hyopneumoniae* colonization at weaning,[2,3] as it has been shown that piglets become colonized during lactation. Colonized suckling piglets are considered initiators of the spread of *M. hyopneumoniae* infections during growing and finishing periods, infecting susceptible pen mates by horizontal transmission.[4] Considering the high replacement rate of breeding farms and the great impact in the endemic status that it causes when gilts negative to *M. hyopneumoniae* are placed in positive farms, controlled exposure to knowingly shedding animals in order to properly acclimate naïve gilts is a method that needs further investigation. Therefore, in this study we investigated various seeder-to-naïve gilt ratios to identify the most successful scenario to safely expose young gilts with *M. hyopneumoniae* in a 4-week exposure period. Our overall goal is to develop a standard protocol for gilt acclimation to *M. hyopneumoniae* prior to entering the breeding herd.

Materials and methods

Sixty gilts were divided in two groups: 21 2-week old seeder gilts that were intra-tracheally inoculated with *M. hyopneumoniae* strain 232 and followed during 28 days post inoculation (dpi) to confirm development of disease, and 39 aged-matched naïve gilts that were exposed to seeders during a 4-week period. At 28 dpi seeders were transferred to clean experimental rooms and placed in contact with naïve gilts for the transmission assessment. Gilts were randomly allocated to one of six rooms based on the following seeder-to-naïve ratios: 1:9, 2:8, 3:7, 4:6, 5:5 and 6:4, for a total of 6 rooms and 10 gilts per room. Laryngeal swabs, oral fluids, and blood samples were collected from all gilts prior inoculation and exposure, and every 14 days thereafter. At the end of the study all gilts were humanly euthanized and bronchial swabs, and lung tissue samples were collected. Infection in seeders was confirmed by development of clinical signs, seroconversion post-inoculation, and detection of *M. hyopneumoniae* genetic material. Naïve gilts were classified as positive after 4 weeks if *M. hyopneumoniae* was detected by IHC on fixed lung tissue or its genetic material by real time PCR of bronchial swabs.

Results and discussion

After 4 weeks of exposure to seeders 3/9 (30%) naïve gilts were positive in the 1:9 ratio, 6/8 (75%) in 2:8, 2/7 (28%) in 3:7, 2/6 (33%) in 4:6, 4/5 (80%) in 5:5 and 4/4 (100%) in the 6:4 ratio. The estimated transmission rate (β) and expected probability of infection (ψ) were 1.28 per pig/week and 0.6, respectively, based on a SI model, which implies that, within a week, expected 0.77 (1.28×0.6) susceptible pigs will become infected in experimental conditions. Under the conditions of this study, six seeders within a group of 10 gilts resulted in a successful exposure of all naïve gilts in the group to *M. hyopneumoniae* in a 4-week period. The main applicability of the results of this study could be translated into the exposure of naïve gilts to adequate proportions of knowingly shedding animals in the breeding herd. Further aspects to be evaluated for appropriate gilt acclimation to *M. hyopneumoniae* are the size of groups for proper exposure and the age to begging exposure. Since a situation where gilts negative for *M. hyopneumoniae* are introduced in positive farms is still forceful, we consider the results of this study to be a valuable first step to develop a standard protocol for gilt acclimation to *M. hyopneumoniae* prior to entering the reproductive herd in order to minimize potential risk factors associated with sow herd instability.

References

1. Pieters M et al. An assessment of the duration of *Mycoplasma hyopneumoniae* infection in an experimentally infected population of pigs. Vet Microbiol, 2009;134.

2. Pieters M et al. Intra-farm risk factors for *Mycoplasma hyopneumoniae* colonization at weaning age. Vet Microbiol, 2014;172.

3. Nathues H et al. Herd specific risk factors for *Mycoplasma hyopneumoniae* infections in suckling pigs at the age of weaning. Acta Vet. Scand, 213:55.

4. Meyns T et al. Quantification of the spread of *Mycoplasma hyopneumoniae* in nursery using transmission experiments. Prev. Vet. Med, 2005:66.

Lactogenic protection against porcine epidemic diarrhea virus in piglets following homologous challenge

A. Schelkopf[1]; D. Magstadt[2]; B. Arruda[2]; P. Arruda[2]; J. Zimmerman[2]; T. Wetzell[3]; S. Dee[1]; D. Madson[2]
[1]Pipestone Veterinary Services, Pipestone, Minnesota; [2]Iowa State University, Ames, Iowa;
[3]Boehringer Ingelheim, Saint Joseph, Missouri

Introduction

Immunity to Porcine Epidemic Diarrhea virus (PEDV), whether active or transferred, is poorly understood. The objectives of this study were to 1) characterize the immune response in naïve sows following exposure to PEDV utilizing different serological assays, and 2) assess lactogenic immunity and protection in piglets subsequently challenged with the homologous strain of PEDV.

Materials and methods

In April 2014, a commercial sow farm infected with a prototype stain of PEDV. Thirty sows were randomly enrolled in a long-term study. Sows were monitored for 1) PEDV serum antibodies by florescent focus neutralization (FFN) assay, indirect fluorescent assay (IFA), IgG and IgA enzyme-linked immunosorbent assays (ELISA), and 2) fecal shedding by RT-PCR on individual fecal samples. Samples were collected at weeks post infection (WPI): 1,2,3,4,6,8,11,15, and 22.

Eight sows from this group were bred and relocated to Iowa State University in September 2014 for homologous PEDV challenge. Sows were randomly allocated into groups: negative control (NC; n = 3) and treatment (T; n = 5). Three additional sows were sourced from a PEDV-naïve farm as positive controls (PC; n = 3). Treatment groups were separated by room.

At 110-111 days of gestation, sows in groups T and PC were challenged via gavage with a 1 mL dose of 1×10^3 $TCID_{50}$/mL of PEDV. At 24-48 hours post parturition, all piglets in groups T and PC were inoculated with the same dose as the sows by gavage. The PEDV used for challenge was propagated from an isolate obtained from the source farm at initial infection.

Individual fecal, serum, and colostrum/milk samples were obtained from sows; serum and feces were collected from piglets, along with individual piglet weights. Sample collection days post inoculation (DPI) included: 1) sow fecal and serum samples on 0, 3-6, 11, 14, 18, 21, and 25 DPI; 2) sow colostrum or milk at farrowing 11, 14, 18, 21, and 25 DPI; 3) piglet serum at birth, 7, 14, and 18 days of age (DOA); and 4) piglet fecal swab at birth, 4, 7, 11, 14, and 18 DOA. Individual piglet weights were determined at birth, 7, 14, and 18 DOA. Study termination was 25 DPI; piglets were 18 DOA.

All serum, colostrum, and milk samples were assayed using the Iowa State University Veterinary Diagnostic Laboratory IFA test, IgA and IgG ELISAs, and the South Dakota State University Veterinary Diagnostic Laboratory FFN test.

Results

At the time of the initial outbreak, all sow fecal samples were RT-PCR positive for PEDV; serum antibodies were not detected, all assays. By 6 weeks post-infection, all sow fecal samples were PCR negative, signifying viral shedding had ceased. Serum antibodies were detected 14 days post-break and throughout the sampling period at the source farm, all assays.

All sows and piglets in group NC were fecal PCR negative for the study duration. All litters of the PC group displayed clinical signs consistent with PEDV infection, with 100% mortality by 4 DOA. Sows in this group also had diarrhea. All litters in group T had RT-PCR positive fecal samples; only one of the five litters demonstrated clinical disease consistent with PEDV infection. Piglet mortality in the T group was 12.9%. Average litter weight did not differ between treatment groups at all DPI ($P > 0.05$). PEDV serum antibody was detected in previously exposed sows and their piglets by IFA, FFN, and ELISA. Piglet FFN.

Conclusion and discussion

From infection in a naïve sow population, fecal shedding of PEDV ceased by 6 weeks post infection. PEDV antibodies were detected in sows for at least a 5 month duration using the IFA test, FFN test, and IgG and IgA ELISA assays. Following homologous PEDV challenge, PEDV antibody was present in colostrum/milk and protective to piglets born to previously exposed sows. Clinical disease was only seen in 1/5 litters and 0/5 sows in the previously exposed T group; viral shedding occurred in all litters. At all sample points, average piglet weights did not differ. This study provides strong evidence of the duration of and extent of lactogenic immunity to a homologous strain of PEDV. Diagnostic assays with FFN, IgA, and IgG ELISAs all had a high degree of agreement on piglet serum and should be considered potential indicators of immunity.

A stochastic, mathematical model of influenza A virus within swine breeding herds: Implications of possible management interventions

L. A. White[1], BS; M. Torremorell[2], DVM, PhD; M. E. Craft[2], PhD
[1]Department of Ecology, Evolution & Behavior, University of Minnesota, Saint Paul, Minnesota;
[2]Department of Veterinary Population Medicine, University of Minnesota, Saint Paul, Minnesota

Introduction

Influenza A virus (IAV) is a globally endemic infection in swine herds that causes significant morbidity in the swine industry and poses a substantial public health risk.[1,2] Initial modeling work suggests that vaccination (especially heterologous) does not effectively remove influenza from breeding herds,[3] and recent empirical findings suggest that IAV is most likely maintained endemically through gilt introductions and in piglet subpopulations.[4] The goal of developing this model is to test intervention strategies that swine producers could feasibly employ to control IAV in their herds.

Materials and methods

Building on the work of Reynolds et al.,[4] we have developed a stochastic Susceptible-Exposed-Infectious-Recovered simulation of IAV dynamics in a swine breeding herd. Using Gillespie's Direct Method,[5] this model allows for stochastic births, deaths, direct and indirect infection, and maternally derived immunity in piglets. The construction of this model reflects the spatial organization of a standard breeding herd and accounts for the different classes of pigs therein including gilts, sows, and piglets in various production and/or immune stages. The interventions tested include: serial routine heterologous mass vaccination with waning immunity, pre-farrow vaccination with and without piglet early weaning, and varying the timing of gilt purchases.

Results

Our preliminary results suggest that none of the interventions can effectively eliminate infection from a medium sized herd (~1500 sows) when implemented individually. In concert, pre-farrow vaccination, early weaning of piglets (removal 7-14 days after birth), and longer periods between introductions of gilts (intervals greater than one month) were the most effective at reducing prevalence, but did not result in elimination of IAV.

Conclusions and discussion

Our findings support other modeling and empirical studies that suggest that piglets maintain IAV in breeding herds.[3,4,6] Our next steps will be to evaluate the sensitivity of results to epidemiological parameters, vaccine efficacy, and farm size. We will also test the potential role of seasonality in transmission rates of IAV.[4] We hypothesize that one or more of these interventions will prove effective at reducing the incidence and prevalence of IAV when timed appropriately with seasonality in flu transmission.

References

1. Torremorell M et al. Transmission of influenza A virus in pigs. *Transbound. Emerg. Dis.* 2012; 59, 68–84.

2. Vincent A et al. Review of influenza A virus in swine worldwide: A call for increased surveillance and research. *Zoonoses Public Health* 2014; 61, 4–17.

3. Reynolds JJH et al. Mathematical modeling of influenza A virus dynamics within swine farms and the effects of vaccination. *PLoS One* 2014; 9, e106177.

4. Diaz A et al. Association between influenza A virus infection and pigs subpopulations in endemically infected breeding herds. *PLoS One* 2015; 10, e0129213.

5. Keeling M and Rohani P. Modeling of infectious diseases in humans and animals. Princeton University Press, Princeton, New Jersey, 2008.

6. Allerson M. Infection dynamics of pandemic 2009 H1N1 influenza virus in a two-site swine herd. *Transbound. Emerg. Dis.* 2013; 61, 490–499.

Field experiences with 119 filtered breeding farms in the United States

S. Tousignant[1,2], DVM, PhD; D. Reicks[2], DVM; P. Davies[1], DVM, PhD;
A. Perez[1], DVM, PhD; R. Morrison[1], DVM, PhD
[1]Veterinary Population Medicine, University of Minnesota, Saint Paul, Minnesota;
[2]Swine Vet Center, Saint Peter, Minnesota

Introduction

Since 2005, there has been rapid adoption of aerosol filtration technology on swine breeding farms.[1] As the early adopter began to gain experience, the industry quickly began to notice the apparent success.[2] Since then, aerosol filtration on swine breeding herds has proven to be a cost effective means of reducing the incidence of at the farm level.[3] Unfortunately, some filtered herds report new porcine reproductive and respiratory syndrome virus (PRRSV) breaks every year and until now, only anecdotal reports of what factors might be associated with these events exists. With this in mind, the intent of this study was to document changes in the improvements made on filtered farms during the past decade.

Materials and methods

A convenience sample of filtered breeding herds within the Swine Vet Center practice was selected. An annual survey, dating back to 2005, was completed for each herd that documented the type of construction, filtration operation, as well as building modifications such as positive pressure, manure pumpouts, and back drafting devices utilized. Additionally, each herd reported if they had experienced an incident case of PRRSV during that year.

Results and discussion

The number of filtered herds increased from 10 in 2005, to 119 in 2014. Additionally, incidence of PRRSv ranged from 0% to 17% on these herds. From the survey, there was a general increase in the use of positive pressure ventilation in the loadout areas and offices (Figures 1a and b). Additionally, there has been a general trend toward continuous filtration since 2005 (Figure 1c). Finally, many filtered farms are employing internal auditing personnel, who are tasked with reviewing biosecurity, filter integrity, as well as continuously monitoring and repairing any damage and air leaks (Figure 1d). The foundation of any filtered farm is a set of strong biosecurity protocols that are designed to prevent the introduction of PRRSV via incoming supplies, animals or people. It appears that once those possible routes of entry are rigorously controlled, adding filtration to a farm may prove to be a crucial final step in reducing the incidence of PRRSV.

References

1. Dee S, Batista L, Deen J, et al. Evaluation of an air-filtration system for preventing aerosol transmission of Porcine reproductive and respiratory syndrome virus. *Can J Vet Res* 2005;69:293-298.

2. Spronk G, Otake S, Dee S. Prevention of PRRSV infection in large breeding herds using air filtration. *Vet Rec* 2010;166:758-759.

3. Reicks D. Field experiences with air filtration: Results and Costs. Allen D Leman Swine Conference 2008; 42 - 43.

Figures 1A-D: Trends since 2005 in the various filter farm modifications documented in this study

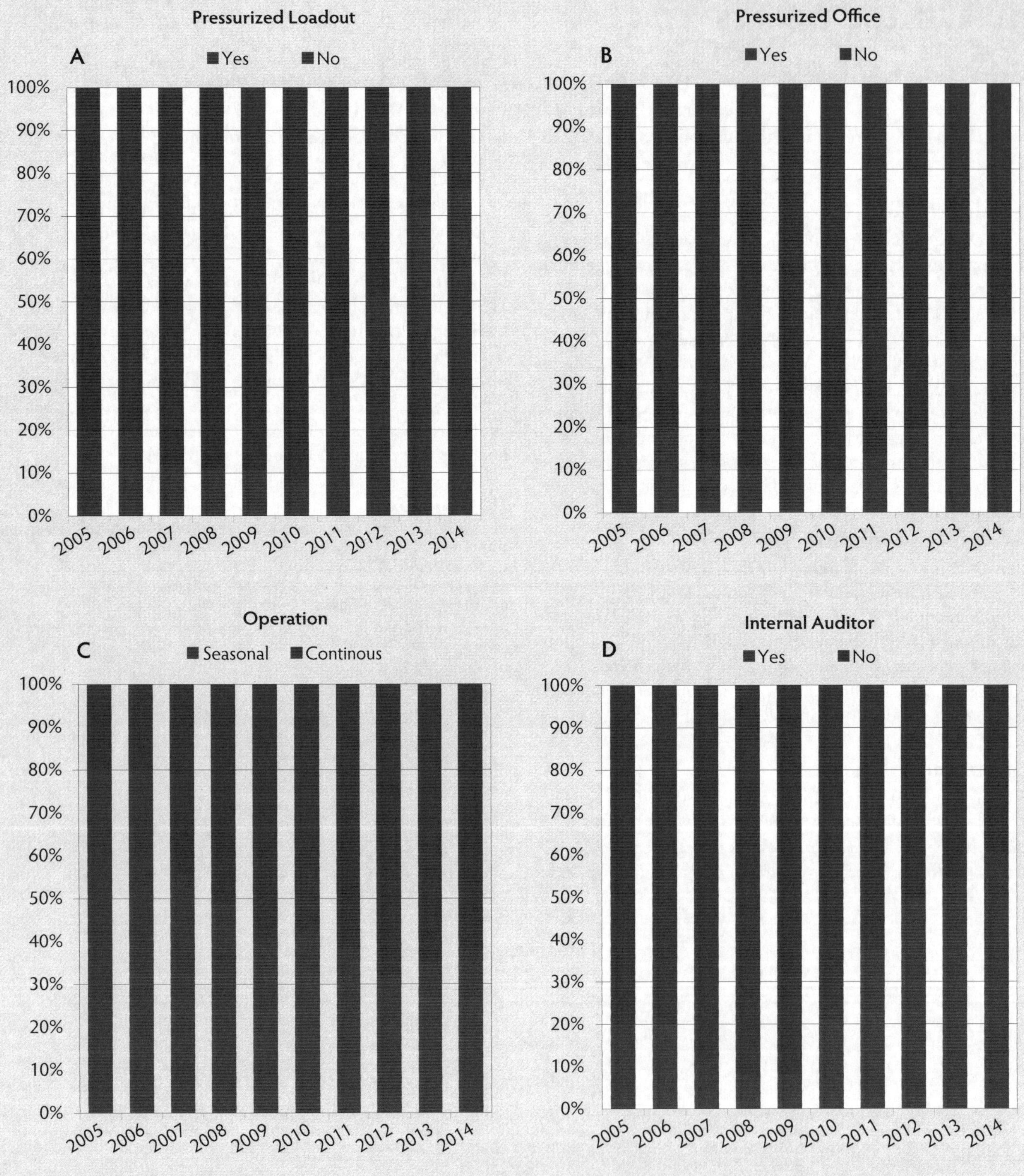

PRRSV and PEDV infection patterns and risk factors in sow farms in the United States

S. Tousignant[1,2], DVM, PhD; P. Valdes-Donoso[1], DVM, MPVM; Andres M. Perez[1], DVM, PhD;
Robert B. Morrison[1], DVM, PhD
[1]Veterinary Population Medicine, University of Minnesota, Saint Paul, Minnesota;
[2]Swine Vet Center, Saint Peter, Minnesota

Introduction

Porcine reproductive and respiratory syndrome virus (PRRSV) and porcine epidemic diarrhea virus (PRRSV) are two economically important diseases of swine production in the United States of America.[1,2] The objective of this study was to assess patterns of co-infection of these two pathogens using data from a group of voluntary participants enrolled in the Swine Health Monitoring Project in the Midwest region of the United States of America.

Materials and methods

Spatial temporal clusters of PRRSV and PEDV were identified using a multinomial space-time scan statistic. A multinomial, multivariate regression model was fit to quantify associations between farm level management factors and disease outcomes including if the farm was located in a disease cluster. A second multinomial, multivariate regression model was fit to identify management behaviors more common among diseased farms within clusters than diseased farms outside of clusters.

Results and discussion

109 herds from five production companies were enrolled in this study. Five significant space-time clusters with two or more herds were identified as being infected with PRRSV and/or PEDV or neither disease between the fall of 2013 to the spring of 2014 (Figure 1). County density was significantly associated with higher odds of being infected with either disease, whereas high biosecurity including bio-aerosol filtration was significantly associated with lower odds of being infected with either disease. Additionally, contracted trucking was more common among diseased farms within clusters versus not in clusters. Efforts should be directed at implementing strict biosecurity practices that are common among filtered farms, especially in swine dense areas and if contracted trucking is employed as a means of mitigating the incidence of these diseases.

References

1. Holtkamp DJ, Kliebenstein JB, Neumann EJ. Assessment of the economic impact of porcine reproductive and respiratory syndrome virus on United States pork producers. *Journal of Swine Health and Production* 2013;21:72-84.

2. Stevenson GW, Hoang H, Schwartz KJ, et al. Emergence of Porcine epidemic diarrhea virus in the United States: clinical signs, lesions, and viral genomic sequences. *J Vet Diagn Invest* 2013;25:649-654.

Figure 1: Locations of time-space clusters of PRRSV and PEDV in a cohort of 109 herds. Footnotes provide details on each time space cluster

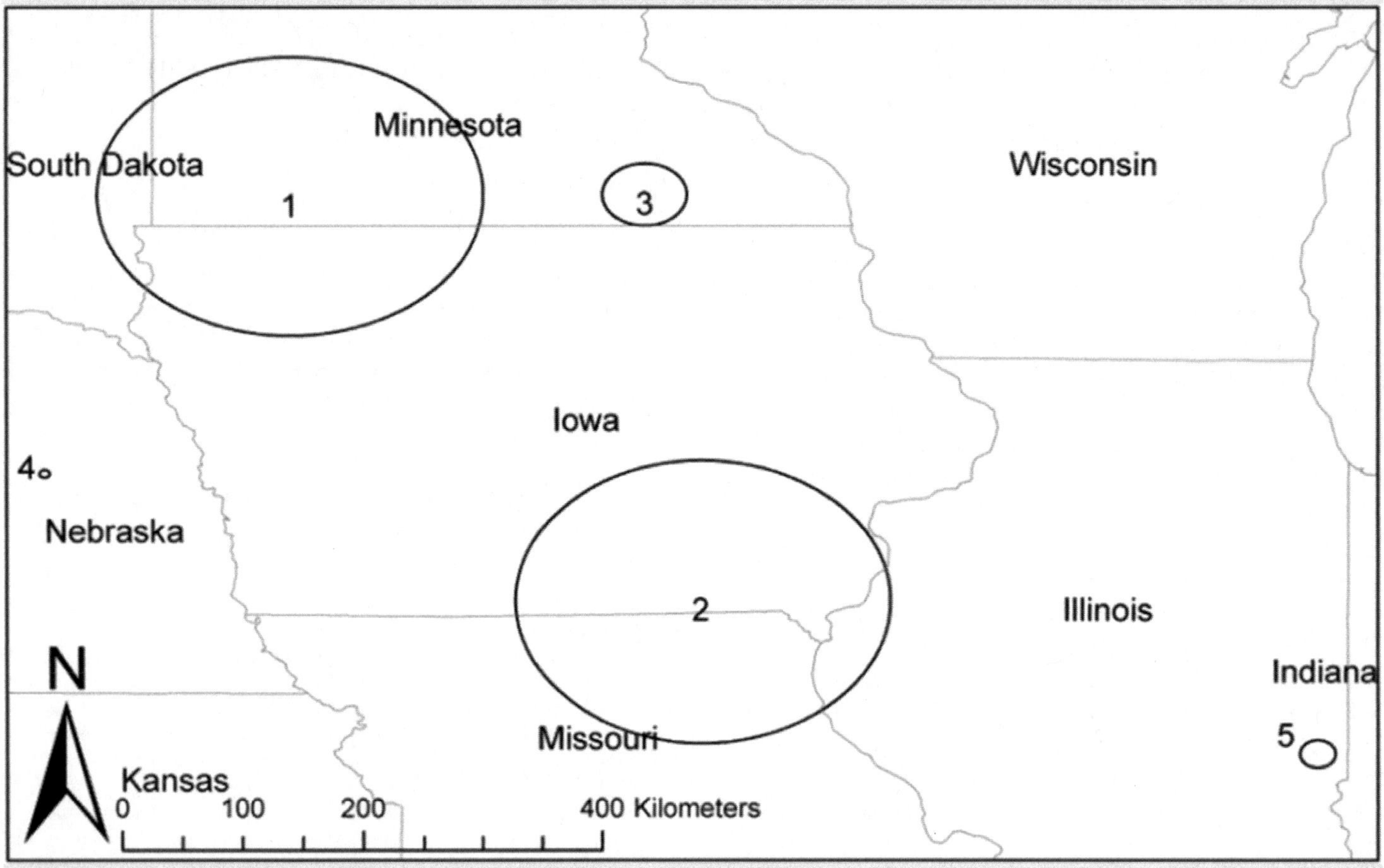

[1] PRRS+/PED+, observed to expected ratio = 4.65, *P*-value = 0.001 (most likely cluster)
[2] PRRS-/PED+, observed to expected ratio = 8.29, *P*-value = 0.001
[3] PRRS+/PED-, observed to expected ratio = 20.88, *P*-value = 0.001
[4] PRRS-/PED+, observed to expected ratio = 9.67, *P*-value = 0.001
[5] PRRS-/PED-, observed to expected ratio = 1.22, *P*-value = 0.001

Measuring progress on the control of porcine reproductive and respiratory syndrome (PRRS) at a regional level: The Minnesota N212 regional control project (RCP) as a working example

Pablo Valdes-Donoso[1,2]; Lovell S. Jarvis[2]; Dave Wright[1]; Julio Alvarez[1]; Andres M. Perez[1]
[1]Department of Veterinary Population Medicine, College of Veterinary Medicine, University of Minnesota, Saint Paul, Minnesota;
[2]Department of Agriculture and Resource Economics, University of California Davis, Davis, California

Abstract

Due to the highly transmissible nature of porcine reproductive and respiratory syndrome (PRRS), implementation of regional programs to control the disease may be critical. Because PRRS is not reported in the US, numerous voluntary regional control projects (RCPs) have been established. However, the effect of RCPs on PRRS control has not been assessed yet. This study aims to quantify the extent to which RCPs contribute to PRRS control by proposing a methodological framework to evaluate the progress of RCPs. Information collected between July 2012 and June 2015 from the Minnesota Voluntary Regional PRRS Elimination Project (RCP-N212) was used. Demography of premises (e.g. composition of farms with sows = SS and without sows = NSS) was assessed by a repeated analysis of variance. By using general linear mixed-effects models, active participation of farms enrolled in the RCP-N212, defined as the decision to share (or not to share) PRRS status, was evaluated and used as a predictor, along with other variables, to assess the PRRS trend over time. Additionally, spatial and temporal patterns of farmers' participation and the disease dynamics were investigated. The number of farms enrolled in RCP-N212 and its geographical coverage increased, but the proportion of SS and NSS did not vary significantly over time. A significant increasing ($P < 0.001$) trend in farmers' decision to share PRRS status was observed, but with NSS producers less willing to report and a large variability between counties. The incidence of PRRS significantly ($P < 0.001$) decreased, showing a negative correlation between degree of participation and occurrence of PRRS ($P < 0.001$) and a positive correlation with farm density at the county level ($P = 0.02$). Despite a noted decrease in PRRS, significant spatio-temporal patterns of incidence of the disease over 3-weeks and 3-kms during the entire study period were identified. This study established a systematic approach to quantify the effect of RCPs on PRRS control. Despite an increase in number of farms enrolled in the RCP-N212, active participation is not ensured. By evaluating the effect of participation on the occurrence of PRRS, the value of sharing information among producers may be demonstrated, in turn justifying the existence of RCPs.

Reduction of antibiotic use to treat post-weaning diarrhoea following oral vaccination with *Escherichia coli* F4 : Case report

Frédéric Vangroenweghe
Elanco Animal Health, Brussels, Belgium

Introduction

Post-weaning *Escherichia coli* diarrhoea (PWD), also called post-weaning enteric colibacillosis, in pigs remains a major cause of economic losses for the pig industry, due to either piglet death, or poor weight gain in surviving piglets.[1,2] PWD typically causes mild to severe watery diarrhoea between 5 and 10 days after weaning and is caused primarily by enterotoxigenic *Escherichia coli* (ETEC). The most common adhesins found on ETEC from PWD in pigs are associated with fimbriae F4 (previously called K88) and F18, while the predominant enterotoxins are heat-labile toxin (LT), heat-stable toxin a (STa), and heat-stable toxin b (STb). In addition to F4 and F18, other fimbrial adhesins, such as F5 (K99), F6 (987P), and F7 (F41), have been associated with PWD, but less frequently.[3-6] The presented case report shows the impact of vaccination on overall post-weaning piglet health following vaccination with a live non-pathogenic *E. coli* F4 vaccine.

Case description

Since more than 2 years, a closed 400-sow farm in a 3-week batch management system suffered from extreme problems with post-weaning diarrhoea. The clinical phase started already within 3 days post-weaning with high levels of mortality and massive use of in-feed and injectable antibiotics (Table 1). During that period, several preventive options have been applied, such as in-feed zinc oxide inclusion, treatment with colistin, lincomycin and enrofloxacin (all based on bacteriology and subsequent antimicrobial resistance patterns). Each weaning batch consisted of approximately 600 piglets, weaned at around 24-26 days of age. For facilitating the weaning process, 500 kg of weaning feed is applied for each weaning batch. Depending on the feed consumption throughout the first post-weaning days, the switch to the post-weaning diet is performed earlier or later. Following diagnosis of an ETEC F4, responsible for post-weaning diarrhoea, an oral vaccination with a live non-pathogenic strain of *E. coli* (Coliprotec® F4; Prevtec Microbia) 7 days before the clinical outbreak is performed at a dose of 2 ml per piglet, containing $1.3\text{-}8.0 \times 10^8$ colony-forming units. An antibiotic-free window of 3 days before and 3 days after vaccination is necessary for optimal immunological response following bacterial growth in the intestines. The following parameters were recorded for each group: # weaned piglets, # dead piglets (% mortality) d1-d10, # antibiotic injections, treatment incidence (TI; treatments per 100 piglets), % pens treated on total, day of switch to post-weaning diet.

Results

The main results of the case are given in Table 1.

Discussion

This case report shows that following a correct diagnosis to confirm the presence of ETEC F4 in post-weaning diarrhoea problems, vaccination with an oral live non-pathogenic *E. coli* F4 vaccine shows improvement of several economically important parameters. There is a major decrease in piglet mortality (4.7% to 2.0-0.5%), combined with a reduced treatment incidence, a higher early post-weaning feed intake and less labour to the swine farmer. In conclusion, vaccination with an oral live non-pathogenic *E. coli* F4 vaccine resulted in lower antibiotic use (90-93% reduction in TI) and no zinc oxide inclusion in combination with lower piglet mortality and improved financial results.

References

1. van Beers-Schreurs et al., 1992. *Veterinary Quarterly* 14(1), 29-34.
2. Fairbrother et al., 2005. *Animal Health Research Reviews* 6(1), 17-39.
3. Kwon et al., 2002. *The Veterinary Record* 150, 35-37.
4. Frydendahl, 2002. *Veterinary Microbiology* 85, 169-182.
5. Chen et al., 2004. *Veterinary Microbiology* 103, 13-20.
6. Vu Khac et al., 2006. *BMC Veterinary Research* 2, 10

Table 1: Measured parameters of case of chronic problems with post-weaning diarrhoea before and after vaccination with an oral, live, non-pathogenic *E. coli* F4 vaccine.

	Treatment group		
Parameters	**Pre-vaccination batch**	**Batch 1 – post-vaccination**	**Batch 2 – post-vaccination**
# weaned piglets	600	550	600
# dead (% mortality) d1-d10	28 (4.7%)	11 (2.0%)	3 (0.5%)
# antibiotic injections	1800	190	150
TI (treatments/100 piglets)	300	34.5	25
% pens injected on total	100%	75%	37.5%
Day of feed change	D8	D5	D6
# h treatment administration	6.0 h	2.0 h	1.5 h
# h vaccination	-	2.75 h	3.0 h

Effect of azaperone (Stresnil®) administration after farrowing on maternally derived immunoglobulin level in suckling piglets using the Immunocrit® method

Frédéric Vangroenweghe[1]; Hinnerk Kranenburg[2]; Johan Kamp[3]; Rutger Jansen[4]
[1]Elanco Animal Health, Brussels, Belgium; [2]De Varkenspraktijk – Overlaet, Oss, The Netherlands; [3]De Oosthof Samenwerkende Dierenartsen – DK Hellendoorn, Hellendoorn, The Netherlands; [4]ForFarmers, Lochem, The Netherlands

Introduction

The dramatic change in sow prolificacy observed in the modern swine industry over the last decade was also accompanied by a significant increase in pre-weaning piglet mortality. This emphasizes the importance to guarantee maximum colostrum intake within the first hours after birth as a readily available energy source and for passive immunity, increasing the piglets chances for survival and optimal performance.[1] The decision on the practical approaches to maximize colostrum intake is a great management challenge.[2] Recent studies show that the use of azaperone in sows at the end of parturition may improve piglet performances.[3,4] Recently, a new assessment technique for the piglet's maternally derived immunoglobulin level has been developed[6] and evaluated.[5] The objective of the current trial was to investigate the effect of azaperone treatment of the sow at parturition on the maternally derived immunoglobulin level using the Immunocrit® method.

Materials and methods

The trial was performed in a 400 sow farrow-to-finish herd with a 1-week batch management system based in The Netherlands. Upon parturition, sows ($n = 74$) were randomly assigned to one of both trial groups, namely control group (C-group) or azaperone group (A-group). The sows in the A-group were injected with 8 ml of azaperone (Stresnil®, Elanco Animal Health) upon the end of parturition. For analytical purposes, 6 piglets in each litter were selected, based on their life weight on day 2, and subsequently bled. Blood was further processed as described previously[5] with the increase of centrifugation to 10 minutes instead of 5 and subsequent immunocrit analysis was performed. Results were statistically analysed using SAS for significant effects between treatment groups.

Results

Results of the immunocrit immunoglobulin analysis showed a numerical non-significant ($P > 0.05$) increase in IgG levels (35.0 ± 0.5 mg/mL in C-group *vs.* 36.1 ± 0.5 mg/mL in A-group) and weight-corrected IgG levels (40.6 ± 0.6 mg/mL in C-group *vs.* 41.4 ± 0.6 mg/mL in A-group) in piglets born from sows injected with azaperone following the end of parturition. Mortality during the lactation period improved significantly ($P < 0.05$) between C-group (17.4%) and A-group (13.7%). Table 1 shows the significant correlation ($P < 0.0001$) between mortality and the level of IgG within 48 h after birth.

Discussion

Using the new immunoglobulin analysis technique, Immunocrit®, a clear idea could be obtained of the colostrum distribution in treated and untreated litters. The obtained results are in accordance with previous studies[3,4] showing an improved colostrum IgG distribution in azaperone (Stresnil®) treated litters. Although a non-significant increase in IgG levels in the azaperone treated group, a significant impact on mortality during the lactation could be observed. Based on these and earlier research,[3,4] the decision could be taken to treat sows at risk with azaperone, in order to obtain a more evenly distributed colostrum IgG level among the piglets in the litter.

References

1. Vasdal, G. et al. (2010) Applied Animal Behaviour Science 125, 96-102.
2. Baxter, E.M. et al. (2009) Livestock Science 124: 266-276
3. Schulze M. et al. (2012) Journal of Animal Science 90, 2331-2336.
4. Jourquin J. et al. (2011) Proc. 3rd European Symposium on Porcine Health Management.
5. Jansen R. et al. (2014) Proc. 6th European Symposium on Porcine Health Management.
6. Vallet, J.L. et al (2013) The veterinary Journal 195, 91-97.

Table 1: Correlation between mortality and the level of IgG within 48 h after birth.

	IgG blood level		
	IgG < 20 mg/mL	**IgG > 20 mg/mL**	**Total**
Survival	59	764	823
Dead	26	49	75
Total	85	813	898

Detection of classical swine fever virus (CSFV) using serum-based assays

Y. Panyasing[1], DVM, PhD; R. Kedkovid[1], DVM, MS; Y. Woonwong[1], DVM; N. Sopipan[1], DVM, MS; C. Sirisereewan[1], DVM; J. Arunorat[1], DVM; P. Makhum[2], BS; P. Choojai[2], BS; R. Tantilertchareon[2], DVM, MS; P.Ninvilai[3], DVM, MS; R. Thanawongnuwech[1], DVM, PhD; J. Zimmerman[4], DVM, PhD, DACVPM

[1]Department of Pathology, Faculty of Veterinary Science, Chulalongkorn University, Bangkok, Thailand; [2]Chulalongkorn University-Veterinary Diagnostic Laboratory, Bangkok, Thailand; [3]Animal Health and Technical Service Office, Bangkok, Thailand; [4]College of Veterinary Medicine, Iowa State University, Ames, Iowa

Introduction

CSFV is endemic in Asia,[1] many parts of South and Central America, the Caribbean, and parts of Europe and Africa.[2] Accurate and rapid diagnostic assays are need to detect, control, and/or eliminate CSFV. The purpose of this study was to compare the diagnostic characteristics of universally-accessible CSFV tests using serum samples from pigs of known CSFV status.

Materials and methods

30 unvaccinated, CSFV antibody-negative pigs (35-40 kg) were individually housed from day post inoculation (DPI) -14 to 28 in research facilities at Chulalongkorn University (Bangkok, Thailand). At DPI 0, pigs were IN inoculated with CSFV (ALD strain, 1×10^5 $TCID_{50}$/ml). Serum samples (n = 276) were collected on DPI -14, -7, 0, 1, 2, 3, 4, 5, 6, 7, 10, 14, 17, 21, 28 and tested by virus isolation (VI), rRT-PCR (3 commercial assays), antigen-capture ELISAs (2 commercial assays), neutralizing antibody (SN), and antibody ELISA (3 commercial assays).

Results

Serum samples were positive by rRT-PCR on DPI 2 (10% of pigs) and peaked at DPI 5 (100% of pigs). Serum was VI-positive on DPI 3 (20% of pigs) and peaked at DPI 6 (100% of pigs). Samples were antigen-capture ELISA positive on DPI 6 (60%) and peaked at DPI 7 (82.1%). The last positives were detected by rRT-PCR, VI, and antigen-capture ELISA on DPI 28 (26.7%), DPI 21 (13.3%), and DPI 17 (21.1%), respectively. SN antibody was detected as early as DPI 7 (21.4% of pigs). Antibody was detected by commercial antibody ELISAs as early as DPI 10 (14.8 - 22% of pigs). By DPI 17, there was no statistically significant difference in the rate of antibody detection between the SN (94.7%) and antibody ELISAs (84.2-89.5%).

Conclusions

The rRT-PCRs provided earlier and longer detection of CSFV versus other assays designed for direct detection (VI and antigen-capture ELISAs). There was no statistically significant difference in the detection of antibody by the SN and commercial antibody ELISAs ~3 weeks after infection. Therefore, commercial CSFV rRT-PCR and antibody ELISAs are suitable for large scale screening whereas VI and SN should be used as confirmatory assays.

References

1. Paton DJ et al. Genetic typing of classical swine fever virus. *Vet Microbiol.* 2000;73:137-157.

2. Kirkland PD et al. Pestiviruses. In: Zimmerman J, Karriker L, Ramirez A, Schwartz KJ, Stevenson G (Eds), Diseases of Swine, 2012;538-553. Wiley-Blackwell Publishing Press, Hoboken, NJ.

Detection of classical swine fever virus (CSFV) in oral fluid samples by RT-PCR

Y. Panyasing[1], DVM, PhD; K. Rungprasert[1], DVM, MS; N. Anakkul[2], DVM, PhD; R. Kedkovid[1], DVM, MS; R. Thanawongnuwech[1], DVM, PhD; A. Kittawornrat[3], DVM, PhD; J. Zimmerman[4], DVM, PhD, DACVPM
[1]Department of Pathology, Faculty of Veterinary Science, Chulalongkorn University, Bangkok, Thailand; [2]Center of Animal Laboratory, Chulalongkorn University, Bangkok, Thailand; [3]CPF (Thailand) Public Company Limited, Bangkok, Thailand; [4]College of Veterinary Medicine, Iowa State University, Ames, Iowa

Introduction

The ability of CSFV to spread over long distances to areas free of CSFV is well-documented.[1] Accurate and rapid surveillance/testing procedures are mandatory for detection, control, and/or elimination because (1) CSFV infections cannot be differentiated from endemic pathogens, e.g., PRRSV, salmonellosis, on the basis of clinical signs and because (2) CSFV can be shed throughout a pig's lifetime without overt clinical signs. The purpose of this study was to evaluate the detection CSFV by RT-PCR using oral fluid samples.

Materials and methods

30 unvaccinated, CSFV antibody-negative pigs (35-40 kg) were individually housed from day post inoculation (DPI) -14 to 28 in research facilities at Chulalongkorn University (Bangkok, Thailand). At DPI 0, pigs were IN inoculated with CSFV (ALD strain, 1×10^5 $TCID_{50}$/ml). Individual oral fluid samples were collected daily from DPI -14 to 28. For comparative purposes, serum samples were collected at DPI -14, -7, 0-, 1, 2, 3, 4, 5, 6, 7, 10, 14, 17, 21, and 28. Oral fluid (n = 504) and serum (n = 276) samples were tested for CSFV RNA using a commercial kit (LSI VetMax™ CSFV (PPC), Life Technologies, Le Bois Dieu, Lissieu, France).

Results

No false positive oral fluid CSFV rRT-PCR results were observed, i.e., 100% specificity was observed among samples collected prior to inoculation (n = 90). CSFV RNA was detected in oral fluid samples from DPI 4 (2/29 pigs) through DPI 28 (5/15 pigs). The highest proportion of positive results was observed on DPI 12 (12/12 pigs). An analysis of serum and oral fluid RT-PCR results (n = 147) showed a statistically significant association ($P = 0.0001$).

Conclusions

CSFV RNA was detected in oral fluid specimens collected from individual CSFV-inoculated pigs before the appearance of clinical signs, during the course of clinical signs, and after recovery. Thus, the data suggest that CSFV surveillance could be done using oral fluid samples. This approach could be particularly useful in for the detection of inapparent CSFV infections.

References

1. David D et al. Emergence of classical swine fever virus in Israel in 2009. *Vet J.* 2011;190:e146-e149.

BioChek software, a 24/7 access link between sample submission and laboratory results

Maartje Wilhelm[1], DVM; Eric van Esch[1], DVM; Tim Goode[2]; Alex Eggen[1], DVM
[1]BioChek BV, Reeuwijk, Netherlands; [2]Tim Goode, BioChek USA

Introduction

For monitoring the health status of a swine herd data has to be collected. Data collection starts with submitting samples to a diagnostic laboratory. This process involves administrative procedures that are time consuming and are often repeated. Mistakes will lead to confusion and extra work. BioChek software makes this process much easier. The BioChek software includes a "sample submitter" and a "sample processor" part. The system allows that the submitter and the processor share the same data. A unique barcode on the submission form assures digital transfer of information. After processing and approval of the submitted samples, the results are stored in the Cloud with 24/7 access on any internet linked device. Historical data can be re-used to generate trend reports.

Materials and methods

Samples are taken by the (veterinary-) staff and the variable data (details on the animals and their location) is entered in the BioChek software. The software-generated-form contains the data related to the submitter, which is the same for all submissions by this user. Samples are sent together with this barcoded form to the in-house or external diagnostic laboratory. The sample processor has the BioChek software and must have a barcode reader. By scanning the barcode the data present on the submission form is entered in the laboratory software. When the test results are available and released by the laboratory, the results are stored in the Cloud. The submitter of the samples can access the data on any internet device by using the (BioChek-) sample submitter software. The laboratory and submitter should use the same Cloud database. The user friendly interface presents the data as is desired by the user. This BioChek software system was tested on a farm where the consulting veterinarian was considering the implementation of a PCV2 vaccination in the breeding stock. He wanted to compare the immune response before and after vaccination including the influence of the vaccination on the Coefficient of Variation (%CV). His aim was to create uniformity in PCV2 antibody levels in the breeding stock.

Results

Serum samples were taken from the non-vaccinated sows before farrowing and presented in Figure 1A.

Sows were vaccinated against PCV2V according to leaflet instructions of the vaccine used. Results are presented in figure 1B.

All samples were analyzed with the BioChek PCV2 ELISA and results are presented in a quantitative manner and as was desired by the submitter (individual sow data in a bar chart format) showing the titer level and the calculated %CV, using the BioChek Software.

Discussion

Sow vaccination provided uniform titers in the sows. With a good colostrum intake management system, these vaccinated sows will raise piglets with uniform MDA titers against PCV2V. The BioChek software provided the graphic view as was requested by the submitter, with the relevant data (titer level and %CV in sows).

Conclusion

The BioChek software, including a sample submitter and a sample processor part provides the veterinarian with a user friendly, easy access system. The unique barcode avoids mis-communications between submitter and processor. Storage of the test results in the Cloud, provides the submitter with 24/7 access to both the latest and historical data, on any internet linked device. In an easy to generate and read manner, a graphical report to analyze the health status of the herd over certain period of time is just a mouse click away. BioChek software is available on request.

Figure 1A: PCV2 antibody (ELISA) titers in the sows before vaccination.

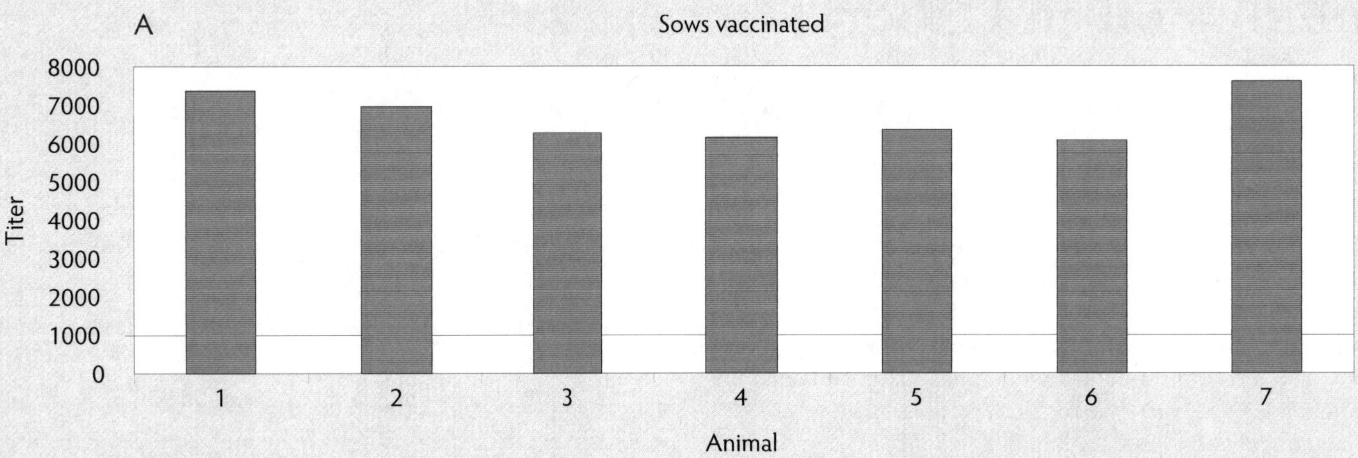

Figure 1B: PCV2 antibody (ELISA) titers in the sows after vaccination, samples taken before farrowing.

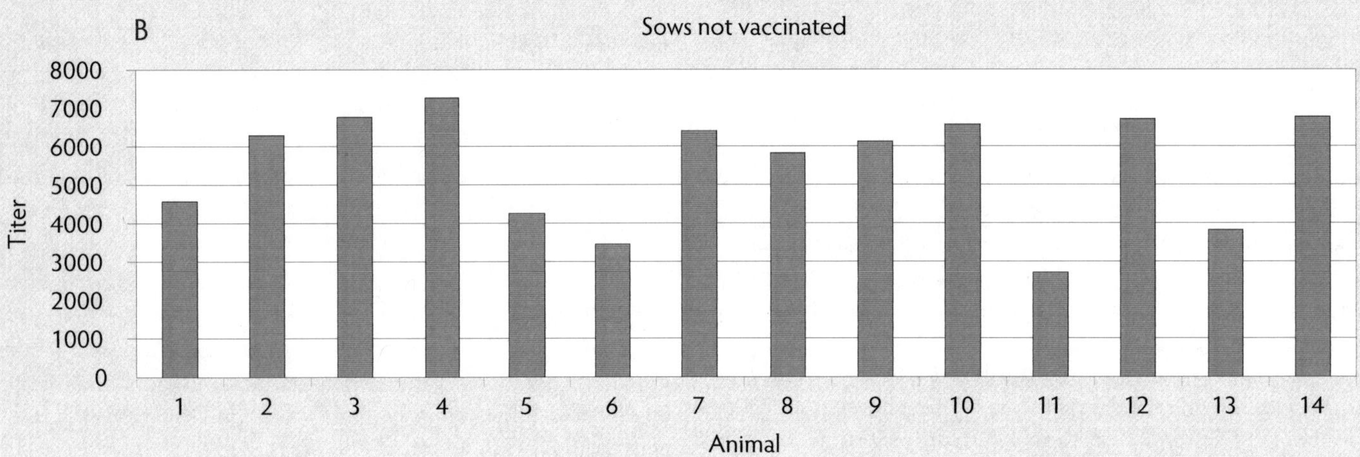

Multiplex testing for PRRSV, SIV, PCV2, and APP antibodies using multiplexed fluorometric immuno assays (MFIA)

I. Caya, BSc; M. Bertrand, DEC; A. Broes, DVM, PhD
Biovet, Saint-Hyacinthe, Quebec, Canada

Porcine respiratory disease complex (PRDC) is a significant economic problem for the swine industry. This syndrome is caused by the interaction of multiple non-infectious and infectious factors, including porcine reproductive and respiratory syndrome virus (PRRSV), swine influenza virus (SIV), porcine circovirus (PCV2), and *Actinobacillus pleuropneumoniae* (APP). Measurement of antibodies to these viruses is routinely performed for monitoring herd status or optimizing vaccination protocols. Testing for all these agents at the same time in a single assay (single well) could potentially save a lot of labor, time, and cost compared to traditional methods. This is now possible thanks to a new immunoassay platform called multiplexed fluorometric immunoassays (MFIA).

MFIA rely on the use of microbead sets (there are several dozen). Each bead set has a unique fluorescent marker that differentiates the various bead sets. The beads are coated with antigens (a different antigen for each bead set). Each serum to be tested is mixed with the relevant bead sets in the well of a microtiter plate. During incubation, the antibodies eventually present react with the corresponding antigen. Antigen-antibody complexes are then revealed by incubating the beads with secondary antibodies conjugated to a fluorochrome. Finally the microplates are examined using a special reader equipped with lasers and driven by a data acquisition and processing software. The reader is able to differentiate the various bead sets (thus the various antigens) and to quantify the intensity of the reporter fixed to each of them. It is thus possible to evaluate the quantity of the relevant antibodies present in the samples. Theoretically this technique allows detection of the antibodies, directed against dozens of different antigens in the same reaction (same well).

Over the last months Biovet has developed several bead sets for several agents involved in PRDC, and we are progressively developing more of them. We are currently ready to offer MFIA testing and reagents for PRRSV, SIV, PCV2, APP1-9-11, APP2, APP3-6-8-15, APP4-7, APP5, APP10, and APP12. The MFIA using these beads have diagnostic performances (sensitivity, specificity, reproducibility) similar to those of corresponding traditional assays.

References

1. Caya, I., Bertrand, M., Broes, A., 2014. A Multiplexed Fluorometric Immunoassay (MFIA) for the Detection of Antibodies to *Actinobacillus pleuropneumoniae* 1-9-11, 2, 3-6-8-15, 4-7, 5, 10 and 12. Proc. of the American Association of Veterinary Laboratory Diagnosticians Annual Conference 57, 188.

2. Caya, I., Bertrand, M., Broes, A., 2014. A Multiplexed Fluorometric Immunoassay for the detection of antibodies to PorcineReproductive and Respiratory Syndrome Virus (PRRSV), Swine Influenza Virus (SIV), and Porcine Circovirus (PCV2). Proc. of the American Association of Veterinary Laboratory Diagnosticians Annual Conference 57, 190.

Observations of a licensed trivalent PRDC vaccine for PCV2, M. hyo and PRRS

Brian Payne[1], DVM; John Kolb[1], DVM; Roy Edler[2], MS, MS; Hayley Oswald[1], MS
[1]Boehringer Ingelheim Vetmedica, Inc. (BIVI), Saint Joseph, Missouri; [2]Health Management Center (HMC), Field Research Services, BIVI, Ames, Iowa

Introduction

Porcine respiratory disease complex (PRDC) is a multi-pathogen disease that can cost owners > $10/pig.[1] Immune protection against clinical disease associated with these pathogens is one control method. A trivalent porcine circovirus type 2 (PCV2), Mycoplasma hyopneumoniae (M.hyo) and porcine reproductive and respiratory syndrome (PRRS) vaccine, 3FLEX® (Boehringer Ingelheim Vetmedica, Inc., St. Joseph, MO, USA), has been available in the US marketplace since 2010 and used to vaccinate hundreds of thousands of pigs. The purpose of this field study was to evaluate pigs vaccinated with 3FLEX® compared to a non-vaccinated group following challenge with PCV2 and PRRS.

Materials and methods

Study pigs (n = 675) were weaned from two typical commercial sow units that were demonstrated PRRS stable (herd category II-A)[2] on weaned pig PRRS PCR (HMC, Ames, IA, USA) testing. To prevent shedding of vaccine from the vaccinated group to the control group, pigs receiving 3FLEX® (2ml, n = 387) were housed approximately 1.0 km north/downwind from non-vaccinated control (NVC) group (n = 288) for 48 days. No PRRS virus (serum PRRS PCR) was detected in NVC pigs 12 hours prior to transport to the challenge site. Following the vaccination period, pigs were transported by treatment group, and assigned to pens based on pig size (small, medium, large) and balanced by treatment. One pen per barn held pigs from NVC and 3FLEX® groups to be challenged with a proven oral PRRS (type 2, 4000 $TCID_{50}$) and PCV2 (8 logs) positive tissue homogenate (5 gallon feed, top-dressed with 500 mL of tissue homogenate/pen) one week post-placement. An additional intramuscular challenge with PRRS virus (2mL, 2 logs) was made 10 days following the oral exposure. A subset of pigs were euthanized and necropsied to confirm a valid challenge by observing lesion development at five weeks post challenge.

Results and discussion

The results of this trial suggest that 3FLEX® provides protection to pigs challenged with PRRS and PCV2 as compared unvaccinated cohorts. To the authors' knowledge, this is the first large scale field study to successfully demonstrate individual pig PRRS and PCV2 challenge following 3FLEX® vaccination. In the face of a PRRS and PCV2 challenge, use of a trivalent vaccine mixture is an option for managing the economic losses associated with PRDC and the single injection reduces time and labor required for vaccination against these costly diseases compared to multiple, single-antigen vaccine or two dose protocols.

References

1. Haden et al., Assessing production parameters and economic impact of swine influenza, PRRS and *Mycoplasma hyopneumoniae* on finishing pigs in a large production system. AASV. 2012;75-6.

2. Holtkamp et al., Standard herd classification system for describing the PRRSV status of herds. Leman.2010; 44-6.

Table 1: Differences in performance variables in 3FLEX® vaccinated versus non-vaccinated control (NVC) then PRRS/PCV2 challenged pigs

Performance variable	3FLEX®	Non-vaccinated control (NVC)	*P*-value
Initial BW (lbs)	61.6	61.7	0.81
Final BW (lbs)	262.2	248.4	< 0.0001
ADG (lbs)	1.93	1.80	< 0.0001
Mortality (%)	5.1	28.2	< 0.0001
Culls (%)	11.8	27.5	< 0.0001

BW = Body weight; ADG = Average daily gain

Birth weight impact on survivability and days to market

Jan Jourquin[1], DVM; Joaquin Morales[2], DVM; Courtney D. Bokenkroger[3]
[1]Elanco Animal Health, Antwerp, Belgium; [2]PigChamp Pro Europa, Segovia, Spain;
[3]Elanco Knowledge Solutions, Greenfield, Indiana

Introduction

A strong focus on litter size selection and management has resulted in a steady increase of the number of piglets weaned per sow per year. High prolificacy in sows and increased fetal survival however lead to intra uterine crowding (IUC) and growth retardation (IUGR).[1] As a result the piglet birth weight is decreasing and its variability increasing.[2] For every extra pig in the litter, the average piglet birth weight drops with 25 to 35 gram.[2,3,4] Low birth weight has been defined as piglets beneath a certain arbitrary weight cut off but there is no uniformity between the studies.[2,5,6] Many authors have described how low birth weight pigs have higher mortality rates, reduced daily gain, reduced pork quality and increased feed conversion rate compared to their heavier litter mates.[1,5,7,8]

The objective of this study was to define a criterion for low birth weight by investigating the relation between the individual birth weight of the piglet, its survival chances and weight gain up to slaughter. Pigs below a predicted threshold are pigs at risk.

Materials and methods

178 litters from 3 farms were followed from birth to slaughter or moment of death. Litter and sow parameters were collected. The pigs were weighed at birth, weaning (26 days), end of nursery period (67 days), end of growing period (110 days) and the day before the first pigs of the batch went to slaughter (157 days). Days to 100 kg life weight was calculated based on life daily gain. If the pig died, date, weight and cause were recorded. Of the 2331 piglets born alive, not all could be followed until death or slaughter; 195 piglets were fostered away and out of the study before weaning. A second group of 254 piglets could not be followed beyond weaning and a third group of 177 pigs was sold at end of nursery. The pigs were categorized in increasing birth weight classes of 100 g. For each class, frequency, mortality and days to 100 kg was calculated. Descriptive statistics and a logistic regression model were applied to evaluate the impact of birth weight on piglet survivability and weight gain. A mixed effects logistic regression model was fit to estimate the probability of pre-wean mortality based on birth weight. A piece-wise linear predictor was selected to best represent the drastic decrease in pre-wean mortality found as birth weights increase in the range of 0.5 to 1.0kg and then the less extreme change in pre-wean mortality observed for changes in weight above 1.0kg. The change point of the linear predictor was found by computing the model fit for birth weights ranging from 0.5kg to 2.0kg and then selecting the point corresponding to the maximum likelihood for the model.

Results and discussion

Litter size was 14.3 total born piglets and 13.1 life born piglets. The average birth weight was 1.46 kg. Farm and litter size had an impact on birth weight. Litter size also impacted the within litter birth weight coefficient of variation. Pre-weaning mortality was 14.2%. A further 1.3, 1.1 and 0.9 % of the pigs died during nursery, growing and finishing phase respectively. 71.2 % of the mortality occurred the first 5 days of life. These piglets averaged 1.13 kg at birth and they lost weight indicating they did not nurse well. Low viability and crushing were the main mortality reasons. As birth weight increased, the probability for pre weaning mortality decreased. This decrease had a very steep slope with a change point at 1.13 kg, after which the mortality rate flattened out (see Figure 1). Below 1.13 kg (17.5% of total) the chance of survival until weaning was 58%. Pigs from 1.13 kg up had a 92% chance of survival. After weaning the relation between birth weight and mortality disappeared. Of the 1338 pigs that were followed until slaughter, pigs below 1.13 kg at birth (9.9 % of total) had an average daily gain of 203, 293, 601 and 805 g for the lactation, nursery, growing and finishing period respectively and the time to reach 100 kg was 191 days. Pigs of 1.13kg and higher (90.1 % of total) at birth had an average daily gain of 228, 359, 659, 845 g for the lactation, nursery, growing and finishing period respectively and the time to reach 100 kg was 177 days. The daily gain of the low weight piglets was constantly lower during each growing phase, demonstrating no compensatory growth potential.

Conclusion

A threshold birth weight of 1.13 kg was defined as the weight under which piglets are “at risk”. Apart from low survival chances, pigs that do survive from this “at risk” group have a negative impact on days to market and on the uniformity of the batches. This leads to a decrease of efficient use of the farm facilities.

References

1. Berard et al. Effect of litter size and birth weight on growth, carcass and pork quality, and their relationship to postmortem proteolysis. *J. Anim. Sci.* 2008. 86:2357–2368.

2. Quiniou et al. Variation of piglets' birth weight and consequences on subsequent performance, *Livestock Production Science 78* (2002) 63–70.

3. Devillers et al. Variability of colostrum yield and colostrum intake in pigs. *Animal* (2007), 1:7, pp 1033–1041.

4. Abell et al. Effect of litter size on piglet birth weight and birth weight as an indicator of pig quality. *Midwest ASAS,* 2015.

5.Gondret et al. Influence of piglet birth weight on postnatal growth performance, tissue lipogenic capacity and muscle histological traits at market weight. *Livestock Production Science 93* (2005) 137–146.

Figure 1: Predicted pre-wean mortality by birth weight

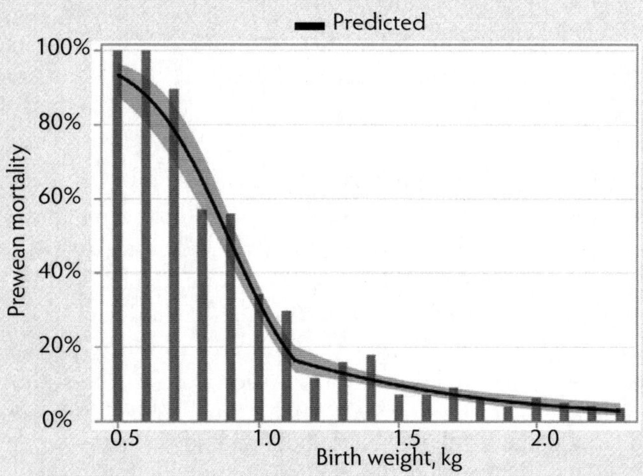

6. Milligan et al. Within-litter birth weight variation in the domestic pig and its relation to pre-weaning survival, weight gain, and variation in weaning weights. *Livestock Production Science 76* (2002) 181–191.

7. Kohler et al. The effect of piglet individual birth weight on profitability. *AASV Annual Meeting* 2014.

8. Fix JS, Relationship of Piglet Birth Weight with Growth, Efficiency, Composition and Mortality. Dissertation North Carolina State University 2010.

The use of antibody monitoring for endemic diseases in two commercial sow farm on-site GDUs and a barn of high health commercial pigs – Field study

Christa Goodell[1], DVM, MS, PhD; John Prickett[2], DVM, PhD; David Baum[3], DVM, PhD; Dyneah Classen[2], DVM; Nicole Eddy[2], MS; Paulette Barter[1]; T. Gard[3]; Joe Connor[2], DVM, MS
[1]IDEXX, Westbrook, Maine; [2]Carthage Veterinary Services, Carthage, Texas; [3]Iowa State University, Ames, Iowa

Introduction

Endemic diseases are costly. While their impact on mortality and growth performance is seldom quantified, they are believed to be significant and particularly notable in times of low hog prices or high feed costs.[1] Monitoring endemic diseases is a valuable tool to permit better understanding of disease dynamics in pig populations. This better understanding might be facilitated by analyzing and graphing routine antemortem diagnostic data and thus improve production and health management decision-making.[2] This study was performed to monitor PRRSV, *M. hyopneumoniae* (M. hyo) and influenza A virus (IAV) antibodies in growing commercial and replacement animals.

Material and methods

A commercial wean-to-finish barn and 2 gilt developer units (GDU's) monitored antibodies to PRRSV, M. hyo and IAV using oral fluids (IDEXX PRRS OF) and sera (IDEXX PRRS X3, IDEXX M. hyo, IDEXX Swine Influenza Virus). Routine (approximately biweekly) pen-based oral fluid collections were performed in the growing pigs or gilts using the same pens at each time point, and sera was collected monthly from a subset of animals from each of these pens.

Results

W-F site: OF was collected 13 times, and sera, 7 times, during the grow-out period. Both sera and OF results remained negative (S/P < 0.40) for PRRSV antibodies over this time. Monthly *M. hyo* serology remained negative (S/P < 0.40) as well. IAV serology showed (likely) maternal antibodies declined by 12 weeks of age and remained negative (S/N > 0.60) through the end of the growing period.

GDU 1: Two groups of gilts were followed in a similar sampling pattern as the commercial study. In weaned pigs, PRRS X3 ELISAs were positive, but by the following sampling period, both sera and OF were negative for PRRSV antibodies. By 10 weeks of age the PRRS OF Test detected seroconversion, subsequently noted in the monthly sera sampling. M. hyo test results demonstrated young piglets negative for M. hyo antibodies; however by 10 weeks of age, there was indication of seroconversion. Finally, similar to pigs on the W-F site, maternal antibodies to IAV were identified in nursery age gilts; but these declined in both groups, and the groups remained IAV antibody negative in finishing age females.

GDU 2: Like GDU 1, two groups of gilts were followed, with a similar sampling pattern as the commercial site. These gilts started and remained both PRRS and M. hyo antibody negative (both by OF and sera) for the duration of GDU testing. Both groups were positive for IAV antibodies early in the nursery phase due to maternal antibodies, and these antibodies declined by ~10 weeks of age. However, one group showed IAV exposure by seroconversion at 14wks of age (S/N ≤ 0.60).

Conclusions

Endemic diseases antibody monitoring is a useful tool to demonstrate the continued freedom from exposure in high health pigs. Antibody monitoring is also an excellent tool for confirming proper implementation of exposure programs, and provides insight into the immune status of growing females acclimated in GDUs prior to sow farm entry.

References

1. Holtcamp D et al., The economic cost of major health challenges in large US swine production systems. *Proceedings of the American Association of Swine Veterinarians.* 2007: 85-89.

2. Alvarez et al., Impact of endemic and epidemic pathogens in performance of growing pigs. *National Hog Farmer* Dec. 28, 2015; 24(2): 262-269.

Improving sow farm profitability by reducing unnecessary culling and non-productive days and improving reproductive management using advanced technologies

Christine Pelland, DVM, MS

As the swine industry intensifies, more pressure is placed on the individual sow to exhibit optimal reproductive performance for the entirety of her life. If she does not fit this mold, she will be culled from the population and replaced with a new replacement gilt. We select ideal candidates to enter into the herd as replacements based on genetic merit and external physical attributes. On the other hand, we are relatively blind as to what is occurring under the surface and therefore uncertain about the biological and hormonal contributions that ultimately determine her overall performance.

Technological advancements in ultrasonography have proven themselves an excellent tool for diagnostic purposes looking at organs in the sow such as the uterus and ovaries.[1] Ultrasound provides us with a real-time tool to asses an individual sow's biological and hormonal status. Utilizing currently available equipment, we are able to successfully use transcutaneous scanning as a quick and non-invasive means to detect puberty in gilts and to characterize ovaries and uteri in reproductively failed sows.[2]

Transcutaneous scanning has the ability to detect pregnancy as early as 18 days,[3] follicular dynamics around the time of estrus (Figure 1A) and ovulation,[4] examination of the gilt and determination of puberty status,[5] and examination of the non-productive female (diagnosis of cystic ovarian disease – Figure 1B).[6]

Using this technology has many economic benefits. One major benefit is the use of transcutaneous scanning in gilts to determine if they are pubertal. This has the advantage of helping to tailor gilt development protocols and weed out poor performers early on. Another benefit is to observe ovarian structures in non-pregnant animals. This allows for more timely individual approaches such as immediate culling or informed hormonal treatments to get them back to optimal reproductive performance most efficiently. All of these benefits are inevitably driving the end result of having the most reproductively sound animals in your herd and being as efficient as possible by reducing the number of non-productive days.

References

1.Weitze KF, Habeck O, Willmen T, Rath D. Detection of ovulation in the sow using ultrasound transcutaneous sonography. Zuchthlygiene 1989;24:40-2.

2. Kauffold J, Rautenberg T, Hoffman H, Beynon N, Schellenberg I, Sobiraj A. A field study into the appropriateness of transcutaneous ultrasonography in the diagnosis of uterine disorders in reproductively failed pigs. Theriogenology 2005; 64:1546-58.

3.Miller GM, Breen SM, Roth SL, Willenburg KL, Rodriquez-Zas A, Knox RV. Characterization of image and labor requirements for positive pregnancy diagnosis in swine using two methods of real-time ultrasound. J Swine Health Prod 2003;11:233-9.

Figure 1: A) Ovary in a sow depicting multiple follicles as she approaches ovulation. B) Ovary in a sow depicting a large ovarian cyst.

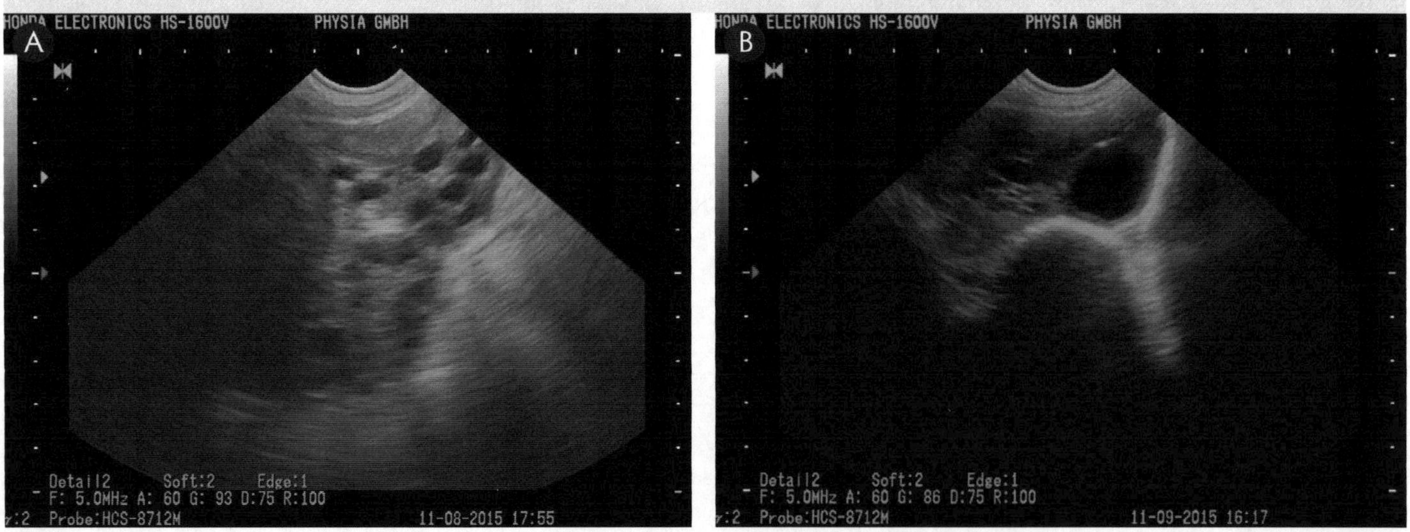

4.Soede NM, Hazeleger W, Kemp B. Follicle size and the process of ovulation in sows as studied with ultrasound. Reprod Domest Anim 1998;33:239-44.

5.Kauffold J, Rautenberg T, Richter A, Waehner M, Sobiraj A. Ultrasonographic characterization of the ovaries and the uterus in prepubertal and pubertal gilts. Theriogenology 2004;61:1635-48.

6.Knox RV, Althouse GC, Visualizing the reproductive tract of the female pig using real-time ultrasonography. J Swine Health Prod 1999; 7:207-15.

Determining viability of porcine mycoplasma pure cultures using a redox indicator

Boh Chang Lin, DVM, MS, PhD; Jeff Kula, DVM, MAM; Terry Marco, BS
MVP Laboratories, Inc. Omaha, Nebraska

Introduction

Mycoplasma hyopneumoniae (Mhp), *Mycoplasma hyorhinis* (Mhr), and *Mycoplasma hyosynoviae* (Mhs) have caused a big economic loss in the pig industry worldwide. In order to prevent mycoplasma infection, pure cultures were submitted from university veterinary diagnostic laboratories to federal licensed biological companies for autogenous vaccine production. PCR assays are used to identify mycoplasma species. However, PCR results cannot tell the difference between viable and dead cultures. In order to determine viability of the submitted cultures, a non-toxic redox (oxidation-reduction) indicator (alamarBlue) was used. The alamarBlue (life technologies) assay is designed to monitor the reducing environment of proliferating mycoplasma culture.[1] If the tested culture is alive, innate metabolic activity of mycoplasma culture will make alamarBlue in reduced form (pink in color). If the tested culture is dead, alamarBlue will be in the oxidized form (blue in color).

Materials and methods

Determining viability of mycoplasma cultures: A sample was removed from a vaccine production jar right before inactivation and another sample was removed from the same vaccine serial right after completion of the inactivation test. A ten-fold serial dilution was made for the live and inactivated culture and incubated at 37°C until the live cultures reached 10^8 color change unit (CCU) per ml and inactivated culture showed no color change. Each of the three sterile snap-top tubes was added with 0.9 ml of modified Friis broth and either 0.1 ml of sterile PBS or 0.1 ml of live Mhp culture or 0.1 ml of inactivated Mhp culture respectively. Each tube was added with 0.1 ml of alamarBlue and incubated at 37°C for 3 days. The color change and OD_{630} of each sample were recorded with an ELISA reader. Aseptic procedure was used in order to avoid bacterial contamination. Each tested sample was also inoculated into a sheep blood agar plate and incubated at 37°C for three days to rule out any bacterial contamination. The test procedure for Mhr and Mhs was the same as above except that a bacterial plate count was used and positive sample must have 10^8 colony forming unit (CFU) per ml.

Results and discussion

So far various cell viability and proliferation assays have been used to find out the effect of a chemical on cells propagated in vitro. Some indirect techniques that use dye exclusion or preferential dye uptake have been found to destroy or interfere with a cell's function. A recent report even indicated that most test methods to determine viable versus dead cells cannot easily be used for quantitative differentiation of mycoplasma cultures.[2] In order to solve this limitation we have adopted a new concept that uses an oxidation-reduction reagent called alamarBlue to detect the metabolic activity of mycoplasma in the culture.

In this study Mhp live culture always turns pink with an OD_{630} reading of 0.070 to 0.210 ($OD_{630} < 0.210$), while inactivated Mhp culture turns blue with an OD_{630} above 0.300. Some cultures may turn purple with an OD_{630} reading between 0.210 and 0.299, which are identified as partial viable cultures. Some partial viable cultures may turn viable after long term incubation with enriched medium. Table 1 shows color change and OD_{630} readings of some tested samples. In this study it was also found that viability of Mhr can be detected at around 10^3 CFU/ml and no color change can be obtained below 10^2 CFU/ml.

References

1. Rampersad, S. N. 2012. Multiple applications of AlamarBlue as an indirect indicator of metabolic function and cellular health in cell viability bioassays. Sensors 2012, 12, 12347-12360; doi:10.3390/s120912347.

2. Anderson, A.; A. Rovira; Pieters, M. 2015. Determining *Mycoplasma hyopneumoniae* viability through the use of ethidium monoazide. Pro AASV Ann Meet. Orlando, Florida. 2015:289-290.

Table 1: Using color change and OD_{630} absorbance value to indicate the oxidized (dead) or reduced (live) environment of fresh and frozen Mhp cultures

Test sample	Freezing	Color	Viability (OD_{630})	True viability*
Sterile PBS	NA	blue	-- (0.420)	--
Live Mhp	NA	pink	+ (0.075)	+
Inactivated Mhp	NA	blue	-- (0.408)	--
Mhp200006	5.5 years	pink	+ (0.100)	+
Mhp200199	5.0 years	blue	-- (0.539)	--
Mhp200579	5.0 years	blue	-- (0.349)	--
Mhp200022	4.0 years	blue	-- (0.387)	--
Mhp200942	4.5 years	pink	+ (0.083)	+
Mhp300934	3.5 years	purple	+/-- (0.295)	+
Mhp300197	3.0 years	pink	+ (0.100)	+
Mhp200934	3.0 years	blue	-- (0.564)	--
Mhp200197	3.0 years	pink	+ (0.100)	+
Mhp300320	2.0 years	pink	+ (0.099)	+
Mhp300335	2.0 years	blue	-- (0.356)	--
Mhp300545	1.0 year	pink	+ (0.098)	+
Mhp300027	0.6 year	purple	+/-- (0.290)	--

* True viability was obtained by incubating tested culture in an enriched medium at 37°C for six weeks to see any color change in liquid medium or any growth on solid agar. NA=not applicable.

Flexible adjuvants for combined live Aujesky's disease and inactivated swine influenza vaccines

Sophia Xu[1], BVM, PhD; Juliette Ben Arous[2], PhD; Sébastien Deville[2]; Laurent Dupuis[2], PhD
[1]SEPPIC Inc, Fairfield, New Jersey; [2]SEPPIC, Paris, France

Introduction

Live attenuated and inactivated antigens are both widely used for pig vaccination. In the field, animals receive multiple vaccines against diverse diseases in a short time, and there is a strong demand for compatible and mixable vaccine formulations comprising diverse types of antigens. However, as inactivated vaccines usually contain vaccine adjuvants, whereas live vaccines are usually not adjuvanted and not compatible with adjuvants, these types of vaccines are not mixable. Diverse adjuvants technologies are used for swine inactivated vaccines, such as oil emulsion adjuvants, aluminium hydroxide and polymeric adjuvants. We have shown previously that Montanide™ adjuvants based either on polymer technology or oil in water emulsions were compatible with a PRRS live vaccine, and enhanced the protection to challenge of vaccinated animals. These adjuvants also allowed the reduction of the antigenic load in a live PRRS vaccine.[1] Here we show that these Montanide™ adjuvants can be used to formulate highly efficient inactivated swine influenza vaccines, and are also compatible with an Aujesky's disease live vaccine. They can therefore allow the formulation of efficient combined inactivated/live vaccines for swine against swine influenza and Aujesky's disease.

Materials and methods

Inactivated vaccines against Swine influenza virus (SIV) were formulated with either the polymer adjuvant Montanide™ Gel 01 (Gel), an oil in water emulsion adjuvant Montanide™ ISA 15A VG (ISA 15A) or without adjuvant. All vaccines contained the same antigenic dose. A non vaccinated group was included as a negative control. At day 0 and day 21, 10 seronegative pigs (15 kg) were vaccinated in each group intramuscularly in the neck simultaneously with 2ml of the corresponding inactivated SIV vaccine and 2ml of Aujesky's disease attenuated live vaccine.

Safety properties of the vaccines (systemic and local reactions) were assessed at vaccination, during the trial and at slaughter. Efficacy was followed both by serological analysis and application of a challenge procedure. Antigen specific ELISA analyses against Aujesky's virus and SIV were performed at day 0 (before vaccination), day 21 (before revaccination), day 42 (before virulent challenge) and day 56 (at slaughter). At day 42, 5 animals in each group were infected with Aujesky's virus, and 5 other animals were infected with SIV. Clinical signs and hyperthermia after challenge were measured for 14 days, and nasal viral loads were scored. Bacterial over-infections of the lungs were scored at slaughter at day 56.

Results

All vaccines tested were safe. Antibody titers against swine influenza virus were significantly superior for Gel or ISA 15A adjuvanted formulations compared to the non adjuvanted vaccine. Immune response after vaccination with adjuvanted vaccines reached positive protective threshold at 21 days after only one injection, whereas in the non adjuvanted group, protective antibody level was not reached before the infective challenge. Protection against SIV challenge was also improved, as hyperthermia after infection, lung lesions at slaughter and nasal viral shedding were reduced compared both to non vaccinated and non adjuvanted groups. Moreover, both adjuvants were compatible with Aujesky's disease live vaccine. Protection conferred by the attenuated Aujesky's disease part of the vaccine was not reduced by the presence of either adjuvant. Moreover, viral shedding after Aujesky's challenge was reduced in the Gel group.

Conclusions and discussion

These results show that relevant aqueous adjuvants such as Montanide™ Gel 01 or Montanide™ ISA 15A VG are compatible with both inactivated and attenuated viral vaccines for swine. These adjuvants are highly effective in improving SIV vaccine protective efficacy and do not impair the efficacy of co-administered live vaccine. Therefore, such adjuvants can allow the formulation of multivalent combined inactivated/live vaccines. Such combined vaccines could lead to the reduction of the number of injections given to pigs in the field.

References

1. Deville et al. 2012. Procedia in Vaccinology 6:134-140.

The prospects of selection for social genetic effects to improve welfare and productivity in pigs

Naomi Duijvesteijn[1], PhD; John. M. Eggert[2], MS, MBA, PhD, Dipl. ACAG
[1]Topigs Norsvin Research Center B.V., Beuningen, The Netherlands; [2]Topigs Norsvin, Burnsville, Minnesota

Introduction[1]

Livestock production is a world-wide industry. In developed countries such as the USA and Europe, the market share of animal proteins is stagnating and animal production systems are focusing more on efficient and sustainable production.[2] The demand for higher efficiency has resulted in 1% to 3% genetic progress per generation in commercial breeding programs due to single or multi-trait models.[3] Breeding organisations are continuously adapting their breeding goals to support the needs of farmers and society. There is a world-wide demand for better feed efficiency due to the competition for land between food and feed production, but there is also more attention on some of the undesirable correlated responses such as increase in disease susceptibility and more observed mal-behaviour such as aggression and tail biting.

Breeding for improved behaviour[1]

A very promising selection method with the potential to improve both animal welfare and economic output is the use of social genetic effects (SGEs) in breeding programs. A social genetic effect is a heritable effect of one individual on the trait value of another individual.[4] The breeding approach using SGEs incorporates both the direct genetic effect due to the focal individual, and the genetic effect an animal has on its pen mates into the trait value of the focal individual.[4] While traditional methods focused on individual performance only, this strategy could improve growth in pigs, as well as the behaviour of pigs which are housed in groups.[5,6] A one generation selection experiment where pigs (gilts and castrates) were grouped based on a high or low SGE for growth was conducted to investigate underlying behavioural differences and confirm previous results (N = 480). Aggression measured by skin lesions and fighting during regrouping did not differ between high and low SGE pigs. However, pigs with a high SGE showed less aggression after reunion with familiar pigs and also had less non-reciprocal biting in the week after regrouping.[5] During the finishing phase, high SGE pigs showed systematically less biting behaviour; 40% less aggressive biting and 27% less oral manipulation of pen mates. High SGE pigs were also chewing 40% less on distraction material and consumed 30% less of the jute sacks provided. These differences were also expressed in the tail damage, where high SGE pigs had a better tail score (less damage) compared to low SGE pigs.[6]

One remark on this one-generation selection experiment should be made. The contrast on growth, by selecting sires and dams with high and low SGEs, did not result in significant differences in growth in their offspring.[7] The authors argue that control measures (jute sacks, sawdust) to limit harmful behaviour might have reduced the expression on SGEs on growth and that research under commercial circumstances needs to be conducted.

Conclusions

Social genetic effects offer the opportunity to breed for improved behaviour while maintaining performance, although results are sometimes conflicting and accurate pen registration is essential. Since group housing is standard practice in finishing pigs, the pen is the production unit and the pen is of great importance for the production, welfare and health of the pig.

References

1. Duijvesteijn, N. Sociable swine; prospects of indirect genetic effects for the improvement of productivity, welfare and quality. *Thesis; general discussion*. 2014; 119-148.

2. Thornton, P. K. Livestock production: recent trends, future prospects. *Philosophical Transactions of the Royal Society B: Biological Sciences*. 2010; 365; 2853-2867.

3. Merks, J. W. One century of genetic changes in pigs and the future needs. *BSAS Occasional Publication*. 2000; 8-19.

4. Griffing, B. Selection in reference to biological groups I. Individual and group selection applied to populations of unordered groups. *Aust J Biol Sci*. 1967; 20; 127-140.

5. Camerlink, I., S. P. Turner, P. Bijma, and J. E. Bolhuis. Indirect genetic effects and housing conditions in relation to aggressive behaviour in pigs. *PloS one*. 2013; 8; e65136.

6. Camerlink, I., W. W. Ursinus, P. Bijma, B. Kemp, and J. E. Bolhuis. Indirect genetic effects for growth rate in domestic pigs alter aggressive and manipulative biting behaviour. *Behav Genet*. 2014; 45: 117-126.

7. Camerlink, I., J. Bolhuis, N. Duijvesteijn, J. van Arendonk, and P. Bijma. Growth performance and carcass traits in pigs selected for indirect genetic effects on growth rate in two environments. *J of Anim. Sci*. 2014; 92: 2612-2619.

Nutritional requirements of immunologically castrated (IC) barrows

D. A. Nelson[1], PhD; M. Bertram[2], PhD; S. Pollmann[3], PhD; M. Tokach[4], PhD
[1]Zoetis, Florham Park, New Jersey; [2]First Choice Livestock, LLC, Urbandale, Iowa; [3]DSP Consulting Inc., Alpine, Utah; [4]Kansas State University, Manhattan, Kansas

Introduction

Male pigs are given IMPROVEST® (*gonadotropin releasing factor analog–diphtheria toxoid conjugate*) in the finishing phase, eliminating the need for physical castration early in life. IMPROVEST involves a two-dose regimen that requires different nutrient requirements prior to and after the second injection to optimize IC barrow performance and to give pork producers full advantage of the economic benefits that it offers. In 2013, Zoetis released the North American Nutritional Guidelines for Immunologically Castrated Male Pigs (Bertram et al., 2013). These guidelines were derived from a committee of academic and industry nutritionists using data generated from 12 US trials,[1] and incorporated the 2012 NRC modeling software. Basis the 12-trial data, the committee recommended lysine concentrations of 104, 108, 112 and 98% of the dietary SID lysine concentration of contemporary gilt diets for IC barrows for the periods from 25 to 50 kg, 50 to 75 kg, 75 to 100 kg and 100 to 130 kg, respectively. The average improvement in ADG and F/G of IC barrows over physical castrates (PC) in these 12 trials was 4.3% and 8.4%, respectively.

Recent research

New research at the University of Illinois[2-4] and Iowa State University[5] offers insights into the performance of IC barrows based on further analysis of their dietary lysine requirements. Specifically, dietary concentrations of lysine and phosphorus should be increased to optimize average daily gain (ADG) and feed efficiency (F/G). Lysine levels for IC barrows should be formulated to provide 125 to 128% of a production system's current physical castrate (PC) barrow diets starting at approximately 50 pounds body weight through two weeks after the second dose of IMPROVEST (Table 1). Phosphorus levels should provide 111% of a production system's current PC barrow diets for the same time period. Beginning two weeks after the second dose of IMPROVEST and until slaughter, IC barrows should be fed the same diets used for PC barrows. After the second dose of IMPROVEST, IC barrows can maintain their ADG advantage over PC barrows under a wide range of marketing strategies. When incorporating the new nutritional requirements and marketing at optimum times post-second dose (Barn empty by 7 weeks after the second dose), IC barrows can exceed a 10% improvement in both ADG and F/G over PC barrows. This improved performance is dramatically greater than earlier reported results. To maximize the economic return from IC barrows, producers must feed IC barrows separately from gilts.

References

1.Cowles B, D. Meeuwse, J. Bradford. 2013. Data summary of immunological castration impact in grow-finish performance of male pigs in USA. Am. Assoc. Swine Vet Proc.; (Abstr. #57) 375-376.

2.Puls, C.L. 2013. Growth performance and carcass characteristics of immunologically-castrated barrows in comparison with to intact barrows, physically-castrated barrows, and gilts. Thesis (Ph.D.), University of Illinois at Urbana-Champaign.

3.Puls, C.L., A. Rojo, M. Ellis, D.D. Boler, F.K. McKeith, J. Killefer, A.M. Gaines, P.D. Matzat, and A.L. Schroeder. 2014. Growth performance of immunologically castrated (with Improvest) barrows (with or without ractopamine) compared to gilt, physically castrated barrow, and intact male pigs. J. Anim. Sci. 92:2289-2295.

4.Puls, C.L., M. Ellis, F.K. McKeith, A.M. Gaines, and A.L. Schroeder. 2014.Effects of ractopamine on growth performance and carcass characteristics of immunologically and physically castrated barrows and gilts J Anim. Sci. 92:4725-4732.

5.Elsbernd, A. J., C.F.M. de Lange, K.J. Stalder, L.A. Karriker, and J.F. Patience. 2014. Determining the SID lysine requirement of immunologically and physically castrated male pigs during three growth phases and feed intake pattern characterization in immunologically castrated pigs following their second injection. Thesis (M.S.), Iowa State University.

Table 1: Key nutritional recommendations of IC barrows

Body Weight, kg	20-30	31-45	46-60	61-75	76-90	91-122[a]	123+[b]
SID Lysine level relative to PC barrows, %	125	125	127	128	128	128	100
SID Lysine level, % of diet[c]	1.58	1.45	1.21	1.09	.95	.82	.62
SID Lysine level, g/day	17.1	22.3	23.0	23.0	22.3	22.0	26.0
SID Lysine level, g/kg of gain	22.60	21.00	20.00	19.20	19.00	18.00	18.00
SID Lysine, g:MCal ME	4.70	4.29	3.56	3.19	2.75	2.39	1.81
Available Phosphorus level relative to PC barrows, %	111	111	111	111	111	111	100
Available Phosphorus level, % of diet[d]	.50	.41	.39	.35	.31	.21	.16
Available Phosphorus level, g/day	5.50	6.05	7.22	7.41	7.22	6.24	5.46

a Assuming 2nd injection at 105 kg to account for 14 day transition period.
b Assumes no ractopamine-HCL. With ractopamine-HCL follow manufactures recommendation and increase 35-45%.
c Amino acid ratios remain consistent as listed in 2012 NRC.
d Ca:P ratio remain consistent as listed in 2012 NRC

Neonatal immunology and vaccinology: Timing is everything

Dick Hesse, MS, PhD
Kansas State University, Manhattan, Kansas

Neonatal immunology

Management is the key to neonatal immunity—housing environment, diet, pre-conditioning/acclamation of the dam, and subsequent environment microbiota colonization of the piglet have dramatic effects on the successful development of a fully functional neonatal immune system. Standard management practices and useful information for the sow herd manager are provided by Chiba[1] and Reese.[2] Hurley[3] summarized and discussed general management concepts regarding the development of the piglet immune system. Mixed beneficial results of sow dietary supplementation[4,5] with various oils or extracts have been observed in colostrum and milk composition and also on passively acquired immunity of the piglet; however, further research is needed to determine the overall effectiveness of this, and other, supplementation practices. Control and potential management of microbiota colonization of neonates[6-8] has dramatic effects on colonization resistance as well as development of the mucosal immunity; however, once again further research is needed to determine the best strategies to use this potential management tool.

Successful pig production is dependent upon the dam. Providing proper housing and diet pre- and post-parturition as well as solid acclimation—exposure or vaccination to a variety of production pathogens and development of good immune response to them—are critical for establishing protective levels of maternally derived immunity (MDI) to the baby pig. Colostrum contains large amounts of antibody, a large number of leukocytes, and innate immune factors (cytokines among others) that need to be delivered prior to gut closure in order to successfully protect the neonatal pig. Lactogenic immunity—maternally derived antibody (MDA) from colostrum and milk—has long been recognized as a critical element for growing healthy pigs.[9,10] Maternally derived cellular immunity (MDCI), also acquired in colostrum prior to gut closure, has not been as well characterized as MDA, but it is clearly essential for protection of the neonate.[11]Porcine neonatal immune system development—what do we really know or think we know? Two excellent contemporary reviews of passively and actively acquired neonatal immunity are summarized by Bandrick and Butler.[12,13]

One important key to manipulating the neonatal immune system is understanding how to manipulate the sow's immune system prior to parturition. The goal of dam "pre-conditioning/acclimation" is the purposeful priming/directing of their immune system to provide the protective levels of MDI to the piglet. Pathogen exposure or vaccination are the time honored "management tools" for gilt/sow acclimation. Monitoring and understanding the magnitude of the dam's MDI response to the these procedures allows the practitioner to successfully manage piglet response so that timing of natural pathogen exposure and/or vaccination against production site pathogens results in a predictable induction of disease resistance or protective active immunity. Knowing specific antibody titers to a given pathogen allows the practitioner to manipulate sow herd immunity and subsequently "baby pig herd" immunity. The following information bullets are intended to reaffirm the things that we have been taught, but also to expose field veterinarians to new findings that are directly related to understanding and manipulating the neonatal immune system.

Key components/concepts of MDI

Understanding gut closure is critical to MDI. Colostrum contains large amounts of antibody, a large number of leukocytes, and innate immune factors-cytokines/antimicrobial peptides that need to be delivered prior to gut closure[9,10] in order to successfully protect the neonatal pig. Once gut closure occurs, passive transfer of these maternally derived protective elements stops. First come, first served – first born pigs get the highest level of MDI which includes MDA-antibody and MDCI-cellular and cytokines. After gut closure occurs, usually 24-48 hours post parturition (assuming the pig suckles), colostrum secretion stops and milk secretion begins. MDA and MDCI are still present in milk; however, it contains significantly higher levels of IgA than IgG and fewer leucocytes than is found in colostrum. The protective effect of MDI in milk is local, within the lumen of the intestine.

As part of her PhD dissertation, Brandrick[12] conducted an in-depth review of MDI literature and investigated the relationship of MDA and MDCI on the development and manipulation of the neonatal pig. Interesting points brought out in her thesis include:

- Cross fostering has important implications on colostral transfer of MDI.
- MDA is not dam/offspring specific. Antibody present in colostrum sources other than the piglet's dam will result in effective transfer and the protective effect of these antibodies will benefit the neonate.
- Colostral transfer of MDCI (immune cells) and subsequent functionality appears to have a genetic factor associated with it. Passive acquired MDCI are functional in the offspring of a dam's own piglets, but not functional in piglets

other than her own. Cross fostering clearly can have beneficial effects, but in order to ensure the highest levels of MDA and MDCI, it should not be done after gut closure. Timing is everything.

- High levels of colostral antibody can interfere with a piglet's active antibody response, but it does not interfere with their active CMI response.

Re-secretion of colostral derived antibodies has been described in calves and a similar mechanism appears to be functional in piglets where colostral IgG is re-secreted back into the intestinal lumen and that the re-secreted antibody can result in reduction of rotaviral clinical disease.[14]

Immunity of a piglet – when does it become as functional as the sow's? We all know that young pigs start to develop a functioning immune system by 80 days (or less) of gestation, but did you know that their immune system is really not as functional as their dam's until they are around 7 weeks of age?[15,16]

Neonatal vaccinology

As already discussed, management is the key to neonatal immunology and also to vaccinology. Passive protection due to MDA and, to some extent MDCI, will not be a significant part of the following discussion; rather, emphasis will be on successful manipulation of the neonatal immune system through vaccination and how it is dependent upon understanding how the sow's immune system was managed prior to parturition.

There is no single vaccination strategy or vaccine that covers all diseases in all production sites. But there are some things that we can consider as neonatal vaccination programs are discussed and designed. As previously stated, a key baby pig management tool is pre-farrow vaccination/exposure which results in pathogen specific transfer of MDA and MDCI to the neonate. Knowing specific antibody titers to a given pathogen allows the practitioner to manipulate sow herd immunity and subsequently "baby pig herd" immunity. Monitoring, understanding, and vaccinating based on the magnitude of the dam/piglet MDI response results in vaccine responses that generate a solid "vaccine take" which provides disease protection prior to pathogen exposure during the production cycle. Profiling specific antibody titers can be a useful management tool.

Key components/concepts of neonatal vaccinology

A consistent vaccine response in baby pigs is dependent upon stable/consistent levels of immunity in the sow herd. Important things to consider are: what is the overall immune level of the breeding herd and when does a pathogen of concern circulate in that particular herd. From a management perspective is it better to rely on intensive vaccination of the breeding herd or vaccination of the neonates. There are pros and cons either way – the only thing that counts is keeping ahead of the site/region specific disease(s).

MDA passive interference with vaccine induced antibody response in the neonate is very real and has been well documented for decades. Bandrick reviewed the literature regarding MDA inhibition on the generation of CMI responses in baby pigs and concluded that, in the case of swine influenza virus (SIV), it did not inhibit a CMI-specific response to SIV.[12] Recent studies investigating antigen specific colostral derived cellular immunity have demonstrated both *in vivo* delayed type hypersensitivity responses and *in vitro* lymphoproliferative responses in piglets from *M. hyopneumoniae* vaccinated dams, but not in piglets from unvaccinated dams.[17] This observation has real implications for defining disease intervention strategies – clearly additional studies in this area are needed.

Vaccination verses immunization – what is the difference? Protection!!!! Vaccination in the face of high levels of maternal antibody usually results in inadequate levels of protection and a short duration of immunity. In order to get the most from vaccines (humoral and CMI) responses, it is necessary to vaccinate when MDA levels are low, and prior to pathogen circulation. If the baby pig herd does not have consistent levels of immunity, then multiple vaccinations are usually required in order to actually immunize the herd. The time period between vaccinations is also an important component of successful immunization.

Modified live vaccines (MLV) are equally safe and efficacious in all pigs at all ages, right? WRONG. First of all, live vaccines may have the same name, but rarely do they have the same characteristics. Differences in attenuation, clearance/duration of viremia, kinetics of neutralizing antibody response, duration of immunity, and interaction with herd specific pathogens all play into when and how live vaccines can safely be used. An example of unintended consequences as a result of using a MLV to control PRRS infection in conjunction with porcine circovirus2b (PCV2b) exposure was observed by Niederwerder while conducting experiments for her PhD studies. PRRS vaccination reduced clinical disease following wild type PRRS exposure, but the PRRS vaccinated animals experienced more severe clinical disease than the animals exposed only to wild type PRRS when PCV2b was used as a co-factor. The study showed that PRRS vaccination may offer control for subsequent PRRS exposure, but increased disease potentiation for PCV2b.

In a perfect world, we would give a single neonatal vaccination at the time of processing that would last through grow/finish – but do you know anyone who lives in that world? The point is, we need to understand immunity in the breeding herd in order to properly time vaccination of the baby pig herd, and no single strategy works for every herd. Sow herd titer profiling and timing of baby pig herd vaccination is critical.

References

1. Chiba, L. (2004). "Section 4: Baby Pigs (Birth to Weaning)". In Swine Production Handbook (13th Revision: 44-55). http://www.ag.auburn.edu/~chibale/swineproduction.html.

2. Reese, D., Hartsock, T., & Morrow, M. (2010). Baby Pig Management - Birth to Weaning - eXtension. Retrieved December 4, 2015, from http://articles.extension.org/pages/27050/baby-pig-management-birth-to-weaning.

3. Hurley, David J. (2004). "Neonatal immune development in swine management". Presented at 2004 AASV Annual Meeting, Des Moines, IA.

4. Leonard, S. G., Sweeney, T., Bahar, B., Lynch, B. P., & O'Doherty, J. V. (2010). "Effect of maternal fish oil and seaweed extract supplementation on colostrum and milk composition, humoral immune response, and performance of suckled piglets." *Journal of animal science* 88 (9): 2988-2997.

5. Ariza-Nieto, C., M. Bandrick, S. K. Baidoo, L. Anil, T. W. Molitor, and M. R. Hathaway. (2011). "Effect of dietary supplementation of oregano essential oils to sows on colostrum and milk composition, growth pattern and immune status of suckling pigs." *Journal of animal science* 89 (4): 1079-1089.

6. Inman, C. F., K. Haverson, S. R. Konstantinov, P. H. Jones, C. Harris, H. Smidt, B. Miller, M. Bailey, and C. Stokes. (2010). "Rearing environment affects development of the immune system in neonates." *Clinical & Experimental Immunology* 160 (3): 431-439.

7. Inman, Charlotte F., Georgina M. Laycock, Louisa Mitchard, Ross Harley, James Warwick, Rachel Burt, Pauline M. Van Diemen, Mark Stevens, and Mick Bailey. (2012). "Neonatal colonisation expands a specific intestinal antigen-presenting cell subset prior to CD4 T-cell expansion, without altering T-cell repertoire." *PloS one* 7 (3): e33707.

8. Bauer, E., Williams, B. A., Smidt, H., Verstegen, M. W., & Mosenthin, R. (2006). "Influence of the gastrointestinal microbiota on development of the immune system in young animals". *Curr Issues Intest Microbiol* 7 (2): 35-51.

9. Chase, Christopher C., & Lunney, Joan K. (2012). "Immune System". In Diseases of Swine, 10th Ed. 222-249. Wiley-Blackwell.

10. Martineau, Guy-Pierre., Farmer, Chantal, & Peltoniem, Olli. (2012). "Mammary System". In Diseases of Swine, 10th Ed. 270-293. Wiley-Blackwell.

11. Hlavova, K., Stepanova, H., & Faldyna, M. (2014). "The phenotype and activation status of T and NK cells in porcine colostrum suggest these are central/effector memory cells." *The Veterinary Journal* 202 (3): 477-482.

12. Bandrick, Meggan M. (2010). "Maternal influences on neonatal immune development in pigs". PhD Dissertation, Retrieved from the University of Minnesota Digital Conservancy, http://purl.umn.edu/95904.

13. Butler, J. E., K. M. Lager, I. Splichal, D. Francis, I. Kacskovics, M. Sinkora, N. Wertz et al. (2009). "The piglet as a model for B cell and immune system development." *Veterinary immunology and immunopathology* 128 (1): 147-170.

14. Ward, L. A., Rich, E. D., & Besser, T. E. (1996). "Role of maternally derived circulating antibodies in protection of neonatal swine against porcine group A rotavirus". *J Infect Dis* 174 (2), 276-282.

15. Christopher Stokes, Michael Bailey, Karin Haverson, Cecilla Harris, Philip Jones, et al. (2004). "Post-natal development of intestinal immune system in piglets: implications for the process of weaning". *Animal Research*, EDP Sciences 53 (4): 325-334. doi: 10.1051/animres:2004020.

16. Potočnjak, D., Kezić, D., Popović, M., Zdolec, N., Valpotić, H., Benković, V., Valpotić, I. (2012). "Age-related changes in porcine humoral and cellular immune parameters." *Veterinarski Arhiv 82*(2): 167-181.

17. Bandrick, M., Pieters, M., Pijoan, C., & Molitor, T. W. (2008). "Passive transfer of maternal Mycoplasma hyopneumoniae-specific cellular immunity to piglets". *Clin Vaccine Immunol* 15(3): 540-543. doi: 10.1128/cvi.00466-07

18. Niederwerder, Megan C. (2015). "Clinical Disease and Host Response of Nursery Pigs Following Challenge with Emerging and Re-emerging Swine Viruses". PhD Diss. Kansas State University. Retrieved from K-Rex. http://krex.k-state.edu.

19. Niederwerder, Megan C., Bawa, Bhupinder., Serão, Nick V.L., Trible, Benjamin R., Kerrigan, Maureen A., Lunney, Joan K., Dekkers, Jack C. M., Rowland, Raymond R. R. (2015). "Vaccination with a porcine reproductive and respiratory syndrome (PRRS) modifies live virus vaccine followed by challenge with PRRS virus and porcine circovirus type 2 (PCV2) protects against PRRS but enhances PCV2 replication and pathogenesis compared to results for nonvaccinated cochallenged controls". *Clin Vaccine Immunol* 22: 1244-1254.

Fair Oaks Pig Adventure: Public perception of what we do

M. Inskeep, DVM
Rensselaer Swine Services, Rensselaer, Indiana

Introduction

The Fair Oaks Pig Adventure is a one-of-a-kind, modern pig farm and tour facility that was designed and built to share with the world the story of modern pig farming. As part of Fair Oaks Farms located on I-65 between Chicago and Indianapolis, the Fair Oaks Pig Adventure opened its doors in the summer of 2013 to help educate the general public and inform consumers about the truths and realities of the pork industry. The pig farm portion of this experience, called Legacy Farms, is a self-sustaining 2600 breed-to-wean farm with an attached nursery/finisher gilt developer unit (GDU), which allows tourists to view all stages of swine production. The Pork Education Center opened in the spring of 2015 on the main campus of Fair Oaks Farms to educate the public about pork as a nutritious food. Legacy Farms is owned by Belstra Milling Company (BMC) along with family investors, and is operated by BMC of DeMotte, Indiana. The Pig and Pork Education Project, which is known as the Fair Oaks Pig Adventure, is a "not-for-profit" owned by the industry and managed by a board of directors. This portion of the adventure received $2 million from the National Pork Board and $2 million from private companies and stake holders, and is involved in fundraising to support the project with the ultimate goal of creating an endowment to fund the current and future growth of the project. Fair Oaks Farms manages the logistics of the tourism portion of the adventure, including marketing, management, and labor.

The tour

For the year of 2015, there will be an estimated 110,000 people visit Fair Oaks Farms with close to 80,000 of those people visiting The Pig Adventure at Fair Oaks Farms. Visitors come from right down the road to all over the world and include school groups, civic groups, families, walk-ins, etc. As visitors arrive at Fair Oaks Farms, they check in and pay admission at the Dairy Adventure Center. From there, they have the option to experience various aspects on main campus, with one aspect being the Pork Education Center. The Pork Education Center is a building designed to provide tourists with knowledge on pork as a final product, the nutritional aspects of pork, and porcine medical uses. There is also a ropes course that tourists have the option in participating in for entertainment. At the Pork Education Center, people have the option to get on the shuttle that takes them one and a half miles down the road to the Pig Adventure. As the shuttle pulls into the farm, it enters a garage where the door closes behind it before people can get off the bus. This helps with biosecurity to make sure that visitors coming through the farm do not walk or enter through the same parking lot or doors as the farm employees. Once inside the Pig Adventure, visitors are exposed to facts about historic and modern pig production on touch screen displays. A tour guide welcomes them to the adventure and starts the tour by taking them though a simulated shower and explaining the purpose of showering in and out. Once they are through the "shower," the rest of the tour is self-guided, allowing visitors to spend as much time as they desire in each wing of the adventure with tour guides along the way to offer information and answer questions. Most often, visitors begin in the farrowing wing where they can look down through the windows into all ten of the farrowing rooms as well as the holding nursery. Visitors can observe the farrowing process, piglets nursing, and farm employees performing daily chores, as well as explore touch screens with facts about piglets and farrowing. Processes such as tail docking, castration, clipping needle teeth, and euthanasia are not performed visibly for the public, but are freely talked about through the tour when those conversations arise. Every hour as needed, Pig Windows are available for visitors, which is when a farm employee brings a newborn piglet upstairs for visitors to view at their level through a window and for a picture opportunity. There is a two-way intercom that allows the farm employee and/or tour guide to explain what happens in the farrowing room on a modern sow farm. The middle wing of the adventure allows visitors to view the breeding and gestation barn with electronic sow feeding stations. Visitors are able to observe breeding, heat checking, sows eating, and every day chores that are performed in the breeding barn. Sows are bred in stalls, and then moved into large static or dynamic pens for gestation. Upstairs on the tour, there are more touch screens that provide information about breeding and gestation at a modern sow farm, a simulated pregnancy detection station for visitors to pretend to ultrasound, and information to get to know the farmers. The final wing of the adventure is the GDU where visitors can view two nursery rooms and five finishing rooms of various ages of pigs. Tour guides do not always have farm production experience, but are scheduled to spend time working downstairs on the farm with the animals as part of their training.

Experience, perception, and challenges

The adventures at Fair Oaks offer a one-of-a-kind experience to open the eyes of people of all ages to a large, modern pig farm who otherwise would have no accurate knowledge or exposure to livestock. There is wide variety of visitors that tour the Pig Adventure from fellow pork producers to kids and adults that have never seen a pig before, and everything in between. Because of the location of Fair Oaks, the adventure is able to draw in visitors from not only nearby rural areas, but also from large cities including Chicago and Indianapolis. While there is a small percentage of visitors who arrive at the Pig Adventure with negative, pre-conceived ideas and minds that they are unwilling to change, the majority of the visitors are fascinated with the experience and leave with a very positive view of raising pigs. Most visitors arrive with such little knowledge of how pigs are raised that they are amazed when their eyes are opened to the facts. Visitors ask a wide variety of questions, including sensitive topics as well as general questions about pigs and the farm itself. One of the most common questions pertaining to sensitive issues that is asked by the visitors is "Why are they in cages?" when referring to the farrowing stalls. Visitors, including adults and children, occasionally express a concern that the pigs never get to go outside, that piglets don't get to see their mother again after weaning, the sham chewing/chomping on the bars and salivation in gestation, and what happens to the animals after they leave the farm. However, the majority of the questions (from children and/or adults) are geared towards general knowledge about the farm and about pigs, such as, are pigs born with their eyes opened, how many pigs can one mom have, what is the pig's name, what do pigs eat, how do you tell a boy from a girl, and what are the different color marks for on the pigs? After going through the tour, visitors are amazed at the number of pigs born each year and are pleasantly surprised at how clean the farm is. Aside from the concerns that the farm expected from the visitors (such as the ones mentioned previously) there have also been other concerns expressed that were not anticipated. Surprisingly, some visitors have expressed concerns about the use of rattle paddles and sort boards, so it was explained that these tools were used for movement/noise to help direct pig movement without causing harm to the animals. Also, children especially are very disturbed by the site of blood that they see during the farrowing process. The last several years of the Pig Adventure being opened has been successful in enlightening many visitors to the truths about raising pigs and earning their trust. At the same time, it has also opened our eyes that we are virtually living in a completely separate world from the average urban consumer with no frame of reference for modern livestock production. With a destination like the Fair Oaks Pig Adventure, we are able to explain to visitors, engage them in conversation, and open their minds to what the pork industry is all about.

Common swine industry audit – Here's what is coming

Chris Rademacher[1], DVM; Sherrie Webb[2]

[1]Veterinary Diagnostic and Production Animal Medicine, College of Veterinary Medicine, Iowa State University, Ames, Iowa; [2] National Pork Board, Clive, Iowa

In October of 2014, the National Pork Board announced that a new standardized common swine industry audit on animal welfare had been created. This program would allow pork producers, packers and processors to utilize one standardized audit tool, rather than having each packer creating their own audit standards on animal welfare and pre-harvest pork safety. Multiple audit standards had created an extraordinary administrative burden on pork producers, particularly to those who sold pigs to multiple packers. The primary complaints were that they were costly, duplicative and it was discovered that the different auditors may judge the same item with different criteria. The Professional Animal Auditor Certification Organization (PAACO) will be training independent, third party auditors to do these audits of the sites selected at random by the packer. The primary reason for this was to create a standardized set of criteria for each question that the auditors could follow in an attempt to minimize auditor to auditor variation. A key understanding is that the auditors work for the packer who commissioned them to come to the site and perform the Common Swine Industry Audit (CSIA). The results will go back to the packer as well as the producer getting a copy.

The audit tool expands upon the current PQA Plus Site Assessment tool in that there are animal benchmarks, facilities, animal handling, and records sections. In addition to these, there will be assessments of the site's Standard Operating Procedures (SOP), evidence of caretaker training as well as some additional questions in the areas listed in Figure 1. A breakout of how the points are allocated is provided in Table 1. While the breadth of items to be measured has vastly increased, it is important to note that from a score standpoint, 50% of the total points are still based on the animal assessments. One of the primary differences is that this is an audit and not an assessment, meaning there will be scores assigned to each question in an effort to quantify the status of animal welfare and animal care practices employed on the site. There are also 5 questions on the audit that, if failed, would result in an automatic failure of the audit. These 5 questions are all critical animal welfare questions dealing with animal abuse, timely euthanasia, and confirming insensibility and death for euthanized animals. Questions 6-89 are all point questions dealing with various topics with different weighting of points on each question. Once the audit is completed, a final score will be tabulated. Right now,

Figure 1: Audit standards*

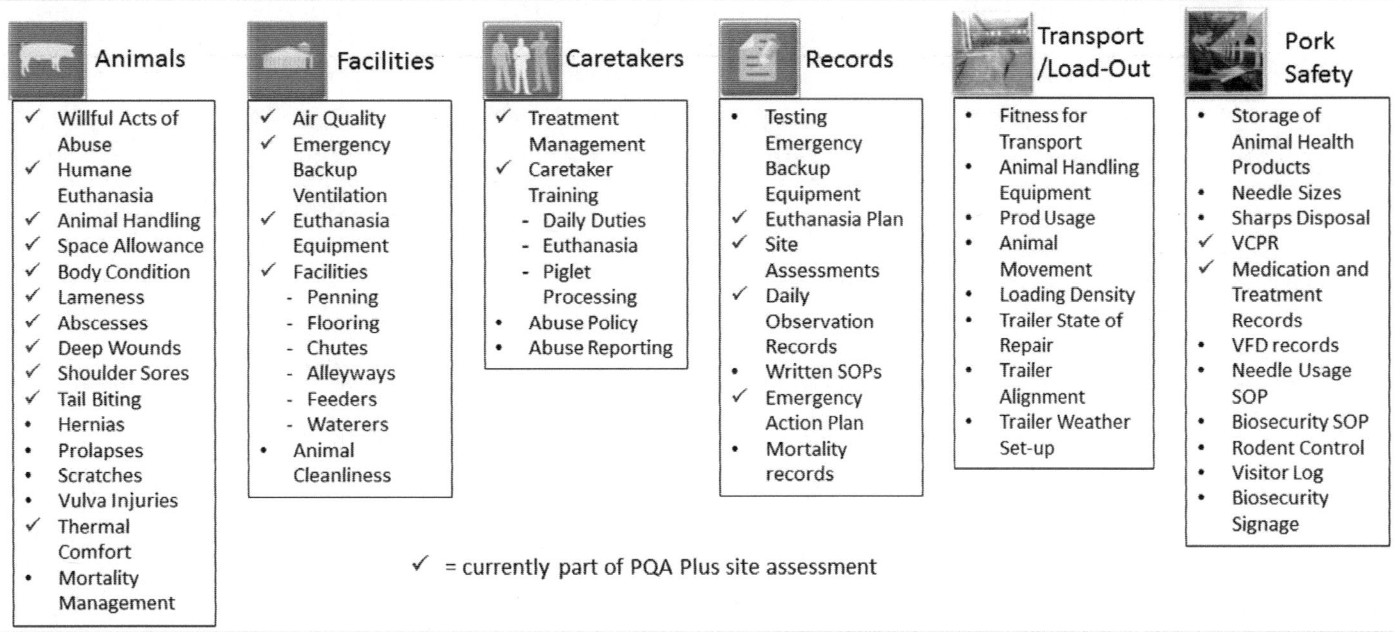

* Figure courtesy of National Pork Board

Table 1: Distribution of points by category for the Common Swine Industry Audit*

Category	No. questions	Non-breeding possible points	No. questions	Breeding possible points	Percent of total points
Animal welfare		372		424	
Critical	5		5		
Animal/benchmark	22	185	28	230	50.3%
Caretaker	6	30	7	35	7.7%
Facilities	10	50	10	50	10.9%
Records	16	32	17	34	7.4%
Transport/load-out	9	75	9	75	16.4%
Food safety	13	33	13	33	7.2%
Total	81	405	89	457	

* Table courtesy of National Pork Board

there is no pass or fail score that has been determined (other than any a failure in any of the 5 critical questions). Rather the intent is to benchmark and strive for continuous improvement.

Some of the biggest changes involve expanding sections on records, having standard operating procedures and caretaker interviews to see if the caretakers are following SOP's on certain questions relating to animal welfare principles. There is also a section that deals exclusively with transportation as well. For each question that is not awarded points, a corrective action form will have to be filled out that describes why the points were not given and how the producer is going to come into compliance. These corrective action forms will ultimately go to the packers at some time period after the audit to be reviewed with the producer.

There is a vast amount of resources on the National Pork Board's website (`http://www.pork.org/common-industry-audit/`), including the audit tool, SOP, policy and record templates to be used by producer if they don't already have one. The Iowa Pork Industry Center (IPIC) in partnership with the Iowa Pork Producer Association (IPPA) put together a series of workshops around the state this past summer to assist producers on how to prepare for an audit. The workshops were designed to ensure producers are prepared for all different facets of the audit. They are particularly focused on making sure they have all the SOP's that are in the audit, ensuring they have all the required records as well as how to prepare caretakers for the interviews from the auditors. More audit training sessions will be planned for the coming year. Veterinarians can serve a key role in helping to get their producers educated and prepared for the CSIA.

It's not about you or me, or them or us, it's about *US* all, together, living the four "C's"

D. Polson[1], DVM, MS, PhD; B. Martinez-Lopez[2], DMV, MPVM, PhD; R. Main[3], DVM, PhD; D. Holtkamp[3], DVM, MS; A. Perez[4], DMV, PhD; E. Lowe[1], DVM; E. Mondaca[1], DMV, PhD; Z. Whedbee[2]; K. Mueller[3], DVM; B. Crim[3]; J. Kraft[3]; C. Mowrer[3]

[1]Boehringer-Ingelheim Vetmedica Inc., Saint Joseph, Minnesota; [2]University of California-Davis, Davis, California; [3]Iowa State University, Ames, Iowa; [4]University of Minnesota, Saint Paul, Minnesota

"Coming together is a beginning, staying together is progress, and working together is success."
- Henry Ford

What we can learn from physicists that has nothing to do with physics

As veterinarians, as an organization of swine veterinarians, as part of the pork industry, and as part of the global animal protein production community; why would we care about the world of physics?

Whether we do or don't care; if there are principles and applications that we can learn, understand and apply to our world; there are compelling reasons that we *should* care – not just care in the sense of a spectator being aware of an event playing out before their eyes, but care very deeply and intensely as active and committed participants in the event with a lot on the line, with a lot to lose as well as gain – directly for the protein producers we are in business to serve, and indirectly for the protein consumers we are all in business to provide for.

Sir Isaac Newton was, to say the least, a brilliant mind and highly influential scientist – responsible for some of the most significant discoveries in all of human history.[1] Isaac Newton was a physicist and mathematician who developed and described classical (Newtonian) mechanics (e.g., the laws of motion and gravitation), and is credited with developing calculus (independently from and in parallel to Gottfried Leibniz). Interestingly, before pursuing his education and academic career in physics and mathematics at the University of Cambridge, Newton was a Lincolnshire farmer who failed at farming.

So what can we learn from Isaac Newton, failed farmer but astronomically successful physicist? Well, let's look at a few things he has been attributed with saying[2] and find those nuggets:

- *"Truth is to be found in simplicity, and not in multiplicity and confusion of things."* Newton appreciated the reality that truth discoverable in knowledge becomes evident when distilled down to its elegant simplicity.
- *"My powers are ordinary. Only my application brings me success."* Newton worked hard to make the most of his potential.
- *"We build too many walls and not enough bridges."* and *"Tact is the knack of making a point without making an enemy."* Newton recognized the importance of *cohesive* communication, cooperation, coordination and collaboration among peers and with constituencies.
- *"If I have seen further than others, it is by standing on the shoulders of giants."* Newton gave credit where credit was due – Galileo, Copernicus, Kepler, Descartes. He became educated on the current state of knowledge generated by those who came before him, and understood that the further knowledge he built was constructed on the solid foundation of that which they discovered.
- *"To explain all nature is too difficult a task for any one man or even for any one age. 'Tis much better to do a little with certainty and leave the rest for others that come after you."* Newton also understood that the knowledge he was building with his peers would be built upon further by those who came after him.

Tim Berners-Lee is a physicist at the Massachusetts Institute of Technology (MIT) in Cambridge, MA. In a March 1989 proposal "*Information Management: A Proposal*", submitted to the CERN (*Conseil Européen pour la Recherche Nucléaire*, or European Council for Nuclear Research), Berners-Lee outlined a plan to create something that did not exist at that time.[3] In his words:

> *"This proposal concerns the management of general information about accelerators and experiments at CERN. It discusses the problems of loss of information about complex evolving systems and derives a solution based on a distributed hypertext system.*
>
> *We should work toward a universal linked information system, in which generality and portability are more important than fancy graphics techniques and complex extra facilities.*
>
> *The aim would be to allow a place to be found for any information or reference which one felt was important, and a way of finding it afterwards. The result should be sufficiently attractive to use that it the information contained would grow past a critical threshold, so that the usefulness the scheme would in turn encourage its increased use.*
>
> *The passing of this threshold accelerated by allowing large existing databases to be linked together and with new ones."*

Yes, Tim Berners-Lee, a physicist, is credited with conceiving and birthing what today we know as the 'World Wide Web' while at CERN...what virtually the entire world, top to bottom and side to side, touches via 'URL', 'http:', 'www.', and 'HTML' virtually every day many times a day; and many of us, in some way, all day long for work and every evening / weekend at home for personal enjoyment.

But Berners-Lee and colleagues could not have created the 'World Wide Web' without the prior existence of the 'Internet'.

For clarity, the 'Internet' and 'World Wide Web' are not the same thing[4], but they are related – one would not exist nor would have exploded across the entire face of the earth without the other, i.e., they are interdependent.

- The 'Internet' is the global aggregate hardware infrastructure network (servers, modems, routers, switches, wired and wireless transmission and reception hardware – like cables, antennas, satellites and satellite dishes)...the digital equivalent of a global transportation system and infrastructure.
- The 'World Wide Web' is the aggregate web software and database global network (web pages, databases, hyperlinks and URL's) ...in essence, all of the data living at their home addresses, and all of the data that is moving between one data home and another (i.e., data warehouses).

So, since Berners-Lee's 'World Wide Web' and the 'Internet' are not synonymous, but represent two different but related 'things' (for lack of a better term), who is responsible for originating the 'Internet'? The answer is incrementally complex and a bit tangled, but, in relatively distilled terms: [5]

- The 'Internet' was spawned out of a concept penned in 1962 by J.C.R. Licklider, a Washington University (St. Louis, MO) educated physicist, mathematician and psychologist – which he referred to as the 'Intergalactic Computer Network'.[6] In Licklider's words:[7]

  "We need to substitute for the book a device that will make it easy to transmit information without transporting material."

  "It should be possible...to type 'clr' on the keyboard and have 'The Council on Library Resources, Inc.' appear on the display."

  "It seems reasonable to envision, for a time 10 or 15 years hence, a 'thinking center' that will incorporate the functions of present-day libraries together with anticipated advances in information storage and retrieval and ... a network of such centers, connected to one another by wide-band communication lines and to individual users by leased-wire services."

  Referred to as the 'Johnny Appleseed' of computing, it has been said of Licklider ('Lick') that, *"...most of the significant advances in computer technology...were simply extrapolations of Lick's vision. They were not really new visions of their own. So he was really the father of it all."*

- Over the following 12 years, a cascade of work on various aspects that, individually would not stand alone, but collectively represented the basic components and raw material needed to construct what would become the 'Internet', were conceived, developed and integrated into / interfaced with the evolving system.
- Vinton Cerf (a mathematician and computer scientist educated at Stanford and UCLA) and Robert Kahn (an electrical engineer educated at City College in New York) are the two persons generally regarded jointly as 'Fathers of the Internet' (and co-developers of TCP/IP). In 1974 they coined the term 'Internet' in their published paper, *"A Protocol for Packet Network Intercommunication"*.[8]
- From that point in 1974, the 'Internet' continued to develop, take shape and forge connections to Asia and Europe.

Between 1984-1989, Ben Segal, a CERN PhD physicist and mathematician out of the Imperial College London, convinced CERN that TCP/IP was the key for CERN to connect the computer systems of their physicist collaborators and their institutions to enable more timely communication, better coordination and more effective collaboration.[9]

Of particular interest from Segal's Internet Hall-Of-Fame Inductee Bio is this statement:

> *"He played an important role as an Internet promoter, spearheading the introduction of IP into a hostile Europe when it was not politically correct or career-friendly to do so there. European Postal Telegraph and Telecommunications Administrations and industry were opposed to these standards, and their use outside the laboratory was forbidden."*

Also Segal, as a mentor:

> *"...helped Tim Berners-Lee, who went on to invent the World Wide Web, with design decisions and pointed him to the RFC series and existing protocols like the File Transfer Protocol and Network News Transfer Protocol, which could guide his invention of the Hypertext Transfer Protocol."*

So, if you use and appreciate the 'Internet' and the 'World Wide Web', thank a physicist.

But beyond appreciating who we should be thankful to; what can we learn from Licklider, Cerf, Roberts, Sagen and Berners-Lee and the rest of their pioneering colleagues and collaborators? Here are several possible learnings (there are likely many more):

- Much in the same fashion as Newton did to accomplish what he is noted for centuries earlier; to *"see further"*, Berners-Lee and collaborators stood on the shoulders of giants – computer scientists, engineers, as well as fellow physicists and CERN colleagues.
- These pioneers – physicists, engineers, computer scientists, psychologists – were highly collaborative inter-institutionally, across disciplines and on a global scale. Why? They all had professional goals as individuals, and within their respective institutions and companies. And they also each had personal egos as well as operated within environments

with institutional egos. No doubt these egos fed professional competition for publications, funding and recognition at the individual and institutional levels. Why would they and their institutions invest the time, resources and professional energy in collaborating with their academic/research competitors in conceiving, gestating, birthing and rearing these 'things' for which the structure is unclear at best and the value is, at best, uncertain? Certainly they all hoped that, once up and running, these 'things' could help them solve mutually experienced gaps/problems/needs (e.g., communication hurdles and information loss/access limitations) whose value were directly experienced by they and their teams, as well as their departments and institutions. But also, they no doubt perceived, intuitively, that these 'things' would continue to develop in ways that would solve problems that had not yet even been recognized as problems, and add value to each of them and their respective institutions as well as collectively for all of them and across all of their institutions in ways that they could not have envisioned at the time.

- They were able to navigate the negative individual psychological and institutional philosophical hurdles; effectively neutralizing resistance to change where it needed to be, even converting resistance to support and advocacy of the dynamic innovations being created and developed.
- And as or even more importantly, it was (and is) through the commitment to and initiation of long-term communication, cooperation, coordination and collaboration among many persons and institutions that drove incredibly good *big* things that provide far-reaching long-lasting benefit, far beyond even the imaginations of the creators. In the words of Vinton Cerf:[10]

  "At some point, you can't lift the boulder with just your own strength. And if you find that you need to move bigger and bigger boulders up hills, you will need more and more help."

To create and sustain the necessary components of the internet and world wide web, and integrate them into a cohesive whole, collaboration among the core advocates of these integrated technological platforms/systems was not just an important option, it was a required imperative for them to establish even the most basic functionality they envisioned.

But beyond the core advocates, why would the members of CERN – a meta-institution, made up of personnel from all over the globe and supported jointly by many different nations and their respective governments – want to better connect *with*, rather than compete *against* each from inside the box of their own in-country infrastructure? Why would they want to better communicate, cooperate, coordinate and collaborate *with* each other as individuals and as institutions? An even more seminal question is, given this same competitive-collaborative dichotomy, why does CERN exist at all?

CERN[11,12] came into being in 1954 thanks to the efforts of the founding 12 member nations, and continues to exist (now with 21 member nations) because there was not then and is not now any other viable way for individual institutions or nations to build and maintain the fixed infrastructure necessary to enable discovery of answers to the biggest questions of high-energy particle physics – answers that would otherwise be far too difficult and costly, and take far too long (if not be effectively impossible) to uncover for any one of them on their own. For CERN to happen at all took (and continues to take) an immense initial (and ongoing) collaborative spirit pursuing mutually beneficial work amongst the nations and governments that make up its members, associate members, observers and non-member cooperators.

Examples of two of the research collaborations that utilize the CERN infrastructure, like the Large Hadron Collider (LHC), are ATLAS (A Toroidal LHC ApparatuS) and the CMS (Compact Muon Solenoid) Collaboration.[13,14] The ATLAS 'experiment' is a global collaboration of over 3,000 physicists and other scientists, technicians and students from 174 institutions in 38 countries. The CMS Collaboration is a global collaboration of over 4,300 physicists and other scientists, technicians and students from 182 institutions in 42 countries.

Let's look a bit closer at one of these collaborations – the CMS Collaboration.

Why would the CMS Collaboration – a decentralized meta-organization, made up of a diffuse global network of institutionally-based physics research departments and their scientists, technicians and students – want to better connect *with* rather than compete *against* each other for research funding, research execution, publications and related recognition? Why would they want to better communicate, cooperate, coordinate and collaborate *with* each other? And, not unlike CERN itself, an even more seminal question is, given this same competitive-collaborative dichotomy, why does the CMS Collaboration (or ATLAS) exist at all?

Nuclear and high-energy particle physics research is hard, really *really* hard. And it is extremely complex as well as very *very* resource intense. Significant progress takes a lot of people a long *long* time. Any one person, no matter how brilliant or passionate, has neither the capacity nor the competence to go it alone with any chance of sustained success. Neither does any one institution, no matter how large the endowment and no matter how well staffed, have the resources or capacity to go it alone and see sustained success. And no one country has the mountains of resources at its disposal and oceans of capabilities within its borders that it would take to go it alone in any sort of sustained manner.

The creation of the CMS Collaboration and their utilization of the CERN infrastructure enabled the pooling and far *far* better leveraging of institutional resources, capacities and competencies; along with enabling access to the fixed (and extremely costly and complex) infrastructure they required to do their work at all – leading to the discovery of solid answers to giant questions about very very *very* tiny particles. CERN and the global research collaborations like CMS that use its infrastructure understand that they would not exist if the really big questions (and the needs and opportunities they represent) did not exist; as do they also understand that their existence would not be necessary if those questions were smaller questions answerable with far less effort and resources.

One only has to find a single published paper to get a sense of the size and scope of the CMS Collaboration. Looking at just one example, the text of the paper "*Search for Dijet Resonances in 7 TeV pp Collisions at CMS*" is only nine pages including the 25 references, but has 2,263 authors from 158 institutions listed.[15] There are currently 487 *peer-reviewed* papers originating from the CMS Collaboration published between 2010 and 2015.[16]

To appreciate an even larger and broader context, since the inception of the CERN Large Hadron Collider (LHC) becoming operational in 2010, for six of the experimental collaborations using the CERN facility (CMS, ATLAS, ALICE, TOTEM, LHCb, LHCf) there have been 1,439 *peer-reviewed* papers published in the same six year period by physicists and other allied scientists within these collaborations from all over the world based on data originating from the LHC.[17] In addition, nearly ten thousand conference proceedings papers and abstracts have been published-presented in the area of high-energy particle physics by these same six collaborations.

Physicists really know how to collaborate. We, each of us and all of us, could learn a lot from them about how they collaborate so effectively.

Facing the "*So what*?"

"*The secret is to gang up on the problem, rather than each other.*" – Thomas Stallkamp

Let's revisit the question we started with: As veterinarians, as an organization of swine veterinarians, as part of the pork industry, and as part of the global animal protein production community; why would we care about any of this coming from the academic world of physics? At this point, it should be inescapably evident that there *are*, in fact, principles and applications that we can learn, understand and apply to *our* world coming from how physicists think and operate – *these* are the compelling reasons we should (must) care.

It bears repeating verbatim – we should not just care in the sense of a spectator being aware of an event playing out before their eyes, but care very deeply and intensely as active and committed participants in the event with a lot on the line – a lot to lose as well as gain – directly for the protein producers we are in business to serve, and indirectly for the protein consumers we are all in business to provide for.

Through all of the efforts leading to the design, formation and sustained operation of CERN and large global research collaborations like ATLAS and CMS; what prevents (or at least mitigates) individual and institutional parochialism – 'me vs. you', 'us vs. them'?

The unprecedented large and sustained collaboration among physicists in the establishment of CERN and, for example, ATLAS and CMS, did not go unnoticed by others in the broader science and philosophy communities. Over the last few decades, behavioral scientists and those from various other disciplines have actively and directly studied the culture of and environment at CERN and, relatedly, within the research collaborations that utilize CERN infrastructure.

One example of that observational behavioral research was published in *Nature* in 2010 – "*Physics: The Large Human Collider*".[18] Although it has not been without challenges – there are no Utopias on earth when human beings and our human nature are involved – a main learning from their work was that those involved with CERN and collaborations like the CMS embrace a culture of and discipline for collaborating to achieve the scientific 'greater good' – a purpose larger than any one of them; and live with a willingly shared sense of community that transcends the individuals, the institutions, and the corresponding nations.

So, it appears that a high level of communication, cooperation, coordination and collaboration is not only possible but a practiced reality in the world of physics. But, is this same type of mindset and behavior even possible in our pig-pork production and veterinary world, much less in a broader competitive business context?

Can we, how can (*will*) we improve our own collaboration capabilitie?

"*I never learned from a man who agreed with me.*" – Robert A. Heinlein

In our own profession and in the broader pork industry, there is certainly a lot of room for improvement in how we communicate, cooperate, coordinate and collaborate in every area we engage – including that of information management platforms and systems we develop and utilize.

In the medical and behavioral sciences there are well run and effective models of broad, large-scale collaboration – the Cochrane Collaboration and the Campbell Collaboration are two such examples that we can and should learn from.[19,20] Larger, broader and more diffuse collaborations that include veterinary medicine and swine within their scope exist as well, such as the One Health Initiative, whose mission statement explicitly refers to cooperation and collaboration, and for whom all seven points in their vision statement aspire to "*Joint efforts...*".[21]

There is no doubt that these collaborations, just as with CERN and CMS, have experienced and still experience 'growing pains' as they cycle through Tuckman's Developmental Sequence of '*form, storm, norm and perform*'.[22] After all, no matter how intelligent or intellectual we think we may be, we are, all of us, simultaneously and inescapably a highly complex mix of the logical, the rational, and the emotional. Linda Elder (1996) said it quite well:[23]

"If we are concerned with developing our rationality in order to improve our lives, we must understand the powerful role that both emotions and thoughts play in our minds. We must understand the ways in which affect and cognition influence one another in determining both our outlook on life and our behavior. Most importantly, we must come to terms with those truths about the human mind that enable us to begin the process of taking charge of our minds: that thoughts and emotions are inextricably bound, that we have both egocentric and rational tendencies, that our inner conflicts are never best understood as a simple matter between emotion and reason, that self-command of mind takes both extended education and self-discipline, that our fullest rational development is dependent on the development of rational affect, that to bring intelligence to bear upon emotions we must take charge of the thinking underlying those emotions."

As it relates to (but clearly is not limited to) information management platforms and systems: improvements in our ability to communicate, cooperate, coordinate and collaborate are needed most in at least four distinct areas:

- Design and Development: The design and development of widely integrated information systems (and their components) which enable us to better manage business, network and area operations for live production and all related up and down-stream aspects; as well as better leverage and learn from the resulting data those operations generate. This would encompass:
  - Front-end data capture and transfer
  - Internal system data management
  - Back-end data analytics and reporting, notification and communication
- Operations: We have dabbled over the past 8-10 years with overlaying an area-based model[24] on top of the legacy structure of live production business operations. While this basic layered approach can be (and has been) effective, it needs to be further adapted to:
  - accommodate all relevant diseases (both domestic and foreign) for swine,
  - be broadened to include other production animal species,
  - be further oriented to encompass (and integrate) both area-coordinated and network-coordinated disease control, and
  - be scaled to a level that optimizes the spatial and network opportunities for maximally effective health management – e.g., to an intra-national level to be sure, but even more worthwhile to an inter-national level, e.g., all of North America. From there it is obvious that a global scaling would be the logical extension of continent-level area-network coordinated efforts.
- Research: Extracting the knowledge (and applying the corresponding wisdom) that lies buried throughout the massive (and growing) data oceans that our live- and post-production operations generate every minute of every day is a tremendous opportunity as well as a major challenge. However, there are opportunities to extract highly useful knowledge from those data oceans – not just on a static/periodic basis, but on a dynamic/continuous and real-time basis – particularly with the greater availability of machine learning and artificial intelligence.
- Learning: Effective cooperation, coordination and collaboration are entirely dependent on effective communication, and effective communication has everything to do with learning. Data communicates its knowledge treasures via the interpretive vehicle of analysis. In turn, knowledge is communicated via the distribution vehicles of education and training. And all elements of learning are of value only insofar as they support continually better decision making – be it at the operational level or with new discovery and synthesis of new innovation.

To be most effective for all of the constituencies we serve, and to most effectively leverage what the physicists (and, of course, many others – engineers and computer scientists, to name only two) have brought us by way of the 'Internet' and 'World Wide Web', our data and information management system infrastructure must have:

- Definitions – Clear definitions and terminology to enable effective communication and understanding of information
- Standards – Standards for data, messaging, protocols, metrics and algorithms to enable effective analysis, reporting and interpretation of integrated data
- Security – Accommodate learning and action while preserving appropriate levels of security, confidentiality and anonymity for participants and their data
- Access – Direct and indirect secure access to all necessary data streams needed for integration, analysis, interpretation and leverage

However, *what* we do is actually not our most meaningful hurdle – the basic areas of need tend to reveal themselves quite well, and the mechanical elements to best meet those needs are typically straight-forward to identify. Rather, it is *how* we go about meeting those needs that represents our greatest challenge. It is within the *how* that resides not only what we have the most to learn, but is where can be found that which represents what we also have the most to learn from the motivations, mindset, methods and mechanics of physicists.

As individuals, as teams and as businesses/institutions we are generally quite good at promoting and applying the four "C's" at an intra-business/intra-institutional/intra-organizational level. We are also generally rather good at fostering the four C's among the pork industry participants whom we enroll in applied research projects and ongoing service initiatives. However, we have room for improvement (and substantially so) in engaging the four C's at the inter-business/inter-institutional/inter-organizational level.

This is not to say that the inherent construct of competition among businesses, institutions and/or organizations is not healthy – it generally can be and often is, and as such drives degrees of progress and innovation. This is also not to say that businesses, institutions and organizations don't already communicate and collaborate reasonably well at various levels – they (we) clearly do. To observe that, one has only to look at how our industry organizations, professional organizations, universities and allied industry have worked and continue to work together on initiatives of value to us all. We all, to varying degrees, practice some level of *positive interdependence*.[25]

It is rather to say that we, all of us, can do far better with practicing the four C's in a more cohesive way to tackle the more massively scaled and complex challenges that we face as the greater pork protein production community, challenges that none of us could hope to solve by depending on more conventionally scaled forms of collaboration. And we can do so to the even greater collective benefit of us all – not just in terms of sustaining that which we depend on to support every one of us and our families; but also in terms of thriving and growing to more effectively meet the needs of those to whom we all contribute in various ways to provide the pork protein we produce.

We all, individually and collectively, have a lot of room to grow, and *social interdependence theory* would suggest that doing far better at living the four C's is eminently possible *with practice*.[26] In the words of Rusbult and Van Lange[27] (and, with no apology, to inject one final reference to the world of physics):

> "*Analogous to contemporary physics – where the relations between particles are as meaningful as the particles themselves – in interdependence theory, between-person relations are as meaningful as the individuals themselves.*"
> – Caryl Rusbult and Paul Van Lange

In as similar a fashion as it applies to individuals and groups of individuals, there is little (no) doubt that such a particle physics analogy applies very well in increasingly higher levels to teams and organizations (to include businesses and institutions) made up of individuals and teams; as well as organizations and industry-based communities made up of those organizations. After all, we and everything around us that exists in the natural universe are constructed of the very same fundamental particles which physicists so effectively collaborate to discover and understand. And it is precisely the acceleration of discovery and understanding of the 'physics' of our shared pork protein production world that should drive us all to greater degrees of cohesive collaboration.

The overriding purpose of this paper and its related presentation is not to offer 'the answer' nor 'the solution' to those of us engaged in pork protein production. Rather, it is to offer up observations on how others have faced and made progress in overcoming the same fundamental barriers and constraints to achieve more effective and efficient collective progress dealing with pan-industry challenges and capitalizing on pan-industry opportunities that transcended their (and transcend our) respective businesses, institutions and organizations.

And to that end, this should only serve to help us intensify and accelerate our efforts to craft better, more sustainable and more collaborative ways to pursue and ultimately capture greater value for those we serve (and, in turn, for ourselves) – actively, purposefully, deliberately, systematically and continuously living all four C's.

"*Life is trying things to see if they work.*" – Ray Bradbury

References

1. http://www.biography.com/people/isaac-newton-9422656 (accessed November 30, 2015).

2. http://www.brainyquote.com/quotes/authors/i/isaac_newton.html (accessed November 30, 2015).

3. http://www.w3.org/History/1989/proposal.html (accessed November 30, 2015).

4. http://www.diffen.com/difference/Internet_vs_World_Wide_Web (accessed November 30, 2015).

5. http://www.internethalloffame.org/brief-history-internet (accessed November 30, 2015).

6. https://en.wikipedia.org/wiki/J._C._R._Licklider (accessed November 30, 2015).

7. http://www.azquotes.com/author/28667-J_C_R_Licklider (accessed November 30, 2015).

8. https://www.cs.princeton.edu/courses/archive/fall06/cos561/papers/cerf74.pdf (accessed November 30, 2015).

9. http://www.internethalloffame.org/inductees/ben-segal (accessed November 30, 2015).

10. http://www.brainyquote.com/quotes/authors/v/vint_cerf.html (accessed November 30, 2015).

11. http://home.cern/about (accessed November 30, 2015).

12. https://en.wikipedia.org/wiki/CERN (accessed November 30, 2015).

13. http://home.cern/about/experiments/atlas (accessed November 30, 2015).

14. http://home.cern/about/experiments/cms (accessed November 30, 2015).

15. http://arxiv.org/pdf/1010.0203.pdf (accessed November 30, 2015).

16. http://cds.cern.ch/collection/CMS (accessed December 6, 2015).

17. http://cds.cern.ch/collection/LHC%20Experiments?ln=en (accessed December 6, 2015).

18. http://www.nature.com/news/2010/100324/full/464482a.html (accessed November 30, 2015).

19. http://www.cochrane.org/about-us (accessed November 30, 2015).

20. http://www.campbellcollaboration.org/about_us/index.php (accessed November 30, 2015).

21. http://www.onehealthinitiative.com/mission.php (accessed December 7, 2015).

22. http://openvce.net/sites/default/files/Tuckman1965DevelopmentalSequence.pdf (accessed December 7, 2015).

23. http://www.criticalthinking.org/pages/cognition-and-affect-critical-thinking-and-emotional-intelligence/485 (accessed December 7, 2015).

24. https://www.aasv.org/shap/issues/v22n2/v22n2p84.pdf (accessed November 30, 2015).

25. http://www.sciencedirect.com/science/article/pii/S1877042813035039 (accessed December 7, 2015).

26. http://www.sfu.ca/~jcnesbit/EDUC220/ThinkPaper/Johnson2003.pdf (accessed December 7, 2015).

27. Rusbult, C.E., Van Lange, P.A.M. (2008); "Why we need interdependence theory." *Social and Personality Psychology Compass*, 2, 2049-2070. (Cited in: "Interdependence Theory and Related Theories", Reis H.T. and Arriaga X.B., 2013 – http://web.ics.purdue.edu/~arriaga/PDFs/2014%20Chap%20Reis%20&%20Arriaga%20Author%20Accepted.pdf; accessed December 7, 2015).

Review of PEDV disinfectant research

Derald Holtkamp[1], MS, DVM; Paul Thomas[2], MS, DVM; Kimberlee Gerardy[1];
Locke A. Karriker[1,3], MS, DVM, DACVPM; Alejandro Ramirez[1], DVM, M.P.H., PhD, DACVPM;
Jianqiang Zhang[1], MS, MD, PhD; Jessica L. Bates[1,3], MS, DVM
[1]Department of Veterinary Diagnostic and Production Animal Medicine, Iowa State University, Ames, Iowa;
[2]AMVC Management Services, Audubon, Iowa; [3]Swine Medicine Education Center, Iowa State University, Ames, Iowa

Introduction

Trailer-to-facility and facility-to-trailer contamination of porcine epidemic diarrhea virus (PEDV) has been demonstrated to occur at swine harvest facilities.[1] Much of the early research on trailer sanitation and disinfection was done for porcine reproductive and respiratory virus (PRRSV), a single stranded enveloped RNA virus like PEDV. This research demonstrated that to successfully inactivate PRRSV, procedures involving a combination of trailer washing, disinfection, and drying using either a thermo-assisted drying and decontamination (TADD) system or natural drying overnight may be used.[2-5] Consequently, the gold standard trailer sanitation today consists of washing, disinfection, and drying either naturally or with a TADD system.

A complete wash, disinfection, and dry; however, takes time and specialized facilities. Unfortunately, there are not enough of these facilities in the United States to serve the large volume of livestock trailers transporting pigs across the country every day. Whether procedures involving scraping the organic material from the trailer followed by disinfection, natural drying, TADD, or a combination of these might be effective at inactivating PEDV when a traditional wash is unavailable had not previously been studied.

The purpose of this paper is to describe the findings and implications from five studies conducted from September 2013 through March of 2015 to evaluate the efficacy of different aspects of commonly used trailer sanitation protocols for inactivating PEDV. One of the studies evaluated trailer washing, disinfection, and drying either naturally or with a TADD system. The other four evaluated disinfection, natural drying and TADD procedures in the presence of feces to represent conditions in livestock trailers that were mechanically scraped but not washed.

Materials and methods

The same experimental model was used in all five studies. A fixed volume (5 or 10 ml) of undiluted PEDV-positive feces from pigs intentionally infected with wild-type PEDV (or PEDV-negative feces for the negative control groups) was spread evenly on the floor of a 6 by 6 inch aluminum coupon with 1 inch sides. The coupons were made to replicate the floor of a livestock trailer. Following treatment as outlined in Tables 1 to 5 for each study, the feces were re-collected from the coupons, and given to 3 to 4-week old PEDV-naïve pigs via gastric tube. These pigs served as a bioassay to detect the presence of infectious PEDV remaining in the treated coupons. Fecal swabs were collected from the pigs on days 3 and 7 post-inoculation. The fecal swabs were tested for the presence of PEDV by PCR to determine if the pigs were infected by the material collected from each coupon. Each treatment group contained 4 replicates (4 coupons and 4 pigs; 1 coupon per pig). All 4 pigs in a treatment group were housed separately in raised tubs but in the same room. Eight replicates were included for the study evaluating Stalosan F.

Objective, treatment groups and results for each study

Evaluation of high pressure wash with detergent followed by disinfection with combination quaternary ammonium/glutaraldehyde with and without additional heat and time

The objective of this study was to investigate a range of relevant time and temperature combinations following wash with detergent (Barnstorm®, Preserve International, Zepher Cove, Nevada) and disinfection with a combination quaternary ammonium/glutaraldehyde product (Synergize®, Preserve International, Reno, NV) to determine if they are sufficient to inactivate PEDV in swine feces on metal surfaces similar to those found in livestock trailers. These sanitation practices together are representative of industry standard practices for swine livestock trailer washing and disinfection. The treatment groups evaluated are in Table 1.

Bioassays were PEDV-negative in 100% (4/4) of the pigs in the Neg control group and in treatment groups WD-68C-10M, WD-66C-10M, WD-60C-20M, WD-49C-20M, WD-20C-12H, WD60M, and WD10M. Bioassays were PEDV-positive in 100% (4/4) of the pigs in the Pos control group. Conclusions: All treatment groups (WD-68C-10M, WD-66C-10M, WD-60C-20M, WD-49C-20M, WD-20C-12H, WD60M, and WD10M) were 100% effective at inactivating PEDV.

Evaluation of time and temperature combinations required to inactivate PEDV in feces

The objective of this study was to investigate a range of time and temperature combinations sufficient to inactivate PEDV in the presence of swine feces on metal surfaces similar to those found in livestock trailers without washing or disinfection. The combinations of time and temperature evaluated are presented in Table 2.

Bioassays were PEDV-negative in 100% (4/4) of the pigs in the Neg control and treatment groups 71C-10M and 20C-7D. Bioassays were PEDV-positive in 25% (1/4) of the pigs in groups 63C-10M, 54C-10M, and 20C-10M. Bioassays were PEDV-positive in 50% (2/4) of the pigs in group 38C-12H and in 100% (4/4) of the pigs in the Pos control group. Conclusions: The 71C-10M

Table 1: Description of treatment groups for evaluation of high pressure wash with detergent followed by disinfection with combination quaternary ammonium / glutaraldehyde, with and without additional heat and time.

Treatment group	Description of treatment
Neg control	5 ml PEDV-negative feces, no treatment
Pos control	5 ml PEDV- positive feces, no treatment
WD-68C-10M	5 ml PEDV-positive feces, pressure washed with detergent, Synergize® disinfectant at 1:256 concentration for 10 minutes, heated to 68° C (155°F) and held at this temperature for 10 minutes
WD-66C-10M	5 ml PEDV-positive feces, pressure washed with detergent, Synergize® disinfectant at 1:256 concentration for 10 minutes, heated to 66° C (150°F) and held at this temperature for 10 minutes
WD-60C-20M	5 ml PEDV-positive feces, pressure washed with detergent, Synergize® disinfectant at 1:256 concentration for 10 minutes, heated to 60° C (140°F and held at this temperature for 20 minutes
WD-49C-20M	5 ml PEDV-positive feces, pressure washed with detergent, Synergize® disinfectant at 1:256 concentration for 10 minutes, heated to 49° C (120°F) and held at this temperature for 20 minutes
WD-20C-12H	5 ml PEDV-positive feces, pressure washed with detergent, Synergize® disinfectant at 1:256 concentration for 10 minutes, left at 20°C (room temperature) for 12 hours
WD60M	5 ml PEDV-positive feces, pressure washed with detergent, Synergize® disinfectant at 1:256 concentration for 60 minutes
WD10M	5 ml PEDV-positive feces, pressure washed with detergent, Synergize® disinfectant at 1:256 concentration for 10 minutes

Table 2: Description of treatment groups for the evaluation of time and temperature combinations to inactivate PEDV in feces.

Treatment group	Description of treatment
Neg control	5 ml PEDV-negative feces, no treatment
Pos control	5 ml PEDV-positive feces, no treatment
71C-10M	5 ml PEDV-positive feces heated to 71° C (160°F) and held at this temperature for 10 minutes
63C-10M	5 ml PEDV-positive feces heated to 63° C (145°F) and held at this temperature for 10 minutes
54C-10M	5 ml PEDV-positive feces heated to 54° C (130°F) and held at this temperature for 10 minutes
38C-12H	5 ml PEDV-positive feces heated to 38° C (100°F) and held at this temperature for 12 hours
20C-24H	5 ml PEDV-positive feces left at 20° C (room temperature) for 24 hours
20C-7D	5 ml PEDV-positive feces left at 20° C (room temperature) for 7 days

and 20C-7D treatment groups were each found to be 100% effective at inactivating PEDV. No other groups were found to be effective.

Evaluation of a dry phosphate-based disinfectant powder disinfectant powder to inactivate PEDV in feces

The objective of this study was to investigate the efficacy of a dry phosphate-based disinfectant powder (Stalosan® F, Vitfoss, Graasten, Denmark) applied with a leaf blower to inactivate PEDV in the presence of swine feces on metal surfaces similar to what is found in livestock trailers. The dry disinfectant powder was evaluated at 20° C in the presence of feces with one hour of contact time. The treatment groups evaluated are in Table 3. Bioassays were PEDV-positive in 100% (8/8) of the pigs for the Pos control and Stalosan® F groups. Bioassays were PEDV-negative in 100% (8/8) of the pigs in the Neg control group. Conclusions: Under the conditions evaluated, the dry phosphate-based disinfectant powder was not effective at inactivating PEDV in the presence of feces.

Evaluation of an accelerated hydrogen peroxide® disinfectant to inactivate PEDV in feces at 20° C

The objective of this study was to test the efficacy of an accelerated hydrogen peroxide® (AHP®) disinfectant (Accel®, Virox Technologies Inc., Oakville, Ontario, Canada) applied as a foam for inactivating PEDV at 20° C in the presence of feces on metal surfaces like those found in livestock trailers. Treatment groups evaluated are in Table 4.

Bioassays were PEDV-negative in 100% (4/4) of the pigs in the Neg control and all treatment groups (5ml-1:16M; 10ml-1:16M; 5ml-1:32M; and 10ml-1:32M). Bioassays were PEDV-positive in 100% (4/4) of the pigs in the Pos control group. Conclusions: All AHP treatment groups (5ml-1:16M; 10ml-1:16M; 5ml-1:32M; and 10ml-1:32M) were 100% effective at inactivating PEDV at 20° C.

Evaluation of an AHP disinfectant mixed in a 10 percent propylene glycol solution to inactivate PEDV in feces at -10° C

The objective of this study was to evaluate the efficacy of an AHP disinfectant (Accel®, Virox Technologies Inc., Oakville, Ontario, Canada) mixed in a 10% propylene glycol (PG) solution and applied as a foam under freezing conditions (-10°C) for inactivating PEDV in the presence of feces on metal surfaces found in livestock trailers. Treatment groups evaluated are in Table 5.

Bioassays were PEDV-negative in 100% (4/4) of the pigs in the Neg control and all treatment groups (5ml-1:32-40M; 10ml-1:32-40M; 5ml-1:16-40M; 10ml-1:16-40M; 5ml-1:32-60M; 10ml-1:32-60M; 5ml-1:16-60M; and 10ml-1:16-60M). Bioassays were PEDV-positive in 100% (4/4) of the pigs in the Pos control group. Conclusions: All AHP treatment groups (5ml-1:32-40M; 10ml-1:32-40M; 5ml-1:16-40M; 10ml-1:16-40M; 5ml-1:32-60l; 10ml-1:32-60M; 5ml-1:16-60M; and 10ml-1:16-60M) were 100% effective at inactivating PEDV at -10° C.

Summary of conclusions

Procedures that successfully inactivated PEDV when washing was possible

- High pressure wash with Barnstorm® detergent followed by disinfection with Synergize® with or without additional natural drying or TADD procedures

Procedures that successfully inactivated PEDV in the presence of feces when washing was not possible

- Combinations of heat and time
  - 160°F (71° C) for 10 minutes
  - 68°F (20° C) (room temperature) for 1 week
- AHP foaming disinfectant at room temperature (20° C) in the presence of feces with 30 minutes of contact time
  - Worked with both 1:16 and 1:32 concentrations of AHP
  - Worked in the presence of both 5 ml (light) and 10 ml (heavy) volumes of feces

Table 3: Description of treatment groups for the evaluation of a dry phosphate-based disinfectant powder to inactivate PEDV in feces.

Treatment group	Description of treatment
Neg control	5 ml PEDV-negative feces, no treatment
Pos control	5 ml PEDV-positive feces, no treatment
Stalosan® F	5 ml PEDV-positive feces, Stalosan® F with one hour of contact time at 20° C (room temperature)

Table 4: Description of treatment groups for evaluation of an accelerated hydrogen peroxide (AHP) disinfectant to inactivate PEDV in feces at 20° C.

Treatment group	Description of treatment
Neg control	5 ml PEDV-negative feces, no treatment
Pos control	5 ml PEDV- positive feces, no treatment
5ml-1:16M	5 ml PEDV positive feces, 1:16 concentration of AHP disinfectant at 20° C for 30 minutes of contact time
10ml-1:16M	10 ml PEDV positive feces, 1:16 concentration of AHP disinfectant at 20° C for 30 minutes of contact time
5ml-1:32M	5 ml PEDV positive feces, 1:32 concentration of AHP disinfectant at 20° C for 30 minutes of contact time
10ml-1:32M	10 ml PEDV positive feces, 1:32 concentration of AHP disinfectant at 20° C for 30 minutes of contact time

Table 5: Description of treatment groups for evaluation of an accelerated hydrogen peroxide (AHP) disinfectant mixed in a 10% propylene glycol solution at -10° C to inactivate PEDV in feces.

Treatment group	Description of treatment
Neg control	5 ml PEDV-negative feces, no treatment
Pos control	5 ml PEDV- positive feces, no treatment
5ml-1:32-40M	5 ml PEDV positive feces, 1:32 AHP disinfectant at -10° C for 40 minutes of contact time
10ml-1:32-40M	10 ml PEDV positive feces, 1:32 AHP disinfectant at -10° C for 40 minutes of contact time
5ml-1:16-40M	5 ml PEDV positive feces, 1:16 AHP disinfectant at -10° C for 40 minutes of contact time
10ml-1:16-40M	10 ml PEDV positive feces, 1:16 AHP disinfectant at -10° C for 40 minutes of contact time
5ml-1:32-60M	5 ml PEDV positive feces, 1:32 AHP disinfectant at -10° C for 60 minutes of contact time
10ml-1:32-60M	10 ml PEDV positive feces, 1:32 AHP disinfectant at -10° C for 60 minutes of contact time
5ml-1:16-60M	5 ml PEDV positive feces, 1:16 AHP disinfectant at -10° C for 60 minutes of contact time
10ml-1:16-60M	10 ml PEDV positive feces, 1:16 AHP disinfectant at -10° C for 60 minutes of contact time

- AHP foaming disinfectant at freezing temperatures mixed in a 10% propylene glycol solution as an antifreeze in the presence of feces
  - Worked with both 1:16 and 1:32 concentrations of AHP
  - Worked with both 40 and 60 minute contact times
  - Worked in the presence of both 5 ml (light) and 10 ml (heavy) volumes of feces

Procedures that failed to inactivate PEDV in the presence of feces when washing was not possible

- Combinations of heat and time
  - < = 145°F (63° C) for 10 minutes
  - 100°F (38° C) for 12 hours
  - 68°F (20° C) for 24 hours
- Dry phosphate-based disinfectant powder applied with a leaf blower with 30 minutes of contact time at room temperature (20° C) in the presence of feces

Acknowledgements

The studies were funded in whole or part by National Pork Board (NPB Project #13-227, NPB Project #13-248 and NPB Project #14-272), PIC North America, Iowa Select Farms, Protekta Inc., Vitfoss and Ogena Solutions.

References

1. Lowe J., Gauger P., Harmon K., Zhang J., Connor J., Yeske P., Loula T., levis I., Dufresne L., Main R. 2014. Role of transportation in spread of porcine epidemic diarrhea virus infection, United States. Emerging Infectious Diseases. 20(5):872-874.

2. Dee, S.A., Deen, J., Otake, S., Pijoan, C., 2004a. An experimental model to evaluate the role of transport vehicles as a source of transmission of porcine reproductive and respiratory syndrome virus to susceptible pigs. Can. J. Vet. Res. 68(2):128-33.

3. Dee, S., Deen, J., Burns, D., Douthit, G., Pijoan, C. 2004b. An assessment of sanitation protocols for commercial transport vehicles contaminated with porcine reproductive and respiratory syndrome virus. Can. J. Vet. Res. 68(3):208-14.

4. Dee, S., Torremorell, M., Thompson, B., Deen, J., Pijoan, C. 2005a. An evaluation of thermo-assisted drying and decontamination for the elimination of porcine reproductive and respiratory syndrome virus from contaminated livestock transport vehicles. Can. J. Vet. Res. 69(1):58-63.

5. Schneider P.T., Zhang J., Ramirez A., Holtkamp D.J. 2015. An evaluation of disinfection protocols to reduce virus transmission via livestock transport vehicles. J. Swine Health Prod. J Swine Health Prod. 23(6):306–316.

Feed biosecurity update

Roger A. Cochrane, MS; Loni L. Schumacher, DVM; Anne R. Huss, MS, PhD; Cassandra Jones, MS, PhD; Jason C. Woodworth, MS, PhD; Charles R. Stark, MS, PhD; Steve S. Dritz, DVM, PhD
Departments of Grain Science and Industry, Animal Science and Industry, Diagnostic Medicine Pathobiology, Kansas State University

Most feed mill biosecurity practices in the past have been focused on preventing the introduction of bacterial biological hazards, specifically the bacterium *Salmonella,* from entering finished feeds. Multiple points-of-entry of biological hazards into a feed mill exist, and can include raw ingredients, transport vehicles, and people. While total prevention of the introduction of *Salmonella* to a facility cannot always be achieved, current good manufacturing practices (cGMP) can help reduce the outgrowth and risk of finished feed contamination. With the introduction of porcine epidemic diarrhea virus (PEDV) to the United States, current feed mill biosecurity practices were not enough to prevent PEDV transmission. The objective of this paper is to highlight some of the recent research we have conducted, along with our collaborators at Iowa State University, evaluating PEDV mechanisms of transmission in feed and contamination within feed mills. Additionally, we have focused efforts on demonstrating the effectiveness of various mitigation strategies to inactivate PEDV in feed.

We certainly acknowledge that this is not a comprehensive list of biosecurity practices that can be employed at the feed mill level. We are at the early stages of understanding the role of feed manufacturing facilities in the prevention of pathogen introduction onto farms. Nevertheless, we have used PEDV as a model for understanding how feed and feed mill practices can be a portal-of-entry for biological hazards, like *Salmonella* and PEDV, onto swine farms.

What has PEDV taught us about infectious dose in feed?

The minimum infectious dose of PEDV through feed is quite low, compared to the infectious dose of *Salmonella*. Using PEDV as a model, a known quantity of PEDV virus was intentionally inoculated to feed. With our initial study, varying concentrations of PEDV was inoculated into feed and infectivity was determined using a 10-day-old pig bioassay. The pig bioassay resulted in a minimum infectious dose in feed of 5.6×10^1 $TCID_{50}/g$. This translates to the potential of 200 infectious PEDV particles to induce clinical signs of infection. In contrast, the infectious dose of *Salmonella* is variable due to host-health, but is generally accepted as 10^3 viable cells.

The practical implications of these results, implicate very small quantities of fecal matter or contaminated dust exhausted from an acutely infected premise could potentially be carried back to a feed manufacturing facility and lead to contamination of subsequent finished feeds and/or ingredients. Based on this, practical recommendations include prohibiting foot traffic by non-mill personnel (delivery drivers) into the mill and monitoring foot traffic within the mill to ensure foot traffic is excluded from incoming ingredient pit grates or hand add ports that are at floor level with direct access to the mixer or finished product areas.

What has PEDV taught us about feed mill contamination after processing PEDV inoculated feed?

Kansas State University is home to the Feed Safety Research Center (FSRC), which was specifically designed to facilitate research related to the impact feed processing has on biological hazards, like *Salmonella* and PEDV. The FSRC is a pilot-scale feed mill with biosafety level-2 (BSL-2) containment operations allowing known levels of pathogens to be introduced into feed and monitored throughout processing. In the FSRC we are able to monitor facility contamination with PEDV after the manufacture of contaminated feed. To replicate commercial feed processing simplified feed mill with a ribbon mixer, bucket elevator, and pilot-scale pellet mill was used in the FSRC.

During operation of the FSRC in BSL-2 containment mode, a 110 pound (50 kg) batch of PEDV-contaminated feed was mixed in the ribbon mixer, discharged into the bucket elevator and the feed collected. The mixer and bucket elevator were discharged and deemed empty when no more feed was discharged, mimicking commercial feed production. Subsequently, four 110 pound (50 kg) batches of PEDV-negative feed were sequenced through the system. Following processing of the contaminated batch and each subsequent sequence diet, environmental swabs were collected throughout the facility. Environmental collection points included direct-feed contact surfaces, surfaces immediately adjacent to direct-feed contact surfaces, and other areas throughout the facility, including floors, drains, and walls. Surface types included metal, concrete, rubber, and plastic.

The direct-feed contact surfaces were all positive for PEDV RNA via RT-qPCR analysis following manufacture of the PEDV-contaminated feed. Although detectable PEDV RNA decreased with each subsequent PEDV-negative diet processed on the direct-feed contact surfaces, more than 80% of the samples remained PEDV-positive. This would indicate diet sequence could potentially reduce PEDV contamination within equipment but does not completely eliminate the hazard. Surprisingly, 100% of samples collected from surfaces directly adjacent to the direct-feed contact surfaces contained PEDV RNA after the production of the PEDV-positive diet and all subsequent sequence diets.

Practical implications from this portion of the experiment tend to suggest that once a contaminated batch of feed is processed, the feed mill environment becomes highly contaminated. Presumably, the non-direct-feed contact surfaces are being contaminated by PEDV viral particles accumulated in dust particles generated during the feed manufacturing process. These results coincide with previous research done evaluating the *Salmonella*-surrogate, *Enterococcus faecium*, during feed processing, in which dust generated during processing led to facility contamination. These data seem to support that once a feed mill environment is contaminated, dust and foot traffic through the mill may provide a route of cross-contamination to other areas of the mill and other diets. This also suggests feed mill sanitation programs and management of dust collection systems are a critical control point in preventing cross-contamination of feed with PEDV RNA. For example, recycling of dust from dust collection systems should be discouraged.

In addition to monitoring environmental contamination, the manufactured feed was directly evaluated for PEDV RNA by RT-qPCR. Feed samples were collected after discharge from the mixer and discharged from the bucket elevator during the manufacture of the contaminated diet and the first two sequence diets. Results from the RT-qPCR analysis indicated PEDV RNA was only detectable in the feed from the contaminated batch and the first sequence diet collected from the mixer. Interestingly, samples collected during the 2nd sequence diet at the bucket elevator discharge had detectable PEDV RNA in 2 of 9 samples. Again, these feed samples were evaluated for infectivity through a 10-day-old pig bioassay. Results from the pig bioassay confirmed infectivity of the inoculated diet. Additionally, infectivity was detected in 1 of 3 replicates from both the 1st and 2nd sequence diets from samples collected from the mixer discharge prior to the bucket elevator. Of these results, it is concerning since samples with no detectable PEDV RNA via RT-qPCR resulted in infectivity during the pig bioassay. Therefore, sequencing of diets through a feed mill may help manage risk, but it does not eliminate the risk.

What have we learned about feed mitigation strategies?

Thermal processing

We conducted two experiments to evaluate thermal processing on PEDV infectivity using various conditioner time and temperature combinations using a pilot-scale pellet mill. In the first experiment, PEDV was inoculated to feed at a low dose (1×10^2 $TCID_{50}$/g) and high dose (1×10^4 $TCID_{50}$/g). Within each dose, batches of the inoculated feed were processed at 1 of 3 conditioning times (45, 90, 180 sec) and 1 of 3 processing temperatures [155 (68.3), 175 (79.4), and 195°F (90.6°C)]. Interestingly, RT-qPCR resulted in detectable PEDV RNA in feed processed at all time and temperature combinations, although none were found to be infective during the 10-day-old pig bioassay. These results indicate that thermal processing does not eliminate all PEDV genetic material but can mitigate infectivity.

Subsequently, a 2nd experiment was performed that evaluated processing temperatures below 155°F (68.3°C). Although target conditioning temperatures below 155°F (68.3°C) are uncommon in commercial facilities, it is possible for feed to be manufactured without reaching higher target temperatures. Production of feed at lower temperatures commonly occurs during start-up of the pellet mill or if the die of the pellet mill becomes plugged. Unplugging of the die is often resolved without manual intervention by turning off steam to the conditioner, decreasing the processing temperature. This can result in significant quantities of pelleted feed to be produced without reaching the target conditioning temperature. Feed produced below the targeted temperature, may no effectively mitigate PEDV infectivity, resulting in contamination of subsequent processing equipment, including the pellet cooler, pelleted feed handling equipment, storage bins at the feed mill, feed transport trucks, and on-farm storage bins, feed lines, and feeders. Therefore, in this experiment we evaluated feed processing temperatures at 100 (37.8), 115 (46.1), 130 (54.4), 145 (62.8), and 160°F (71.1°C).

Similar to the previous results, PEDV RNA was detectable at all processing temperatures via RT-qPCR. However, infectivity was demonstrated in pig bioassay of only feed processed at the 2 lowest temperatures, 100 (37.8) and 115°F (46.1°C). No infectivity was detected for the higher conditioning temperatures [130 (54.4), 145 (62.8), and 160°F (71.1°C)]. Therefore, these results indicate the minimum temperature to inactive PEDV in feed during conditioning is between 115 (46.1), and 130°F (54.4°C). While thermal processing has been shown to effectively mitigate the risk of PEDV, it is only a point-in-time strategy. Thermally processed feeds are still susceptible to post-processing contamination from within the facility or on the farm.

Chemical mitigation

Addition of chemical additives to feed are an attractive strategy to mitigate PEDV, primarily due to the potential residual activity to mitigate post-processing contamination. Therefore, in an initial experiment we evaluated 6 different chemical additives, which were applied to a finished feed and feed ingredients, followed by direct inoculation with PEDV. Subsamples were collected over time to monitor PEDV degradation using RT-qPCR. The chemical additives included a commercial formaldehyde product, sodium bisulfate, sodium chlorate, an organic acid blend, an essential oil blend, and a medium chain fatty acid blend. The chemical additives were evaluated in 4 different matrices, including: 1) a complete swine diet, 2) blood meal, 3) porcine meat and bone meal, and 4) spray-dried porcine plasma.

The results indicated that feed or feed ingredients could be treated with the different chemical additives as a means to mitigate subsequent PEDV contamination. The medium chain fatty acid blend, essential oils blend, and commercial formaldehyde were found to be particularly effective. Importantly, the success of the various chemical additives were dependent upon matrix and the stability of PEDV over time was also matrix-dependent. PEDV was found to be most stable in porcine meat and bone meal and spray-dried porcine plasma compared to blood meal and a complete swine diet. While the effectiveness of the medium chain fatty acid blend to degrade PEDV RNA has been confirmed through RT-qPCR and shows promise as an alternative mitigation strategy, although elimination of infectivity must be confirmed through pig bioassay.

Conclusion

Introduction of PEDV in the United States and the subsequent research to identify and control transmission has emphasized the need for continued improvement in feed biosecurity. An effective biosecurity plan requires the identification and evaluation of hazards, as well as procedures to control significant hazards that are reasonably foreseeable to occur within a particular facility. These control procedures can include the prevention of entry of the pathogen during ingredient receiving, monitoring and exclusion of foot traffic by high risk individuals, prevention of cross-contamination, and proactive mitigation using thermal processing or chemical additives. Many of these preventative strategies are included in cGMPs and quality control programs, but a separate feed biosecurity plan may be helpful to concentrate efforts. An assessment strategy, such as a self-assessment, may help facilities to identify gaps in their biosecurity plans. Additionally, future research is needed to continue to quantify the relative risk of pathogens in various feeds and ingredients, as well as research on improved control methods.

Appreciation is expressed to the National Pork Board for providing funding support for the majority of this research. Appreciation is also expressed to Q. Chen, R. G. Main, J. Zhang, and P. C. Gauger (Iowa State University) for their collaboration providing the PEDV and bioassay.

PEDV prevalence

R. B. Morrison, DVM, MBA, PhD; C. Betlach BSc
College of Veterinary Medicine, University of Minnesota

Overview of Swine Health Monitoring Project

The Swine Health Monitoring Project (SHMP) was started in 2011.[1] The target population is sow herds in the United States and we started with a convenience sample of systems and herds. Participation is voluntary and anonymous unless the system chooses to make its participation known publicly by including its logo on the weekly report. First participants provided retrospective data from July 2009 to the date of enrollment and then prospective data thereafter. Within each production system, eligible sow herds wean piglets into offsite locations and are requested to report weekly health status for PRRSv, PEDv and Seneca Valley Virus. All participants are required to sign participation and confidentiality agreements.

Beginning July 1 of each year, cumulative incidence is computed as the proportion of susceptible herds (epidemiological units) with new cases. The weekly incidence is monitored with an EWMA. The approximated EWMA (smoothed) number of PRRSV infected herds is estimated per week and an epidemic threshold was defined. In this way, we have an objective indicator if and when an epidemic has started and ended for our sample of herds. The SHMP has gradually expanded to include 27 systems composed of 3 regional PRRS projects and 24 companies or coordinated systems. As of today, 996 herds (2.53 million sows) participate for PEDV.

Figure 1: Cumulative incidence of PEDV

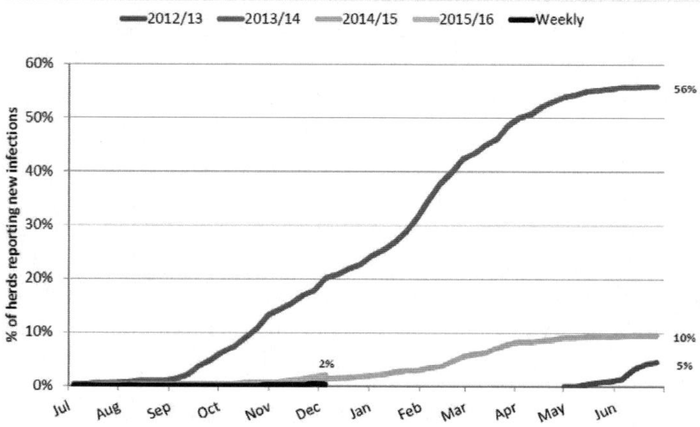

Current projects

Monitoring incidence – One can't help but be impressed by the decreased incidence since the initial epidemic in 2013/14 (Figure 1, 2). Cumulative incidence decreased from 56% of our SHMP herds experiencing an outbreak to 10% just one year later. Thus far, we are on target for a similarly low incidence this year. We can attribute this decrease to what we have learned in this short time:

- Extremely low infectious dose required to infect pigs,[2]
- Time and temperature required to inactivate,[3]
- Importance of transportation biosecurity,[4]
- Feed biosecurity,[5-9]
- Immune memory in previously infected sows,[10,11]
- Importance of spread within a region,[12] and
- Improved methods for eliminating virus from sow herds.[13]

Production impact – This study involved participants and data from the Swine Health Monitoring Project. Veterinarians were invited to share production data from herds infected with PEDV. These data, in conjunction with diagnostic reports, were used to estimate the time required for the herd to produce PEDV PCR negative pigs and the production loss. Of 183 infected herds that shared production records, a median of 2.7 piglets / inventoried sow were not weaned and the average time required to recover to baseline production was 10 weeks. Herd infected in quarters 3 or 4 of the year had approximately twice the negative impact.[14]

Time to stability – Of the 429 infected herds that achieved the stable state of weaning PEDv PCR negative pigs, the median time was 28 weeks, ranging from 7 – 64 weeks.

Both production loss and time to stability have been reduced since conducting these analyses by veterinarians having developed and adopted improved methods for control and elimination.

Going forward – We will continue to monitor incidence and use the SHMP data to conduct and report in-depth analyses. We continue to develop improved methods of data collection to reduce chance of error and burden on the veterinarians. Finally, we continue to listen to our participants for ways by which we can continue to deliver value.

Acknowledgements

- SHMP participating producers and veterinarians who share their health status data (see logos on weekly report).
- SHMP advisory group - The following individuals provide input as needed and requested for the SHMP project. Their role is valuable and appreciated. Paul Yeske, Matthew Turner, Joe Connor, Gordon Spronk, Luc Dufresne, Clayton Johnson; Paul Sundberg (ad hoc).
- Funding
  - Swine Health Information Center

Figure 2: Exponential weighted moving average of PEDV incidence

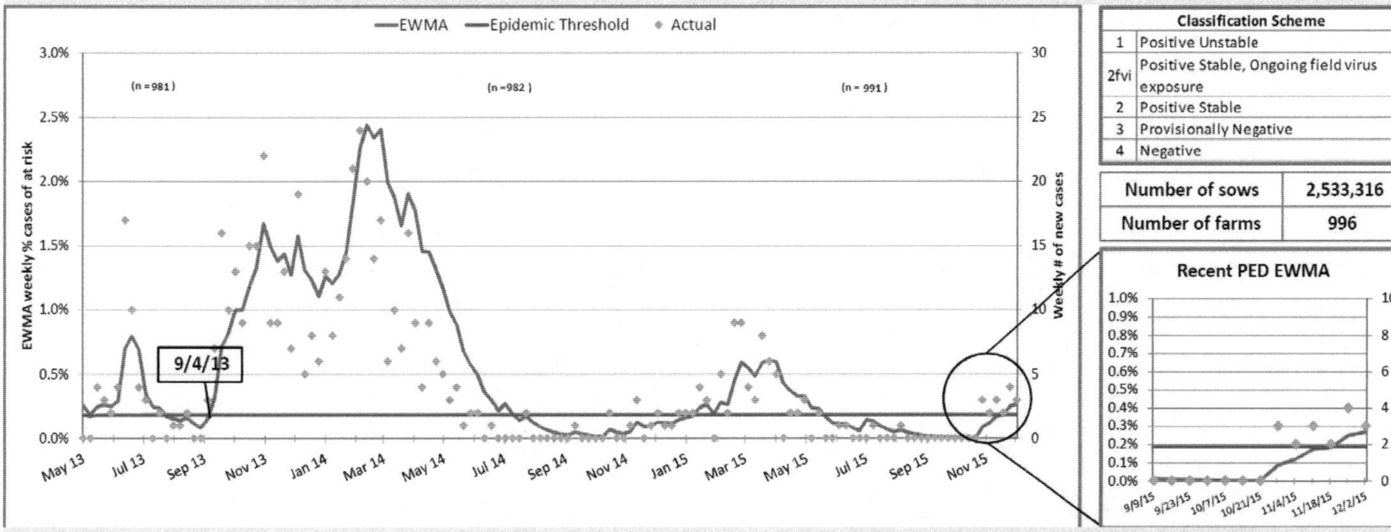

References

1. SJP Tousignant et al. Temporal and spatial dynamics of porcine reproductive and respiratory syndrome virus infection in the United States. *Am J Vet Res* 2015; 76:70–76.

2. X. Liu et al. Determination of the infectious titer and virulence of an original US porcine epidemic diarrhea virus PC22A strain. Veterinary Research (2015) 46:109.

3. P. Thomas et al. Evaluation of time and temperature sufficient to inactivate porcine epidemic diarrhea virus in swine feces on metal surfaces. J Swine Health Prod. 2015;23(2):84–90.

4. J. Lowe et. al. Role of Transportation in Spread of Porcine Epidemic Diarrhea Virus Infection, United States Emerging Infectious Diseases, www.cdc.gov/eid, Vol. 20, No. 5, May 2014.

5. S. Dee et al. An evaluation of contaminated complete feed as a vehicle for porcine epidemic diarrhea virus infection of naïve pigs following consumption via natural feeding behavior: proof of concept. BMC Vet Res. 2014; 10: 176.

6. H.S. Kochlar et al. Porcine epidemic diarrhea in Canada: An emerging disease case study. Can Vet J. 2014 Nov; 55(11): 1048–1049.

7. J. Pasick et al. Investigation into the role of potentially contaminated feed as a source of the first-detected outbreaks of porcine epidemic diarrhea in Canada. Transbound Emerg Dis. 2014 Oct;61(5):397-410.

8. S. Dee et. al. An evaluation of a liquid antimicrobial (Sal CURB®) for reducing the risk of porcine epidemic diarrhea virus infection of naïve pigs during consumption of contaminated feed. BMC Veterinary Research 2014, 10:220.

9. P. Davies et al. The dilemma of rare events: Porcine epidemic diarrhea virus in North America. http://dx.doi.org/10.1016/j.prevetmed.2015.08.006.

10. D. Goede et al. Previous infection of sows with a "mild" strain of porcine epidemic diarrhea virus confers protection against infection with a "severe" strain. Vet Micrbiol 76; 2015. 161-164.

11. A. Schelkopf et al. Lactogenic protection against PEDv in piglets following homologous challenge. 2015. Proc Leman Swine Conference abstracts. 28.

12. Alvarez et al. Spatial and temporal epidemiology of porcine epidemic diarrhea (PED) in the Midwest and Southeast regions of the United States. http://dx.doi.org/10.1016/j.prevetmed.2015.11.003.

13. L. Dufresne. Managing acute and chronic PEDv in sow herds: what have we learned so far. 2015. Proc Leman Swine Conference 2015.

14. D. Goede et. al. Production impact & time to stability in sow herds infected with Porcine Epidemic Diarrhea virus. 2015. Prev Vet Med – accepted.

Experience with PEDV in North Carolina

Sara Hough, DVM
Prestage Farms, Inc., Clinton, North Carolina

In the fall of 2013 and spring of 2014, 89% of our sow farms broke with PEDV. The following year, in the fall of 2014 and spring of 2015, 57% of our herds broke, with 94% of those being clinical re-breaks on farms that had been previously infected. How did we become so unlucky? Living in pig-dense North Carolina, we measure distance between farms in feet, not miles. Many of the farms were constructed when air transmission, was thought to be impossible. We are considered 3-site-production, but finishers and nurseries surround many of the sow farms. When PEDV arrived to North Carolina in the summer of 2013, it did not take long to wreak havoc on Sampson and Duplin County, two of the most pig-populated counties in the United States. There were a couple of sites that dodged a bullet, but the majority of our farms (sow, nursery and finisher) in dense areas broke with PEDV within a 6-8 month time frame. Pig losses were horrendous and continued beyond 10-12 weeks post break on some farms. The plan was to eliminate PEDV from our sow farms and nursery sites, with eventual elimination from finishing. We verified negative status of our sow farms by utilizing testing procedures outlined in the Swine Health Producer Guide provided by the Pork Board.[1]

We were successful in elimination from sow farms and nursery sites the first time around. We continued to have some finishing circulation through the summer of 2014. We associated that circulation with continuous flow sites and poor sanitation between the plant and farms. During the summer of 2014, RFLP 1-7-4 PRRSV started to circulate in nursery and finishing, with sow farms eventually breaking late summer. PRRSV circulated through 93% of the farms from then until spring of 2015, with the majority of isolates within 2% homology of one another. During the same time period, farms started to have scours and mortality on gilt litters. The index PEDV farm in our system in 2013 was also the index PEDV farm in 2014. Initially, gilt litters would scour with pretty significant mortality, and then some sow litters would become sick, with increased morbidity, but not significant mortality. Previously exposed sows would scour, but prevalence was low and looseness was transient. Once PEDV was confirmed, a protocol was put in place for re-break, with intentions of sow farm elimination.

When a farm was confirmed positive, pigs nursing naïve sows were weaned down to 10 days and previously exposed sows were vaccinated pre-farrow with PEDV vaccine to booster pre-existing immunity. Gestation was exposed with either piglet intestines or sow feces after wean-down took place. Gilts were loaded into the farm for preparation of up to a 20-week closure. Pigs were euthanized at birth on naïve sows for 18 days following break, knowing that there was very little chance of those pigs surviving and the possibility of further amplification and shedding of more virus. Lactogenic immunity tended to appear in about 2.5 weeks in naïve animals with initial breaks, so we decided to target 18 days for euthanasia of at-risk litters. Cross-fostering was not allowed for at least the first 6 weeks post break to ensure that virus was not moved from crate to crate. Farrowing was never exposed in order to keep as much virus away from neonates as possible, so gestation was re-exposed each week with live PEDV material until all sows had seen at least one exposure. White washing of farrowing began immediately, along with pre-farrow PEDV vaccination, which continued until the farm was deemed negative. Strict biosecurity between gestation and farrowing was also reinforced by restricting movement of people and equipment, and proper sow sanitation before loading into farrowing. After the last feedback, barns were washed and disinfected. We started testing farms at 6 weeks and required 4 negative weekly Swiffer tests from piglets before farm was considered negative. Swiffer testing consisted of 'swiffering' up to 15 litters less than 10 days and 15 pigs/litters at weaning with any loose stool. Samples were collected and tested weekly.

With the original PEDV breaks in 2013-2014 we started testing at 10 weeks post break, about 50% of the farms were still testing positive at 20 weeks. With the PEDV re-breaks, almost 75% had negative tests by 10 weeks post break, only 4 weeks after they started testing, and all were negative by 16 weeks. It is difficult to compare losses from initial breaks and re-breaks due to PRRSV associated losses around the same time of infection. Simply comparing the pigs weaned per sow at 16 weeks post break on the original 51 farms that broke in 2013-early 2014 and 28 farms that re-broke the following year, farms weaned half a pig more with re-break.

We realized that with monthly gilt introductions of naïve animals, yearly circulation and re-breaks could become a permanent reality. We decided to start a gilt exposure plan, which would expose gilts as feeder pigs to material collected from an initial break and amplified through PRRSV naïve, multiplication animals that were less than 10 days old. We would then allow for an extended off-site cool down phase before shipment to sow farms as mature gilts. During this phase, gilts are tested for active circulation and shedding. This acclimation process began in the late spring of 2015, with the first acclimated gilts entering all of our farms in North Carolina late summer. We did have two farms that remained naïve, in very dense locations, during both outbreaks. In the spring of 2015, we decided to utilize an off-site sow farm to take a subset of sows so that we could do a controlled exposure on-farm, as these previously naïve farms would be receiving exposed gilts. We moved 5 groups of sows off-site and loaded the farm with 20 weeks-worth of gilts. We waited for the remaining pigs to be born and weaned off sow farm and then exposed gestation the day the last wean pig left. As sows began to wean pigs at off-site location and return, gestation was re-exposed. This was done with each returning group. Farrowing rooms were white-washed, sows were vaccinated pre-farrow with PEDV vaccine and no cross-fostering was allowed. Farms did have clinical litters once they started to farrow back at the sow farm. One farm had an elevated pre-wean mortality the first week of 60% and then returned to normal level with following group of sows, while the other had an elevated group at 25% before returning to 15%. With our experience, 100% would be likely for the first 3 weeks post break on a naïve farm, with lingering mortality for the next two months or more. We deemed these exposures a success and have since entered PEDV-acclimated gilts without issue.

It is difficult to determine what our plan will be in the next five years. Currently, there is PEDV circulation in grow-finish and surrounding sow farms. As long as we have a threat in our area, we will need to continue to acclimate gilts for PEDV, just as we do for other impending diseases to maintain herd immunity. We hope there are additional tools and technologies developed that will guide us in the near future or even a better understanding of the virus that will help to control or eradicate this disease.

References

1. American Association of Swine Veterinarians and National Pork Producers Council. *Swine Health Producer Guide: How to Verify PEDV-Negative Status after the Virus is Eliminated on a Farm.* Des Moines, Iowa: National Pork Board; 2014. Available at: www.pork.org.

Experiences with porcine epidemic diarrhea virus (PEDV) acclimation of replacement gilts

Pete Thomas, DVM, MS
Smithfield Foods, Hog Production Division, Midwest Region, Algona, Iowa

Introduction

Porcine epidemic diarrhea virus (PEDV) was first detected in the United States in April 2013. Since that time, it has spread widely throughout North America infecting herds in essentially all geographical areas. Despite an industry focus on improved biosecurity, this pathogen has rapidly and entered many herds with excellent previous track records of disease prevention.

Initially, anecdotal reports indicated that experiences and expectations for disease prevention, clinical course of disease, and herd elimination efforts are similar to those for transmissible gastroenteritis virus (TGEV). Over the course of the past 18 months we have found that this disease is much more difficult to manage. Some of the challenges that have been encountered with disease include:

- Very high load of viral shedding and low infectious dose.[3]
- Stability of the virus in the environment.[4]
- Ability of the virus to overwhelm passive immunity.

These challenges have led to an increased time to herd stability – 15-20 weeks on average to return to 100% of normal weaning numbers, with a wide range in variability between herds.[2] Estimates of piglet losses per sow during a break have continued to increase with increased number of herds that became endemically affected by high pre-weaning mortality. The most recent estimate has increased to at least 2.6-4.7 Pigs lost per sow.[2]

Since the original outbreak of PEDV in the United States, some herds have experienced a long-term endemic state of disease or clinically severe re-infections of disease. Due to the continued risk of herd infection, effective development of active immunity in naïve replacement animals and passage of passive transfer of this immunity to offspring is important for many herds. An industry survey of swine veterinarians had previously found that 50% of veterinarians recommended exposure of replacement females with live PEDV to induce immunity in gilts prior to breeding. In addition, this same survey found that 84% of veterinarians planned to start or continue gilt acclimation if a naïve herd was infected with PEDV virus.[5]

Herd background

The operation discussed, includes 62,000 sows located within 16 sow farms ranging between 1,200 - 7,600 sows. These herds are located within a 10 mile radius of each other, in an otherwise low-density area of the Midwest. Despite an excellent track-record of external biosecurity and prevention of economically significant disease introduction, all 16 sow herds became infected with PEDV virus between February 4, 2014 and February 14, 2014. All sequencing conducted in this operation demonstrated that the strain was similar to the original PEDV strain introduced into the United States.

Typical clinical disease associated with PEDV presented. Severe watery diarrhea, vomiting, and mortality approaching 100% in pigs under 7 days of age was consistent among all farms. All animals received oral feed-back 2×, 5-7 days apart. Follow-up testing was conducted at all 16 sow farms. Results indicated that 590/590 (100%) sows were positive for anti-PEDV IgG antibodies by IFA at 4-6 weeks post-exposure. These results indicate that a successful and uniform exposure occurred for sows across the operation. Intense cleaning and disease management protocols were introduced immediately. All herds in this operation were able to eliminate the virus within 5 months of initial infection and wean PEDV negative pigs.

Replacement females are produced within the operation and all multiplication sow farms became infected within the same time period as the commercial farms. All replacement gilts are raised in off-site all-in-all-out nurseries and then moved to two large (8-10 barns each) gilt finishing sites. Both finishing sites are continuous flow and were infected with PEDV as well. Due to the source of gilts, continuous flow nature of the gilt finishing facilities, and the desire to mitigate risk associated with being naïve; It was determined that a controlled exposure program would be implemented as soon as negative feeder pigs were being placed into the gilt finishing sites.

Gilt acclimation

Exposure material

Exposure material was generated using intestines from piglets at multiplication sow farms collected during the original outbreak. This material was homogenized and administered orally to naïve 10 week old replacement gilts at each finishing site. After clinical signs developed; diarrhea was collected, pooled, and stored in 50ml centrifuge tubes. All samples are stored on-site in conventional chest freezers.

Tubes from each finishing site were sent to Iowa State University Diagnostic Laboratory. Each sample was tested for the presence of PEDV by PCR and the CT value was determined to verify a high quantity of PEDV RNA was present. The RNA was also sequenced to verify homology to the original strain detected during the outbreak in February 2014. In addition to the PEDV testing, the samples were tested for swine deltacorona virus, porcine circovirus type 2, porcine reproductive and respiratory virus, TGEV, *Lawsonia intracellularis*, and porcine parvovirus by PCR to rule out the presence of other significant pathogens. All of these additional PCR tests were negative.

Exposure techniques and confirmation

Three different exposure techniques were attempted based upon feedback from others in the industry. This included: 1) Top-dress of exposure material on feed, 2) Administration of exposure material in the drinking water through a water proportioner, and 3) Saturation of cotton ropes in a slurry of exposure material. Verification of infection was determined by observation of clinical signs and oral fluids PCR. In addition, 15 animals from two barns of each exposure method were bled and tested for the presence of PEDV antibodies by ELISA. All exposure techniques appeared to be successful.

All replacement females are exposed to PEDV within 24 hours of arrival at the gilt finishing sites. This is done in order to help assure that we do not have continuous natural exposure occurring, which could hypothetically lead to an increased risk of mutation occurring. All finishing barns are tested for the presence of PEDV by oral fluids PCR one week post-exposure to verify that successful infection has occurred in the population. Success is measured by detection of PEDV RNA at a CT value in the upper teens to lower twenties in addition to visual verification of clinical signs.

Internal biosecurity related to PEDV acclimation

All gilts have approximately 12-14 weeks between the time of planned exposure and movement into gilt development units attached to the breeding herds. Previous research has shown that virus shedding should not exceed this time period. Protective immunity developed via viral infection has been demonstrated to last at least 7 months.[1] However, personal experience in addition to the anecdotal experience of other veterinarians, has shown that animals can shed high amounts of virus and experience clinical disease infrequently late in the finishing period after the group was previously exposed. For these reasons, internal biosecurity protocols have been developed in order to mitigate the risk of live virus introduction into the breeding herd.

All barns are all-in-all-out. In one of the finishing sites where barns are not connected the animals are kept in the same barn where they are originally exposed. In the second finishing site, all barns are connected by internal hallways. At this site there is a "hot side" which consisted of barns where animals are exposed and housed for 8 weeks before being moved to the "cool" side where they are housed for an additional 6 weeks. Order of chores is performed oldest to youngest with appropriate boot change and disinfection protocols in place. In addition, when moving from "hot" groups back into "cool" groups is necessary, coveralls are changed and hands/arms are washed thoroughly. Separate chutes are utilized for incoming and outgoing animals. Separate mortality removal areas are utilized. Hallways are washed and disinfected routinely and always before movement of mature animals out of the site and after movement of animals form the "hot" to the "cool" side.

These practices were not initially implemented. Since implementation we have not seen late clinical symptoms or detected high amounts of virus in older animals.

Breeding herd entry

Prior to movement into the breeding herd, oral fluids are collected and tested for PEDV by PCR for two consecutive weeks. In addition, 30 rectal swabs are collected and tested in pools of five for PEDV by PCR. Animals are allowed to move into breeding herds if all rectal swab pools are negative and oral fluids demonstrate a CT value higher than 32. In my experience, it is difficult to consistently get completely negative results on oral fluid PCR. This is likely due to residual viral RNA in the environment.

Discussion and conclusion

Since the original outbreak of PEDV in this operation, previously exposed gilts have entered our breeding herds. In some of the initial groups there were known occasions at both finishing sites where clinical signs or high virus levels were detected in groups when a finishing barn was partially moved into breeding herds. During these instances, clinical disease was not observed in piglets. Conversely, we had half of the sow farms with minimal to moderate clinical signs develop about one year after the initial outbreak. This occurred at a time when the finishing populations appeared stable clinically and on testing. Symptoms typically started in gilt litters when this occurred. At the time, most of the gilts farrowing came from multiplication sow farms during the endemic phase of disease during the initial outbreak, and therefore first exposed to virus in the farrowing crate or early nursery period.

It is unknown if infection at this younger age was inadequate for optimal immune development, if live virus acclimation at the finishing site, or another factor was the main risk factor involved in these re-breaks. Interestingly, there was no correlation to which finishing site supplied the gilts. Some herds sourced from each site experienced the re-break while other herds did not. Disease was much milder in the re-break herds and all herds cleaned-up much faster than was seen with the original outbreak. Since the introduction of strict internal biosecurity and more rigorous testing herds have maintained stability.

PEDV acclimation of replacement gilts has been a strategy utilized by many in the industry to maintain a consistent supply of gilts through continuous flow gilt development sites, while at the same time reducing the large risks associated with maintaining a PEDV naïve herd. However, there are risks associated with utilizing live pathogenic virus for exposing replacement gilts. In the long-term better tools are needed in the industry in order to safely stimulate effective immunity in naïve animals.

References

1. Goede DP, et al. Previous Infection of Sows With a "Mild" Strain of PED Virus Confers Protection Against Infection with a "Severe" Strain. AASV. 2015;55-57.

2. Goede DP, Morrison RB, et al. PED Incidence, Time to Stability, and Production Impact. Allen D. Leman Swine Conference St. Paul, MN, USA. 2014.

3. Thomas JT, et al. Effect of Porcine Epidemic Diarrhea Virus Infectious Doses on Infection Outcomes in Naïve Conventional Neonatal and Weaned Pigs. *PLoS One*;2015 Oct 6;10(10):e0139266.

4. Thomas PR, Karriker LA, Ramirez A, et al. Evaluation of Time and Temperature Sufficient to Inactivate Porcine Epidemic Diarrhea Virus in Swine Feces on Metal Surfaces. *J Swine Health Prod*. 2015;23(2):84-90.

5. Turner MB. Review of Porcine Epidemic Diarrhea Virus Exposure Techniques. AASV. 2015;417-420.

Evaluation of the effects of PEDV vaccine on PEDV naïve and previously PEDV exposed sows in a challenge model comparing immune response and preweaning mortality

T. J. Schwartz[1]; C. J. Rademacher[2]; L. G. Gimenez-Lirola[2]; Y. Sun[2]; J. J. Zimmerman[2]
[1]Suidae Health and Production, Algona, Iowa; [2]Department of Veterinary Diagnostic and Production Animal Medicine, College of Veterinary Medicine, Iowa State University, Ames, Iowa

Introduction

Porcine epidemic diarrhea virus (PEDV), a coronavirus, was first recognized in the United States in April, 2013.[1] In affected herds it has caused 100% morbidity and 50-100% mortality in suckling piglets.[2] PEDV has proven to be a very difficult disease to treat or control, and no single intervention strategy has been 100% successful. Colostrum from immunized sows has been found to increase survival rates of PEDV challenged piglets.[3] It has been well documented that coronaviruses such as PEDV or transmissible gastroenteritis virus (TGEV) stimulate a strong IgA response in the sow which is passed to pigs via lactogenic immunity.[4] The practice of feeding back feces from affected PEDV-positive piglets to sows is effective in stimulating the immune response and lactogenic immunity in sows; however, this practice does present the possibility of continued presence of virulent virus thus continuing disease outbreaks within a herd, significant economic losses, and uncertainty because of variation in immune response in sows.[4] There is a need for an efficacious vaccine that can stimulate a protective immune response, can be differentiated from a field strain of PEDV, is safe, and does not pose the threat of prolonging disease outbreaks.

PEDV vaccines were in use in Asia at the time of the first US outbreaks, but no vaccines were available for use within the US. Zoetis received a conditional license for a killed PEDV vaccine, accompanied by data suggesting a 2.5-fold increase in serum antibody titer in PEDV-vaccinated sows versus those vaccinated with placebo as well as a 90% relative reduction in the PEDV-associated preweaning mortality rate compared to litters from placebo-vaccinated sows.[5] Zoetis studies demonstrated that PEDV vaccinated sows weaned 20.1% more litters compared to control sows.[5]

There have been few studies to date reporting the efficacy of PEDV vaccination using an experimental challenge model; however, it has been suggested that vaccination increases sow PEDV serum and colostrum antibody titers, increases lactogenic immunity, and decreases pre-weaning mortality of litters undergoing PEDV challenge. Therefore, continued research needs to be examined to further evaluate the efficacy of these vaccines and how they may be implemented in the future. The aim of this study was to compare the effect of vaccination versus no vaccination in litters and sows that were either previously PEDV exposed sows or PEDV naïve.

Materials and methods

Thirty-two 6th-parity sows were obtained at approximately six weeks pre-farrowing. Sixteen of the sows had been previously exposed (P.E.) to PEDV and were negative for PEDV in feces by PCR for the previous nine months prior to the start of the trial. The other 16 sows were naïve to PEDV. Prior to acclimation animals were verified negative for PRRSv and PEDV by PCR on serum and feces, respectively. Both the P.E. and the naïve groups were split into two groups of eight each for allocation to two treatments, vaccinated (Vac) non-vaccinated (Non-Vac). The sows in the P.E. group were allocated to assure an even number of sows with high and low PEDV serum neutralization titers were present in each group while the PEDV-naive sows were randomly assigned to a treatment group (Table 1).

All sows were individually identified and allowed to acclimate for one week in gestation pens. Sows were monitored daily and fed a diet formulated to meet or exceed their daily requirements. Water was available *ad libitum.* After one week of acclimation, the sows received their first intramuscular vaccination by injection at five weeks pre-farrow and received a second dose at 2 weeks pre-farrow. Nonvaccinated sows received a placebo (saline) injection on the same schedule. All sows farrowed naturally within a 4-day period. At farrowing (Day 0), sow serum was collected for IgA and IgG testing. Sow colostrum and milk samples were taken on days 0, 2, and 9 days post farrowing. Sows were observed daily for clinical signs of PEDV. Pig serum samples were collected at 2 and 9 days of age for IgA and IgG testing at Iowa State University Veterinary Diagnostic Laboratory.

At 5 days of age, each piglet was orally inoculated with 1 ml of inoculum containing the original US strain of PEDV. Inoculum was previously prepared by collection of feces from 1-day-old piglets inoculated with PEDV at birth, diluted to a PCR cycle-threshold (Ct) of 19, and stored in aliquots at -80^{0} C. Exposure to PEDV was confirmed by PCR testing of pooled fecal swabs from each pen 1 day after inoculation. Piglets were observed twice daily for clinical signs. Non-ambulatory or moribund piglets were humanely euthanized according to AASV/AVMA guidelines. Observations continued until the end of the trial which was at 21 days post-farrowing.

Summary statistics were calculated for all groups to assess the overall quality of the data sets. A one way ANOVA test was used to test for differences among the four treatment groups for sow S/P IgG and S/P IgA. Differences were considered statistically significant at $P < .05$ (Table 2 & 3).

Results

See attached table and figures.

Discussion

From the results of this study, it is evident that serum IgG is induced by previous exposure to PEDV, this is demonstrated by a significantly higher IgG value for P.E. non vaccinated animals compared to Naïve non-vaccinated animals ($P < .01$) and that PEDV vaccination of naïve sows can stimulate IgG antibody titers of significantly similar magnitude as those previously exposed sows ($P = .02$). Vaccination also numerically increased serum and colostrum IgG levels in previously exposed sows (Figures 1 -3). The serum IgG response was similar in P.E. and Naïve vaccinated groups.

On the other hand, IgA titers did not follow the pattern of IgG response. Both vaccinated and nonvaccinated P.E. groups had significantly higher IgA titers in all serum ($P < .01$) and milk samples. The vaccinated P.E. group had the highest IgA titers, followed by the P.E. Non-vaccinated group. Very low IgA antibody was detected in the vaccinated naïve group with virtually no IgA detected in non-vaccinated naïve group (Figures 1-5). The IgA levels were not significantly different between naive vaccinated and naïve non-vaccinated animals ($P = 0.18$). These results are not unexpected as parenteral vaccines predominately invoke a humoral IgG response. Even though vaccination of naïve sows did induce low but detectable IgA in colostrum and milk when compared to non-vaccinates (Figure 3), it was not sufficient to reduce mortality or statistically significant (Figure 6).

Table 1: Treatment group designation of bred sows.

Previously exposed – Zoetis PEDV vaccinated (P.E.-vac)	Previously exposed Non-vaccinated (saline) (P.E.-non-vac)	Naïve – Zoetis PEDv vaccinated (Naïve-vac)	Naïve non-vaccinated (saline) (Naïve-non-vac)
8 bred sows	8 bred sows	8 bred sows	8 bred sows

Table 2: Statistical comparisons of PEDV IgA ELISA least squared means of sow serum at farrowing.

Statistical comparison of sow serum IgA at farrowing	
Naïve vaccinated vs. Naïve non-vaccinated	$P = .18$
Naïve vaccinated vs. P.E. vaccinated	$P < .01$
Naïve vaccinated vs. P.E. non-vaccinated	$P < .01$
Naïve non-vaccinated vs. P.E. vaccinated	$P < .01$
Naïve non-vaccinated vs. P.E. non-vaccinated	$P < .01$
P.E. vaccinated vs. P.E. non-vaccinated	$P < .01$

Table 3: Statistical comparisons of PEDV IgG ELISA least squared means of sow serum at farrowing.

Statistical comparison of sow serum IgG at farrowing	
Naïve vaccinated vs. naïve non-vaccinated	$P < .01$
Naïve vaccinated vs. P.E. vaccinated	$P < .05$
Naïve vaccinated vs. P.E. non-vaccinated	$P < .05$
Naïve non-vaccinated vs. P.E. vaccinated	$P < .01$
Naïve non-vaccinated vs. P.E. non-vaccinated	$P < .01$
P.E. vaccinated vs. P.E. non-vaccinated	$P < .01$

Figure 1: Comparison of sow serum IgG and IgA levels at farrowing using S/P ratios. A positive level is any value greater than 0.8 designated in this graph by a positive threshold line at 0.8.

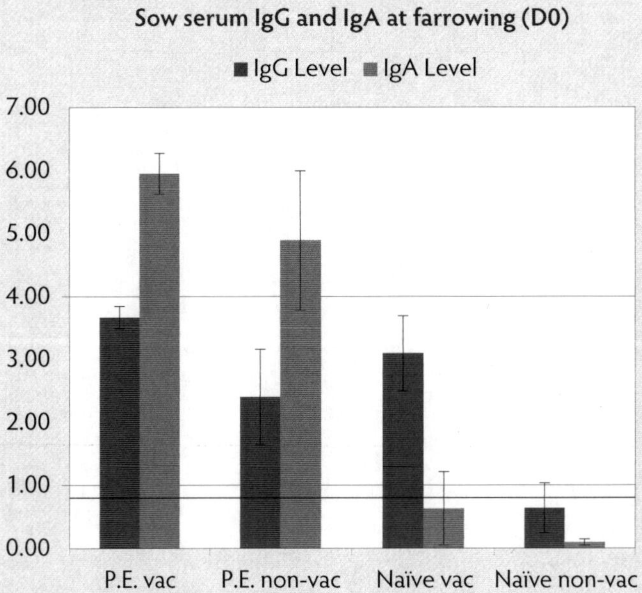

Figure 2: Comparison of sow colostrum and milk IgG levels using S/P ratios. A positive level is any value greater than 0.8 designated in this graph by a positive threshold line at 0.8.

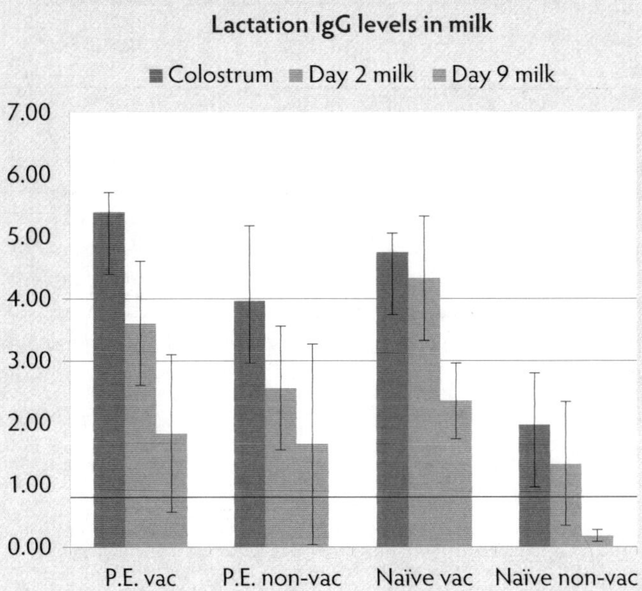

Figure 3: Comparison of sow colostrum and milk IgA levels using S/P ratios. A positive level is any value greater than 0.8 designated in this graph by a positive threshold line at 0.8.

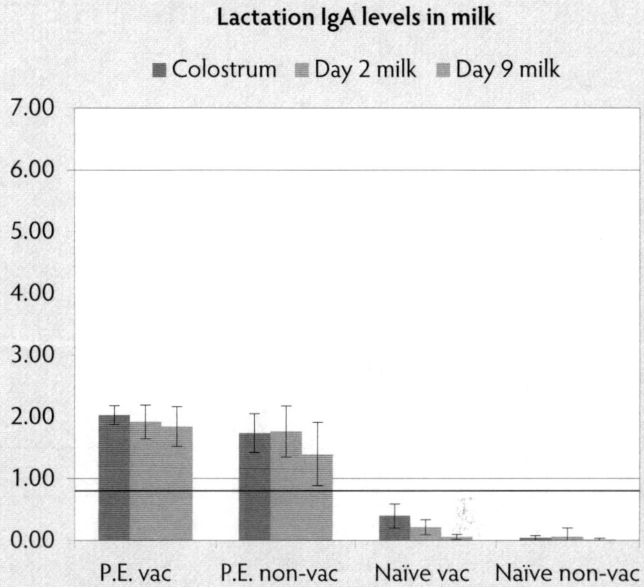

Figure 4: Comparison of piglet serum Day 2 IgG and IgA levels using S/P ratios. A positive level is any value greater than 0.8 designated in this graph by a positive threshold line at 0.8.

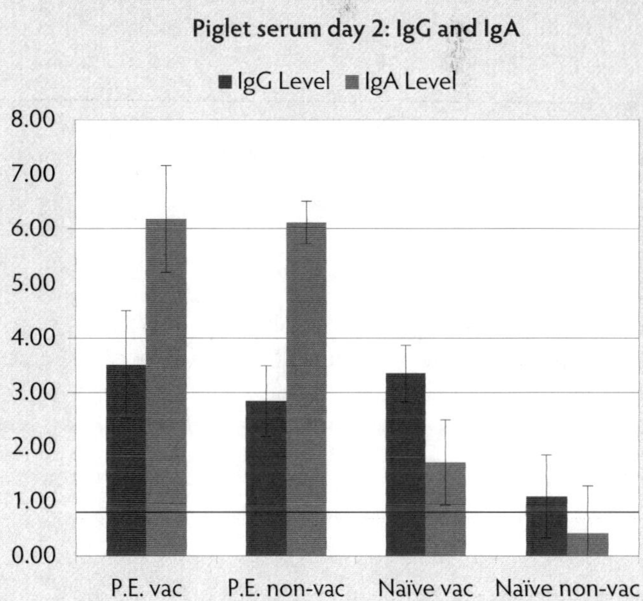

Figure 5: Comparison of piglet serum Day 9 IgG and IgA levels using S/P ratios. A positive level is any value greater than 0.8 designated in this graph by a positive threshold line at 0.8.

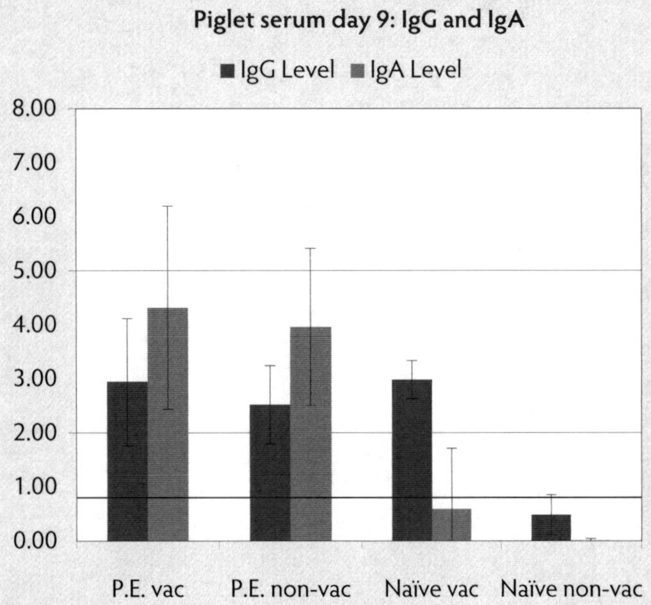

Figure 6: Comparison of the piglet mortality between treatment groups. PEDV inoculation occurred on Day 5 and mortality occurred until 15 days post farrowing. No mortality occurred after Day 15 and the trial concluded on Day 21.

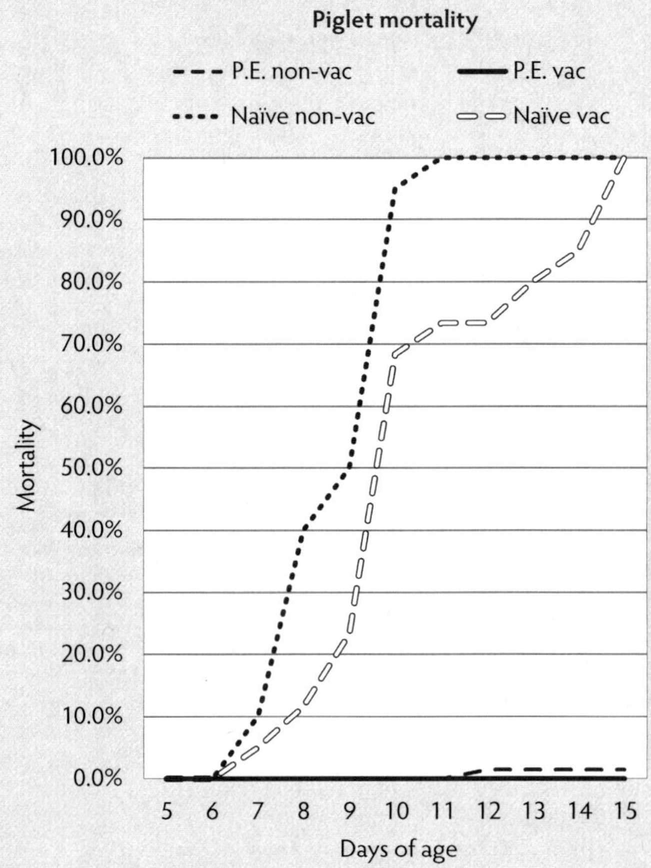

Piglet serum IgG antibody levels trended similarly to sow serum and colostrum, with higher levels found in both P.E. groups and in the vaccinated naïve group which is as expected with normal passive transfer via colostrum. Piglet serum IgG and IgA decreased from D2 to D9, likely the result of expected passive antibody decay. As expected, the titer of IgG decreased in milk over time compared to colostrum since IgG is predominately found and concentrated in colostrum and both immunoglobulin isotypes tend to become more diluted as milk production increases.

The IgA levels are of particular interest as IgA levels are strongly correlated with protection from mucosal pathogens. In this study, piglet mortality was strongly influenced by the level of IgA, implied by results in Figures 4, 5 and 6. Previously exposed sows which had a high IgA titers in colostrum experienced mild clinical signs and had no mortality following challenge with PEDV. In contrast, all pigs from the naïve sows, whether vaccinated or non-vaccinated, succumbed to the effects of PED. Mortality was delayed in the naïve vaccinated group compared to naïve non-vaccinated group (Figure 6); however, mortality was still 100% in both treatment groups in sows previously naïve to PEDV.

Under the parameters of this study, vaccination with Zoetis PEDV vaccine stimulated statistically significant increases in IgG levels in serum of sows and numerical increases in piglets as well as in colostrum and milk from sows irrespective of whether sows were previously exposed to PEDV or were naïve. However, vaccine did not significantly increase IgA levels in previously exposed sows and stimulated very little IgA in naïve sows. In this study, vaccination did not induce sufficient passive immunity in the form of IgA to decrease or eliminate preweaning mortality in vaccinated naïve sows. These findings support the hypothesis that continuing ingestion of fairly high levels of IgA is important for piglet survival when faced with PEDV challenge. Effective PEDV vaccination of sows to protect PEDV challenged piglets in naïve herds will likely require stimulation of higher IgA titers in milk.

References

1. Stevenson, G *et al:* Emergence of porcine epidemic diarrhea virus in the United States: clinical signs, lesions, and viral genomic sequences.

2. United States Department of Agriculture (USDA). Swine Enteric Coronavirus Disease Testing Summary Report. Available at: https://www.aasv.org/PEDV/SECoV_weekly_report_141112.pdf. Nov 24, 2014.

3. I. Shibata, M. ono, M. Mopri, J. vet. Med. Sci./Jpn. Soc. Vet. Sci. 63, 655-658 (2001)

4. Saif, LJ, Pensaert MB, Sestak K, Sang-Geon Y, Kwonil J. Coronaviruses. In: Zimmerman JJ, Karriker LA, Ramirez A, Schwartz KJ, Stevenson GW, eds. *Diseases of Swine.* 10th ed. Hoboken, New Jersey: Wiley-Blackwell Publishing; 2012: 503-514.

5. Data on File, Study Report No. B826R-US-13-258, Zoetis Inc.

PRRS epidemiology: Data from the Swine Health Monitoring Project

R. B. Morrison, DVM, MBA, PhD; C. Betlach BSc; H. Baldry BSc; T. Homann
College of Veterinary Medicine, University of Minnesota

Overview of Swine Health Monitoring Project

The Swine Health Monitoring Project (SHMP) was started in 2011.[1] The target population is sow herds in the United States and we started with a convenience sample of systems and herds. Participation is voluntary and anonymous unless the system chooses to make its participation known publicly by including its logo on the weekly report. First participants provided retrospective data from July 2009 to the date of enrollment and then prospective data thereafter through June 2013. Within each production system, eligible sow herds wean piglets into offsite locations and are requested to test at least 30 piglets of weaning age monthly for PRRSV by means of a PCR assay on the sow farm. All participants are required to sign participation and confidentiality agreements.

Beginning July 1 of each year, cumulative incidence is computed as the proportion of susceptible herds (epidemiological units) with new PRRSV cases. The weekly incidence of new PRRSV infected herds is monitored with an EWMA. The approximated EWMA (smoothed) number of PRRSV infected herds is estimated per week and an epidemic threshold was defined. In this way, we have an objective indicator if and when an epidemic has started and ended for our sample of herds.

During the first 4-year period of this study, weekly incidence was low during spring and summer and high during fall and winter. The EWMA signaled the onset of a PRRSV epidemic during the middle 2 weeks of October each year. Results indicated a striking repeatability in annual PRRSV temporal and spatial patterns across 4 years of data among 371 sow herds from 14 production companies (Figure 1, 2). This suggested that efforts to control PRRSV at a regional level should continue to be supported.

The incidence changed in 2013/14 with significantly fewer PRRSV cases being observed during this study period and the onset of the annual epidemic was delayed approximately 3 wk. Cluster analysis revealed a pattern similar to previous years.[2] This reduction could be due to:

Figure 1: Incidence of PRRS virus as depicted by exponential weighted moving average (EWMA).

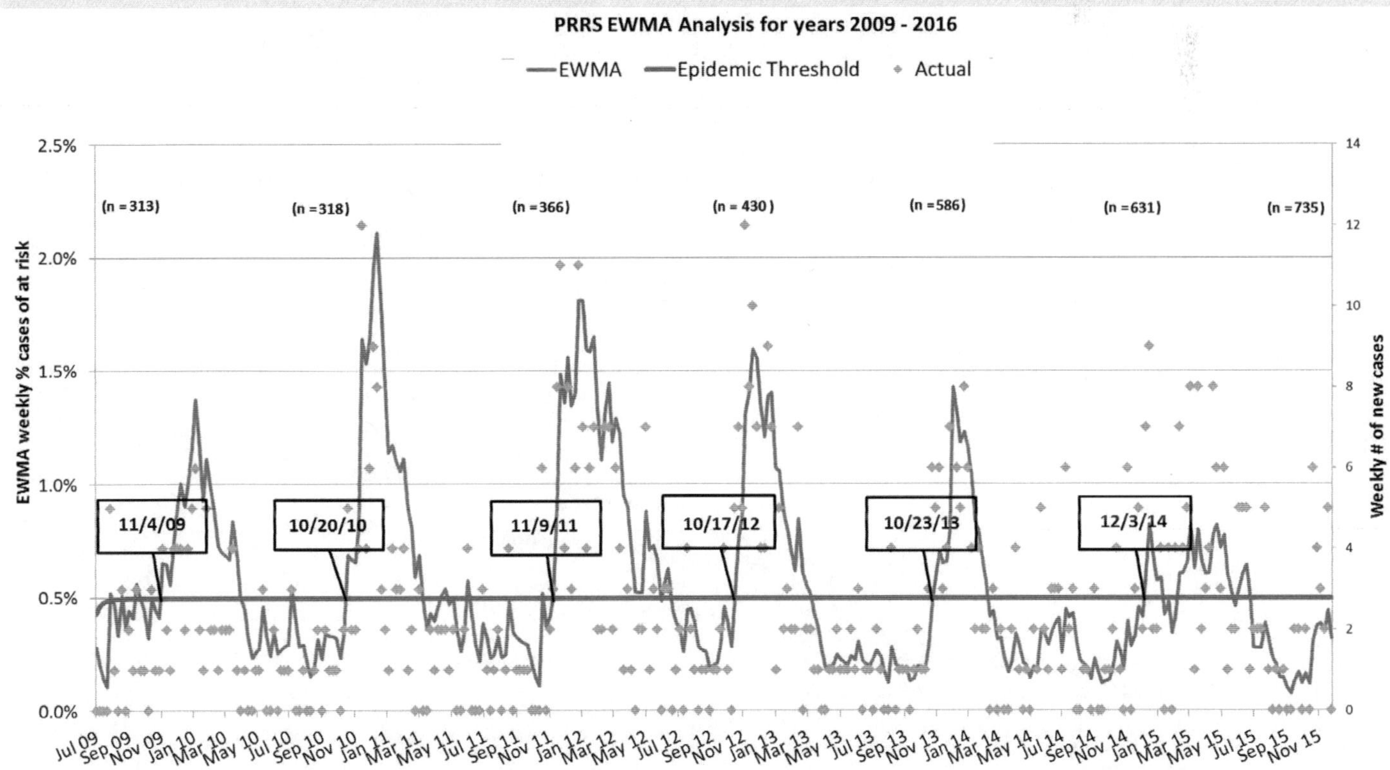

- spurious observations and no real decrease,
- increased awareness of PRRS seasonality and consequently, increased application of biosecurity for PRRSV,
- biosecurity steps to decrease spread of PED virus may have also decreased PRRS virus incidence,
- increased number of herds filtering incoming air, decreased incidence in these filtered herds and therefore less virus in growing pig sites,[3]
- increased use of PRRS virus vaccine which could decrease shedding (Figure 3), and/or
- weather changes.

The SHMP has gradually expanded to include 27 systems composed of 3 regional PRRS projects and 24 companies or coordinated systems. As of today, 735 herds (1.99 million sows) participate for PRRS and 994 (2.53 million sows) report PEDV activity. The PRRS incidence has been similar although was lower again in 2014/15 and promises to be low again in 2015/16. This lower incidence is very encouraging and suggests a fundamental change in biosecurity or risk of PRRSV infection.

Current projects

Production impact - PRRSV is reported to be the single most costly pathogen affecting the North American swine industry by causing reproductive failure and reduced performance in growing pigs. There is considerable variability among infected herds for reproductive impact due to strain virulence, immune status of the sow herd, management interventions and possibly other factors. In one outbreak, we estimated the loss of weaned pigs to be 3.57/inventoried sow with 54.3%, 39.1% and 1.1% of the loss to be caused by reduced born alive, increased prewean mortality and reduced litters born, respectively (plus 5.5% error). We are calculating these losses in a group of herds to gain understanding on the variability among causes of reproductive failure and association with immune status for PRRSV at the time of the outbreak.

Time to stability - The development, publication and adoption of the guidelines for characterizing a sow herd's PRRS status was a breakthrough for understanding and communicating methods for control and elimination of PRRS virus from sow herds.[4] To control and eliminate PRRSV from breeding herds, some veterinarians adopt a strategy called load-close-expose which consists of interrupting replacement pig introduction for several months and may include exposing the pigs to a replicating PRRSV. We performed an important study to determine the time required for sow herds to achieve stability.[5] This was a prospective quasi-experiment that followed 61 breeding herds acutely infected with PRRSV that adopted one of two exposure pro-grams: modified-live virus (MLV) vaccine or live-resident virus inoculation (LVI). Treatment groups (load-close-expose with MLV or LVI) were compared for: (a) time-to-PRRSV stability (TTS), defined as time in weeks it took to produce PRRSV negative pigs at weaning; (b) the time-to-baseline production (TTBP), defined using statistical process control methods to represent time to recover to the number of pigs weaned per week that herds had prior to PRRSV-detection; and (c) the total production loss in terms of number of pigs weaned per week. The median TTS among participating herds was 26.6 weeks. The overall TTBP was 16.5 weeks (range 0-29 weeks). The magnitude of production losses following whole-herd exposure averaged 2217 pigs not

Figure 2: Annual cumulative incidence of PRRS virus

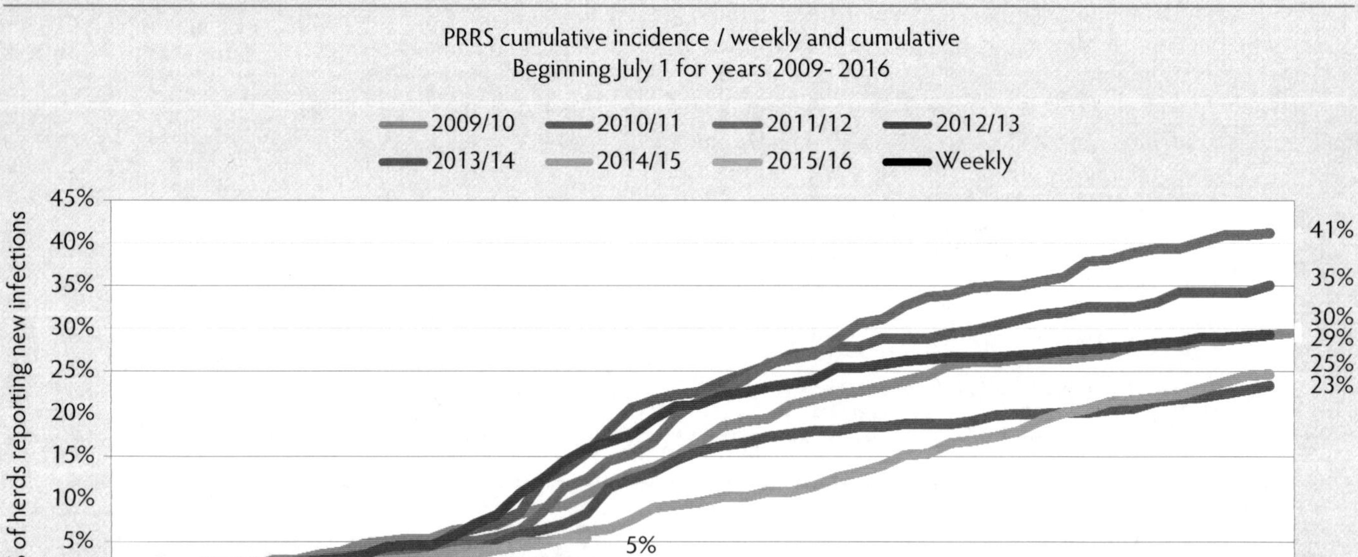

Figure 3: Prevalence of PRRS virus status

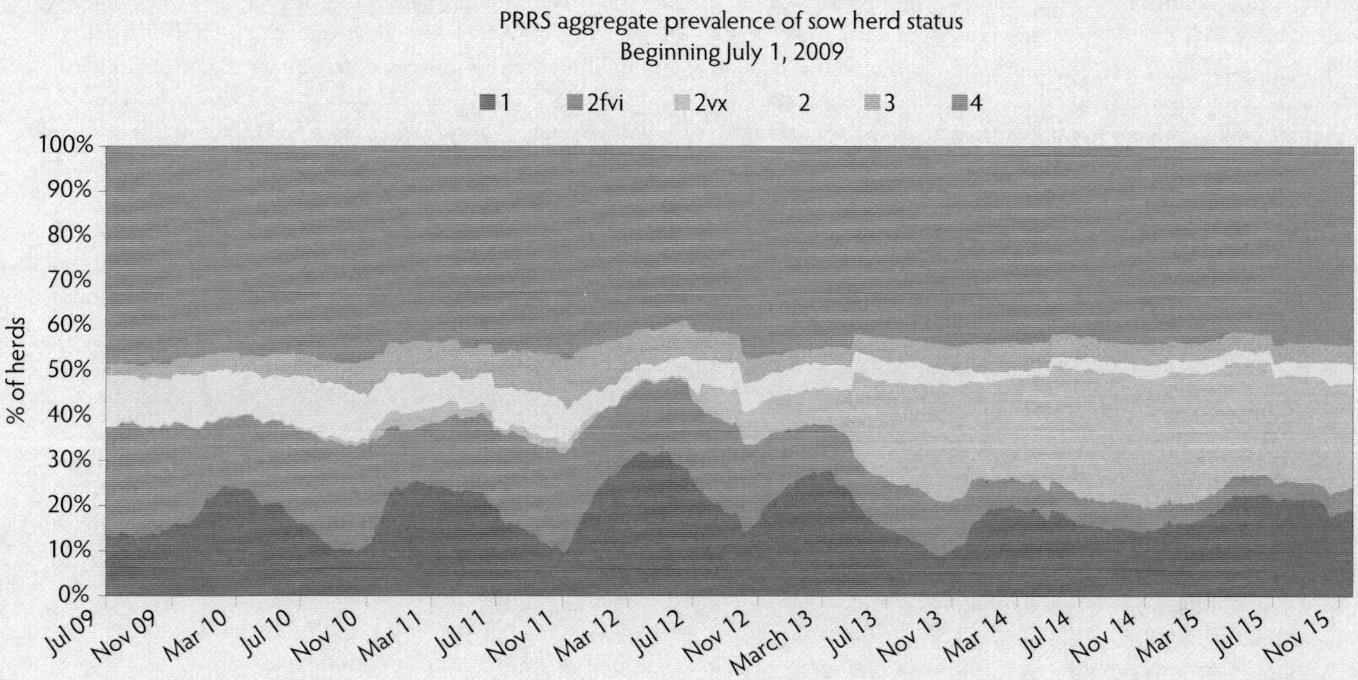

weaned/1000 sows and was correlated with TTBP. Herds in the MLV group recovered production sooner and had less total loss than herds in the LVI group.

The creation of SHMP has given us a database of herds that we can benchmark measures such as TTS in an ongoing fashion. We recently surveyed participants for their testing requirements for establishing stability and learned that most have a testing sensitivity of at least the AASV guidelines; that is, at least 4 consecutive samples of 30 pigs. In the future, we will break out the prevalence chart by systems that adhere to guidelines vs those with a lower standard.

RFLP 1-7-4 - Every few years a new PRRSV strain seems to emerge in the US. PRRSV RFLP 1-7-4, which first began infecting herds in 2014, is one particular strain many veterinarians and producers have claimed to cause severe clinical signs in sow herds. Participants in the Swine Health Monitoring Project (SHMP) were invited to enroll in a study aimed at determining time to stability (TTS), time to baseline production (TTBP) and associated risk factors in sow herds infected with PRRS 1-7-4; a total of 8 participants and 149 sow herds enrolled. Herd inclusion criteria was as follows: farrow-to-wean herd enrolled in SHMP, diagnostic evidence of a PRRSV 1-7-4 infection, no diagnostic evidence that herd was infected with another strain of PRRSV from date of first detection to date of reaching stability, and monthly herd testing of at least 30 pigs starting no later than 12 weeks after last known PRRS virus exposure.

A preliminary analysis of herds (n = 62) was conducted to determine time to first negative test (TT1) and its implications on herd stability. Stability was defined as the date of the last negative test of at least four consecutive tests over a minimum of 90 days post PRRSV detection. An earlier study of 61 herds infected with PRRSV (Linhares et al., 2014) demonstrated that TT1 and TTS were significantly correlated +0.49 ($P = 0.002$). TT1 in the subset of 1-7-4 herds had a median value of 17.5 weeks, 3.9 weeks shorter than that noted in the Linhares et al study ($P = 0.11$) (Figure 1, 2). These results revealed that TT1 was not significantly different from historical data.

To date, 26 of 149 participating 1-7-4 herds have reached stability. From these 26, 20 were selected for analysis based on either electing not to perform whole herd exposure post detection or choosing to expose sows and gilts using live virus vaccine. The frequency distribution of TTS was positively skewed with a median of 29.5 weeks from date of first negative test followed by three consecutive negative tests. This is 9 weeks shorter than historical Linhares et al data. However, the remaining herds will lengthen this TTS duration and diagnostic results will continue to be collected and TTS data will be updated accordingly. In addition, multiple variables will be analyzed to determined potential risk factors of TTS. Phase II of the study will be an analysis of production data to determine time to baseline production.

TTS at farrowing - There is a cost to keeping a herd closed to gilt entries and therefore an incentive to decreasing time for herds to achieve stability. An interesting observation from the Linhares et al study was that the shortest time was just 12 weeks to achieve stability at weaning. Time to stability is composed of 2 time components: time to farrow negative pigs and time to eliminate virus from farrowing area. So, sow herd immunity may play a key role as does management and internal biosecurity. It has been reported that virus can be detected in lymphoid tissues for as long as 225 days.[6] Although it is unknown how commonly this occurs, we did not detect an association between herd size and TTS suggesting that long term persistence may not be a practical concern. We have initiated a study to determine time to stability as measured at farrowing. That is, how long does it take after a herd is infected to farrow PRRSV negative litters? One gestation interval (approximately 16 weeks) would be a conservative time expectation. This assumes that few or no sows are being re-infected. Several researchers have reported that immunity from infection is very good at preventing re-infection with the same virus suggesting that re-infection may be unusual during the closure period.

Once a herd is no longer producing viremic pigs at birth, we need to reduce transmission during lactation. Time for all virus transmission to stop during lactation will be a function of the biosecurity protocol and ability of management and staff to adhere to it and sow herd immunity (which will influence shedding and susceptibility to infection). Once all pigs are virus negative at farrowing, an aggressive biosecurity protocol is indicated. This may even include a wean-down to 7 days in an attempt to eliminate all virus from farrowing.

Proposed projects

Real-time risk scores – We propose to create a dynamic system of risk characterization for PRRS. To accomplish this, we will tie pig movement data to source herd PRRS status and location and estimate PRRSV risk of such movement. A multivariable logistic regression model will be used to quantify the association between variables in the databases (locations, time, strains) and occurrence of outbreaks. The model will be fitted and validated using prospective data and, subsequently, be used to estimate prospectively and on a routine basis the probability that a PRRS outbreak occurs on a given farm. The product of this project will be an algorithm to dynamically estimate risk for PRRS outbreaks in participating farms and systems. Such a system may be an aid to planning pig movement and thereby reduce area spread of the virus.

Detect emerging strains of PRRS - We have proposed to conduct prospective monitoring of PRRS virus sequence evolution and impact. To accomplish this, we will capture diagnostic data from cooperating diagnostic labs, production data from producers and develop analytic methods. Our goal is to detect emerging strains of PRRSV and give producers the opportunity to respond to emerging, highly virulent strains. We propose to initiate this objective with a convenience sample of participants' herds. We will use AgConnect to capture diagnostic data, specifically PRRSV sequences, from diagnostic labs at UMN, ISU and SDSU. As our methods are developed and refined, we intend to expand this objective to include all participants' sequence data. And as potential emerging strains of PRRSV are detected, we will solicit production data from affected farms to ascertain the apparent virulence of the strain. Herd owners will be informed that their PRRS virus appears to be an emerging, virulent strain.

Acknowledgements

- SHMP participating producers and veterinarians who share their health status data
- SHMP Advisory Group - The following individuals provide input as needed and requested for the SHMP project. Their role is valuable and appreciated. Paul Yeske, Matthew Turner, Joe Connor, Gordon Spronk, Luc Dufresne, Clayton Johnson; Paul Sundberg (ad hoc).
- Funding
  - Swine Health Information Center
  - Boehringer Ingelheim Vetmedica for Linhares et al TTS study and proposed study for stability at farrowing

References

1. SJP Tousignant et al. Temporal and spatial dynamics of porcine reproductive and respiratory syndrome virus infection in the United States. *Am J Vet Res* 2015; 76:70-76.

2. SJP Tousignant et al. Comparison between the 2013-2014 and 2009–2012 annual porcine reproductive and respiratory syndrome virus epidemics in a cohort of sow herds in the United States. Can Vet J 2015;56: 1087-1089

3. C Alonso et al. Epidemiological study of air filtration systems for preventing PRRSV infection in large sow herds. Preventive veterinary medicine. , 2013, Vol.112(1), p.109-117.

4. D. Holtkamp et al. Terminology for classifying swine herds by porcine reproductive and respiratory syndrome virus status. J Swine Health Prod. 2011;19(1):44-56.

5. D Linhares et al. Comparison of time to PRRSV-stability and production losses between two exposure programs to control PRRSV in sow herds. Prev Vet Med 116 (2016) 111-119.

6. Wills et al. Duration of Infection and Proportion of Pigs Persistently Infected with Porcine Reproductive and Respiratory Syndrome Virus. J Clin Microbiol, Jan. 2003, p. 58-62.

PRRSV: Emerging isolates

Chad Smith, DVM
Maxwell Foods, LLC. Goldsboro, North Carolina

Introduction

During the spring of 2014, a new PRRSV strain was confirmed in eastern NC and subsequently spread throughout the region. The first Maxwell Foods' sow farm became infected in May of 2014. By May of 2015, 90% of our sow herds, 98% of nursery sites, and 99% of our finishing sites had been confirmed positive with 1-7-4 PRRSV at some point during the previous year. This pattern of spread was nearly identical to PEDV spread that occurred beginning in the summer of 2013. The initial sow farms identified in both PRRSV 1-7-4 and PEDV were the same sites, with additional breaks kept to a minimum until the winter months. Both viruses spread uncontrollably in our system throughout February to April. Because of the limited number of farms and the confounding variables always present in commercial swine production, no statistical analysis was conducted on data from our system. All questions, findings, observations, and unfortunately, more questions are all anecdotal in nature.

1-7-4 PRRSV spread

The percentage and number of farms that were considered positive for 1-7-4 are illustrated in Figures 1 and 2, respectively. The first sow farms were diagnosed in May of 2014 with the next sow farms diagnosed nearly six months later in November. Two nurseries and one finishing site were confirmed positive through lateral spread during this time period. Lateral spread to nurseries and finishing sites was not followed after this point due to the increasing number of farms that were becoming infected and the inability to manage the diagnostics and data. All growing pig sites considered positive after this point are due strictly to movement of infected pigs.

At first glance of figures 1 and 2, it appears nurseries are somewhat responsive and finishing statuses follow changes in the number of sow farms becoming infected. This is somewhat logical as an increased number of positive sow farms will produce more piglets that must be moved to nursery and finishing sites. Sow farms would be considered the point sources of virus for the system. However, the five sow farm breaks that occurred between November and the first of February were geographically close to nursery and finishing sites positive from pig movements. The average distance was 1.7 miles with a minimum and maximum of 0.5 and 3.0 miles, respectively. It is unknown if these sow farms would have become infected without the presence of positive nursery and finishing sites nearby. A video of the progression of 1-7-4 exposure by site will be shown during the presentation, but it appears to indicate once the virus was present in the system, positive nursery and finishing sites may have been the source of virus for subsequent sow farm breaks and further virus propogation.

Dendrogram analysis

Beginning May 2014, Maxwell Foods began sampling 30 poor-doing piglets per month from every sow farm. One PRRSV ORF 5 sequence was obtained from each sampling event per farm. As illustrated in Figure 3, up to 20% system-wide divergence was seen initially, with some individual farms maintaining a "resident" strain, and others with multiple strains of virus. Once farms became exposed to 1-7-4 PRRSV, system-wide divergence was reduced to less than 2% over a year's monitoring (Figure 4). At the time of writing, no sequences other than 1-7-4 have been recovered from any farm in the system, after the farm broke with 1-7-4. Due to the limitations of sampling strategy and utilization of Sanger sequencing[3], it is not possible to determine if other viruses/sequences are present within samples or the system.

Management interventions and time to stability (TTS)

Throughout the spread of 1-7-4 PRRSV, management interventions to improve the health and TTS of individual farms varied greatly. Herd closures, no pig movement, "batch" farrowing, and no interventions were all employed. At the time of writing, the farms where no interventions were employed appear to have stabilized

Figure 1: Percentage of PRRSV 1-7-4 positive farms by stage of production and month of infection. Cumulative % from April 2014 thru September 2015.

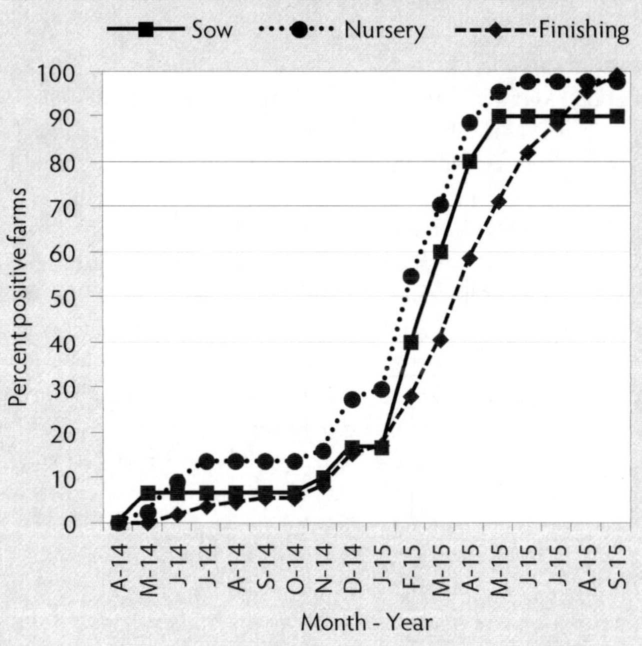

Figure 2: Number of PRRSV 1-7-4 positive farms by stage of production and month of infection. Cumulative count from April 2014 thru September 2015.

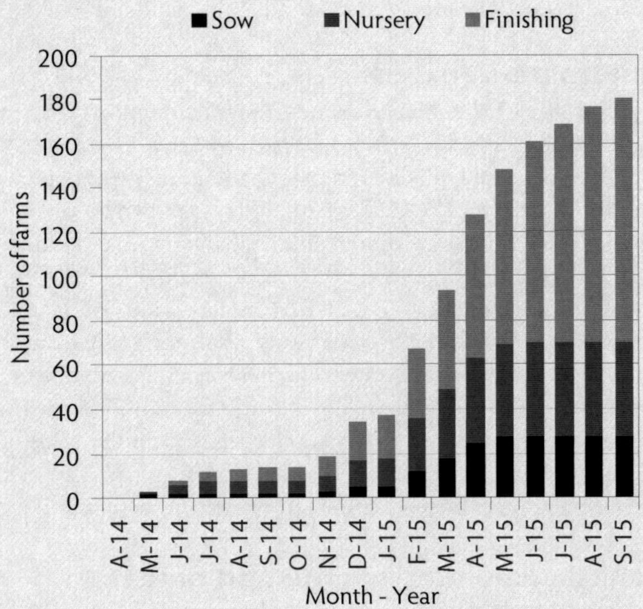

Figure 3: Neighbor-Joining[2] dendrogram constructed in MEGA[1] of PRRSV ORF5 samples from May 2014 thru September 2015. PRRSV 1-7-4 sequences are collapsed to allow visualization of non-1-7-4 terminal nodes.

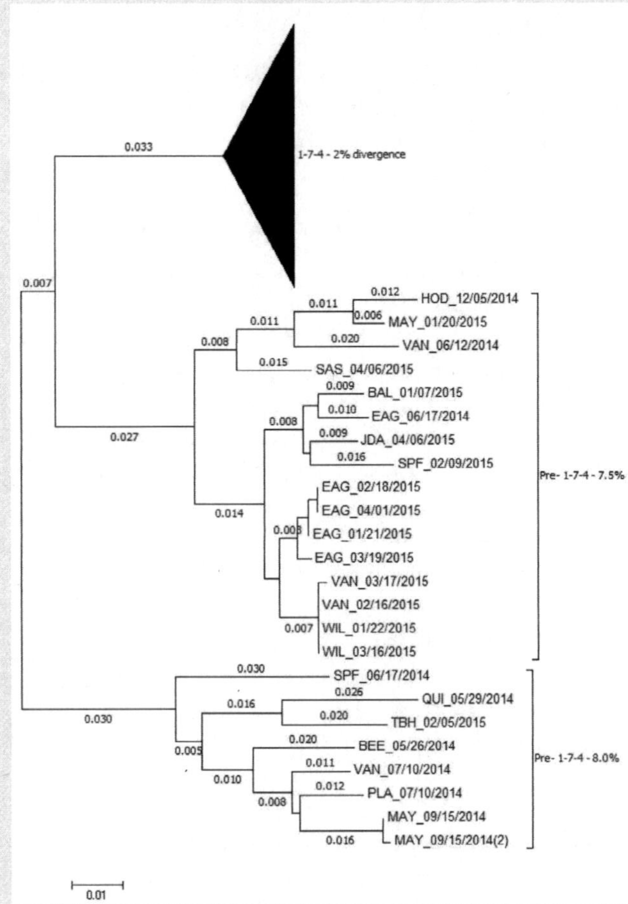

quicker than any of the other sites. However, there may be a seasonal effect to when the herds broke responsible for this observation. Table 1 summarizes the intervention strategies and the time when the first of three negative sampling events occurred. Of the 27 sow farms that broke, only 22 are evaluated. Five sites were either part of a parity segregation system or previously planned depopulation that would be misleading and are therefore excluded. Three sites continue to have positive samples and clinical signs while the majority of the sites appear to be stabilizing.

Seasonal effect

Generally, Maxwell Foods management considers 16-20 weeks or one "turn" of the sow reproductive cycle to be a near magical time period, and usually the minimum amount of time required for any changes in production or health statuses to take full effect. Once the sow inventory has been exposed to an agent with subsequent development of immunity and virus elimination from the individuals, it must be removed from the environment and preventing spread to newborn piglets or new gilt introductions. The sow farms in our system that broke at the end of February through March appear to have achieved stability in a numerically shorter time frame despite no management interventions. Potentially, this is due to the added heat and humidity of NC summers, particularly July and August, reducing the amount of time PRRSV may survive in the environment.

Herd closure

Loading and closing herds to new animal introductions was attempted on farms 1, 2, and 4. Whole herd exposure after loading was not employed. Due to the facility design, a maximum of 14-16 weeks of gilts may be held at each site. After those gilts are exhausted, decisions were made regarding whether to keep the herd closed and miss breeding targets, or attempt another closure. These decisions were heavily influenced by the PCR results from wean pig monitoring programs. Farms 1 and 2 each had three attempts at herd closure, none resulting in stability. Farm 1 has yet to stabilize, and farm 2 stabilized after implementation of a batch farrowing system described below. Farm 4 was deemed close to stability, and that herd remained closed until piglets had three consecutive negative sampling events. Additionally, the farrowing house staff voluntarily implemented rules regarding zero cross-fostering of piglets between sows for the last 10 weeks of the closure period.

Figure 4: Neighbor-Joining[2] dendrogram constructed in MEGA[1] of PRRSV ORF5 samples from May 2014 thru September 2015. Non-PRRSV 1-7-4 sequences are collapsed to allow visualization of 1-7-4 terminal nodes.

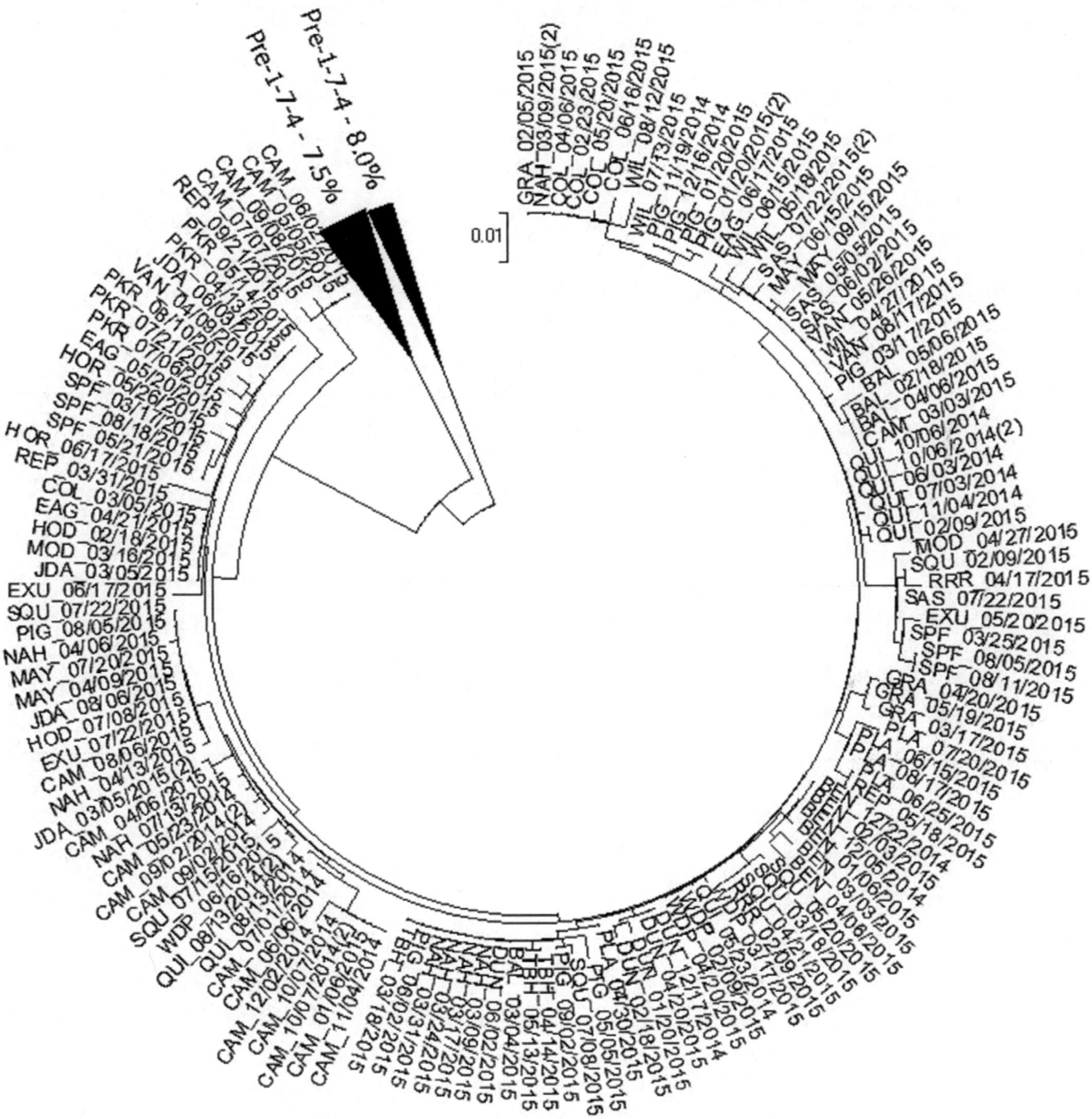

Table 1: PRRSV 1-7-4 sow farms, time to first negative test (TT1), current number of negative sampling events, and intervention strategies by farm.

Farm	Positive	TT1 month	TT1 days	20 wks post-positive date	#Neg. samples	Intervention
1*	May-14			October 10, 2014	1	Herd closed, no pig movement
2	May-14	Apr-15	313	October 10, 2014	4	Herd closed, batch
3	Nov-14	Oct-15	316	April 8, 2015	2	Batch
4	Dec-14	May-15	147	April 24, 2015	4	Herd closed, no pig movement
5	Dec-14	Jul-15	201	May 6, 2015	4	Batch
6	Feb-15	Aug-15	203	June 29, 2015	3	None
7	Feb-15	Aug-15	175	June 29, 2015	3	None
8	Feb-15	Aug-15	189	June 29, 2015	3	None
9	Feb-15	Aug-15	166	July 8, 2015	4	None
10	Feb-15	Jul-15	147	July 13, 2015	3	None
11	Mar-15	Jul-15	137	July 23, 2015	2	None
12	Mar-15	Aug-15	147	August 3, 2015	3	None
13	Mar-15	Jun-15	75	August 5, 2015	4	None
14	Mar-15	Jul-15	97	August 18, 2015	4	None
15*	Apr-15			August 27, 2015		None
16*	Apr-15			August 27, 2015		None
17	Apr-15			August 31, 2015	1	None
18	Apr-15			September 14, 2015	1	None
19	Apr-15			September 17, 2015	2	None
20	Apr-15			September 24, 2015	1	None
21	May-15			October 7, 2015	2	None
22	May-15			October 13, 2015	1	None

* Continues to test positive

Batch farrowing

At the time of infection, farms 3 and 5 were participating in a 10-2 or biweekly farrowing program. The goal of the program was to double the size of the cohort group of wean pigs from each farm to allow filling of nursery buildings/sites with single aged, single source piglets. Clinical signs of PRRSV resolved relatively quickly in these herds, but piglets continued to test positive longer than farms that were infected later in the year. After farm 2 failed to stabilize with herd closures, sows were moved within the farm to allow ½ of the farm to farrow on even weeks and ½ of the farm to farrow on odd weeks. Within 45 days of this intervention, piglets began testing negative and four negative tests have finally been obtained for this site.

No intervention

Due to the lack of response from the first farms that broke with 1-7-4, and the frequency of which farms began to break in February 2015, no interventions were done on farms 6 thru 22. A gentle reminder was given to farm managers regarding limited cross-fostering, but gilt shipments continued on their regular schedule to each farm. At the time of writing, it appears the farms we did not attempt interventions at are stabilizing quicker. We are cautiously optimistic that this is due to the potential seasonal effect of when the herds were infected and not a result of our interventions.

Results and discussion

PRRSV 1-7-4 spread throughout our region and the Maxwell system in a relatively quick fashion. Biosecurity measures appear to have limited the spread until the winter months unless we placed infected pigs in close proximity to a sow herd. During February, farms became infected quickly and subsequent pig movement worsened the problem. The 1-7-4 strain appears to be overwhelming the diagnostics/sequencing results and no historical or new PRRSV strains have been found since 1-7-4 infection. Management interventions to decrease the TTS appeared unsuccessful as employed, however a numerical trend indicated that there was a seasonal effect to TTS based on infection dates. The spread of PRRSV 1-7-4 was similar to the pattern observed with PEDV. Once sites are infected in our system/region with a "new" organism, we appear to have little control over preventing its spread, specifically through the winter months.

References

1. Koichiro Tamura, Glen Stecher, Daniel Peterson, Alan Filipski, and Sudhir Kumar (2013) MEGA6: Molecular Evolutionary Genetics Analysis version 6.0. Molecular Biology and Evolution:30 2725-2729.

2. Saitou N & Nei M (1987) The neighbor-joining method: A new method for reconstructing phylogenetic trees. Molecular Biology and Evolution. 4:406-425.

3.Sanger F, Coulson AR (May 1975). A rapid method for determining sequences in DNA by primed synthesis with DNA polymerase. Journal of Molecular Biology. 94 (3): 441–8.

Mycoplasma hyopneumoniae elimination

P. E. Yeske, DVM, MS
Swine Vet Center Saint Peter, Minnesota

Introduction

Mycoplasma hyopneumoniae (*M. hyopneumoniae*) continues to be one of the most prevalent and economically significant respiratory pathogens in the swine industry.[1,2] *M. hyopneumoniae* is the etiologic agent of enzootic pneumonia, a chronic respiratory disease in swine characterized by a chronic, non-productive cough.[3,4] Economic losses related to *M. hyopneumoniae* are associated with decreased feed efficiency, reduced average daily gain, and increased medication costs.[1] (Table 1) Details the differences in performance average daily gain, feed efficiency, percent mortality, percent marketed, feed grade medication and other medication costs in Mycoplasma negative pigs vs positive pigs (all pigs are similar genetics and nutrition).

One of the industries challenges is to enter mycoplasma negative gilts into positive herds. Today almost all genetic suppliers are *Mycoplasma hyopneumoniae* negative production. This has affected the stability of many positive farms resulting in an increase in colonization and more problems in the finishing phase. The difficulty of exposing *M. hyopneumoniae* negative gilts early enough to develop good immunity and not be shedding high numbers by the time they get to farrowing continues to be a part of the problem. Figure 1 shows the time required to get gilts exposed and immune and reduced shedding before farrowing. Recent research work has shown that to infect gilts need to have at least 60% of the gilts shedding to effectively get the negative gilts exposed.[5] This has been a factor in frustration of being able to get this done on a consistent basis has pushed some farms to elimination just to not have to do this and not have the down stream problems.

The economics of this disease had been the primary driver for producers to look at elimination. Many herds have been able to stay negative for an extended period of time. This has also encouraged more herds to move forward with elimination when looking at the return on investments and the amount of time that herds have been able to stay negative. (Table 2) Outlines the success of elimination programs and the amount of time that they have been able to stay negative.

M. hyopneumoniae elimination protocols

If elimination of *M. hyopneumoniae* from a herd or flow is desired, *M. hyopneumoniae* elimination protocols can be implemented. The following protocols for *M. hyopneumoniae* elimination have been described: 1.) Depopulation and repopulation, 2.) Herd closure and medication, 3.) Whole herd medication without herd closure and 4.) Change of flow in a parity segregated flow.

Depopulation and repopulation is the most direct approach for *M. hyopneumoniae* elimination, as it involves removal of the entire breeding herd and restocking with *M. hyopneumoniae* negative replacements.[6] Advantages of depopulation and repopulation include the ability to eliminate more than one disease at once and the opportunity to improve genetics;[6] however, there is a complete loss of production from the time the breeding herd is liquidated until replacement females begin farrowing. The duration of lost production can be reduced with an offsite breeding project, but is associated with additional costs. Furthermore, total depopulation of the breeding herd may be undesirable on farms with animals that have a high genetic potential (i.e., genetic nucleus or multiplier farms). If herds have had problems with parity structure this also offers an opportunity to get the herd started back over.

Herd closure and medication approach has been adapted from the Swiss model for *M. hyopneumoniae* and from herd closures done for PRRS control and many times in combination with

Table 1: Performance differences between *M. hyopneumoniae* positive and negative groups from 2008-2014.

Per 1000 sows	Myco (+)	Myco (-)	Difference
Finishing mortality	3.8%	3.0%	-0.8%
Finishing culls (underweight MKT)	1.9%	1.6%	-0.3%
Total pigs sold	27081	27406	325
Cost of treatments	$ 3.60	$ 2.39	$ (1.21)
Finish ADG	1.72	1.80	$ 0.08
F/G	2.75	2.73	$ (0.02)
Total cost per pig marketed			**$ 4.44**
Total cost per sow			**$ 121.79**

Figure 1: Time line for negative gilt exposure prior to farrowing in a positive herd to remain stable.

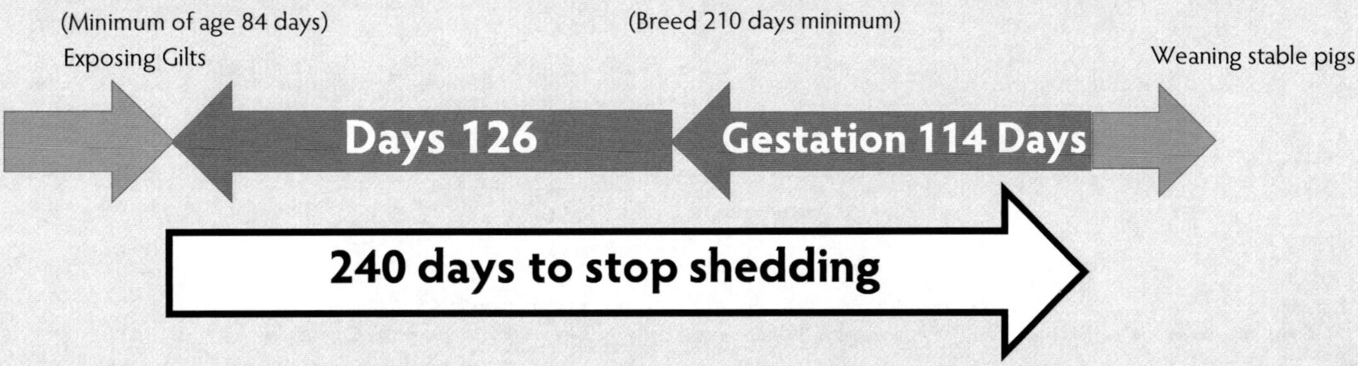

Table 2: Success of *Mycoplasma hyopneumoniae* eliminations.

	Herd Closure	Medication	Total
Number of sows	130088	47450	177638
Number of herds	46	20	66
Percent negative at 1 year	98%	58%	86%
Percent negative to date	89%	53%	80%
Herds negative	40	10	51
Average months negative	44	32	40
Average herd size	2828	2373	2691
Years negative	3.6	2.7	3.4

PRRS elimination from the herd as well.[2] The basis of this procedure is to close the herd once all replacement animals have been exposed to *M. hyopneumoniae* and not make any additions for 240 days.[2] Work done by Pijoan and Peters demonstrated that this was the time that it took for animals to no longer be shedding following natural infection.[7] Using this information gilts are accumulated for the closure into the herd when possible or with an offsite breeding project. For gilts that will be added to the herd in this time they must be infected with *M hyopneumoniae* to start the clock ticking. This can take 1-2 months and is a critical step in the process to ensure success. Once this is completed then herd immunity is boostered every 90 days with whole herd vaccination. At 7 weeks before the negative replacements are scheduled to arrive the whole herd is medicated (sows and piglets). Although different combinations have been done the most common method had been to use Lincomycin in the water for the sows and injection of Draxxin® to the piglets at birth and at 14 days to piglets. (Figure 2)[2] A detailed time line of herd activates including the offsite breeding project. (Table 3) Details a check list that can be used to make sure that all activities are being completed in the process and following the timeline. Herds have had extended closures over 13-14 months and been able to achieve negative status with no medication just longer closures.

Whole herd medication is another method that allows for much faster elimination from the herd but has a lower rate of success. This procedure does not involve a specific herd closure but is generally done when gilts have just entered the farm. Treating all animals on site and then coming back 2 weeks later to treat all sows again. Treating piglets that are born for 4 weeks at birth and 14 days of age with Draxxin® starting at time of first medication. Waiting until all treated piglets are weaned off site and then bringing negative replacements back into the herd.

Combinations of herd closure and whole herd medication have been done in cases where closure for a full 240 days is not possible. Having a closure period before starting the whole herd medication program appears to have increased the odds for success of the medication program. This has varied in closure time frames but it is best to be as long as possible to push towards greater success.

Parity segregated flow offers a unique opportunity to eliminate *M. hyopneumoniae* in taking advantage of the flow to allow for immune sows. Since the older parity sows should not be shedding and not have exposure to younger sows that are shedding, these sows should be negative and allow for flowing the negative replacement animals into these herds, and then flow the system

Figure 2: Time of events in a herd closure for *Mycoplasma hyopneumoniae* including an offsite breeding project.

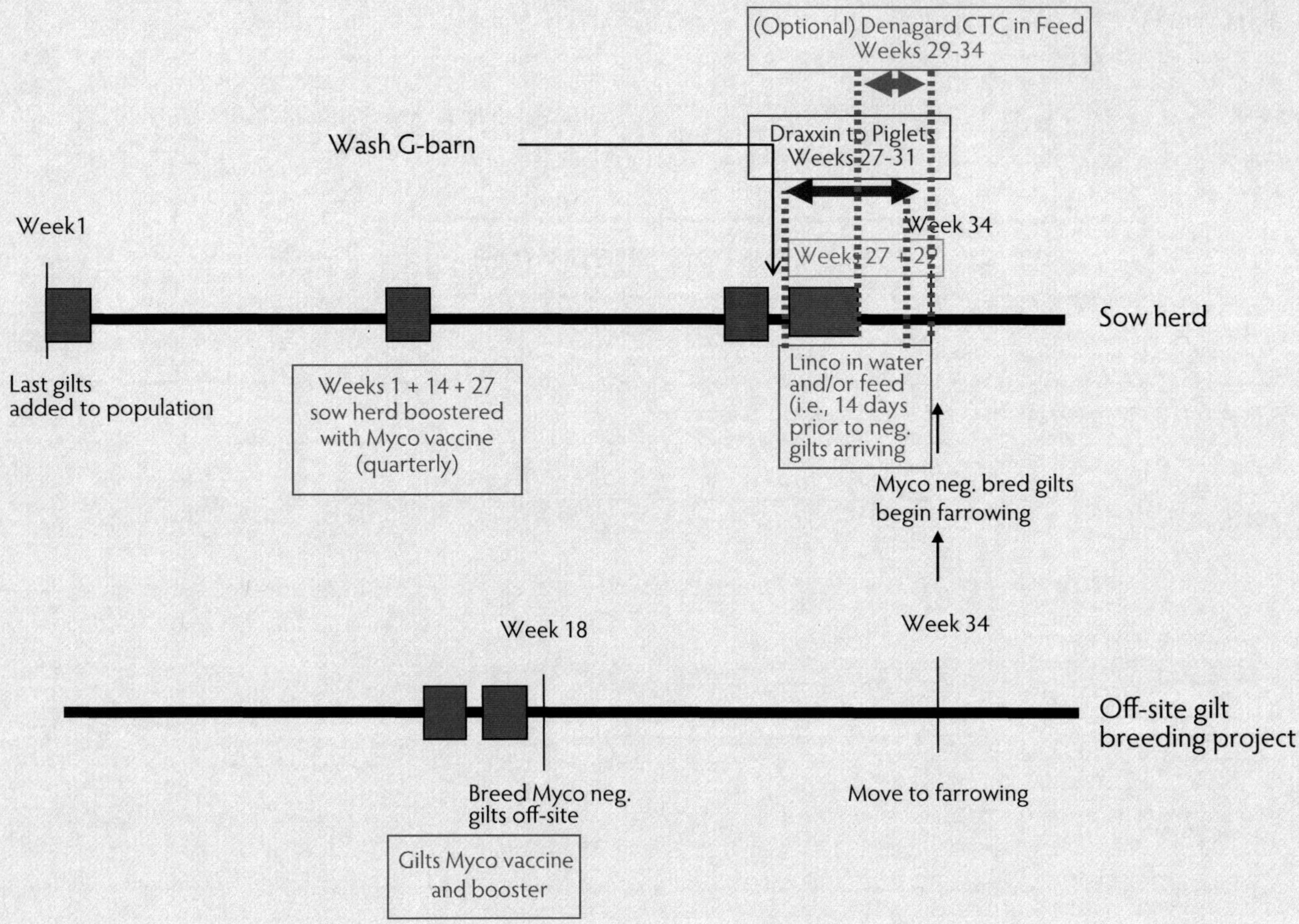

backwards (following older immune sows to parity 1 site) until 240 days have passed on the last positive gilts introduced into the parity 1 herd. Once this is completed the flow can go back to normal. Some herds using this approach have used medication along with the change of flow to drive the odds to higher success.

Table 2 details the success of these methods and the period of time that these herds have been able to stay negative on average. The longest herd in this data base has stayed negative for 12 years following elimination. These procedures have been done in farrow to wean farms of various sizes with the largest herd being 8000 sows. A spreadsheet has been developed to document the cost of *M hyopneumoniae* in a herd as well as the various procedures to eliminate it and how to calculate return on investment.[8,9] This tool has been very useful in detailing multiple scenarios for owners as they look at the possibilities.

Discussion

Mycoplasma hyopneumoniae continues to be a problem for the industry. With essentially all negative replacement stock it is more and more difficult to keep positive herds stable, resulting in more problems in the grow finish population and more cost of disease. There are many different ways to eliminate *M hyopneumoniae* from herds; every herd will be different with the goals for the elimination plan and the risk levels that the owner is willing to take. Hopefully with this presentation one can see some of the values for elimination and different methods to do it. The successes of herds that have followed these procedures is outlined below. Every herd owner will need to decide if there is enough return for doing one of these projects but hopefully with this information they along with the herd veterinarians can make a more informed decision on how to move forward for their specific herd.

Table 3: Checklist of activities for *Mycoplasma hyopneumoniae*.

Mycoplasma Eradication Herd Closure Protocol

Farm		
Date		

Date	WK	Activity	Product	Dosage	Withdrawal	Completion date	INT
5/12/2015	20	Herd closure	N/A	N/A	N/A		
5/12/2015	20	Myco exposure complete (confirmed)	N/A	N/A	N/A		
5/13/2015	20	Whole herd vaccination (sow unit & GDU)	Respisure 1	2 cc	21 days		
8/10/2015	33	Whole herd vaccination (sow unit & GDU)	Respisure 1	2 cc	21 days		
9/4/2015	36	Vaccinate gilts at off-site breeding project	Respisure 1	2 cc	21 days		
9/18/2015	38	Vaccinate gilts at off-site breeding project	Respisure 1	2 cc	21 days		
10/2/2015	40	Begin off-site breeding project	N/A	N/A	N/A		
11/5/2015	45	Wash the gestation	N/A	N/A	N/A		
11/8/2015	46	Whole herd vaccination(sow unit & GDU)	Respisure 1	2 cc	21 days		
5/24/2015	22	Rope test gilts in GDU test all pens and keep track of pens so that they can be retested	N/A	N/A	N/A		
11/19/2015	47	Whole sow herd medication Draxxin	Draxxin	1cc per 88 ibs	33 days		
11/19/2015	47	Piglet treatment all piglets on site (larger pigs will need to adjust dose by weight)	Draxxin	1cc per pig Diluted to 1cc 4.4.lbs	33 days		
11/20/2015	47	Piglet treatment at birth	Draxxin	1cc per pig Diluted to 1cc 4.4.lbs	33 days		
12/3/2015	49	Whole sow herd medication Draxxin	Draxxin	1cc per 88 ibs	33 days		
12/3/2015	49	Begin Denagard CTC in feed lactation and gestation	Denagard CTC	1600 gms ton CTC 35 gms ton Denagard	14 days		
12/3/2015	49	Begin piglet treatment at 14 days of age (treat all piglets that are 14 days and older)	Draxxin	2cc per pig Diluted to 1cc 4.4.lbs	33 days		
12/17/2015	51	Last piglets treated at birth	N/A	N/A	N/A		
12/31/2015	53	Last Piglet treatment at 14 days of age	N/A	N/A	N/A		
1/7/2016	2	Elimination completed	N/A	N/A	N/A		
1/8/2016	2	Negative gilts enter breeding herd	N/A	N/A	N/A		
1/8/2016	2	Date pigs are expected negative	N/A	N/A	N/A		
240		Days since the herd was confirmed (+)					
241		Days since closure					
268		Last gilts birth date	4/15/2015				
304		Gilts 10 months of age	2/13/2016				

References

1. Maes D, Segales J, Meyns T, Sibila M, Pieters M, Haesebrouck F. Control of *Mycoplasma hyopneumoniae* infections in pigs. Rev. *Vet Microbiol.* 2008;126(4):297-309.

2. Holst S, Yeske P, Pieters M Herd closure and medication protocols for elimination of *M hyopneumoniae* Journal of Swine Health & Production 2015; Volume 23, Number 6: 321-330.

3. Mare C, Switzer W. New species: *Mycoplasma hyopneumoniae*. A causative agent of virus pig pneumonia. *Vet Med*. 1965;60:841-846

4. Goodwin RFW, Pomeroy AP, Whittlestone P. Production of enzootic pneumonia in pigs with a mycoplasma. *Vet Rec*. 1965;77:1247-1249.

5. Roos L. M. hyopneumoniae: Investigating the gilt acclimatization; Allen D Leman conference 2015

6. Yeske P. Mycoplasma eradication strategies. *Proc AASV.* Orlando, Florida. 2007;367-370.

7. Pieters M, Pijoan C, Fano E, Dee S. An assessment of the duration of *Mycoplasma hyopneumoniae* infection in an experimentally infected population of pigs. *Vet Microbiol.* 2009;134(3-4):261-266.

8. Yeske P. Cost of eradicating diseases according to method. *Proc AASV.* Omaha, Nebraska. 2010;15-18.

9. Yeske P. Economic impact of *mycoplasma hyopneumoniae* eliminations. Proc 23rd IPVS Mexico 2014: 336.

Gilt acclimation options for *Mycoplasma hyopneumoniae* infected breeding herds

M. Turner, DVM
Prestage Farms, Clinton, North Carolina

Introduction

Mycoplasma hyopneumoniae (*Mhp*) is widely recognized as an economically significant bacterial pathogen of growing pigs, is the primary pathogen of enzootic pneumonia, and an important player in Porcine Respiratory Disease Complex (PRDC).[1] Prevalence of infection with *Mycoplasma hyopneumoniae* is very high worldwide where pigs are commercially produced and most pigs are considered infected unless specific programs were put in place to eliminate the organism. *Mycoplasma hyopneumoniae* infection and clinical disease in growing pigs sourced from *Mhp* positive breeding herds is closely tied to the shedding status of the dam and her propensity to shed *Mhp* and infect her offspring during lactation.[2] Specifically, primiparous sows are more likely to infect their offspring than are multiparous sows. Additionally, it is clear that replacement breeding animals can shed *Mhp* and infect contacts for at least 200 days after infection.[3] Successful attempts to eliminate *Mycoplasma hyopneumoniae* from breeding herds are well documented through various protocols involving combinations of antimicrobial therapy, vaccination, herd closure, depopulation, or partial depopulation.[4] The concept of maintaining a breeding population free of *Mycoplasma hyopneumoniae* and thus producing growing pigs free of enzootic pneumonia is ideal and should be accomplished wherever it is feasible. *Mycoplasma hyopneumoniae* naïve breeding replacements are commonly available today and the utilization of these naïve replacement breeding animals has been both a value and a challenge. *Mhp* naïve gilts used in established *Mhp* infected breeding farms has sometimes led to increased colonization of the growing pig population, poor gilt reproductive performance and poor gilt retention due to chronic respiratory disease. Aerosol transmission of *Mycoplasma hyopneumoniae* is well documented and proximity to infected farms is a significant risk factor for infection or reinfection.[5,6] Viable *Mhp* has been documented out to at least 9.2km from an infected source farm.[7] Unfortunately, not all breeding herds are good candidates for elimination due to the risk of re-infection. Knowing that some breeding farms will be required to maintain a positive stable status due to infection risk, the goal for these farms must be to reduce the rate of colonization with *Mhp* as low as possible during lactation so the growing pig population is minimally infected. A critical component of reducing piglet colonization is proper, early gilt acclimation.

Experiences with *Mhp* instability and acclimation

Case 1

A total of seven farrow to wean farms were populated PRRSV and *Mhp* naïve between 2001 and 2005. Two of these farms were located 5or more miles from other farms, while five had other commercial pigs within 2.5 miles. The two remote farms remain free of PRRSV and *Mhp*, but the five farms with commercial pigs in the region all became infected with PRRSV and *Mhp* within one year. Several iterations of elimination for PRRSV and mycoplasma were attempted but re-infection was too frequent to manage and eventually the decision was made to keep the high risk farms positive for both PRRSV and *Mhp*. Replacement breeding animals continue to be naïve for *Mhp*, leading to episodes of coughing gilts at entry to the breeding barn and episodes of coughing pigs in the nursery and early finisher (confirmed as *Mycoplasma hyopneumoniae* pneumonia). The pattern of gilt respiratory disease was undulating seasonally with some groups of gilts appearing not to be infected until well into gestation while others became sick within weeks of entry. Vaccination of the breeding herd to reduce shedding, medication pulses, and early vaccination of piglets with *Mhp* vaccine was not very rewarding, piglet medication programs helped but ultimately could not achieve the desired results.

The primary problem with this scenario was the fact that gilts were not consistently infected with *Mhp* which led to both gilt clinical disease and high levels of weaned pig colonization- resulting in downstream growing pig disease. Attempts were made to move shedding gilts back to the growing gilt pool to maintain circulation, but lack of on-site growing time and lack of predictable infection resulted in poor control of *Mhp*. Eventually it was decided to attempt a controlled exposure of *Mhp* in the replacement gilt flow and maintain that exposure naturally through flow modifications.

Old Flow (Figure 1)- Gilts were not adequately exposed to *Mhp* before breeding. Numbers represent total age of gilts.

New Flow (Figure 2)- Gilts are infected with *Mhp* in an acclimation site and circulation is maintained by introduction of susceptible animals and direct contact with a *Mhp* shedding population. Serology (Idexx M hyo ELISA Westbrook, ME) is utilized to confirm *Mhp* exposure in the acclimation site. The barrows from this flow are reared in commercial finishers and serve as a comparison to the gilts (Figure 3).

Figure 1: M hyo non-acclimated gilt flow

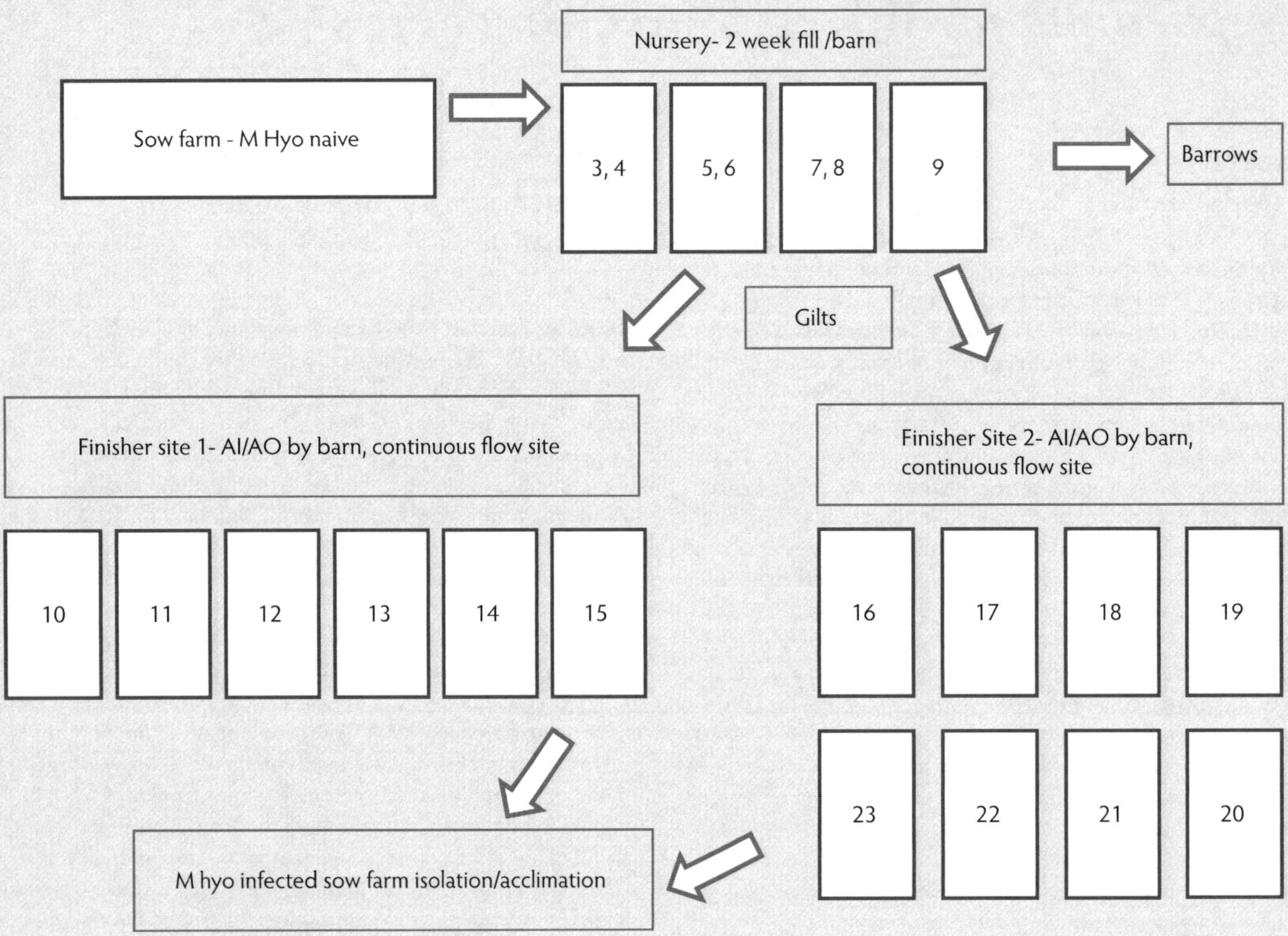

Case 2

A group of *Mhp* positive farms receive MH positive gilts through a GDU system consisting of 4 separate barns. These 4 barns were managed continuous flow by barn but all in/all out by airspace (room). A source change for replacement gilts was made to another *Mhp* positive gilt supplier. The flow of the GDU was changed from continuous flow by barn to all in / all out by barn. Soon it became apparent that not all groups of gilts had the same infection status. Depending on the status of the multiplication farm from week to week, the *Mhp* status of the particular gilt group would be variable- leading to inconsistent exposure status in gilts. Mild respiratory disease appeared in the replacement gilts soon after entry into sow units and downstream pig performance rapidly degraded as those gilts produced offspring. *Mhp* pneumonia was a common diagnostic finding in nursery and finisher pigs.

Since the gilt source was *Mhp* infected, a flow was designed to create an opportunity for *Mhp* to move from group to group and maintain more uniform early exposure and recovery. Incoming feeder gilts were placed in a way to allow fence-line contact with gilts that were 4 and 8 weeks older to facilitate early and more uniform exposure to *Mhp* similar to case 1.

Results and discussion

The goal of any gilt acclimation program is to manage exposure and immunity which hopefully results in improved performance of the terminal market animal. For *Mhp,* the goal is for young sows to be immune and non-infectious at the time of farrowing. Pieters et al, 2009 proved that animals can shed *Mhp* to direct contacts for at least 200 days.[3] Acclimation must be accomplished well before breeding if reduced sow to pig transmission is desired. However, proper *Mhp* acclimation is difficult to accomplish and there are significant costs associated with maintaining *Mhp* infection in the growing gilt population. These case examples resulted in decreased growth rates, increased mortality, decreased selection rates, and ultimately an increased financial investment in replacement breeding stock. *Mycoplasma hyopneumoniae* is a slow spreading pathogen with R_0 rates reported to vary from 0.85 to 1.47 depending on virulence[8] that transmits better with direct

Figure 2: M hyo acclimated gilt flow

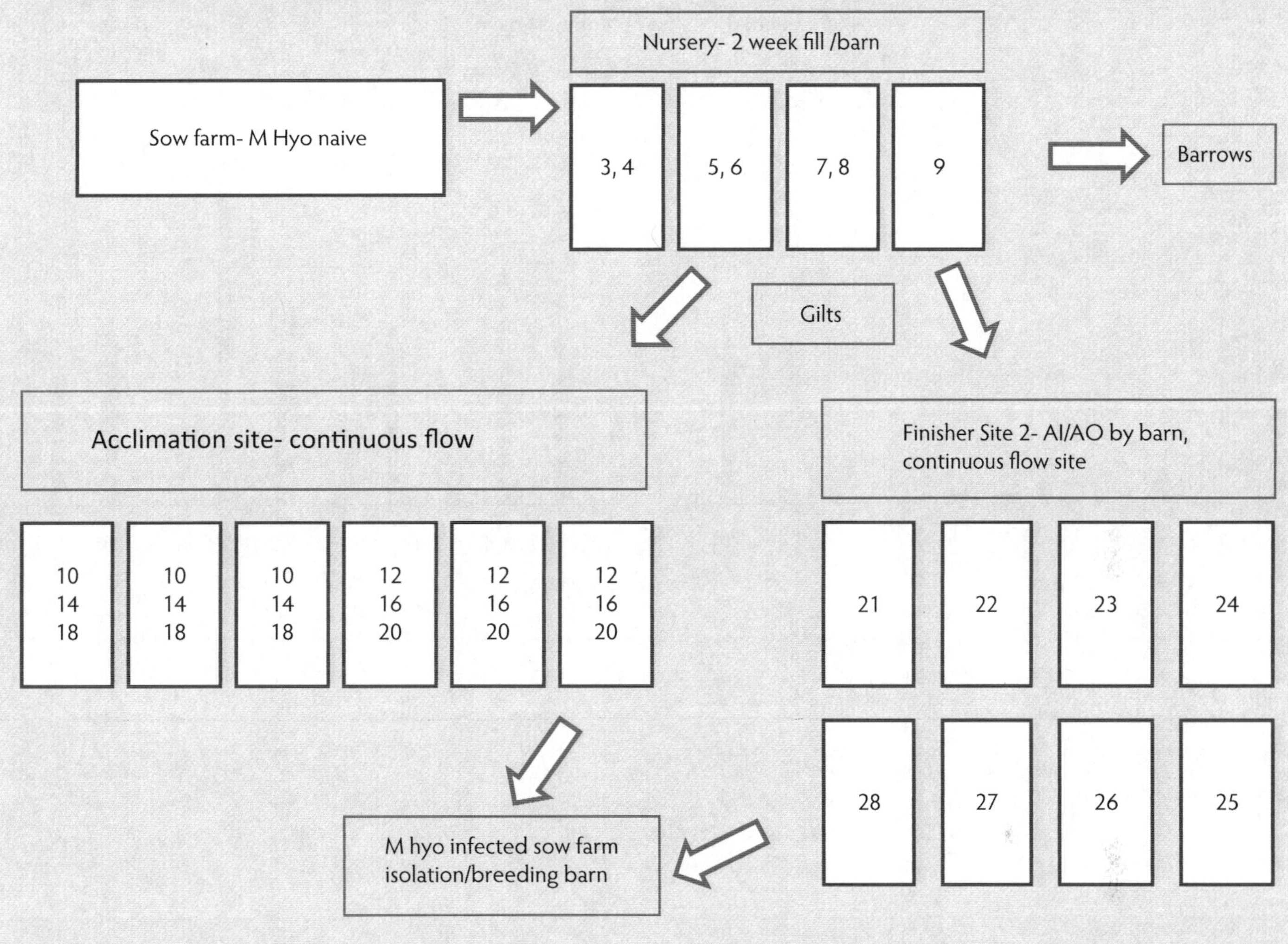

contact between shedding and susceptible animals[9] for an extended period of time compared to the viruses we design our acclimation systems around, yet it moves over long distances and can re-infect farms.[5-7] Breeding farms that continually circulate *Mhp* within the young sow population will result in *Mhp* colonization of piglets during lactation. These groups of pigs that are colonized before weaning have the entire grow-out period available for slow lateral spread within the group, which can result in chronic pneumonia. Repeated introductions of susceptible breeding replacements will continue to maintain colonization of the weaned pig population and respiratory disease in the growing pig population. Gilt acclimation is a useful tool for breeding farms where exclusion of *Mhp* is not practical and reduced clinical disease in the growing pig population is desired. Additionally, for *Mhp* elimination programs, having a source of young replacements with verified exposure to *Mhp* is valuable for successful elimination.

References

1. Maes D. *Mycoplasma hyopneumoniae* infections in pigs: update on epidemiology and control. Proceedings of the 21st IPVS Congress, Vancouver, Canada. 2010; 30-35.

2. Fano E, Pijoan C, Dee S, Deen J. Effect of *Mycoplasma hyopneumoniae* colonization at weaning on disease severity in growing pigs. Canadian Journal of Vet Research 2007;71:195-200.

3. Pieters M, Pijoan C, Fano E, Dee S. An assessment of the duration of *Mycoplasma hyopneumoniae* infection in an experimentally infected population of pigs. Vet Microbiol. 2009;134:261-266.

4. Holst S, Yeske P, Pieters M. Elimination of *Mycoplasma hyopneumoniae* from breed-to-wean farms: A review of current protocols with emphasis on herd closure and medication. J Swine Health Prod. 2015;23(6):321-330.

5. Hege R, Zimmermann W, Scheidegger R, Stark KD. Incidence of reinfections with *Mycoplasma hyopneumoniae* and Actinobacillus pleuropneumoniae in pigs farms located in respiratory disease-free regions of Switzerland-Identification and quantification of risk factors. Acta Vet Scand 2002;43:145-156.

6. Thomsen BL, Jorsal Se, Andersen S, Willeberg P. The Cox regression model applied to risk factor analysis of infections in the breeding and multiplying herds in the Danish SPF system. Prev Vet Med 1992;12:287-297.

7. Dee S, Otake S, Oliveira S, Deen J. Evidence of long distance airborne transport of porcine reproductive and respiratory syndrome virus and *Mycoplasma hyopneumoniae*. Vet Res 40:39.

Figure 3: M hyo antibody comparison of acclimated gilts and non-acclimated barrows by age.

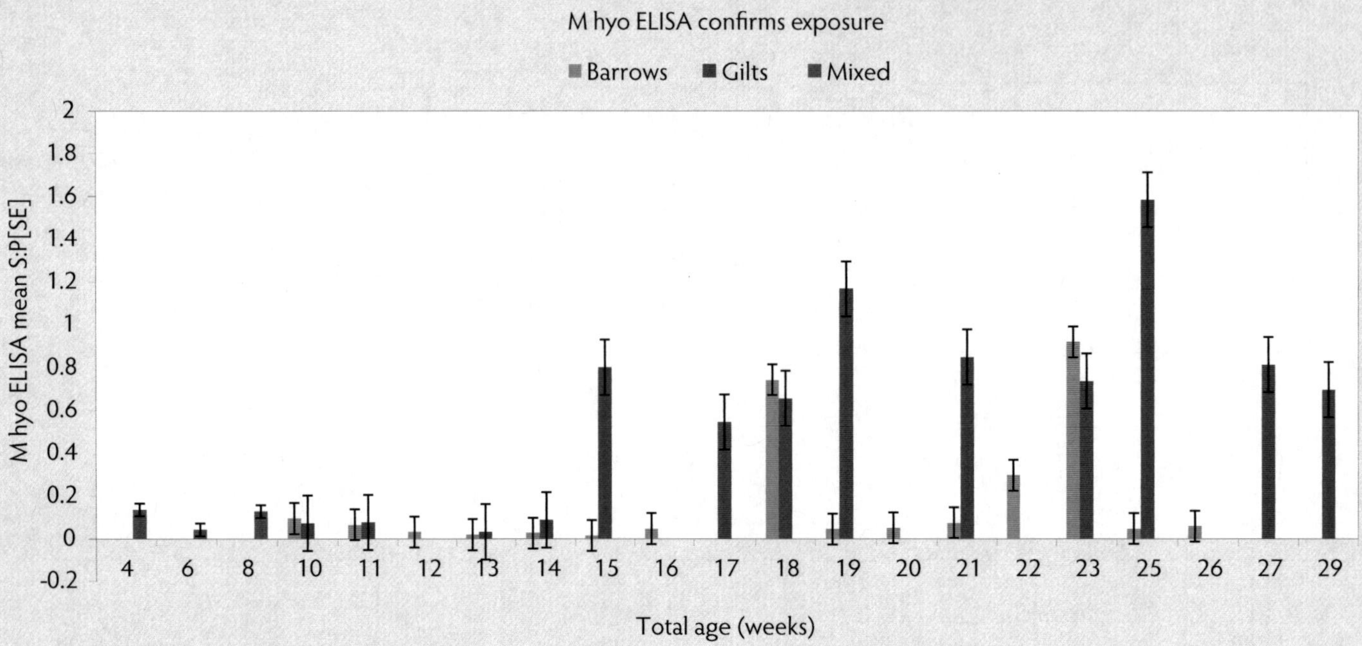

8. Meyns T, Maes D, Dewulf J, Vicca J, Haesebrouck F, de Kruif A. Quantification of the spread of Mycoplasma hyopneumonaie in nursery pigs using transmission experiments. Prev Vet Med 2004;66:265-275.

9. Morris C, Gardner I, Hietala S, Carpenter T, Anderson R, Parker K. Seroepidemiologic study of natural transmission of *Mycoplasma hyopneumoniae* in a swine herd. Preventive Veterinary Medicine 1995;21:323-337.

Mycoplasma hyosynoviae: Retrospective diagnostic case review

Paisley Canning[1], DVM; Kent Schwartz[2], DVM, MS; Locke Karriker[1], DVM, MS, DACVPM
[1]Swine Medicine Education Center, Ames, Iowa; [2]Iowa State University Veterinary Diagnostic Laboratory, Ames, Iowa

Diagnostic challenges of lameness cases

Locomotor disorders, commonly termed lameness, are a significant issue in the growing pig population, affecting both animal well-being and production parameters. Practitioners and caretakers in the field cite an increasing concern about these clinical syndromes for several reasons. Lameness negatively affects a pig's welfare causing pain and reducing the animal's ability to ambulate comfortably. In late finishing, lameness may limit our ability to transport affected pigs and may increase the prevalence of exhausted or down pigs in market loads. Implementing treatment for infectious lameness in market weight pigs also creates tension between drug withdrawal periods and pre-planned marketing logistics. As the use of antimicrobials comes under increased regulation by means of government and customer demand, elucidating the cause of growing pig lameness and applying effective interventions will be more critical than ever.

Growing pig lameness cases rarely involve only one pathogen or primary cause. Visual assessment of lameness alone can be inconsistent between observers and is nonspecific to lameness etiologies. Veterinarians often then rely on diagnostic testing on post-mortem samples. These diagnostic investigations often do not implicate specific etiology and frequently warrant additional diagnostics and examination of herd records to generate useful information on contributing causative factors. Lameness submissions to diagnostic labs commonly lack comprehensive case histories and complete fixed and fresh tissue sets. In cases where practitioners have invested in the submission of several pigs and auxiliary samples (feed, water, blood) a clear or singular diagnostic finding is still not guaranteed; however if samples are representative, some etiologies can be ruled out. The lack of ante mortem sampling tools limits the practitioner's ability to survey large groups economically. Such tools would be useful but would need to be validated to correlate testing results to clinical signs, particularly for pathogens such as *Mycoplasma hyosynoviae* (MHS) which are present in non-clinical carrier animals.

One common cause of growing pig lameness is arthritis and/or synovitis, which often has an infectious cause. The exact field prevalence of this lesion in growing pigs is unknown. However, a recent study described infectious diagnoses in cases from 2003-2010 at the Iowa State University Veterinary Diagnostic Laboratory (ISU VDL). Overall, 69% of these cases had evidence of infectious arthritis, with *Mycoplasma* species involved in up to 37% of cases in 2010. This includes both *Mycoplasma hyorhinis*, seen pigs < 10 weeks of age and MHS in pigs > 10 weeks of age (Neto et al 2012).

Post-mortem samples provide a one-time snapshot of the disease process which, for dynamic disease processes, often correlates poorly to clinical outcomes and predicted treatment efficacy. *Mycoplasma hyosynoviae* is a highly dynamic pathogen. MHS may colonize the joint transiently and may or may not cause disease, often requiring a secondary risk factor to cause arthritis. Depending upon the course of the disease process, different diagnostics tests may be of value. Diagnostic assays for MHS are limited to polymerase chain reaction (PCR) and culture, with some insight also gained by histopathology. A PCR test may identify the primary pathogen (MHS) early in the disease process, but may be negative in a more chronic joint once MHS has been cleared from the joint. Although *Mycoplasma hyosynoviae* (MHS) has been identified as a primary cause of infectious arthritis and clinical lameness, a clear-cut case definition associating clinical signs with diagnostic results has yet to be created.

With the complexity of MHS diagnostic investigations, a case definition is needed to create a standardized way of discussing and interpreting MHS cases. Secondly, a retrospective analysis of MHS diagnostic investigations would provide valuable information on submitter behaviors and case findings in MHS submissions. The objective of the case analysis is to better understand current MHS cases and generate recommendations to improve sampling, submission and the diagnostic value of testing modalities.

Case definition for *Mycoplasma hyosynoviae*

The format for the case definition follows that of the USDA case definition format applied recently to Swine Enteric Coronavirus Disease (SECD) includes the following details: 1) general disease/pathogen information (etiological agent, distribution, clinical presentation); 2) incubation period, differential diagnosis, transmission, reservoir and epidemiology); 3) diagnostic guidelines (agent identification assays, gross lesions and histopathology) and 4) case definition (including criteria for designating positive animals and herds).

The case definition will then be applied to the MHS cases identified and examined in the retrospective case review.

Mycoplasma hyosynoviae case review

A retrospective case series review of MHS cases submitted to the ISUVDL was conducted. Below is an outline of the case review and selection process (Table 1).

The following information was extracted from the MHS case submission sheets and final reports:

1. Clinical history supplied by veterinarian on the diagnostic lab submission form;
2. Herd descriptors – site size, age of pigs, type of site;
3. Type and number of samples submitted (example: joint fluid, joint swab, whole pig, whole leg);
4. Tests requested on submission form;
5. Tests conducted and results of each test.

The case definition will then be applied and findings summarized. The objective of the case series is to generate hypotheses on the importance of co-factors/concurrent infections, summarize sample submission trends, create recommendations on improving lameness submissions and present common and conflicting characteristics between MHS cases and case definition.

Acknowledgements

- Iowa Pork Producers Association, National Pork Board and PIC for funding.
- Dr. Maria Clavijo, Emily Mahan Riggs and Nicole Hershberger for their guidance and assistance.

References

1. Gomes Neto, J. C. (2012). *Diagnostic and field investigations in Mycoplasma hyosynoviae and Mycoplasma hyorhinis*. Master of Science, Iowa State University.

Table 1: Search strategy and total number of cases included for analysis for MHS case review

Inclusion and exclusion criteria	Total number of unique cases
Original search criteria on ISU VDL database. Identify cases that had the following characteristics: 1. May 1st 2010 to April 30th 2015 2. porcine species 3. histopathology performed on submitted tissues 4. one of 24 relevant diagnostic codes related to lameness or locomotor dysfunction **and/or** MHS PCR and/or culture performed	1847 cases
Total unique cases remaining once those failing to meet the following criteria were removed: • May 1st 2010 to April 30th 2015 • Must have histopathology • Must not be research project • > 50 days of age • > 7 weeks of age	1019 cases
Total number of lameness cases Cases that did not involve bones, joint, trauma or deficiency were deleted. Example of case types removed would be serology, systemic *Mycoplasma hyorhinis* and those involving animals failing to meet the age criteria.	668 cases
Total number of MHS cases For these cases, the original submission sheet and final diagnostic report were entered into the case database.	~200 MHS specific cases (case review is ongoing at time of proceedings submission)

Interspecies transmission of influenza A viruses

Marie Culhane, DVM, PhD; Amanda Beaudoin, DVM, PhD, DACVPM
College of Veterinary Medicine, University of Minnesota, Saint Paul, Minnesota

Introduction

Influenza A viruses (IAVs) infect many animal species including human,[1] avian,[2] swine,[3] equine,[4] canine,[5] feline,[6] some marine mammals,[7] and bats.[8] Transmission of avian influenza viruses from the natural reservoir to other wild and domestic species does occur, however, interspecies transmission of influenza viruses is less common than intraspecies transmission (e.g. pig to pig), as viruses become adapted to the species in which they circulate. Over time, influenza A viruses may develop the ability to bind well (adapt) to receptors on host cells of non-reservoir species, including humans, making it a zoonotic pathogen of considerable importance.

Swine are susceptible to IAVs from all species. To use an ugly and overused term, they are suitable "mixing vessels" for influenza A viruses, capable of generating new reassortant, mammalian adapted IAVs of pandemic potential. The mixing vessel term was coined some decades ago, because it was known that the pig had both mammalian α-2,6 Gal and avian α-2,3 Gal influenza viral receptors. While best avoided, the term continues to pop up, especially in the lay press, and cannot be ignored. Rather than deny that interspecies transmission exists, it is best to more fully explore the issue and call to light that interspecies transmission occurs in a variety of species, in several directions, and is not limited solely to pigs.

Pigs and humans

Persons working in pig farms represent the immediate interface of two vastly important networks of influenza virus transmission. Both populations habitually experience infection with IAVs. In pig farms in the Midwest US, transmission occurs year-round, peaking in spring and fall,[9] and human seasonal influenza occurs annually, extending from fall to spring. Influenza infection in both populations causes considerable morbidity and, particularly in humans, severe illness and death. Pre-existing immunity to influenza viruses through past exposure to similar strains or through vaccination is protective for both species. However, introduction of novel viruses into a naïve population can lead to high morbidity and mortality, unfettered transmission, and, potentially, pandemic spread. Such an event occurred in the spring of 2009, when the first cases of a novel influenza virus were identified in Southern California.[10] The Centers for Disease Control and Prevention (CDC) identified the virus as an influenza A H1N1 virus of swine lineage (A/California/2009/H1N1). The virus proceeded to spread globally, and on June 11, 2009, the World Health Organization declared an H1N1 influenza pandemic.

Although considerably less common than seasonal influenza, human infection with animal influenza viruses does occur. During December 2005–December 2009, there were 19 reported human cases of sporadic infection with swine-origin viruses in the United States, compared with an estimated 15-60 million seasonal influenza cases annually (5-20% of the population).[11,12] Human infections with human or swine influenza viruses are clinically indistinguishable.[13] Serologic studies have shown that swine-exposed persons are more likely than persons outside of an agricultural setting to have antibodies specific to influenza viruses of swine origin.[14] Highlighting the critical position of swine workers at the intersection of the swine and human populations is the finding that the spouses of workers also are more likely to have swine-specific influenza antibodies, despite having no direct contact with pigs.

This potential for sporadic transmission of swine viruses to persons in close proximity certainly presents a concern for worker health on an acute basis. Furthermore, the risk of introduction of human influenza viruses into pig populations, reassortment into novel viruses over time, and the potential spillover back into the human population has significant implications for human health.[15,16] Attention must be paid to the fact that the swine worker serves as the first point of contact, and thus the first at risk, for newly emerging viruses.

Pigs and avian

Reports of H5 and H7 avian IAV infections in swine are scarce and typically from Asia.[17] However, in North America, avian IAV have infected swine with whole genomes intact (H4N6) and the predominant H1N1 in Europe, prior to 2009, was a wholly avian H1N1.[17] The incorporation of avian polymerase genes into reassortant IAV found in swine is a driver of genetic diversity (TRIG and H2N3). The possible reassortment of avian IAV and swine IAV is a constant concern; however, previous studies have shown that most avian-adapted IAV replicate very poorly in swine.[17] The highly-pathogenic avian influenza (HPAI) H5NX outbreak in North America in 2015 raised this concern to the highest levels ever because, for the first time in US history, millions of domestic American poultry became infected with HPAI right in the MN/IA cornbelt - amongst thousands of pigs. The 2015 HPAI H5NX viruses were experimentally inoculated into pigs at the USDA Animal Research Service (ARS). The ARS experimental study results demonstrated once again that avian adapted viruses replicate poorly in pigs, with only directly inoculated pigs having detectable virus in the lungs. There was no spread to contact pigs.[18] In conclusion, most avian H5 (and H7 and H9) viruses previously tested in swine failed to sustain efficient replication and transmission. Nevertheless, low levels of infection may facilitate reassortment and as influenza researchers with due diligence, this interspecies interaction will continue to be under investigation.

Avian and humans

Direct infection of humans with non-human-adapted avian influenza viruses has been rarely and only sporadically reported, and the only subtypes known to have directly infected humans are H5, H7 and H9.[19] H7 and H9 viruses have occasionally caused clinical signs (usually conjunctivitis) after occupational exposure.[20-23] Notably, after the 2003 poultry outbreak of H7N7 in the Netherlands, approximately 50% of those with poultry contact had antibodies to the virus, and it was estimated that 1000 people may have been infected with AI.[21] Subclinical seropositivity to H9N2 viruses, which are endemic in Asia, was identified in nearly 10% of poultry workers and 4% of general citizens tested in Guangdong Province, China.[24] More widespread and serious have been the infections caused by highly pathogenic avian influenza (HPAI) H5N1 viruses, beginning in 1997 and continuing to this day as part of the multi-continental epizootic of poultry infection. Efforts and investments in animal IAV research really increased as a result of HPAI H5N1 infections in humans in 1997 in Asia. The direct transmission of avian influenza viruses into humans thwarted the dogma of the pig as a necessary intermediate host for avian to mammalian transmission to occur. As a result of this fascinating interspecies transmission, the Centers for Excellence in Influenza Research and Surveillance (CEIRS) and OFFLU[26] networks were formed and are performing active surveillance and research in the field of animal influenza. OFFLU is the joint OIE-FAO global network of expertise on animal influenzas. OFFLU aims to reduce negative impacts of animal influenza viruses by promoting effective collaboration between animal health experts and with the human health sector.

Human and equine; equine and canine, avian and canine, avian and feline

Equine influenza has a long history, perhaps longer than swine influenza. Dr. David Morens, in a historical medical manuscript, recounted the story of EIV so well and it is best read in its entirety.[27] In summary, Dr. Morens pointed out that at the end of the 19th century, "...horses were ubiquitous in virtually all rural and urban settings, it is thus conceivable that they played a role in influenza virus maintenance and evolution analogous to that of pigs in the modern era...." He further speculates that influenza was originally transmitted from humans to horses in the 1800s, and that the cause of the 1889 influenza pandemic, while unknown, was caused by an H3N8 equine influenza virus that had become adapted to humans, and which then spread pandemically. That same H3N8 virus or some of its gene segments persists in horses today and eventually spread to canines.

Canine IAV was first identified as a factor in canine respiratory disease in 2004.[28] This virus is a genetic variant of the H3N8 equine influenza virus that gained the ability to infect dogs. The key change in the virus was the ability for transmission of the virus from dog to dog. The Asian H3N2 canine influenza virus that caused outbreaks in urban areas of the USA in spring 2015 is derived from an avian strain that also gained the ability to infect dogs and be transmitted from dog to dog.[29,30] As is the case with all influenza viruses, there is the opportunity for changes in the virus that could affect transmission rates and increase or decrease the ability of the virus to cause respiratory illness.

And again, because pigs are constantly under scrutiny as the "mixing vessels," these new canine H3N2 IAV were also experimentally inoculated into pigs by researchers at USDA ARS.[31] An isolate originating from a sick dog from the 2015 spring outbreak of H3N2 canine IAV was intratracheally (IT) and intranasally (IN) inoculated into 5 pigs in an controlled experiment. The inoculated pigs were placed nose-to-nose with contact pigs after 2 days. While there was limited virus detected in lung fluid of the IT/IN inoculated pigs, there were no lung lesions and no transmission to contact pigs.

Conclusions

Animal influenza virus surveillance and research will continue for many years because the reservoirs and transmission of influenza virus continue to change as the virus itself evolves. Influenza A viruses are global and involve multiple species, thus we need comprehensive surveillance programs with extensive virologic and epidemiologic information to detect new influenza strains and determine the most common strains causing disease. Detection of new strains in pigs that may have epidemic or pandemic potential is not only necessary for human health, but is critical to provide better prevention, treatment, and control of IAV to the pigs and clients under our care as veterinarians. Detection of influenza viruses in other mammalian species such as cats, dogs, and horses are considered part of these integrated, One Health, influenza surveillance programs. As veterinarians, we are essential participants in these surveillance programs.

References

1.Bedford T, Riley S, Barr IG, Broor S, Chadha M, Cox NJ, et al. Global circulation patterns of seasonal influenza viruses vary with antigenic drift. Nature. 2015;523(7559):217-U06.

2.Chen RB, Holmes EC. Hitchhiking and the Population Genetic Structure of Avian Influenza Virus. Journal of Molecular Evolution. 2010;70(1):98-105.

3.Torremorell M, Allerson M, Corzo C, Diaz A, Gramer M. Transmission of Influenza A Virus in Pigs. Transboundary and Emerging Diseases. 2012;59:68-84.

4.Murcia PR, Baillie GJ, Stack JC, Jervis C, Elton D, Mumford JA, et al. Evolution of equine influenza virus in vaccinated horses. J Virol. 87. United States2013. p. 4768-71.

5.Hoelzer K, Murcia PR, Baillie GJ, Wood JLN, Metzger SM, Osterrieder N, et al. Intrahost Evolutionary Dynamics of Canine Influenza Virus in Naive and Partially Immune Dogs. Journal of Virology. 2010;84(10):5329-35.

6.Ali A, Daniels JB, Zhang Y, Rodriguez-Palacios A, Hayes-Ozello K, Mathes L, et al. Pandemic and Seasonal Human Influenza Virus Infections in Domestic Cats: Prevalence, Association with Respiratory Disease, and Seasonality Patterns. Journal of Clinical Microbiology. 2011;49(12):4101-5.

7.Harder TC, Siebert U, Wohlsein P, Vahlenkamp T. Influenza A virus infections in marine mammals and terrestrial carnivores. Berliner Und Munchener Tierarztliche Wochenschrift. 2013;126(11-12):500-8.

8.Tong S, Zhu X, Li Y, Shi M, Zhang J, Bourgeois M, et al. New world bats harbor diverse influenza a viruses. PLoS Pathog. 2013;9(10):e1003657.

9.Beaudoin AL, et al. Characterization of influenza A outbreaks in Minnesota swine herds and measures taken to reduce the risk of zoonotic transmission. Zoonoses Public Health. 2012 Mar;59(2):96-106

10.CDC. Swine InfluenzaA(H1N1) Infection in Two Children, Southern California, March-April 2009.

11.CDC. CDC - Seasonal Influenza (Flu) - Key Facts about Swine Influenza (Swine Flu) in Humans. 2009.

12.CDC. CDC - Seasonal Influenza (Flu) - Q & A: Seasonal Influenza (Flu): The Disease. 2011

13.Myers KP, Olsen CW, Gray GC. Cases of Swine Influenza in Humans: A Review of the Literature. Clinical infectious diseases : an official publication of the Infectious Diseases Society of America. 2007;44(8):1084-1088. doi:10.1086/512813.

14.Gray, GC et al. Swine workers and swine influenza virus infections. Emerg Infect Dis. 2007.13(12):1871-1878.

15.Ma W, et al. The role of swine in the generation of novel influenza viruses. Zoonoses Public Health 2009; 56:326-37.

16.Nelson MI, et al. Introductions and evolution of human-origin seasonal influenza A viruses in multinational swine populations. Jrnl. Virol. 2014. 88(17):10110–19.

17. Lipatov AS, Kwon YK, Sarmento LV, Lager KM, Spackman E, Suarez DL, et al. (2008) Domestic Pigs Have Low Susceptibility to H5N1 Highly Pathogenic Avian Influenza Viruses. PLoS Pathog 4(7): e1000102. doi:10.1371/journal.ppat.1000102

18. Vincent, A. Personal Communication. September 22, 2015.

19.Cox, N. J., and T. M. Uyeki, 2008: Public health implications of avian influenza viruses. In: Swayne, D. E. (ed), Avian influenza, 1st Edn , pp. 453-454-484. Blackwell Pub. Ames, Iowa.

20.Koopmans, M., B. Wilbrink, M. Conyn, G. Natrop, H. van der Nat, H. Vennema, A. Meijer, J. van Steenbergen, R. Fouchier, A. Osterhaus, and A. Bosman, 2004: Transmission of H7N7 avian influenza A virus to human beings during a large outbreak in commercial poultry farms in the Netherlands. Lancet 363(9409), 587-593. DOI: 10.1016/S0140-6736(04)15589-X.

21.Bosman, A., A. Meijer, and M. Koopmans, 2005: Final analysis of Netherlands avian influenza outbreaks reveals much higher levels of transmission to humans than previously thought. Euro Surveill 10(1), E050106.2.

22.Du Ry van Beest Holle,M., A. Meijer, M. Koopmans, and C. M. de Jager, 2005: Human-to-human transmission of avian influenza A/H7N7, The Netherlands, 2003. Euro Surveill 10(12), 264-268.

23.Meijer, A., A. Bosman, E. E. van de Kamp, B. Wilbrink, Du Ry van Beest Holle,M., and M. Koopmans, 2006: Measurement of antibodies to avian influenza virus A(H7N7) in humans by hemagglutination inhibition test. J Virol Methods 132 (1-2), 113-120. DOI: 10.1016/j.jviromet.2005.10.001.

24.Lu, C. Y., J. H. Lu, W. Q. Chen, L. F. Jiang, B. Y. Tan, W. H. Ling, B. J. Zheng, and H. Y. Sui, 2008: Potential infections of H5N1 and H9N2 avian influenza do exist in Guangdong populations of China. Chin Med J (Engl) 121(20), 2050-2053.

25.CEIRS. https://www.niaid.nih.gov/labsandresources/resources/ceirs/Pages/introduction.aspx. Accessed 10 December 2015.

26.OFFLU. http://www.offlu.net/fileadmin/home/en/about-us/pdf/OFFLU_brochure_v2-29-May-2015.pdf. Accessed 10 December 2015.

27. Morens, David M., and Jeffery K. Taubenberger. "Historical Thoughts on Influenza Viral Ecosystems, or Behold a Pale Horse, Dead Dogs, Failing Fowl, and Sick Swine." Influenza and Other Respiratory Viruses 4.6 (2010): 327–337. PMC. Web. 10 Dec. 2015.

28.Crawford PC, Dubovi EJ, Castleman WL, Stephenson I, Gibbs EP, Chen L, Smith C, Hill RC, Ferro P, Pompey J, et al. Science. 2005 Oct 21; 310(5747):482-5. Epub 2005 Sep 26.

29. Vincent, A. Personal Communication, December 3, 2015.

29.Park S-J, Kang B-K, Jeoung H-Y, et al. Complete Genome Sequence of a Canine-Origin H3N2 Feline Influenza Virus Isolated from Domestic Cats in South Korea. Genome Announcements. 2013;1(2):e00253-12. doi:10.1128/genomeA.00253-12.

30.Song D, Kang B, Lee C, et al. Transmission of Avian Influenza Virus (H3N2) to Dogs. Emerging Infectious Diseases. 2008;14(5):741-746. doi:10.3201/eid1405.071471.

Considerations for vaccines against influenza A virus in US swine

Amy Vincent[1], DVM, PhD; Daniela Rajao[1], DVM, PhD; Tavis Anderson[1], PhD; Eugenio Abente[1], PhD; Rasna Walia[1], PhD; Nicola Lewis[2], BSc, BVetMed, PhD, MRCVS
[1]National Animal Disease Center, USDA-ARS, Ames, Iowa;
[2]Department of Zoology, University of Cambridge, Cambridge, United Kingdom

Introduction

Swine influenza is an acute respiratory disease caused by influenza A virus (IAV) of the Orthomyxovirus family. The virus decreases health and welfare of pigs and results in an economic loss for the swine industry worldwide. The classical H1N1 IAV was the predominant virus circulating among swine populations in the US for nearly 80 years. In the late 1990s, the triple reassortant H3N2 virus emerged with hemagglutinin (HA), neuraminidase (NA), and polymerase basic 1 (PB1) gene segments derived from seasonal human influenza viruses, polymerase basic 2 (PB2) and polymerase acidic (PA) gene segments derived from an avian IAV, and nucleoprotein (NP), matrix (M), and nonstructural (NS) gene segments from the classical-swine H1N1.[40,66] Subsequently, this triple-reassortant H3N2 reassorted with cH1N1, resulting in new genotypes of H1N1 and H1N2.[23] In addition, newly introduced lineages of H1N2, H1N1 and H3N1 with HA and/or NA genes derived from human seasonal IAV have been detected in North American swine herds over the past 15 years.[45,57]

In 2009, the USDA initiated a surveillance system with the goal of characterizing the genetic diversity of swine IAV in the US.[4] These data revealed that the vast majority of circulating IAV subtypes in swine are H1N1, H1N2, and H3N2, with rare H3N1 detections. The H1 viruses are divided into seven different HA genetic clades; δ-1, δ-2, α, β, γ, γ2, and pandemic. The predominant H1 clades currently circulating in the swine population are γ and δ-1. From July 2014 to July 2015, 49% of isolates were γ-clade and 30% were δ-1. The H3 viruses are divided into eight different HA genetic clades (IV, IV-A, IV-B, IV-C, IV-D, IV-E, IV-F, and human-like H3s), with clade IV-A representing the majority of circulating H3 in production herds (63% of H3 viruses during the last year). The remaining H3 are sporadically detected in swine herds and are classified as IV-B, IV-C, IV-E, and human-like H3s. Viruses from clades IV, IV-C, and IV-F have not been detected since the last quarter of 2013. These HA genetic lineages may be paired with either one of two N2 genes derived from human H3N2 seasonal-lineages from either the 1998 or 2002 human influenza virus seasons,[39] with N1 genes from the classical swine-lineage, or with human H1N1 seasonal-lineage N1 genes.[4] In currently circulating swine IAV, N1s are generally of classical-lineage (94%), and the N2s are generally the 2002-lineage (83%). The 1998-lineage of N2 is most often paired with δ-2 H1 when it is detected, but these represent a smaller component overall. Despite the presence of dominant phylogenetic HA and NA clades, minor HA and NA clades also continue to co-circulate, thus complicating vaccine strategies and strain updates.

The US is divided into five different regions for reporting purposes (see Figure 1) based on USDA-APHIS veterinary services districts, with district 1 and 2 combined into one region (Region 1). Though the general genetic diversity patterns are consistent across these regions, there are subtle and important differences between regions in diversity and in raw abundance of samples provided to the USDA surveillance system. While Regions 1 and 2 both have higher detections of the H1N1 subtype, Region 3 has more of the H1N2 subtype and Region 4 has more of the H3N2 subtype. At present, very little surveillance data is observed from Region 5. Region 2 shows the most diversity in terms of the different HA and NA clades observed, and most of the swine IAV isolates are submitted from this region. Overall, the HA and NA clade distributions suggest that regional decisions on vaccine components may be of value.

For comparison, the World Health Organization (WHO) organizes and hosts vaccine strain consultation meetings twice each year to select human vaccine strains for the Northern or Southern Hemispheres.[3] This consultation is the forum for sharing data from the Global Influenza Surveillance and Response System (GISRS), with over 60 years of coordinating monitoring and response to influenza A and B viruses in the human population. Virologic surveillance is the cornerstone of the WHO vaccine strain selection process, and is carried out through a network

Figure 1: Regional map for IAV surveillance in swine.

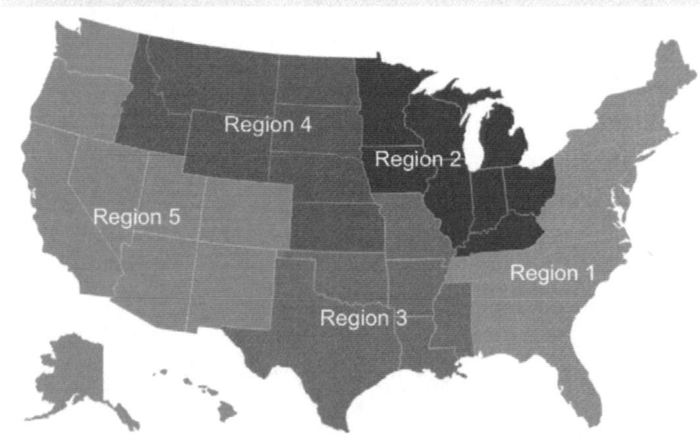

of national influenza centers and collaborating centers, with data representing national, regional, and global levels, as well as climatic or transmission zone data aggregation. Other viral and clinical illness data is also collected and reported to WHO by national and regional centers. Laboratory characterization of virus isolates by hemagglutination inhibition (HI) assays remains the gold standard, with specific ferret antisera as well as pre- and post-vaccination human sera. The antigenic data is combined with sequence, virus growth properties, epidemiological, and clinical information to inform the vaccine strain selection process.[49] Newer strategies, technologies or data management tools have recently made their way into the strain selection process. This includes monitoring animal viruses involved in zoonotic events and assessing their pandemic potential, incorporating next generation whole genome sequencing and sequence analysis pipelines, and antigenic mapping and antibody landscape modeling. Efforts to improve the understanding of NA immunity and NA evolution as a consideration for vaccine candidate selection has also been recently re-emphasized. The WHO network was instrumental in establishing vaccine standards and improving production capacity and is engaged in facilitating the manufacturing and regulatory aspects of vaccine production as well. High growth optimized laboratory backbone strains that accept the HA and NA from the selected seasonal vaccine candidates are provided to vaccine manufacturers. The scope and magnitude of this coordinated and inter-sectoral network for the comparably more simplistic situation with human seasonal influenza viruses accentuates the very basic level from which we approach strain selection for swine vaccines.

Vaccine platforms

Historically, all of the influenza vaccines licensed for swine in the United States were killed whole virus products and several fully licensed products in a number of different formulations are available. Whole inactivated viruses (WIV) are typically combined with potent oil-in-water adjuvants to stimulate robust antibody responses that can be measured by HI assays. Different adjuvants can alter the response to antigens by stimulating different arms of the immune system or stimulating broader cross-reactive antibodies. According to the Center for Veterinary Biologics autogenous vaccines represented at least half of the swine influenza virus (SIV) vaccine doses produced for a period of consecutive years in the United States. However, this trend made a pronounced drop in 2012. This is likely due in part to an alternative source of vaccines that were not captured through full or autogenous licensure, but a real drop in autogenous WIV usage cannot be ruled out. In 2012, a new vaccine platform product was licensed for swine H3N2 virus,[17] along with the manufacturing establishment, using technology based on a non-replicating alphavirus RNA particle.[55,56] A recent review by Sandbulte *et al.* provided guidance for choosing vaccines based on the currently available technology.[52] Although it appears that there are many options for preventing IAV in swine, veterinarians and producers remain frustrated with the available products. Licensed products must be demonstrated to be efficacious in experimental studies with homologous or related viruses, yet they frequently fail or perform sub-optimally in the context of field application.

Research in swine on live-attenuated IAV vaccines (LAIV) has repeatedly demonstrated their general safety and superior efficacy under experimental conditions that are more similar to those in field-settings when directly compared to inactivated vaccines.[35,41,48,54,58] Experimental data have demonstrated that attenuated vaccines are superior in their ability to overcome maternal antibody interference, protect against heterologous antigenically distinct viruses, as well as protect against transmission of heterologous challenge viruses to naïve contact pigs.[22,34] The caveat to the LAIV are the limitations of the HI assay in predicting protection or cross-protection. Subunit and/or vectored vaccines targeting the HA, NA, and/or the more conserved M or NP proteins have also demonstrated efficacy against IAV in swine,[5,61,62] and some have clinical or manufacturing advantages over WIV vaccines. However, subunit vaccines have an even heavier weighted importance of including contemporary antigens to increase antigenic coverage of the circulating viral diversity.

The concept of developing vaccines that induce broadly cross-protective and long lasting immunity has been a focus for human IAV research for many years.[8] The term "universal" vaccine was coined from this field of study. Universal vaccine platforms include viral-like particles expressing multiple HA and NA subtypes, vectored peptides from highly conserved epitopes of HA or NA, as well as several strategies targeting the more conserved HA2 stem or stalk.[63] One stem-targeting approach is to use sequential heterologous exposures to vaccine viruses with chimeric HA proteins.[27,28,37] The HA1 globular head containing the highly variant portion of the mature HA is replaced by HA1 from an irrelevant subtype. Each series of the chimeric vaccines contain the conserved HA2 stem from a relevant IAV strain to drive a more robust antibody response to the conserved HA2 stem. Although antibodies induced from this platform are not neutralizing in the traditional sense, there is experimental evidence that this vaccine strategy can reduce viral shedding and transmission and prevent clinical disease against heterosubtypic viruses with the conserved stem in laboratory animal models. Although these clinical outcomes may have a place in human medicine and pandemic preparedness, it is not clear if this vaccine strategy has an advantage over the efficacy yielded with current swine IAV vaccines. However, studies in swine, especially used in prime-boost strategies, need to be conducted for critical comparison to current technologies.

Sequence analyses for vaccine purposes

Gathering and monitoring sequence data is critical for studying IAV and is the foundation from which to build a scientific basis to vaccine decision making. The pace of swine IAV evolution, as well as its relatively small genome, make it possible to identify the genetic basis linked to differences in several virus or virus-host properties.[36] By studying virus gene sequence diversity, scientists gain an appreciation for the processes that shape the evolution and epidemiology of infection.[16] One of the most important advances in understanding the evolutionary properties of IAV has been the development of coalescent-based methods that estimate rates of viral population growth.[42,43] The phylodynamic toolkit has also been used to address epidemic spread, spatio-temporal dynamics, zoonotic transmission, and antigenic drift.[19,60]

Genetic evolution analyses or investigating the genetic basis of viral properties are conducted using nucleotide sequences, however, protein sequences may be more appropriate to use when asking questions regarding host immune responses to viral proteins, such as the HA of IAV. Given that the nucleotide sequence of genes contains the information to create functional proteins, and the flexibility of nucleotide-based evolutionary models, it is occasionally advocated that nucleotides rather than protein sequences should be used in analysis. However, DNA or RNA consist of only four possible nucleotide bases, whereas the linear sequences of proteins are determined by 20 possible amino acids, resulting in greater resolution in certain analyses. In addition, proteins are the biologic unit at which natural selection operates, such as the process of immune evasion, whereby genetic mutations in the HA gene translate into protein modifications that escape antibodies from previous exposures. Not all genetic mutations lead to mutations in the corresponding proteins, but have one of three possible outcomes. First, mutations may be deleterious, and purifying selection would eliminate viruses with these mutations; second, they may be neutral and have no effect on protein function, and because they are not penalized by selection pressures, neutral mutations may accumulate over time; or third, they may be beneficial, and consequently become favorably selected in the viral population, contributing to such processes as genetic drift in IAV. These processes may be inferred using online resources such as DataMonkey.[9]

A number of factors make HA the most studied of the IAV genes. The HA contains the receptor binding site required for infecting the host and is thus the target of neutralizing antibodies. The HA protein is also the source of the hemaggutinating property of IAV and is therefore critical for readouts of HI tests that use agglutination of red blood cells. Understanding HA gene evolution in in the context of vaccine and natural immunity is critical in informing vaccine choices and strain updates. If possible, HA gene datasets should be analyzed in both ways (DNA and protein). However, nucleic acid sequences will likely have limited phylogenetic signal with distantly related sequences, and for sequences that are closely related, protein based analyses may have problems such as codon bias or saturation of the third position of codons. More generally, both analytical approaches for HA sequences are limited with IAV from swine due to the absence of the epidemiological metadata, such as detailed spatial and temporal sampling information or data related to clinical conditions.[20] Thus, the strengths of the available sequence analysis techniques are weakened since phylogenetic pattern and other biological variables like antigenic variation are not easily integrated with specific host population level or other relevant epidemiologic data with the anonymous nature of the USDA surveillance system.

In addition to monitoring overall changes in the HA amino acid sequences, defined regions of the HA may be targeted to ask specific questions regarding antibody recognition. The antigenic regions of a pandemic human H3 virus from 1968 were deduced using monoclonal antibodies against naturally occurring and laboratory produced antigenic variants. These 'antigenic sites' have long served as a reference for antigenic positions of relevance to antigenic drift on the globular head of H3 HAs (131 positions, referred to as regions A-E).[6,64] More recently, antigenic cartography, a computational analysis of antibody-antigen binding assays such as HI assay data,[29] was used to characterize the antigenic evolution of human, swine and equine H3 IAV strains.[30,31,53] The evolution of human influenza H3N2 viruses circulating from 1968 to 2003 resulted in 11 discrete antigenic clusters – where each cluster consisted of antigenically-similar viruses. The distance between each antigenic cluster required an update in the human influenza virus seasonal vaccine strain, and 67 amino acid positions in the HA protein were associated with the transition from one cluster to the next. Using site-directed mutagenesis, it was then determined that single amino acid changes largely accounted for the phenotypic change that resulted in the emergence of a new human H3 antigenic cluster from previous influenza seasons. Although more than 130 positions on the HA were previously characterized as putative antigenic sites, major substitutions leading to the emergence of antigenically distinct seasonal epidemic viruses occurred at just 7 amino acid positions in the HA.[25] Similarly, antigenic cartography of equine H3 from 1968 to 2007 identified 3 antigenic clusters, and again, single amino acid substitutions were likely responsible for cluster transitions and these cluster-defining substitutions occurred at 2 of the same 7 positions identified for human H3.[31] Analyses of the antigenic evolution of swine H3 field isolates from the United States identified two major antigenic clusters (red and cyan)[11,30] and several distinct outlier groups or strains. By combining antigenic cartography with sequence analysis, it was determined that 6 of the same 7 positions identified for human H3 antigenic cluster transitions[30] were associated with cluster/outlier transition in the swine H3 wild type viruses.

The HA sequences of the vaccine and field strains may be compared to predict vaccine efficacy. However, there can be poor correlation between the sequence similarity and the antigenic phenotype of the viruses depending on the nature and location of the mutations and resulting amino acid substitutions. Relying entirely on phylogenetic similarity to assess the appropriateness of a particular vaccine strain has questionable value. Based on previous experimental data, antigenic phenotype of H3N2 did not strongly correlate with phylogenetic clades.[30] Extending on this analysis, we extrapolated potential antigenic phenotype based on the antigenic sites of contemporary swine H3 sequences that are publically available. According to these data, there are at least three major antigenic clusters: cyan, red, and green. Putative cyan-like viruses were predominant in 2009 and are the most common antigenic motif of clade IV and IV-F. However, detection of viruses with the cyan antigenic motif was greatly reduced after 2009 and has been rarely detected since 2013. Commercial vaccines produced and tailored to the epidemiology of IAV-S prior to 2009 would have likely targeted cyan motif viruses, potentially having reduced efficacy against the red or green cluster viruses that emerged after 2010.

Similar to the antigenic regions identified for H3 viruses, the antigenic structure of the H1 HA was also characterized, in which four immunodominant antigenic regions or sites were determined, designated Sa, Sb, Ca, Cb.[7,15] And as shown for the human and swine H3N2 viruses, major antigenic changes in the recent evolution of seasonal viruses pre-H1N1pdm09 were predominantly caused by single amino acid substitutions near the receptor-binding site and previously determined antigenic sites.[25] Although the fairly recent H1N1pdm09 viruses have not yet undergone a major antigenic transition, substitutions in or near the receptor-binding site (mostly in the 151-159 loop) were shown to influence the antigenic properties of these viruses.[26] In a recent study in our laboratory, we observed that the remarkable genetic diversity greatly influences the antigenic diversity of currently circulating US swine H1 IAV. In contrast to the H3 HA, a few amino acid substitutions in or near the antigenic sites or receptor-binding site of H1 HA seem to act cumulatively as antigenic determinants, but antigenic changes could not be correlated with single substitutions at specific amino acid positions at this time.

On the horizon for new tools to look at IAV HA sequences, screening of sequences of human H3 IAV through a user-friendly website that allows for real-time automated tracking of antigenic motifs from circulating strains is now in development,[38] and could be applicable for swine as we better understand the antigenic determinants of the HA from the swine host. Continued research into better understanding the antigenic evolution of IAV-S is a necessary step to help guide swine vaccine manufacturers to develop more efficacious vaccines. Additionally, structural modeling of the HA and NA genes can provide insights into the mechanism of interaction with host innate or adaptive immune mediators (such as antibodies). Since influenza viruses constantly mutate to evade the immune system, computational modeling of the 3D protein structure of the envelope proteins can serve to guide wet lab experiments into research questions of interest. Such protein modeling approaches were used to show that the HAs of H1N1pdm09 and the 1918 pandemic H1N1 actually shared a significant number of amino acid residues in known antigenic sites, suggesting that common epitopes exist for neutralizing antibodies that are cross-reactive to both HAs.[21]

Antigenic characterization

As indicated above, sequence analysis without concurrent antigenic characterization may provide limited value for choosing existing vaccines or making vaccine strain updates since a very small number of amino acid changes in key positions can dramatically affect cross-reactivity. The HI assay remains the gold standard for antigenic characterization in the WHO vaccine strain selection process, despite years invested in attempts to overcome the limitations of the assay. Neutralizing assays in standard[13] and micro[24] formats also can be used to antigenically characterize IAV and have been used by swine vaccine manufacturers.[18] Serum neutralization (SN) assays tend to be more technically challenging, especially when a wide panel of virus antigens needs to be tested, but the SN is reasonably correlated with HI results. Neuraminidase inhibition (NI) assays are also important to consider when multiple lineages of NA genes co-circulate, as is the case for U.S. swine IAV.[50,51] Unpublished data from our lab indicate that the presence of cross-reacting NI antibodies block the effect of mismatched HA in our vaccine associated enhanced respiratory disease (VAERD) model.

Antigenic cartography is a relatively new computational technique for the analysis of large sets of binding assay data and provides a quantification and visualization of HI assay data into "antigenic maps." This tool is routinely used in the WHO vaccine consultation meetings with cross-HI data between ferret anti-sera and circulating human seasonal and vaccine viruses. Although its utility has been established in both research and public and animal health in the context of IAV, the techniques of antigenic cartography are not specific to influenza, to viruses, or even to pathogens in general. Binding assay data (HI, SN, or NI), on which antigenic cartography operates, is widely used in biology for estimating general receptor-ligand interactions. Antigenic maps for swine H3[30,45] and H1[33] IAV have been generated and continue to be useful for quantitating antigenic distance as it relates to genetic variation and antigenic evolution over time, and as a research tool to measure the antigenic impact of amino acid mutations. A related modeling technique, called antibody landscaping, visualizes antibody responses to influenza viruses in relation to currently and previously circulating strains in a given individual or population over time and also following vaccine or natural exposure.[12]

Why vaccines fail

Although sterilizing immunity would be ideal, the reality is that this type of vaccine immunity is rarely achievable with IAV, at least long term, due to its remarkable ability to evolve to evade immunity. The more realistic goals are to prevent clinical disease and reduce virus replication and shedding to the extent that transmission events are greatly reduced without causing immunopathology. Although traditional IAV vaccines seem fairly basic, there are many mitigating factors that can interfere with IAV vaccines being successful at achieving these goals. At a basic level, the importance of antigenic balance in multivalent vaccines, especially with H3 and H1 components, is well known.[44] Some viruses are poor antigens and induce minimal HI titers even with adjuvants. When implementing a new vaccine formulation or strategy into a herd, pre- and post-vaccination serum should be tested against each of the vaccine antigens in the HI assay at the minimum. HI tests with post-vaccination serum against herd-specific outbreak strains gives further confidence in the vaccine. In addition to considerations with vaccine immunogenicity, WIV used in the field can have diminished performance, even with vaccines that performed well under test conditions due to more complicated host immune factors.

In recent years, several different immunologic profiles against a series of exposures over time have been described or resurrected from older literature and reveal potentially complex immune responses to conserved and heterologous epitopes of IAV. There are a number of examples with IAV to indicate the sequence of exposures is extremely important to each subsequent exposure, especially with vaccines. The concept of original antigenic sin with IAV has been a controversial topic for decades,[10] which described a phenomenon whereby the strain specific response to the first exposure blocked a strain specific response to a second virus of a distant antigenic variant of the same subtype or a different subtype. Generally the second heterologous exposure boosted the antibody response to the initial priming exposure at the expense of the secondary response. This has also been shown to happen with less extreme pairs of heterologous viruses, where the heterologous secondary exposure still boosts the priming exposure, but the secondary exposure is only muted rather than blocked. This phenomenon has been given the name of back-boosting.[12] Additionally, the initial infection with IAV that an individual experiences has been proposed to shape subsequent responses and potential mortality patterns seen with pandemics, particularly the 1918 pandemic, and is termed antigenic imprinting.[65] VAERD has been reported in multiple scenarios, but in general the phenomenon results when pigs are vaccinated with oil-in-water adjuvanted WIV and challenged with a homosubtypic but antigenically distinct IAV and results in extensive lung damage. The immune response elicited from the vaccination consists of cross-reactive antibodies that fail to neutralize the mismatched virus.[14,46] Understanding these phenomena in pigs at the individual and population levels may be used to avoid unwanted responses and even to an advantage, such as the potential for heterologous prime and boost vaccine strategies to broaden the antibody repertoire or intentionally inducing a back-boost with a heterologous boost vaccine to increase the antibody titers to the priming exposure.

Control of IAV infection in swine production systems traditionally relied on the vaccination of breeding females with WIV to result in the passive transfer of antibodies to their piglets via colostrum. Ideally, the titers of the MDA should be high enough to protect the piglets during the first weeks of life before they can mount a protective active immune response. However, these passive antibodies can interfere with active humoral and cellular immune responses to vaccine and may lead to VAERD when pigs are vaccinated in the presence of MDA and then exposed to heterologous virus.[32,59] Furthermore, MDA do not usually protect piglets against infection and transmission but rather against clinical disease, provided the antibodies match the IAV strain to which pigs are exposed.[1,2] Mismatched MDA alone was shown to negatively impact the outcome of disease as MDA positive pigs infected with heterologous virus developed VAERD.[47] These complicated immune interactions with IAV are important to keep in perspective when considering the great diversity of IAV circulating in pigs currently and the widespread use of inactivated products for immunization of pigs.

Conclusions

Key questions are: how can we use current vaccines to improve the ability to control IAV in swine and what is on the horizon in the future? The first question requires going back to basic epidemiology and disease control. Each farm needs to know the virus status and immune profile at the pig level and in different age groups. Do the MDA derived from the sows react against the strains the piglets need to be protected from downstream? Then each flow needs to know the virus status and population immune profiles at farm and flow levels. When do MDA wane and does hyper-immunizing sows with WIV prevent the use of vaccine in growing pigs? What are the consequences for IAV evolution and emergence of novel variants within the pig flow with this strategy? What are the consequences if our current sow vaccination strategies were changed? What are the flu viruses circulating at the regional level, i.e., what IAV do your neighbors and pig sources have? How are IAV transmitted, transported, or otherwise shared within pig flows and between regions and farms? Once these profiles are established, endemic and outbreak strains need to be matched to vaccines. This requires knowledge at the genetic and antigenic levels for the strains contained in the vaccines. Can different vaccination strategies be employed to increase effectiveness by expanding the breadth or quality of immunity? Different vaccine application strategies should be tested in swine to better understand the impact of heterologous prime-boost or the effects of applying different vaccines at different timings based on immune profiles as discussed above.

How can we do better with future technologies and opportunities? New vaccine platforms may be on the market soon that will provide alternative options to what is currently available. Nonetheless, an integrated multi-agency approach is needed to improve IAV vaccine strain selection for use in swine. The USDA IAV swine surveillance system has reached a critical mass and is in a position to provide input to national and regional vaccine strain selection, but surveillance alone doesn't lead to better vaccines. Finer scale resolution of spatio-temporal trends in subtype, genotype, and antigenic diversity will allow the identification of long-term patterns of transmission and evolution. Representative viruses from predominant phylogenetic clusters should be evaluated at an antigenic level for serologic cross-reactivity with anti-sera from vaccinated swine. New swine antisera should be generated with IAV that have drifted antigenically or with newly added vaccine strains. *In vivo* swine studies should be conducted when field strains of IAV with substantial changes have been identified or when vaccine strains have been substantially changed. Following the human vaccine model, IAV backbones could be approved by the Center for Veterinary Biologics (CVB) for ease of quickly generating viruses with new and relevant HA and NA for vaccine seed strains. These backbones should exhibit high yield growth properties as well as attenuating mutations in the case of LAIV. Companies may select viruses based on HA and NA sequences from the surveillance system or from their own internal surveillance data for their customers. Viruses from the USDA IAV repository are available for this purpose.

Doing better requires continued funding for surveillance and research, greater collaboration, and translation of newer methodologies from human and other host species for use in swine. Greater transparency with farm and flow level meta-data to connect with virus sequences to fully understand virus evolution and migration is also needed to build upon the foundation established with the surveillance system thus far. Additionally, knowledge of vaccine strains to match to endemic or outbreak strains is also needed. We gratefully acknowledge the pork producers, swine veterinarians, and laboratories for participating in the USDA IAV Swine Surveillance System, establishing the critical foundation from which to build future strategies to prevent and control IAV in swine.

References

1. Allerson M, Deen J, Detmer SE, Gramer MR, Joo HS, Romagosa A, et al.: The impact of maternally derived immunity on influenza A virus transmission in neonatal pig populations. *Vaccine* 2013:31(3):500-505.

2. Allerson MW, Davies PR, Gramer MR, Torremorell M: Infection Dynamics of Pandemic 2009 H1N1 Influenza Virus in a Two-Site Swine Herd. *Transbound Emerg Dis* 2013.

3. Ampofo WK, Azziz-Baumgartner E, Bashir U, Cox NJ, Fasce R, Giovanni M, et al.: Strengthening the influenza vaccine virus selection and development process: Report of the 3rd WHO Informal Consultation for Improving Influenza Vaccine Virus Selection held at WHO headquarters, Geneva, Switzerland, 1-3 April 2014. *Vaccine* 2015:33(36):4368-4382.

4. Anderson TK, Nelson MI, Kitikoon P, Swenson SL, Korslund JA, Vincent AL: Population dynamics of cocirculating swine influenza A viruses in the United States from 2009 to 2012. *Influenza Other Respir Viruses* 2013:7 Suppl 4:42-51.

5. Braucher DR, Henningson JN, Loving CL, Vincent AL, Kim E, Steitz J, et al.: Intranasal vaccination with replication-defective adenovirus type 5 encoding influenza virus hemagglutinin elicits protective immunity to homologous challenge and partial protection to heterologous challenge in pigs. *Clinical and vaccine immunology: CVI* 2012:19(11):1722-1729.

6. Bush RM, Bender CA, Subbarao K, Cox NJ, Fitch WM: Predicting the evolution of human influenza A. *Science* 1999:286(5446):1921-1925.

7. Caton AJ, Brownlee GG, Yewdell JW, Gerhard W: The antigenic structure of the influenza virus A/PR/8/34 hemagglutinin (H1 subtype). *Cell* 1982:31(2 Pt 1):417-427.

8. Cox NJ, Hickling J, Jones R, Rimmelzwaan GF, Lambert LC, Boslego J, et al.: Report on the second WHO integrated meeting on development and clinical trials of influenza vaccines that induce broadly protective and long-lasting immune responses: Geneva, Switzerland, 5-7 May 2014. *Vaccine* 2015.

9. Delport W, Poon AF, Frost SD, Kosakovsky Pond SL: Datamonkey 2010: a suite of phylogenetic analysis tools for evolutionary biology. *Bioinformatics* 2010:26(19):2455-2457.

10. Fazekas de St G, Webster RG: Disquisitions of Original Antigenic Sin. I. Evidence in man. *J Exp Med* 1966:124(3):331-345.

11. Feng Z, Gomez J, Bowman AS, Ye J, Long LP, Nelson SW, et al.: Antigenic characterization of H3N2 influenza A viruses from Ohio agricultural fairs. *J Virol* 2013:87(13):7655-7667.

12. Fonville JM, Fraaij PL, de Mutsert G, Wilks SH, van Beek R, Fouchier RA, et al.: Antigenic Maps of Influenza A(H3N2) Produced With Human Antisera Obtained After Primary Infection. *J Infect Dis* 2015.

13. Gauger PC, Vincent AL: Serum virus neutralization assay for detection and quantitation of serum-neutralizing antibodies to influenza a virus in Swine. *Methods Mol Biol* 2014:1161:313-324.

14. Gauger PC, Vincent AL, Loving CL, Lager KM, Janke BH, Kehrli ME, Jr., et al.: Enhanced pneumonia and disease in pigs vaccinated with an inactivated human-like (delta-cluster) H1N2 vaccine and challenged with pandemic 2009 H1N1 influenza virus. *Vaccine* 2011:29(15):2712-2719.

15. Gerhard W, Yewdell J, Frankel ME, Webster R: Antigenic structure of influenza virus haemagglutinin defined by hybridoma antibodies. *Nature* 1981:290(5808):713-717.

16. Grenfell BT, Pybus OG, Gog JR, Wood JL, Daly JM, Mumford JA, et al.: Unifying the epidemiological and evolutionary dynamics of pathogens. *Science* 2004:303(5656):327-332.

17. Harrisvaccines: Harrisvaccines awarded vaccine product licensure; 2012. Available at: `http://www.harrisvaccines.com/blog/posts/2012/september/harrisvaccines-awarded-vaccine-product-licensure/`. Accessed 9 November 2015.

18. Hause BM, Oleson TA, Bey RF, Stine DL, Simonson RR: Antigenic categorization of contemporary H3N2 Swine influenza virus isolates using a high-throughput serum neutralization assay. *J Vet Diagn Invest* 2010:22(3):352-359.

19. Holmes EC: The phylogeography of human viruses. *Mol Ecol* 2004:13(4):745-756.

20. Holmes EC: Viral evolution in the genomic age. *PLoS Biol* 2007:5(10):e278.

21. Igarashi M, Ito K, Yoshida R, Tomabechi D, Kida H, Takada A: Predicting the antigenic structure of the pandemic (H1N1) 2009 influenza virus hemagglutinin. *PLoS One* 2010:5(1):e8553.

22. Kappes MA, Sandbulte MR, Platt R, Wang C, Lager KM, Henningson JN, et al.: Vaccination with NS1-truncated H3N2 swine influenza virus primes T cells and confers cross-protection against an H1N1 heterosubtypic challenge in pigs. *Vaccine* 2012:30(2):280-288.

23. Karasin AI, Landgraf J, Swenson S, Erickson G, Goyal S, Woodruff M, et al.: Genetic characterization of H1N2 influenza A viruses isolated from pigs throughout the United States. *J Clin Microbiol* 2002:40(3):1073-1079.

24. Kitikoon P, Vincent AL: Microneutralization assay for Swine influenza virus in Swine serum. *Methods Mol Biol* 2014:1161:325-335.

25. Koel BF, Burke DF, Bestebroer TM, van der Vliet S, Zondag GC, Vervaet G, et al.: Substitutions near the receptor binding site determine major antigenic change during influenza virus evolution. *Science* 2013:342(6161):976-979.

26. Koel BF, Mogling R, Chutinimitkul S, Fraaij PL, Burke DF, van der Vliet S, et al.: Identification of amino acid substitutions supporting antigenic change of influenza A(H1N1)pdm09 viruses. *J Virol* 2015:89(7):3763-3775.

27. Krammer F, Hai R, Yondola M, Tan GS, Leyva-Grado VH, Ryder AB, et al.: Assessment of influenza virus hemagglutinin stalk-based immunity in ferrets. *J Virol* 2014:88(6):3432-3442.

28. Krammer F, Margine I, Hai R, Flood A, Hirsh A, Tsvetnitsky V, et al.: H3 stalk-based chimeric hemagglutinin influenza virus constructs protect mice from H7N9 challenge. *J Virol* 2014:88(4):2340-2343.

29. Lapedes A, Farber R: The geometry of shape space: application to influenza. *J Theor Biol* 2001:212(1):57-69.

30. Lewis NS, Anderson TK, Kitikoon P, Skepner E, Burke DF, Vincent AL: Substitutions near the hemagglutinin receptor-binding site determine the antigenic evolution of influenza A H3N2 viruses in U.S. swine. *J Virol* 2014:88(9):4752-4763.

31. Lewis NS, Daly JM, Russell CA, Horton DL, Skepner E, Bryant NA, et al.: Antigenic and genetic evolution of equine influenza A (H3N8) virus from 1968 to 2007. *J Virol* 2011:85(23):12742-12749.

32. Loeffen WL, Heinen PP, Bianchi AT, Hunneman WA, Verheijden JH: Effect of maternally derived antibodies on the clinical signs and immune response in pigs after primary and secondary infection with an influenza H1N1 virus. *Vet Immunol Immunopathol* 2003:92(1-2):23-35.

33. Lorusso A, Vincent AL, Harland ML, Alt D, Bayles DO, Swenson SL, et al.: Genetic and antigenic characterization of H1 influenza viruses from United States swine from 2008. *J Gen Virol* 2011:92(4):919-930.

34. Loving CL, Lager KM, Vincent AL, Brockmeier SL, Gauger PC, Anderson TK, et al.: Efficacy in pigs of inactivated and live attenuated influenza virus vaccines against infection and transmission of an emerging H3N2 similar to the 2011-2012 H3N2v. *J Virol* 2013:87(17):9895-9903.

35. Masic A, Booth JS, Mutwiri GK, Babiuk LA, Zhou Y: Elastase-dependent live attenuated swine influenza A viruses are immunogenic and confer protection against swine influenza A virus infection in pigs. *J Virol* 2009:83(19):10198-10210.

36. Moya A, Elena SF, Bracho A, Miralles R, Barrio E: The evolution of RNA viruses: A population genetics view. *Proc Natl Acad Sci U S A* 2000:97(13):6967-6973.

37. Nachbagauer R, Wohlbold TJ, Hirsh A, Hai R, Sjursen H, Palese P, et al.: Induction of broadly reactive anti-hemagglutinin stalk antibodies by an H5N1 vaccine in humans. *J Virol* 2014:88(22):13260-13268.

38. Neher RA, Bedford T: nextflu: real-time tracking of seasonal influenza virus evolution in humans. *Bioinformatics* 2015.

39. Nelson MI, Vincent AL, Kitikoon P, Holmes EC, Gramer MR: Evolution of novel reassortant A/H3N2 influenza viruses in North American swine and humans, 2009-2011. *J Virol* 2012:86(16):8872-8878.

40. Olsen CW: The emergence of novel swine influenza viruses in North America. *Virus Res* 2002:85(2):199-210.

41. Pena L, Vincent AL, Ye J, Ciacci-Zanella JR, Angel M, Lorusso A, et al.: Modifications in the polymerase genes of a swine-like triple-reassortant influenza virus to generate live attenuated vaccines against 2009 pandemic H1N1 viruses. *J Virol* 2011:85(1):456-469.

42. Pybus OG, Charleston MA, Gupta S, Rambaut A, Holmes EC, Harvey PH: The epidemic behavior of the hepatitis C virus. *Science* 2001:292(5525):2323-2325.

43. Pybus OG, Rambaut A, Harvey PH: An integrated framework for the inference of viral population history from reconstructed genealogies. *Genetics* 2000:155(3):1429-1437.

44. Quinnan GV, Schooley R, Dolin R, Ennis FA, Gross P, Gwaltney JM: Serologic responses and systemic reactions in adults after vaccination with monovalent A/USSR/77 and trivalent A/USSR/77, A/Texas/77, B/Hong Kong/72 influenza vaccines. *Rev Infect Dis* 1983:5(4):748-757.

45. Rajao DS, Gauger PC, Anderson TK, Lewis NS, Abente EJ, Killian ML, et al.: Novel Reassortant Human-Like H3N2 and H3N1 Influenza A Viruses Detected in Pigs Are Virulent and Antigenically Distinct from Swine Viruses Endemic to the United States. *J Virol* 2015:89(22):11213-11222.

46 Rajao DS, Loving CL, Gauger PC, Kitikoon P, Vincent AL: Influenza A virus hemagglutinin protein subunit vaccine elicits vaccine-associated enhanced respiratory disease in pigs. *Vaccine* 2014:32(40):5170-5176.

47. Rajao DS, Vincent AL, Loving CL, Gauger PC, Sandbulte MR, Kitikoon P: Maternally derived antibodies induce vaccine-associated enhanced respiratory disease in weaned pigs challenged with heterologous virus. *American Association of Swine Veterinarians Annual Meeting*. Dallas; 2014: 51-54.

48. Richt JA, Lekcharoensuk P, Lager KM, Vincent AL, Loiacono CM, Janke BH, et al.: Vaccination of Pigs Against Swine Influenza Viruses Using a NS1-Truncated Modified Live Virus Vaccine. *J Virol* 2006:80(22):11009-11018.

49. Russell CA, Jones TC, Barr IG, Cox NJ, Garten RJ, Gregory V, et al.: Influenza vaccine strain selection and recent studies on the global migration of seasonal influenza viruses. *Vaccine* 2008:26 Suppl 4:D31-34.

50. Sandbulte MR, Eichelberger MC: Analyzing swine sera for functional antibody titers against influenza A neuraminidase proteins using an enzyme-linked lectin assay (ELLA). *Methods Mol Biol* 2014:1161:337-345.

51. Sandbulte MR, Gao J, Straight TM, Eichelberger MC: A miniaturized assay for influenza neuraminidase-inhibiting antibodies utilizing reverse genetics-derived antigens. *Influenza Other Respir Viruses* 2009:3(5):233-240.

52. Sandbulte MR, Spickler AR, Zaabel PK, Roth JA: Optimal Use of Vaccines for Control of Influenza A Virus in Swine. *Vaccines (Basel)* 2015:3(1):22-73.

53. Smith DJ, Lapedes AS, de Jong JC, Bestebroer TM, Rimmelzwaan GF, Osterhaus AD, et al.: Mapping the antigenic and genetic evolution of influenza virus. *Science* 2004:305(5682):371-376.

54. Solorzano A, Webby RJ, Lager KM, Janke BH, Garcia-Sastre A, Richt JA: Mutations in the NS1 protein of swine influenza virus impair anti-interferon activity and confer attenuation in pigs. *J Virol* 2005:79(12):7535-7543.

55. Vander Veen RL, Loynachan AT, Mogler MA, Russell BJ, Harris DL, Kamrud KI: Safety, immunogenicity, and efficacy of an alphavirus replicon-based swine influenza virus hemagglutinin vaccine. *Vaccine* 2012:30(11):1944-1950.

56. Vander Veen RL, Mogler MA, Russell BJ, Loynachan AT, Harris DL, Kamrud KI: Haemagglutinin and nucleoprotein replicon particle vaccination of swine protects against the pandemic H1N1 2009 virus. *Vet Rec* 2013:173(14):344.

57. Vincent AL, Ma W, Lager KM, Gramer MR, Richt JA, Janke BH: Characterization of a newly emerged genetic cluster of H1N1 and H1N2 swine influenza virus in the United States. *Virus Genes* 2009:39:176-185.

58. Vincent AL, Ma W, Lager KM, Janke BH, Webby RJ, Garcia-Sastre A, et al.: Efficacy of intranasal administration of a truncated NS1 modified live influenza virus vaccine in swine. *Vaccine* 2007:25(47):7999-8009.

59. Vincent AL, Ma W, Lager KM, Richt JA, Janke BH, Sandbulte MR, et al.: Live attenuated influenza vaccine provides superior protection from heterologous infection in pigs with maternal antibodies without inducing vaccine-associated enhanced respiratory disease. *J Virol* 2012:86(19):10597-10605.

60. Volz EM, Koelle K, Bedford T: Viral phylodynamics. *PLoS Comput Biol* 2013:9(3):e1002947.

61. Wesley RD, Lager KM: Overcoming maternal antibody interference by vaccination with human adenovirus 5 recombinant viruses expressing the hemagglutinin and the nucleoprotein of swine influenza virus. *Vet Microbiol* 2006:118(1-2):67-75.

62. Wesley RD, Tang M, Lager KM: Protection of weaned pigs by vaccination with human adenovirus 5 recombinant viruses expressing the hemagglutinin and the nucleoprotein of H3N2 swine influenza virus. *Vaccine* 2004:22(25-26):3427-3434.

63. Wiersma LC, Rimmelzwaan GF, de Vries RD: Developing Universal Influenza Vaccines: Hitting the Nail, Not Just on the Head. *Vaccines (Basel)* 2015:3(2):239-262.

64. Wiley DC, Wilson IA, Skehel JJ: Structural identification of the antibody-binding sites of Hong Kong influenza haemagglutinin and their involvement in antigenic variation. *Nature* 1981:289(5796):373-378.

65. Worobey M, Han GZ, Rambaut A: Genesis and pathogenesis of the 1918 pandemic H1N1 influenza A virus. *Proc Natl Acad Sci U S A* 2014:111(22):8107-8112.

66. Zhou NN, Senne DA, Landgraf JS, Swenson SL, Erickson G, Rossow K, et al.: Genetic reassortment of avian, swine, and human influenza A viruses in American pigs. *J Virol* 1999:73(10):8851-8856.

FDA's antibiotic regulatory changes and NPB's pork producer educational outreach

J. Koeman, MSc, MPH, DVM, DACVPM
National Pork Board, Clive, Iowa

Introduction

The US Food and Drug Administration (FDA) is taking a new approach on the judicious use of medically important (to treat human illness) antibiotics in food-animal production that involves voluntary cooperation by all sectors.

On Jan. 1, 2017, the FDA's new regulations addressing on-farm antibiotic use in food-animal production will take effect. The agency's effort is aimed at eliminating the use of medically important (to human illness) antibiotics for growth-promotion purposes in food-animal production and bringing therapeutic use in feed and water – to treat, control or prevent spe cific disease – under additional veterinary oversight. Producers, veterinarians, feed mills and suppliers will all face new requirements.

FDA's new regulations defined

All sectors of animal agriculture, from drug companies to producers and veterinarians to feed mills, are collaborating to implement the new FDA antibiotic regulations:

Guidance 209

In 2010, FDA outlined its intent and recommendations regarding growth promotion uses of medically important antibiotics in food-animal production. FDA also specified that veterinarian oversight will increase for the remaining therapeutic applications (prevention, treatment and control) of medically important antibiotics. This action applies to both feed-grade and water-based antibiotics.

Guidance 213

On Dec. 11, 2013, FDA initiated a three-year transition process to complete its food-animal antibiotic strategy. This action requests animal-health companies to outline intentions to voluntarily remove any production/growth-promotion uses from product labels of medically important antibiotics. The guidance also eliminates over-the-counter status of these medications and increases veterinary oversight for on-farm therapeutic use by requiring a veterinary feed directive (VFD) for feed applications and a prescription for water treatments. Jan. 1, 2017, is when implementation must be completed.

Final Veterinary Feed Directive (VFD) Rule

On June 3, 2015, FDA announced the final VFD rule. The final rule outlines specific requirements of the VFD process for medically important (to treat human infections) feed-grade antibiotics.

USCARE: Producers checklist for antibiotic success

The Pork Checkoff offers USCARE as an easy way to remember the six key steps producers need to complete before January 2017 to ensure successful compliance with FDA's new regulations for on-farm use of medically important (to human health) antibiotics.

Understand the new feed (Veterinary Feed Directive) and water (prescription) rules: As of Jan. 1, 2017, growth promotion use of medically important (to human health) antibiotics will not be allowed. Only therapeutic use (treatment, control, prevention) for a specific animal health condition will be allowed under the direction of a veterinarian.

Strengthen your veterinary-client-patient relationship (VCPR): Schedule periodic herd visits with your veterinarian and review health monitoring and herd health strategies.

Communicate with your feed mill: Ensure the mill personnel understand and are prepared to implement new VFD record-keeping procedures for antibiotics and that stringent feed delivery protocols are in place.

Assess your herd health and welfare strategies: Sit down with your veterinarian to outline production and management changes to maximize animal health and minimize antibiotic use.

Renew your commitment to responsible antibiotic use: Stay up to date with Pork Quality Assurance Plus certification and make it part of your worker education program.

Ensure your record-keeping compliance: For producers, this means keeping copies (print or electronic) of VFD for two years and prescription records for one year.

Impacts on the farm

The implementation of Guidance 213 and the revised VFD rule will change how antimicrobials are used in food animal production on farms of all sizes. Previously, only three products required a VFD, and all are labeled for swine use. Under the new regulations, most antibiotics labeled for use in swine will be affected by the VFD rule; however, a few compounds of interest will not be affected. These include bacitracin, tiamulin, carbadox, ionophores and bambermycin. For a list of affected drug classes, see FDA's Guidance for Industry 152: Appendix A.

While it is expected that the revisions to the VFD rule will make the process more user friendly, the new FDA rule will require additional time and effort on the part of producers and their veterinarians. Veterinarians and producers will need to sit down together to discuss how to apply the new VFD and prescription requirements and other herd health strategies. The veterinarian can help design herd health and production plans to decrease disease risks, enhance performance and produce safe, wholesome pork. Each producer should review these on-farm protocols to further minimize the need for antibiotic use. During the transition period there will be many situations and scenarios that will need to be clarified to allow the veterinarian, feed mill and producer to properly comply with the rule.

Some feel that the overall success of these changes should be measured by a decrease in pounds of antibiotics sold. However, this is a far too simplistic method of assessment. It is uncertain how much clinical disease is currently being prevented by the uses of antibiotics that will require treatment following the removal of growth promotion antibiotics. The pork industry, along with the other species groups, is currently assessing the feasibility and obstacles to collection and analysis of privately held antibiotic use data for the purpose of monitoring trends in antibiotic use in their respective industries.

Conclusions

The FDA Guidance for Industry #213 and the changes to the VFD rule will have substantial impact on antimicrobial use decisions on farm. Producers and veterinarians should plan for increased time, effort and cost to ensure access to needed medications and to comply with documentation of the judicious use of therapeutic antimicrobials. A close working relationship between veterinarians and producers is key to complying with the new regulatory changes and to identifying other herd health strategies to minimize the need for antibiotics and to maintain optimal animal health.

Implementing the VFD rule: Feed company perspective

Ronny Moser, PhD; Scott Ringger
JBS United, Inc.

Introduction

The evolution of the veterinary feed directive (VFD) began with the United States Food and Drug Administration's efforts to ensure appropriate use of antimicrobials provided in feeds since the 1960's. The VFD process began with the Animal Drug Availability Act (ADAA) of 1996 from which a "judicious use" policy was established in lieu of prescriptions governed by state or federal law. Within 70 days of enactment of ADAA, the first VFD, Tilmicosin (Pulmotil®) was approved by FDA and was the only VFD drug in the marketplace for many years. Currently, there are three VFD drugs approved for use in swine, beef and fish. There will potentially be greater than 280 VFD labeled drugs after January 1, 2017 when FDA's judicious use of antimicrobials is fully implemented.

Compliance to the VFD process will be dependent on cooperation of the drug sponsors, veterinarians, the feed industry and our collective clients. The feed industry's response has largely been driven by the actions of the American Feed Industry Association (AFIA) which is committed to representing the total feed industry and member companies' interest on matters involving federal and state legislation and regulation. AFIA has been actively involved with the FDA in the interpretation of the regulations, and with its membership to educate and formulate appropriate responses. Even though AFIA is committed to one industry leadership voice, the following is JBS United's specific response to compliance with VFD regulation.

JBS United, Inc., based in Sheridan, Indiana, develops and markets animal nutrition and health solutions that drive profit for our clients. We have 6 feed mills throughout the Midwestern USA that manufacture blended products containing drugs that fall under the VFD regulations. Fostering a culture of quality service that exceeds our clients' expectations requires us to become heavily invested in assuring the success of the VFD process. For the sake of this discussion and to be consistent with terminology of the VFD regulations, JBS United feed mills will be designated as "Distributors".

Regarding VFD's, JBS United's roles and responsibilities for each of our distributors are: 1) receive an electronic or handwritten VFD, 2) review it for accuracy and completeness, 3) store and access submitted VFD prior to distributing VFD products, 4) supply the client with a high quality VFD products within the requested timeframe and 5) ensure that all JBS United distributors comply with regulatory requirements.

Past challenges

As JBS United complied with the VFD regulation, we faced several challenges. Disruptions occurred in supplying clients with VFD products due to our distributors having received incomplete VFD's or not receiving original VFD within 5 days as required by the original regulations. In some cases, the need to deliver product to meet a client's urgent need resulted in nonconformities observed during FDA inspections. In addition, VFD compliance dramatically increased recording-keeping by the distributor. In 2014, JBS United's 6 distributors received and processed 240 VFD's, involving only one drug. We received 70% of the VFD's electronically, the rest were received via mail or fax. Even the electronic records had to be printed as a paper copy. All records had to be retained for two years, a greater retention time than is required for Current Good Manufacturing Practices (CGMPs) for medicated feeds. For example, medicated feed assay records have to be retained for only one year.

To address these initial challenges and to prepare for increased effort starting in January, 2017, JBS United developed procedures to manage VFD's at each of our distributors. Upon receipt of a VFD, the distributor's customer service representative will review and approve each VFD. The customer service rep will access SharePoint, a web application platform in the Microsoft office server suite that is used as a secure place to store, organize, share and access information. The valid VFD will be stored prior to distributing a VFD product and the JBS United account manager is alerted via email that the VFD has been received and approved. A client's feed order is submitted with a valid VFD for manufacturing, loading and delivery to the client. To further manage a specific VFD, SharePoint is programmed to send email alerts to the distributor and the client's account manager on dates 2 weeks prior, 1 week prior and upon date of expiration.

As the collective industry gained experience, sought further clarity in interpretation and addressed challenges, the FDA approved changes to the VFD process in June, 2015. Some items previously required were deleted from the VFD submission form. Maximum VFD expiration was extended from 3 to 6 months. Distributors are no longer required to maintain the original VFD. And generic substitutions are allowed unless indicated by the veterinarian. The feed industry, including JBS United embraced these changes viewed as improvement to the VFD process.

® Registered Trademark of Elanco Animal Health

Future concerns

As we approach January 1, 2017, when the judicious use of antimicrobials will be fully implemented, JBS United and the feed industry have concerns for FDA to consider. Their interpretation of the two-year record retention rule needs clarity given the amount of paper that could be required to keep on file. Some from FDA have stated that retention of paper copies will be required while others have stated that electronic copies will be sufficient. Not all feed industry distributors can comply with the electronic signature rule (21 CFR part 11) required for distributors to store electronic records. Therefore, feed industry companies and their distributors have been advised to keep paper copies until the rule is clarified. As stated earlier, we prefer a one-year retention time because it is consistent with CGMPs for medicated feeds required for the feed industry.

The feed industry must receive label changes from the drug sponsors in time to make changes. FDA has stated that it will "bank or hold" all changes submitted by drug sponsors until their release on one day in December, 2016, making it very difficult for the feed industry to make any changes in a timely manner.

During this transition, adequate supply of old, approved drugs from drug sponsors is required as the new ones come onto the marketplace. Some drug sponsors have stated that it may take many months to move the newly labeled drugs into the marketplace. In addition, adequate time to exhaust existing supply of drugs with old labels is needed. Otherwise, there may be requests to return unused drug to drug sponsors. If all options are exhausted, disposal of drugs will be required at considerable cost to the collective industry.

The collective industry needs to formulate a plan for unused premix product with old drug labels located at the distributor and/or client. FDA has stated that on January 1, 2017, all drug premix products labeled with growth promotion claims will be allowed to be utilized with a valid VFD for only therapeutic/preventive purposes. These premix products will likely be labeled with stickers indicating that a VFD is required and that no growth promotion claims will be allowed on finished feed products (type C medicated feed) after January 1, 2017.

It is vital that all veterinarians issuing VFDs be trained and knowledgeable of feed drugs, approved combinations and different approvals by different drug sponsors for the same compounds. We recognize that most veterinarians are knowledgeable about the VFD process, but the feed industry is concerned that many veterinarians not knowledgeable about feed drugs will likely become involved in the process. We recommend the development of online training that describes the VFD process and stores a list of veterinarians who have participated in the training.

Cooperation between the drug sponsors, veterinarians, the feed industry and our clients is paramount to address all major issues and ensure clients receive VFD products in a timely manner to prevent and treat disease and promote animal welfare. The feed industry is committed to this cooperation.

Implementing the VFD rule: Practitioner perspective

Paul D. Ruen, DVM
Fairmont Veterinary Clinic, LLP, Fairmont, Minnesota

Introduction

The Food and Drug Administration (FDA), due to increased concerns and pressures surrounding the use of antimicrobials in livestock, implemented a new Veterinary Feed Directive (VFD) rule in June 2015.[1] The result is that food animal veterinarians will now have direct oversight of all antibiotic use in feed for prevention, control and therapy. The VFD rule followed release in December 2013 of Guidance 213,[2] which outlined a 3-year plan to transition the industry away from all growth promoter use of medically important antibiotics, eliminate over-the-counter status of these medications, and establish veterinary oversight for on-farm use of therapeutic antibiotics in feed (VFD) and water (prescription).

In most cases swine veterinarians are already engaged with pork producers in the decisions related to the use of feed-grade antibiotics. The Pork Quality Assurance®[3] program, initiated by the National Pork Producers Association in 1989 and later updated to the Pork Quality Assurance Plus® program, has relied on a close partnership between producers and their veterinarians and has achieved wide acceptance in the US swine industry. Regardless of changing regulations, it is important that swine veterinarians measure up to the standards of our veterinary oath, working hand-in-hand with our clients in support of the best care of the pig, stewardship of natural resources, and the promotion of public health. The diagnostic tools and resources available at universities, commercial laboratories, and in our own practices must be used to direct health strategies that minimize the impact of disease in pig populations. The purpose of this presentation is to outline components of the new VFD and provide direction for practical implementation in the field.

Discussion

Unlike other food animal veterinarians, swine practitioners have some familiarity with the use of VFDs because of the previous VFD requirement to use feed-grade Pulmotil® and Nuflor®. Requiring VFDs for all other feed antibiotics listed as "important for humans" and approved for prevention, control, and treatment will mean additional responsibilities, complications, and costs to manage the health care of swine.[1] Guidance for Industry #120,[4] released by the FDA in September 2015, provides some clarification to the details of the new VFD rule. Feed medications unaffected by the rule change are bacitracin, carbadox, bambermycin, tiamulin, and ionophores. VFDs for affected medications will be required as labels are changed. Sponsors and the FDA are working to synchronize this in December 2016. Practical implementation of VFDs for our clients will require us to meet 4 objectives: legal, accurate, efficient, and on-time.

Legal:

The following are legal responsibilities of veterinarians writing VFDs:[5]

- License to practice veterinary medicine in the state where pigs are to be fed
- Valid veterinary client patient relationship (VCPR)
- Use of VFD in compliance with laws
- Prepared written VFD, with all required information and signature
- Provide client and distributor a copy of VFD
- Retain VFD records for 2 years
- Provide VFD orders for inspection by FDA upon request

In the new VFD rule, there is no requirement to list a veterinary license number or State of issuance. However, it is required that the issuing veterinarian be licensed in the State where the pigs are to be fed, and is operating in compliance with all applicable veterinary practice requirements. The FDA defers to individual States' veterinary practice acts for defining veterinary-client-patient-relationship (VCPR) except in those cases where no appropriate State VCPR exists, in which case the Federal VCPR will apply. This is meant to be consistent with antibiotic use via other routes where prescription is required and appropriate veterinary oversight is determined by State licensing regulations and boards. The Model Practice Act, Federal definition of VCPR,[6] and most state definitions do not define documentation requirements. And there are no specific legal requirements for diagnostics, bacterial culture, and sensitivities. The key elements of the federally-defined VCPR include: (1) engage with the client to assume responsibility for making clinical judgments about patient health, (2) have sufficient knowledge of the patient by virtue of examination and/or visits to the facility where patient is managed, and (3) provide for any necessary follow-up evaluation or care. Tables 1 and 2 list states that follow Federal v. State VCPR,[7] respectively.

Veterinary practices and systems are continuing on the path toward medical records systems that promote timely and accurate decisions in the field. When I first joined the practice in Fairmont in 1992, we visited the majority of our swine clients' farrow-to-finish farms on a weekly or biweekly schedule. "Boots on the ground" ensured we were in the loop for most treatment decisions. Today the amount of contact swine veterinarians have with individual groups of pigs varies considerably depending on the type of practice and system of production. While site visits are still an important component to a working VCPR, quality diagnostics, accurate production reports, and a valuable team of skilled swine technicians and managers are now part of the new models of swine health care. By bringing all antibiotic medication

Table 1: States following Federal VCPR

Alabama	Hawaii	Pennsylvania
Alaska	Kansas	Rhode Island
Arkansas	Maryland	South Dakota
Connecticut	Massachusetts	Vermont
Delaware	Michigan	Washington
District of Columbia	Montana	West Virginia
Florida	New Jersey	Wisconsin
Georgia	New York	

Table 2: States following State VCPR

Arizona	Maine	Ohio
California	Minnesota	Oklahoma
Colorado	Mississippi	Oregon
Idaho	Missouri	South Carolina
Illinois	Nebraska	Tennessee
Indiana	Nevada	Texas
Iowa	New Hampshire	Utah
Kentucky	North Carolina	Virginia
Louisiana	North Dakota	Wyoming

use under the oversight of veterinarians, the FDA expects us to deliver on our veterinary training to work closely with clients to promote animal care, better health, and food safety.

Accurate:

The key components necessary for completing a VFD include:[5]

- Veterinarian's name, address, and phone number
- Client's name, business or home address, and phone number
- Premise(s) where the animals are located and will be eating the medicated feed
- VFD issuance date and VFD expiration date (maximum 6 months) based on calendar date
- Name of drug to be used, level of drug in the feed, and duration of use to the animals
- Instructions and usage information if intended to be fed in combination with other VFD or OTC drug
- Species and production class of animals receiving medication
- Approximate number of animals to receive the medicated feed before the VFD expires
- Withdrawal time, and special instructions and/or precautions
- Number of reorders (refills) that are authorized (if permitted by drug approval)
- Statement declaring that extra label use is not permitted

Accuracy will be in proportion to the investment of time and effort spent by veterinarians to communicate with clients and mills. Communication flow will need to address two categories of VFDs, one targeted to a farm's normal disease challenges and a second one that addresses disease outbreaks or emergency. We need to ensure in all cases that VFDs help get the right feed medication to the right pig at the right time. With enforced oversight, comes a responsibility on our part to make the agreed plan easy to understand and follow. Fortunately, swine producers have a good record of achieving proper antibiotic withdrawals. Data from USDA FSIS[8] clearly demonstrate the success of pig producers in delivering antibiotic residue-free pigs to harvest. Preliminary results of multi-drug KIS testing on roaster swine and sows in 2015 found 0.26% (29 carcasses out of 10,959 sampled) with any violative residue. Because current feed grade antibiotics have largely existed as over-the-counter (OTC) products, this demonstrates the safety and inclination of producers and feed mills to follow the label directions.

Additional information may be included or attached to the VFD to improve accuracy of the VFD such as a more specific description of animal location(s), approximate age range of animals, approximate weight range of animals, diet(s) that should include the listed medication, or any other information the veterinarian determines to be appropriate in identifying the animals.

Efficient:

During the course of developing the new VFD, the FDA considered and implemented some changes to the rule that improve the practicality of VFD orders. One such change extends the maximum allowable expiration time for a VFD to 6 months instead of the old rule's 90 days. For farmers, feed mills, and veterinarians, this longer window of time fits closely to a feeding cycle for the wean-to-finish herd. But costs anticipated by the FDA related to the implementation of the new VFD rule likely underestimate the full and ongoing expenses we will incur to managing the new process for use of antibiotics in feed. The FDA estimates a one-time industry cost of $1,411,000 to cover review of the rule and development of compliance plans. However they go on to predict savings of $13,000 per year due to improved efficiency of the new VFD process and another $7.87 million per year savings in veterinary labor.[1] Only the government can add extensive regulatory and oversight requirements yet claim to have created a cost savings! Obviously the comparison is between implementation of the old VFD to the new VFD rule. Time estimates based on personal experience indicate that it will conservatively run 7 to 10 minutes per VFD. That time will be spread across veterinarian and office staff, and does not include investments for software and records systems, or people hours to update information on clients, animals, sites, flows, and feed mills. A few examples of anticipated VFD use in my practice illustrates that the time involved to comply with the new VFD rule will be substantial.

Table 3 describes a swine producer that gets a share of weaned pigs from a producer-owned sow farm every 9 weeks. His feed originates from a local commercial mill, and all the pigs are on a single farm site, so it's a simple 1 nursery and 2 finisher barn flow. As you will note from the table, this producer requires 5 VFDs to accomplish the current feed medication strategy.

Table 4 is the flow of pigs from another 3000 sow farm that weans to a common nursery and then member-owners buy the feeder pigs in rotation to be delivered to their sites for finishing. The 9 owners, 20 sites, 43 barns, and 8 feed mills results in the writing of 42 VFDs.

Table 5 lists the sites and barns for a farmer that owns shares in several sow farms. Pigs arrive as weaners and feeders, entering both owned and contract barns. Two feed mills are utilized in feeding the pigs. Altogether this producer needs 30 VFDs to accomplish the feed medication plans.

All distributors must submit a one-time notification to the FDA of their intent to distribute medicated feed containing a VFD drug.[5] The term distributor means any person who distributes a medicated feed containing a VFD drug to another person, whether that is to another distributor or a recipient of a VFD such as a client making feed for their own pigs. The distributor notification must include the distributor's name and business address, signature of the distributor or their authorized agent, and the date notice was signed.

The FDA improved the efficiency of the VFD process when they removed the requirement for veterinarians to list the amount of feed that will be consumed during the VFD feeding period. That cumbersome step was often inaccurate and to most of us seemed unnecessary because only labeled feeding rates and durations of use are legal. An estimate of the number of pigs to be fed under the VFD is required, but the flexibility to include multiple sites under the same VFD will save considerable time in many farm situations. Multiple groups of animals with the same ownership and fed from the same mill can now be listed together on one VFD when the veterinarian deems that appropriate. This provides for both judicious antibiotic use and improved process efficiency.

On-time:

Extra paperwork should not hinder animals from getting the medication they need when they need it. Therefore veterinarians

Table 3: Single pig flow, single pig site

Client	Flow	Sites	Barns	# Pigs	Feed mills	Diets w/ AB	Antibiotic(s)	# VFDs
Ole	A	1	1 Nur	625	1 com E	Nursery 1	Neo/Terra 600g	1
			2 Fin	1200		Nursery 2	Neo/Terra 400g	1
						Nursery 3	CTC 400g, Den 35g	1
						Nursery 4	No med	
						Finish 1	CTC 400g, Den 35g	1
						Finish 2	BMD 30g	0
						Finish 3	BMD 30g	0
						Finish 4	CTC 600g	1
						Finish 5	BMD 30g	0
						Finish 6	BMD 30g	0
Total			**3**		**1**			**5 VFDs**

Table 4: Single pig flow, multiple owners, pig sites, and mills

Client	Flow	Sites	Barns	# Pigs	Feed mills	# AB Diets	Antibiotic(s)	# VFDs
Nurs	RD	1	6 N	7800	1 com T	3	CTC/Den, Tylan	3
					1 farm			5
Prod 1	RD	8	14 F	13800	1 com LR	6	CTC/Den, Tylan, CTC, BMD	5
Prod 2	RD	1	2	4800	1 farm	7	CTC/Den, Tylan, CTC, BMD	4
Prod 3	RD	1	5	3600	1 farm	6	CTC/Den, Tylan, CTC	6
Prod 4	RD	2	2	1200	1 com J	5	CTC/Den, BMD, CTC	2
Prod 5	RD	1	1	600	1 com L	5	CTC/Den, Tylan, CTC	5
Prod 6	RD	1	2	1200	1 farm	3	CTC/Den, Tylan, CTC, Skycis	3
Prod 7	RD	3	7	4200	1 com J	5	Pulm, CTC/Den, BMD, CTC	3
Prod 8	RD	1	2	1200	1 com J	5	CTC/Den, BMD, CTC	3
Prod 9	RD	1	2	2400	1 com T	3	CTC/Den, Tylan, CTC, Skycis	3
Totals		**20**	**43**	**40800**	**8**			**42 VFDs**

Table 5: Single owner, multiple pig flows and sites

Client	Flow	Sites	Barns	Feed mills	# AB diets	Antibiotic(s)	# VFDs
Sven	CC	6	6 WF	1 farm	8	CTC/Den, Tylan, CTC, BMD	6
	CC	2	8 WF	1 com LR	8	CTC/Den, Tylan, CTC, BMD	6
	RD	7	12 F	1 farm	6	CTC/Den, Tylan, CTC, BMD	5
	RD	1	2 F	1 com LR	6	CTC/Den, Tylan, CTC, BMD	5
	PC	2	2 N	1 farm	3	CTC/Den, Tylan	3
	PC	3	4 F	1 farm	6	CTC/Den, Tylan, CTC, BMD	5
Total	**3**	**20**	**34**	**1**			**30 VFDs**

will want to work closely with clients and feed mills to share required information in preparation for the advent of new medicated feed labels and VFDs. The FDA requires all VFDs to be communicated non-verbally. As long as the correct and necessary information is included in the VFD, forms can be custom made or used from a commercial template. Paper and electronic options are accepfigu. GlobalVetLINK®[9] is a company that provides a web-based service called FeedLINK® for completing, sending, and storing electronic VFDs. It is anticipated that others may also develop electronic VFD services, including drug sponsors and information management companies. The new rule requires all parties to keep a copy of the VFD for 2 years, including the original by the veterinarian. Going forward, electronic VFDs will be used in a high percentage of cases in order to maximize the efficiency of veterinary time and resources, and to improve the time lag between pig need and pig feed in the bin. Consideration for using a commercial service versus an internally developed system will hinge on costs, staffing, and privacy of data.

It is just as critical that veterinarians, producers, and feed mills recognize that medicated feed covered under a VFD cannot be delivered, stored, or remain in the feeder after the VFD has expired. It will be desirable to work within a process that notifies the veterinarian automatically when a VFD is set to expire.

Summary

The joint efforts of veterinarians, producers, and distributors will be important if we are to accomplish the best care for the pig and maintain and build upon the successes our industry has achieved in producing safe, quality, and affordable pork protein for the consumer. Regulated use of antibiotics in livestock will include veterinary oversight for all routes of administration in the future, including feed drugs via the VFD. Our job requires us to manage the process for our clients so that we meet all legal requirements, ensure accuracy in the documentation, be efficient with our time, yet ensure medications are targeted and on-time. How our profession responds to the challenges and opportunities in the new VFD rule will have a bearing on the future availability of antibiotics for food animals.

References

1. Federal Register. Final Rule: Veterinary Feed Directive. Available at: https://www.federalregister.gov/articles/2015/06/03/2015-13393/veterinary-feed-driective

2. US Food and Drug Administration. Guidance for Industry #213, December 2013. Available at: http://www.fda.gov/downloads/AnimalVeterinary/GuidanceComplianceEnforcement/GuidanceforIndustry/UCM299624.pdf

3. National Pork Board. PQA Plus®. www.pork.org

4. US Food and Drug Administration. Guidance for Industry #120, September 2015. Available at: http://www.fda.gov/downloads/AnimalVeterinary/GuidanceComplianceEnforcement/GuidanceforIndustry/ucm052660.pdf

5. 21CFR558.6 http://www.gpo.gov/fdsys/pkg/CFR-2011-title21-vol6/pdf/CFR-2011-title21-vol6-part558.pdf

6. Federal definition of VCPR, 21 CFR530.3(i). Available at: http://www.accessdata.fda.gov/scripts/cdrh/cfdocs/cfcfr/CFRSearch.cfm?fr=530.3

7. FDA list of State v. Federal VCPR. Available at: http://www.fda.gov/AnimalVeterinary/DevelopmentApprovalProcess/ucm460406.htm

8. C. Pixley, USDA FSIS U.S. Report on National Residue Program Lab Methods. 2015.

9. GlobalVetLINK® https://www.globalvetlink.com/

Pork production without antibiotics

J. Risser, DVM; K. Thornburg
Country View Family Farms, Middletown, Pennsylvania

Country View Family Farms is a system of 270 team members, 60,000 sows on 16 sow farms, and the supporting nursery and finisher contract growers. The majority of the farms are located in Pennsylvania; while others are in Indiana, Ohio, New York, and Virginia. All hogs return to Pennsylvania for harvest at the Clemens Food Group in Hatfield, Pennsylvania. Our vision is "To be a provider of wholesome, sustainable food for our families and customers with a world class team practicing Ethics, Integrity, and Stewardship."

This has been a long journey for Country View Family Farms. Our ability to raise pigs without antibiotics was not a one-year planning process. This was a program shaped by our core values and vision established by senior leadership almost 15 years ago. The planning started with awareness of conversations around the world regarding animal production for food, and we then continued those discussions with our customers. While the majority of customer demand is centered on a safe and affordable food supply, there is growing demand for alternative produced food products, including proteins from animals raised without antibiotics. If we can provide those niche products while still maintaining, or in some cases improving, animal care it is our responsibility to do so.

The entire system flows through three production phases to utilize specialized facilities and skill sets of the animal caretakers in each stage. All of the sow farms are managed by Country View Family Farms' team members. We evaluate contract growers in relation to the phase of production based on personalities and their other business endeavors. It is essential for us to have very dedicated detail oriented growers for the nursery phase to start the pigs off right. While it is still important for finishing growers to be detail oriented, we also know that these barns are often managed by individuals who have alternative primary sources of income, on or off the farm, and will not have has much time to offer the hog barn. It is also critical to commit to barns in strategic locations for biosecurity. We frequently review the Pennsylvania Area Regional Control Map during discussions of barn placement or flow logistics. This control program has increased communication between pig producers in an effort to improve herd health across the state. The maps are easily accessible online and contain swine site locations throughout Pennsylvania, their type, and PRRSV and PED status. Knowing what health challenges are in the area aids in the systematic placement of pig flows with the goal to move the pigs closer to the harvest facility, while adhering to the overall strategy of single stocking, sourcing, and all – in- all – out production. The whole system birth to market is built to minimize the variation in pig age within the same barn. All of these factors will enhance the overall health program long term, reducing administration of antibiotics.

Realizing we have only mentioned antibiotics once up to this point, let us emphasize it's not about raising pigs without antibiotics; but how to prevent the health challenges we face and the need for antibiotics. As noted above, for Country View Family Farms, we feel that our success starts with the team and grower selection; but specifically for veterinarians it starts by being involved in the early decisions regarding facility design, barn locations, flow placement, stocking density, and nutrition. These decisions, whose effects are not immediately quantifiable, will have an impact on the continuous improvement of animal care in the system. Our attention then turned to the preventative health program. We focus the vaccinations in the sow and young pig with all injectable vaccinations administered by weaning. The only other vaccine administered is for *Lawsonia intraceullularis* via the water. We were very early adopters of responsibly using feed and water antibiotics only for prevention and treatment of disease. Instead we focus our team members and growers on individual pig care principles of early detection and individual treatment of health challenged animals. As shown in Figure 1, our vaccination program is over 60% of total system medication and vaccination cost per market hog; while feed medication makes up only 10-20%.

Once we were successful with these strategies in our entire system, we started exploring the possibilities of raising pigs without antibiotics from birth to market to bring solutions to our customers. Today we have 16,800 sows in our Farm Promise® Program, which includes a never ever antibiotics component. The standards of our USDA process verified program are outlined in Table 1. This program does allow the flexibility to treat sows and individual pigs. However, the treated pigs must be individually identified, segregated, and harvested under traditional market channels. Our system identifies treated piglets pre-weaning by ear notching; and those and any pigs treated post-weaning are tattooed with digits known to represent treated animals by all program growers and team members at the harvest facility.

Another key relationship to reducing antibiotic usage is with a nutritionist. The feed components and quality are imperative to making the program successful. We believe in preparing the pig for health challenges starts before conception with the nutrition program for the sows. Direct fed microbials are utilized to obtain the best quality of colustrum, and then are included in the nursery diets following the pig through the process. At every phase, instead of focusing on vaccine inputs, we focus on the nutritional inputs for the long term success of the pig. There are a multitude of products available, mainly feed additives that offer promising yet unproven enhancements for improving health in place of antibiotics. Many of these vendors have experience in other species but limited research to support their claims. It becomes

Figure 1: Percent of total system medication & vaccination cost for vaccine, antibiotic, other, and feed medication.

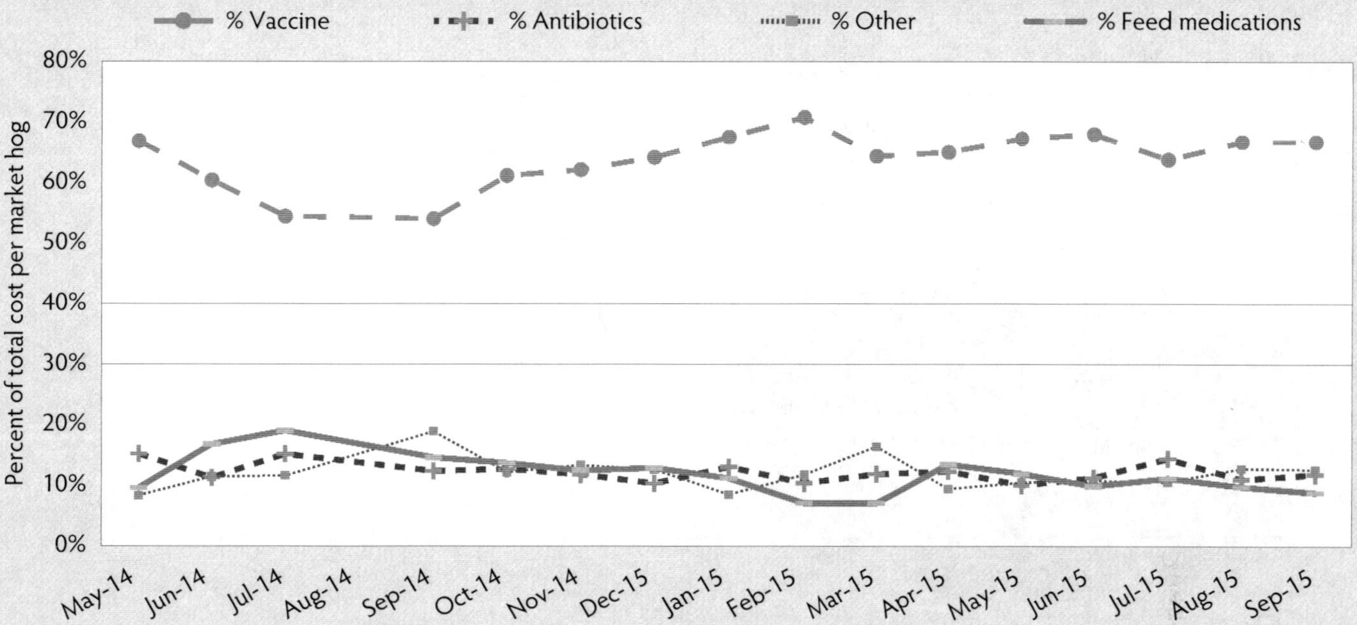

Table 1: Outline of the Farm Promise® program standards.

Free to roam	Sows housed in open pen environment for gestation
Source assured	Traceability of product back to farm
Certified standard of care	Animals raised with proper and humane care at every stage of life
Never ever given antibiotics	Birth to market, pigs are never administered antibiotics
Never ever given growth promotants	Birth to market, pigs are not administered growth promotants; such as ractopampine or ionophores
Never ever given animal by-products	Birth to market, pigs are not fed diets containing animal by-products

important for the nutritionist and veterinarian to review the research data jointly and make a decision for the betterment of the program and system.

When we first embarked on this new journey, our strategy was to simply remove all feed and water antibiotics from the program flows. However, we quickly realized this type of systemic change is not that simple. The health program was re-evaluated to determine if there were any additional commercially available or scientifically proven autogenous products to add to the prevention strategy. Also, production practices that we may take for granted had to be re-evaluated; such as the ventilation, stocking density, and water quality. We also worked with the Food Safety Quality Assurance department at Clemens Food Group to ensure that by removing antibiotics we were not creating a subclinical impact on food safety. As we have learned, there is a cost to raising a niche product; but the customer must understand and be willing to utilize the entire carcass to maximize the return. We have been successful in minimizing the increase in the Farm Promise® cost of production over traditionally raised hogs. The program pigs must provide the return on investment for themselves, and the pigs removed from the program due to individual treatments. This additional cost is very dependent on the program cost of production and the percent of pigs removed for treatment. Figure 2 shows the percent allocation of the increased cost of Farm Promise® program pigs to traditional pigs with the largest percentage due to the impact of feed conversion.

To truly minimize the use of antibiotics throughout a system, it takes a large team from the sow farm technician attending farrowing, to the contract nursery and finisher grower, the veterinarian, the production manager, the program coordinator, the food safety quality assurance technician, the harvest operations technician, the marketing team, and sales team. Success is not defined as pork production without antibiotics; it is created by focusing our strategies and resources on long term solutions to prevent health challenges, in order to reduce the administration of antibiotics to a select few individual hogs.

Figure 2: Percent of cost of Farm Promise® over traditional.

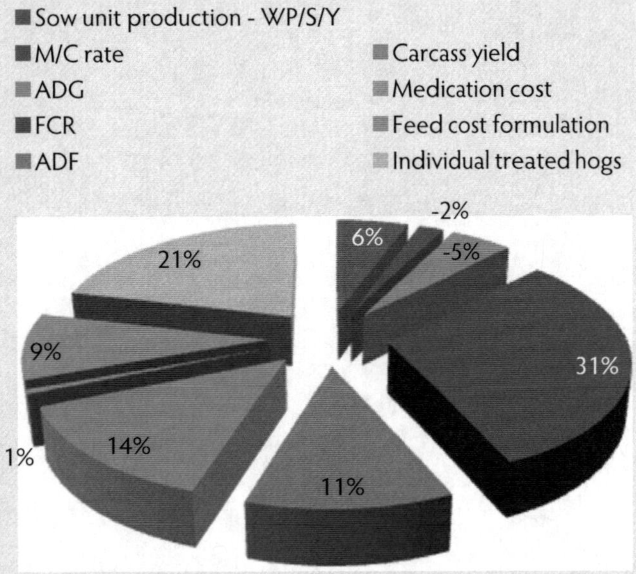

Strategies to reduce antibiotic use in practice

Laura Bruner, DVM
Swine Vet Center, Saint Peter, Minnesota

Antibiotic use has become one of the major issues facing livestock producers and food animal veterinarians today. In the eve of increased veterinary oversight through guidance 213 and veterinary feed directives, producers and veterinarians alike are searching for insight on what antibiotic use will and should look like in the future. As you evaluate treatment protocols for your clients, it's important to ask yourself what can be done to reduce the need for antibiotics in a herd. I think one of the first steps in reducing antibiotic use is working on:

1. Eradicating a disease
2. Stabilizing a disease
3. Preventing a disease from occurring

Eradicating a disease

PRRS

The first step in a health management program is controlling what you can control. Vertical transmission of PRRS virus into nurseries out of PRRS positive sow farms in our experience increases the need for antibiotic intervention even if there are only a few pigs positive at weaning. Herd closures for the elimination of PRRS virus within a herd have proven successful for many veterinarians. The term "Load, Expose, and Close" has been used frequently in the swine industry. This involves loading the farm with as many replacement gilts as possible, exposing the herd to the PRRS strain affecting the herd, commercial PRRS vaccine, or both, and closing the farm to any new replacements until PRRS virus is absent in the farm. This typically happens around 210 days after herd exposure of animals to the PRRS virus or vaccine.

Mycoplasma

Many of our clients have also chosen to eradicate *Mycoplasma hyopneumoniae* at the same time they are eradicating PRRS from a sow farm. We have successfully eradicated Mycoplasma from 50 farms in the last 10 years. In our experience, successful Mycoplasma eradication programs will decrease the amount of antibiotics needed both in the nursery and finishing. Producers that have had successful Mycoplasma closures have shown a decrease in medication costs as much as $0.33 per head marketed.

Stabilizing a disease

Influenza A virus (IAV)

Some diseases we just haven't found ways to consistently eradicate from herds yet. One of these is IAV in swine (IAV-S). Influenza can be one of the most frustrating pathogens to deal with in that there are no clear cut answers on how to eliminate it from a population. In many cases, if stabilization does occur, soon, a new strain of IAV-S is found in the population. IAV-S causes a bulk of challenges in the nursery. As maternal antibodies begin to wane, the virus spreads through the nursery population. At this time many pigs succumb to bacterial infections such as *Streptococcus suis* and *Haemophilus parasuis* which typically require implementation of some antibiotic therapy. The most common strategies we see in our practice to help control or stabilize within a sow herd is to vaccinate breeding groups either 4 to 6 weeks prefarrow, go through herd with mass vaccinations 1 month apart, or quarterly/bi-annual mass vaccination.

PRRS virus

If the choice of a producer is to not eradicate PRRS from a sow farm, then stabilization of the farm is imperative to decrease the need for antibiotic use in the downstream pigs. This can be accomplished by ensuring all gilts entering the farm are exposed to the farm virus and are free from shedding virus upon entry into the sow farm. Protocols must be in place in the farrowing house to minimize the spread of PRRS in the farrowing rooms.

Rotavirus

Rotavirus is a common cause of baby pig scour on the sow farm. I believe that we seriously underestimate the role people play in prolonging scours in the farrowing house. Crate to crate biosecurity is essential in the control/stabilization of rotavirus on the sow farm. Implementing procedures that we use commonly in "PRRS protocols" or "McRebel"(**M**anagement **C**hanges to **Re**duce **E**xposure to **B**acteria to **E**liminate **L**osses)[1] will decrease the incidence of scour and the need for antibiotics to treat secondary *E coli* and Clostridium infections. Boostering the immunity in the sow to pass along in colostrum to piglets is essential. Farms I work with have implemented "rota cubes" as a way of having continuous and known rotavirus in the feedback for sows prefarrow.

Circovirus

Although we have good vaccines for circovirus, there can still be some vertical transmission of circovirus to piglets which in turn will cause early circulation of the virus in the nursery. Depending on your vaccination strategy, vaccine may or may not be on board at the time this happens. Controlling circovirus in the sow herd by vaccination of replacement animals and boostering those gilts with circovirus prior to breeding can help keep sow farm stabilization.

Prevention

Biosecurity

Biosecurity is paramount in preventing disease in all phases from farrow to finish. Auditing the big risk procedures within a farm should be done as frequent as necessary to address those risks. Testing of incoming replacement animals, transportation, people entry, supply entry, removal of mortalities, garbage removal, maintenance and repair procedures are all primary areas to focus on in a biosecurity audit.

Filtration

Filtration of incoming air into a farm is one of the greatest tools we have in prevention of some major diseases swine producer face. Most commonly sow farms are filtered to prevent the introduction of PRRS. Although filtration does not completely prevent farms from breaking with PRRS it has proven to decrease the rate of PRRS breaks by 43%.[2] In the process of preventing PRRS through filtration, it appears anecdotally that they have also had less Influenza and Mycoplasma hyopneumoniae breaks as well.

Summary

Antibiotic use will continue to be at the forefront for the consumer and therefore the packer. Eradication, stabilization, and vaccination strategies have the potential to help us decrease the need for antibiotics.

It is in our best interest as an industry to figure out ways to reduce the need for antibiotics and use them judiciously so that we can preserve the right to medicate pigs the way we see fit as veterinarians.

References

1. McCaw, M. PRRS Control: Whole herd management concepts and research update. North Carolina Healthy hog seminar, 1995. North Carolina State University.

2. Feder, J, Reicks, D. Filtration update, 10 years and 106 farms later. Ames poultry meeting, July 2015.

Treatment decision trees

Locke Karriker, DVM, MS, Diplomate ACVPM
Swine Medicine Education Center, Iowa State University, Ames, Iowa

Uses a treatment context

In the context of animal treatment, there are three main objectives for simple decision trees: determine if treatment of animal or population is warranted, determine if a chosen treatment meets a certain standard (such as legality in food animals), or select a best treatment from a list of options. Decision trees are stepwise graphical representations of a decision process. They are especially useful in the following situations: 1) the situation being considered is complex and has many steps to reach a solution, 2) There are many potential conclusions or categories that answer the decision to be made, 3) the decision can be broken down into discrete parts that have "yes or no" answers, 4) when a complex situation occurs infrequently and important considerations or inputs might be forgotten in the interim, 5) to illustrate and simplify a new concept to employees or peers, 6) to help an individual check the quality of their own decision by reviewing the parts for consistency and agreement, 7) when answers to previous questions suggest additional or different criteria that should be considered in subsequent steps, 8) to provide consistent and predictable management decisions among many different operators. These circumstances are frequently encountered when making decisions about treatment or when selecting appropriate treatment from a list of options.

Software needed

Decision trees can be made using several common software packages including Microsoft PowerPoint, most word processors, and many drawing programs. Essentially any program that allows you to link boxes and arrows can be used to communicate the decision tree. A specialized program from Microsoft called Visio is specifically designed to create charts and diagrams that can be saved and shared in a wide number of formats including other Microsoft programs. The examples that follow were built using Visio and PowerPoint.

Steps

1. Brainstorm the list of input information that is relevant to the decision being made and formulate these as questions.
2. Consider the potential answers to these questions.
3. Organize questions in order of importance.
4. Order questions so that more general questions are asked earlier in the process than very specific questions.
5. Use one of the software programs to present the questions or decision steps in graphical manner. Software programs that allow "drag and drop" construction from menu of shapes and arrows are easiest and quickest to use.
6. Link dependent questions to previous questions so that the reader is directed to the next important consideration in the process based on the answer given to the previous step.
7. Test the resulting decision tree with someone that has little knowledge of the likely answer for clarity and completeness.
8. Refine the decision tree as new information is found that can simplify the decision or has an impact on the correct conclusion.

Examples of each of the objectives

Of the three broad objectives for which a decision tree can be useful, the first is determining if treatment of animal or population is warranted, or more specifically, empowering caretakers to make this determination as soon as a problem is detected. This serves the purposes of addressing welfare concerns immediately, positioning employees to have an impact, and conserving veterinary resources for solving health issues. An example of this type of situation would be the observation of vaginal discharges in sows. A small amount of discharge at certain times of the reproductive cycle is normal while at other times, the appearance of any discharge is reason for concern. Instead of applying veterinary resources at each observation of a discharge, a decision tree might help caretakers to "triage" them when observed. Figure 1 is an example of a decision tree employed to help caretakers determine if treatment is warranted or veterinary intervention is warranted.

Figure 2 illustrates a similar objective, in this case for evaluating problems first observed as lactational insufficiency. There are a variety of potential sow, piglet, environmental, management and infectious causes that might create lactational insufficiency. In this particular case, the farm had specific resources that could be implemented to treat animals if it were appropriate and the decision tree references these sources when they are appropriate. Ultimately, if correcting the non-infectious causes and applying the previously approved treatment interventions are completed without success, the tree terminates with the involvement of the veterinarian. A benefit for the progressive veterinarian of this is that they get to focus on high quality challenges because the easiest solutions have already been implemented.

The best example of using decision trees to determine if a chosen treatment meets a certain standard (such as legality in food animals) is the AMDUCA algorithm published by the AVMA and available as a PDF in brochure format.[1] It helps veterinarians to navigate the complex decision making associated with using a drug in an extra-label manner.

Figure 1: Troubleshooting vaginal discharges – Treat or not to treat?

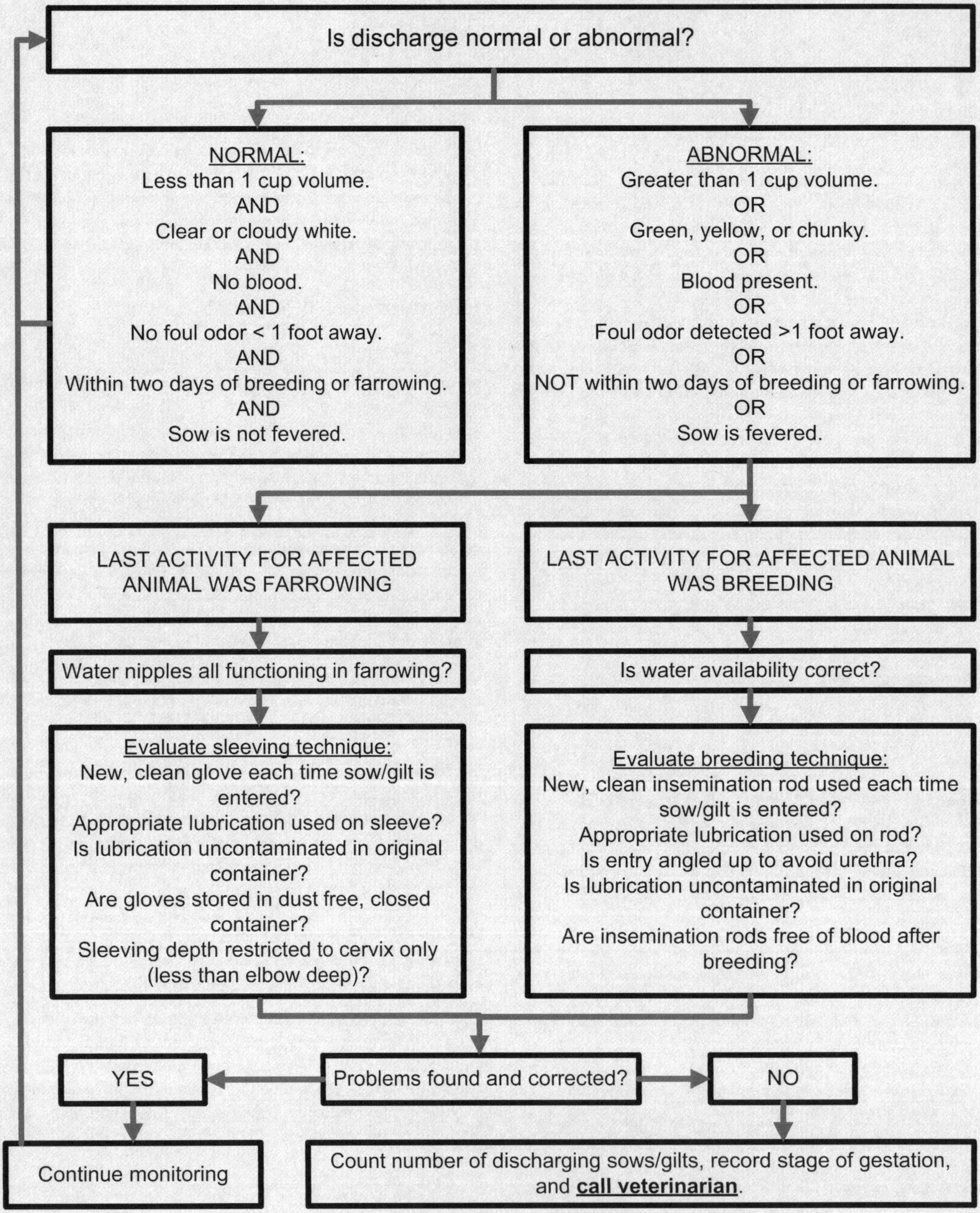

Figure 2: Response plan - Sow lactation problems

When to use this plan:

- When a litter of pigs is found that appear to be losing weight, or are dehydrated.

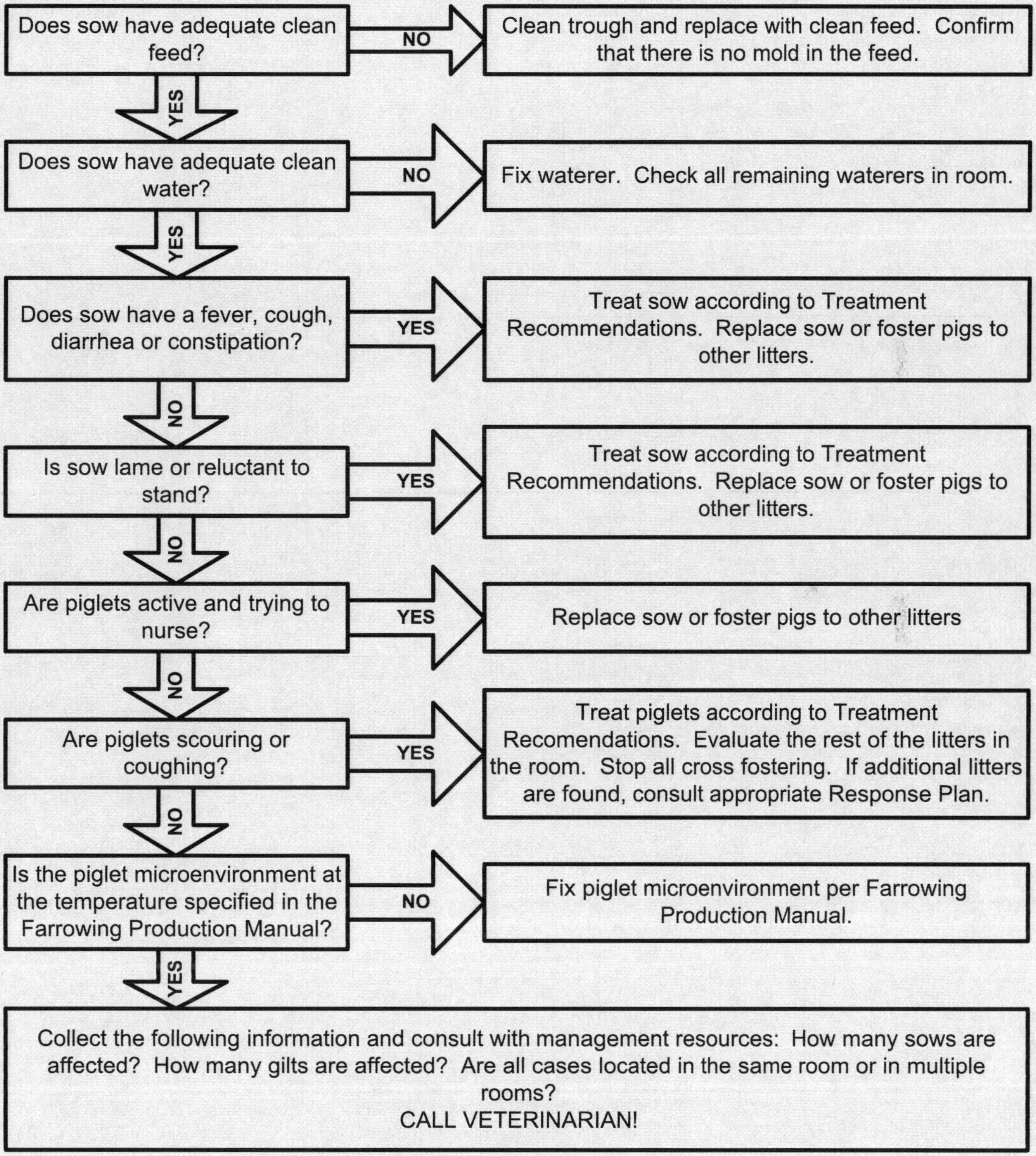

Selecting a best treatment from a list of options is more specific to particular cases or production settings and decision trees have the most utility here to help consolidate resources and make sure that important considerations are not skipped. Given the specific nature of these types of trees and the ramifications of recommending treatments without a valid veterinary-client-patient relationship, an example is not published here but is presented in the session.

References

1. Extralabel Drug Use (ELDU): An Informational Outline of the Animal Medicinal Drug Use Clarification Act (AMDUCA). Website of the American Veterinary Medical Association. https://ebusiness.avma.org/files/productdownloads/extralabel_brochure.pdf (accessed on December 16, 2015).

Considerations for preserving antibiotic use

Michael D. Apley, DVM, PhD, DACVCP
Kansas State University, Manhattan, Kansas

Introduction

There are three general categories for us to consider related to preserving the ability to use antibiotics to prevent, control, or treat infectious disease in swine. The first is consideration of any effects our current use may have on the ability to treat infections in the future due to antibiotic resistance. The second category includes forces of law due to legislation and regulation. The third concerns access to certain market channels due to "retaliation", food retailer and quick service restaurant (QSR) initiated requirements. All of these categories will have an effect on which antibiotics we use for what in the future.

Antibiotic resistance in swine pathogens

Antimicrobial resistance is a real issue. The severity of the challenge in humans was recently highlighted in a 2013 CDC report which described the pathogens and resistance patterns which threaten human health.[1] Do we have a similar situation developing for pathogens in swine, regardless of any real or perceived relationship to human health?

First, how would we identify significant trends in swine pathogen susceptibility profiles if they are occurring? One of our challenges is that some of our antibiotics have breakpoints for at least one pathogen in swine respiratory disease (e.g., ceftiofur, florfenicol, enrofloxacin, penicillin, injectable oxytetracycline, tiamulin, tulathromycin), and some have breakpoints adapted from human medicine or for other routes of administration which may have little or no relationship to clinical efficacy (e.g., oral tetracyclines, trimethoprim-sulfa, lincomycin). In the case of enteric disease pathogens, no breakpoints have been established for clinical efficacy in any veterinary species. Therefore we are best informed by looking for shifts in MIC distributions for many of these applications. Significant information may be gained by these shifts, even if the changes do not cross a breakpoint. However, many of the available publications and data focus on "% resistant", with some including the intermediate category.

The complexity of identifying significant trends was addressed in a recent evaluation of four temporal data-analysis options for susceptibility of *Pasteurella multocida* and *Streptococcus suis* to ampicillin, tetracycline, tiamulin, and trimethoprim-sulfamethoxazole.[2] In evaluation of isolates collected from 1998 to 2010 at the University of Guelph Animal Health Laboratory, there was a trend towards increased *Pasteurella multocida* resistance to ampicillin. Temporal clusters of specific pathogen-antibiotic resistance patterns were identified throughout the study, after which the resistance patterns declined to previous levels. Seasonal trends were also identified for some pathogen-antibiotic combinations. Recent trends in increasing resistance of *Pasteurella multocida* to trimethoprim-sulfamethoxazole were noted, although this breakpoint has an unknown relationship to clinical outcome in swine; however, these trends would indicate a move in population MIC distribution.

A similar study was conducted in Europe, focusing on isolates of *Pasteurella multocida, Actinobacillus pleuropneumoniae, and Streptococcus suis* from 11 countries. Isolates were evaluated from 2002 through 2006, with reporting of both % resistant and MIC distributions. For APP, there was increasing resistance over the time period to tetracycline, gentamicin, and TMP/sulfa. For most antibiotic-pathogen combinations, resistance was low and stable.[3]

A recent study in Australia indicated that antibiotic resistance in swine respiratory pathogens exhibited low levels of resistance to commonly used antibiotics.[4]

A report of increased Brachyspira-associated disease in US swine herds also evaluated the species identity of 79 isolates from disease incidents.[5] Of the 79 isolates, 38% were identified as *B. hyodysenteriae*, 8% as *B. pilosicoli*, 25% as *B. murdochii*, and 14% could not be identified as to species. Using extended range dilution MIC testing, it was noted that the latter two categories had significantly higher MIC distributions than the other species categories, and the authors hypothesized that these findings might explain the increased incidence of *Brachypira*-associated disease in recent years.

A report of other enteric pathogen susceptibility profiles has been published for *Escherichia* coli and *Salmonella* spp, from submissions to the Minnesota Veterinary Diagnostic Laboratory from 1995-to 2004.[6] In *E. coli* isolated from enteric and septicemia cases, ceftiofur resistance rose from 4 to 22%, characterized as an MIC $\geq$ 2 µg/ml, which is the respiratory disease resistance breakpoint for swine. Resistant percentages for TMP/sulfa rose from 13% in 1995 to as high as 33% during the testing period. As discussed above, the relation of the breakpoint to clinical efficacy is unknown. The tetracyclines consistently showed 95 to 100% resistance. For *Salmonella*, ceftiofur resistance was consistently in the 5-11% range, while ampicillin increased from 34% resistant in 1995 to 74% in 2004. The tetracyclines were consistently in the 80% to 100% resistant range. Salmonella resistance to gentamicin remained steady in the 12-15% range, while neomycin resistance decreased from 32% to 6%. These aminoglycoside results are difficult to interpret due to the questionable application of these breakpoints adapted from human medicine.

A look at both enteric and respiratory pathogens in the Iowa State University Diagnostic Laboratory Bacterial Susceptibility Profiles indicates minor numerical changes, both more and less susceptible, from 2003 through 2014.[7] The susceptibility profiles in the Iowa State data closely resemble the Minnesota data discussed above.

This is not an exhaustive review of swine antimicrobial susceptibility trends. While there could be the impression that since there is little moving over a 10-20 year period that resistance will not be a problem, discussions I have had with practitioners about problematic *Strep. suis*, *E. coli*, and *Salmonella* spp. infections displaying significant resistance would indicate that there are still problem areas. The trends in other mammalian species, including humans, would indicate that resistance is a real and persistent problem across all uses of antibiotics, and that using them only when necessary, and paying close attention to resistance trends, is a prudent approach.

Legislation

The recent legislation in California, and the lack of activity on repeatedly introduced bills in the national legislature, would suggest that any effects of legislation on our ability to use antibiotics in food animals will most likely come from state assemblies. California Senate Bill 27, signed into law on October 10, 2015, moves beyond the requirements of Food and Drug Administration Center for Veterinary Medicine (FDA/CVM) Guidance for Industry #209 in requiring a veterinary prescription for all medically important antimicrobials used in food animals, including those labels currently holding an over-the-counter status and that will continue to do so after January 1, 2017.[8] The bill contains the following language in relation to control and prophylaxis uses of an antibiotic.

> **"14401.**
> Beginning January 1, 2018, a medically important antimicrobial drug shall not be administered to livestock unless ordered by a licensed veterinarian through a prescription or veterinary feed directive, pursuant to a veterinarian-client-patient relationship that meets the requirements of Section 2032.1 of Title 16 of the California Code of Regulations.
> **14402.**
> (a) Beginning January 1, 2018, a medically important antimicrobial drug may be used when, in the professional judgment of a licensed veterinarian, the medically important antimicrobial drug is any of the following:
> (1) Necessary to treat a disease or infection.
> (2) Necessary to control the spread of a disease or infection.
> (3) Necessary in relation to surgery or a medical procedure.
> (b) A medically important antimicrobial drug may also be used when, in the professional judgment of a licensed veterinarian, it is needed for prophylaxis to address an elevated risk of contraction of a particular disease or infection.
> (c) A person shall not administer a medically important antimicrobial drug to livestock solely for purposes of promoting weight gain or improving feed efficiency.
> (d) Unless the administration is consistent with subdivision (a), a person shall not administer a medically important antimicrobial drug in a regular pattern."

This legislation mirrors actions being taken on the regulatory front to evaluate and address uses of medically important antibiotics in food animals where the label does not have a specified period of administration.

National legislation has been introduced in an attempt to bring about more detailed reporting. Senator Diane Feinstein put a hold on the Animal Drug User Fee Act (ADUFA) in 2013 as an attempt to force inclusion of increased reporting requirements, which was not successful. Representative Henry Waxman first introduced the "Delivering Antimicrobial Transparency in Animals (DART) Act of 2013" as HR 820, which did not make it out of the Subcommittee on Health. This bill would have required increased reporting of antimicrobial sales for all food animal antimicrobials, and required reporting by end users of antimicrobials in the feed. In 2015, the bill was sponsored by Rep Louise Slaughter (D-NY) as H.R. 2459.[9]

Bills which purpose to drive the evaluation of prevention and control uses, but which in fact would result in their removal for at least a protracted period of time, continue to be introduced. Representative Louise Slaughter again introduced the latest edition of the PAMTA act in 2015, "Preservation of Antibiotics for Medical Treatment Act of 2015" (HR 1552).[10] This bill had 39 cosponsors. This bill does not outright prohibit the use of antimicrobials in food animals for anything other than individual animal therapeutic use and non-routine preventive use, but sets a very high bar with a very short timeline to retain their use, clearly with the intention of establishing unattainable requirements. On the senate side, Senator Dianne Feinstein (D-CA) again introduced the "Preventing Antibiotic Resistance Act of 2015" (S 621). This bill had 5 cosponsors and is very similar to PAMTA.[11] These bills have typically not made it out of committee to the floor, and have been repeatedly introduced over the last decade.

Regulation

Regulations have had a dramatic effect on our uses of antibiotics in animal agriculture, and will continue to do so in the future. The Animal Medicinal Use Clarification Act Regulations give the FDA/CVM the legal authority to limit or prohibit the extralabel use of drugs in food animals (ELDU). This authority has been used to prohibit the extralabel use of fluoroquinolones, glycopeptides, and cephalosporins in recent years.

The Cephalosporin ELDU prohibition is a precedent suggesting that hard data indicating a problem with ELDU is not required to use this avenue to address medically important antibiotic use in food animals.[12] The primary concern is that even though there was absolutely no evidence to separate concerns regarding label and extralabel use, the action taken was directed at extralabel use. The use of the Animal Medicinal Drug Use Clarification Act (AMDUCA) regulations as a lower-resistance regulatory pathway is troublesome to those who invested considerable effort in both the AMDUCA and the regulation development process, resulting in legalizing extralabel use in veterinary medicine under specified conditions.

There is also a concern over species inclusion. Regardless of the lack of evidence to indicate a concern for swine, this species is included in the cephalosporin ELDU prohibition. In the evidence cited for cattle, the authors of two of the cited papers state in their discussions that you really can't make the conclusion from the paper for which they were used in the FDA decision, which in the opinion of this author was obvious from reading the articles.[13,14] The FDA also left out key articles related to cephalosporin use in cattle that were not supportive of their stance on the issue.[15,16] A published systematic review suggests that the finding of multidrug-resistant bacteria on organic and "conventional" dairies is much more complicated that just comparing antimicrobial use.[17] The key evidence which really supported the ELDU ban was for the injection of chicken eggs and the resulting change in susceptibility profiles of surviving *Salmonella*. The evidence for concern in cattle was inconsistent, and nonexistent for swine.

The prohibition allows the use of ceftiofur for extralabel indications but not with an extralabel regimen. The result of having the ability to use an antimicrobial for off-label indications but not the ability to adjust the dosage appropriately is completely nonsensical, and is likely to contribute to selection for antimicrobial resistance.

The most telling direct quote from the order of prohibition was from the section refuting the allegation that the FDA/CVM was relying on the precautionary principle. *"In the preamble to the final rule, FDA addressed the question of what type of evidence would be necessary by saying that the risk determinations that would lead to prohibition of an extralabel use typically will involve documented scientific information. However, the Agency believes that it is not limited to making risk determinations based solely on documented scientific information, but may use other suitable information as appropriate."*

While the current FDA/CVM leadership is committed to prevention and control uses being classified as judicious therapeutic uses of medically important antimicrobials, future leadership may not share this view.

The future direction of regulatory action concerning antibiotics in food animals is being clearly communicated to the industries. If you go to the Fact Sheet: Veterinary Feed Directive Final Rule and Next Steps on the FDA/CVM website, you can read some very telling language as to the next initiative to follow GFIs 209 and 213.[18]

> *"The FDA acknowledges the important role medically important antimicrobials play in treating, controlling, and preventing disease in food-producing animals. However, the agency has been actively engaging veterinary organizations, animal producer organizations and other stakeholders to express our position that medically important antibiotics labeled for continuous or undefined durations of use is not consistent with judicious use principles, as outlined in previously-released guidance documents.*
>
> *In the case of disease prevention, the FDA believes it is important such use is appropriately targeted to animals at risk for a specific disease and the use duration is limited and risk-based... Long-term or open-ended prevention uses are not covered by the phase-out process for production uses described in Guidance #213. However, the* ***National Action Plan for Combating Antibiotic-Resistant Bacteria*** *calls for the identification and implementation of measures to foster stewardship of antibiotics in animals. The FDA believes long-term or open-ended use of medically important antibiotics is a significant stewardship issue and intends to seek broad public input on this issue in the summer of 2015."*

It could not be clearer that the next focus of regulatory activity will be continuous feeding of medically important antibiotics for the purpose of disease prevention or control.

Retailation

Our use of antibiotics in food animals has achieved the status of a marketing pawn. In my conversations with food retailers and quick service restaurants, it is quite clear that the consumer is king, and if product meeting certain specifications leaves the shelves or goes over the counters as opposed to other options (especially from competitors), they will continue to try and supply this product until the demand is satisfied; that is if supply and cost are consistent with financial reality. Actually that may be misstated, social media drives a lot of decisions, and I am left wondering if it is really clear if the social media messages reflect actual customers. In my opinion the restraint from going completely antibiotic free is about 1 part concern about animal welfare and 5 parts supply.

The initial large scale antibiotic free initiatives have involved the poultry industry, with organic and natural never-ever programs in almost every food animal industry. There are obvious differences in lifespan and consequently time of disease exposure. This isn't really understood by the public, and frankly many retailers and QSRs, in my experience.

The use of antibiotic use status as a marketing tool has become a convoluted swirl of one-upmanship, anti-food animal agendas, and sometimes shameless misdirection as to the real dangers to consumers. One thing is clear, science is only a minor driver, if that, in many of the decisions.

Summary

Resistance is a real issue and in my opinion is having, and will continue to have, an impact on our ability to prevent, control, and treat disease in food animals. We are just beginning to understand the ways that resistance genes move, both through bacteria that we have come to now intimately, and through the less well characterized populations of bacteria which we cannot work with in the lab. To think that we have reached some kind of resistance plateau from which there will be no more deterioration of antibiotic efficacy is nonsensical. The biggest concerns are relegated to a few pathogen/antibiotic combinations, but they are occurring. Any management practices which can economically replace the need for antibiotics should be well received.

We will continue to see pressure on food animal antibiotic use for prevention and control from legislative and regulatory entities, as well as through supply chains. Transparency will become an industry standard, although the final form of monitoring is not yet clear. Within that transparency, we must be seen to be using well thought through written treatment protocols, using best-evidence approaches to constructing these protocols, and relying on accurate treatment and outcome records to guide evolution of these protocols. We will be asked hard questions about the nature of systems which require antibiotics for the prevention or control of diseases, and we should expect pressure on prevention and control uses of antibiotics. More hard questions will be asked about the relationship of prescribers to profits from the antibiotics prescribed.

1. Centers for Disease Control and Prevention. Antibiotic Resistance Threats in the United States, 2013. Accessed 1/5/2016 at http://www.cdc.gov/drugresistance/pdf/ar-threats-2013-508.pdf.

2. Glass-Kaastra SK, Pearl DL, Reid-Smith J, et al. Surveillance of Antimicrobial Resistance in Clinical Isolates of Pasteurella multocida and Streptococcus suis form Ontario Swine. Can J Vet Res 78(4);241-249,2014.

3. De Jong A, Thomas V, Simjee S, et al. Antimicrobial Susceptibility Monitoring of Respiratory Tract Pathogens Isolated from Diseased Cattle and Pigs across Europe: The VetPath Study. Vet Microbiol 172:202-215, 2014.

4. Dayao DA, Gibson JS, Blackall PJ, et. al. Antimicrobial Resistance in Bacteria Associated with Porcine Respiratory Disease in Australia. Vet Microbiol 171:232-235, 2014.

5. Clothier KA, Kinyon JM, Frana TS, et. al. Species Characterization and Minimum Inhibitory Concentration patterns of Brachyspira Species Isolates from Swine with Clinical Disease. J Vet Diagn Invest 23:1140-1145, 2011.

6. Malik YS, Chander Y, Olsen K, Goyal SM. Antimicrobial Resistance in Enteric Pathogens Isolated from Minnesota Pigs from 1995 to 2004. Can J Vet Res 75:117-121, 2011.

7. Iowa State University Veterinary Diagnostic and Production Animal Medicine Bacterial Susceptibility Profiles. Accessed 1/5/2016 at http://vetmed.iastate.edu/vdpam/disease-topics/bacterial-susceptibility-profiles.

8. California Senate Bill 27: SB-27 Livestock: Use of Antimicrobial Drugs. Accessed 1/6/2016 at http://leginfo.legislature.ca.gov/faces/billTextClient.xhtml?bill_id=201520160SB27.

9. H.R. 2459 114th Congress, 1st session. Accessed 7/12/2015 at https://www.congress.gov/bill/114th-congress/house-bill/2459/text.

10. Preservation of Antibiotics for Medical Treatment Act of 2015. Accessed 7/12/2015 at https://www.congress.gov/bill/114th-congress/house-bill/1552.

11. Preventing Antibiotic Resistance Act of 2015. Accessed 7/12/2015 at https://www.congress.gov/bill/114th-congress/senate-bill/621.

12. New Animal Drugs; Cephalosporin Drugs; Extralabel Animal Drug Use; Order of Prohibition. 21 CFR Part 530. Final rule. Accessed 10/15/2013 at http://www.gpo.gov/fdsys/pkg/FR-2012-01-06/pdf/2012-35.pdf.

13. Berge AC, Moore DA, Sischo WM. Prevalence and antimicrobial resistance patterns of Salmonella enterica in preweaned calves from dairies and calf ranches. Am J Vet Res 67:1580-1588, 2006.

14. Warnick LD, Kanistanon PL, McDonough LP. Effect of previous antimicrobial treatment on fecal shedding of Salmonella enterica subsp. enterica serogroup B in New York dairy herds with recent clinical salmonellosis. Prev Vet Med 56:285-297, 2003.

15. Heider LC, Funk JA, Hoet AE, et al. Identification of Escherichia coli and Salmonella enteric organisms with reduced susceptibility to ceftriaxone from fecal samples of cows in dairy herds. Am J Vet Res 70:389-993, 2009.

16. Daniels JB, Call DR, Hancock D, et al. Role of ceftiofur in selection and dissemination of bla-CMY-2 mediated cephalosporin resistance in Salmonella enterica and commensal Escherichia coli isolates from cattle. Appl Environ Microbiol 75:3648-3655, 2009.

17. Wilhelm B, Rajic A, Waddell L, et al. Prevalence of zoonotic or potentially zoonotic bacteria, antimicrobial resistance, and somatic cell counts in organic dairy production: current knowledge and research gaps. Foodborne Pathog Dis 6:525-539, 2009.

18. FDA/CVM FACT SHEET: Veterinary Feed Directive Final Rule and Next Steps. Accessed 7/12/2015 at http://www.fda.gov/AnimalVeterinary/DevelopmentApprovalProcess/ucm449019.htm.

Emerging disease response: Global roadmap but local drivers

Corrie Brown, DVM, PhD, DACVP
College of Veterinary Medicine, University of Georgia, Athens, Georgia

Recent years have seen great progress in planning for and controlling certain animal health emergencies. Governmental and intergovernmental efforts to contain outbreaks caused by agents such as highly pathogenic avian influenza, African swine fever, and MERS-CoV, are three examples that have received widespread attention, and often accolades for the rapid and coordinated responses. So the question can reasonably be asked, why was there not a similar response to the North American introduction and widespread devastation caused by porcine epidemic diarrhea virus (PED)? Well, unfortunately, not all animal health emergencies are created equal, and much depends on the missions and abilities of the emergency management systems operated by the governmental and intergovernmental organizations. This requires some backgrounding in how national animal health systems are constructed and how they operate in today's interconnected and globalized world.

World Trade Organization, the OIE, and notifiable diseases

Globalization is a phenomenon that has changed how animals and animal products are moved around the world. The advent of the World Trade Organization (WTO) 20 years ago has regularized international trade and allowed products to move more freely and abundantly, in a relatively sanction-less manner. The World Organization for Animal Health (also known as OIE, according to the previous name of the organization, Office of International Epizootics) is the intergovernmental body charged with interpreting the WTO agreements, creating standards by which trade can occur safely. All countries which participate in any kind of trade involving animals and animal products belong to the OIE and have to abide by the OIE standards written in support of the WTO agreements.

The OIE maintains a list of "notifiable" diseases and all member countries must report the occurrence of these diseases every 6 months. Additionally, if a country has previously been considered free of one of these notifiable diseases, and the disease makes an incursion, the country's Chief Veterinary Officer is required to notify the OIE of the presence of this disease within 24 hours of diagnosis. Prior to 2004, notifiable diseases were categorized into List A and List B, with differing reporting requirements, but then the lists were combined, with one reporting regimen, and this combined list initially totaled 134 different diseases.

Since the inception of the combined "notifiable disease" list in 2004, each year the OIE membership examines the list and continues to refine how it is populated. For 2015, there were 117 notifiable diseases. Notifying the international community of the occurrence of any of these diseases is required of all OIE member nations. Failure to report is considered a serious breach of transparency and therefore renders the country at risk of appearing obfuscating, which will endanger overall abilities to trade. Of the 117 notifiable diseases, six are swine specific - African swine fever, classical swine fever, Nipah virus infection, porcine cysticercosis, porcine reproductive and respiratory syndrome, and transmissible gastroenteritis. There are several other notifiable diseases of pigs that require notification, but are not totally swine-specific and so are included in the multi-species category – *Brucella suis, Echinococcus granulosus*, Aujeszky's disease, rabies, trichinellosis, Japanese encephalitis, and foot-and-mouth disease. As spelled out in the OIE regulations, it is the responsibility of the Chief Veterinary Officer to report to the OIE, for general dissemination to the membership, the occurrence of any of these notifiable diseases.

National animal health system and reporting/response

The Chief Veterinary Officer, as the lead government veterinarian in each country, oversees the national animal health system in his/her nation. This national animal health system is, as defined by the OIE, responsible for diseases of "public good." As such, the government assumes responsibility and facilitation of control for those diseases which could harm the national economy or the public health. All of the notifiable diseases fall in this category of "public good." For example, each of the 13 diseases of swine listed above could cause trade restrictions or human disease. And in the US, for some of the diseases which would have massive negative trade implications, such as classical swine fever, African swine fever, or foot-and-mouth disease, the national animal health system maintains full emergency response plans and contingencies for requesting funds from Congress to eliminate infected animals and pay indemnities.

Emergency response, global efforts

In addition to OIE, there is a second large intergovernmental organization for animal health, the United Nations Food and Agriculture Organization (UN-FAO). This group functions largely to help control transboundary and many other notifiable diseases. The FAO has been very active in promoting Emergency Prevention Systems and maintains a training program for developing nations, Good Emergency Management Practice, which helps countries to develop preparedness and response capacities for the diseases of public good.

So, there is no doubt that over the last two decades, both developed and developing nations have built excellent programs, on a general and shared framework, for responding to emergency situations. But all these emergency preparations and response plans are for diseases which, by consensus or regulations, are considered to belong in the "public good" category.

Porcine epidemic diarrhea – response?

So, why wasn't massive governmental control applied for PED, which entered the US in 2013, and created major disruptions for swine producers? This disease is considered endemic in many European and Asian countries, and can cause serious losses for producers. But because the disease is not notifiable, it is not required that it be reported to OIE, and so it cannot legally affect trade issues. The national animal health systems of the world do not consider this disease to be a disease of "public good," as it is not on the list of transboundary animal diseases that would impair trade, and it has no negative public health implications. Basically, this is a disease that internationally is considered to be a "private good" disease, and therefore not necessitating major government or intergovernmental efforts to contain and control.

Our US systems of intensive production and geographic concentrations of swine rearing, allowed for severe spread of PED and massive negative impacts for farming communities across the country. Even though the disease was first noted in the US in May of 2013, it wasn't until June 2014 that the USDA issued a Federal Order requiring the reporting of the disease, and made federal funding available for combating the outbreak. This initial funding totaled $26.2M and included cost-sharing for development of biosecurity plans, diagnostic testing, and research, but no monies for indemnities.

In comparison, the highly pathogenic avian influenza (HPAI) outbreak in 2015 consisted of a full-fledged federally-directed outbreak response, complete with incident command systems, massive depopulation of affected birds, and more than $950M paid for control efforts and indemnities. This is because HPAI is a notifiable disease, with major trade impacts and potential public health considerations, and emergency response plans for this disease were already fully developed and ready to be implemented. In contrast, with PED, the federal government stepped in only when damage was sufficient to be of greater economic impact, far beyond individual producers. And still, the funding levels were meager in comparison to what would be seen with a bona fide transboundary animal disease.

Summary

As Napoleon the pig stated in the classic novel *Animal Farm* by George Orwell, "All animals are equal, but some animals are more equal than others." The same is true for diseases. Our regulatory systems, both national and international, are designed to deal with diseases of established public good. For those diseases that are traditionally considered within the private good realm, it falls to local actors and the industry to do the assessment and control. Only when the impacts of these private good diseases grow to demonstrate significant damage to the economy do they assume a position garnering sufficient attention for federal control. Understanding the national animal health system's construction and its inherent constraints could be a good starting point for beginning dialogue about policy changes in the national animal health system and/or enhanced preparedness in the industry for the next private good disease that emerges in full force.

References

1. Anonymous. FAO in Emergencies, http://www.fao.org/emergencies/en/, accessed Nov 12, 2015.

2. Anonymous. International standard setting activities – OIE, USDA-APHIS, https://www.aphis.usda.gov/wps/portal/aphis/ourfocus/animalhealth/sa_import_into_us/!ut/p/a1/rZFLb8IwEIR_C4ceI2-clzmGVxJIS1XUFnKx3IQQt4kdEoPKv8cgL1UFpVJ9W-vb0cwOStESpYLt-YYpLgWrTnPq0uk8xOYAcBQs-mOIH18nCYk9ax7aG1hpYBj4oe3FAGATDNFoEI68fgIQub_tv6EUpZ1QjSrRijU172gmhVoLRSv-3rL28AAdo3LX0kJmu-48McFrVtFyzSpVnn943chWUS6UpLvupN1kPNeKxbooXNMxXGDMsI1nG8TBuVFgTFw3M3Mzsy4Z4Mrz4a4M35BgPNDIJH7yZ1MMM-cC3DrTGbjhYaVNeldd9G20-GPq6R3V4jYZJhsty1RpcFFItPx5bM3xj-029XWRp-q-FFr-W5NN_VIT62B8PhOwnGof-73eESGmBhA!/?urile=wcm:path:/aphis_content_library/sa_our_focus/sa_animal_health/sa_export_from_us/sa_live_animals/sa_international_regulations/ct_international_standard_setting_activities_oie, accessed Nov 12, 2015.

3. Anonymous. OIE, Notifiable diseases, http://www.oie.int/animal-health-in-the-world/oie-listed-diseases-2015/, accessed Nov 12, 2015

4. Anonymous. Questions and Answers: Swine enteric coronavirus diseases monitoring and control program. APHIS-USDA Factsheet, May 2014, accessed https://www.aphis.usda.gov/wps/portal/aphis/ourfocus/animalhealth?urile=wcm%3Apath%3A%2Faphis_content_library%2Fsa_our_focus%2Fsa_animal_health%2Fsa_animal_disease_information%2Fsa_swine_health%2Fct_ped_info.

5. Anonymous. Technical Note: Porcine Epidemic Diarrhea USDA-APHIS, https://www.aasv.org/aasv%20website/Resources/Diseases/PorcineEpidemicDiarrhea.php, accessed Nov 12, 2015.

6. Domenech J, Lubroth J Eddi C, Martin V, Roger F. Regional and international approaches on prevention and control of animal transboundary and emerging diseases, Ann NY Acad Sci 1081:90-107, 2006, doi: 10.1196/annals.1373.010.

7. Hill C, Raizman E, Snider T, Goyal S, Torremorell M, Perez AM. Focus on emergence of porcine epidemic diarrhea in North America. FAO EMPRES No. 9, July 2014.

Modeling the transboundary risk of feed ingredients contaminated with porcine epidemic diarrhea virus

Scott Dee[1]; Casey Neill[1]; Aaron Singrey[2]; Travis Clement[2]; Roger Cochrane[3]; Cassandra Jones[3]; Gilbert Patterson[4]; Gordon Spronk[1]; Jane Christopher-Hennings[2]; Eric Nelson[2]

[1]Pipestone Applied Research, Pipestone Veterinary Services, Pipestone, Minnesota; [2]Animal Disease Research and Diagnostic Laboratory, South Dakota State University, Brookings, South Dakota; [3]Department of Grain Science, Kansas State University, Manhattan, Kansas; [4]Center for Animal Health and Food Safety, University of Minnesota, Saint Paul, Minnesota

Background

This study describes a model developed to evaluate the transboundary risk of PEDV-contaminated swine feed ingredients and the effect of two mitigation strategies during a simulated transport event from China to the US.

Results

Ingredients imported to the USA from China, including organic & conventional soybeans and meal, lysine hydrochloride, D-L methionine, tryptophan, Vitamins A, D & E, choline, carriers (rice hulls, corn cobs) and feed grade tetracycline, were inoculated with PEDV. Control ingredients, and treatments (ingredients plus a liquid antimicrobial (SalCURB, Kemin Industries (LA) or a 2% custom medium chain fatty acid blend (MCFA)) were tested. The model ran for 37 days, simulating transport of cargo from Beijing, China to Des Moines, IA, US from December 23, 2012 to January 28, 2013. To mimic conditions on land and sea, historical temperature and percent relative humidity (% RH) data were programmed into an environmental chamber which stored all containers. To evaluate PEDV viability over time, ingredients were organized into 1 of 4 batches of samples, each batch representing a specific segment of transport. Batch 1 (segment 1) simulated transport of contaminated ingredients from manufacturing plants in Beijing (day 1 post-contamination (PC)). Batch 2 (segments 1 and 2) simulated manufacturing and delivery to Shanghai, including time in Anquing terminal awaiting shipment (days 1-8 PC). Batch 3 (segments 1, 2 and 3) represented time in China, the crossing of the Pacific and entry to the US at the San Francisco, CA terminal (day 1-27 PC). Batch 4 (segments 1-4) represented the previous events, including transport to Des Moines, IA (days 1-37 PC). Across control (non-treated) ingredients, viable PEDV was detected in soybean meal (organic and conventional), Vitamin D, lysine hydrochloride and choline chloride. In contrast, viable PEDV was not detected in any samples treated with LA or MCFA.

Conclusions

These results demonstrate the ability of PEDV to survive in a subset of feed ingredients using a model simulating shipment from China to the US. This is proof of concept suggesting that contaminated feed ingredients could serve as transboundary risk factors for PEDV, along with the identification of effective mitigation options.

The next one! A Canadian vision for emerging disease preparedness*

Doug MacDougald, DVM

Introduction

The most recent emerging disease in Canada, Porcine Epidemic Diarrhea (PED), highlighted our lack of cohesive preparedness for dealing with current important endemic and future emerging diseases, in spite of much better containment than in the US This paper has three parts: A) a review of the past and why we have arrived at this point without seriously addressing emerging disease preparedness, B) our current position on PED and the recent changes in disease preparedness that have occurred to date, and C) a plan for a comprehensive and sustainable swine disease strategy for Canada.

I will reference extensively from Dr. John Harding's paper, *Accountability in the age of transboundary and emerging porcine diseases*, IPVS 2013[1] and from Dr. Robert Desrosiers's paper, *Emerging diseases: The past and the future*, AASV 2015.[2] These two veterinarians have challenged all of us in the swine industry to change how we think and act and I would like to thank them.

Part A: Past

The cost of current endemic disease is high. Porcine Reproductive Respiratory Syndrome (PRRS) is the most expensive in our industry with a cost of 15 billion dollars since first being detected in the US industry and current estimates at 664 million dollars per year and 130 million dollars per year for the Canadian industry. Yet as Desrosier[2] states, "it took more than two decades to identify and accept one of the most important transmission means (aerosol) of that organism." The US and Canada export 15 and 46 % of total pork production, respectively, and thus any trade limiting disease will have a material to catastrophic impact on our industry, depending on the disease.[3] Dr. Patrick Webb of the National Pork Board (NPB) reported the cost to the swine sector alone would be 57 billion dollars over ten years.[4]

As Harding[1] so succinctly stated "never has the risk of transboundary and emerging diseases been so real. The "perfect storm" caused by swine industry intensification, globalization, societal and environmental encroachment is having an insidious impact on the global swine industry." Desrosiers's[2] summary points pertaining to the past include; a) we don't know exactly how any of the main emerging pathogens of the last 40 years were introduced, or why they emerged; this equates to persistent vulnerability, b) we have not gotten rid of any emerging pathogens of the last 40 years; adding diseases without eliminating is not a good long term strategy, c) some emerging pathogens of the future are already present in our herds, d) it can take years to determine the cause of an emerging disease.

Abelseth[5] concluded that the use of Specific Pathogen Free (SPF) swine will influence the hog industry in two main ways. The first is that it will result in the absence of some of the diseases which commonly plague swine herds. It has been demonstrated that atrophic rhinitis (AR) and virus pneumonia (VPP) can be eliminated...thus improving on financial return. The second is to achieve maximum results without resorting to treatment levels of antibiotics. Since approximately 2,000,000 tons of feed (90% medicated) is fed to hogs in Canada annually, considerable savings would be realized if medication were unnecessary. Fifty four years ago the link between better health and antibiotic use was being highlighted and it seems not much has changed!

So why aren't we better prepared? There is currently a gap in who does what, depending on if a disease is reportable to regulatory agencies or not. In the case of PED, it was not reportable in the US until June 2014 with the epidemic essentially over by then. In Canada, PED did not fall under the responsibility of the Canadian Food Inspection Agency (CFIA) and each province had a differing response to this transboundary disease being reportable, notifiable or nothing. The Food and Agriculture Organization of the United States defines transboundary animal diseases as (paraphrased): *"Those epidemic diseases which are highly contagious or transmissible and have the potential for very rapid spread, irrespective of national borders, causing serious socio-economic and possibly public health consequences. These diseases which cause a high morbidity and mortality in susceptible animal populations, constitute a constant threat to the livelihood of livestock farmers. Furthermore, their potential consequences are of such a magnitude that their occurrence may also have a significant detrimental effect on national economies."*[1]

It is just not clear who does what within our federal government, provincial government and industry regarding emerging diseases. Even though we export a significant amount of our product, the provincial pork boards and Canadian Pork Council (CPC) in conjunction with governments have not created a "single voice" organization with the support of all involved to shape swine health strategy in Canada. The reason this has not occurred is, in my opinion, complacency driven by the long history of no Foreign Animal Disease (FAD), the "regulatory black hole" that transboundary diseases fall into and provincial organizations maintaining a "silo mentality," protecting their turf without true focus on delivering emerging disease preparedness on a coordinated national basis. A new approach with different outcomes is needed.

Lastly and especially pertinent to current endemic disease and future emerging disease is the issue of antimicrobial use in agriculture. The World Health Organization in 2011 lists the critically important antimicrobials for human medicine[6] and states:

* Paper was published in *Advances in Pork Production* Vol 27; 2016 Banff Pork Seminar

"Antimicrobials are used widely in agriculture. This includes non-therapeutic use such as for growth promotion. It also includes use as prophylaxis to try to prevent infections developing in food animals and therapeutic use to treat sick animals. However, this use also includes using agents defined by WHO as "critically important" for human medicine".

"Resistant Gram negative bacteria (e.g., E. coli) have become a major and rapidly increasing problem. There are no new classes of antimicrobials in the pipeline and so it is unlikely that any new classes of effective antimicrobials will be available for 10 years or more to treat infections caused by resistant Gram negative bacteria". "The end of the age of the miracle drug may indeed be upon us." "On a global scale, E. coli is the most important human pathogen and causes substantially many more infections than Salmonella and Campylobacter combined. Thus, the importance of resistance in E. coli, typically considered a benign commensal, should not be underestimated."

The uncontrolled use of critically important antibiotics for human medicine in agriculture is ending. Thus, it seems clear we are in need of a different model for sustainable swine health.

Part B: Current

The summary of PED cases in Ontario is: 63 primary cases from January, 2014 to July, 2014 with voluntary elimination at the producer and site level. No cases from July, 2014 to November, 2014 and then 21 primary cases from November, 2014 to October, 2015. Ontario Swine Health Advisory Board (OSHAB) implemented a PED Area Regional Control & Elimination (ARC&E) component into an already functioning PRRS ARC&E and this through 2015 has become province wide with over 1,100 of the estimated 1,800 targeted swine sites participating (61%) with a target of achieving over 80% by 2016.

Pig sites contaminated by pig movement from the primary sites are not tracked in Ontario thus total positive sites are not accurately known. However, of sites that are registered in the ARC&E program, 127 PED infected sites (54 primary sites reported by OMAFRA and the remainder are herds infected by pig flow), 77% (98) are now negative with 23% (29) in the elimination process but not yet negative.[7]

In a survey of swine veterinarians in this province in October 2015 there were six primary cases with an estimated <25 contaminated sites and three of the six are in process of elimination. The summary of PED cases in the Maritimes, Western Canada (Manitoba) and Quebec[8] is:

Maritimes: The farm in Prince Edward Island, among the original group in Canada with feed as the source of infection, has tested negative on all animal and environmental tests since October 2014 and can be presumed negative.

Manitoba: Of the five infected herds, one has depopulated and not yet repopulated, another was destroyed by fire and a third was reported as presumed negative. The remaining positive farms are in the process of becoming 'confirmed' negative, possibly within two to three months.

Quebec had 16 sites infected to date with PEDV and, at the time of writing this paper, every site is presumed negative with confirmatory testing pending for two sites.

What organizations/regions have done in the face of PED?

The Quebec industry formed Equipe Quebecoise de sante procine (EQSP) in 2008 to develop a Foreign Animal Disease preparedness plan. The EQSP mission is to work in collaboration with government and industry stakeholders on prevention, planning and intervention against targeted swine diseases to minimize their potential impact on the Quebec swine sector.[9] The specific plan is shown in Figure 1 which shows a government command centre and an EQSP command centre essentially bridging the gap of government and industry.

This organization and project was extended in 2012 with support from a broad base of industry including; Les Éleveurs de porcs, AQINAC and packing plants (Figure 2).

In June 2013, with PED emerging in the US, EQSP was given the mandate to develop a two part strategy to address this threat:

1. Quickly identify actions to be taken to prevent the introduction of PED in Quebec's swine herds.
2. Establish an action plan to eliminate PED in the event the virus was introduced into Quebec (Figure 3).

This strategic plan led to surveillance of

- Packing plants; testing unloading docks at least once a day.
- Depending on the plant, either testing all trailers from Ontario or receiving Ontario pigs at a specified time and then wash and disinfect and test the dock.
- Random sampling of:
  - Quebec trailers
  - Assembly yards
  - Feed ingredients.
- Recommendations on buying hogs to prevent spread of porcine epidemic diarrhea virus (PEDV) and porcine delta coronavirus (PDCoV).

The elimination plan has been successful as detailed above.

Western Canada to date has responded to PED with the provincial pork boards and the Chief Veterinary Officers (CVO) collaborating in what would appear to be an effective manner. The newly formed Western Canadian Swine Health Alliance (WCSHA), representing a majority of production and processing, is now collaborating with the above organizations and supporting Canada- West Swine Health Intelligence Network, a member of the Canadian Swine Health Intelligence Network (CSHIN).

Ontario was the least coordinated province to date with the CVO, Ontario Pork, OSHAB and other organizations lacking a clear operational or communication plan for prevention and elimination but in spite of this the containment and elimination of PED has been a success story albeit as detailed above, work still

Figure 1: EQSP Plan 2008

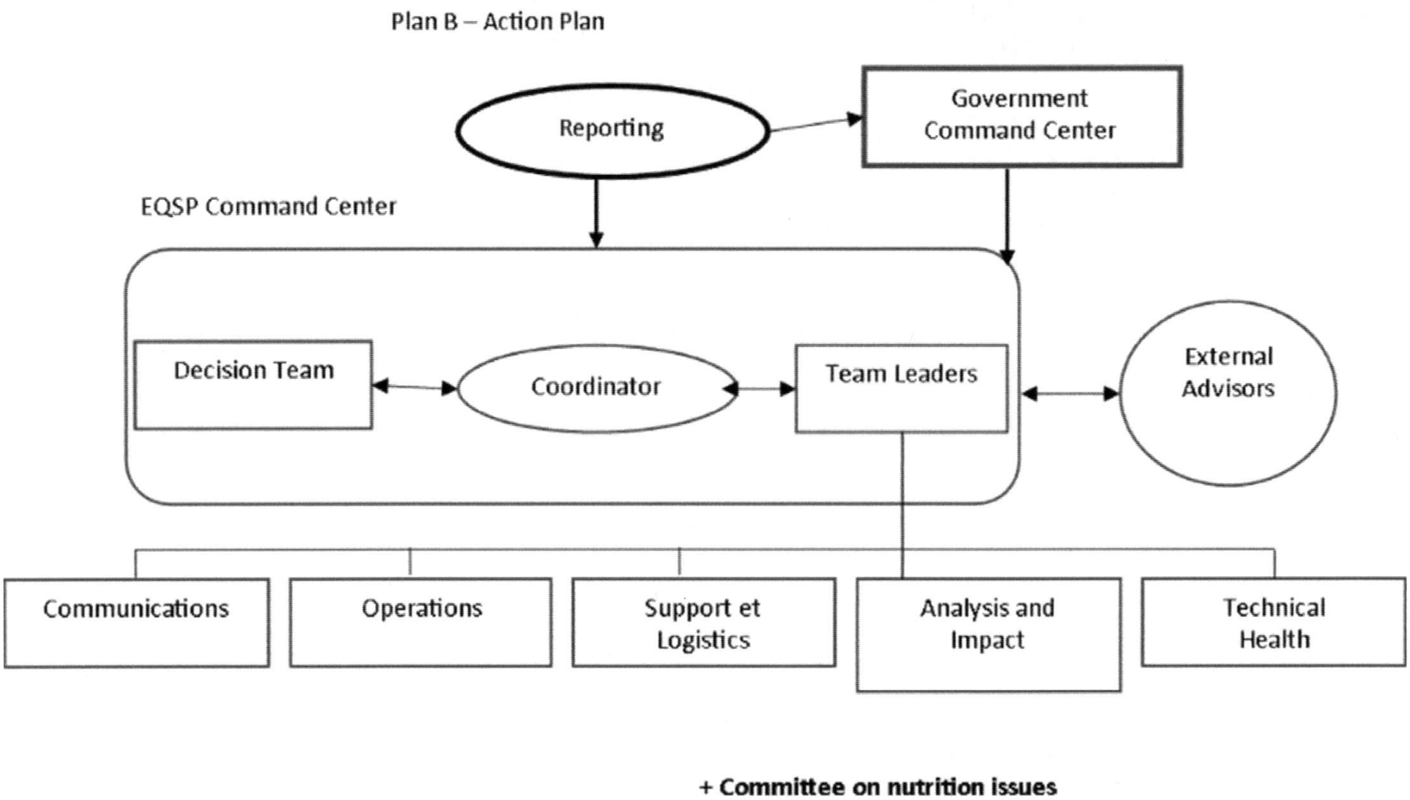

to be done with a small number of producers with endemic PED and with another high risk season coming – if province wide elimination is to be the goal.

OSHAB issued a position statement in the fall of 2015; "PEDV and PDCoV can and should be eradicated from Ontario and the rest of Canada."[10] It states these viruses are the perfect model for assessing the robustness of a swine health plan for control, containment and elimination in preparing for the next emerging disease. Key requirements to achieve the goal of elimination:

Regulatory oversight is required to market virus-positive pigs efficiently and for all-site elimination of the virus.

1. A plan for virus-positive assembly yard elimination is implemented in the low-risk summer season. Funding support for undue expenses must be made available.
2. Livestock trucks returning from the United States must have sanitation solutions.
3. A surveillance plan for early detection and intervention in new cases and in critical control points are in place. This plan should include surveillance to allow for declaration of freedom from PEDV and PDCoV infection on a provincial and national level.

   Ontario is rolling out its Ontario Animal Health Network (OAHN) surveillance plan which is a program new to the swine sector. The goal of the program is to improve swine health in Ontario through communication, surveillance, and sharing information. The idea is to improve emergency preparedness, and carefully share what diseases are being seen across Ontario so that veterinarians and producers can make better herd decisions. It also allows for follow up funds for disease outbreaks and projects to address these issues. The network was assembled in October 2014, but was previously the Ontario node of CSHIN. This plan is to integrate with other provinces and species in Canada with Quebec's existing surveillance plan RAIZO as an example. See Figure 4 & 5 for the high level strategy.[11]

On a national level the Canadian Pork Council (CPC) with a levy of 1¢ per hog from the provincial pork boards has created a health team to support national issues such as, surveillance (CSHIN), antimicrobial use, international border issues and support of provincial issues and initiatives with knowledge transfer and communication including a monthly "health meeting" with provincial boards.

Part C: A plan for a comprehensive and sustainable swine disease strategy for Canada

This would be easy if we were creating an industry from zero, but we are not, and any vision of a sustainable plan has to take into account, history, culture, relations and be able to deliver a

Figure 2: EQSP Plan 2012

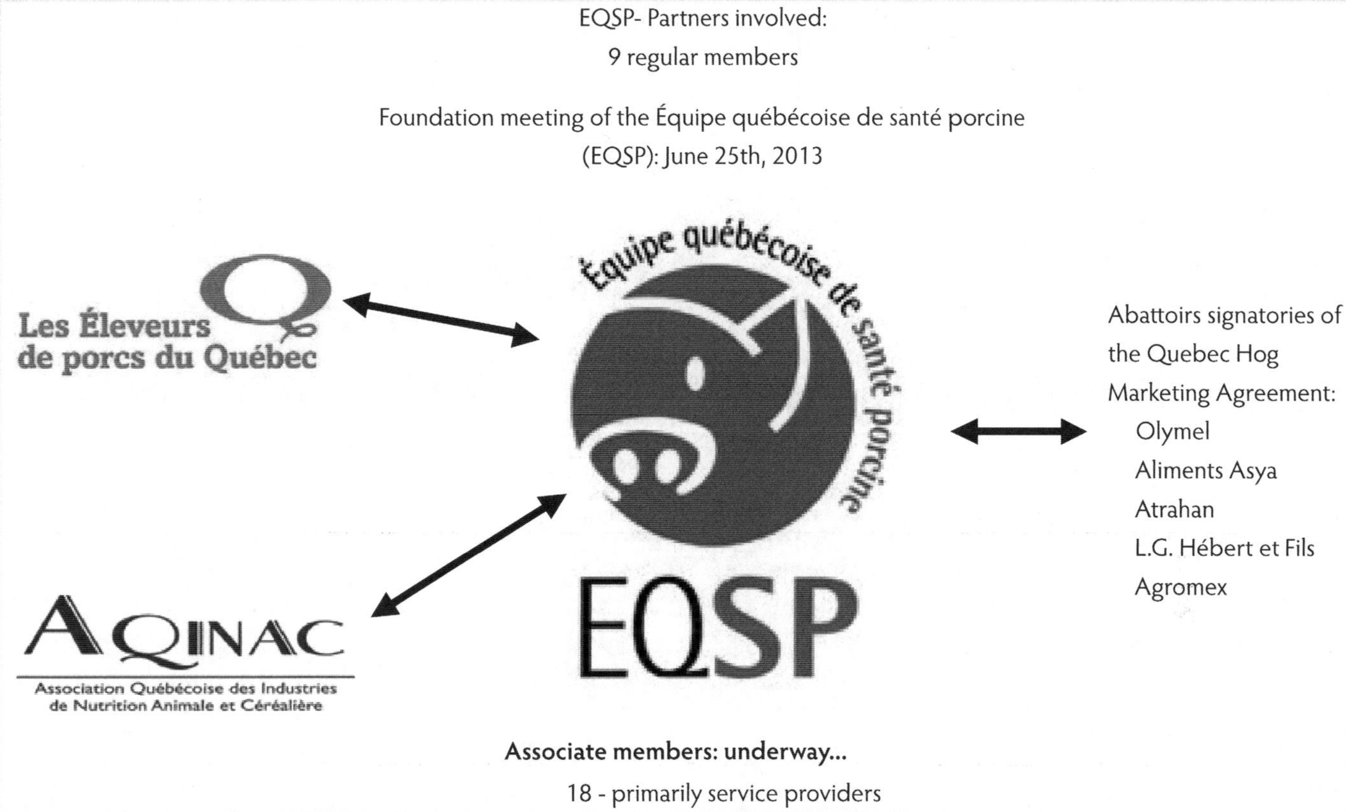

Figure 3: Case management process

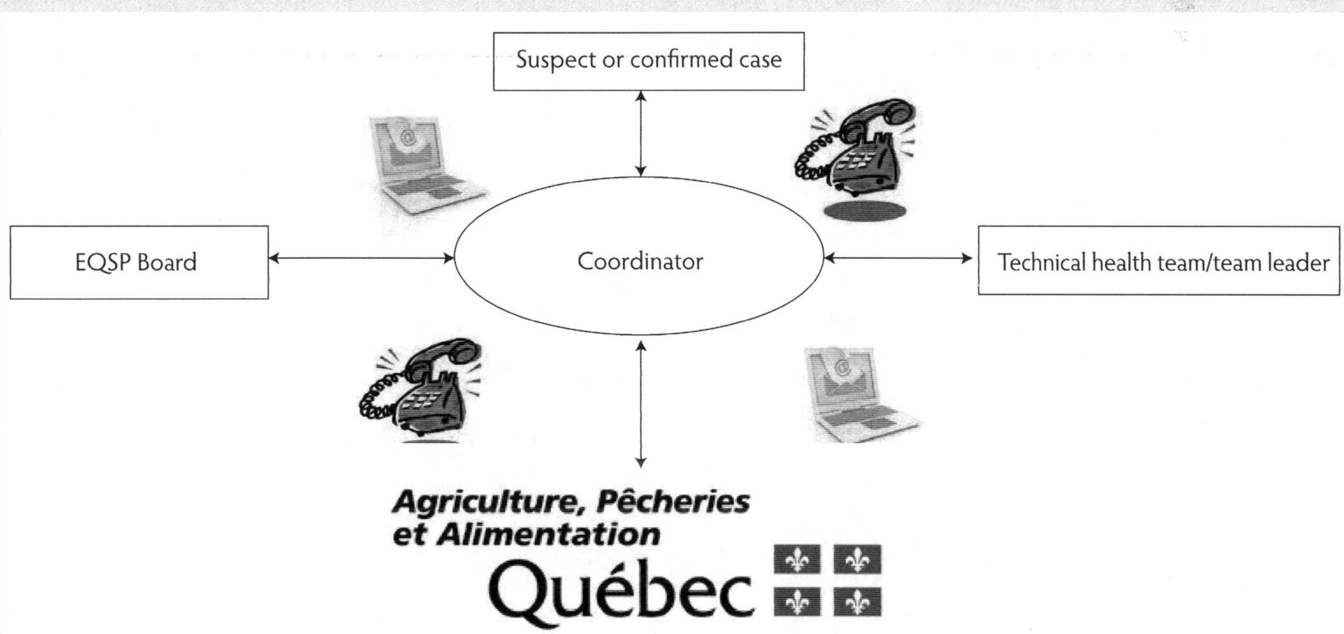

Figure 4: OAHN Strategy

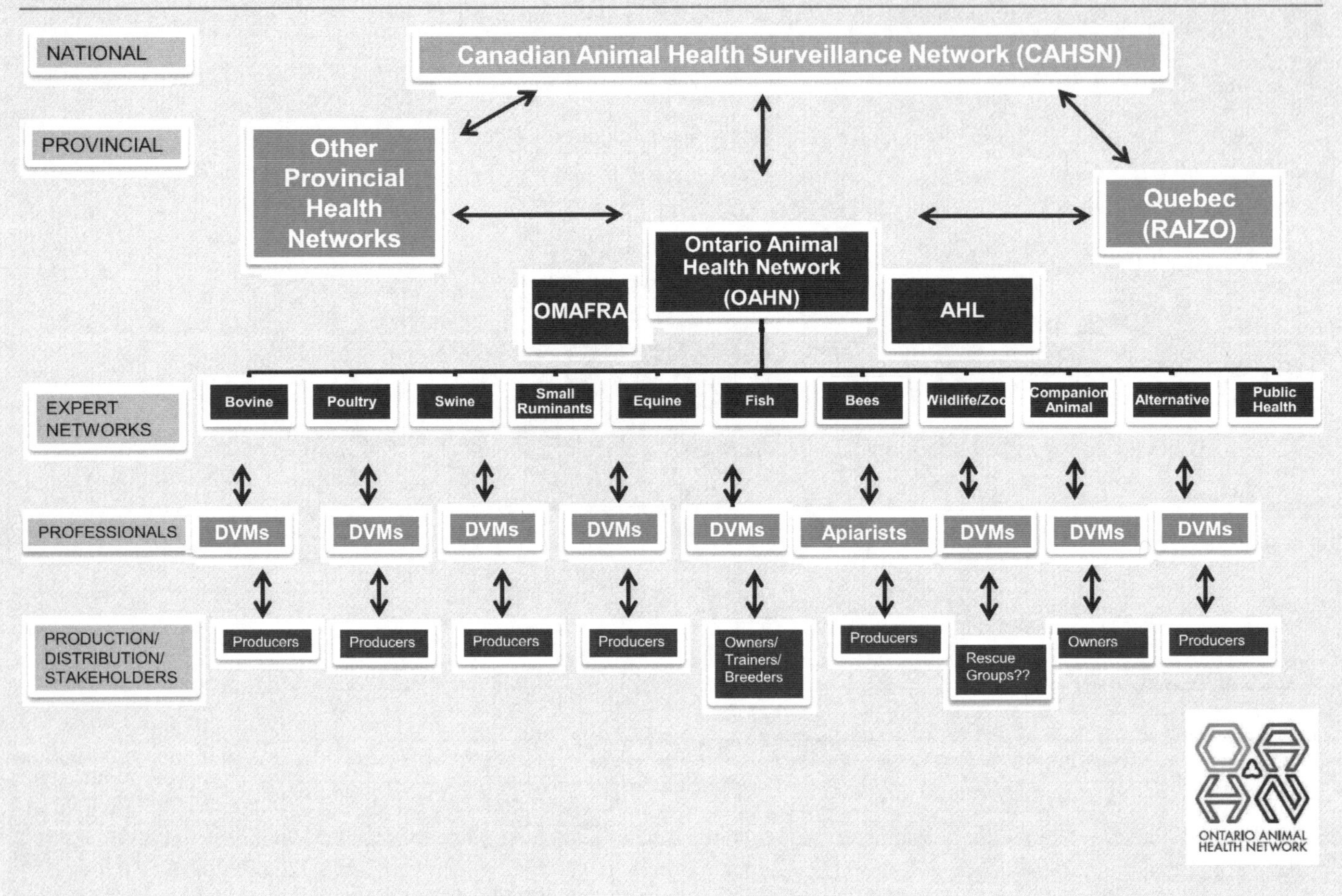

good solution that will work long term. Wishful thinking doesn't work!

First some history of Canada from William A. Macdonald's essay on Canada's history and role in the world.[12]

"Use words, not force. Make railways, not war. These overly simple ideas capture a national story – the Canadian one – that differs from those of most other countries.

Canada's story has increasingly been driven by persuasion...since its beginnings – first Quebec in 1608 and then Confederation in 1867 – Canada has had three very big achievements. First, it has survived – not just as a nation but as one that includes the distinctive province of Quebec. Second, it made itself coast-to-coast. Finally, despite its divisions of nationality, culture, language, religion and class, it has developed a political and socio-cultural outlook that works. All these achievements have been based on mutual accommodation. Today's Canada is the product of its capacity for mutual accommodation and a belief in an underlying shared order."

It seems clear from history that to accomplish Desrosiers's[1] conclusion, "for pathogens, reportable or not and emerging or not, that are easily transmitted indirectly, a collective approach coordinated by a central entity is needed to effectively control spread and losses", we need to build a Canadian plan from the ground - provincial industry – up and with mutual accommodation. As shown in Part B there are good working parts as highlighted by Quebec's EQSP model[8] and this, in my opinion, needs to be replicated albeit with different versions in other regions as well as nationally.

Ontario is forming an independent health leadership team called Swine Health Ontario (SHO). This is being created by four founding members – Ontario Pork (OP), Ontario Pork Industry Council (OPIC), OSHAB and Ontario Ministry of Agriculture Food and Rural Affairs (OMAFRA).

The impetus is "while Ontario has been successful at managing swine health issues, there is room for further improvement in terms of co-ordination, communication and the need for a broad, proactive strategy that includes all aspects of swine health".

The role thus of the SHO is to create and oversee a broad proactive, industry-wide, long term strategy for swine health operating within the mission of "continuous improvement in swine health management through a proactive, practical and responsive plan

Figure 5: The disease surveillance tartan

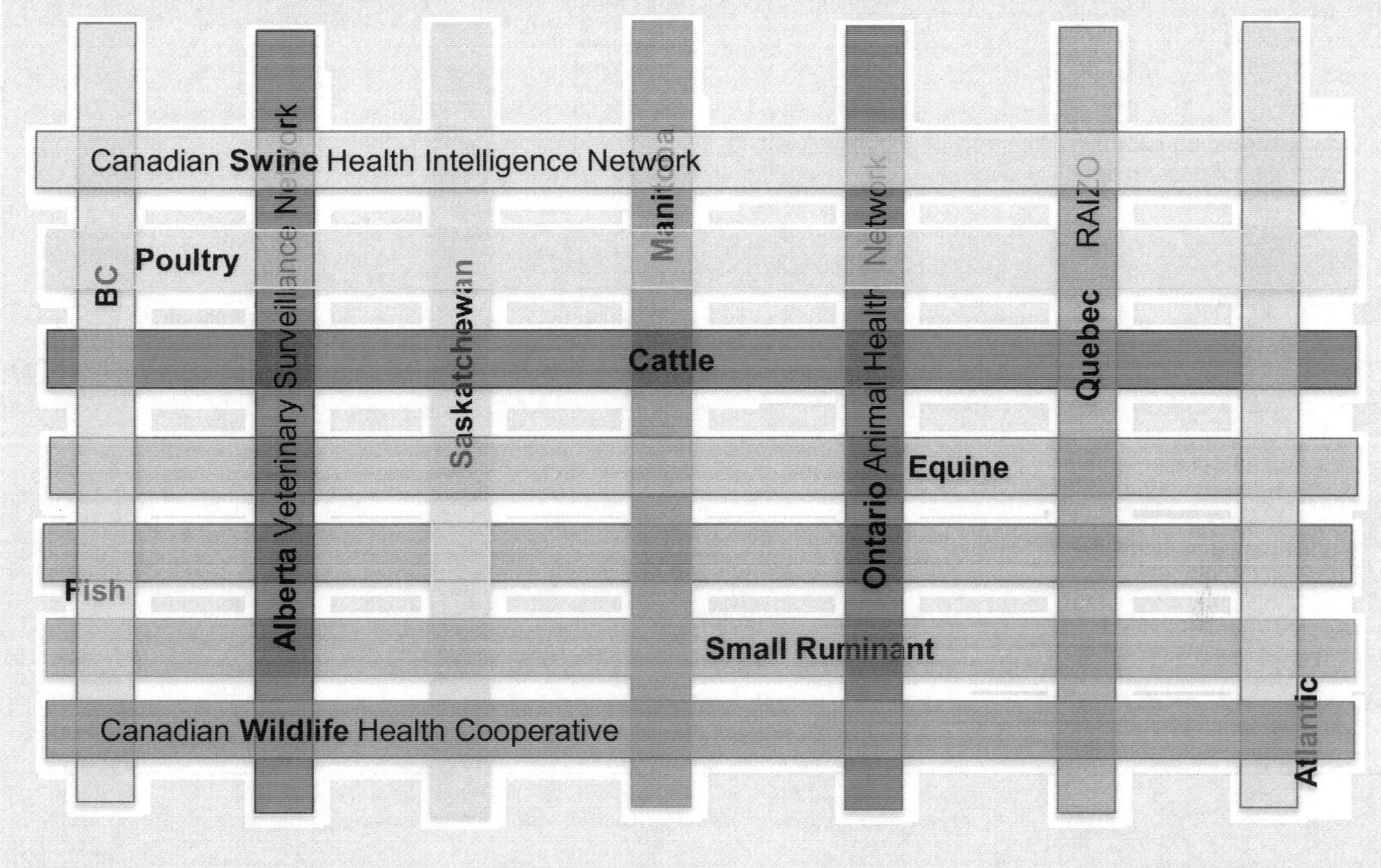

that engages the entire swine industry."[11]

Thus as you can see, Ontario is building with broad industry support, a central coordinating entity similar to Quebec's EQSP and, in my opinion, this works.

Western Canada, in spite of the current coordinated approach, needs to create a similar model for the simple reason that leadership at the provincial pork boards and the CVO's change and because leadership is working well together today does not mean it works well in the future.

The Maritime region, with a very small number of herds can be organized by the current organizations, but needs to be plugged into a national entity.

What about nationally?

In the US, the Swine Health Information Center (SHIC) was formed by AASV, NPD and NPPC because of PEDV introduction with the lessons including:

1. Pathway of pathogen introduction is difficult – at best. Be prepared for the next one because it is coming.
2. Don't expect USDA alone to respond to emerging diseases.
3. Industry needs to take more responsibility for non-regula tory diseases.
4. Better state-federal-industry response coordination is essential.
5. Agreed upon response plan before the next disease.

SHIC is a tool to:

1. Implement industry preparedness.
2. Improve swine health management.
3. Improve non-regulatory disease response.
4. Enhance AASV, NPPC and NPB acting for producers.

I believe we have to have a national entity following the model of EQSP and SHO to coordinate, collaborate and communicate with the regional organizations within Canada and with international organizations such as SHIC.

The structure of the Canadian Vision aside what does it have to deliver? Harding summarized this best and I am paraphrasing some of his key points that pertain to this paper[1]:

1. Clearly define what our national and provincial governments will do in the response to FAD and transboundary disease with insistence of commonality of response.
2. Industry needs to take more responsibility for all emerging disease preparedness.
3. Nationally deliver a practical and mandatory surveillance system that will detect early emerging disease and that will

work transparently with the key organizations on containment and elimination.

4. Veterinarians need to deliver a lot more to solve the complex issues of the day thinking globally and leading the swine industry to do the same.

5. Industry needs to develop a more long-term sustainable health focused pork industry with a clearly stated and deliverable responsible antimicrobial plan.

Conclusion

"It is no use saying, 'We are doing our best.' You have got to succeed in doing what is necessary." ~Winston Churchill

References

1. Harding, J.C.S. (2014) Accountabilities in the age of transboundary and emerging porcine diseases. 23rd IPVS Congress.
2. Desrosiers, Robert (2014) Emerging diseases: The past and the future. Annual meeting 2014 AASV.
3. US Meat Export Federation (2015). Available at www.usmef.org.
4. Webb, P. (2012) Economic Impacts of FMD in the United States.
5. Abelseth, M.K. (1962) The Application of Specific Pathogen Free Animals to Research and Production.
6. World Health Organization (2011) Critically Important Antimicrobials for Human Medicine 3rd Revision.
7. OSHAB ARC&E Progress Update (2015).
8. Byra, C. (2015) CSHIN Porcine Epidemic Diarrhea Update – Sept 3. 2015.
9. Équip Québécoise de santé porcine (2015) List of EQSP recommendations and documents on PED and SDCV.
10. OSHAB Position Statement (2015).
11. Swine Health Ontario Leadership Team (2015) Swine Health Strategy Proposal.
12. Macdonald, W.A. (2015) The Globe and Mail Metro (Ontario Edition).

SHIC: Enhancing our preparedness

Paul Sundberg, DVM, PhD, DACVPM
Executive Director, Swine Health Information Center

Organization

The Swine Health Information Center (SHIC) began operation as a 501(c)(3) corporation on July 4, 2015. Its mission is to protect and enhance the health of the US swine herd through coordinated global disease monitoring, analysis of swine health data and targeted research investments that minimize the impact of future disease threats.

The National Pork Board, National Pork Producers Council and the American Association of Swine Veterinarians have each appointed two representatives to the SHIC Board of Directors. Three at-large producer representatives are also members of the Board. The Board has approved an initial operating budget and a plan for federally insured investments of the money in excess of the operating budget that is modeled after that of the National Pork Board.

A Monitoring and Analysis Working Group and a Response Working Group have been formed to provide program oversight and decision-making. Each are actively meeting to fulfill its objectives. The Swine Disease Matrix content and prioritization has been updated through work of the SHIC Monitoring and Analysis Working Group. The SHIC Response Working Group developed the SHIC Seneca Valley Virus Research Priorities and selected projects for funding.

Outreach and communication

Twenty-eight meetings separate meetings with producers, veterinarians, colleges of veterinary medicine, academics, USDA and other stakeholders has increased industry understanding of SHIC and its mission and the feedback has helped to focus and refine SHIC responsibilities, activities and programs. Written, radio, TV and web-based communications (`www.swinehealth.org`) has offered information and resources about SHIC and its roles and responses to Seneca Valley Virus emergence.

Swine Disease Matrix Project

Organizing and funding the research needed for diagnostic and informational preparedness for prioritized, potential production diseases will be a core activity of the Center.

The SHIC Monitoring and Analysis Working Group developed their system for prioritizing the Swine Disease Matrix and has completed that work. The Matrix is a living document that will be periodically or as needed reviewed and updated. Input into the priorities and process of Matrix research was gained through a meeting with multiple USDA agencies and the major veterinary diagnostic laboratories with swine disease research expertise.

With industry encouragement USDA-NVSL has studied the capabilities of our current diagnostic tests to detect an emerging, highly pathogenic strain of pseudorabies virus now circulating in pigs in China and they are conducting a vaccine efficacy trial to understand our capabilities to respond to this disease with the current, commercially available vaccines should it enter the US.

Global swine health and issues identification

Through global disease monitoring, the Center will manage the Swine Disease Matrix and heighten awareness and actions about emerging diseases.

Draft Fact Sheets with what is known about pathogenesis, epidemiology, disinfection, diagnostics and routes of transmission for 18 viruses presenting risk to the US swine herd are completed and out for external review.

The Center is forming a focus group to help develop a national system for implementing and coordinating Rapid Response Teams in response to new or emerging diseases. The Rapid Response Teams will be made up of veterinarians and epidemiologic subject matter experts that will be charged with rapid investigation of new or emerging diseases to, as quickly as possible, discover pathways of entry and transmission. The information will be used to inform the industry's response to these disease outbreaks.

Swine health data analysis and monitoring for trends

The Center will enable producers to get information about emerging and endemic disease trends. It will review and provide support to improve active disease introduction risk assessments, which will help producers improve their farms' biosecurity.

SHIC is funding technology tools and projects to support epidemiological analysis of producer-offered swine health data. And for this data to be analyzed it must first be in a standardized format used by the veterinary diagnostic laboratories. The Center is coordinating a work plan and identifying the resources needed so these laboratories can communicate their data to other laboratories and enable access and analysis of the data.

Seneca Valley virus as an example of SHIC response to an emerging disease

Coordinating and partnering with the National Pork Board, National Pork Producers Council, the American Association of Swine Veterinarians and a variety of state and federal animal health officials, other regulatory agencies and swine health researchers

SHIC helped to assess the needs, fund appropriate research and provide answers about the increase in Seneca Valley virus clinical infections during the summer and fall of 2015. During the approximately 2 months after the increase in clinical cases was reported, SHIC coordinated multiple stakeholder conference calls, developed research and response coordination priorities and funded over $375,000 of research into the most important gaps in the ability to respond to the disease.

Lessons learned from highly pathogenic avian influenza (HPAI)

C. Rowles, DVM

Introduction

The United States animal agricultural industry has been impacted by two major disease introductions in the past three years. First, porcine epidemic diarrhea virus (PEDV) was first confirmed in the domestic swine population on May 17,2013.[1] Increased mortality associated with this disease reduced overall market hog slaughter from 112 million swine in 2013 to 106 million in 2014.[2]

Since the index case in December 2014 in the state of Washington, highly pathogenic avian influenza (HPAI) has been detected in commercial and backyard poultry, wild birds, or captive wild birds in 21 states. To date, a total of 211 commercial and 21 backyard poultry premises have been affected, particularly affecting the states in the upper Midwest (Iowa, Minnesota). This devastating outbreak resulted in the destruction of nearly 50 million poultry, including 7.5 million turkeys and 42.1 million table-egg layer and pullet chickens. The cost to federal taxpayers is over $950 million. Iowa alone spent over $600million, and over 400 people lost their jobs due to this catastrophic event.[3-5]

The purpose of this paper will be to elaborate on the lessons learned in the HPAI outbreak and provide insight for how those lessons might be applied to the swine industry.

Discussion

Biosecurity

While biosecurity discussions usually begin with the application of the policies and procedures at the farm level, HPAI and PEDV have shown the need for discussions of biosecurity at the national and company level as well. While specific routes of introduction have not been confirmed for PEDV, there is evidence to suggest a potential route of introduction through the use of flexible intermediate bulk containers (aka tote bags).[6] Evidence suggests the original HPAI infection was introduced through migratory waterfowl. Subsequent epidemiological investigations also suggest evidence of transmission through other fomites. In the majority of cases in the study, people, feed trucks, live haul loaders, pre-loaders, and other items were shared by multiple farms. While equipment sharing makes economical and logistical sense, it also increases the risk of lateral spread of HPAI between farms.[7] The important lesson here is that as we have driven decisions based on continual incentive to drive costs lower, we have incurred the potential for increasing risk of disease introduction. Future managerial decisions need to factor biosecurity risk analysis as part of the cost analysis.

At the farm level, many poultry farms identified gaps in their policies and procedures. Many are adopting the same "line of separation" and "perimeter buffer zone" concepts as are used currently in swine operations. Many operations are investing capital in either Danish entry systems or shower in/shower out facilities.

Aphis has also recommended the establishment of a biosecurity officer for farms. This person would be responsible for the implementation of the policies. More importantly, this person would also be responsible for ongoing monitoring of biosecurity. Biosecurity on paper is one thing. Daily implementation and adherence is another.

Depopulation

The first concept to address is that depopulation is not synonymous with euthanasia. Euthanasia is a term described as ending the life of an individual animal in a way that minimizes or eliminates pain and distress. By contrast, depopulation is defined as destruction of animals in a quick and efficient method with consideration given to the welfare of the animals as practical.[3,8]

During the early phases of the HPAI outbreak there were significant delays in prompt depopulation of flocks. These delays are thought to have contributed to the rapid spread of the virus in the Midwest. USDA-Aphis has stated that it is important to depopulate a flock within 24 hours after initial diagnosis. The reason for this is to reduce shedding of virus in order to protect neighboring flocks. Over time the USDA and the poultry industry worked hard to develop farm specific methodologies to advance this goal.

One methodology incorporated was ventilation shut down. (VSD) VSD involves shutting off the ventilation and allowing the temperature to elevate to 104 degrees within 30 minutes and to hold this temperature for at least 3 hours. This allows for rapid hyperthermic death.[3] While this is not considered to be an ideal method of euthanasia, the USDA, avian pathologists, and industry stakeholders agreed that the method could be used provided that 1) all other more humane methods had been considered, and 2) it had been determined that the time taken for other methods would allow the amount of virus to become excessively high, resulting in undue spread of the disease.[9] This particular issue highlights an area where all stakeholders came together to cooperate and collaborate in a way to deal with a difficult issue and arrive at a solution.

Disposal of carcasses

One area that was particularly problematic was the disposal of carcasses. The typical options of burial, incineration, composting, landfill, and rendering were considered. However, questions regarding public health concerns and environmental impacts often times delayed the proper implementation of a solution which only compounded the problems. One important lesson for the swine industry is to have thought through and preplanned for a mass disposal event.

Cleaning and disinfection

Cleaning and disinfection of facilities was another difficult step for HPAI. In particular, layer facilities are typically wood frame structures that are rarely "wet" cleaned. Early on in the outbreak, USDA attempted to use contractors to wet clean and disinfect. This was very quickly determined to be expensive and impractical. In this instance, the poultry developed protocols to dry clean the barns followed by heat treatments to inactivate the virus. In this case it was determined that sealing the barns and heating to temperatures of 120 degrees for 7 days would render the virus inactive.[10]

Secure egg supply

It is important to remember that not all flocks were affected by HPAI. Despite some early difficulties in implementation, the USDA issued over 8,000 permits for movement of birds and eggs without a single documented transmission of virus from any of those permitted movements.[11] Currently, the swine industry is working to develop a secure pork supply plan that is patterned after the secure egg supply plan. Today's modern swine industry is highly dependent on the ability to transport pigs for housing and marketing. It behooves the industry to continue to develop plans with USDA to enable a permitted movement process in the face of a foreign animal disease.

Disease risk and economic mitigation

HPAI had a significant economic impact on the producer, the industry, and the country. Hundreds of millions of taxpayer dollars were expended in an effort to indemnify the producer for losses associated with depopulation and cleanup. However, those indemnification dollars only covered a portion of the economic impact. Currently the poultry industry is exploring disease related insurance programs to help mitigate those additional expenses. The swine industry is highly dependent upon our export markets for sales and marketing of pork products. The potential economic impact of a foreign animal disease in the United States would do significant damage to the producer and the industry. The swine industry should continue to explore alternative solutions to this potential problem.

Summary

HPAI was a significant event for animal agriculture. In this case it was a defined foreign animal disease in which the USDA had a proscribed role and obligation. However, there were a myriad of other state, county, and city governmental agencies, as well as associations and producers involved. At the lessons learned conference in Des Moines Iowa in July of 2015 a clear theme of improved communication and collaboration was echoed by all speakers. The USDA learned that producers often times have the most practical and cost effective solutions to problems. Producers learned they need to be willing to communicate with other producers to help stop the spread of disease. All expressed the need for clear lines of communication between all agencies and producer groups and the public in order to provide for consistency in messaging.

The swine industry is also vulnerable to the introduction of a foreign animal disease. The introduction of a transboundary disease like PEDV is a wakeup call for the industry. In particular, the swine industry needs to further enhance biosecurity at all levels. It needs to continue to monitor disease activity in all areas of the world. The national pork board has taken an important step in the direction with the formation of the Swine Health Information Center. The industry needs to prepare for the potential introduction of a foreign animal disease by further developing the secure pork supply. And finally, the industry needs to continue to collaborate and cooperate with the USDA to prepare for the potential of a foreign animal disease outbreak.

References

1. Iowa pork industry center PEDV fact sheet. http://vetmed.iastate.edu/sites/default/files/vdpam/Disease_Topics/PED-virus-fact-sheet-5-17-13.pdf.
2. M. Haley, USDA ERS, personal communication. November 2015.
3. Sato Y. Highly Pathogenic Avian Influenza (HPAI) Lessons Learned. 2015 ISU James D McKean swine Disease Conference. 2015; 62-67.
4. USDA-Aphis, Veterinary Services. Fall 2015 HPAI Preparedness and Response Plan. https://www.aphis.usda.gov/animal_health/downloads/animal_diseases/ai/hpai-preparedness-and-response-plan-2015.pdf.
5. US poultry. Proceedings and notes from Avian Influenza Outbreak...Lessons Learned Conference in Des Moines, Iowa. July 28-29, 2015. https://www.aphis.usda.gov/animal_health/downloads/animal_diseases/ai/hpai-lessons-learned-conference-iowa.pdf.
6. USDA-Aphis. Swine enteric coronavirus introduction to the United States: root cause investigation report. www.aphis.usda.gov/animal_health/animal_dis_spec/swine/downloads/secd_final_report.pdf.
7. USDA-Aphis. Epidemiologic and other analysis of HPAI-affected poultry flocks: July 15, 2015 report. www.aphis.usda.gov/animal_health/animal_dis_spec/poultry/downloads/Epidemiologic-Analysis-July-15-2015.pdf.
8. Thornber PM, Rubira RJ, and Styles DK. 2014. Humane killing of animals for disease control purposes. Rev. sci. tech. Off. int. Epiz. 33(1):303–310.
9. American Association of Avian Pathologists. 2015. Position Statement of Ventilation Shutdown. http://www.aaap.info.
10. USDA. 2015. Cleaning & Disinfection Basics: Virus Elimination. https://www.aphis.usda.gov/animal_health/emergency_management/downloads/hpai/cleaning_disinfection.pdf.
11. B. McCluskey: personal communication. November 2014.

Keep your feet on the ground but stick your head in the "cloud"

Web-based tools and services being developed for producer group specific and larger-scale (aggregate data) spatiotemporal swine health monitoring applications

Rodger Main[1], DVM, PhD; Jordan Bjustrom-Kraft[1], BS; Bret Crim[1], BS; Erin Lowe[2], DVM; Zack Whedbee[3], BS; Enrique Mondaca[2], DVM, PhD; Kate Mueller[1], DVM; Dale Polson[2], DVM, PhD; Beatriz Martínez-López[3], DVM, MPVM, PhD

[1]Iowa State University, Ames, Iowa; [2]Boehringer-Ingelheim, Saint Joseph, Missouri; [3]University of California-Davis, Davis, California

Abstract

Managing, monitoring, and maintaining the health status of US swine has become increasingly complex. The continued expanse of today's multisite pig production systems; extensive pig movement within and across broad geographic regions of US and Canada; a growing list of emerging and reemerging diseases of local, regional, and national significance; diagnostic data being derived from any number of veterinarians, veterinary clinics, production systems, and veterinary diagnostic laboratories; and a growing appreciation for the effect that the health status of a region, pig-flow, or production system has on the biological and financial performance of pork production operations have each contributed to a need for improved systems (or intercommunicative networks) capable of effectively collating, managing, summarizing, and securely communicating information pertaining to the health status of swine herds. A primary aim of our highly collaborative efforts over the past few years has been to develop a broadly applicable and streamlined system for linking veterinary diagnostic laboratory submissions, corresponding test results, attending veterinarian insight, and an interpreted health status of farm sites to an exceptionally capable spatiotemporal disease management tool (Disease BioPortal®, University of California – Davis, `http://bioportal.ucdavis.edu/`) for use in area-regional, veterinary clinic, or production system specific swine health monitoring and disease control initiatives. Partnering with three different producer groups (pilot participants) enabled access to the premises-specific information and corresponding flow of diagnostic information needed to evaluate and trouble-shoot the performance and functionality of the systems and web-based animal health information management tools being developed. These efforts have led to the development of a suite of complementary methodologies and web-based tools (Animal Health Information Management Network, Figure 1) that will provide producers a new system for monitoring and maintaining the health status of swine farms in their practice, region, or production system over time. While the focus of this animal health information management network development project has initially concentrated on porcine diseases, specifically PEDV, PDCoV, and PRRSV; the tools developed can be adapted to and used across any number of pathogens, livestock species, and laboratories in the National Animal Health Laboratory Network. Moving forward, the Iowa State University Department of Veterinary Diagnostic and Production Animal Medicine (in collaboration with the University of California-Davis, participating VDLs, and GlobalVetLINK) plans to continue to enhance the capabilities of the network and pilot a service whose aim is to support the efforts of veterinarians, veterinary clinics, and/or production systems interested in using this suite of web-based applications to monitor and maintain the health status of swine farms in their practice, region, or production system over time. Additionally, through collaborations with the USDA National Animal Health Laboratory Network and peer veterinary diagnostic laboratories in the National Animal Health Laboratory Network (e.g., Iowa State University Veterinary Diagnostic Laboratory, Kansas State University Veterinary Diagnostic Laboratory, South Dakota State University Animal Disease Research and Diagnostic Laboratory, and the University of Minnesota Veterinary Diagnostic Laboratory); the tools and systems of inter-laboratory connectivity being further developed in the Animal Health Information Management Network were used to establish a web-based dashboard of time and space sensitive graphics in Disease BioPortal® that depict the national and state level trends in case level PEDV and PDCoV PCR test results observed in case submissions made to VDLs and reported to the USDA. These efforts may well represent the first working example in US history whereby aggregate diagnostic data has been electronically captured from all of the VDLs located throughout the US and seamlessly integrated into an interactive web-based summary of national and state level diagnostic trends. Collectively, this suite of novel methods, web-based tools, and systems of inter-laboratory connectivity being advanced and further developed hold promise for helping create a step-wise improvement for both producer group specific and larger-scale (aggregate data) swine health monitoring applications. The continued development and use of the next generation of animal health information management tools for routine private enterprise applications are

Figure 1: Animal Health Information Management Network

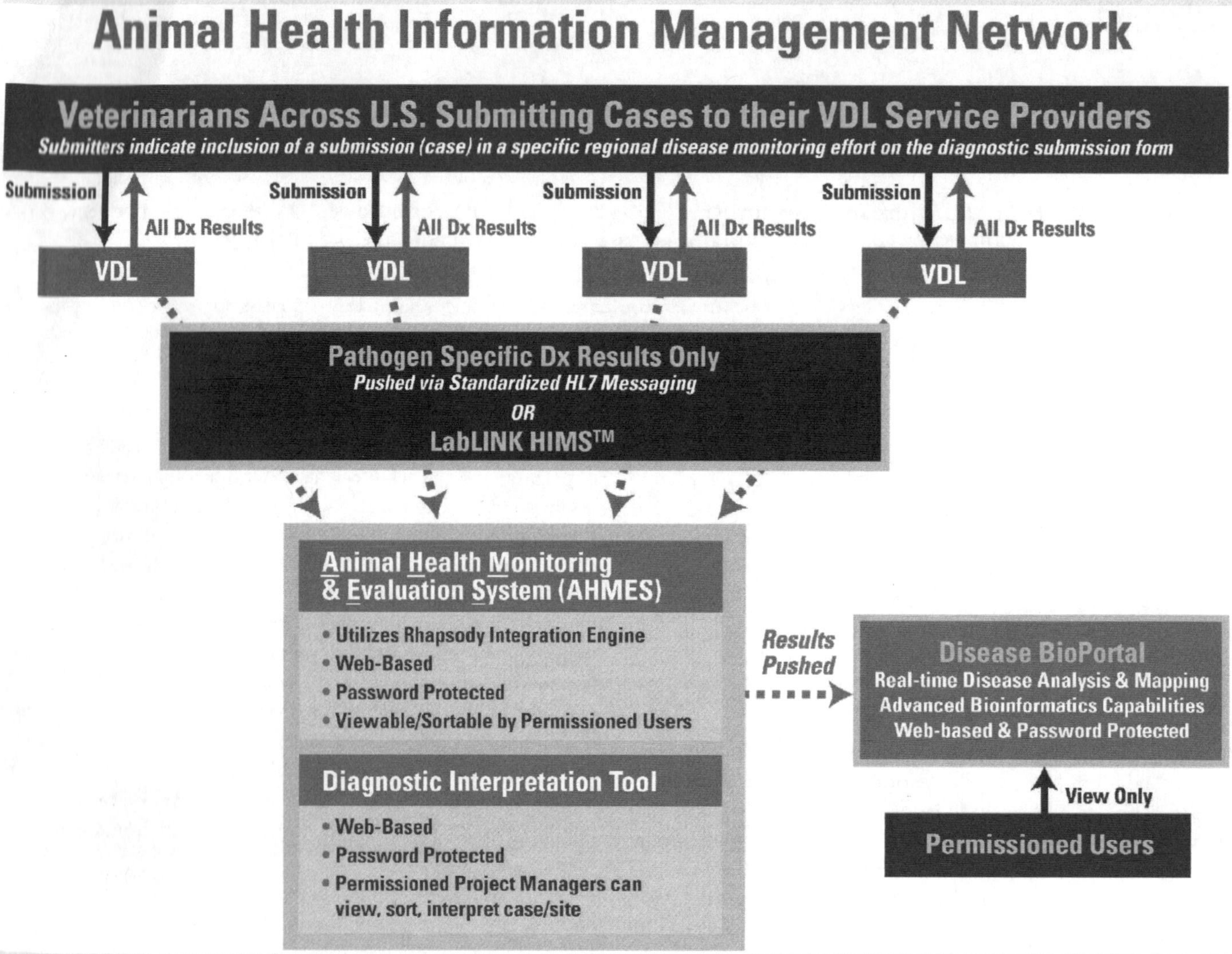

unquestionably necessary to create the systems of connectivity and web-based analytics needed to enhance the US pork industry's overall capabilities and preparedness for managing endemic, emerging, and/or transboundary diseases of high consequence.